Volkhard May, Oliver Kühn

# Charge and Energy Transfer Dynamics in Molecular Systems

Volkhard May, Oliver Kühn

# Charge and Energy Transfer Dynamics in Molecular Systems

*Second, Revised and Enlarged Edition*

WILEY-VCH Verlag GmbH & Co. KGaA

**Authors**

*Dr. Volkhard May*
Humboldt-Universität Berlin, Germany
e-mail: may@physik.hu-berlin.de

*Dr. Oliver Kühn*
Freie Universität Berlin, Germany
e-mail: ok@chemie.fu-berlin.de



**Library of Congress Card No.: applied for**
**British Library Cataloging-in-Publication Data:**
A catalogue record for this book is available from the British Library

**Bibliographic information published by Die Deutsche Bibliothek**
Die Deutsche Bibliothek lists this publication in the Deutsche Nationalbibliografie; detailed bibliographic data is available in the Internet at <http://dnb.ddb.de>.



Printed in the Federal Republic of Germany
Printed on acid-free paper

**Printing** Strauss Offsetdruck GmbH, Mörlenbach
**Bookbinding** Litges & Dopf Buchbinderei GmbH, Heppenheim

**ISBN** 3-527-40396-5

# Foreword

Our understanding of the elementary processes of charge and energy transfer in molecular systems has developed at an enormous pace during the last years. Time–resolved spectroscopy has opened a real–time look at the microscopic details of molecular dynamics not only in the gas, but also the condensed phase. Atomic scale structures are available for a virtually uncountable number of biological systems which in turn triggers spectroscopic investigations like in the case of photosynthetic complexes or photoactive proteins. The emerging combination of structural and temporal resolution in time–resolved X–ray crystallography bears an unprecedented potential for the understanding of the interrelation between molecular structure and function. On the theoretical side, accurate electronic structure methods are becoming available for systems with hundreds of atoms, thus providing valuable information about interaction potentials governing molecular motions. The combination of quantum and molecular mechanics offers a way to condensed phase systems. Quantum dynamics methods, on the other hand, suffer from exponential scaling. Fortunately, the detailed information contained in the full wave function is quite often not needed and effective model simulations based on quantum chemical, classical molecular dynamical, but also experimental input are appropriate.

"Charge and Energy Transfer Dynamics in Molecular Systems" has been successful in providing an advanced level introduction into modern theoretical concepts of a very active area of research. Here, the quantum statistical density operator approach reveals its full flexibility by facilitating an integrative description of such diverse topics as there are vibrational relaxation, optical excitation, or electron, proton, and exciton transfer. It served the goal set by the authors to contribute to the bridging of the communication gap between researchers with different backgrounds. In addition having this self–contained source proved invaluable for the education of graduate students.

With the enlarged Second Edition V. May and O. Kühn incorporate many of the recent developments in the field. The scope of the introduction into condensed phase dynamics theory has been broadened considerably. It includes a discussion of quantum–classical concepts which emerged with the prospect of being able to describe the approximate quantum time evolution of hundreds of degrees of freedom. The detection of transfer processes by means of ultrafast nonlinear spectroscopy has received a greater emphasis. The timely topic of utilizing tailored laser fields for the active control of charge and energy transfer is introduced in a new chapter. Throughout new illustrative examples have been added which will enhance the appreciation of the mathematical formalism.

I am happy to recommend this Second Edition to an interdisciplinary audience.

*Klaus Schulten*

Urbana, Illinois, September 2003

# Preface

The positive response to the First Edition of this text has encouraged us to prepare the present Revised and Enlarged Second Edition. All chapters have been expanded to include new examples and figures, but also to cover more recent developments in the field. The reader of the First Edition will notice that many of the topics which were addressed in its "Concluding Remarks" section have now been integrated into the different chapters.

The introduction to dissipative quantum dynamics in Chapter 3 now gives a broader view on the subject. Particularly, we elaborated on the discussion of hybrid quantum–classical techniques which promise to be able to incorporate microscopic information about the interaction of some quantum system with a classical bath beyond the weak coupling limit. In Chapter 4 we give a brief account on the state–space approach to intramolecular vibrational energy and the models for treating the intermediate time scale dynamics, where the decay of the survival probability is nonexponential. Chapter 5 now compares different methodologies to compute the linear absorption spectrum of a molecule in a condensed phase environment. Furthermore, basic aspects of nonlinear optical spectroscopy have been included to characterize a primary tool for the experimental investigation of molecular transfer processes. Bridge–mediated electron transfer is now described in detail in Chapter 6 including also a number of new examples. Chapter 7 on proton transfer has been supplemented by a discussion of the tunneling splitting and its modification due to the strong coupling between the proton transfer coordinate and other intramolecular vibrational modes. Chapter 8 dealing with exciton dynamics has been considerably rearranged and includes now a discussion of two–exciton states.

Finally, we have added a new Chapter 9 which introduces some of the fundamental concepts of laser field control of transfer processes. This is a rapidly developing field which is stimulated mostly by the possibility to generate ultrafast laser pulse of almost any shape and spectral content. Although there are only few studies on molecular transfer processes so far, this research field has an enormous potential not only for a more detailed investigation of the dynamics but also with respect to applications, for instance, in molecular based electronics.

Following the lines of the First Edition we avoided to make extensive use of abbreviations. Nevertheless, the following abbreviations are occasionally used: DOF (degrees of freedom), ET (electron transfer), IVR (intramolecular vibrational redistribution), PES (potential energy surface), PT (proton transfer), QME (quantum master equation), RDM (reduced density matrix), RDO (reduced density operator), VER (vibrational energy relaxation) and XT (exciton transfer).

We have also expanded the "Suggested Reading" section which should give a systematic starting point to explore the original literature, but also to become familiar with alternative views on the topics. Additionally, at the end of each Chapter, the reader will find a brief list of references. Here, we included the information about the sources of the given examples and refer to the origin of those fundamental concepts and theoretical approaches which have been directly integrated into the text. We would like to emphasize, however, that these lists are by no means exhaustive. In fact, given the broad scope of this text, a complete list of references would have expanded the book's volume enormously, without necessarily serving its envisaged purpose.

It is our pleasure to express sincere thanks to the colleagues and students N. Boeijenga, B. Brüggemann, A. Kaiser, J. Manz, E. Petrov, and B. Schmidt, which read different parts of the manuscript and made various suggestions for an improvement. While working on the manuscript of this Second Edition we enjoyed the inspiring atmosphere, many seminars, and colloquia held within the framework of the Berlin Collaborative Research Center (Sfb450) "Analysis and Control of Ultrafast Photoinduced Reactions". This contributed essentially to our understanding of charge and energy transfer phenomena in molecular systems. Finally, we would like to acknowledge financial support from the Deutsche Forschungsgemeinschaft and the Fonds der Chemischen Industrie (O.K.).

*Volkhard May and Oliver Kühn*

Berlin, September 2003

# Preface to the First Edition

The investigation of the stationary and dynamical properties of molecular systems has a long history extending over the whole century. Considering the last decade only, one observes two tendencies: First, it became possible to study molecules on their natural scales, that is, with a spatial resolution of some ångström ($10^{-10}$ meters) and on a time scale down to some femtoseconds ($10^{-15}$ seconds). And second, one is able to detect and to manipulate the properties of single molecules. This progress comes along with a steadily growing number of theoretical and experimental efforts crossing the traditional borderlines between chemistry, biology, and physics. In particular the study of molecular transfer processes involving the motion of electrons, protons, small molecules, and intramolecular excitation energy, resulted in a deeper understanding of such diverse phenomena as the photoinduced dynamics in large molecules showing vibrational energy redistribution or conformational changes, the catalysis at surfaces, and the microscopic mechanisms of charge and energy transfer in biological systems. The latter are of considerable importance for unraveling the functionality of proteins and all related processes like the primary steps of photosynthesis, the enzymatic activity, or the details of the repair mechanisms in DNA strands, to mention just a few examples. In a more general context also molecular electronics, that is, the storage and processing of information in molecular structures on a nanometer length scale, has triggered enormous efforts. Finally, with the increasing sophistication of laser sources, first steps towards the control of chemical reaction dynamics have been taken.

The ever growing precision of the experiments requires on the theoretical side to have microscopic models for simulating the measured data. For example, the interpretation of optical spectroscopies in a time region of some tenths of femtoseconds, demands for an appropriate simulation of the molecular dynamics for the considered system. Or, understanding the characteristics of the current flowing through a single molecule in the context of scanning tunneling microscopy, needs detailed knowledge of the electronic level structure of the molecule as well as of the role of its vibrational degrees of freedom. These few example already demonstrate, that advanced theoretical concepts and numerical simulation techniques are required, which are the combination of methods known from general quantum mechanics, quantum chemistry, molecular reaction dynamics, solid state theory, nonlinear optics, and nonequilibrium statistical physics.

Such a broad approach is usually beyond the theoretical education of chemists and biologists. On the other hand, quantum chemistry and chemical reaction dynamics are quite often not on the curriculum of physics students. We believe that this discrepancy quite naturally does not facilitate communication between scientists having different backgrounds. Therefore it is one of the main intentions of the present book to provide a common language for bridging this gap.

The book starts with an introduction and general overview about different concepts in Chapter 1. The essentials of theoretical chemical physics are then covered in Chapter 2. For the chemistry student this will be mostly a repetition of quantum chemistry and in particular the theory of electronic and vibrational spectra. It is by no means a complete introduction into this subject, but intended to provide some background mainly for physics students. The prerequisites from theoretical physics for the description of dynamical phenomena in molecular systems are presented in Chapter 3. Here we give a detailed discussion of some general aspects of the dynamics in open and closed quantum systems, focusing on transfer processes in the condensed phase.

The combination of qualitative arguments, simple rate equations, and the powerful formalism of the reduced statistical operator constitutes the backbone of the second part of the book. We start in Chapter 4 with a discussion of intramolecular transfer of vibrational energy which takes place in a given adiabatic electronic state. Here we cover the limits of isolated large polyatomic molecules, small molecules in a matrix environment, up to polyatomics in solution. In Chapter 5 we then turn to processes which involve a transition between different electronic states. Special emphasis is put on the discussion of optical absorption, which is considered to be a reference example for more involved electron–vibrational transfer phenomena such as internal conversion which is also presented in this chapter. Chapter 6 then outlines the theoretical frame of electron transfer reactions focusing mainly on intramolecular processes. Here, we will develop the well–known Marcus theory of electron transfer, describe nuclear tunneling and superexchange electron transfer, and discuss the influence of polar solvents. In Chapter 7 it will be shown that, even though proton transfer has many unique aspects, it can be described by adapting various concepts from electron transfer theory. The intermolecular excitation energy transfer in molecular aggregates is considered in Chapter 8. In particular the motion of Frenkel excitons coupled to vibrational modes of the aggregate will be discussed. In the limit of ordinary rate equations this leads us to the well–known Förster expression for the transfer rate in terms of emission and absorption characteristics of the donor and acceptor molecules, respectively.

By presenting a variety of theoretical models which exist for different types of transfer processes on a common formal background, we hope that the underlying fundamental concepts are becoming visible. This insight may prepare the reader to take up one of the many challenging problems provided by this fascinating field of research. Some personal reflections on current and possible future developments are given in Chapter 9.

The idea for writing this book emerged from lectures given by the authors at the Humboldt University Berlin, the Free University Berlin, and at the Johannes Gutenberg University Mainz during the last decade. These courses have been addressed to theoretically and experimentally oriented undergraduate and graduate students of Molecular Physics, Theoretical Chemistry, Physical Chemistry, and Biophysics, being interested in the fast developing field of transfer phenomena. The book is self–contained and includes detailed derivations of the most important results. However, the reader is expected to be familiar with basic quantum mechanics. Most of the chapters contain a supplementary part where more involved derivations as well as special topics are presented. At the end of the main text we also give some comments on selected literature which should complement the study of this book.

Of course this book would not have been possible without the help, the critical comments, and the fruitful discussions with many students and colleagues. In this respect it is a pleasure for us to thank I. Barvik, N. P. Ernsting, W. Gans, L. González, O. Linden, H. Naundorf, J. Manz, S. Mukamel, A. E. Orel, T. Pullerits, R. Scheller, and D. Schirrmeister. We also are grateful for continuous financial support which has been provided by the Deutsche Forschungsgemeinschaft, in particular through the Sonderforschungsbereich 450 "Analysis and Control of Ultrafast Photoinduced Reactions".

*Volkhard May and Oliver Kühn*

Berlin, September 1999

# Contents

# 1 Introduction

The understanding of molecular transfer phenomena requires a unified theoretical treatment which should have its foundations in a microscopic definition of the considered molecular system. There are three questions which need to be answered in this respect: First, what is the appropriate theoretical description of the molecular system, second, what is the form of the dynamical equations which describe the transfer process, and third, how can the computed results be related to experimental observations.

From a general point of view, quantum mechanics gives the framework for all phenomena occurring in *molecular systems*. In the following the term "molecular system" shall cover single molecules, simple molecular aggregates, but also larger arrangements of molecules like supra–molecular complexes and, in particular, molecules embedded in different types of environments will be of interest. The definition even encompasses biological macromolecules such as membrane–bound protein complexes. The common link between these molecular systems is that they show *transfer processes*. By "transfer process" we understand the flow of vibrational energy, the dynamics of electrons, protons, and electronic excitation energy. The nature of these processes is intimately related to the kind of preparation of the initial conditions, for instance, by the interaction with an electromagnetic field. In view of this broad scope it is clear that an exact quantum mechanical treatment is impossible if we go beyond the level of simple model systems.

Therefore, we will start in Chapter 2 with a discussion of the steps which lead us from the formally exact to some approximate molecular Hamilton operator. Given a molecule in the gas phase (vacuum) as shown in the upper part of Fig. 1.1, the *Born–Oppenheimer separation* of nuclear and electronic motions can be performed. Here, the molecular wave function is split up into an electronic and a nuclear part, a procedure which is justified by the large mass–difference between both types of particles. This results in a Schrödinger equation for the electronic wave function alone, for given fixed positions of the nuclei. Calculating the electronic energy spectrum for different positions of the nuclei one obtains *potential energy surfaces* which govern the motion of the nuclei. These potential energy surfaces are at the heart of the understanding of stationary molecular spectra and chemical reaction dynamics. If nuclear and electronic motion are adiabatically separable, that is, the coupling between different electronic states is negligible and one can carry out the Born–Oppenheimer approximation. Under certain conditions, however, so–called nonadiabatic transitions between different electronic states as a consequence of the nuclear motions are to be expected.

If we step from the gas to the condensed phase, for example, by considering a molecule in solution as shown in the lower part of Fig. 1.1, the effect of the molecule–environment interaction has to be taken into account. The simplest way to do this is to add an additional

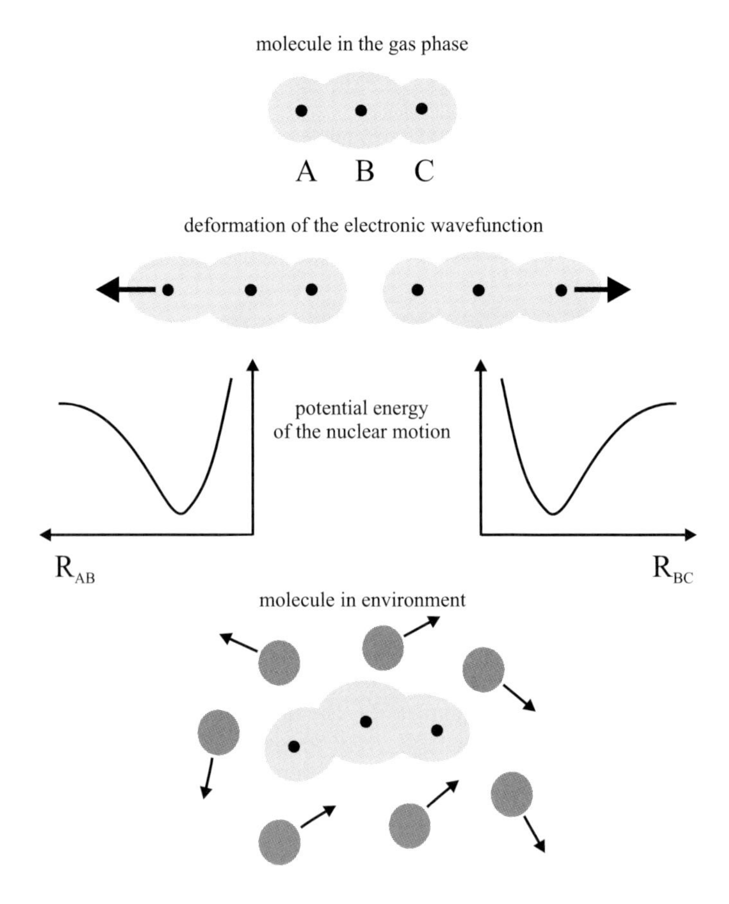


**Figure 1.1:** The complicated problem of the interaction between electrons and nuclei is reduced to some tractable level by employing the Born–Oppenheimer separation of their motions. Upper panel: Three–atomic molecule with nuclei labelled by A, B, and C. The electronic wave function is indicated by a grey area. Middle panel: The bond length between atom A and B (left part) as well as atom B and C (right part) is increased accompanied by an instantaneous deformation of the electronic wave function. As a result, a potential energy curve is formed determining the dynamics of the bond length. Lower panel: If the molecule is taken from the gas to the condensed phase its stationary properties cannot be calculated without invoking further approximations.

external potential to the molecular Hamiltonian. Often the environment can be described as a macroscopic dielectric and its influence can be judged from its dielectric properties.

Having discussed the stationary molecular properties we turn to the *molecular dynamics* in Chapter 3. Here, the reader will become familiar with concepts ranging from incoherent to coherent transfer events. The connection between these limits is provided by the relevant time scales; of particular importance is the relation between intramolecular relaxation and intermolecular transfer times. In view of experimental advances in ultrafast optical spectroscopy, our treatment reflects the historical evolution of knowledge on molecular dynamics.

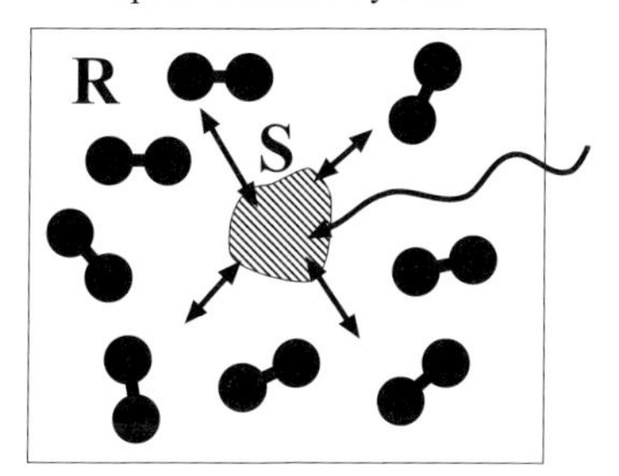


**Figure 1.2:** Open molecular system $S$ interacting with its environment (reservoir) $R$. In addition the system may be influenced by external fields (wiggly line).

The essential ingredient for the theoretical modelling is the concept of an *open molecular system* $S$ interacting with its *environment* (reservoir) $R$ by collision processes or via other means of energy exchange. A schematic illustration of this situation is given in Fig. 1.2. The *relevant* system $S$ may represent any type of molecule, but it may also comprise selected so–called *active* degrees of freedom of a particular molecule. In order to study the dynamics of the relevant system in an actual experiment, its response to some external field such as an electromagnetic field is measured.

The most general description of the total system, $S$ plus $R$, is given by the quantum statistical operator $\hat{W}$ as indicated in the left part of Fig. 1.3. This operator is based on the concept of a *mixed* quantum state formed by $S$ and its macroscopic environment. However, the operator $\hat{W}$ contains much more information than we will ever need, for instance, to simulate a particular experiment. Indeed, it is the relevant system $S$ we are interested in. Making use of a reduction procedure we obtain a *reduced statistical operator* $\hat{\rho}$ which contains the information on the dynamics of $S$ only, but including the influence of the environment $R$ (right–hand part of Fig. 1.3). When deriving equations of motion for the reduced statistical operator, the so–called *Quantum Master Equations*, a number of approximations have to be invoked. Most fundamental in this respect will be the assumption of a weak interaction between the system

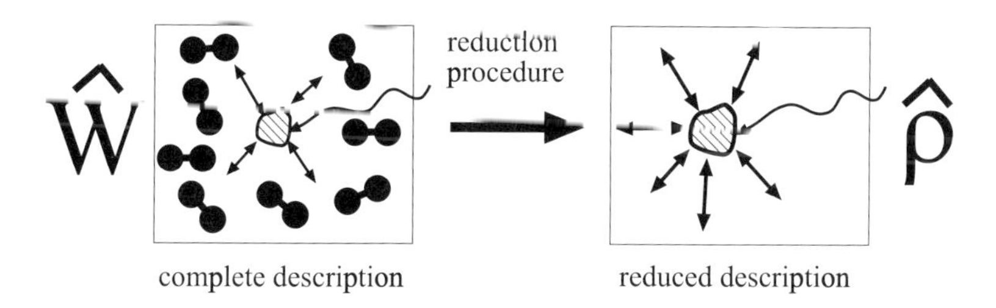


**Figure 1.3:** The total system $S + R$ is completely described by the quantum statistical operator $\hat{W}$. By means of a reduction procedure one can focus on the relevant system using the reduced statistical operator $\hat{\rho}$.

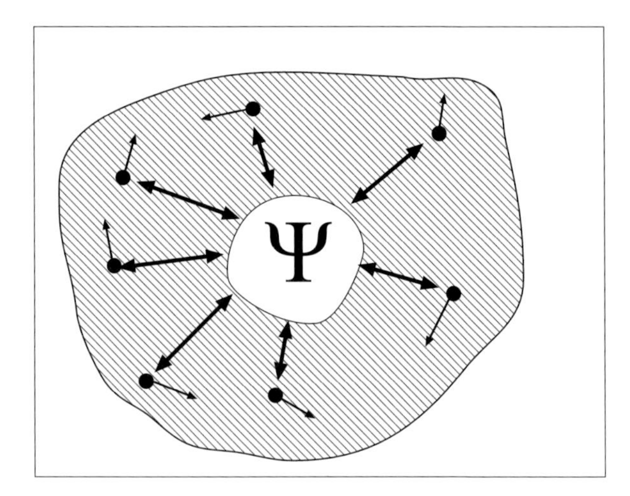


**Figure 1.4:** Mixed quantum–classical description of condensed phase dynamics. The classical particles move in the mean field generated by the quantum particle described by the wave function $\Psi$.

$S$ and the reservoir $R$, which in practice requires a proper separation into relevant and environmental coordinates for the molecular system at hand. If there is no interaction at all, the Quantum Master Equation would be equivalent to the time–dependent Schrödinger equation. This is the regime of *coherent* dynamics. If the interaction is not negligible, however, the system dynamics gradually changes with increasing coupling strength from a *partially coherent* one to an *incoherent* one. The incoherent motion of a quantum system is commonly described using ordinary rate equations which are based on the *Golden Rule* rate expression of quantum mechanics.

The Quantum Master Equation, however, affords a more general frame since it comprises the various dynamical limits. Therefore, it gives a unified description of quantum molecular dynamics of a relevant system and its interaction with some environment.

The concept of the statistical operator provides a *quantum–statistical* description of $S$ and $R$. However, in many situations it is sufficient to describe $R$ by means of classical mechanics. Then, $S$ can be characterized by a wave function $\Psi$ instead of a statistical operator and the dynamics of the environmental degrees of freedom is governed by Newton's equations. Often the dynamics is split up in such a way that the classical particles move in the mean field of the quantum particle. This situation is visualized in Fig. 1.4.

The formal concepts developed in Chapters 2 and 3 are then applied to describe different transfer phenomena. In principle, the different transfer processes can be classified according to the type of transferred particle. In addition, one can distinguish between intramolecular and intermolecular particle transfer. The common frame is provided by the molecular Schrödinger equation together with the Born–Oppenheimer separation of electronic and nuclear motions as mentioned above.

The coupled nuclear dynamics in polyatomic molecules which might be immersed in some condensed phase environment is treated in Chapter 4. We will show how an initially prepared vibrational state decays while its excitation energy is distributed over all possible environmental modes, as illustrated in the left–hand part of Fig. 1.5. For small polyatomic molecules the reversible energy flow out of the initial state is called *intramolecular vibrational energy redistribution*. For condensed phase situations the irreversible dissipation of energy into the environment is called *vibrational energy relaxation*. In both cases the transferred objects are the quanta of vibrational energy.

As sketched in Fig. 1.5 the preparation of the initial state can be due to an optical transition between two electronic states as a consequence of the interaction between the molecular system and an external electromagnetic field. In Chapter 5 we will discuss the processes of photon absorption and emission sketched in Fig. 1.5. It will be shown that the coupled electron–vibrational dynamics which is responsible for the absorption lineshape can be described by a combined density of states which is the Fourier transform of some correlation function. This theoretical result will turn out to be quite general. In particular we will show that different types of transfer processes can be accommodated into such a framework. For example, the *internal conversion* dynamics of nonadiabatically coupled electronic states (see right–hand part of Fig. 1.5) can, in the incoherent limit, be described by a combined density of states.

The external field interaction, on the other hand, provides the means for preparing nonequilibrium initial states which can act as a donor in a photoinduced electron transfer reaction which is discussed in Chapter 6. The concerted electron–vibrational dynamics accompanying electron transfer reactions can often be modelled in the so–called *curve–crossing* picture of two coupled potential energy surfaces representing two electronic states along a *reaction*

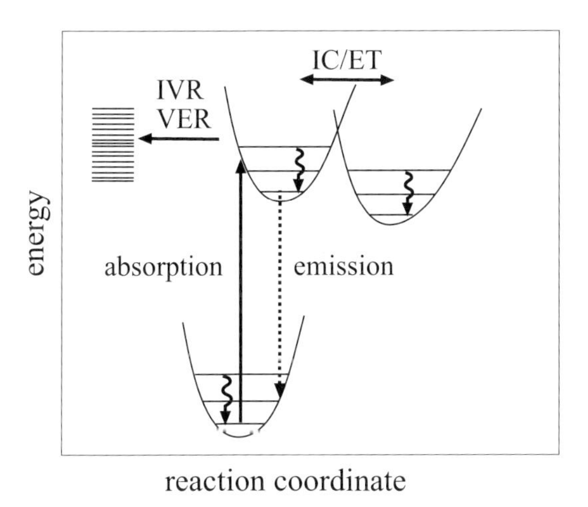


**Figure 1.5:** After optical preparation of an electronically and vibrationally excited initial state (absorption), different transfer processes can occur. If the electronic state is not changed, but there is a coupling to some manifold of vibrational states, intramolecular energy redistribution (IVR) or vibrational energy relaxation (VER) can be observed. If there is some coupling to another electronic state, intramolecular internal conversion (IC) or electron transfer (ET) take place. At the same time, one has VER as indicated by the wiggly lines. In addition the system may return to the ground state by emitting a photon.

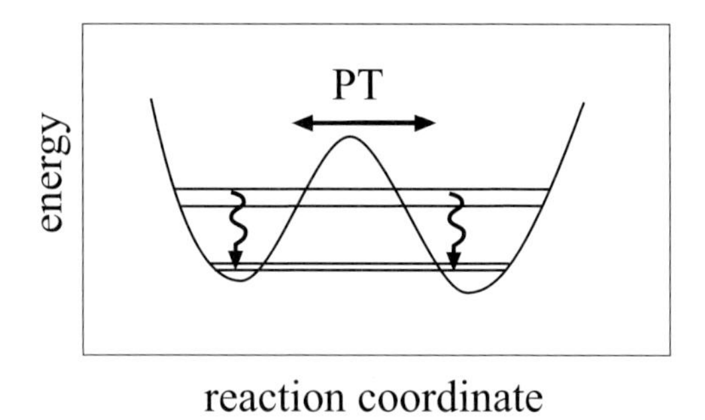


**Figure 1.6:** Hydrogen bonding which governs the proton transfer (PT) dynamics often leads to a double minimum potential along a reaction coordinate. The interaction between the proton and some environment may cause vibrational relaxation (wiggly lines).

*coordinate* (see right–hand part Fig. 1.5).

In contrast, the proton or hydrogen atom transfer investigated in Chapter 7 usually does not involve electronic transitions. In Fig. 1.6 we have sketched a typical situation for intramolecular proton transfer which is realized as an isomerization reaction in the adiabatic electronic ground state. Since the proton has a rather small mass, tunneling processes may play an important role for proton transfer. The small mass ratio between the proton and the other heavy atoms provides the background for the introduction of a second Born–Oppenheimer separation. This will enable us to adapt most of the concepts of electron transfer theory to the case of proton transfer.

In Chapter 8 we discuss excitation energy transfer or so–called exciton transfer in molecular aggregates as another example for coupled electron–vibrational motion. In Fig. 1.7 the mechanism of exciton transfer in the limit of localized excitations is shown. The donor (left) is initially excited, for example, by an external field. As a consequence of the Coulomb interaction between the excited molecule and surrounding molecules, excitation energy is transferred to some acceptor (right). Due to the large spatial separation, donor and acceptor are usually described by different sets of nuclear (reaction) coordinates. The process can formally be understood in a picture where the donor emits radiation energy which is in turn absorbed by the acceptor.

A successful analysis of molecular transfer processes triggers the desire to take *active control* of the dynamics. For example, it would be intriguing to have a means for depositing energy into specific bonds or reaction coordinates such as to dissociate a polyatomic molecule into desired products. To utilize electron transfer processes as part of an ultrafast switch is of tremendous importance in the emerging area of molecular electronic. Or controlling the fate of an exciton, for instance, in a photosynthetic light–harvesting complex as well as guiding the motion of protons in enzymatic catalysis would reveal a microscopic picture of biological functions.

Fortunately, many of the theoretical concepts for laser control developed over the last two decades eventually may turn into working schemes because recent years have witnessed an impressive evolution of laser pulse shaping techniques. But the exploration of the potential

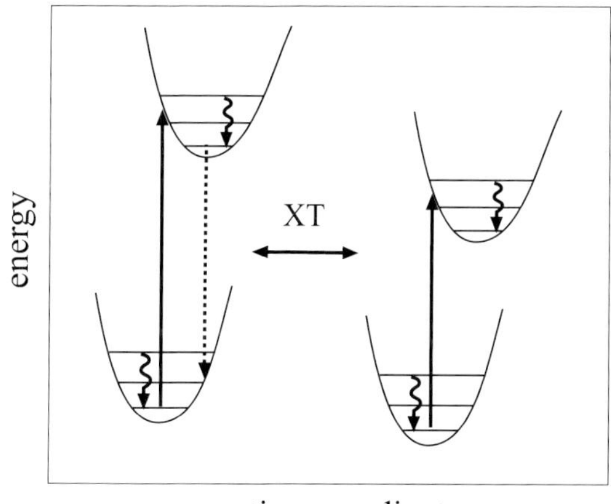


**Figure 1.7:** Excitation energy transfer (XT) which occurs after optical preparation of an electronically and vibrationally excited initial state (donor, left). The Coulomb interaction is responsible for de–excitation of the donor and excitation of the acceptor (right). The nuclear dynamics may be subject to relaxation processes (wiggly lines). Often two independent nuclear (reaction) coordinates are used for the donor and the acceptor site.

of laser pulse control in the context of molecular transfer processes is still in its infancy. The active control of transfer processes will be discussed in Chapter 9.

# 2 Electronic and Vibrational Molecular States

*This chapter provides the background material for the subsequent development of a microscopic description of charge and energy transfer processes in the condensed phase. After introducing the molecular Hamiltonian operator we discuss the Born–Oppenheimer separation of electronic and nuclear motions as the key to the solution of the molecular Schrödinger equation. The Hartree–Fock method which is a simple yet very successful approach to the solution of the ground state electronic structure problem is explained next. It enables us to obtain, for instance, the potential energy surface for nuclear motions. To prepare for the treatment of condensed phase situations we further introduce the dielectric continuum model as a means for incorporating static solvent polarization effects into the electronic structure calculations.*

*The topology of the potential energy surface can be explored by calculating the first and second derivatives with respect to the nuclear coordinates. Of particular interest are the stationary points on a potential energy surface which may correspond to stable conformations of the molecule. In the vicinity of a local minimum it is often possible to analyze nuclear motions in terms of small amplitude normal mode vibrations. If one wants to model chemical reaction dynamics, however, the shape of the potential energy surface away from the stationary points is required as an input. We present two different approaches in this respect: The minimum energy reaction path and the Cartesian reaction surface model. Particularly the latter will provide the microscopic justification for the generic Hamiltonians used later on to simulate small molecular systems embedded in some environment. Finally, we discuss the diabatic and the adiabatic representation of the molecular Hamiltonian.*

## 2.1 Introduction

The development of quantum theory in the 1920's was to a considerable extent triggered by the desire to understand the properties of atoms and molecules. It was soon appreciated that the Schrödinger equation together with the probabilistic interpretation of its solutions provides a powerful tool for tackling a variety of questions in physics and chemistry. The mathematical description of the hydrogen atom's spectral lines could be given and developed to a textbook example of the success of quantum mechanics. Stepping into the molecular realm one faces a complicated many–body problem involving the coordinates of all electrons and all nuclei of the considered molecule. Its solution can be approached using the fact that nuclei and electrons have quite different masses allowing their motion to be adiabatically separated. This concept was first introduced by Born and Oppenheimer in 1927. Within the Born–Oppenheimer adiabatic approximation the simplest molecule, the hydrogen molecule ion, $H_2^+$, can be treated.

From the electronic point of view the appearance of one more electron, for instance, in $H_2$, necessitates the incorporation of the repulsive electronic interaction. Moreover, since one deals with two identical electrons care has to be taken that the wave function has the proper symmetry with respect to an exchange of any two particle labels. In a straightforward way this is accomplished by the *self–consistent field method* according to Hartree, Fock, and Slater. Despite its deficiencies Hartree–Fock theory has played an enormous role in the process of exploring the electronic structure of molecules during the last decades. It still serves as the basis for many of the more advanced approaches used nowadays.

However, it is not only the electronic structure at the equilibrium configuration of the nuclei which is of interest. The form of the potential energy hypersurfaces obtained upon varying the positions of the nuclei proves crucial for an understanding of the vibrational and rotational structure of molecular spectra. Moreover it provides the key to chemical reaction dynamics. While the adiabatic Born–Oppenheimer ansatz is an excellent approximation in the vicinity of the ground state equilibrium configuration, nonadiabatic couplings leading to transitions between electronic states become an ubiquitous phenomenon if the nuclei are exploring their potential surface in processes such as photodissociation and electron transfer reactions, for example.

This chapter introduces the concepts behind the keywords given so far and sets up the stage for the following chapters. Having this intention it is obvious that we present a rather selective discussion of a broad field. We first introduce the molecular Hamiltonian and the respective solutions of the stationary Schrödinger equation in Section 2.2. This leads us directly to the Born–Oppenheimer separation of electronic and nuclear motions in Section 2.3. A brief account of electronic structure theory for polyatomic molecules is given next (Section 2.4). This is followed by a short summary of the dielectric continuum model in Section 2.5 which allows for incorporation of solvent effects into electronic structure calculations. On this basis we continue in Section 2.6 to discuss potential energy surfaces and the related concepts of harmonic vibrations and reaction paths. In Section 2.7 we focus attention to the problem of nonadiabatic couplings which are neglected in the Born–Oppenheimer adiabatic approximation. Finally, the issue of diabatic versus adiabatic pictures which emerges from this discussion is explained and alternative representations of the molecular Hamiltonian are given.

## 2.2 Molecular Schrödinger Equation

In the following we will be interested in situations where atoms made of point–like nuclei and electrons are spatially close such that their mutual interaction leads to formation of stable molecules. Let us consider such a molecule composed of $N_{\rm nuc}$ atoms having atomic numbers $z_1, \ldots, z_{N_{\rm nuc}}$. The Cartesian coordinates and conjugate momenta for the $N_{\rm el}$ electrons are denoted $\mathbf{r}_j$ and $\mathbf{p}_j$, respectively. For the $N_{\rm nuc}$ nuclei we use $\mathbf{R}_n$ and $\mathbf{P}_n$ (see Fig. 2.1). The Hamiltonian operator of the molecule has the general form

$$H_{\rm mol} = T_{\rm el} + V_{\rm el-nuc} + V_{\rm el-el} + T_{\rm nuc} + V_{\rm nuc-nuc} \, . \tag{2.1}$$

Here the kinetic energy of the electrons is given by ($m_{\rm el}$ is the electron mass)

$$T_{\rm el} = \sum_{j=1}^{N_{\rm el}} \frac{\mathbf{p}_j^2}{2m_{\rm el}} \, , \tag{2.2}$$

and for the nuclei it is

$$T_{\rm nuc} = \sum_{n=1}^{N_{\rm nuc}} \frac{\mathbf{P}_n^2}{2M_n} \, , \tag{2.3}$$

with $M_n$ being the mass of the $n$th nucleus. Since both kinds of particles are charged they interact via Coulomb forces. The repulsive Coulomb pair interaction between electrons is

$$V_{\rm el-el} = \frac{1}{2} \sum_{i \neq j} \frac{e^2}{|\mathbf{r}_i - \mathbf{r}_j|} \, , \tag{2.4}$$

and for the nuclei we have

$$V_{\rm nuc-nuc} = \frac{1}{2} \sum_{m \neq n} \frac{z_m z_n e^2}{|\mathbf{R}_m - \mathbf{R}_n|} \, . \tag{2.5}$$

(Note that the factor $1/2$ compensates for double counting.) The attractive interaction between electrons and nuclei is given by

$$V_{\rm el-nuc} = -\sum_{j,n} \frac{z_n e^2}{|\mathbf{r}_j - \mathbf{R}_n|} \, . \tag{2.6}$$

Since there are $N_{\rm el}$ electrons and $N_{\rm nuc}$ nuclei, the molecule has $3(N_{\rm el} + N_{\rm nuc})$ spatial degrees of freedom (DOF). Each electron is assigned an additional quantum number $\sigma_j$ to account for its spin. The purely quantum mechanical concept of electron spin was introduced to explain the fine structure of certain atomic spectra by Uhlenbeck and Goudsmit in 1925. Later its theoretical foundation was laid in the relativistic extension of quantum mechanics developed by Dirac in 1928. When using the nonrelativistic Hamiltonian Eq. (2.1) we have no means to rigorously introduce spin operators and to derive the interaction potential between

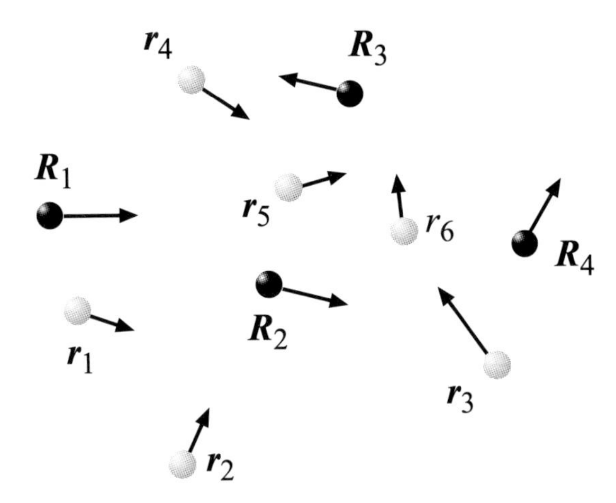


**Figure 2.1:** Arrangement of electrons and nuclei forming a molecule by virtue of their mutual interaction. Note that both types of particles are assumed to be point–like throughout the book. The arrows indicate the classical momentum vector of the particle at a certain instant in time.

coordinate and spin variables (spin–orbit coupling). Therefore, the existence of linear Hermitian spin operators is usually postulated and their action on spin functions defined. We will not consider relativistic effects in this text and therefore carry the spin variable along with the electron coordinate only in the formal considerations of Section 2.4.

All quantum mechanical information about the stationary properties of the molecular system defined so far is contained in the solutions of the time–independent nonrelativistic Schrödinger equation

$$H_{\rm mol}\Psi(r,\sigma;R) = \mathcal{E}\Psi(r,\sigma;R)\ . \tag{2.7}$$

Here and in the following we will combine the set of electronic Cartesian coordinates in the multi–indices $r = (\mathbf{r}_1, \mathbf{r}_2, \dots, \mathbf{r}_{N_{\rm el}})$. A similar notation is introduced for the nuclear Cartesian coordinates, $R = (\mathbf{R_1}, \mathbf{R_2}, \dots, \mathbf{R_{N_{\rm nuc}}})$. In addition we will frequently use the more convenient notation $(\mathbf{R_1}, \mathbf{R_2}, \dots, \mathbf{R_{N_{\rm nuc}}}) \to (R_1, \dots, R_{3N_{\rm nuc}}) = R$. Momenta and masses of the nuclei will be written in the same way. (In this notation $M_1 = M_2 = M_3$ is the mass of nucleus number one etc.) For the spin we use the notation $\sigma = (\sigma_1, \sigma_2, \dots, \sigma_{\rm N_{el}})$.

As it stands Eq. (2.7) does not tell much about what we are aiming at, namely electronic excitation spectra, equilibrium geometries etc. However, some general points can be made immediately: First, the solution of Eq. (2.7) will provide us with an energy spectrum $\mathcal{E}_\lambda$ and corresponding eigenfunctions, $\Psi_\lambda(r,\sigma;R)$. The energetically lowest state $\mathcal{E}_0$ is called the ground state. If $\mathcal{E}_\lambda$ is negative the molecule is in a stable bound state. Note that in the following we will also make use of the more formal notation where the eigenstates of the molecular Hamiltonian are denoted by the state vector $|\Psi_\lambda\rangle$. The wave function is obtained by switching to the $(r,\sigma;R)$ representation: $\Psi_\lambda(r,\sigma;R) = \langle r,\sigma;R|\Psi_\lambda\rangle$.

Second, the probability distribution, $|\Psi_\lambda(r,\sigma;R)|^2$, contains the information on the distribution of electrons as well as on the arrangement of the nuclei. Having this quantity at hand one can calculate, for example, the charge density distribution $\mathbf{x}$, $\rho_\lambda(\mathbf{x})$ for a particular molecular state at some spatial point. The classical expression

$$\rho(\mathbf{x}) = -e\sum_{j=1}^{N_{\text{el}}}\delta(\mathbf{r}_j - \mathbf{x}) + e\sum_{n=1}^{N_{\text{nuc}}} z_n\delta(\mathbf{R}_n - \mathbf{x}) \,, \tag{2.8}$$

is quantized by replacing the coordinates by the respective operators. Taking the matrix elements of the resulting charge density operator with respect to the state $\Psi_\lambda(r, \sigma; R)$ we get

$$\begin{aligned}\rho_\lambda(\mathbf{x}) &= -eN_{\text{el}}\sum_\sigma \int d^3\mathbf{r}_2 \ldots d^3\mathbf{r}_{N_{\text{el}}}\, dR\, |\Psi_\lambda(\mathbf{x}, \mathbf{r}_2, \ldots, \mathbf{r}_{N_{\text{el}}}, \sigma; R)|^2 \\ &\quad + \sum_{n=1}^{N_{\text{nuc}}}\sum_\sigma ez_n \int dr\, dR\, \delta(\mathbf{R}_n - \mathbf{x})|\Psi_\lambda(r, \sigma; R)|^2 \,. \end{aligned} \tag{2.9}$$

Third, since the Hamiltonian does not depend on spin, the solution of Eq. (2.7) can be separated according to

$$\Psi(r, \sigma; R) = \psi(r; R)\,\zeta(\sigma) \,. \tag{2.10}$$

Here, $\zeta(\sigma)$ is the electronic spin function which is obtained by projecting the molecule's spin state vector $|\zeta\rangle$ onto the spin states of the individual electrons, $\zeta(\sigma) = (\langle\sigma_1|\langle\sigma_2|\ldots\langle\sigma_{N_{\text{el}}}|)|\zeta\rangle$. The individual spin states, $|\sigma_i\rangle$, describe electrons whose spin is parallel (spin up) or antiparallel (spin down) with respect to some direction in coordinate space.

Finally, owing to the *Pauli principle* which states that the wave function of a system of electrons has to be antisymmetric with respect to the interchange of any two electronic indices, $\Psi(r, \sigma; R)$ will be antisymmetric in electronic Cartesian plus spin coordinates. The fact that there can be identical nuclei as well is frequently neglected when setting up the exchange symmetry of the total wave function. This is justified since the nuclear wave function is usually much more localized as compared with the electronic wave function and the indistinguishability is not an issue. Exceptions may occur in systems containing, for example, several hydrogen atoms.

## 2.3 Born–Oppenheimer Separation

The practical solution of Eq. (2.7) makes use of the fact that due to the large mass difference ($m_{\text{el}}/M_n < 10^{-3}$), on average electrons can be expected to move much faster than nuclei. Therefore, in many situations the electronic degrees of freedom can be considered to respond instantaneously to any changes in the nuclear configuration, i.e., their wave function corresponds always to a stationary state. In other words, the interaction between nuclei and electrons, $V_{\text{el-nuc}}$, is modified due to the motion of the nuclei only *adiabatically* and does not cause transitions between different stationary electronic states. Thus, it is reasonable to define an electronic Hamiltonian which carries a parametric dependence on the nuclear coordinates:

$$H_{\text{el}}(R) = T_{\text{el}} + V_{\text{el-nuc}} + V_{\text{el-el}} \,. \tag{2.11}$$

As a consequence the solutions of the time–independent electronic Schrödinger equation describing the motion of the electrons in the electrostatic field of the stationary nuclei (leaving aside the electron's spin)

$$H_{\rm el}(R)\,\phi_a(r;R) = E_a(R)\,\phi_a(r;R)\;, \tag{2.12}$$

will parametrically depend on the set of nuclear coordinates as well. Here, the index $a$ labels the different electronic states. The *adiabatic* electronic wave functions $\phi_a(r;R) = \langle r;R|\phi_a\rangle$ define a complete basis in the electronic Hilbert space. Hence, given the solutions to Eq. (2.12) the molecular wave function can be expanded in this basis set as follows

$$\psi(r;R) = \sum_a \chi_a(R)\,\phi_a(r;R)\;. \tag{2.13}$$

The expansion coefficients in Eq. (2.13), $\chi_a(R)$, depend on the configuration of the nuclei. It is possible to derive an equation for their determination after inserting Eq. (2.13) into Eq. (2.7). One obtains

$$\begin{aligned} H_{\rm mol}\psi(r;R) &= (H_{\rm el}(R) + T_{\rm nuc} + V_{\rm nuc-nuc})\sum_a \chi_a(R)\,\phi_a(r;R) \\ &= \sum_a [E_a(R) + V_{\rm nuc-nuc}]\,\chi_a(R)\,\phi_a(r;R) \\ &\quad + \sum_a T_{\rm nuc}\chi_a(R)\,\phi_a(r;R) \\ &= \mathcal{E}\sum_a \chi_a(R)\,\phi_a(r;R)\;. \end{aligned} \tag{2.14}$$

Multiplication of Eq. (2.14) by $\phi_b^*(r;R)$ from the left and integration over all electronic coordinates yields the following equation for the expansion coefficients $\chi_a(R)$ (using the orthogonality of the adiabatic basis)

$$\begin{aligned} \int dr\;\phi_b^*(r;R)\,H_{\rm mol}\,\psi(r;R) &= [E_b(R) + V_{\rm nuc-nuc}]\,\chi_b(R) \\ &\quad + \sum_a \int dr\;\phi_b^*(r;R)\,T_{\rm nuc}\,\phi_a(r;R)\chi_a(R) \\ &= \mathcal{E}\,\chi_b(R)\;. \end{aligned} \tag{2.15}$$

Since the electronic wave functions depend on the nuclear coordinates we have with $\mathbf{P}_n = -i\hbar\nabla_n$ and using the product rule for differentiation

$$\begin{aligned} T_{\rm nuc}\,\phi_a(r;R)\,\chi_a(R) &= \sum_n \frac{1}{2M_n}\Big\{\mathbf{P}_n^2\phi_a(r;R)]\,\chi_a(R) \\ &\quad +2\,[\mathbf{P}_n\phi_a(r;R)]\;\mathbf{P}_n\chi_a(R) \\ &\quad +\phi_a(r;R)\,\mathbf{P}_n^2\,\chi_a(R)\Big\}\;. \end{aligned} \tag{2.16}$$

The last term is simply the kinetic energy operator acting on $\chi_a(R)$. The other terms can be comprised into the so–called *nonadiabaticity* operator

$$\Theta_{ab} = \int dr\ \phi_a(r;R)\, T_{\text{nuc}}\, \phi_b(r;R) + \sum_n \frac{1}{M_n}\left[\int dr\ \phi_a(r;R)\mathbf{P}_n\phi_b(r;R)\right]\mathbf{P}_n\ . \tag{2.17}$$

Thus, we obtain from Eq. (2.15) an equation for the coefficients $\chi_a(R)$ which reads

$$(T_{\text{nuc}} + E_a(R) + V_{\text{nuc-nuc}} + \Theta_{aa} - \mathcal{E})\ \chi_a(R) = -\sum_{b\neq a}\Theta_{ab}\chi_b(R)\ . \tag{2.18}$$

This result can be interpreted as the stationary Schrödinger equation for the motion of the nuclei with the $\chi_a(R)$ being the respective wave functions. The solution of Eq. (2.18), which is still exact, requires knowledge of the electronic spectrum for all configurations of the nuclei which are covered during their motion. Transitions between individual adiabatic electronic states become possible due to the electronic *nonadiabatic coupling*, $\Theta_{ab}$. This is a consequence of the motion of the nuclei as expressed by the fact that their momentum enters Eq. (2.17). The diagonal part of the nonadiabaticity operator, $\Theta_{aa}$, is usually only a small perturbation to the nuclear dynamics in a given electronic state.

Looking at Eq. (2.18) we realize that it will be convenient to introduce the following effective potential for nuclear motion if the electronic system is in its adiabatic state $|\phi_a\rangle$

$$U_a(R) = E_a(R) + V_{\text{nuc-nuc}}(R) + \Theta_{aa}\ . \tag{2.19}$$

This function defines a hypersurface in the space of nuclear coordinates, the *potential energy surface* (PES) which will be discussed in Section 2.6 in more detail. Its exceptional importance for a microscopic understanding of molecular transfer phenomena will become evident in Chapters 4–7.

The solution to Eq. (2.18) is given by $\chi_{aM}(R) = \langle R|\chi_{aM}\rangle$. The index $M$ denotes the (set of) *vibrational* quantum numbers. The molecular wave function is

$$\psi_M(r,R) = \sum_a \chi_{aM}(R)\,\phi_a(rR)\ . \tag{2.20}$$

By virtue of the expansion (2.20) it is clear that the vibrational quantum number $M$ in general is related to the *total* electronic spectrum and not to an individual electronic state.

### Born–Oppenheimer Approximation

Solving the coupled equations (2.18) for the expansion coefficients in Eq. (2.20) appears to be a formidable task. However, in practice it is often possible to neglect the nonadiabatic couplings altogether or take into account the couplings between certain adiabatic electronic states only. In order to investigate this possibility let us consider Fig. 2.2. Here we have plotted different adiabatic electronic states for a simple diatomic molecule as a function of the bond distance. Without going further into the details of the different states we realize that

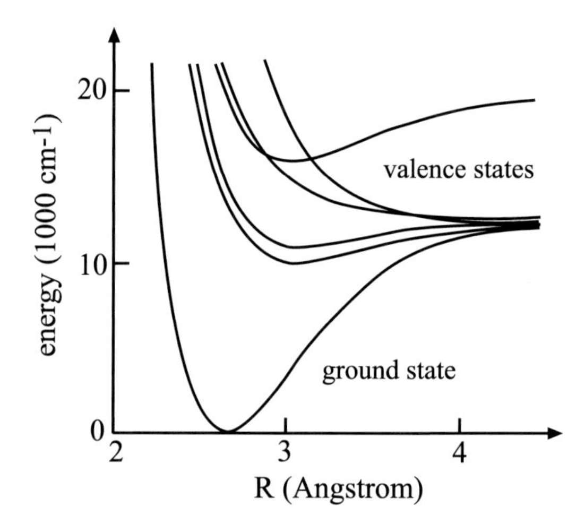


**Figure 2.2:** Potential energy curves $U_a(R)$ for different adiabatic electronic states $|\phi_a\rangle$ along the bond distance $R$ of a diatomic molecule (ground and valence states of $I_2$).

there is one state, the electronic ground state $|\phi_0\rangle$, which particularly close to its minimum is well separated from the other states $|\phi_{a>0}\rangle$. Intuitively we would expect the nonadiabatic couplings, $\Theta_{0a}$, to be rather small in this region. In such situations it might be well justified to neglect the nonadiabatic couplings, i.e., we can safely set $\Theta_{0a} = 0$ in Eq. (2.18). The nuclear Schrödinger equation then simplifies considerably. For $\Theta_{ab} = 0$ we have

$$H_a(R)\chi_a(R) = (T_{\text{nuc}} + U_a(R))\,\chi_a(R) = \mathcal{E}\chi_a(R)\ , \tag{2.21}$$

where $H_a(R)$ defines the *nuclear Hamiltonian* for the state $|\phi_a\rangle$. Thus, the nuclei can be considered to move in an effective potential $U_a(R)$ generated by their mutual Coulomb interaction and the interaction with the electronic charge distribution corresponding to the actual configuration $R$. The solutions of Eq. (2.21) are again labeled by $M$, but this quantum number is now related to the *individual* adiabatic electronic states. The total *adiabatic* wave function becomes

$$\psi_{aM}^{(\text{adia})}(r;R) = \chi_{aM}(R)\phi_a(r;R)\ . \tag{2.22}$$

The neglect of the nonadiabatic couplings leading to the wave function (2.22) is called the *Born–Oppenheimer approximation.*

Going back to Fig. 2.2 it is clear, however, that in particular for excited electronic states one might encounter situations where different potential curves are very close to each other. If $\Theta_{ab}$ does not vanish for symmetry reasons it can no longer be neglected. The physical picture is that electronic and nuclear motions are no longer adiabatically separable, i.e., the change of the nuclear configuration from $R$ to some $R + \Delta R$ causes an electronic transition.

In order to estimate this effect we consider a perturbation expansion of the energy with respect to the nonadiabaticity operator. The second–order correction to the adiabatic energies

$\mathcal{E}_{aM}^{(\text{adia})}$ is obtained as

$$\mathcal{E}_{aM}^{(2)} = \mathcal{E}_{aM}^{(\text{adia})} + \sum_{bN} \frac{|\langle \chi_{aM}|\Theta_{ab}|\chi_{bN}\rangle|^2}{\mathcal{E}_{aM}^{(\text{adia})} - \mathcal{E}_{bN}^{(\text{adia})}}, \tag{2.23}$$

where the $\chi_{aM}(R) = \langle R|\chi_{aM}\rangle$ are the Born–Oppenheimer nuclear wave functions. Apparently, the matrix elements $\langle \chi_{aM}|\Theta_{ab}|\chi_{bN}\rangle$ have to be small compared to the energy difference $|\mathcal{E}_{aM}^{(\text{adia})} - \mathcal{E}_{bN}^{(\text{adia})}|$ in order to validate the adiabatic Born–Oppenheimer approximation. Looking at the definition of $\Theta_{ab}$ is clear that this operator will be a small perturbation whenever the character of the electronic wave function does not change appreciably with $R$. On the other hand, the denominator in Eq. (2.23) will become small if two electronic states approach each other. Thus, knowledge about the adiabatic states is necessary to estimate the effect of nonadiabatic couplings. The actual calculation of $\Theta_{ab}$ is a rather complicated issue and an alternative representation will be discussed in Section 2.7.

**Some Estimates**

We complete our qualitative discussion by considering the dynamical aspect of the problem. For simplicity let us take a diatomic molecule in the vicinity of the potential minimum where the potential is harmonic, i.e., $U_a(R) = \kappa R^2/2$. Here $\kappa$ is the harmonic "spring" constant which is calculated from the second derivative of the potential with respect to $R$ (see below). The frequency of harmonic vibration is obtained from $\omega = \sqrt{\kappa/M_{\text{nuc}}}$, with $M_{\text{nuc}}$ being the reduced mass of the vibration. If $\langle v\rangle$ and $\langle \Delta R\rangle$ denote the average velocity and deviation from the minimum configuration, respectively, the virial theorem tells us that $M_{\text{nuc}}\langle v\rangle^2/2 = \kappa\langle \Delta R\rangle^2/2$. According to quantum mechanics this will also be proportional to $\hbar\omega = \hbar\sqrt{\kappa/M_{\text{nuc}}}$. Now consider the electrons: Let us assume that the most important contribution to the potential comes from the electrostatic electronic interaction. If $d_{\text{el}}$ is some typical length scale for the electronic system, for example, the radius of the electron cloud, its potential energy will be proportional to $e^2/d_{\text{el}}$. Further, the average electronic velocity is $\langle v_{\text{el}}\rangle \approx \hbar/m_{\text{el}}d_{\text{el}}$. Applying a reasoning similar to the virial theorem gives $e^2/d_{\text{el}} = \hbar^2/m_{\text{el}}d_{\text{el}}^2$ for the electronic subsystem. Eq. (2.21) tells us that the average electronic energy is of the order of the potential energy for nuclear motion $\hbar\sqrt{\kappa/M_{\text{nuc}}}$. This gives for the spring constant $\kappa \approx \hbar^2 M_{\text{nuc}}/m_{\text{el}}^2 d_{\text{el}}^4$. Using this result we obtain the relations

$$\frac{\langle v\rangle}{\langle v_{\text{el}}\rangle} \propto \left(\frac{m_{\text{el}}}{M_{\text{nuc}}}\right)^{3/4} \tag{2.24}$$

and

$$\frac{\langle \Delta R\rangle}{d_{\text{el}}} \propto \left(\frac{m_{\text{el}}}{M_{\text{nuc}}}\right)^{1/4}. \tag{2.25}$$

Since $m_{\text{el}} \ll M_{\text{nuc}}$ the nuclei move on average much slower than the electrons and explore a smaller region of the configuration space.

With $d_{\text{el}}$ and $\langle v_{\text{el}}\rangle$ at hand we can estimate the period for the bound electronic motion as $T_{\text{el}} \approx d_{\text{el}}/\langle v_{\text{el}}\rangle \approx m_{\text{el}}d_{\text{el}}^2/\hbar$. Thus, the average energy gap between electronic states is of the order of $\langle \Delta E \rangle_{\rm el} \approx \hbar/T_{\rm el} \approx \hbar^2/m_{\rm el}d_{\rm el}^2$. Comparing this result with the vibrational frequency given above we obtain

$$\frac{\langle \Delta E \rangle_{\rm el}}{\hbar\omega} \propto \sqrt{\frac{M_{\rm nuc}}{m_{\rm el}}} \,. \tag{2.26}$$

Thus, it is the large mass difference which makes the gap for vibrational transitions much smaller than for electronic transitions in the vicinity of a potential minimum. Therefore, the denominator in Eq. (2.23) is likely to be rather large and the second–order correction to the adiabatic energy becomes negligible in this case.

## 2.4 Electronic Structure Methods

Our knowledge about the microscopic origin of spectral properties of molecules, their stable configurations as well as their ability to break and make chemical bonds derives to a large extent from the progress made in electronic structure theory during the last decades. Nowadays modern quantum chemical methods routinely achieve almost quantitative agreement with experimental data, for example, for transition energies between the lowest electronic states of small and medium–size molecules. With increasing number of electrons the computational resources limit the applicability of the so–called ab initio (i.e., based on fundamental principles and not on experimental data) methods and alternatives have to be exploited. Semiempirical methods, such as the Hückel or the Pariser–Parr–Pople method, simplify the exact ab initio procedure in a way that gives results consistent with experimental data. On the other hand, recent developments in Density Functional Theory shift the attention to this more accurate method. Stepping to situations of molecules in the condensed phase, for example, in solution, requires more approximate methods as given, for example, by the reduction of the solvent to a dielectric continuum surrounding the solute[1] (see Section 2.5).

In the following we will outline a tool for the practical solution of the electronic Schrödinger equation (2.12) for fixed nuclei. For simplicity our discussion will mostly be restricted to the electronic ground state $E_0(R)$. Specifically, we will discuss the Hartree–Fock self–consistent field procedure in some detail. It is the working horse of most more advanced ab initio methods which also include the effect of electronic correlations missing in the Hartree–Fock approach. Whereas these methods are based on the electronic wavefunction, Density functional Theory (discussed afterwards) builds on the electron density function. We note in caution that this section by no means presents a complete treatment of the field of electronic structure theory. The intention is rather to provide a background for the following discussions. The reader interested in a more comprehensive overview of the state of the art is referred to the literature quoted at the end of the book.

Let us start with the situation in which the Coulomb interaction between the electrons is switched off. Then the electronic Hamiltonian (2.11) becomes a sum of *single–particle* Hamiltonians, $H_{\rm el}(R) = \sum_{j=1}^{N_{\rm el}} h_{\rm el}(\mathbf{r}_j)$, containing the kinetic energy of the $j$th electron and the Coulomb energy due to its interaction with the static nuclei. Note that we will drop the parametric dependence on the nuclear coordinates in the following. The stationary Schrödinger

[1] Throughout we will use the terms solute and solvent to describe a molecule (solute) embedded in a medium (solvent), no matter whether the latter is really a solvent in the usual sense or, for instance, a solid state matrix.

equation for $h_{\rm el}(\mathbf{r}_i)$ is solved by the single–particle wave function $\varphi_{\alpha_j}(\mathbf{r}_j, \sigma_j)$,

$$h_{\rm el}(\mathbf{r}_j)\varphi_{\alpha_j}(\mathbf{r}_j, \sigma_j) = [T_{\rm el}(j) + V_{\rm el-nuc}(\mathbf{r}_j)]\varphi_{\alpha_j}(\mathbf{r}_j, \sigma_j) = \epsilon_{\alpha_j}\varphi_{\alpha_j}(\mathbf{r}_j, \sigma_j) \,. \tag{2.27}$$

Here the index $\alpha_i$ runs over all possible single–particle states (including spin) of the $N_{\rm el}$–electron system which have the energy $\epsilon_{\alpha_j}$. The single–particle functions $\varphi_{\alpha_j}(\mathbf{r}_j, \sigma_j)$ are called *spin orbitals*.

There are several points to remark concerning the solutions of Eq. (2.27): First, since we are dealing with identical particles the single–particle spectrum $\epsilon_{\alpha_j}$ is the same for all electrons. Second, for the spin–independent Hamiltonian we use here, the spin function can be separated from the spatial orbital in the single–particle wave function according to $\varphi_{\alpha_j}(\mathbf{r}_j, \sigma_j) = \varphi_{a_j}(\mathbf{r}_j)\zeta_{a_j}(\sigma_j)$ and $\alpha_j = (a_j, \sigma_j)$. As mentioned above the orthogonal spin functions $\zeta_{a_j}(\sigma_j)$ describe spin–up or spin–down electrons. Therefore for $N_{\rm el}$ spatial orbitals $\varphi_{a_j}(\mathbf{r}_j)$ there will be $2N_{\rm el}$ possible spin orbitals $\varphi_{\alpha_j}(\mathbf{j}_j, \sigma_j)$. Thus, given $N_{\rm el}$ electrons, the electronic ground state would correspond to the situation where we fill in electrons in the different spin orbitals starting from the one with the lowest energy. Of course, we have to take care of the *Pauli principle*, i.e., each electron must have a distinct set of quantum numbers. In the present case this implies that each spatial orbital may be occupied by two electrons having spin up and spin down, respectively. The result of the distribution of electrons over the available spin orbitals is referred to as an *electronic configuration*.

Depending on whether there is an even number of electrons in the ground state (*closed shell* configuration) or an odd number (*open shell* configuration) all electrons will be paired or not, respectively. For simplicity we will focus in the following on the electronic ground state of closed shell systems only. Here $N_{\rm el}$ spin orbitals are occupied. One can further require the spatial orbitals to be identical for spin–up and spin–down electrons so that there will be $N_{\rm el}/2$ doubly occupied spatial orbitals in the ground state. Needless to say that the total spin of this many–electron system is zero. A closed shell situation is shown for the water molecule in Fig. 2.3.

The Pauli principle which we invoked above can be traced back to a fundamental property of the total wave function of a many–electron system. First, we observe that in contrast to classical mechanics, in quantum mechanics the electrons described by a wave function are not distinguishable. This means that the total probability distribution, $|\phi(r, \sigma)|^2$, should be invariant with respect to the exchange of any two particle indices. The permutation of the particle indices is conveniently written using a permutation operator $\mathcal{P}$ which, when acting on a many–particle wave function, exchanges the indices of any two particles. After the application of $\mathcal{P}$ the wave function can change at most by a constant factor $\xi$ (of modulus 1). Therefore, applying $\mathcal{P}$ twice one should recover the original wave function, i.e., we have $\xi^2 = 1$ or $\xi = \pm 1$. For spin $1/2$ particles like electrons it turns out that $\zeta = -1$ and therefore the total wave function has to be antisymmetric with respect to the exchange of any two electron indices.

If we go back to the single–particle spin orbitals defined by Eq. (2.27) it is clear now that even in the absence of the electron interaction the so–called *Hartree product* ansatz

$$\phi^{\rm HP}_{\{\alpha_j\}}(r, \sigma) = \prod_{j=1}^{N_{\rm el}} \varphi_{\alpha_j}(\mathbf{r}_j, \sigma_j) \tag{2.28}$$

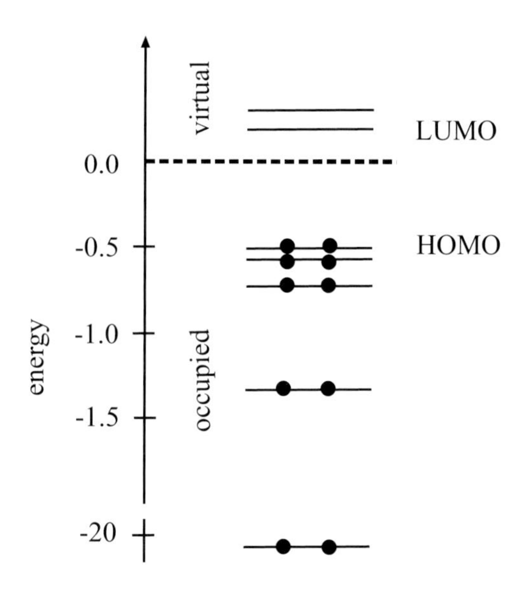


**Figure 2.3:** Orbital diagram for water calculated using Hartree–Fock theory (the energy is given in atomic units ($\hbar^2/(ea_0^2)$). There are $N_{\rm el}/2 = 5$ doubly occupied orbitals; the empty orbitals are called virtual. The highest occupied molecular orbital (HOMO) and the lowest unoccupied molecular orbital (LUMO) are assigned. Notice that Hartree–Fock theory predicts the LUMO energies to be positive implying that additional electrons cannot bind and the negative ion does not exist.

cannot be correct since it does not have the required antisymmetry ($\{\alpha_j\}$ comprises the set of quantum numbers $\alpha_j$). However, Eq. (2.28) can be used to generate an antisymmetric wave function. To this end we make use of the permutation operator $\mathcal{P}$. Keeping track of the number of permutations, $p$, which have been performed one obtains an antisymmetric wave function by the prescription

$$\phi(r,\sigma) = \frac{1}{\sqrt{N_{\rm el}!}} \sum_{\rm perm} (-1)^p \mathcal{P} \left[\phi^{\rm HP}_{\{\alpha_j\}}(r,\sigma)\right] . \tag{2.29}$$

Here the summation is carried out over all $N_{\rm el}!$ possible permutations of the electron indices $(\mathbf{r}_j, \sigma_j)$ ($j = 1, \ldots, N_{\rm el}$) in the Hartree product. Alternatively, Eq. (2.29) can be written in form of a determinant, the so–called *Slater determinant*, where the rows contain the single–particle spin orbitals for a given state and all possible electron coordinates and in the columns the different electronic states for a given coordinate are written down. The elementary properties of determinants then guarantee the antisymmetry of the ansatz for the total electronic wave function.

### The Hartree–Fock Equations

So far we have not considered the effect of the Coulomb interaction between the electrons. Within Hartree–Fock theory this is usually done by starting from the correct antisymmetric ansatz (2.29) for the wave function. Then the goal is to optimize the single–particle spin orbitals such that the total energy is minimized. This can be achieved by invoking the calculus of variation. Consider a Slater determinant $\phi(r,\sigma)$ which shall be a function of some parameters. In practice the spatial orbitals are expanded in terms of some fixed basis set and the expansion coefficients then take the role of the parameters. The basis set is usually chosen to consist of functions which are centered at the different atoms in the molecule (linear combination of atomic orbitals, LCAO).

The expectation value of the energy is then given by

$$\langle H_{\rm el}\rangle = \int dr \sum_{\sigma} \phi^*(r,\sigma) \left[ \sum_{j=1}^{N_{\rm el}} h_{\rm el}(\mathbf{r}_j) + \frac{1}{2} \sum_{i,j=1}^{N_{\rm el}} V_{\rm el-el}(\mathbf{r}_i,\mathbf{r}_j) \right] \phi(r,\sigma) \,. \qquad (2.30)$$

Here, the first term denotes the single–particle Hamiltonian including the electron–nuclei Coulomb interaction, Eq. (2.27), and the second term describes the electron–electron repulsion, Eq. (2.4). In Section 2.8.1 it is shown that variational optimization of Eq. (2.30) leads to the following so–called Hartree–Fock integro–differential equation for determination of the optimal orbitals for a closed shell configuration

$$\left[ h_{\rm el}(\mathbf{x}) + \sum_{b}^{N_{\rm el}/2} \left[2J_b(\mathbf{x}) - K_b(\mathbf{x})\right] \right] \varphi_a(\mathbf{x}) = \varepsilon_a \varphi_a(\mathbf{x}) \,. \qquad (2.31)$$

Here $\varepsilon_a$ is the energy associated with the spatial orbital $\varphi_a(\mathbf{x})$. Further the operator on the left hand side is called the *Fock operator*; it is an *effective* one–electron operator.

Without the electron–electron interaction and wavefunction antisymmetrization the Fock operator reduces to the single electron Hamiltonian, $h_{\rm el}(\mathbf{x})$. Different spatial orbitals are coupled by means of the Coulomb operator $J_b(\mathbf{x})$ (see Eq. (2.140)) and the exchange operator $K_b(\mathbf{x})$ (see Eq. (2.141)). The Coulomb operator represents the average local potential of an electron in orbital $\varphi_b(\mathbf{x})$ felt by the electron in $\varphi_a(\mathbf{x})$. Thus, the exact two–particle Coulomb interaction is replaced by an effective one–electron potential. The fact that each electron only sees the *mean field* generated by all other electrons is a basic characteristic of the Hartree–Fock approach. Of course, in this way the interaction between electrons becomes blurred and correlations between their individual motions are lost.

It has been discussed above that for electrons having parallel spins there is some particular correlation introduced by the antisymmetric ansatz for the wave function. This effect is contained in the exchange operator. However, the action of $K_b(\mathbf{x})$ on the orbital $\varphi_a(\mathbf{x})$ obviously cannot be viewed in terms of a local potential for the electron in $\varphi_a(\mathbf{x})$. In fact it is the exchange operator which makes the Fock operator *nonlocal* in space.

The Hartree–Fock equations are nonlinear, since the Fock–operator itself depends on the orbitals $\varphi_a(\mathbf{x})$. Hence the solution can only be obtained by iteration. Starting from some trial orbitals one first constructs the Fock operator and then uses it to obtain improved orbitals which are the input for a new Fock operator. This iterative procedure is continued until the potentials $J_a(\mathbf{x})$ and $K_a(\mathbf{x})$ are consistent with the solutions for the orbitals. Therefore the approach is usually termed Hartree–Fock *self–consistent field* method.

Given the solution of the Hartree–Fock equations one has at hand the ground state energy as well as the ground state adiabatic electronic wave function which follows from a *single* Slater determinant built up by the optimal orbitals. Both quantities are functions of the nuclear coordinates; by exploring possible nuclear configurations the ground state Hartree–Fock potential energy surfaces can be constructed according to Eq. (2.19). However, if, for instance, the bond in a diatomic molecule is stretched towards dissociation the character of the electronic state will change considerably, e.g., from a closed shell to an open shell system. This effect of having contributions from different electronic configurations cannot be described by a single Slater–determinant, Eq. (2.29); the predicted potential energy curve will be qualitatively incorrect. The effect of the simultaneous presence of different electronic configurations, which is also an ubiquitous phenomenon for electronically excited states in the region where potential curves intersect (cf. Fig. 2.2), is called static correlation. It has to be distinguished from dynamic correlations which are related to that part of the electron–electron interaction which is not accounted for by the mean–field approximation based on a single Slater–determinant.

Conceptually the simplest approach to account for such correlations is the configuration interaction (CI) method. Here one starts with the Hartree–Fock ground state and generates a basis for expanding the total electronic wavefunction by forming all possible Slater–determinants which result from promoting different numbers of electrons from the occupied to the unoccupied orbitals, i.e.,

$$|\phi^{(\mathrm{CI})}\rangle = C_0|\phi^{(0)}\rangle + C_1|\phi^{(1)}\rangle + C_2|\phi^{(2)}\rangle + \dots \,. \tag{2.32}$$

Here $|\phi^{(0)}\rangle$ stands for the Hartree–Fock ground state, $|\phi^{(1)}\rangle$ and $|\phi^{(2)}\rangle$ comprises all possible single and double excitations, respectively, starting from the ground state. The coefficients $C_i$ give the weight for these configurations. Upon diagonalization of the electronic Hamiltonian in this basis set the expansion coefficients are obtained and the problem of electron correlations is solved in principle. In practice the number of possible excitations increases rapidly[2] and the approach has to be restricted, for instance, to include at most double excitations. Several alternatives to the configuration interaction method have been developed and the reader is referred to the literature list at the end of the book for more details.

### Density Functional Theory

The methods discussed so far have been based on the electronic wavefunction, i.e., the Hartree–Fock ground state energy was assumed to be a functional of the wavefunction and variational minimization has been applied (cf. Section 2.8.1). A different strategy is followed in Density Functional Theory where the one–electron probability density

$$\rho(r) = N_{\mathrm{el}} \sum_{\sigma} \int d^3\mathbf{r}_2 \dots d^3\mathbf{r}_{N_{\mathrm{el}}} \; |\phi(\mathbf{r}, \mathbf{r}_2, \dots, \mathbf{r}_{N_{\mathrm{el}}}, \sigma)|^2 \tag{2.33}$$

is the central object of interest. The foundation of Density Functional Theory is laid by the Hohenberg–Kohn theorems. They state that for a given electron–nuclear interaction potential[3] the full many–particle ground state energy, $E_0$, is a unique functional of the electronic density

[2] Given $M$ spin–orbitals there are $\begin{pmatrix} M \\ N_{\mathrm{el}} \end{pmatrix}$ possibilities for the distribution of $N_{\mathrm{el}}$ electrons.

[3] In fact the first Hohenberg–Kohn theorem holds for an arbitrary external potential for the electron motion.

and that any density $\rho(r)$ other than the ground state density $\rho_0(r)$ will give an energy higher than the ground state energy, i.e., $E[\rho] \geq E[\rho_0] \equiv E_0$ implying that a variational principle can be applied.

The energy functional can be decomposed as follows

$$E[\rho] = \int dr\, V_{\text{el-nuc}}(r)\, \rho(r) + T_{\text{el}}[\rho] + \frac{1}{2}\int dr dr' \,\frac{\rho(r)\rho(r')}{|r-r'|} + E_{\text{XC}}[\rho]\,. \tag{2.34}$$

The different terms correspond to the interaction between electrons and nuclei, the kinetic energy of the electrons[4] , the classical electron–electron interaction energy, and the non–classical contribution from the electron–electron interaction due to exchange and correlation effects. It should be noted that apart from the first term all contributions to the energy functional (2.34) are universal, i.e., not molecule specific. They are comprised in what is called the Hohenberg–Kohn functional and depend only on the properties of the electronic degrees of freedom.

The practical calculation of the electron density starts from the variational principle. Here the stationarity condition for the energy $\delta E[\rho]/\delta\rho = 0$ has to be fulfilled subject to the constraint that the system must contain a fixed number of electrons. This leads to the so–called Kohn–Sham equations

$$\left[T_{\text{el}} + V_{\text{el-nuc}} + \int dr' \,\frac{\rho(r')}{|r-r'|} + V_{\text{XC}}(r)\right]\varphi_a^{\text{KS}}(r) = \varepsilon_a^{\text{KS}}\varphi_a^{\text{KS}}(r)\,. \tag{2.35}$$

From this equation the Kohn–Sham orbitals $\varphi_a^{\text{KS}}(r)$, and thus the electron density $\rho(r) = \sum_a |\varphi_a^{\text{KS}}(r)|^2$, as well as the respective orbital energies $\varepsilon_a^{\text{KS}}$ are determined in a self–consistent manner. Apart from the exchange–correlation potential, here $V_{\text{XC}}(r) = \delta E_{\text{XC}}[\rho(r)]/\delta\rho(r)$, Eq. (2.35) resembles the Hartree–Fock equations (2.31). However, it is important to emphasize that upon adding $V_{\text{XC}}$ the Kohn–Sham equations become formally exact and they are still local in space. This has to be contrasted with the Hartree–Fock equations where the exchange operator introduces a nonlocal spatial dependence of the orbitals.

But, unfortunately the form of the exchange–correlation functional is not specified by the Hohenberg–Kohn theorems and in fact it is not known. In practice this problem is approached by developing approximate functionals which may incorporate sum–rules, asymptotic properties of the electron density, information from approximations to the electron density, and fits to exact numerical results available for some test systems. A simple form for the exchange–correlation energy is given, e.g., by the so–called local density approximation

$$E_{\text{XC}}^{\text{LDA}}[\rho] = \int dr\, \rho(r)\, \varepsilon_{\text{XC}}[\rho(r)] \tag{2.36}$$

where $\varepsilon_{\text{XC}}[\rho(r)]$ is the known exchange–correlation energy per particle for a homogeneous electron gas moving on a positive background charge density. This model works rather well, e.g., for perfect metals. For more complicated functionals we refer the reader to the literature list given at the end of this book.

[4]Note that $T_{\text{el}}[\rho]$ refers to the kinetic energy of some noninteracting reference system which has the some density as the real system. The difference between real and reference kinetic energy is assumed to be part of the unknown exchange–correlation energy.

Despite this fundamental deficiency of an unknown $E_{\rm XC}[\rho]$, in practical applications modern Density Functional Theory often outperforms the Hartree–Fock method, e.g., when predicting barrier heights for chemical reactions, because it includes correlation effects at least approximately. Compared to high level wavefunction based methods it is numerically much less expensive making it a tool for studying larger molecules.

## 2.5 Dielectric Continuum Model

In the previous sections we have been concerned with the electronic structure of polyatomic molecules and their parametric dependence on the positions of the nuclei. The numerical effort for calculating ground state energies or reaction surfaces clearly prohibits an application to systems of hundreds of interacting molecules or to macroscopic systems such as molecules in solution.

A straightforward but approximate solution of this problem is the inclusion of a few solvent molecules or if possible even the first solvation shell into the quantum chemical calculation. This so–called supermolecule approach has the advantage that short–range interactions between solute and solvent molecules are reasonably accounted for. Thus, one can learn about the local structure of the solvent around the solute. Such a treatment is necessary, for instance, to describe the formation of hydrogen bonds which may occur if the solvent is water.

The long–range electrostatic interactions are, of course, not included in the supermolecule approach. They are, however, accounted for in the so–called continuum models which are in turn applicable whenever short–range interactions are negligible. The model implies that we discard the discrete nature of the solvent and treat it as a homogeneous entity fully characterized by its macroscopic properties. This approach will be discussed in the present section. Indeed, it is flexible enough to accommodate the supermolecule approach yielding a mixed description which may distinguish between the first solvation shell and the rest of the solvent.[5]

In the next section we give a brief summary of some concepts of classical electrostatics. The selection shall provide a background for the reaction field approach discussed in the Section 2.5.2 as well as for the elaboration of electron transfer theory in Chapter 6.

### 2.5.1 Medium Electrostatics

Consider a solvent in a container whose dimension is such that effects due to the walls can be neglected. If there are no free charges the solvent is a *dielectric*. The $m$th solvent molecule can be characterized by its charge density distribution $\rho_m(\mathbf{x})$. Using the definition of Section 2.2 the classical expression for $\rho_m(\mathbf{x})$ reads

$$\rho_m(\mathbf{x}) = -e \sum_{j=1}^{N_m^{(\rm el)}} \delta(\mathbf{r}_j(m) - \mathbf{x}) + e \sum_{n=1}^{N_m^{(\rm nuc)}} z_n(m)\delta(\mathbf{R}_n(m) - \mathbf{x}) \,, \tag{2.37}$$

[5]Another strategy is followed in the so–called Quantum Mechanics/Molecular Mechanics (QM/MM) approach, where a quantum chemical calculation of the solute is combined with point charges resulting from a classical but atomistic treatment of the environment.

where the additional index $m$ is used to label the respective molecule here and in the following.[6]

The stationary version of Maxwell's equations $\nabla \mathbf{E}(\mathbf{x}) = 4\pi\rho(\mathbf{x})$ and $\nabla \times \mathbf{E}(\mathbf{x}) = 0$ enables one to compute the electric field $\mathbf{E}(\mathbf{x})$ induced by the complete molecular charge distribution

$$\rho(\mathbf{x}) = \sum_m \rho_m(\mathbf{x}) . \tag{2.38}$$

The field is related to the scalar potential by $\mathbf{E}(\mathbf{x}) = -\nabla\Phi(\mathbf{x})$. The scalar potential can be obtained from the Poisson equation $\Delta\Phi(\mathbf{x}) = -4\pi\rho(\mathbf{x})$, which gives as

$$\Phi(\mathbf{x}) = \int d^3\mathbf{x}' \frac{\rho(\mathbf{x}')}{|\mathbf{x} - \mathbf{x}'|} . \tag{2.39}$$

Often the complete information on the microscopic electric field contained in these expressions is of little practical use. In many experiments one is only interested in macroscopic quantities which are *averaged* with respect to their microscopic contributions. This averaging is equivalent to the elimination of the short–range part of the field from all expressions.

In order to explore this point further let us assume that we have divided the macroscopic probe volume into smaller volumes $\Delta V(\mathbf{x}_s)$ which still contain a large number of molecules. Here $\mathbf{x}_s$ is a vector pointing to the $s$th small volume (see Fig. 2.4). Replacing the total integration of Eq. (2.39) by integrations with respect to the $\Delta V(\mathbf{x}_s)$ we get

$$\Phi(\mathbf{x}) = \sum_s \int\limits_{\Delta V(\mathbf{x}_s)} d^3\mathbf{x}' \frac{\rho(\mathbf{x}')}{|\mathbf{x} - \mathbf{x}'|} . \tag{2.40}$$

We are only interested in the long–range contributions of the charges located in $\Delta V(\mathbf{x}_s)$ to the potential. Therefore, we take $\mathbf{x}$ to be far away from $\mathbf{x}_s$ such that $|\mathbf{x} - \mathbf{x}_s| \gg |\mathbf{x}' - \mathbf{x}_s|$. This inequality enables us to expand the factor $|\mathbf{x} - \mathbf{x}'|^{-1}$ into a Taylor series with respect to $\mathbf{x}' - \mathbf{x}_s$. Keeping only the first two terms we get

$$\frac{1}{|\mathbf{x} - \mathbf{x}_s - (\mathbf{x}' - \mathbf{x}_s)|} \approx \frac{1}{|\mathbf{x} - \mathbf{x}_s|} - (\mathbf{x}' - \mathbf{x}_s)\nabla_{\mathbf{x}} \frac{1}{|\mathbf{x} - \mathbf{x}_s|} . \tag{2.41}$$

Inserting this into Eq. (2.40) one obtains the first two contributions of the so–called *multipole expansion* of $\Phi(\mathbf{x}_s)$. The monopole term

$$\Phi_{\text{mp}}(\mathbf{x}) = \sum_s \frac{1}{|\mathbf{x} - \mathbf{x}_s|} \int\limits_{\Delta V(\mathbf{x}_s)} d^3\mathbf{x}' \rho(\mathbf{x}') \tag{2.42}$$

corresponds to the potential of a point charge located at $\mathbf{x} = \mathbf{x}_s$. If there is no net charge in $\Delta V(\mathbf{x}_s)$ this contribution vanishes. Introducing the *dipole moment* of $\Delta V(\mathbf{x}_s)$ as

$$\mathbf{d}_s = \int\limits_{\Delta V(\mathbf{x}_s)} d^3\mathbf{x}' (\mathbf{x}' - \mathbf{x}_s)\rho(\mathbf{x}') \tag{2.43}$$

[6]If the operator of the charge density is needed, the electronic and nuclear coordinates in Eq. (2.37) have to be understood as quantum mechanical operators. To obtain in this case the charge density which enters Maxwell's equations one has to take the expectation value of the charge density operator with respect to the molecular wave function.

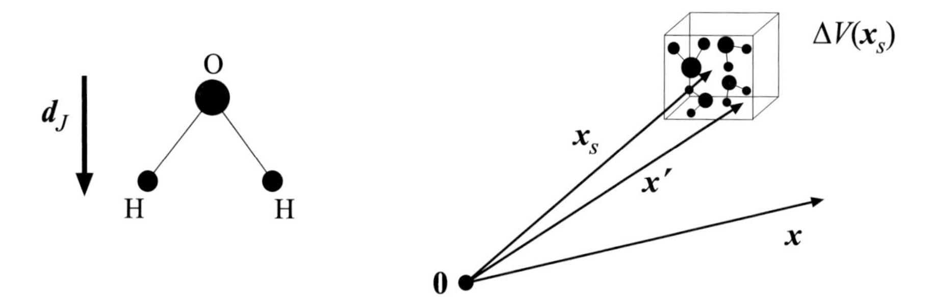


**Figure 2.4:** Dipole moment of $H_2O$ (left). Macroscopic electrostatic quantities are obtained by averaging over the volume elements $\Delta V(\mathbf{x}_s)$. The $\Delta V(\mathbf{x}_s)$ contain a large number of individual molecules but have a dimension small enough to neglect the discrete nature of the vector $\mathbf{x}_s$ pointing to it (right). The vector $\mathbf{x}$ should be positioned far away from $\mathbf{x}_s$.

the second term in the above expansion can be written as

$$\Phi_{\rm dp}(\mathbf{x}) = \sum_s \mathbf{d}_s \frac{\mathbf{x} - \mathbf{x}_s}{|\mathbf{x} - \mathbf{x}_s|^3} . \tag{2.44}$$

The dipole moment is the quantity we will be concerned with in the following discussion of dielectric media. In the spirit of the Taylor expansion (2.41) the contribution of higher–order multipole moments is usually small compared to the dipole term. An important exception occurs if the dipole moment vanishes for symmetry reasons.

The dipole moment of the small volume element $\mathbf{d}_s$ can of course be traced to the individual molecular dipole moments. We have

$$\mathbf{d}_s = \sum_{m \in \Delta V(\mathbf{x}_s)} \mathbf{d}_m \tag{2.45}$$

with

$$\mathbf{d}_m = \int d^3\mathbf{x}' \rho_m(\mathbf{x}')\mathbf{x}' . \tag{2.46}$$

Apparently, whether a molecule has a permanent dipole moment or not is determined by its symmetry. Systems like $CCl_4$ or diatomics like $H_2$ or $N_2$ do not have a permanent dipole; the dielectric is nonpolar. However, application of an external field can lead to a distortion of the molecular charge density and in this way induce a dipole moment. On the other hand, $H_2O$ or $NH_3$, for instance, do have a permanent dipole and form polar dielectrics, (see Fig. 2.4).

For the description of the behavior of the dielectric in some external field, for example, it is customary to introduce the dipole density or the *polarization* which is defined as

$$\mathbf{P}(\mathbf{x}_s) = \frac{\mathbf{d}_s}{\Delta V(\mathbf{x}_s)} . \tag{2.47}$$

Suppose that the discrete nature of our subdivision into the $\Delta V(\mathbf{x}_s)$ can be neglected. Then, $\mathbf{x}_s$ becomes a continuous quantity and we can write the macroscopic potential in dipole approximation and under the assumption of charge neutrality as

$$\begin{aligned}\Phi_{\mathrm{mac}}(\mathbf{x}) &= \sum_s \mathbf{d}_s \frac{\mathbf{x}-\mathbf{x}_s}{|\mathbf{x}-\mathbf{x}_s|^3} \approx \int d^3\mathbf{x}'\,\mathbf{P}(\mathbf{x}')\frac{\mathbf{x}-\mathbf{x}'}{|\mathbf{x}-\mathbf{x}'|^3} \\ &= \int d^3\mathbf{x}'\,\mathbf{P}(\mathbf{x}')\nabla_{\mathbf{x}'}\frac{1}{|\mathbf{x}-\mathbf{x}'|} = -\int d^3\mathbf{x}'\,\frac{\nabla_{\mathbf{x}'}\mathbf{P}(\mathbf{x}')}{|\mathbf{x}-\mathbf{x}'|}. \end{aligned} \tag{2.48}$$

Here the integration is with respect to the entire probe volume. Furthermore the last line has been obtained by making use of the Gauss theorem.

Comparison of this expression with Eq. (2.40) suggests the interpretation of $-\nabla\mathbf{P}$ as a charge density. Specifically, we can define the (macroscopic) dipole density

$$\rho_{\mathrm{P}}(\mathbf{x}) = -\nabla\mathbf{P}(\mathbf{x}) . \tag{2.49}$$

Besides the polarization charge density an additional externally controlled charge density $\rho_{\mathrm{ext}}$ may be present. By this we mean, for example, the charge density introduced in a dielectric if a solute molecule is placed into it (see below). Note that we are only interested in the long–range contribution of the solute to the electric field. The equation for the macroscopic electric field in the medium is then given by

$$\nabla\mathbf{E}_{\mathrm{mac}}(\mathbf{x}) = 4\pi(\rho_{\mathrm{ext}}(\mathbf{x}) + \rho_{\mathrm{P}}(\mathbf{x})) . \tag{2.50}$$

Defining the dielectric displacement vector as

$$\mathbf{D} = \mathbf{E}_{\mathrm{mac}} + 4\pi\mathbf{P} , \tag{2.51}$$

the macroscopic source equation becomes

$$\nabla\mathbf{D}(\mathbf{x}) = 4\pi\rho_{\mathrm{ext}}(\mathbf{x}) . \tag{2.52}$$

According to this relation the dielectric displacement field can be interpreted as the *external* field.

So far we discussed how a given charge distribution of the medium results in an electric field. But one can also ask the question how an external field leads to a change of the medium charge distribution. Within the present approach the answer to this question is that the polarization of the medium will be a complicated functional of the electric field, $\mathbf{P} = \mathbf{P}(\mathbf{E}_{\mathrm{mac}})$. If we assume that the perturbation of the medium due to the electric field is weak, a Taylor expansion of the polarization in terms of $\mathbf{E}_{\mathrm{mac}}$ is justified. In linear approximation the relation between the electric field and the polarization is expressed in terms of the so–called linear *susceptibility* $\chi$ as

$$\mathbf{P}(\mathbf{x}) = \chi\mathbf{E}_{\mathrm{mac}}(\mathbf{x}) . \tag{2.53}$$

Here we assumed that the medium is homogeneous and isotropic. In general, however, the susceptibility is a tensor, i.e., the vectors of the polarization and the electric field do not have to be parallel. Further, for an inhomogeneous medium the relation between polarization and electric field may be nonlocal in space. One can introduce the dielectric constant

$$\varepsilon = 1 + 4\pi\chi \tag{2.54}$$

and write

$$\mathbf{E}_{\mathrm{mac}}(\mathbf{x}) = \varepsilon^{-1}\mathbf{D}(\mathbf{x}) \,. \tag{2.55}$$

This expression shows that the total macroscopic field $\mathbf{E}_{\mathrm{mac}}(\mathbf{x})$ results from the response of the medium to the external field. The response properties of the medium are contained in the inverse dielectric function.

Finally, we give the expression for the potential energy of a charge distribution:

$$W = \frac{1}{2}\int d^3\mathbf{x}\, \rho(\mathbf{x})\Phi(\mathbf{x}) \,. \tag{2.56}$$

### 2.5.2 Reaction Field Model

In this section we address the influence a continuously distributed solvent has on the solute's electronic properties. In principle we expect the following behavior: The solute's electrons and nuclei feel the charge of the solvent molecules and vice versa. As a result, the charge distribution in the solute changes and, consequently, its electronic spectrum. But at the same time the charge distribution of the solvent is rearranged too.

In the following this situation will be discussed using a model where the solute is treated by ab initio quantum chemistry and the solvent enters through its macroscopic dielectric properties. The solute is supposed to reside inside a *cavity* ($V_{\mathrm{cav}}$ with dielectric constant equal to one (vacuum)) within the dielectric ($V_{\mathrm{sol}}$). We will assume for simplicity that the solvent is homogeneous and isotropic, i.e., we can characterize it by a dielectric constant $\varepsilon_{\mathrm{sol}}$. This neglects, for instance, effects coming from a locally inhomogeneous distribution of the solvent molecules in the first solvation shell.

The first important step is the definition of the size and the shape of the cavity. Various cavity shapes are possible which should in the ideal case give a reasonable approximation to the molecular charge distribution. The simplest and most approximate model is that of a spherical cavity. More elaborate calculations could be based, for instance, on the union of overlapping spheres centered at the different nuclei (see Fig. 2.5). The size of the cavity should is also an important parameter. In particular one must be aware that serious errors can be expected if the cavity size is too small to accommodate most of the charge distribution as described by the molecular wave function. Thus, we will assume that the solute's charge distribution, $\rho_{\mathrm{mol}}(\mathbf{x})$, is confined inside $V_{\mathrm{cav}}$.

Provided that the molecular charge distribution $\rho_{\mathrm{mol}}(\mathbf{x})$ is given and the size and the shape of the cavity have been defined, we still have to account for the coupling between solvent and solute. In the spirit of the dielectric continuum description of the solvent the exact microscopic Coulomb interaction (cf. Eq. (2.1)) is approximated by the respective expressions for a dielectric discussed in the previous section.

We first note that the solute's charge distribution generates an electrostatic potential which is obtained from $\Delta\Phi(\mathbf{x}) = -4\pi\rho_{\mathrm{mol}}(\mathbf{x})$ inside the cavity and from $\Delta\Phi(\mathbf{x}) = 0$ within $V_{\mathrm{sol}}$.

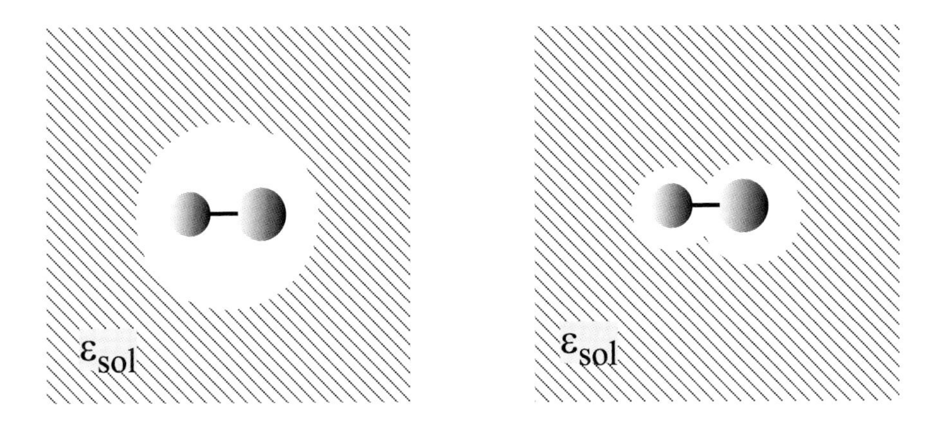


**Figure 2.5:** Different shapes of the cavity for accommodation of the solute molecule within the dielectric continuum model.

The boundary conditions at the cavity surface are given by $\Phi(\mathbf{x} \in V_{\rm cav}) = \Phi(\mathbf{x} \in V_{\rm sol})$ and $\partial\Phi(\mathbf{x} \in V_{\rm cav})/\partial\mathbf{n} = \varepsilon_{\rm sol}\partial\Phi(\mathbf{x} \in V_{\rm sol})/\partial\mathbf{n}$. Here, $\mathbf{n}$ is a unit vector on cavity surface pointing outwards.

The potential of the solute's charge density induces a polarization of the dielectric. This polarization gives rise to a potential $\Phi_{\rm pol}(\mathbf{x})$. In the present case $\Phi_{\rm pol}(\mathbf{x})$ depends on the polarization charge densities which are induced at the cavity surface. The total electrostatic potential inside $V_{\rm cav}$ is therefore $\Phi_{\rm pol}(\mathbf{x}) + \Phi(\mathbf{x})$. According to Eq. (2.56) we can calculate the interaction energy (polarization energy) between the solute's charge distribution and the induced so–called *reaction field* as follows

$$W_{\rm pol} = \frac{1}{2}\int d^3\mathbf{x}\,\rho_{\rm mol}(\mathbf{x})\Phi_{\rm pol}(\mathbf{x}) \tag{2.57}$$

In a next step the electrostatic problem has to be linked to the quantum mechanical treatment of the solute molecule. This is straightforwardly done by replacing the discrete classical charge distribution in Eq. (2.57) by the quantum mechanical expectation value of the respective charge density operator: $\rho_{\rm mol}(\mathbf{x}) \to \langle\hat{\rho}_{\rm mol}(\mathbf{x})\rangle$. It is customary to stay with a classical description of the nuclei such that $\langle\hat{\rho}_{\rm mol}(\mathbf{x})\rangle = \rho_{\rm nuc}(\mathbf{x}) + \langle\hat{\rho}_{\rm el}(\mathbf{x})\rangle$. Here the nuclear and the electronic parts are given by the second term in Eq. (2.8) and the first term in Eq. (2.9), respectively. Using the Born–Oppenheimer separation of electronic and nuclear motions the averaging in (2.9) is performed with respect to the adiabatic electronic states for a fixed nuclear configuration. In order to incorporate the effect of the continuous dielectric on the solute's electronic properties we have to interpret $W_{\rm pol}$ as the expectation value of the single–particle operator

$$\hat{V}_{\rm int} = \frac{1}{2}\int d^3\mathbf{x}\hat{\rho}_{\rm mol}(\mathbf{x})\Phi_{\rm pol}(\mathbf{x})\ . \tag{2.58}$$

Within Hartree–Fock theory this operator is simply added to the single–particle Hamiltonian in the Fock operator in Eq. (2.31).

At this point it is important to notice that $\Phi_{\rm pol}(\mathbf{x})$ itself depends on the molecular charge distribution. This makes the determination of the electronic states of the solute a nonlinear problem which has to be solved *iteratively*: Starting from some initial guess for the reaction field potential one first calculates the charge distribution of the molecule. The resulting potential is then used to generate a new $\Phi_{\rm pol}(\mathbf{x})$. This procedure is repeated until some convergence criteria are fulfilled. Finally, one obtains the electronic energies and the respective wave functions for the molecule inside the cavity.

The reaction field method has found various applications. In particular one is frequently interested in knowing the energy required to adjust the solvent molecules (in the present case their dipole moments) in response to the introduction of a solute (solvation energy, cf. Section 6.5.2 ). This solvation energy, for example, often is responsible for the stabilization of certain isomers of the solute.

In preparation of the next chapter we point out that the reaction field approach has also a dynamical aspect. In order to appreciate this we have to recall that it is the (quantum mechanically) *averaged* charge distribution of the solute which is "seen" by the solvent. The time scale for electronic motion is typically of the order of $10^{-15}$–$10^{-16}$ s. Thus, for the solvent to experience only the *mean field* due to the solute's electrons, it is necessary to assume that the time scale required for building up a polarization in the solvent is much longer than that of the electronic motion. If we consider, for example, rotational motion of the solvent molecules on a time scale of about $10^{-12}$ s (orientational polarization), this reasoning is certainly valid. However, if the polarization is of electronic character the description in terms of a static dielectric constant is likely to fail.

## 2.6 Potential Energy Surfaces

In the previous sections it has already been indicated that the potential energy hypersurface defined by Eq. (2.19) is the key quantity when it comes to investigate chemical reaction dynamics or more generally nuclear motions. In the following we will consider some properties of the adiabatic Born–Oppenheimer PES ($\Theta_{ab} = 0$) for a particular electronic state,

$$U_a(R) = E_a(R) + V_{\rm nuc-nuc}(R) \,. \tag{2.59}$$

In general $U_a(R)$ is a function of all the $3N_{\rm nuc}$ nuclear coordinates $R$ (remember the notation $R = (R_1, \ldots, R_{3N_{\rm nuc}})$). Since the energy is independent of overall translations and rotations of the molecule, there are actually only $3N_{\rm nuc} - 6$ coordinates necessary to completely specify the energy of the molecule in the configuration space of the nuclear coordinates (for linear molecules there are only $3N_{\rm nuc}$-5 independent coordinates).

Let us assume for the moment that we have obtained $U_a(R)$. Then we are in the position to draw several conclusions, for example, on the nature of the bonding as well as on the dynamical behavior to be expected in the considered system. To this end we define the gradient of the potential as

$$\nabla U_a(R) = \{\partial U_a(R)/\partial R_1, \ldots, \partial U_a(R)/\partial R_{3N_{\rm nuc}}\} \,. \tag{2.60}$$

This vector points along the direction of the steepest rise of the potential and its negative is just the force acting along that particular direction in configuration space. Another quantity

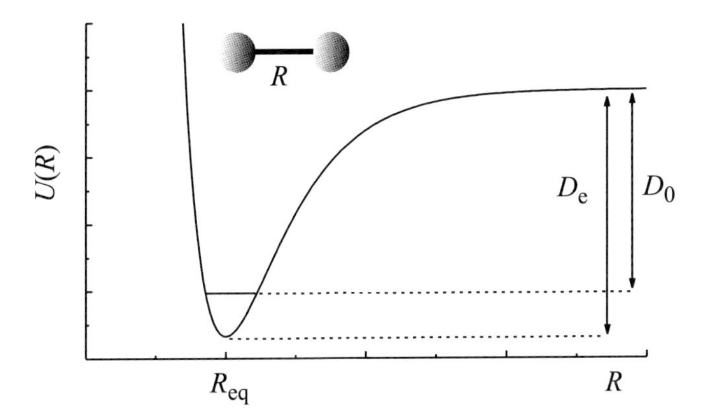


**Figure 2.6:** Schematic view of a typical potential energy curve of a diatomic molecule. Here, $R_{\text{eq}}$ denotes the equilibrium bond length and $D_0$ ($D_e$) the dissociation energy which does (does not) take into account the quantum mechanical zero–point energy.

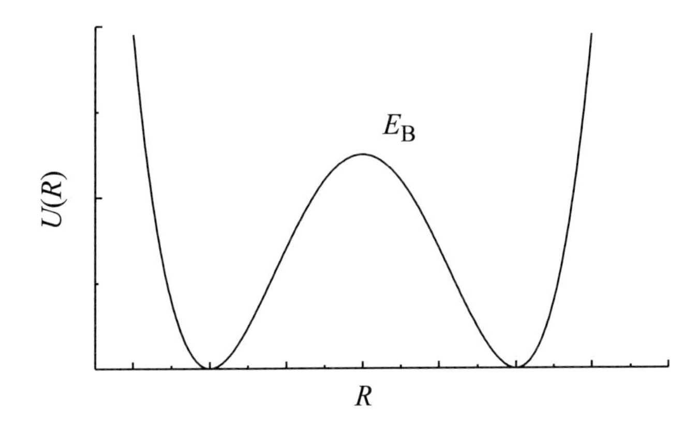


**Figure 2.7:** Schematic view of a potential energy curve typical for isomerization reactions. Reactants and products are separated by a reaction barrier of height $E_{\text{B}}$ along the reaction coordinate $R$.

of great importance is the $3N_{\text{nuc}} \times 3N_{\text{nuc}}$ force constant matrix or *Hessian matrix* whose elements are defined as

$$\kappa_{mn}^{(a)} = \frac{\partial^2 U_a(R)}{\partial R_m \partial R_n} \quad (m, n = 1, \ldots, 3N_{\text{nuc}}) \ . \tag{2.61}$$

The points in configuration space for which the gradient of the potential vanishes,

$$\nabla U_a(R) = 0 \ , \tag{2.62}$$

are called *stationary* points. Suppose we have located a stationary point at the the equilibrium configuration $R^{(a)}$. The nature of the potential energy surface in the vicinity of this stationary

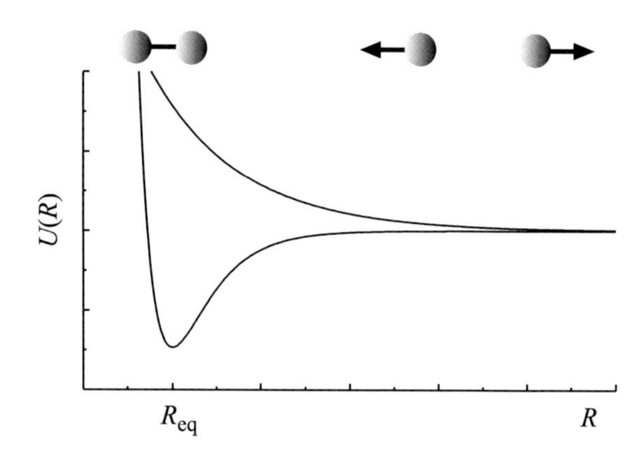


**Figure 2.8:** Schematic view of a typical ground and excited state potential energy curve of diatomic molecules. If the molecule is promoted to the excited state, for example, by means of an external field, dissociation will occur.

point can then be investigated by looking at the eigenvalues of the Hessian matrix. In general there will be 6 eigenvalues equal to zero reflecting the fact that there are only $3N_{\rm nuc}-6$ independent coordinates necessary to determine the energy (see below). If the remaining eigenvalues of the Hessian matrix are all positive we are at a minimum of the potential energy surface. In Fig. 2.6 this situation is plotted for a diatomic molecule where $R$ is the bond length. The minimum of $U(R)$ at $R = R_{\rm eq}$ gives the equilibrium distance between the two atoms. As a consequence of quantum mechanical zero–point motion the lowest possible energy eigenvalue is above the bottom of the potential minimum (solid line in Fig. 2.6). The molecule is said to be stable if the difference between this zero–point energy and the energy it takes to separate the atoms, $U(R \to \infty)$, is finite (dissociation energy, $D_0$ in Fig. 2.6).

Fig. 2.6 corresponds to the situation where $U(R)$ only has a single global minimum. In fact there are many systems which support multiple minima in the potential energy landscape. In Fig. 2.7 we have plotted a potential showing two equivalent minima. These minima in $U(R)$ correspond to different isomers of the molecule. Such situations occur, for example, in systems showing intramolecular hydrogen transfer. Another standard example is the umbrella vibration of $NH_3$. In the course of isomerization the system has to pass a maximum of the potential curve which corresponds to a saddle point of $U(R)$. At such a simple saddle point the Hessian matrix will have one negative eigenvalue.

Finally, we consider a case one typically encounters in excited states. In Fig. 2.8 we plotted potential energy curves for the adiabatic ground and excited states of a diatomic molecule. Apparently, the excited state potential has no minimum. This implies that an electronically excited molecule will experience a force, $-\partial U/\partial R$, leading to dissociation as indicated in the figure.

For larger molecules it is no longer possible to plot the potential energy as a function of all coordinates. It goes without saying that in addition the calculation of these PES becomes computationally very demanding. Fortunately, quite often one has to deal with situations where only few coordinates are important for a reaction. Then it becomes possible to describe this reaction by taking into account only the motion along a single so–called *reaction coordinate* while keeping the remaining coordinates fixed a their equilibrium positions. Consider, for instance, the dissociation of the A–B bond of a triatomic molecule ABC. If the internal excitation of the BC fragment during the cleavage of the A–B bond is negligible, BC can be treated as an entity characterized by its center of mass. Before discussing more advanced concepts (applicable for polyatomics) in Section 2.6.3, we will focus on the nuclear dynamics in the vicinity of stationary points.

### 2.6.1 Harmonic Approximation and Normal Mode Analysis

After having discussed some general aspects of adiabatic potential energy surfaces we turn to the problem of nuclear dynamics. Let us assume that we have located a stationary point $R^{(a)}$ in configuration space corresponding to a global minimum of $U_a(R)$. Restricting our discussion to small deviations, $\Delta R_n^{(a)} = R_n^{(a)} - R_n$ $(n = 1, \ldots, 3N_{\text{nuc}})$, from the stationary point the potential can be approximated by a second order Taylor expansion with respect to $R^{(a)}$,

$$U_a(R) = U_a(R^{(a)}) + \sum_{m,n=1}^{3N_{\text{nuc}}} \frac{1}{2} \kappa_{mn}^{(a)} \Delta R_m^{(a)} \Delta R_n^{(a)} . \tag{2.63}$$

Here, the Hessian matrix has to be taken at the point $R^{(a)}$. Note that at the stationary point the first derivatives vanish because of the condition (2.62). According to Eq. (2.21) the Hamiltonian for the nuclear degrees of freedom in the adiabatic approximation reads

$$H_a = U_a(R^{(a)}) + \sum_{n=1}^{3N_{\text{nuc}}} \frac{P_n^2}{2M_n} + \sum_{m,n=1}^{3N_{\text{nuc}}} \frac{1}{2} \kappa_{mn}^{(a)} \Delta R_m^{(a)} \Delta R_n^{(a)} . \tag{2.64}$$

The linear transformation

$$\Delta R_n^{(a)} = \sum_{\xi} M_n^{-1/2} A_{n\xi}^{(a)} q_{a,\xi} \tag{2.65}$$

can be used to diagonalize the kinetic and potential energy operators. Expressed in the so–called *normal mode coordinates* $q_{a,\xi}$, Eq. (2.64) becomes (note that the normal mode coordinates are mass–weighted)

$$H_a = U_a(q_{a,\xi} = 0) + H_a^{(\text{nm})} \tag{2.66}$$

with the normal mode Hamiltonian defined as

$$H_a^{(\text{nm})} = \frac{1}{2} \sum_{\xi} \left( p_\xi^2 + \omega_{a,\xi}^2 q_{a,\xi}^2 \right) . \tag{2.67}$$

Here the normal mode frequencies $\omega_{a,\xi}$ have been introduced.

The nuclear motions according to Eq. (2.67) can be understood as a superposition of independent harmonic vibrations around the equilibrium configuration $R^{(a)}$ which corresponds

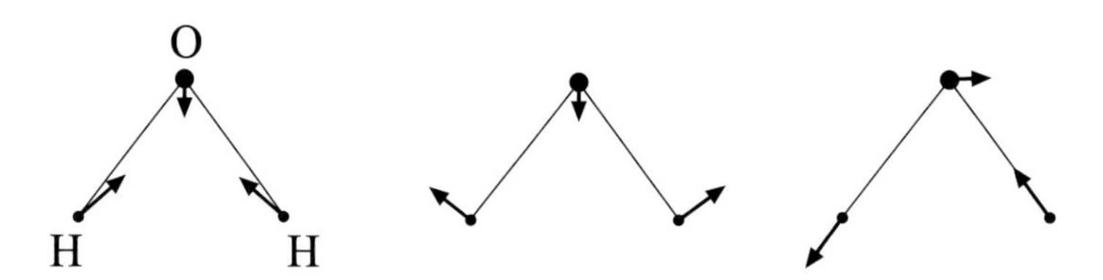


**Figure 2.9:** The displacement vectors for the three normal modes of water. The different amplitudes are determined by the atomic masses (cf. Eq. (2.65)).

to $q_{a,\xi} = 0$. It is noteworthy that the harmonic oscillations of the *individual* atoms within a normal mode have all the same frequency, $\omega_{a,\xi}$, but different amplitudes determined by their masses (cf. Eq. (2.65)). In Fig. 2.9 we show as an example the three normal modes of water. The different amplitudes are represented by arrows of different length. It should be noted that the normal mode vibrations do not lead to any translations or rotations of the molecules as a whole, i.e., linear and angular momentum are conserved. In addition to the $3N_{\text{nuc}} - 6$ normal mode frequencies the diagonalization of the Hessian will result in 6 eigenvalues which are equal to zero. In terms of the potential energy surface this means that there is no restoring force along these zero–frequency normal mode displacements. Thus, it is clear that the eigenvectors obtained for the zero–eigenvalues must correspond to the free translation and rotation of the molecule.

Having specified the vibrational Hamiltonian for the adiabatic electronic state $|\phi_a\rangle$ in Eq. (2.67) the nuclear Schrödinger equation can be solved by making a factorization ansatz with respect to the normal modes for the wave function. Using the standard textbook solution for harmonic oscillators we have ($q$ comprises all normal mode coordinates)

$$H_a^{(\text{nm})} \chi_{aN}^{(\text{adia})}(q) = \mathcal{E}_{aN} \chi_{aN}^{(\text{adia})}(q) \tag{2.68}$$

with

$$\chi_{aN}^{(\text{adia})}(q) = \prod_{\xi} \chi_{aN_\xi}(q_{a,\xi}) \; . \tag{2.69}$$

Here, the set of quantum numbers is written as $N = \{N_1, N_2, \ldots\}$ and the eigenfunctions for mode $\xi$ are given by

$$\chi_{aN_\xi}(q_{a,\xi}) = \frac{\lambda_{a,\xi}}{\sqrt{\sqrt{\pi}\, 2^{N_\xi}\, N_\xi!}} \exp\left( -\frac{1}{2} \lambda_{a,\xi}^2 q_{a,\xi}^2 \right) H_{N_\xi}(\lambda_{a,\xi}\, q_{a,\xi}) \; , \tag{2.70}$$

with $\lambda_{a,\xi}^2 = \omega_{a,\xi}/\hbar$. The $H_{N_\xi}$ in Eq. (2.70) are the Hermite polynomials. The eigenenergies in Eq. (2.68) read

$$\mathcal{E}_{aN} = \sum_{\xi} \hbar\omega_{a,\xi}(N_\xi + \frac{1}{2}) \; , \tag{2.71}$$

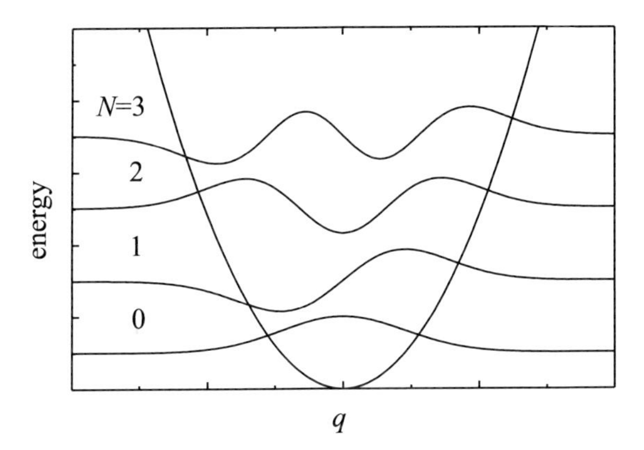


**Figure 2.10:** Harmonic oscillator potential together with the eigenfunctions for the lowest energy eigenstates along the normal mode coordinate $q$.

with the vibrational quantum numbers for mode $\xi$ being $N_\xi = 0, 1, 2, \ldots$

In Figure 2.10 we have plotted the oscillator potential for a single mode together with the eigenfunctions corresponding to the lowest eigenenergies. Note that in contrast to classical mechanics the lowest possible state has finite energy due to quantum mechanical zero–point motion (see Eq. (2.71)). Having solved the electronic and the nuclear problem separately, we are in a position to give the solutions, $\Psi_N^{(\text{adia})}(r, \sigma; R)$ (Eq. (2.22)), to the molecular Schrödinger equation (2.7) within the adiabatic Born–Oppenheimer approximation.

In preparation of the following chapters we now address the issue of the relation between normal modes belonging to different electronic states. Suppose that we have made a normal mode analysis for the electronic ground state PES $U_{a=g}(R)$ which had a stationary point at $R^{(g)}$. We then proceed by searching for the minima in some excited state PES $U_{a=e}(R)$. This excited state shall be selected, for instance, because it is accessible from the ground state via an optical transition (see Chapter 5). Let us assume that we found a stationary point for the configuration $R^{(e)}$. Assuming further the harmonic approximation to the potential surface in the vicinity of $R^{(e)}$ to be valid we can write

$$U_e(R) = U_e(R^{(e)}) + \sum_{m,n=1}^{3N_{\text{nuc}}} \frac{1}{2} \kappa_{mn}^{(e)} \, \Delta R_m^{(e)} \, \Delta R_n^{(e)} \,. \tag{2.72}$$

According to Eq. (2.65) the normal modes are obtained by a linear transformation of the Cartesian displacements. We can relate the displacement vectors for the excited state to those for the ground state via

$$\Delta R_n^{(e)} = R_n - R_n^{(g)} - (R_n^{(e)} - R_n^{(g)}) = \sum_\xi M_n^{-1/2} A_{n\xi}^{(g)} (q_{g,\xi} - \Delta q_{e,\xi}) \,. \tag{2.73}$$

Here the $\Delta q_{e,\xi}$ are defined by taking the deviations $\Delta R_n^{(e)}$ with respect to $R_n = R_n^{(e)}$. This situation is illustrated in Fig. 2.11 for a single normal mode.

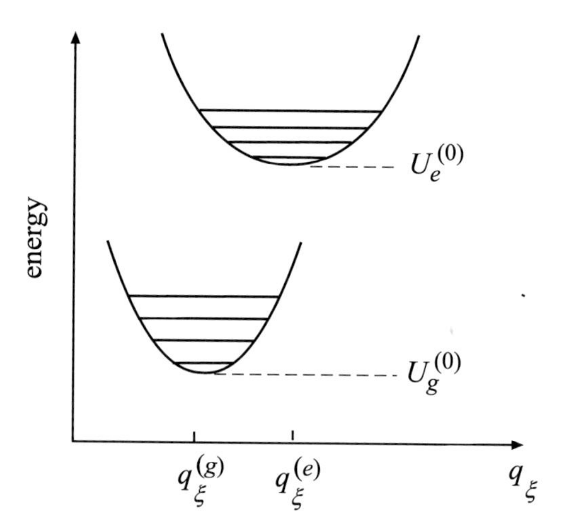


**Figure 2.11:** Shifted harmonic oscillator potential surfaces for two electronic states which are described by the same normal coordinate $q_\xi$.

In the general case the geometry of the molecule may be different in different electronic states. This would imply that the normal mode transformation does not bring the Hamiltonian for the ground and the excited state into diagonal form simultaneously. Thus, the Hessian $\kappa_{mn}^{(e)}$ is not diagonalized by the transformation matrix of the ground state, $A_{n\xi}^{(g)}$. In the following we will assume for simplicity that ground and excited states can be described by the *same* normal mode coordinates. We will allow, however, for state dependent normal mode frequencies, $\omega_{a,\xi}$. With this restriction we can write the Hamiltonian for the excited state as

$$H_e = U_e(q_\xi = q_\xi^{(e)}) + \frac{1}{2}\sum_\xi \left(p_\xi^2 + \omega_{e,\xi}^2(q_\xi - q_\xi^{(e)})^2\right). \tag{2.74}$$

Here and in the following, we will drop the electronic state index at the normal coordinates, $q_{g,\xi} = q_{e,\xi} = q_\xi$ and introduce the abbreviation $q_\xi^{(e)} = \Delta q_{e,\xi}$. Typical PES along some normal coordinate valid for the ground and the excited state are plotted in Fig. 2.11. The solutions of the stationary Schrödinger equation for the excited state Hamiltonian, (2.74), are now shifted oscillator states which read for the mode $\xi$

$$\chi_{e,N_\xi}(q_\xi) = \frac{\lambda_{e,\xi}}{\sqrt{\sqrt{\pi}\,2^{N_\xi}\,N_\xi!}}\exp\left(-\frac{1}{2}\lambda_{e,\xi}^2(q_\xi - q_\xi^{(e)})^2\right)\,H_{N_\xi}(\lambda_{e,\xi}(q_\xi - q_\xi^{(e)}))\,. \tag{2.75}$$

This procedure is easily generalized to incorporate any excited electronic state which can be described by the normal modes of the electronic ground state.

### 2.6.2 Operator Representation of the Normal Mode Hamiltonian

The properties of harmonic oscillators are conveniently derived using so–called creation and annihilation operators. We define the operator (dropping the electronic state index which is unnecessary if the frequency is state–independent as will be assumed in the following).[7]

$$C_\xi = \sqrt{\frac{\omega_\xi}{2\hbar}}\,\hat{q}_\xi + i\frac{1}{\sqrt{2\hbar\omega_\xi}}\,\hat{p}_\xi \tag{2.76}$$

which acts on some oscillator state $|\chi_{N_\xi}\rangle = |N_\xi\rangle$ according to

$$C_\xi\,|N_\xi\rangle = \sqrt{N_\xi}\,|N_\xi - 1\rangle \tag{2.77}$$

and

$$C_\xi\,|0_\xi\rangle = 0\,. \tag{2.78}$$

Thus, the number of vibrational quanta is decreased by one; the operator $C_\xi$ is called annihilation operator. Its hermitian conjugate, the creation operator $C_\xi^+$, increases the number of vibrational quanta in mode $\xi$ by one

$$C_\xi^+\,|N_\xi\rangle = \sqrt{N_\xi + 1}\,|N_\xi + 1\rangle\,. \tag{2.79}$$

These operators obey the boson commutation relation

$$[C_\xi, C_{\xi'}^+] = \delta_{\xi\xi'}\,. \tag{2.80}$$

The coordinate and momentum operators can be expressed by means of these operators as

$$\hat{q}_\xi = \sqrt{\frac{\hbar}{2\omega_\xi}}(C_\xi + C_\xi^+) \tag{2.81}$$

and

$$\hat{p}_\xi = -i\sqrt{\frac{\hbar\omega_\xi}{2}}(C_\xi - C_\xi^+)\,. \tag{2.82}$$

Using these relations the normal mode Hamiltonian (2.67) takes the simple form

$$H^{(\mathrm{nm})} = \sum_\xi \hbar\omega_\xi\left(C_\xi^+ C_\xi + \frac{1}{2}\right)\,. \tag{2.83}$$

The operator $C_\xi^+ C_\xi = \hat{N}_\xi$ is the so–called occupation number operator whose eigenvalue equation is $\hat{N}_\xi|N_\xi\rangle = N_\xi|N_\xi\rangle$. All eigenstates $|N_\xi\rangle$ can be obtained by successive application of the creation operator $C_\xi^+$ on the ground state $|0_\xi\rangle$

$$|N_\xi\rangle = \frac{1}{\sqrt{N_\xi!}}(C_\xi^+)^{N_\xi}|0_\xi\rangle\,. \tag{2.84}$$

Of course the eigenenergies do not change, i.e., they are given by Eq. (2.71).

[7] In the following $\hat{p}_\xi$ and $\hat{q}_\xi$ denote abstract operators in Hilbert space spanned by the vectors $|\chi_N\rangle$.

In the previous section we learned that the nuclear motion in two different electronic states can – under certain conditions – be described using the same normal mode coordinates. The different equilibrium positions of the normal mode oscillators are then accounted for by shifting the equilibrium position of the potential and the respective oscillator wave function by $q_\xi^{(a)}$. Introducing dimensionless coordinates according to $\hat{q}_\xi \sqrt{2\omega_\xi/\hbar} = (C_\xi + C_\xi^+)$ the Hamiltonian for the shifted oscillator, (2.74), becomes

$$
\begin{aligned}
H_a^{(\mathrm{nm})} &= U_a^{(0)} + \sum_\xi \hbar\omega_\xi \left( C_\xi^+ C_\xi + \frac{1}{2} \right) \\
&\quad + \sum_\xi \hbar\omega_\xi [g_a(\xi)(C_\xi^+ + C_\xi) + g_a^2(\xi)] \,. \qquad (2.85)
\end{aligned}
$$

Here we introduced the dimensionless shift of the potential energy surface belonging to state $a$ as

$$
g_a(\xi) = -\sqrt{\frac{\omega_\xi}{2\hbar}}\, q_\xi^{(a)} \,. \qquad (2.86)
$$

The respective energy offset has been abbreviated as $U_a^{(0)}$. In order to transform the shifted oscillator functions (2.75) to an operator form similar to Eq. (2.84) we will introduce the so–called displacement operator. Suppose we expand the wave function $\chi_{aN_\xi}(q_\xi - q_\xi^{(a)})$ (Eq. (2.75)) in powers of the displacement according to

$$
\begin{aligned}
\chi_{aN_\xi}(q_\xi - q_\xi^{(a)}) &= \sum_{n=0}^{\infty} \frac{(-q_\xi^{(a)})^n}{n!} \frac{d^n}{dq_\xi^n} \chi_{aN_\xi}(q_\xi) \\
&= \exp\left\{ -\frac{i}{\hbar} q_\xi^{(a)} p_\xi \right\} \chi_{aN_\xi}(q_\xi) \,, \qquad (2.87)
\end{aligned}
$$

where we have used the coordinate representation of the momentum operator for mode $\xi$, $\hat{p}_\xi = -i\hbar d/dq_\xi$. Using Eqs. (2.82) and (2.86) the exponent can be written in the operator form

$$
-\frac{i}{\hbar} q_\xi^{(a)} \hat{p}_\xi = g_a(\xi)(C_\xi - C_\xi^+) \,. \qquad (2.88)
$$

This suggests the introduction of the displacement operator according to

$$
D^+(g_a(\xi)) = \exp\left\{ g_a(\xi)(C_\xi - C_\xi^+) \right\} \,. \qquad (2.89)
$$

Thus, if $|N_\xi\rangle$ corresponds to an eigenstate of some non–shifted reference oscillator Hamiltonian the eigenstates of the shifted oscillator Hamiltonian, $|\chi_{aN_\xi}\rangle = |N_\xi^{(a)}\rangle$, can be generated as follows

$$
\begin{aligned}
|N_\xi^{(a)}\rangle &= \frac{1}{\sqrt{N_\xi!}} D^+(g_a(\xi))\, (C_\xi^+)^{N_\xi}\, |0_\xi\rangle \\
&= D^+(g_a(\xi))\, |N_\xi\rangle \,, \qquad (2.90)
\end{aligned}
$$

and the corresponding wave functions (2.87) are obtained from $\langle q_\xi|\chi_{aN_\xi}\rangle$.

The displacement operator is unitary, i.e.,

$$D^+(g_a(\xi)) = D(-g_a(\xi)) = D^{-1}(g_a(\xi)) \,. \tag{2.91}$$

Further the following useful property can be derived by expanding the displacement operator in a power series

$$D(g_a(\xi))\, C_\xi^+\, D^+(g_a(\xi)) = [D^+(g_a(\xi))\, C_\xi\, D(g_a(\xi))]^+ = C_\xi^+ - g_a(\xi) \,. \tag{2.92}$$

We can rewrite the vibrational Hamiltonian, (2.74), in the form

$$\begin{aligned} H_a^{(\text{nm})} &= U_a^{(0)} + \sum_\xi \hbar\omega_\xi \left[ (C_\xi^+ + g_a(\xi))\,(C_\xi + g_a(\xi)) + \frac{1}{2} \right] \\ &= U_a^{(0)} + \sum_\xi \hbar\omega_\xi \left[ D^+(g_a(\xi))\, C_\xi^+ C_\xi\, D(g_a(\xi)) + \frac{1}{2} \right] , \end{aligned} \tag{2.93}$$

where we used the unitarity of the displacement operator.

Comparing Eqs. (2.93) and (2.74) we realize that the introduction of the displacement operator yields a very compact notation for the Hamiltonian of a set of harmonic oscillators whose equilibrium positions are displaced with respect to each other. This situation we will encounter in Chapters 5 and 6. There the overlap integral between two shifted oscillator states will play an important role. Assuming $|\chi_{aM}\rangle$ and $|\chi_{bN}\rangle$ to be two normal mode eigenstates for a particular mode belonging to the electronic states $a$ and $b$, respectively, the overlap integral can be written as (skipping the mode index)

$$\langle\chi_{aM}|\chi_{bN}\rangle = \langle M|D(g_a)D^+(g_b)|N\rangle \,, \tag{2.94}$$

where $|N\rangle$ and $|M\rangle$ are the non–shifted states. In order to rewrite the product of the two displacement operators we make use of the operator identity

$$e^{\alpha(A+B)} = e^{\alpha A} e^{\alpha B} e^{-\alpha^2[A,B]/2} \,, \tag{2.95}$$

which holds if $[A,B]$ commutes with $A$ and $B$. Here, $\alpha$ is some parameter. For the displacement operators we obtain with the help of Eq. (2.80)

$$\begin{aligned} D(g_a)D^+(g_b) &= D(\Delta g_{ab}) \\ &= e^{\Delta g_{ab} C^+} e^{-\Delta g_{ab} C} e^{-\Delta g_{ab}^2/2} \end{aligned} \tag{2.96}$$

with $\Delta g_{ab} = g_a - g_b$. The action of the exponential operator on the oscillator states is calculated using a Taylor expansion

$$\begin{aligned} e^{-\Delta g_{ab} C}|N\rangle &= \sum_{n=0}^{N} \frac{(-\Delta g_{ab})^n}{n!} C^n |N\rangle \\ &= \sum_{n=0}^{N} \frac{(-\Delta g_{ab})^n}{n!} \sqrt{\frac{N!}{(N-n)!}}\, |N-n\rangle \,. \end{aligned} \tag{2.97}$$

where we made use of the properties (2.77) and (2.78). Applying the same expansion to the bra vector we obtain for the matrix elements

$$\begin{aligned}\langle \chi_{aM} | \chi_{bN} \rangle &= e^{-(\Delta g_{ab})^2/2} \sum_{m=0}^{M} \sum_{n=0}^{N} \frac{(-1)^n (\Delta g_{ab})^{m+n}}{m!n!} \\ &\times \sqrt{\frac{M!N!}{(M-m)!(N-n)!}} \delta_{M-m,N-n} \, . \end{aligned} \tag{2.98}$$

This expression is called the *Franck–Condon factor* (see Chapter 5). The most apparent property of the overlap integral Eq. (2.98) is certainly the fact that due to the exponential prefactor for any given pair of states the overlap decreases upon increasing the shift between the two PES . The elements of Eq. (2.98) which are diagonal in the vibrational quantum number can be further simplified. Since $\delta_{N-m,N-n} = \delta_{mn}$ we have

$$\begin{aligned}\langle \chi_{aN} | \chi_{bN} \rangle &= e^{-(\Delta g_{ab})^2/2} \sum_{n=0}^{N} \frac{(-1)^n (\Delta g_{ab})^{2n}}{n!^2} \frac{N!}{(N-n)!} \\ &= e^{-(\Delta g_{ab})^2/2} L_N((\Delta g_{ab})^2) \, , \end{aligned} \tag{2.99}$$

where $L_N(x)$ is a Laguerre polynomial. Expressions of this type we will meet in Chapter 5 and a generalization to the case of different frequencies as well as a numerical recipe for an efficient calculation is given in Section 2.8.2.

### 2.6.3 Reaction Paths

Chemical reaction dynamics can be understood in terms of the adiabatic Born–Oppenheimer PES for nuclear motion.[8] Let us consider the simple example of a PES for an isomerization reaction shown in Fig. 2.7. Suppose that initially the nuclei are in some reactant configuration corresponding to the left minimum. The properties of nuclear motion in the vicinity of this minimum (equilibrium configuration) have been considered in the previous section. In order to understand how the nuclei move to the right minimum corresponding to the product state, it is necessary to explore the properties of the PES away from the stationary points. For this purpose we return to the general Hamiltonian

$$H_{\text{nuc}} = \sum_{n=1}^{3N_{\text{nuc}}} \frac{P_n^2}{2M_n} + U(R_1, \ldots, R_{3N_{\text{nuc}}}) \, . \tag{2.100}$$

This expression poses a serious problem for polyatomic molecules since the numerical calculation of a full $3N_{\text{nuc}}$–6 dimensional potential energy surface becomes prohibitive with increasing $N_{\text{nuc}}$. In practice, however, the case that all DOF move appreciably during a reaction is rather unlikely. This observation suggests to separate all DOF into *active* and *spectator* or *substrate* coordinates. This concept can be realized in several ways which differ in the way the substrate DOF are treated and in the choice of the coordinate system.

[8] As discussed in Section 2.3 in the general case it might be necessary to include the nonadiabatic coupling between PES belonging to different electronic states.

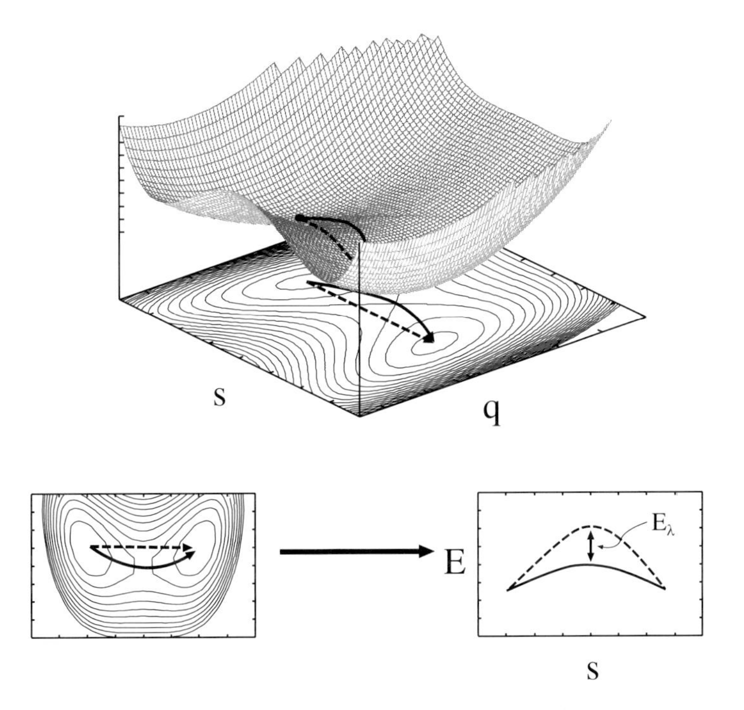


**Figure 2.12:** Upper panel: Schematic plot of a two–dimensional PES. The coordinate $s$ is a reaction coordinate while $q$ describes a harmonic vibration orthogonal to the reaction coordinate. Also shown is the minimum energy path (solid line) as well as a straight line path (dashed line) connecting reactant and product wells. In the lower panel we show the contour view (left) together with a cut along the straight line path where $q = 0$ (right). The energetic difference between both paths is the reorganization energy of the oscillator coordinate (see, Eq. (2.110)). (Figure courtesy of H. Naundorf)

First, let us consider the standard approach of quantum chemistry. Suppose we have performed a search for stationary points and transitions states on the multi–dimensional PES (geometry optimization). For simplicity we assume that there are two minima separated by a single transition state as shown in Fig. 2.12.

This situation may correspond to an isomerization reaction occurring, for example, in the course of intramolecular proton transfer (cf. Chapter7). In order to learn more about the way the reaction takes from the reactant to the product well via the transition state, one can follow the so–called *minimum energy path*. This path is obtained by starting from the transition state configuration[9] and following the steepest descent path to the reactant and product well minima (see solid line in Fig. 2.12).[10]

The $3N_{\rm nuc}$–dimensional vector $R^{(p)}$ which points to the minimum energy path defines a

[9]In principle one could also start at a minimum and follow the shallowest ascent path to the transition states. However, it is numerically very difficult to reach to transition state this way, because at a minimum the potential energy increases in *all* directions; at a saddle point there is only one downward path.

[10]In practice one follows the steepest descent path defined in mass–weighted coordinates.

curve in the $3N_{\text{nuc}}$–dimensional space of the nuclear coordinates. This curve $s = s(R^{(p)})$, which is the arc length along the minimum energy path, can be considered as the one–dimensional *reaction coordinate*. This one–dimensional description provides a valuable framework for the understanding of many reaction mechanisms. Looking at Fig. 2.12 it becomes clear, however, that restricting the reaction *dynamics* to take place on the minimum energy path only, may be a rather crude approximation. In many cases the minimum energy path will be considerably curved if in full $3N_{\text{nuc}}$–dimensional space. Let us imagine a (classical) ball starting at the transition state with some very small velocity. It is clear that unless the ball moves very slowly down into the reactant/product valley, the trajectory of the ball will not follow the minimum energy path if this path is curved. This implies that a one–dimensional description of the dynamics is not adequate.

There are several ways to account for the motion away from the minimum energy path. In the following we will outline a strategy leading to a Hamiltonian which is particularly suited for large molecules or condensed phase situations as will be encountered in later chapters.[11] The first step consists in the identification of those Cartesian coordinates which describe atoms undergoing arbitrarily large displacements, $\mathbf{s} = (s_1, \dots s_{N_{\text{rc}}})$. These are the *active coordinates*. They are separated from the remaining $3N_{\text{nuc}} - N_{\text{rc}}$ *substrate coordinates* $\mathbf{Z}$. The key assumptions is that the substrate coordinates stay close to their equilibrium configurations $\mathbf{Z}^{(0)}(\mathbf{s})$ during the reaction. As indicated this equilibrium configuration may depend on the positions of the reaction coordinates $\mathbf{s}$. As an example consider the transfer of a light atom A between two heavy fragments B and C, i.e., $\text{B}-\text{A} \cdots \text{C} \rightarrow \text{B} \cdots \text{A}-\text{C}$. Such a situation is typical for intramolecular hydrogen transfer reactions, for instance. Proper choice of the coordinate system allows a one–dimensional treatment of the A atom motion along the coordinate $s_1$. The coordinates describing the fragments are then comprised to the vector $\mathbf{Z}$.

Since the substrate atoms perform only small–amplitude motion around their equilibrium positions $U(R) = U(\mathbf{s}, \mathbf{Z})$ can be expanded in term of the deviations $\Delta\mathbf{Z}(\mathbf{s}) = (\mathbf{Z} - \mathbf{Z}^{(0)}(\mathbf{s}))$ as follows

$$
\begin{aligned}
U(R) \approx\; & U(\mathbf{s}, \mathbf{Z}^{(0)}(\mathbf{s})) + \left(\frac{\partial U(\mathbf{s}, \mathbf{Z})}{\partial \mathbf{Z}}\right)_{\mathbf{Z}=\mathbf{Z}^{(0)}(\mathbf{s})} \Delta\mathbf{Z}(\mathbf{s}) \\
& + \frac{1}{2}\Delta\mathbf{Z}(\mathbf{s}) \left(\frac{\partial^2 U(\mathbf{s}, \mathbf{Z})}{\partial \mathbf{Z} \partial \mathbf{Z}}\right)_{\mathbf{Z}=\mathbf{Z}^{(0)}(\mathbf{s})} \Delta\mathbf{Z}(\mathbf{s})\,. 
\end{aligned}
\tag{2.101}
$$

The different terms have a straightforward interpretation: $U(\mathbf{s}, \mathbf{Z}^{(0)}(\mathbf{s}))$ is the potential energy on the (in general multi–dimensional) Cartesian reaction surface, with the spectator degrees of freedom frozen at some reference geometry. This can be, for example, the equilibrium geometry of the spectator atoms at a given value of the reaction coordinates. The second term in Eq. (2.101) contains the forces exerted on the substrate atoms due to the motion of the important DOF away from their equilibrium positions:

$$
\mathbf{f}(\mathbf{s}) = -\left(\frac{\partial U(\mathbf{s}, \mathbf{Z})}{\partial \mathbf{Z}}\right)_{\mathbf{Z}=\mathbf{Z}^{(0)}(\mathbf{s})}\,. \tag{2.102}
$$

[11] For an alternative formulation, which is based on the minimum energy path and harmonic vibrations perpendicular to it, see [Mil80].

Finally, the third term describes the change of the Hessian matrix

$$\kappa(\mathbf{s}) = \left(\frac{\partial^2 U(\mathbf{s},\mathbf{Z})}{\partial \mathbf{Z}\partial \mathbf{Z}}\right)_{\mathbf{Z}=\mathbf{Z}^{(0)}(\mathbf{s})} \tag{2.103}$$

(and thus of the vibrational frequencies) due to the motion along $\mathbf{s}$.

Since the substrate atoms are assumed to perform small–amplitude harmonic motions we can introduce normal modes. Note that the normal modes have to be defined with respect to some fixed reference configuration $\mathbf{Z}^{(0)}(\mathbf{s}_{\text{ref}})$ to preserve the decoupling from the external motions. Thus we have

$$\begin{aligned}\Delta\mathbf{Z}(\mathbf{s}) &= \mathbf{Z} - \mathbf{Z}^{(0)}(\mathbf{s}_{\text{ref}}) + \mathbf{Z}^{(0)}(\mathbf{s}_{\text{ref}}) - \mathbf{Z}^{(0)}(\mathbf{s}) \\ &= \mathbf{M}^{-1/2}\mathbf{A}\mathbf{q} + \mathbf{Z}^{(0)}(\mathbf{s}_{\text{ref}}) - \mathbf{Z}^{(0)}(\mathbf{s}),\end{aligned} \tag{2.104}$$

where $\mathbf{M}$ is the diagonal matrix containing the atom masses and $\mathbf{A}$ is the normal mode transformation matrix (see also Eq. (2.65)). Straightforward application of this transformation to the Hamiltonian with the potential Eq. (2.101) gives the all-Cartesian form [12]

$$H = \mathbf{T_s} + U(\mathbf{s},\mathbf{Z}^{(0)}(\mathbf{s})) + U_{\text{add}}(\mathbf{s},\mathbf{Z}^{(0)}(\mathbf{s})) + \mathbf{T_q} + \frac{1}{2}\mathbf{q}\mathbf{K}(\mathbf{s})\mathbf{q} - \mathbf{F}(\mathbf{s})\mathbf{q}. \tag{2.105}$$

Here, $\mathbf{T_s}$ and $\mathbf{T_q}$ is the diagonal kinetic energy operator for the reaction coordinates and the substrate modes, respectively, and $\mathbf{K}(\mathbf{s}) = \mathbf{A}^{+}\mathbf{M}^{-1/2}\kappa(\mathbf{s})\mathbf{M}^{-1/2}\mathbf{A}$ is the transformed Hessian. Note that it includes a coupling between different substrate modes due to the motion of the reaction coordinates away from the reference configuration $\mathbf{s}_{\text{ref}}$. Since this motion is not restricted to some minimum energy path, there is also a force acting on the substrate modes

$$\mathbf{F}(\mathbf{s}) = \left[\mathbf{f}(\mathbf{s}) - (\mathbf{Z}^{(0)}(\mathbf{s}_{\text{ref}}) - \mathbf{Z}^{(0)}(\mathbf{s}))\kappa(\mathbf{s})\right]\mathbf{M}^{-1/2}\mathbf{A} \tag{2.106}$$

Finally, the special choice of the reference configuration for the definition of the normal modes leads to an additional potential defined by

$$\begin{aligned}U_{\text{add}}(\mathbf{s},\mathbf{Z}^{(0)}(\mathbf{s})) &= -\mathbf{f}(\mathbf{s})(\mathbf{Z}^{(0)}(\mathbf{s}_{\text{ref}}) - \mathbf{Z}^{(0)}(\mathbf{s})) \\ &+\frac{1}{2}(\mathbf{Z}^{(0)}(\mathbf{s}_{\text{ref}}) - \mathbf{Z}^{(0)}(\mathbf{s}))\kappa(\mathbf{s})(\mathbf{Z}^{(0)}(\mathbf{s}_{\text{ref}}) - \mathbf{Z}^{(0)}(\mathbf{s})).\end{aligned} \tag{2.107}$$

Of course, not all substrate modes will couple strongly to the reaction coordinates. A convenient measure for this coupling is the substrate oscillator's displacement from their equilibrium value of zero taken at the reference geometry $\mathbf{Z}(\mathbf{s}_{\text{ref}})$, that is,

$$\mathbf{q}^{(0)}(\mathbf{s}) = -[\mathbf{K}(\mathbf{s})]^{-1}\,\mathbf{F}(\mathbf{s})\,. \tag{2.108}$$

[12] Note that an arbitrary displacement of some active atom in general does not conserve linear and angular momentum of the total system. Strictly speaking, a rigorous treatment of the molecule's rotation would require the use of curvilinear coordinates and therefore destroy the all–Cartesian character of the Hamiltonian. However, since we focus on a description of large molecules or even condensed phase reactions, rotation/translation does not play an important role. In the numerical implementation of this approach it is accounted for approximately by projecting out infinitesimal rotations and translations of the substrate atoms from the Hessian before performing the normal mode transformation (for details see [Ruf88]).

Introducing this quantity into Eq. (2.105) yields after some rearrangement

$$H = \mathbf{T_s} + U(\mathbf{s}, \mathbf{Z}^{(0)}(\mathbf{s})) + U_{\text{add}}(\mathbf{s}, \mathbf{Z}^{(0)}(\mathbf{s})) - E_\lambda(\mathbf{s})$$
$$+\mathbf{T_q} + \frac{1}{2}(\mathbf{q} - \mathbf{q}^{(0)}(\mathbf{s}))\mathbf{K}(\mathbf{s})(\mathbf{q} - \mathbf{q}^{(0)}(\mathbf{s})) \,. \tag{2.109}$$

Here, we introduced the so–called *reorganization energy* defined as

$$E_\lambda(\mathbf{s}) = \frac{1}{2}\mathbf{q}^{(0)}(\mathbf{s})\mathbf{K}(\mathbf{s})\mathbf{q}^{(0)}(\mathbf{s}) \,. \tag{2.110}$$

The interpretation of the substrate mode part of Eq. (2.109) (second line) is straightforward. It is the Hamiltonian for a set of shifted oscillators, whose equilibrium positions depend on the coupling to the reaction coordinates. In our considerations of PES for different electronic excited states we already met this type of Hamiltonian. There the shift of the PES was due to different electronic charge distributions in the considered electronic states. In the present case the shift is a consequence of the motion of the reaction coordinates (s) away from a stationary point on the PES. This can be rationalized by looking at the two–dimensional case shown in Fig. 2.12. Let us further assume that the configuration of the left minimum has served as a fixed reference for the expansion in Eq. (2.101). Therefore, at this minimum the force on the substrate oscillator is zero. Now we move the reaction coordinate on a straight line toward the right potential well (dashed line). This force is trying to push the oscillator back to the minimum energy path (solid line). Restoring the equilibrium position of the substrate oscillator requires the reorganization energy $E_\lambda(\mathbf{s})$ as indicated in the lower right panel of Fig. 2.12.

Keeping track of the dependence of the reference geometry for the spectator atoms on the value of the reaction coordinates is important whenever one wishes to describe a reaction where reactants and products have quite different geometries. If this is not the case, one might simplify the expansion (2.101) and consider the spectator atoms remain frozen at some suitable reference geometry $\mathbf{Z}^{(0)}(\mathbf{s}) = \mathbf{Z}^{(0)}(\mathbf{s}_{\text{ref}}) = \mathbf{Z}^{(0)}$. In this case the additional potential as well as the second term in the force Eq. (2.106) vanishes.

Let us simplify the reaction surface Hamiltonian Eq. (2.109) to establish the contact with a widely used system–bath Hamiltonian. To this end we neglect the change of the reference geometry as well as the coupling between different substrate modes. Furthermore, the normal mode frequencies are assumed to be independent of the reaction coordinate, i.e. we have $K_{\xi\xi'}(\mathbf{s}) \approx \delta_{\xi\xi'}\omega_\xi^2$. Then the Hamiltonian can be written as $H = H_\text{S} + H_\text{R} + H_\text{S-R}$ with $H_\text{S}$ and $H_\text{R}$ describing the motion of the system (s) and bath ($q_\xi$) DOF, respectively. $H_\text{S-R}$ contains the interaction between both subsystems:

$$H_\text{R} + H_\text{S-R} = \frac{1}{2}\sum_{\xi=1}^{3N_\text{nuc}-N_\text{rc}-6}\left[p_\xi^2 + \omega_\xi^2\left(q_\xi - \frac{F_\xi(\mathbf{s})}{\omega_\xi^2}\right)^2\right] \,. \tag{2.111}$$

where we used $q_\xi^{(0)}(\mathbf{s}) = -F_\xi(\mathbf{s})/\omega_\xi^2$. With the reorganization energy given by $E_\lambda(\mathbf{s}) = \sum_{\xi=1}^{3N_\text{nuc}-N_\text{rc}-6} F_\xi^2(\mathbf{s})/2\omega_\xi^2$ the renormalized system Hamiltonian becomes

$$H_\text{S} = \sum_{n=1}^{N_\text{rc}} \frac{p_n^2}{2M_n} + U(\mathbf{s}, \mathbf{Z}^{(0)}) - E_\lambda(\mathbf{s}) \,. \tag{2.112}$$

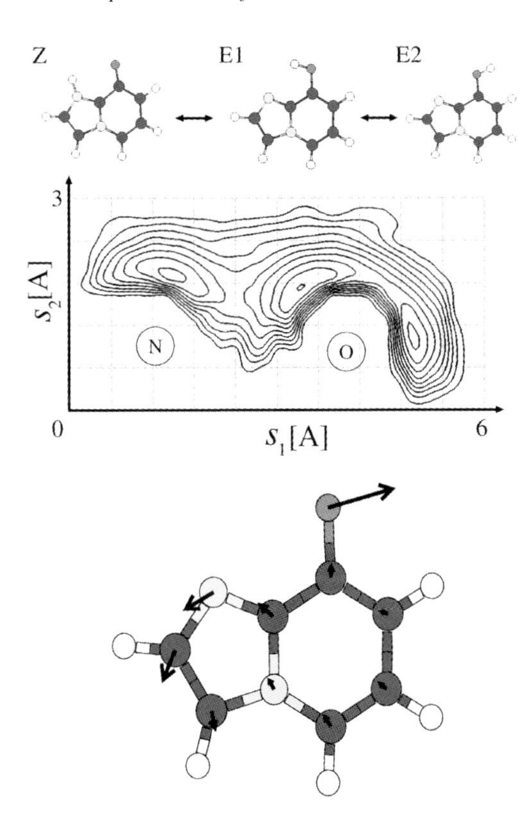


**Figure 2.13:** Two–dimensional Cartesian reaction surface for the motion of the proton in 8–hydroxyimidazo[1,2–a]pyridine (The positions of the oxygen and nitrogen atom are indicated.). There are actually three isomers shown in the upper part of the figure. The most stable one is **E1**. In the lower part of the figure we give the displacement vectors of that substrate mode which couples most strongly to the proton transfer at the transition state between **E1** and **Z**. (Grey code: C, O, N, H (from dark to light); figure courtesy of H. Naundorf.).

An example for the calculation of a Cartesian reaction surface Hamiltonian as defined in Eq. (2.109) is given in Fig. 2.13.

## 2.7 Diabatic versus Adiabatic Representation of the Molecular Hamiltonian

In Section 2.3 we have given the general form of the molecular wave function as (cf. Eq. (2.20))

$$\psi_M(r;R) = \sum_a \chi_{aM}(R)\, \phi_a(r;R)\ . \tag{2.113}$$

In principle, the summation has to be carried out over the complete set of adiabatic electronic states. These states are possibly coupled through the nonadiabaticity operator (Eq. (2.17)). Fortunately, in practice reasonable results are often obtained by including only a finite number of states in the actual calculation. This happens, for example, if one is interested in the electronic excitation spectrum of a molecule or if one wants to model photodissociation dynamics occurring upon irradiation by a laser having a certain fixed wavelength (see, for example, Fig. 2.2).

Let us suppose we have obtained the adiabatic electronic wave function $\phi_a(r;R) = \langle r;R|\phi_a\rangle$. The representation of the molecular Hamiltonian in this electronic basis is then obtained as (using the definitions (2.17) and (2.21))

$$H_{\text{mol}} = \sum_{ab} \left(\delta_{ab}\, H_a(R) + (1-\delta_{ab})\Theta_{ab}\right) |\phi_a\rangle\langle\phi_b| \,. \tag{2.114}$$

Note that $H_a(R)$ and $\Theta_{ab}$ are still operators concerning the nuclear coordinates. We can go one step further and write down the molecular Hamiltonian in the matrix representation of the adiabatic states $|\psi_{aM}\rangle = |\phi_a\rangle|\chi_{aM}\rangle$ which define the Born–Oppenheimer wave function (2.22), $\psi_{aM}^{(\text{adiab})}(r;R) = \langle r;R|\psi_{aM}\rangle$. We have

$$H_{\text{mol}} = \sum_{aM} \mathcal{E}_{aM}|\psi_{aM}\rangle\langle\psi_{aN}| + \sum_{aM,bN} \Theta_{aM,bN}|\psi_{aM}\rangle\langle\psi_{bN}| \,, \tag{2.115}$$

where we introduced

$$\Theta_{aM,bN} = \int dR\, \chi_{aM}^{*}(R) \left[\langle\phi_a|T_{\text{nuc}}|\phi_b\rangle - \sum_n \frac{\hbar^2}{M_n}\langle\phi_a|\nabla_n|\phi_b\rangle\nabla_n\right]\chi_{bN}(R) \,. \tag{2.116}$$

We note that the coupling is mediated by the momentum operator $\mathbf{P}_n = -i\hbar\nabla_n$. It is therefore referred to as *dynamic coupling* and its calculation requires knowledge of the first and second derivatives of the electronic wave function. This poses a serious computational problem especially for polyatomics. Further, the second term in $\Theta_{ab}$ is often rather sharply peaked if not singular indicating that the character of the electronic wave function changes rapidly within a narrow range of configuration space (see Fig. 2.14). Such a behavior of the coupling is likely to cause numerical problems, for example, in a quantum dynamical calculation based on the Hamiltonian (2.115) and using the methods which will be introduced in Chapter 2. On the other hand, since the adiabatic electronic states contain information on the instantaneous nuclear configuration it can be expected that they will lead to a very compact representation of the molecular wave function.

However, the question arises whether there is an alternative to the adiabatic representation of the Hamiltonian. In order to investigate this point consider an electronic basis $\phi_a(r;R^{(0)})$, where the positions of the nuclei are *fixed* at some point $R^{(0)}$ in configuration space. A typical choice for $R^{(0)}$ could be, for instance, some local minimum of the potential energy surface in the electronic state $a$. Of course, $\phi_a(r;R^{(0)})$ is no longer an eigenfunction of $H_{\text{el}}$ except at $R^{(0)}$. Defining $H_{\text{el}}(R^{(0)}) = H^{(0)}$ the electronic Hamiltonian can be written as

$$H_{\text{el}}(R) = H^{(0)}(R^{(0)}) + V(R, R^{(0)}) \tag{2.117}$$

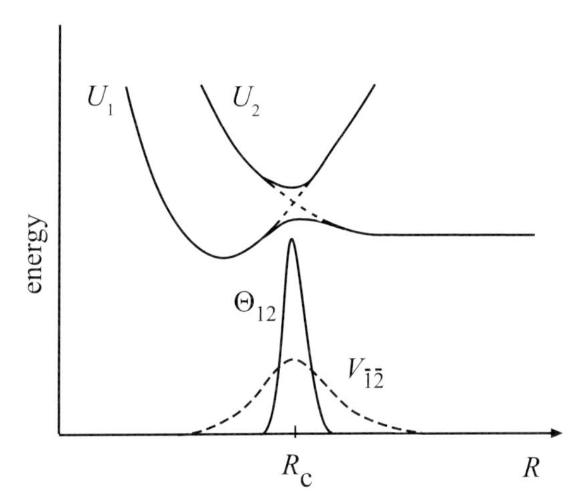


**Figure 2.14:** Schematic view of adiabatic (solid) and diabatic (dashed) potential energy curves along a nuclear coordinate. For $R \ll R_c$ both potential curves are well separated; the lower and upper diabatic states belong to a bound and repulsive electronic state, respectively, and so do the adiabatic potentials $U_1$ and $_2$. For $R \gg R_c$ the character of the potential curves changes; $U_2$ corresponds to a bound state and $U_1$ is repulsive now. This is reflected in the electronic wave functions and therefore in the state couplings shown in the lower part.

with the potential coupling given by

$$V(R, R^{(0)}) = H_{\rm el}(R) - H^{(0)}(R^{(0)}) \, . \tag{2.118}$$

The molecular wave function expanded in this so–called *diabatic* basis set[13] reads

$$\psi(r; R) = \sum_{\bar{a}} \chi_{\bar{a}}(R) \, \phi_{\bar{a}}(r; R^{(0)}) \, , \tag{2.119}$$

where we have used the quantum number $\bar{a}$ to distinguish diabatic states from adiabatic ones.

Suppose the diabatic basis is complete and the summations in Eqs. (2.113) and (2.119) are carried out with respect to the whole set of quantum numbers, both representations shall give identical results. In practice, however, one is interested only in a certain subset of the electronic state manifold in order to model some property of the molecule. Since $\phi_{\bar{a}}(r; R^{(0)})$ does not account for the change in nuclear configuration it can in general be assumed that the diabatic representation is not as compact as the adiabatic one. Thus, more terms in the expansion (2.119) may be needed to represent some feature of the molecular wave function. On the other hand, all matrix elements of the nonadiabaticity operator vanish identically because the diabatic basis functions are not $R$–dependent. The coupling between different electronic states is now due to $V(R; R^{(0)})$ defined in Eq. (2.118); the respective matrix elements are $V_{\bar{a}\bar{b}}(R; R^{(0)}) = \langle \phi_{\bar{a}} | V(R, R^{(0)}) | \phi_{\bar{b}} \rangle$. Thus, the representation of the molecular Hamiltonian

[13]Note that the special choice of $\phi_a(r; R^{(0)})$ as a diabatic basis is sometimes also called crude adiabatic basis.

in terms of the diabatic electronic states is

$$H_{\text{mol}} = \sum_{\bar{a}\bar{b}} (\delta_{ab} H_{\bar{a}} + (1 - \delta_{\bar{a}\bar{b}}) V_{\bar{a}\bar{b}}) \, |\phi_{\bar{a}}\rangle\langle\phi_{\bar{b}}| \; . \tag{2.120}$$

Here we introduced the Hamiltonian for the motion of the nuclei in the diabatic electronic state $|\phi_{\bar{a}}\rangle$ as

$$H_{\bar{a}}(R) = T_{\text{nuc}} + U_{\bar{a}}(R) \; , \tag{2.121}$$

with

$$U_{\bar{a}}(R) = E_{\bar{a}}(R^{(0)}) + V_{\text{nuc-nuc}} + V_{\bar{a}\bar{a}}(R, R^{(0)}) \tag{2.122}$$

being the diabatic potential energy surface. The $E_{\bar{a}}(R^{(0)})$ are the diabatic electronic energies according to $H^{(0)}$. The shift of the electronic state coupling from the kinetic to the potential energy operator is the general feature for a diabatic basis as compared to the adiabatic basis.

Having such a broad definition it should be clear that the crude adiabatic basis, $\phi_{\bar{a}}(r; R^{(0)})$, is not the only possible choice of a diabatic basis. In general one can argue that any complete basis set is suited which solves the stationary Schrödinger equation for a part of the Hamiltonian and yields negligibly small matrix elements of the nonadiabaticity operator. The potential coupling term has to be properly adjusted for each case. A typical situation we will be encountered in Chapter 6 where electron transfer in donor–acceptor complexes is considered. There one can define local electronic states with respect to donor and acceptor groups.

Finally, we would like to focus on those configurations of the nuclei for which different electronic states approach each other. The situation for a diatomic molecule is shown in Fig. 2.14. The diabatic potential energy curves (dashed line) intersect at $R = R_c$. This can be rationalized by looking at the Hamiltonian (2.120). The crossing condition $U_{\bar{a}}(R) = U_{\bar{b}}(R)$ can in principle be fulfilled for any $R$. However, if one would incorporate the diabatic state coupling $V_{\bar{a}\bar{b}}(R)$, i.e., by plotting the potential curves of the diagonalized diabatic Hamiltonian (cf. Eq. (2.131) below), the crossing is replaced by an *avoided crossing*. This behavior (solid lines in Fig. 2.14) is due to the fact that the conditions $U_{\bar{a}}(R) = U_{\bar{b}}(R)$ *and* $V_{\bar{a}\bar{b}}(R) = 0$ cannot simultaneously be fulfilled. An exception occurs if $V_{\bar{a}\bar{b}}(R)$ vanishes due to the symmetry of the diabatic wave functions. This will usually be the case if the symmetry of the two diabatic states is different. Since diabatic and adiabatic Hamiltonians describe the same molecular system, it is clear that adiabatic potential curves of states having the same symmetry will not cross as well. This behavior of potential energy curves is called the *non–crossing rule*.

The situation is different in polyatomic molecules. Here the crossing conditions $U_{\bar{a}}(R) = U_{\bar{b}}(R)$ and $V_{\bar{a}\bar{b}}(R) = 0$ can be simultaneously fulfilled even for states having the same symmetry. However, for an $N$–dimensional PES these are only 2 conditions, i.e., the crossing will only be in $N-2$ dimensions (if $V_{\bar{a}\bar{b}}(R) = 0$ due to symmetry the crossing is in $N-1$ dimensions only). For instance, in the two–dimensional case the PES of two electronic states of the same symmetry will intersect in a single point ($N-2=0$). The form of the PES in the vicinity of this point is usually called a *conical intersection* (this was first discussed by E. Teller in 1937). A numerical example is shown in Fig. 2.15(A).

It is straightforward to derive the equation for the expansion coefficients in (2.119), i.e., the diabatic nuclear wave functions, along the lines outlined in Section 2.3. One obtains

$$(H_{\bar{a}}(R) - \mathcal{E})\chi_{\bar{a}}(R) = -\sum_{\bar{b}\neq\bar{a}} V_{\bar{a}\bar{b}}(R, R^{(0)})\chi_{\bar{b}}(R) \,, \tag{2.123}$$

Neglecting the coupling between different states we get

$$H_{\bar{a}}(R)\chi_{\bar{a}M}(R) = \mathcal{E}_{\bar{a}M}\chi_{\bar{a}M}(R) \,. \tag{2.124}$$

The solutions of this equation, $\chi_{\bar{a}M}(R)$, together with the diabatic electronic states can be used to define the molecular Hamiltonian in the diabatic representation ($|\psi_{\bar{a}M}\rangle = |\phi_{\bar{a}}\rangle|\chi_{\bar{a}M}\rangle$)

$$H_{\text{mol}} = \sum_{\bar{a}M} \mathcal{E}_{\bar{a}M}|\psi_{\bar{a}M}\rangle\langle\psi_{\bar{a}M}| + \sum_{\bar{a}M,\bar{b}N} V_{\bar{a}M,\bar{b}N}|\psi_{\bar{a}M}\rangle\langle\psi_{\bar{b}N}| \,. \tag{2.125}$$

Here, $E_{\bar{a}M}$ are the eigenvalues following from Eq. (2.124) and

$$V_{\bar{a}M,\bar{b}N} = \int dR\, \chi^*_{\bar{a}M}(R)V_{\bar{a}\bar{b}}(R, R^{(0)})\chi_{\bar{b}N}(R) \,. \tag{2.126}$$

In contrast to the adiabatic representation the Hamiltonian matrix contains only coupling terms between different electronic states which stem from the potential energy operator. To distinguish this from the dynamic coupling, the potential coupling is called *static*. Static couplings are usually not as sharply peaked as dynamic ones and in general easier to treat in numerical applications (see Figs. 2.14 and 2.15). But, as already pointed out for a choice such as the crude adiabatic basis it may be required to take into account many terms in the expansion of the total wave function. Thus, the dimension of the diabatic Hamiltonian matrix in this case is likely to be higher than that of the adiabatic matrix.

Quantum chemical ab initio calculations usually provide adiabatic potential energy surfaces and wave functions. But as already emphasized, nonadiabatic couplings are not very convenient for dynamical calculations, for example. Thus, the question arises, whether it is possible to construct a diabatic basis which provides a compact representation of the molecular wave function.

Suppose we know the adiabatic coupling matrix for two electronic states $(a = 1, 2)$ which are of interest say for a dynamical simulation (see Fig. 2.14). Then we can make use of the fact that the general coupled two–state problem can be solved exactly as shown in Section 2.8.3. To this end one expands the diabatic states in terms of the adiabatic states:

$$\phi_{\bar{a}}(r; R) = \sum_{a=1,2} C_{\bar{a}}(a)\,\phi_a(r; R) \,. \tag{2.127}$$

The coefficients $C_{\bar{a}}(a)$ have been determined in Section 2.8.3 (cf. Eqs. (2.174) and (2.175), for instance). They depend on the so–called *mixing angle* which in the present context is assumed to be some unknown function of the coordinates $\tilde{\gamma}(R)$. Without referring to any particular model for the Hamiltonian, it is clear that the asymptotic ( $R \gg R_c$ and $R \ll R_c$) diabatic and adiabatic wave functions should coincide. Further, it is reasonable to assume that

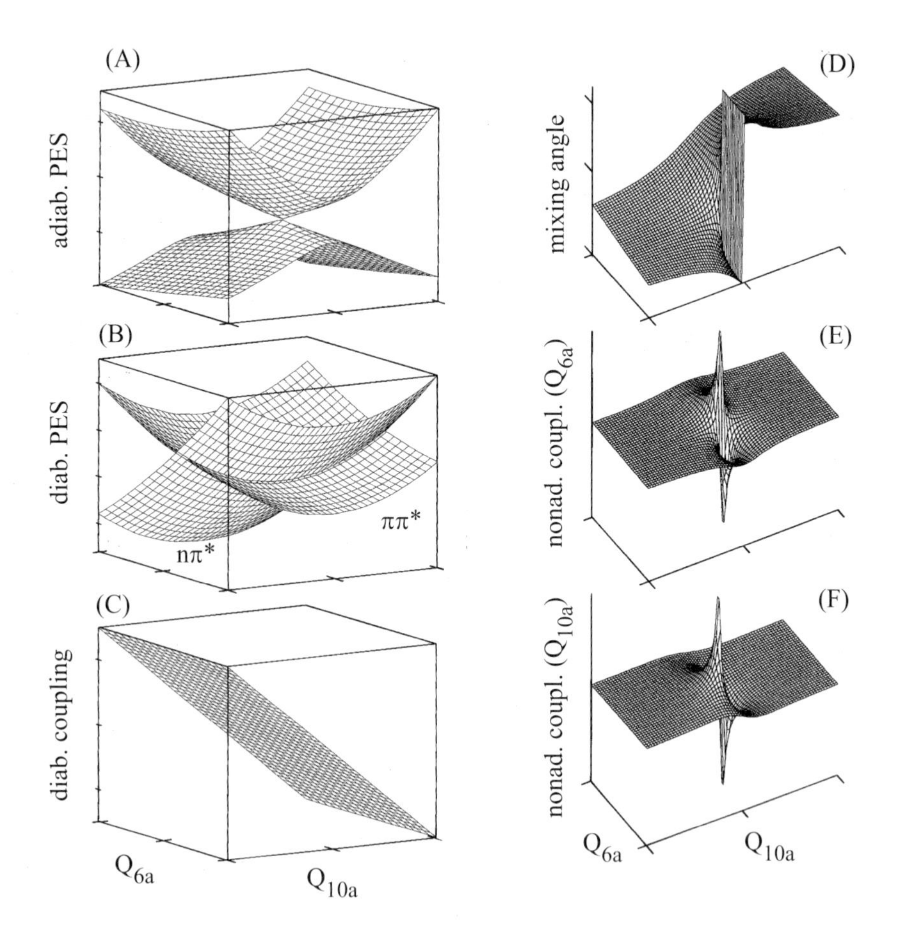


**Figure 2.15:** Results of quantum chemical calculations of the potential energy surfaces (PES) for the $S_1$ $(n\pi^*)$ and $S_2$ $(\pi\pi^*)$ electronic states of pyrazine along the $Q_{6a}$ and $Q_{10a}$ normal mode coordinates of the electronic ground state. (A) adiabatic PES showing a *conical intersection*, (B) diabatic PES, (C) coupling between diabatic electronic states ($V_{\bar{1}\bar{2}}$), (D) mixing angle (Eq. (2.130)), (E) and (F) singular part of the nonadiabatic coupling (second term in (2.17)) given as the derivative of the mixing angle with respect to the two normal mode coordinates (figure courtesy of G. Stock, for details see [Dom97]).

the state coupling is rather localized in the vicinity of $R_c$, i.e., any appreciable change of the mixing angle should occur in this region (cf. Fig. 2.15(D)).

The goal is to determine the mixing angle in such a way that the dynamic coupling is as small as possible. In particular we would like to eliminate the troublesome second term in (2.17), i.e., we demand that

$$\sum_n \langle \phi_{\bar{a}} | \nabla_n | \phi_{\bar{b}} \rangle = 0 \, . \tag{2.128}$$

Inserting the transformation (2.127) into this expression yields the following set of equations for the mixing angle

$$\sum_n \nabla_n \tilde{\gamma}(R) = -\sum_n \langle \phi_a | \nabla_n | \phi_b \rangle \,. \tag{2.129}$$

It can be shown that this equation has a unique solution only for the case of a single nuclear degree of freedom (see, for instance [Bae85]). Besides, the numerical calculation of the required derivative couplings for polyatomic systems presents a difficult task.

An alternative can be developed by starting from a diabatic basis which is constructed in a way that certain properties such as the dipole moments of the molecule behave smoothly. A related constraint is that the electronic wave function should not change appreciably when moving in the configuration space of the nuclear coordinates. Thus, diabatic wave functions for neighboring points should overlap considerably. The simplest approach in this respect is certainly to use some parameterized form for the diabatic potential surface, the static coupling, and if necessary also for other quantities such as the dipole moment. The parameters are then chosen to make observables, for example, those relevant for dynamic processes, agree with experiment.

In the following we will mainly be concerned with the situation of two electronic states and a single nuclear coordinate. In Fig. 2.14 we show typical diabatic and adiabatic potential curves for this case. Also the diabatic coupling $V_{\bar{1}\bar{2}}$ is plotted, which is not as sharply peaked as the nonadiabatic one. Given the diabatic Hamiltonian, the related mixing angle $\gamma(R)$ of the transformation, which has to be distinguished from $\tilde{\gamma}(R)$, is well–defined and given by (cf. Eq. (2.173))

$$\gamma(R) = \frac{1}{2} \arctan\left( \frac{2|V_{\bar{1}\bar{2}}(R)|}{|U_{\bar{1}}(R) - U_{\bar{2}}(R)|} \right) \,. \tag{2.130}$$

It guarantees that the diabatic to adiabatic back transformation (inverse of Eq. (2.127)) brings the potential energy operator into diagonal form. The adiabatic potential then reads

$$U_{1/2}(R) = \frac{1}{2} \left( U_{\bar{1}}(R) + U_{\bar{2}}(R) + \sqrt{[U_{\bar{1}}(R) - U_{\bar{2}}(R)]^2 + 4|V_{\bar{1}\bar{2}}(R)|^2} \right) \,. \tag{2.131}$$

Of course, Eqs. (2.130) and (2.131) are only of practical value if the diabatic potential matrix is known, which is generally not the case unless a parameterized analytical form is used.

In Fig. 2.15 we show the results of quantum chemical calculations for the coupled $S_1$–$S_2$ potential energy surfaces of pyrazine along two normal mode coordinates. Diabatic and adiabatic potentials are plotted together with the respective state couplings. Note that the diabatic potentials as well as the diabatic state coupling are rather smooth functions of the coordinates (panels (B) and (C)). On the other hand, the nonadiabatic potential energy surfaces form a conical intersection (panel (A)). In the vicinity of this intersection the nonadiabatic couplings (panels (E) and (F)) change rapidly (in fact they become singular).

# 2.8 Supplement

## 2.8.1 The Hartree–Fock Equations

In the following we will given some details on the variational optimization of the expectation value of the electronic Hamiltonian Eq. (2.30) which leads to the Hartree–Fock equations (2.31). For a stationary point such as a minimum we require that a linear variation of the wave function, $\phi = \phi + \delta\phi$ should not affect the energy. Here $\delta\phi$ stands for a small change of the parameters in $\phi$. Thus, from

$$\int dr \sum_{\sigma} \left[\phi^*(r,\sigma) + \delta\phi^*(r,\sigma)\right] H_{\rm el} \left[\phi(r,\sigma) + \delta\phi(r,\sigma)\right] \approx \langle H_{\rm el}\rangle + \delta\langle H_{\rm el}\rangle \qquad (2.132)$$

follows that $\delta\langle H_{\rm el}\rangle = 0$. When applying this condition to find the optimum spin orbitals we will impose an additional restriction, namely that the latter shall be orthonormal

$$\int d^3\mathbf{r}_j \sum_{\sigma_j} \varphi^*_{\alpha_j}(\mathbf{r}_j,\sigma_j)\varphi_{\alpha_i}(\mathbf{r}_j,\sigma_j) = \delta_{\alpha_i\alpha_j} \,. \qquad (2.133)$$

The minimization of the energy using the flexibility of the single–particle orbitals under the constraint (2.133) can be performed by solving

$$\delta\left[\langle H_{\rm el}\rangle - \sum_{\alpha_j} \varepsilon_{\alpha_j} \int d\mathbf{r}_j \sum_{\sigma_j} \varphi^*_{\alpha_j}(\mathbf{r}_j,\sigma_j)\varphi_{\alpha_j}(\mathbf{r}_j,\sigma_j)\right] = 0 \,. \qquad (2.134)$$

Here the constraint has been accounted for using the method of Lagrange multipliers; the latter are denoted $\varepsilon_{\alpha_j}$.

Next we need to know the expectation value (2.30) in terms of the spin orbitals. The single–electron part is readily obtained due to the construction of the Slater determinant from single–particle spin orbitals. These orbitals imply that the two Slater determinants in (2.30) must contain the same spin orbitals and that only those terms contribute which have identical permutations in $\phi^*(r,\sigma)$ and $\phi(r,\sigma)$. The contributions of all $N_{\rm el}!$ such permutations are equal thus canceling the prefactor $1/N_{\rm el}!$. One obtains

$$\begin{aligned}
\sum_{j=1}^{N_{\rm el}} \langle h_{\rm el}(\mathbf{r}_j)\rangle &= \sum_{j=1}^{N_{\rm el}} \int d^3\mathbf{r}_j \sum_{\sigma_j} \varphi^*_{\alpha_j}(\mathbf{r}_j,\sigma_j) h_{\rm el}(\mathbf{r}_j)\varphi_{\alpha_j}(\mathbf{r}_j,\sigma_j) \\
&= 2\sum_{a}^{N_{\rm el}/2} \int d^3\mathbf{x}\, \varphi^*_a(\mathbf{x}) h_{\rm el}(\mathbf{x})\varphi_a(\mathbf{x}) \\
&\equiv 2\sum_{a}^{N_{\rm el}/2} h_{aa} \,. \qquad (2.135)
\end{aligned}$$

To arrive at this result we have used the fact that $h_{\rm el}(\mathbf{r}_j)$ carries no spin dependence. Thus, the summation over the two possible spin states results in a factor of 2. Further the sum over the electrons has been written as a sum over the $N_{\rm el}/2$ doubly occupied spatial orbitals.

We now turn to the calculation of the matrix elements of the Coulomb pair interaction in Eq. (2.30). Apparently, due to the indices of the Coulomb potential $V_{\mathrm{el-el}}(\mathbf{r}_i, \mathbf{r}_j)$ only those spin orbitals will be affected which have the same particle indices, $(i, j)$, in the two determinants. One can distinguish the following two cases: First, the order of the spin orbitals is identical in both determinants and second, the ordering in one determinant has been changed. Both cases will differ in sign since they differ in a single permutation. Putting this into formulas we have

$$\begin{aligned}\langle V_{\mathrm{el-el}}(\mathbf{r}_i, \mathbf{r}_j)\rangle &= \int d^3\mathbf{r}_i d^3\mathbf{r}_j \sum_{\sigma_i,\sigma_j} \\ &\times \Big[\varphi^*_{\alpha_i}(\mathbf{r}_i,\sigma_i)\varphi^*_{\alpha_j}(\mathbf{r}_j,\sigma_j)V_{\mathrm{el-el}}(\mathbf{r}_i,\mathbf{r}_j)\varphi_{\alpha_j}(\mathbf{r}_j,\sigma_j)\varphi_{\alpha_i}(\mathbf{r}_i,\sigma_i) \\ &- \varphi^*_{\alpha_j}(\mathbf{r}_i,\sigma_i)\varphi^*_{\alpha_i}(\mathbf{r}_j,\sigma_j)V_{\mathrm{el-el}}(\mathbf{r}_i,\mathbf{r}_j)\varphi_{\alpha_i}(\mathbf{r}_i,\sigma_i)\varphi_{\alpha_j}(\mathbf{r}_j,\sigma_j)\Big] \,. \end{aligned} \tag{2.136}$$

Before giving an interpretation we specify Eq. (2.136) to the case of closed shell configurations and perform the summation with respect to the spin functions. Since $V_{\mathrm{el-el}}(\mathbf{r}_i, \mathbf{r}_j)$ does not depend on spin the first term in Eq. (2.136) gives for the normalized spin functions $\sum_{\sigma_i} \zeta_{a_i}(\sigma_i)\zeta_{a_i}(\sigma_i) \sum_{\sigma_j} \zeta_{a_j}(\sigma_j)\zeta_{a_j}(\sigma_j) = 4$. For the second term one has $\sum_{\sigma_i} \zeta_{a_i}(\sigma_i)\zeta_{a_j}(\sigma_i)$ $\times \sum_{\sigma_j} \zeta_{a_i}(\sigma_j)\zeta_{a_j}(\sigma_j) = 2$, if both electrons have the *same* spin; otherwise the summation gives zero.

The respective contribution to the expectation value of the energy is according to (2.30) obtained by summing over all indices $i$ and $j$ and multiplying the result by 1/2. Instead of summing over $(i, j)$ we can equivalently take the summation with respect to all $N_{\mathrm{el}}/2$ occupied spatial orbitals. This finally gives for the ground state energy to be minimized

$$\langle H_{\mathrm{el}}\rangle = 2\sum_a^{N_{\mathrm{el}}/2} h_{aa} + \sum_{ab}^{N_{\mathrm{el}}/2} [2J_{ab} - K_{ab}] \,. \tag{2.137}$$

From the first term in Eq. (2.136) we have the matrix element $J_{ab}$ defined as

$$J_{ab} = \int d^3\mathbf{x}\, d^3\mathbf{x}'\, |\varphi_a(\mathbf{x})|^2 V(\mathbf{x},\mathbf{x}')|\varphi_b(\mathbf{x}')|^2 \,. \tag{2.138}$$

This matrix describes the classical Coulomb interaction between charge densities due to electrons occupying the orbitals $\varphi_a(\mathbf{x})$ and $\varphi_b(\mathbf{x})$. The second matrix

$$K_{ab} = \int d^3\mathbf{x}\, d^3\mathbf{x}'\, \varphi_a(\mathbf{x})\varphi^*_a(\mathbf{x}')\, V(\mathbf{x},\mathbf{x}')\, \varphi^*_b(\mathbf{x})\varphi_b(\mathbf{x}') \tag{2.139}$$

has no classical analogue. Compared with (2.138) in (2.139) the electron "labels" $\mathbf{x}$ and $\mathbf{x}'$ of the spatial orbitals are exchanged. Further this term only gives a contribution if the electrons in the two considered spin orbitals have the same spin. Thus, two electrons having the same spin experience an extra potential and they do not move independently. This effect is usually called *exchange correlation* and it is a direct consequence of the choice of an antisymmetric ansatz for the total wave function.[14]

[14]Note that for $a = b$ in (2.137) the sum describes the interaction between electrons in the same spatial orbital but with different spin. Of course, this is pure Coulomb interaction since $K_{aa} = J_{aa}$.

The matrices $J_{ab}$ and $K_{ab}$ can also be understood as the matrix elements of the Coulomb operator

$$J_a(\mathbf{x}) = \int d^3\mathbf{x}' |\varphi_a(\mathbf{x}')|^2 \, V(\mathbf{x}', \mathbf{x}) \tag{2.140}$$

and the exchange operator

$$K_b(\mathbf{x})\varphi_a(\mathbf{x}) = \Big[\int d^3\mathbf{x}' \varphi_b^*(\mathbf{x}') \, V(\mathbf{x}, \mathbf{x}') \, \varphi_a(\mathbf{x}')\Big] \varphi_b(\mathbf{x}) \, , \tag{2.141}$$

respectively.

Having specified the expectation value of the electronic energy and using

$$\sum_{\alpha_i} \varepsilon_{\alpha_i} \int d\mathbf{r}_i \sum_{\sigma_i} \varphi^*_{\alpha_i}(\mathbf{r}_i, \sigma_i) \varphi_{\alpha_i}(\mathbf{r}_i, \sigma_i) = 2 \sum_a^{N_{\mathrm{el}}/2} \varepsilon_a \int d^3\mathbf{x} \; \varphi_a^*(\mathbf{x}) \varphi_a(\mathbf{x}) \, , \tag{2.142}$$

the variational determination of the spatial orbitals according to Eq. (2.134) can be performed by introducing $\varphi_a = \varphi_a + \delta\varphi_a$ as mentioned above. After some straightforward algebra one obtains the Hartree–Fock equation (2.31) for the determination of the optimal orbitals for a closed shell configuration.

### 2.8.2 Franck–Condon Factors

In Section 2.6.2 we have discussed the Franck–Condon factors which describe the overlap between wave function of different potential energy surface. The expression Eq. (2.98) is limited to the case of two harmonic potentials of *equal* curvature. If the curvatures are different the resulting expressions become more complicated. In terms of the numerical implementation, however, it is much more convenient to express the Franck–Condon factors via recursion relations. Their derivation for the general case of different curvatures will be outlined in the following.

Using the operator notation introduced in Section 2.6.2 the Franck–Condon factor reads

$$\langle \chi_{aM} | \chi_{bN} \rangle = \langle 0_a | \frac{(C_a)^M}{\sqrt{M!}} D(g_a) D^+(g_b) \frac{(C_b^+)^N}{\sqrt{N!}} | 0_b \rangle \, , \tag{2.143}$$

where we skipped the normal mode index but accounted for the fact that the operators and the vacuum states depend on the index of the potential energy surface because of the different frequencies. To proceed we have to reformulate Eq. (2.143) into a state vector product which only contains one type of oscillator operator, e.g., $C_a$ and one type of vacuum $| \, 0_a \rangle$. This is possible if we use the so–called squeezing operator

$$S_b^+(z) = \exp(z(C_b^2 - C_b^{+\,2})/2) \tag{2.144}$$

to write

$$C_a = S_b^+(z_{ab}) \, C_b \, S_b(z_{ab}) \tag{2.145}$$

with $z_{ab} = \ln(\omega_a/\omega_b)/2$. After some algebra one finds the following expression for the Franck–Condon factor

$$\langle\chi_{aM}|\chi_{bN}\rangle = \langle 0 \mid \frac{C^M}{\sqrt{M!}} D(g)\, S(z)\, \frac{C^{+\,N}}{\sqrt{N!}} \mid 0\rangle\;. \tag{2.146}$$

Here, we have introduced $g = g_a - g_b\sqrt{\epsilon}$, $\epsilon = \omega_a/\omega_b$, $C = C_a$, $|0\rangle = |0_a\rangle$, and $z = z_{ab} = (\ln\epsilon)/2$. holds. Starting with the interchange of one annihilation operator from the left to the right in Eq. (2.146) a recursion relations for the Franck–Condon factor can be derived. One obtains

$$\begin{aligned}\langle\chi_{aM}|\chi_{bN}\rangle &= \sqrt{\frac{N-1}{N}}\frac{c2-\epsilon}{1+\epsilon}\langle\chi_{aM}|\chi_{bN-1}\rangle - \frac{2g\sqrt{\epsilon}}{\sqrt{N}(1+\epsilon)}\langle\chi_{aM}|\chi_{bN-1}\rangle \\ &+\sqrt{\frac{M\epsilon}{N}}\frac{2}{1+\epsilon}\langle\chi_{aM-1}|\chi_{bN-1}\rangle\end{aligned} \tag{2.147}$$

and

$$\begin{aligned}\langle\chi_{aM}|\chi_{bN}\rangle &= -\sqrt{\frac{M-1}{M}}\frac{c2-\epsilon}{1+\epsilon}\langle\chi_{aM-2}|\chi_{bN}\rangle + \frac{2g}{\sqrt{M}(1+\epsilon)}\langle\chi_{aM-1}|\chi_{bN}\rangle \\ &+\sqrt{\frac{N\epsilon}{M}}\frac{2}{1+\epsilon}\langle\chi_{aM-1}|\chi_{bN-1}\rangle\;.\end{aligned} \tag{2.148}$$

Notice that terms with "negative" quantum numbers have to be set equal to zero. The initial value for the recursion relations can be simply calculated in the coordinate representation which gives

$$\langle\chi_{a0}|\chi_{b0}\rangle = \frac{\sqrt{2\sqrt{\epsilon}}}{\sqrt{1+\epsilon}}\exp\left(-\frac{g^2}{1+\epsilon}\right)\;. \tag{2.149}$$

Eqs. (2.147)-(2.149) together with the relation $\langle\chi_{aM}|\chi_{bN}\rangle = (\langle\chi_{bN}|\chi_{aM}\rangle)^*$ allow for a numerically stable determination of the Franck–Condon overlap integrals.

### 2.8.3 The Two–Level System

There are many situations where the relevant molecular system can be modeled as an effective two–level system. A prominent example is given by the one–dimensional double minimum potential shown in Fig. 2.16. This type of potential is relevant for isomerization reactions such as intramolecular proton transfer. Provided the temperature is low enough such that thermal occupation of higher states is negligible the dynamics for the situation of Fig. 2.16 is readily described in terms of the lowest two states. In the following we will study the eigenstates as well as the population dynamics of a generic two–level system. This exactly solvable model will provide a reference case for the subsequent discussions.

The Hamiltonian for a two–level system can be written in two alternative ways. First, we can assume that we know the eigenstates $|\pm\rangle$ and eigenenergies $\mathcal{E}_\pm$, for instance, of the model potential shown in Fig. 2.16. Then we can write

$$H = \sum_{\kappa=\pm}\mathcal{E}_\kappa|\kappa\rangle\langle\kappa|\;. \tag{2.150}$$

If we do not know the eigenstates but some zeroth–order states $|1\rangle$ and $|2\rangle$ which correspond to a situation where, for instance, the coupling between the left and the right well in Fig. 2.16 is switched off, the Hamiltonian reads

$$\bar{H} = \varepsilon_1 |1\rangle\langle 1| + \varepsilon_2 |2\rangle\langle 2| + V|1\rangle\langle 2| + V^*|2\rangle\langle 1| \tag{2.151}$$

Here, the level energies of the zeroth–order states are denoted $\varepsilon_{a=1,2}$ and the coupling between these states is given by $V$. Independent of the specific situation the Hamiltonian (2.151) can be transformed to take the form (2.150). In the following we will outline how this diagonalization of (2.151) is achieved.

In a first step we determine the eigenvalues and eigenstates which follow from the stationary Schrödinger equation

$$H|\Psi\rangle = \mathcal{E}|\Psi\rangle \,. \tag{2.152}$$

We expand the state vector with respect to the states $|a = 1, 2\rangle$

$$|\Psi\rangle = C(1)|1\rangle + C(2)|2\rangle \,, \tag{2.153}$$

which leads to a matrix equation for the expansion coefficients $C(a = 1, 2)$,

$$\begin{pmatrix} \varepsilon_1 & V \\ V^* & \varepsilon_2 \end{pmatrix} \begin{pmatrix} C(1) \\ C(2) \end{pmatrix} = \mathcal{E} \begin{pmatrix} C(1) \\ C(2) \end{pmatrix} . \tag{2.154}$$

The eigenvalues are obtained from the secular equation

$$(\mathcal{E} - \varepsilon_1)(\mathcal{E} - \varepsilon_2) - |V|^2 = 0 \,. \tag{2.155}$$

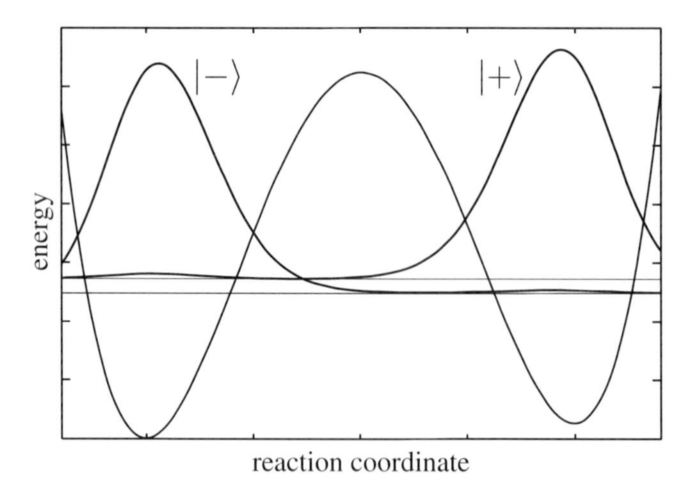


**Figure 2.16:** Potential energy surface along a reaction coordinate describing the intramolecular proton transfer in asymmetrically substituted malonaldehyde. The potential supports two below barrier states whose probability density is plotted here with a vertical offset corresponding to the respective eigenenergies (figure courtesy of H. Naundorf).

Solving this quadratic equation gives

$$\mathcal{E}_{\kappa=\pm} = \frac{1}{2}\left\{\varepsilon_1 + \varepsilon_2 \pm \sqrt{(\varepsilon_1 - \varepsilon_2)^2 + 4|V|^2}\right\} . \tag{2.156}$$

To determine the expansion coefficients and thus the eigenstates the $\mathcal{E}_{\kappa=\pm}$ are inserted into the eigenvalue equation (2.154),

$$\begin{pmatrix} \mathcal{E}_\kappa - \varepsilon_1 & -V \\ -V^* & \mathcal{E}_\kappa - \varepsilon_2 \end{pmatrix} \begin{pmatrix} C_\kappa(1) \\ C_\kappa(2) \end{pmatrix} = 0 . \tag{2.157}$$

Note, that the expansion coefficients $C(m)$ have been labeled by the quantum numbers $\kappa = \pm$. If we make use of the normalization condition

$$\sum_m |C_\kappa(m)|^2 = 1 \tag{2.158}$$

we obtain

$$|C_\kappa(1)|^2 = \frac{(\mathcal{E}_\kappa - \varepsilon_2)^2}{(\mathcal{E}_\kappa - \varepsilon_2)^2 + |V|^2} . \tag{2.159}$$

From Eq. (2.156) we get the relations

$$(\mathcal{E}_\kappa - \varepsilon_1)(\mathcal{E}_\kappa - \varepsilon_2) = |V|^2 \tag{2.160}$$

and

$$\mathcal{E}_+ + \mathcal{E}_- = \varepsilon_1 + \varepsilon_2 , \tag{2.161}$$

which, if inserted into (2.159), gives

$$\begin{aligned} |C_\kappa(1)|^2 &= \frac{(\mathcal{E}_\kappa - \varepsilon_2)^2}{(\mathcal{E}_\kappa - \varepsilon_2)^2 + (\mathcal{E}_\kappa - \varepsilon_1)(\mathcal{E}_\kappa - \varepsilon_2)} = \frac{\mathcal{E}_\kappa - \varepsilon_2}{\mathcal{E}_\kappa - \varepsilon_2 + \mathcal{E}_\kappa - \varepsilon_1} \\ &= \frac{\mathcal{E}_\kappa - \varepsilon_2}{\mathcal{E}_\kappa - \mathcal{E}_{\bar\kappa}} . \end{aligned} \tag{2.162}$$

To have compact notation we introduced $\bar\kappa = \pm$, if $\kappa = \mp$. The complex expansion coefficient itself reads

$$C_\kappa(1) = \sqrt{\frac{\mathcal{E}_\kappa - \varepsilon_2}{\mathcal{E}_\kappa - \mathcal{E}_{\bar\kappa}}}\, e^{i\chi_1(\kappa)} , \tag{2.163}$$

where the phase $\chi_1(\kappa)$ remains open at this point. In a similar manner we can derive

$$|C_\kappa(2)| = \sqrt{\frac{\mathcal{E}_\kappa - \varepsilon_1}{\mathcal{E}_\kappa - \mathcal{E}_{\bar\kappa}}} . \tag{2.164}$$

However, the phase of $C_\kappa(2)$ is not free but has to be determined from

$$C_\kappa(2) = |C_\kappa(2)|e^{i\chi_2(\kappa)} = \frac{|V|e^{-i\arg(V)}}{\mathcal{E}_\kappa - \varepsilon_2}\sqrt{\frac{\mathcal{E}_\kappa - \varepsilon_2}{\mathcal{E}_\kappa - \mathcal{E}_{\bar\kappa}}}\, e^{i\chi_1(\kappa)} . \tag{2.165}$$

We note that for $\kappa = +$, it is $\mathcal{E}_\kappa > \varepsilon_2$ and for $\kappa = -$, one has $\mathcal{E}_\kappa < \varepsilon_2$. Consequently, the phase $\chi_2(\kappa)$ is given by $\chi_2(+) = \chi_1(+) - \arg(V)$ and $\chi_2(-) = \chi_1(-) - \arg(V) + \pi$.

There exist alternative formulas for $|C_\kappa(1)|^2$ and $|C_\kappa(2)|^2$. Before presenting them we note that $|C_{\bar{\kappa}}(1)| = |C_\kappa(2)|$, what is easily demonstrated using, for example Eq. (2.161). To get the first alternative to the Eqs. (2.163) and (2.163) one introduces

$$\Delta E = \mathcal{E}_- - \varepsilon_2 \equiv \frac{1}{2}\left\{\varepsilon_1 - \varepsilon_2 + \sqrt{(\varepsilon_1 - \varepsilon_2)^2 + 4|V|^2}\right\} . \tag{2.166}$$

Using the abbreviation

$$\eta = \frac{\Delta E}{|V|} , \tag{2.167}$$

it follows that

$$|C_+(1)|^2 = |C_-(2)|^2 = \frac{\eta^2}{1+\eta^2} , \tag{2.168}$$

and

$$|C_-(1)|^2 = |C_+(2)|^2 = \frac{1}{1+\eta^2} . \tag{2.169}$$

To arrive at another alternative notation one defines the ratio

$$\lambda = \frac{2|V|}{|\Delta\varepsilon|} , \tag{2.170}$$

with $\Delta\varepsilon = \varepsilon_1 - \varepsilon_2$. We obtain for the expansion coefficients ("sgn" is the sign function)

$$|C_\kappa(1)|^2 = \frac{1}{2}\left(1 + \kappa\frac{\text{sgn}(\Delta\varepsilon)}{\sqrt{1+\lambda^2}}\right) . \tag{2.171}$$

Next we use the trigonometric relation

$$\cos^2\gamma = \frac{1}{2}\left(1 + \frac{1}{\sqrt{1+\tan^2(2\gamma)}}\right) , \tag{2.172}$$

which is also valid for the sine function after replacing the plus sign in the bracket on the right–hand side by a minus sign. We identify

$$\gamma = \frac{1}{2}\arctan\frac{2|V|}{|\Delta\varepsilon|} , \tag{2.173}$$

and obtain from Eq. (2.171) the expressions

$$|C_+(1)|^2 = |C_-(2)|^2 = \cos^2\left(\gamma + \frac{\pi}{4}[1 - \text{sgn}(\Delta\varepsilon)]\right) , \tag{2.174}$$

and

$$|C_-(1)|^2 = |C_+(2)|^2 = \sin^2\left(\gamma - \frac{\pi}{4}[1 - \text{sgn}(\Delta\varepsilon)]\right) . \tag{2.175}$$

The quantity $\gamma$ is the so–called *mixing angle*. Finally, we point out that the coefficients fulfill the condition

$$\sum_\kappa C_\kappa^*(m)C_\kappa(n) = \delta_{mn} , \tag{2.176}$$

which is obtained by expanding the orthogonal zeroth–order states in terms of the eigenstates.

### 2.8.4 The Linear Molecular Chain and the Molecular Ring

The linear molecular chain represents a simple model system for studying transfer phenomena as well as the behavior of energy spectra in dependence on the system size. In different contexts it is also known as the *tight–binding* or the *Hückel* model. We will encounter this model when discussing electron and excitation energy transfer in Chapter 6 and 8, respectively. In the present section we will focus on the most simple setup consisting of an arrangement of $N$ identical quantum levels at energy $\varepsilon_0$ and being coupled via the matrix element $V_{m,m+1} = V_{m-1,m} = V$, i.e., only nearest neighbor couplings are assumed. This situation might describe, for example, the diabatic states $|\varphi_m\rangle$ of different parts of an electron transfer system (donor, bridge, acceptor, see Chapter 6). The potential coupling between these diabatic electronic states is then given by $V$ (see also Eq. (2.120)).

This results in the following Hamiltonian

$$H_{\text{chain}} = \sum_{m=1}^{N} \varepsilon_0 |\varphi_m\rangle\langle\varphi_m| + \sum_{m=1}^{N-1} (V|\varphi_{m+1}\rangle\langle\varphi_m| + \text{h.c.}) . \tag{2.177}$$

In a first step we determine the eigenstates $|\Psi_a\rangle$ of the chain solving the stationary Schrödinger equation

$$H_{\text{chain}}|\Psi_a\rangle = \mathcal{E}_a|\Psi_a\rangle . \tag{2.178}$$

Since the states $|\varphi_m\rangle$ are supposed to be known we can expand the $|\Psi_a\rangle$ in this basis:

$$|\Psi_a\rangle = \sum_m C_a(m)|\varphi_m\rangle . \tag{2.179}$$

Inserting (2.179) into (2.178) and using (2.177) we obtain the equation for the expansion coefficients $C_a(m)$

$$(\mathcal{E}_a - \varepsilon_0)C_a(m) = V[C_a(m+1) + C_a(m-1)] , \tag{2.180}$$

which is valid for $1 < m < N$. For $m = 1$ and $m = N$ we have to take into account the finite structure of the chain. This gives two additional equations

$$(\mathcal{E}_a - \varepsilon_0)C_a(1) = VC_a(2) , \tag{2.181}$$

and

$$(\mathcal{E}_a - \varepsilon_0)C_a(N) = VC_a(N-1) . \tag{2.182}$$

The set of equations (2.180), (2.181), and (2.182) can be solved using the following ansatz

$$C_a(m) = C \sin(am) , \tag{2.183}$$

where $C$ is a real constant. Inserting (2.183) into (2.180) gives

$$(\mathcal{E}_a - \varepsilon_0)\sin(am) = V\left(\sin(a[m+1]) + \sin(a[m-1])\right) . \tag{2.184}$$

With the help of some theorems for trigonometric functions this equation can be transformed into

$$\mathcal{E}_a = \varepsilon_0 + 2V\cos a . \tag{2.185}$$

This expression tells us how the energy spectrum depends on the yet unknown quantum number $a$; the same result is obtained from Eq. (2.181). Eq. (2.182), however, gives the condition

$$(\mathcal{E}_a - \varepsilon_0)\sin(aN) = V\sin(a[N-1]) , \tag{2.186}$$

which can be rewritten as

$$\begin{aligned}((\mathcal{E}_a - \varepsilon_0) - 2V\cos a)\sin(aN) &= -V\left(\sin(aN)\cos a + \cos(aN)\sin a\right) \\ &= 0 ,\end{aligned} \tag{2.187}$$

where the second line follows from Eq. (2.185). Rearranging the right–hand side of Eq. (2.187) gives the condition for the eigenvalues

$$\sin\left(a(N+1)\right) = 0 , \tag{2.188}$$

which is solved by

$$a = \frac{\pi j}{N+1} \quad (j = 0, \pm 1, \pm 2, \ldots) . \tag{2.189}$$

Thus the energy spectrum becomes

$$\mathcal{E}_a = \varepsilon_0 + 2V\cos\left(\frac{\pi j}{N+1}\right) . \tag{2.190}$$

The normalization constant $C$ appearing in Eq. (2.183) is obtained from the relation

$$\sum_{m=1}^{N} |C_a(m)|^2 = C^2 \sum_m \sin^2(am) = 1 . \tag{2.191}$$

Using the tabulated result for the sum one arrives at

$$C = \sqrt{\frac{2}{N+1}} . \tag{2.192}$$

If one considers the expansion coefficients, Eq. (2.183) it is obvious, that they are identical to zero for $j = 0$ and for multiples of $N + 1$. Furthermore, an inspection of Eq. (2.190) and of Eq (2.183) shows that identical results are obtained for $j$ being in the interval $1, \ldots, N$, and for all other $\bar{j}$ which differ from $j$ by multiples of $N$. Therefore, $j$ has to be restricted to the interval $1, \ldots, N$.

In the remaining part of this section we will discuss the model of a molecular ring. Such a system we will encounter, for instance, in Chapter 8 (see, Fig. 8.4). To arrive at a Hamiltonian for a molecular ring, i.e. a circular and regular arrangement of identical molecules the following specification of the model for the chain becomes necessary. The first molecule of the chain is connected with the last one in a way that the coupling strength between both takes the value $V$ which is the strength of the nearest neighbor couplings between all other molecules, too. For this model we may use the ansatz, Eq. (2.179), but the expansion coefficient have to fulfill $C_a(m) = C_a(m + \nu N)$ (where $\nu$ is an integer). This requirement can be satisfied by choosing

$$C_a(m) = C \exp(iam) \tag{2.193}$$

with $a = 2\pi j/N$, $(j = 0, \ldots, N_{\text{mol}} - 1)$. Since $|C_a(m)|^2 = |C|^2$ one easily verifies that $C = 1/\sqrt{N}$. Inserting the expansion coefficients into Eq. (2.180) (the Eqs. (2.181), and (2.180) are dispensable) it again follows Eq. (2.190) for the eigenvalues $E_a$, but now with a modified definition of the quantum numbers $a$ as given above

---

# Notes


[Bae85] M. Baer, in *Theory of Chemical Reaction Dynamics*, M. Baer (ed.), (CRC Press, Bocca Raton, 1985), Vol. II, p. 219.

[Dom97] W. Domcke and G. Stock, Adv. Chem. Phys. **100**, 1 (1997).

[Mil80] W. H. Miller, N. C. Handy, J. E. Adams, J. Chem. Phys. **72** , 99 (1980).

[Ruf88] B. A. Ruf and W. H. Miller, J. Chem. Soc. Faraday Trans. 2 **84**, 1523 (1988).

# 3 Dynamics of Isolated and Open Quantum Systems

*A quantum mechanical description of time–dependent phenomena in two types of molecular systems is given. First, we consider small systems which are isolated from the surroundings. This situation can be modelled using the time–dependent Schrödinger equation. Some basic properties of the time–evolution operator are discussed and the concept of the scattering operator is introduced which can serve as a starting point of a perturbation expansion. It is shown that with increasing dimensionality of the considered system the treatment of transitions between different manifolds of quantum states can be replaced by a rate description based on the Golden Rule of quantum mechanics.*
*To go beyond a description of the system by a single wave function the density operator (statistical operator) is introduced. This concept when specified to the reduced density operator (reduced statistical operator) is used in the second part to treat the dynamics of the system when interacting with some macroscopic environment. The interaction can be systematically incorporated using the projection operator formalism. The latter is shown to provide a means to develop a perturbation theory in line with a reduction scheme onto the state space of the small system. Restricting ourselves to the second order of the perturbation expansion, we derive a Generalized Master Equation which is the basic equation for the considered system–reservoir situation. As a consequence of the reduction procedure this equation contains memory effects. The conditions for neglecting the latter are discussed and the Markovian Quantum Master Equation is obtained. The key quantity providing the link between both types of equations is shown to be the reservoir correlation function. As an example the dynamics of a coupled two–level system is discussed in detail.*
*Finally, we give a brief introduction into nonperturbative methods for dealing with condensed phase dynamics. In particular we discuss the path integral representation of the reduced density operator and the quantum–classical hybrid approach.*

## 3.1 Introduction

In the development of quantum theory the pioneers in this field concentrated on simple systems like the harmonic oscillator or the hydrogen atom assuming them to be isolated from the rest of the universe. The dynamics of such isolated quantum systems is completely described by the time–dependent Schrödinger equation for the wave function $\Psi(x,t)$,

$$i\hbar\frac{\partial}{\partial t}\Psi(x,t) = H\Psi(x,t) \,. \tag{3.1}$$

Here, $x$ comprises some set of degrees of freedom (DOF). An unambiguous solution of this first–order differential equation is obtained by fixing an initial wave function $\Psi(x,t_0)$. Provided Eq. (3.1) has been solved for a particular Hamilton operator $H$, the time dependence of physical observables of the system is given by the expectation values of the associated hermitian operators, $\hat{A}$, with respect to the time–dependent wave function, $A(t) = \langle\Psi(t)|\hat{A}|\Psi(t)\rangle$.[1]

However, the model of an isolated system is an oversimplification and different perturbations from the environment have to be taken into account. One may ask the question how the dynamics of the quantum system of interest (the system $S$) is influenced by some environment. Of course, the answer depends on the actual type of environment, and in particular on its coupling strength to the system. If the environment comprises only a small number of DOF, one can attempt to solve the time–dependent Schrödinger equation, but now for the system plus the small environment. A typical example are small clusters embedding a diatomic molecule. Such an approach is impossible if the environment is large and forms a macroscopic system $R$ (see Fig. 3.1). If the environment stays in thermal equilibrium at temperature $T$ as it is the case for many applications it represents a *heat bath* for the system $S$ and one has to resort to statistical methods as we will see below.

Any coupling to external DOF results in energy exchange between the system $S$ and its environment. If initially energy is deposited into $S$, it will be transferred to the reservoir in the course of time. The DOF of the reservoir accept the energy and distribute it among themselves. If the environment is a macroscopic system the energy is distributed over its huge number of DOF. At the end of this process the environment does not "feel" this negligibly small increase of its internal energy. If the environment stays in thermal equilibrium, $S$ will eventually relax into a state of thermal equilibrium with $R$. The situation is different for the case of a small environment. Here, all DOF may become noticeably excited and it may be possible that the energy moves back into the system $S$. This phenomenon is known as *recurrence*. The energy transfer from $S$ to its surroundings (possibly followed by a recurrence) is termed *relaxation*. If there is no chance for the energy to move back into $S$ the unidirectional energy flow into the environment is called *dissipation*.[2] Obviously, on short time scales the distinction between relaxation and dissipation is likely to be blurred. Hence, there is often no strict discrimination between the two terms in literature.

In the short–time limit it is possible that one enters a regime where the interaction of $S$ with its surroundings is negligible. An upper limit for this time scale would be given, for example, by the mean time between two scattering events of the molecule of interest with

[1] Note that whenever the context requires to distinguish operators from observables we will use a "hat" to mark the operator.

[2] If $S$ is "cold" compared to the macroscopic $R$ the reverse energy flow appears, but without any recurrence.

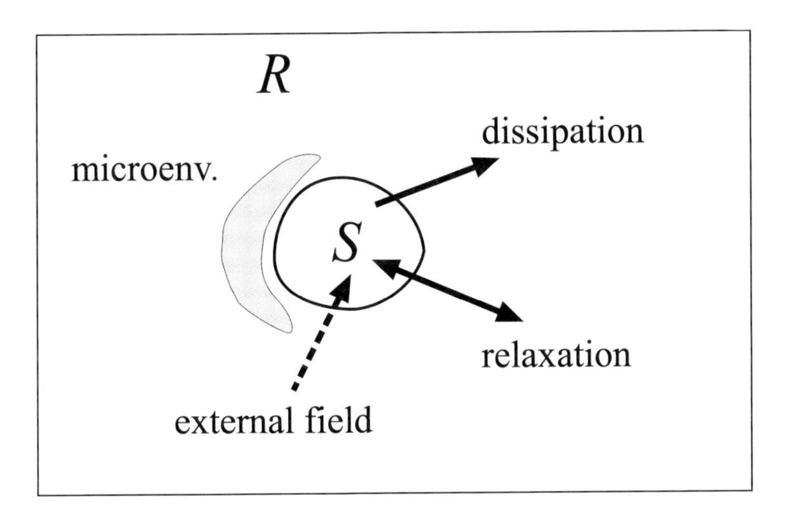


**Figure 3.1:** Schematic view of a typical situation encountered in condensed phase chemistry. A small system interacting with its surroundings (thermal reservoir or microenvironment) is investigated by means of an externally applied field. The system–reservoir interaction leads to unidirectional dissipation or bidirectional relaxation of energy initially disposed into the system.

the surrounding molecules. In this time region the time–dependent Schrödinger equation for the systems $S$ alone may provide an adequate description. This means that for a short time there exists a time–dependent wave function which, however, will be strongly disturbed in its evolution at later times. To indicate the existence of a quantum mechanical wave function during this early state of the time evolution of $S$ the motion is called *coherent*. If the coupling to the environment becomes predominant the motion changes to an *incoherent* one.

The incoherent motion can be described by time–dependent occupation probabilities, $P_a(t)$, of certain quantum states of the system, $|a\rangle$. The $P_a(t)$ are obtained as the solution of rate equations of the type

$$\frac{\partial}{\partial t}P_a = -\sum_b (k_{ab}P_a - k_{ba}P_b) \ . \tag{3.2}$$

This equation contains the rates (of probability transfer per unit time), $k_{ab}$, for the transition from $|a\rangle$ to $|b\rangle$. In the first term of the right–hand side the decrease of $P_a$ with time due to probability transfer from $|a\rangle$ to all other states is given. The reverse process is described by the second term which contains the transfer from all other states $|b\rangle$ into the state $|a\rangle$. Eq. (3.2) was "intuitively derived" by W. Pauli in 1928. It is frequently called *Pauli Master Equation* or just *Master Equation*. It is already obvious at this point that a method is required which allows to connect the description of *coherent* and *incoherent* motion. Before dealing with this problem we will give a more general characterization of the quantum system interacting with an environment.

There is an impressive number of different experiments which are of the type that rather small systems are studied under the influence of a thermal environment using external electromagnetic fields. A typical example is optical spectroscopy of dye molecules (the system $S$) in solution (the reservoir $R$). Studying electronic and vibrational transitions in these molecules, which are induced by the absorption of photons, one simultaneously detects the influence of the solvent molecules. This influence is often given by a random sequence of scattering events between the dye and the solvent molecules.[3] As a consequence there is a stochastic modulation of the initial and final states involved in the optical transition. A closer look at this example provides us with some general aspects of condensed phase dynamics.

First, experimentalists seek to arrange their setup in such a manner that the external field exclusively acts on the dye molecule (solute), without directly influencing the solvent molecules. Indeed, to carry out a good experiment it is necessary to minimize any disturbances of the medium the molecular system of interest is contained in. This situation demands for a theoretical description which focused on the DOF of the molecular system but does not a priori neglect the influence of the environment. In terms of the probabilistic aspect inherent in quantum mechanics this means that *reduced probabilities* valid only for the molecular system $S$ have to be introduced.

Second, if molecular properties are sensitive to the so–called *microenvironment*, i.e., if energy levels and other intramolecular quantities change their values with a change of the molecular structure in close proximity to the studied molecule, a careful description of the system–reservoir coupling has to be carried out or $S$ is supplemented by the microenvironment. An example for a microenvironment is the first solvation shell of molecules in solution (see Fig. 3.1).

Third, experiments, although possible, usually do not concentrate on the spectroscopy of a single molecule. Instead, a large number of molecules is excited simultaneously by the external field. Therefore, an *averaging* with respect to this *ensemble* has to be carried out also in the theoretical description. According to statistical physics the ensemble average can be replaced by an average taken with respect to the possible states of the environment $R$, provided all molecules are identical. The standard example is a thermal environment where this averaging is done using the canonical distribution function for a given temperature $T$.

Fourth, one often studies a system of identical molecules. But it is very likely that every molecule feels a somewhat different environment and as a result molecular properties like the electronic energy spectrum, vibrational frequencies, dipole moments etc. may differ from molecule to molecule. In this case, we have some static *disorder* in the system and an additional averaging over the different possible values of, for example, the transition frequencies is necessary. This particular situation may lead to a broad absorption band in the linear optical spectroscopy of the respective transition. Since this broadening is caused by different values of the transition frequency found for different molecules located at different points in the probed sample, it is called *inhomogeneous broadening*. In contrast the line broadening caused by the rapid stochastic fluctuations of the molecular properties is called *homogeneous*.

And finally, if the reservoir is noticeably disturbed by the dynamics of the molecular system the state of the microenvironment may be driven away from equilibrium and a description in terms of a thermal equilibrium distribution of the whole reservoir becomes invalid.

The *density matrix* formalism is the key to the theoretical description of condensed phase experiments. It was introduced by L. Landau and J. von Neumann in 1927. Before looking

[3]Note that a polar solvent may also act on the dye via long–range electrostatic forces.

at the concept of density matrices in more detail we shall introduce some useful definitions. In the following we will refer to the molecular system of interest or more specifically to all those DOF of a molecule which actively participate in a particular experiment as the *relevant system* or *active system* $S$. All other DOF form the irrelevant part of the system. For nearly all applications discussed below this irrelevant part forms a macroscopic *reservoir* $R$ and is assumed to stay in thermal equilibrium at some temperature $T$, i.e., it can be considered as a *heat bath* (see Fig. 3.1).

Usually the relevant quantum system $S$ consists of a small number of DOF ($< 5$) and has a relatively simple energy spectrum. It is the aim of the theory explained in the following to study the dynamic properties of $S$ on a microscopic basis. In contrast the reservoir $R$ consists of a large number of DOF ($10^2 \dots 10^{23}$) and may form a macroscopic system. Since the reservoir does not participate in an active manner in the dynamics initiated, for example, by an externally applied field, we do not aim at its detailed description. As a matter of fact, statistical physics tells us that such a detailed knowledge is not only impossible but useless as well. Instead, a formulation in terms of quantum statistics, classical statistics or of stochastic concepts is appropriate. Here, the choice of the approach is dictated by the problem at hand. For instance, most liquid phase environments are very likely to behave classically.

One basic question which will be answered by the theory introduced in this chapter is: How do the equilibrated reservoir DOF influence the externally induced dynamics of the relevant system? Our starting point for developing the formalism is the general Hamiltonian

$$H = H_{\rm S} + H_{\rm S-R} + H_{\rm R} \ . \tag{3.3}$$

It is composed of the Hamiltonian $H_{\rm S}$ of the relevant system, the Hamiltonian $H_{\rm R}$ of the reservoir and the interaction $H_{\rm S-R}$ between them. For the moment let the system be characterized by the set of coordinates $s = \{s_j\}$ and their conjugate momenta $p = \{p_j\}$. The reservoir coordinates and momenta are $Z = \{Z_\xi\}$ and $P = \{P_\xi\}$, respectively. Note that this type of Hamiltonian has already been considered in Section 2.6.3. There, however, isolated polyatomics were discussed. We shall see in Chapter 4 how the concept of a Taylor expansion of the global PES around some stable equilibrium configurations leads to Hamiltonians of the type (3.3) even in the context of condensed phase problems.

Since $S$ and $R$ are coupled by means of $H_{\rm S-R}$, it is impossible to introduce a wave function of the system or the reservoir alone. There only exists the total wave function, $\Psi(s, Z)$, which does not factorize into a system part $\Phi_{\rm S}(s)$ and a reservoir part $\chi_{\rm R}(Z)$,

$$\Psi(s, Z) \neq \Phi_{\rm S}(s)\, \chi_{\rm R}(Z) \ , \tag{3.4}$$

unless the coupling between $S$ and $R$ vanishes.

To accomplish the aim of the present approach, i.e., to treat the system dynamics without an explicit consideration of the reservoir dynamics, one could attempt to reduce the wave function $\Psi(s, Z)$ to a part depending on the system coordinates $s$ alone. But, in quantum mechanics we have a probabilistic interpretation of the *square* of the wave function. Thus, the only reduced quantity which can be introduced is the reduced probability density following from an integration of $|\Psi(s, Z)|^2$ with respect to all reservoir coordinates $Z$.

We encounter a generalization of this reduced probability distribution if we try to define the expectation value of an observable described by the hermitian operator $\hat{O}$ which acts in the state space of $S$ only, i.e., we have $\hat{O} = O(s)$. (A dependence on the momenta $p$ is possible but does not change any conclusion given below.) The expectation value reads

$$\langle \hat{O} \rangle = \int ds\, dZ\, \Psi^*(s, Z) O(s) \Psi(s, Z) \; . \tag{3.5}$$

If we introduce

$$\rho(s, \bar{s}) = \int dZ\, \Psi^*(s, Z) \Psi(\bar{s}, Z) \; , \tag{3.6}$$

Eq. (3.5) can be rewritten as

$$\langle \hat{O} \rangle = \int ds\, [O(\bar{s}) \rho(s, \bar{s})]_{s=\bar{s}} \; . \tag{3.7}$$

In this notation the averaging with respect to the large number of reservoir coordinates is absorbed in the definition of $\rho(s, \bar{s})$. Changing from the coordinate representation to a representation with respect to some discrete system quantum numbers $a$, $b$, ... the name *density matrix* introduced for $\rho_{ab}$ becomes obvious. The density matrix $\rho(s, \bar{s})$ or $\rho_{ab}$ shall be more precisely called *reduced density matrix*, since it is the result of a reduction of the total probability density onto the state space of the relevant system.

If there is no coupling between the system and the reservoir, i.e., if $H_{\mathrm{S-R}} = 0$, the density matrix is given as a product of wave functions

$$\rho(s, \bar{s}) = \Phi^*_{\mathrm{S}}(s) \Phi_{\mathrm{S}}(\bar{s}) \; . \tag{3.8}$$

Since this expression contains no more information than the wave function itself, it should be clear that in the case of a quantum system isolated from its environment the characterization by a wave function should be sufficient.

Besides the convenience of notation, density matrices offer a systematic way to describe the dynamics of the reduced quantum system embedded in a thermal reservoir. This theme will explored in the remainder of this chapter. In Section 3.2 we start with reviewing some fundamental aspects of time–dependent quantum mechanics as based on the Schrödinger equation. This will lead us to the important result of the Golden Rule description of quantum transitions in the relevant system. In Section 3.4 the density matrix formalism will be introduced in detail. Equations of motion for the reduced density operator are derived whose approximate treatment is considered in Sections 3.5–3.8. Further methods for describing the quantum dynamics in a molecular system are offered in the Sections 3.9 – 3.11.

## 3.2 Time–Dependent Schrödinger Equation

### 3.2.1 The Time–Evolution Operator

The time–dependent Schrödinger equation, which has been given in Eq. (3.1) in the coordinate representation, will be discussed without using a particular representation in the following. To this end the state vector $|\Psi\rangle$ is introduced which is related to the wave function $\Psi(x)$ through

$\langle x|\Psi\rangle$ ($|x\rangle$ comprises the eigenstates of the system coordinate operators). Using the state vector notation Eq. (3.1) becomes

$$i\hbar\frac{\partial}{\partial t}|\Psi(t)\rangle = H|\Psi(t)\rangle \,, \tag{3.9}$$

and the initial value of the state vector is $|\Psi_0\rangle \equiv |\Psi(t_0)\rangle$. If the Hamiltonian is time–independent a formal solution of Eq. (3.9) is given by

$$|\Psi(t)\rangle = e^{-iH(t-t_0)/\hbar}|\Psi_0\rangle \,. \tag{3.10}$$

The exponential function which contains the Hamiltonian is defined via a Taylor expansion: $\exp\{-iHt/\hbar\} = 1 - iHt/\hbar + \dots$. This expression is conveniently written by introducing the time–evolution operator

$$U(t,t_0) \equiv U(t-t_0) = e^{-iH(t-t_0)/\hbar} \,. \tag{3.11}$$

Note that in the case of a time–dependent Hamiltonian $U(t,t_0) \neq U(t-t_0)$ (see below). The operator $U(t,t_0)$ is unitary and obeys the following equation of motion

$$i\hbar\frac{\partial}{\partial t}U(t,t_0) = HU(t,t_0) \,, \tag{3.12}$$

with the initial condition $U(t_0,t_0) = 1$. The time–evolution operator has the important property that it can be decomposed as

$$U(t,t_0) = U(t,t_{N-1})U(t_{N-1},t_{N-2})\dots U(t_2,t_1)U(t_1,t_0) \tag{3.13}$$

where $t_1 \leq t_2 \dots \leq t_{N-1}$ are arbitrary times in the interval $[t_0,t]$. Note that Eqs. (3.12) – (3.13) are also valid if the Hamiltonian depends explicitly on time (see below).

If the solution of the stationary Schrödinger equation

$$H|a\rangle = E_a|a\rangle \tag{3.14}$$

with eigenstates $|a\rangle$ and eigenvalues $E_a$ is known, it is straightforward to solve the time–dependent Schrödinger equation (3.9). To do this we expand the state vector with respect to the states $|a\rangle$ which form a complete basis. We have

$$|\Psi(t)\rangle = \sum_a c_a(t)|a\rangle \,. \tag{3.15}$$

Since the state vector is time–dependent, the expansion coefficients $c_a(t) = \langle a|\Psi(t)\rangle$ are time–dependent as well. Using Eq. (3.10) and the eigenvalue equation (3.14) we may write

$$c_a(t) = \langle a|e^{-iE_a(t-t_0)/\hbar}|\Psi_0\rangle = e^{-iE_a(t-t_0)/\hbar}c_a(t_0) \,, \tag{3.16}$$

and the solution of the time dependent Schrödinger equation is obtained as a superposition of oscillatory terms[4]

$$|\Psi(t)\rangle = \sum_a c_a(t_0)e^{-iE_a(t-t_0)/\hbar}|a\rangle \,. \tag{3.17}$$

[4]If the Hamiltonian has also a continuous spectrum the sum over the states has to be replaced by an integral with respect to the continuous energy.

Which oscillations are present is determined by the expansion coefficients $c_a(t_0) = \langle a|\Psi(t_0)\rangle$ of the initial value of the state vector. In Section 3.8.7 we give a detailed discussion of the dynamics of a simple yet nontrivial system, the coupled two–level system.

The superposition state Eq. (3.17) is known as a *wave packet* from standard textbooks on quantum mechanics. This name has its origin in the fact that such a superposition of state vectors may correspond to a localized probability distribution if it is transformed into the coordinate representation. Since the state vector $|\Psi(t)\rangle$ is given here as a superposition of (time–dependent) states $c_a(t)|a\rangle$, it is alternatively called *coherent* superposition state. This coherent superposition is phase sensitive and so called *quantum beats* in the time evolution of the occupation probability of eigenstates can occur (see Fig. 3.2 below).

If we choose the initial state for the solution of the time–dependent Schrödinger equation according to $|\Psi_0\rangle = |a\rangle$ we get $|\Psi(t)\rangle = \exp(-iE_\alpha(t-t_0)/\hbar)|a\rangle$. Here the initial state is multiplied by a time–dependent phase factor which cancels when looking at probabilities, $|\Psi(t)|^2$. Hence, we can state that time–dependent phenomena such as quantum beats in an isolated quantum system can only be expected if a non–eigenstate, i.e., a superposition of eigenstates, is initially prepared.

Let us calculate the time–dependent expectation value of the operator $\hat{O}$:

$$O(t) = \langle\Psi(t)|\hat{O}|\Psi(t)\rangle = \sum_{a,b} c_b^*(t_0)c_a(t_0)\langle b|\hat{O}|a\rangle e^{-i(E_a-E_b)(t-t_0)/\hbar} . \quad (3.18)$$

The different time–dependent contributions are determined by *transition frequencies* $\omega_{ab} = (E_a - E_b)/\hbar$ which follow from combinations of the eigenvalues of the Hamiltonian $H$. The time–dependent expectation value, Eq. (3.18), can be rewritten using the time–evolution operator, Eq. (3.11), as

$$O(t) = \langle\Psi(t_0)|U^+(t,t_0)\hat{O}U(t,t_0)|\Psi(t_0)\rangle . \quad (3.19)$$

By means of this relation the time–dependence of the state vector can be transferred to the operator. This yields the so–called *Heisenberg* picture where time–dependent operators are defined as

$$\hat{O}^{(\mathrm{H})}(t) = U^+(t,t_0)\hat{O}U(t,t_0) \quad (3.20)$$

and the state vector is time–independent.

In the case where the states $|a\rangle$ are also eigenstates of $\hat{O}$ with eigenvalues $o_a$, Eq. (3.18) simplifies to

$$O(t) = \sum_a |c_a(t_0)|^2 o_a , \quad (3.21)$$

i.e., the expectation value becomes time–independent. If $\hat{O}$ is the Hamiltonian itself this relation reflects energy conservation during the time evolution of a wave function which is not an eigenstate of the system Hamiltonian.

If $\hat{O}$ is the projector $|\Psi_0\rangle\langle\Psi_0|$ on the initial state we obtain (note $t_0 = 0$)

$$P_{\mathrm{surv}}(t) = \langle\Psi(t)|\Psi_0\rangle\langle\Psi_0|\Psi(t)\rangle = \sum_{a,b} |c_a(0)\,c_b(0)|^2 e^{-i\omega_{ab}t} . \quad (3.22)$$

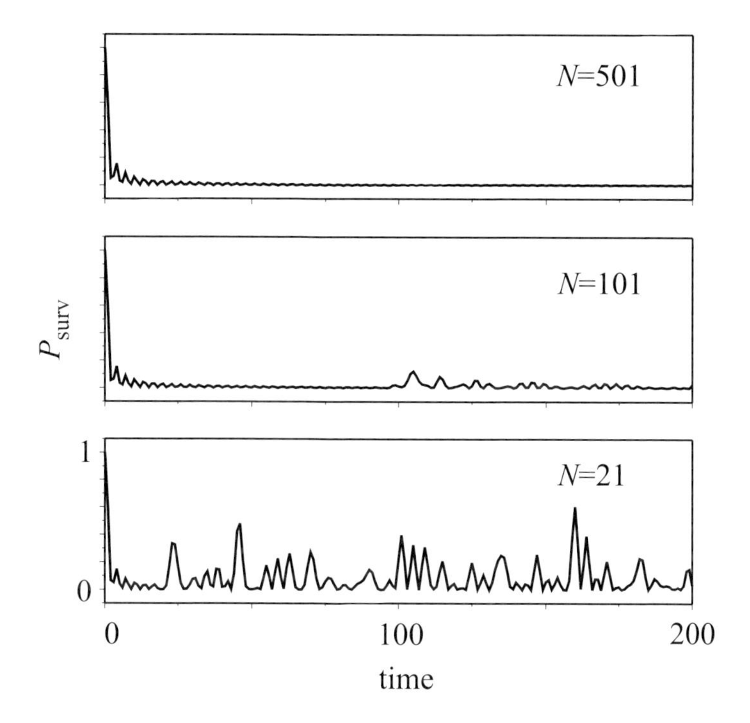


**Figure 3.2:** Survival probability for a system with $N$ eigenstates. The energy spectrum is that of a linear regular chain according to Eq. (2.185), the time is given in units of $\hbar/2V$, and the initial values $c_a(0)$ have been set equal to $1/\sqrt{N}$.

The expression is called *survival probability* since $\langle\Psi_0|\Psi(t)\rangle$ gives the probability amplitude for the initial state to be present in the actual state $|\Psi(t)\rangle$ at time $t$. $P_{\text{surv}}(t)$ has a time–independent part given by all terms with $a = b$. The summation over the different terms with $a \neq b$ which oscillate with time give rise to a decay of survival probability. Since this is due to the fact that the different terms are running out of phase one speaks about a *dephasing* at this point. Depending on the number of eigenstates a *rephasing* triggering a recurrence peak in $P_{\text{surv}}(t)$ may occur during a later stage of the evolution.

In order to illustrate dephasing we show in Fig. 3.2 the survival probability for a system with $N$ eigenstates whose energy spectrum is given by that of a linear molecular chain (cf. Section 2.8.4). To get a pronounced behavior we take as the initial state an equal distribution of probability ($c_a(0) = 1/\sqrt{N}$). It is evident from this figure that with increasing $N$ the structured behavior of $P_{\text{surv}}(t)$ seen for $N = 21$ disappears in the considered time interval. Note that for $N = 101$ there is some indication of a partial rephasing in the middle of the interval.

We notice that even when there is complete dephasing the survival amplitude does not decay to zero, but to the time–independent limit $P_{\text{surv}}(t \gg 0) = \sum_a |c_a(0)|^4$. Since the coefficients are normalized the asymptotic value $P_{\text{surv}}(t \gg 0)$ will be proportional to the inverse number of basis states $|a\rangle$ present in the initial state $|\Psi_0\rangle$. Thus only in the case of an infinite number of eigenstates participating in the dynamics is it possible that the survival amplitude vanishes completely.

### 3.2.2 The Interaction Representation

If the Hamiltonian $H$ of the system under consideration can be decomposed as $H = H_0 + V$ where $V$ represents a small perturbation of the dynamics given by $H_0$, a perturbation expansion with respect to $V$ can be performed. Usually one will attempt to separate $H$ such that the eigenvalue problem of $H_0$ can be solved analytically or by means of numerical diagonalization. Provided such a separation can be made the time–dependent state vector

$$|\Psi(t)\rangle = U(t, t_0)|\Psi(t_0)\rangle \tag{3.23}$$

is conveniently written as

$$|\Psi(t)\rangle = U_0(t, t_0)|\Psi^{(\mathrm{I})}(t)\rangle \; . \tag{3.24}$$

This representation makes use of the formal solution which is available for the unperturbed time–dependent Schrödinger equation for $H_0$ (Eq. (3.10)), i.e.

$$U_0(t, t_0) = e^{-iH_0(t-t_0)/\hbar} \; . \tag{3.25}$$

The new state vector $|\Psi^{(\mathrm{I})}(t)\rangle$ is called the state vector in the *interaction representation*. Since $U(t_0, t_0) = 1$ we have

$$|\Psi^{(\mathrm{I})}(t_0)\rangle = |\Psi(t_0)\rangle \; . \tag{3.26}$$

The equation of motion for the state vector in the interaction representation follows directly from the original time–dependent Schrödinger equation,

$$i\hbar\frac{\partial}{\partial t}|\Psi(t)\rangle = U_0(t, t_0)\left(H_0|\Psi^{(\mathrm{I})}(t)\rangle + i\hbar\frac{\partial}{\partial t}|\Psi^{(\mathrm{I})}(t)\rangle\right) = H|\Psi(t)\rangle \; . \tag{3.27}$$

After some rearrangement we get (note that $U^{-1} = U^{+}$)

$$i\hbar\frac{\partial}{\partial t}|\Psi^{(\mathrm{I})}(t)\rangle = U_0^{+}(t, t_0)VU_0(t, t_0)|\Psi^{(\mathrm{I})}(t)\rangle \equiv V^{(\mathrm{I})}(t)|\Psi^{(\mathrm{I})}(t)\rangle \; . \tag{3.28}$$

The quantity $V^{(\mathrm{I})}(t)$ is the interaction representation of the perturbational part of the Hamiltonian. This representation is defined for an arbitrary operator $\hat{O}$ as

$$\hat{O}^{(\mathrm{I})}(t) = U_0^{+}(t, t_0)\hat{O}U_0(t, t_0) \tag{3.29}$$

The formal solution of Eq. (3.28) is obtained by introducing the so–called *S–operator* (the scattering matrix) defined via

$$|\Psi^{(\mathrm{I})}(t)\rangle = S(t, t_0)|\Psi^{(\mathrm{I})}(t_0)\rangle \equiv S(t, t_0)|\Psi(t_0)\rangle \; , \tag{3.30}$$

where we made use of Eq. (3.26). Comparison with Eq. (3.24) yields

$$U(t,t_0) = U_0(t,t_0)S(t,t_0) \,. \tag{3.31}$$

The $S$–operator can be determined by the iterative solution of the equation of motion (3.28) for $|\Psi^{(\mathrm{I})}\rangle$. Formal time–integration gives

$$|\Psi^{(\mathrm{I})}(t)\rangle = |\Psi^{(\mathrm{I})}(t_0)\rangle - \frac{i}{\hbar}\int_{t_0}^{t} d\tau V^{(\mathrm{I})}(\tau)|\Psi^{(\mathrm{I})}(\tau)\rangle \,. \tag{3.32}$$

This equation is suited to develop a perturbation expansion with respect to $V^{(\mathrm{I})}$. If there is no interaction one gets

$$|\Psi^{(\mathrm{I},0)}(t)\rangle = |\Psi^{(\mathrm{I})}(t_0)\rangle \,. \tag{3.33}$$

If we insert this result into the right–hand side of Eq. (3.32) we get the state vector in the interaction representation, which is the first–order correction to $|\Psi^{(\mathrm{I},0)}(t)\rangle$ in the presence of a perturbation,

$$|\Psi^{(\mathrm{I},1)}(t)\rangle = -\frac{i}{\hbar}\int_{t_0}^{t} d\tau_1 V^{(\mathrm{I})}(\tau_1)|\Psi^{(\mathrm{I},0)}(\tau_1)\rangle \,. \tag{3.34}$$

Upon further iteration of this procedure one obtains the $n$th–order correction as

$$|\Psi^{(\mathrm{I},n)}(t)\rangle = -\frac{i}{\hbar}\int_{t_0}^{t} d\tau_n V^{(\mathrm{I})}(\tau_n)|\Psi^{(\mathrm{I},n-1)}(\tau_n)\rangle \,. \tag{3.35}$$

Thus the total formally exact state vector in the interaction representation is

$$|\Psi^{(\mathrm{I})}(t)\rangle = \sum_{n=0}^{\infty}|\Psi^{(\mathrm{I},n)}(t)\rangle \,. \tag{3.36}$$

Let us consider the total wave function containing the effect of the interaction up to the order $n$. This function is obtained by explicit insertion of all orders into the right–hand side of Eq. (3.35)

$$\begin{aligned}
|\Psi^{(\mathrm{I},n)}(t)\rangle &= \left(-\frac{i}{\hbar}\right)^n \int_{t_0}^{t} d\tau_n V^{(\mathrm{I})}(\tau_n)\int_{t_0}^{\tau_n} d\tau_{n-1} V^{(\mathrm{I})}(\tau_{n-1}) \times \dots \\
&\quad \dots \times \int_{t_0}^{\tau_2} d\tau_1 V^{(\mathrm{I})}(\tau_1)|\Psi^{(\mathrm{I})}(t_0)\rangle \\
&= \left(-\frac{i}{\hbar}\right)^n \frac{1}{n!}\hat{T}\int_{t_0}^{t} d\tau_n \dots d\tau_1 V^{(\mathrm{I})}(\tau_n)\dots V^{(\mathrm{I})}(\tau_1)|\Psi^{(\mathrm{I})}(t_0)\rangle \,.
\end{aligned} \tag{3.37}$$

In the last part of this expression all integrals are carried out to the upper limit $t$. Double counting is compensated for by the factor $1/n!$. In order to account for the fact that the time–dependent operators $V^{(\mathrm{I})}$ do not commute for different time arguments the *time ordering operator* $\hat{T}$ has been introduced. It orders time–dependent operators from the right to the left with increasing time arguments, i.e., if $t_1 > t_2$, $\hat{T}[V^{(\mathrm{I})}(t_2)V^{(\mathrm{I})}(t_1)] = V^{(\mathrm{I})}(t_1)V^{(\mathrm{I})}(t_2)$. This formal rearrangement enables us to write for the exact state vector in the interaction representation

$$|\Psi^{(\mathrm{I})}(t)\rangle = \hat{T}\sum_{n=0}^{\infty}\frac{1}{n!}\prod_{k=1}^{n}\left(-\frac{i}{\hbar}\int_{t_0}^{t} d\tau_k V^{(\mathrm{I})}(\tau_k)\right)|\Psi^{(\mathrm{I})}(t_0)\rangle\,. \tag{3.38}$$

The summation on the right–hand side is formally identical to the expansion of the exponential function. Comparing this expression with Eq. (3.30) we see that the $S$–operator can be written as a *time–ordered exponential function*

$$S(t,t_0) = \hat{T}\,\exp\left\{-\frac{i}{\hbar}\int_{t_0}^{t} d\tau V^{(\mathrm{I})}(\tau)\right\}\,. \tag{3.39}$$

This expression is an example for a compact notation of a resumed perturbation expansion which is very useful when doing formal manipulations with the time–evolution operator. Nevertheless, for any specific calculation it is necessary to go back to the expansion Eq. (3.37).

## 3.3 The Golden Rule of Quantum Mechanics

The Golden Rule rate formula is certainly one of the most important and widely used expressions of quantum mechanics. It offers a simple way to determine the transition rate between different quantum states of some zeroth–order Hamiltonians in the presence of a small coupling. Therefore, the formula enables one to calculate the change of probability of some initial state due to transition events as a function of time. The basic assumption is that these transitions are irreversible. As discussed earlier (cf. Sections 3.1 and 3.2) such a behavior can be found whenever the transition proceeds into a macroscopic number of final states forming an energetic continuum. In such a situation the mutual interferences among the final states and with the initial state preclude any recurrence of probability back into the initial state. The recurrences are additionally suppressed when the coupling between the initial and final states is sufficiently weak. Such an irreversible transition can also be found if a fast relaxation from the final state to further additional states is possible. Here, the final state itself may be discrete, but there is a coupling to another continuum of states.

There exist different situations which lead to a description by the Golden Rule formula. In the following we will present two alternatives before we embed the formalism into a more general framework in Section 3.4.5.

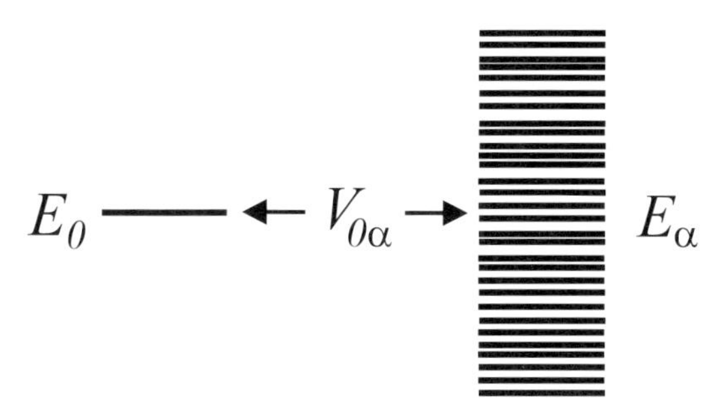


**Figure 3.3:** Coupling of the single state $|0\rangle$ to the manifold of states $|\alpha\rangle$ as described by the Hamiltonian (3.40).

### 3.3.1 Transition from a Single State into a Continuum

Let us consider quantum transitions between some state $|0\rangle$ with energy $E_0$ and a continuum of states $|\alpha\rangle$ with energies $E_\alpha$. The state $|0\rangle$ is supposed to be initially populated and the transitions into the states $|\alpha\rangle$ are due to some inter–state coupling expressed by the matrix $V_{0\alpha}$. The situation is sketched in Fig. 3.3. The total system is described by the Hamiltonian[5]

$$H = E_0|0\rangle\langle 0| + \sum_\alpha \left(E_\alpha|\alpha\rangle\langle\alpha| + V_{0\alpha}|0\rangle\langle\alpha| + V_{\alpha 0}|\alpha\rangle\langle 0|\right) . \tag{3.40}$$

Our goal is to obtain an expression which tells us how the initially prepared state $|0\rangle$ decays into the set of states $|\alpha\rangle$. This transfer of occupation probability can be characterized by looking at the population of state $|0\rangle$ which reads

$$P_0(t) = |\langle 0|U(t)|0\rangle|^2 . \tag{3.41}$$

$U(t)$ is the time–evolution operator, already introduced in Eq. (3.11), and defined here by the Hamiltonian Eq. (3.40). Note, that $P_0(t)$ is a survival probability as introduced in Eq. (3.22).

Provided that one would know the eigenstates of the Hamiltonian (3.40), the survival amplitude would take the form (3.22). The initial values of the coefficients in (3.22) in the present case are determined by the projection of the eigenstates of (3.40) onto the initial state $|0\rangle$. Since we are considering a continuum of states it should be clear from the discussion of Section 3.2 that the recurrence time is very long and the survival amplitude itself is characterized by a rapid decay. Of course, introducing the eigenstates of (3.40) is not the appropriate way to compute, for example, the decay time since a large Hamiltonian has to be diagonalized.

As an alternative let us derive equations of motion for the matrix elements of the time–evolution operator

$$A_{\nu\mu}(t) = \Theta(t)\langle\nu|U(t)|\mu\rangle . \tag{3.42}$$

The quantum numbers $\mu$ and $\nu$ represent the states $|0\rangle$ and $|\alpha\rangle$, and the unit–step function $\Theta(t)$ has been introduced to restrict the definition of $A_{\nu\mu}(t)$ to times larger than zero. The

[5]This setup is similar to a system of two diabatic states (see Section 2.7), $|0\rangle$ denotes an initial electron–vibrational state and the set $|\alpha\rangle$ contains the vibrational states belonging to the final electronic state.

quantity $A_{\nu\mu}(t)$ is called *transition amplitude* and tells us how the state $|\nu\rangle$ is contained in the propagated state $U(t)|\mu\rangle$ at time $t$ if at time $t = 0$ the system was in the state $|\mu\rangle$. The survival amplitude, $P_0(t)$, is equal to $|A_{00}(t)|^2$.

The equations of motion for the transition amplitudes read [6]

$$i\hbar\frac{\partial}{\partial t}A_{\nu\mu} = i\hbar\,\delta(t)\,\delta_{\nu\mu}\delta_{\mu 0} + \sum_{\kappa}\langle\nu|H|\kappa\rangle A_{\kappa\mu}\;. \tag{3.43}$$

In order to solve Eq. (3.43) we introduce the Fourier transform of the transition amplitudes

$$A_{\nu\mu}(\omega) = \int dt\; e^{i\omega t} A_{\nu\mu}(t)\;. \tag{3.44}$$

Taking the Fourier transform of Eq. (3.43) we obtain for the transition amplitudes the following equations

$$\hbar\omega A_{\nu\mu}(\omega) = i\hbar\,\delta_{\nu\mu}\delta_{\mu 0} + \sum_{\kappa}\langle\nu|H|\kappa\rangle A_{\kappa\mu}(\omega)\;. \tag{3.45}$$

In particular for $\nu = \mu = 0$ this gives

$$\hbar\omega A_{00}(\omega) = i\hbar + E_0 A_{00}(\omega) + \sum_{\alpha} V_{0\alpha} A_{\alpha 0}(\omega)\;. \tag{3.46}$$

The off–diagonal elements, $A_{\alpha 0}(\omega)$, can be obtained from

$$\hbar\omega A_{\alpha 0}(\omega) = E_{\alpha} A_{\alpha 0}(\omega) + V_{\alpha 0} A_{00}(\omega)\;. \tag{3.47}$$

Inserting the solution of this equation into the equation for $A_{00}$ yields a closed equation for this quantity which can be solved to give

$$A_{00}(\omega) = i\hbar\left(\hbar\omega - E_0 - \sum_{\alpha}\frac{|V_{0\alpha}|^2}{\hbar\omega - E_{\alpha} + i\varepsilon} + i\varepsilon\right)^{-1}\;. \tag{3.48}$$

Here, $\varepsilon$ has to be understood as a small and positive number which we will let go to zero at the end of the calculation. It guarantees that $A_{00}(\omega)$ is an analytical function in the upper part of the complex frequency plane and, consequently, that the inverse Fourier transformation becomes proportional to $\Theta(t)$. Carrying out this back transformation into the time domain we obtain the desired occupation probability as $P_0(t) = |A_{00}(t)|^2$.

The contributions in the denominator of $A_{00}(\omega)$, which are proportional to the square of the coupling matrix, result in a complicated frequency dependence of $A_{00}(\omega)$. One effect is apparent: the coupling to the continuum shifts the energy $E_0$ of the initial state to a new value. This shift, which is in general a complex quantity, is commonly called *self–energy*

$$\Sigma(\omega) = \sum_{\alpha}\frac{|V_{0\alpha}|^2}{\hbar\omega - E_{\alpha} + i\varepsilon}\;. \tag{3.49}$$

[6] Note, that Dirac's delta function appears on the right–hand side since the time derivative of the unit–step function is given by $d\Theta(t)/dt = \delta(t)$.

The separation into a real and imaginary part gives[7]

$$\Sigma(\omega) \equiv \hbar\Delta\Omega(\omega) - i\hbar\Gamma(\omega) = \sum_\alpha \mathcal{P}\frac{|V_{0\alpha}|^2}{\hbar\omega - E_\alpha} - i\pi\sum_\alpha |V_{0\alpha}|^2\delta(\hbar\omega - E_\alpha)\ . \quad (3.50)$$

If the energies $E_\alpha$ form a continuum the summation with respect to $\alpha$ has to be replaced by an integration. In this case and provided that the coupling constant has no strong dependence on the quantum number $\alpha$, the variation of the self–energy in the region where $\hbar\omega \approx E_0$ can be expected to be rather weak. This means that the frequency dependence of $A_{00}(\omega)$ is dominated by the resonance at $\hbar\omega = E_0$. Since this will give the major contribution to the inverse Fourier transform we can approximately replace $\hbar\omega$ in $\Sigma(\omega)$ by $E_0$. In contrast, if the levels $E_\alpha$ were discrete, $\Sigma(\omega)$ would go to infinity at $\hbar\omega = E_\alpha$ and the frequency dependence of the self–energy can no longer be neglected.

To carry out the inverse Fourier transformation we replace the quantity $\Sigma(\omega)$ by the frequency–independent value $\Sigma(E_0/\hbar)$ and obtain the desired state population $P_0(t)$ as

$$P_0(t) = \left| \int \frac{d\omega}{2\pi} e^{-i\omega t} \frac{i\hbar}{\hbar\omega - (E_0 + \hbar\Delta\Omega(E_0/\hbar)) + i\hbar\Gamma(E_0/\hbar)} \right|^2 = \Theta(t)\ e^{-2\Gamma(E_0/\hbar)t}\ . \quad (3.51)$$

The integral has been calculated using the residue theorem of the theory of complex functions. As expected the occupation probability of the initially occupied state $|0\rangle$ decreases in time due to transitions into the manifold of states $|\alpha\rangle$. For the time evolution of $P_0$ one gets from Eq. (3.51) the simple equation

$$\frac{\partial}{\partial t} P_0(t) = -2\Gamma P_0(t)\ . \quad (3.52)$$

which is a particular example for Eq. (3.2).[8] Following Eq. (3.2) the rate of change of the survival probability is called $k$. It is defined as

$$k = 2\Gamma = \frac{2\pi}{\hbar}\sum_\alpha |V_{0\alpha}|^2\ \delta(E_0 - E_\alpha)\ . \quad (3.53)$$

This type of expression is known as the *Golden Rule* of quantum mechanics. It was first discussed by P. A. M. Dirac and E. Fermi. According to Eq. (3.53) the Golden Rule allows the determination of the rate for occupation probability transfer from some initial state $|0\rangle$ into the manifold of final states $|\alpha\rangle$. The delta function appearing in the rate expression can be interpreted as the energy conservation law for the transition. Only those transitions from $|0\rangle$ to $|\alpha\rangle$ are possible for which the energy of the initial state $E_0$ matches some energy $E_\alpha$ of the final states. The rate is proportional to the square of the interstate coupling $V_{0\alpha}$. This is a direct consequence of replacing the variable energy argument of the self–energy, $\hbar\omega$, by $E_0$.

[7] Here we used the Dirac identity which states that expressions $\propto 1/(\hbar\omega + i\varepsilon)$ appearing in a frequency integral can be rewritten as $\mathcal{P}\ 1/\hbar\omega\ -\ i\pi\delta(\hbar\omega)$, where $\mathcal{P}$ denotes the principal part evaluation of the integral.

[8] Note that in principal the right–hand side of the equation has to be supplemented by the term $\delta(t)P_0(0)$ which stems form the time–derivate of the unit–step function.

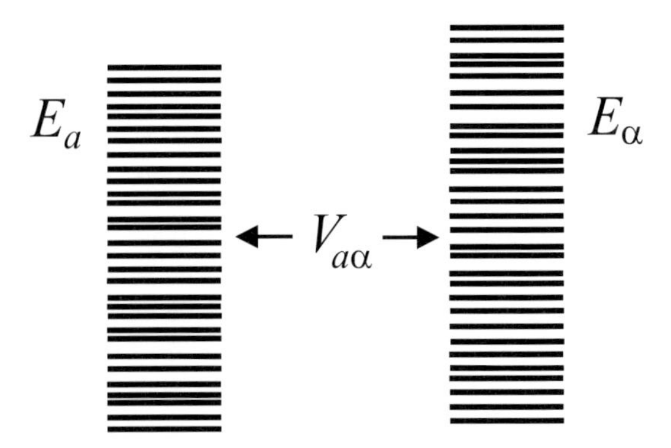


**Figure 3.4:** Coupling of the manifold of initial states $\{|a\rangle\}$ to the manifold of final states $\{|\alpha\rangle\}$ as described by the Hamiltonian (3.54).

Otherwise, higher–order approximations with respect to the coupling matrix elements would have been obtained. Furthermore, it should be taken into account that the derivation assumes an initial population of the discrete state $|0\rangle$ which is not an eigenstate of the complete system. Therefore, Eq. (3.53) is only justified for the case of a weak coupling matrix $V_{0\alpha}$. For cases with stronger coupling the *Golden Rule* expression for the transition rate would be valid at best for times less than the recurrence time.

## 3.3.2 Transition Rate for a Thermal Ensemble

Let us extend the considerations of the preceding section to the case shown in Fig. 3.4. Here, we have a manifold of initial states labeled by $|a\rangle$ which is coupled to some final states $|\alpha\rangle$. The generalization of the Hamiltonian (3.40) reads

$$H = \sum_a E_a|a\rangle\langle a| + \sum_\alpha E_\alpha|\alpha\rangle\langle\alpha| + \sum_{a,\alpha}(V_{a\alpha}|a\rangle\langle\alpha| + \text{h.c.}) \ . \tag{3.54}$$

The situation described by this Hamiltonian is typical, for example, for the nonadiabatic coupling between two electronic states in a molecule where the manifolds $\{|a\rangle\}$ and $\{|\alpha\rangle\}$ take the role of the different vibrational states. Various realizations of this scenario are discussed in the forthcoming chapters.

We are interested in the occupation probabilities $P_a(t)$ of the initially populated states $|a\rangle$. The rate in lowest order with respect to the inter–state coupling $V_{a\alpha}$ can be calculated in close analogy to Section 3.3.1. We obtain for the survival amplitude of the $a$th state

$$P_a(t) = \Theta(t)e^{-k_a t} \ , \tag{3.55}$$

provided that the initial population of this state was equal to unity. The transition rate characterizing the population decay of this state is

$$k_a = \frac{2\pi}{\hbar}\sum_\alpha |V_{a\alpha}|^2\delta(E_a - E_\alpha) \ . \tag{3.56}$$

In a next step let us consider an *ensemble* of $N$ independent but identical systems of the two state manifolds. In the most general case the initially prepared state $|a\rangle$ may be different for each member of the ensemble. The total occupation probability of the initial state is $P(t) = \sum_a n_a \exp(-k_a t)$, where $n_a = N_a/N$ and $N_a$ counts how often the state $|a\rangle$ is initially present in the ensemble. Apparently, the time evolution of every member of the ensemble is independent of the other members and the total population is given as the sum with respect to the single–member populations $P_a$.

If the ensemble stays initially in thermal equilibrium with some environment at temperature $T$ we expect a disturbance of that type of population decay given by Eq. (3.55). This initial equilibrium is the result of a thermalization over all levels, and the quantity $n_a$ has to be replaced by the quantum statistical equilibrium distribution

$$f(E_a) = \frac{\exp(-E_a/k_B T)}{\sum\limits_{a'} \exp(-E_{a'}/k_B T)} . \tag{3.57}$$

To discuss the case where a finite coupling $V_{a\alpha}$ is present, two characteristic times will be introduced. First, we have the time scale $\tau_{S-R}$ which characterizes the coupling of the different members of the ensemble to the thermal reservoir. For example, $\tau_{S-R}$ could be the collision time of the system of interest with the atoms or molecules forming the reservoir. Second, the inter–state coupling introduces a time scale given by $1/k_a$. Now we can distinguish the cases $\tau_{S-R} \gg 1/k_a$ (slow thermalization compared to the transition), $\tau_{S-R} \ll 1/k_a$ (fast thermalization), and $\tau_{S-R} \approx 1/k_a$ (intermediate case).

Case $\tau_{S-R} \gg 1/k_a$:

We suppose that the interaction with an external field promotes the ensemble into the state manifold $\{|a\rangle\}$, where each state occurs $N_a$ times in the ensemble. Since the interaction with the environment is weak compared to the state coupling the population will evolve according to $P(t) = \sum_a n_a \exp(-k_a t)$ as stated above (absence of thermalization on the time–scale of the transfer).

Case $\tau_{S-R} \ll 1/k_a$:

Here, thermalization proceeds at every time–step of the transfer and Eq. (3.55) is no longer applicable. To derive the appropriate equations typical for thermalization let us introduce the time–step $\Delta t \approx \tau_{S-R}$. Usually we will be interested in the time evolution of the system on time scales much longer than $\Delta t$ such that we can consider $\Delta t$ to be a continuous quantity on the time scale of observation. This is, of course, an approximation leading to a *coarse graining* of the time axis. At the initial time the population of the manifold $\{|a\rangle\}$ is thermalized. Starting at $t = 0$, Eq. (3.55) tells us that $P_a(\Delta t) \simeq (1 - k_a \Delta t) P_a(0) = (1 - k_a \Delta t) f(E_a)$. Since $\sum_a f(E_a) = 1 = P(0)$ it follows that $P(\Delta t) \approx 1 - \sum_a k_a f(E_a)\Delta t = P(0) - \sum_a k_a f(E_a) P(0)\Delta t$. Because we are using a course–grained time axis where the state population $P_a(t)$ is thermalized within the time step $\Delta t$ we can generalize this expression and obtain $P(t+\Delta t) \approx P(t) - \sum_a k_a f(E_a) P(t)\Delta t$ for each $t$. Apparently, the occupation probability of state $|a\rangle$ decreased from $f(E_a)$ at $t = 0$ to $f(E_a)P(t)$. Since $\Delta t$ has been assumed to be very small we can rewrite the expression for the total population as

$$\frac{P(t + \Delta t) - P(t)}{\Delta t} \approx \frac{\partial}{\partial t} P(t) = -k P(t) , \tag{3.58}$$

where we introduced the rate for transitions from a thermalized state manifold

$$k = \sum_a f(E_a) k_a = \frac{2\pi}{\hbar} \sum_{a,\alpha} f(E_a) |V_{a\alpha}|^2 \, \delta(E_a - E_\alpha) \,. \tag{3.59}$$

The strong coupling of the system of interest to a thermal reservoir leads to a thermalization which is fast compared to the transfer, i.e., every step of probability transfer from the manifold of initial states to the manifold of final states starts from a thermalized initial state population.
Case $\tau_{\text{S-R}} \approx 1/k_a$:

In this case one can no longer make a separation of time scales and the reasoning used in the previous two cases breaks down. A more general description of the simultaneous influence of the interstate coupling and the coupling to the reservoir is necessary. This more general approach is offered by the *density matrix theory* as will be demonstrated in the forthcoming sections.

Up to now our discussion has been concentrated on the transitions from the states $|a\rangle$ (the initial states) to the states $|\alpha\rangle$ (the final states). Of course, one can consider also the reverse process along the same line of arguments. We expect that there the rate $k_{i\to f} = k_{if} \equiv k$ of the forward transition from the initial to the final state manifold has a counterpart which is the reverse rate $k_{fi} = k_{f\to i}$. The latter follows from Eq. (3.59) by interchanging $E_a$ with $E_\alpha$ in the thermal distribution (transfer starts from the thermalized distribution at the state manifold $|\alpha\rangle$)

$$k_{fi} = \sum_\alpha k_\alpha = \frac{2\pi}{\hbar} \sum_{a,\alpha} f(E_\alpha) |V_{a\alpha}|^2 \delta(E_a - E_\alpha) \,. \tag{3.60}$$

Instead of a single rate equation for $P(t)$ one obtains the *Pauli Master Equations* already discussed in Section 3.1

$$\frac{\partial}{\partial t} P_i(t) = -k_{if} P_i(t) + k_{fi} P_f(t) \,, \qquad \frac{\partial}{\partial t} P_f(t) = -k_{fi} P_f(t) + k_{if} P_i(t) \,. \tag{3.61}$$

The possible transfer forth and back between the state manifolds $\{|a\rangle\}$ and $\{|\alpha\rangle\}$ needs some comments. It seems as if recurrences (as a result of constructive wave function interference) are incorporated. However, on a much shorter time scale and resulting from the coupling to the reservoir any phase relation among the states $|a\rangle$ has been destroyed. We can therefore state that a completely *incoherent* transfer takes place.

It is easy to find the solution of the above given coupled rate equations. Because conservation of probability $P_i(t) + P_f(t) = 1$ holds, the two equations can be transformed to a single one for $P_i(t) - P_f(t)$. Taking as the initial condition $P_i(0) = 1$ the solutions read (note $K = k_{if} + k_{fi}$)

$$P_i(t) = \frac{1}{K} \left( k_{if} e^{-Kt} + k_{fi} \right) \,, \qquad P_f(t) = \frac{k_{if}}{K} \left( 1 - e^{-Kt} \right) \,. \tag{3.62}$$

It is instructive to put both solutions ($\kappa = i, f$) into the form

$$P_\kappa(t) = P_\kappa(\infty) + \big( P_\kappa(0) - P_\kappa(\infty) \big) e^{-Kt} \,. \tag{3.63}$$

with $P_i(\infty) = k_{fi}/K$ and $P_f(\infty) = k_{if}/K$. As it has to be expected the result indicates a complete depletion of the initial state if there is no back–transfer ($k_{fi} = 0$). Otherwise both manifolds remain populated.

A generalization of the Pauli Master Equation to a larger set of different states is straightforward. As an example, one may consider diabatic Born–Oppenheimer states, where each state manifold would represent the vibrational eigenstate for a particular electronic state. To obtain a general solution of the Eqs. (3.2) we denote the right–hand side as $-\sum_b K_{ab}P_b$ with the general rate–matrix $K_{ab} = \delta_{ab}\sum_{c\neq a} k_{a\to c} - (1-\delta_{ab})k_{b\to a}$. Given the eigenvalues $\kappa(\eta)$ and (normalized) eigenvectors $e_a(\eta)$ of $K_{ab}$, the general solution for the population of state $|a\rangle$ reads as $P_a(t) = \sum_\eta c(\eta)e_a(\eta)\exp(-\kappa(\eta)t)$ ($\eta$ counts the rate–matrix eigenvalues). The additional factors $c(\eta)$ are determined from the initial conditions. The decay of the various populations is multi–exponential. Since the smallest $\kappa(\eta)$ equals zero the respective term in $P_a(t)$ fixes $P_a(\infty)$. It is obvious that the given solution (except some special examples) can be only achieved by numerical computations.

The Pauli Master Equation has found numerous applications and we will return to it in the subsequent chapters. However, the basic assumptions are those leading to the Golden Rule for the transition rates which are not always fulfilled (see above). In order to go beyond this level of description a more flexible theory for open quantum systems has to be introduced. This will be done in the next section where we discuss the density matrix approach. It goes without saying that the Pauli Master Equation will be recovered as a limiting case of the more general Quantum Master Equation which is derived below.

## 3.4 The Nonequilibrium Statistical Operator and the Density Matrix

### 3.4.1 The Density Operator

From elementary quantum mechanics we know that a *complete description* of a system is only possible if a set of observables exists from which all physical quantities can be measured simultaneously. Within quantum mechanics this situation is described by a set of commuting operators $\{\hat{A}_\alpha\}$, i.e., the relation

$$\left[\hat{A}_\alpha, \hat{A}_{\alpha'}\right]_- = \hat{A}_\alpha\hat{A}_{\alpha'} - \hat{A}_{\alpha'}\hat{A}_\alpha = 0 \qquad (3.64)$$

has to be fulfilled for all possible pairs of indices. If for the considered system the maximal number of such operators is known, a complete description can be accomplished.

The measurement of some set of observables corresponds to the application of the respective operators $\hat{A}_\alpha$ on the state vector $|\Psi\rangle$. If this exclusively gives the eigenvalues $a_\alpha$, i.e., if

$$\hat{A}_\alpha|\Psi\rangle = a_\alpha|\Psi\rangle\,, \qquad (3.65)$$

the state $|\Psi\rangle$ is called a *pure* state. Alternatively, one can say that a pure state is prepared if a measurement of all observables belonging to the operators $\hat{A}_\alpha$ has been carried out (complete measurement). The expectation value of any operator $\hat{A}$ can be determined as

$$\langle \hat{A} \rangle = \langle \Psi | \hat{A} | \Psi \rangle \ . \tag{3.66}$$

The choice of a complete set of observables is not unequivocal. There may exist another complete set $\{\hat{B}_\beta\}$, independent of the set $\{\hat{A}_\alpha\}$. The respective pure states are denoted $|\Phi_\nu\rangle$. Then, the *superposition principle* of quantum mechanics states that the superposition of all pure states related to the complete set $\{\hat{B}_\beta\}$ reproduces *any* pure state $|\Psi\rangle$:

$$|\Psi\rangle = \sum_\nu c_\nu |\Phi_\nu\rangle \ . \tag{3.67}$$

If the complete measurement of all $\hat{A}_\alpha$ has not been carried out, for example, because the complete set of observables is principally unknown, only an incomplete description of a quantum system is possible (incomplete preparation or measurement of the system). In this case the state of the quantum system has to be described as a *statistical mixture* of pure states $|\Psi_\nu\rangle$. The probability for a single state to be in the mixture will be denoted by $w_\nu$. The states $|\Psi_\nu\rangle$ are assumed to be normalized, and therefore the $w_\nu$ must satisfy the relation

$$\sum_\nu w_\nu = 1 \ . \tag{3.68}$$

Although it is not necessary to demand that the states $|\Psi_\nu\rangle$ form an orthonormal set, it is convenient to do so in the following. Hence, we require in addition that

$$\langle \Psi_\mu | \Psi_\nu \rangle = \delta_{\mu\nu} \ . \tag{3.69}$$

According to this characterization of a mixture of pure states the expectation value of an observable becomes

$$\langle \hat{O} \rangle = \sum_\nu w_\nu \langle \Psi_\nu | \hat{O} | \Psi_\nu \rangle \ . \tag{3.70}$$

This expression provides the idea of the *density operator* (the statistical operator) which will be defined as

$$\hat{W} = \sum_\nu w_\nu |\Psi_\nu\rangle\langle\Psi_\nu| \ . \tag{3.71}$$

It is a summation of projection operators on the states $|\Psi_\nu\rangle$ weighted by the probabilities $w_\nu$. This definition allows a simple notation of the expectation value of any observable using the *trace formula*

$$\langle \hat{O} \rangle = \mathrm{tr}\{\hat{W}\hat{O}\} \ . \tag{3.72}$$

The abbreviation "tr" is defined as the trace with respect to the matrix formed by all matrix elements which are determined in a complete orthonormal basis $|a\rangle$

$$\mathrm{tr}\{\bullet\} = \sum_a \langle a | \bullet | a \rangle \ . \tag{3.73}$$

If $\hat{O}$ and $\hat{P}$ are two operators acting in the Hilbert space spanned by the basis set $|a\rangle$, we have

$$
\begin{aligned}
\mathrm{tr}(\hat{O}\hat{P}) &= \sum_{\alpha} \langle a|\hat{O}\hat{P}|a\rangle = \sum_{a,b} \langle a|\hat{O}|b\rangle\langle b|\hat{P}|a\rangle \\
&= \sum_{a,b} \langle b|\hat{P}|a\rangle\langle a|\hat{O}|b\rangle = \mathrm{tr}(\hat{P}\hat{O})
\end{aligned} \tag{3.74}
$$

This property is called *cyclic invariance* of the operator arrangement in a trace expression (it is also valid if three or more operators are involved). The density operator is normalized such that $\mathrm{tr}\{\hat{W}\} = 1$. If this is not the case it can always be achieved by replacing $\hat{W}$ with $\hat{W}/\mathrm{tr}\{\hat{W}\}$. Furthermore, we mention that the density operator is Hermitian, $\hat{W} = \hat{W}^+$, which follows from Eq. (3.71).

As an example we give the canonical density operator for the thermal equilibrium

$$
\hat{W}_{\rm eq} = \frac{1}{\mathcal{Z}} e^{-H/k_{\rm B}T} = \frac{1}{\mathcal{Z}} \sum_{\alpha} e^{-E_\alpha/k_{\rm B}T} \, |\alpha\rangle\langle\alpha| \,. \tag{3.75}
$$

Here, $\mathcal{Z}$ is the partition function $\mathrm{tr}[\exp\{-H/k_{\rm B}T\}]$ ensuring proper normalization of $\hat{W}_{\rm eq}$. The second part of Eq. (3.75) is obtained using the eigenenergies $E_\alpha$ and eigenstates $|\alpha\rangle$ of the Hamiltonian $H$.

Further, we quote the density operator of a pure state $|\Psi\rangle$ which is defined via the projection operator onto the pure state

$$
\hat{W}_{\rm pure} = |\Psi\rangle\langle\Psi| = \hat{P}_\Psi \,. \tag{3.76}
$$

Comparing this expression with the general definition of the density operator Eq. (3.71) it is obvious that $\hat{W}_{\rm pure}$ corresponds to the special case where all probabilities $w_\nu$ are equal to zero except the one related to the state vector $|\Psi\rangle$.

Suppose we expand the state vector $|\Psi\rangle$ with respect to the complete orthogonal basis $|\alpha\rangle$,

$$
|\Psi\rangle = \sum_{\alpha} c_\alpha |\alpha\rangle \,. \tag{3.77}
$$

Introducing this expansion into the expression for the pure state density operator one obtains

$$
\hat{W}_{\rm pure} = \sum_{\alpha,\bar{\alpha}} c_\alpha c^*_{\bar{\alpha}} |\alpha\rangle\langle\bar{\alpha}| \neq \sum_{\alpha} |c_\alpha|^2 |\alpha\rangle\langle\alpha| \,. \tag{3.78}
$$

The last part of this equation indicates that this expansion does not result in projections onto the basis states $|\alpha\rangle$. Instead, the flip operators $|\alpha\rangle\langle\bar{\alpha}|$ introduce a mixture of states $|\alpha\rangle$ and $|\bar{\alpha}\rangle$ which results in nonvanishing off–diagonal elements of the matrix $(c^*_\alpha \, c_{\bar{\alpha}})$. This is typical for pure states expanded in a particular basis set.

There exists a measure which tells us whether the state is a pure state or not. This measure is called the *degree of coherence* and defined as

$$
C = \mathrm{tr}\{\hat{W}^2\} \,. \tag{3.79}
$$

It takes the value 1 for pure states since the statistical operator in this case is a projector

$$C_{\text{pure}} = \text{tr}\{\hat{W}^2_{\text{pure}}\} = \text{tr}\{\hat{P}^2_\Psi\} = \text{tr}\{\hat{P}_\Psi\} = \text{tr}\{\hat{W}_{\text{pure}}\} = 1 \, , \tag{3.80}$$

where the projector property $P^2_\Psi = P_\Psi$ has been used. For a mixed state it follows that

$$\begin{aligned} C_{\text{mixed}} &= \text{tr}\,\{\hat{W}^2\} = \sum_{\mu,\nu} w_\mu w_\nu \,\text{tr}\left\{\hat{P}_{\Psi_\mu}\hat{P}_{\Psi_\nu}\right\} \\ &= \sum_{\mu,\nu}\sum_\alpha w_\mu w_\nu \langle\alpha|\Psi_\mu\rangle\langle\Psi_\mu|\Psi_\nu\rangle\langle\Psi_\nu|\alpha\rangle \\ &= \sum_\mu\sum_\alpha w^2_\mu \langle\Psi_\mu|\alpha\rangle\langle\alpha|\Psi_\mu\rangle = \sum_\mu w^2_\mu \, < 1 \, . \end{aligned} \tag{3.81}$$

Hence, the degree of coherence becomes less than one. If one studies this quantity for time–dependent density operators the decrease of $C$ indicates the loss of coherence during the time evolution which is caused by interaction of the relevant system with the reservoir.

### 3.4.2 The Density Matrix

In Section 3.1 the concept of the density matrix has been introduced. In order to discuss the density matrix formalism in more detail we consider a complete orthogonal basis of states $|a\rangle, |b\rangle, \ldots$ . Using the completeness relation the density operator can be expanded as

$$\hat{W} = \sum_{a,b} \langle a|\hat{W}|b\rangle \; |a\rangle\langle b| \, . \tag{3.82}$$

The expansion coefficients are called *density matrix* and denoted by

$$\rho_{ab} = \langle a|\hat{W}|b\rangle \, . \tag{3.83}$$

Alternatively, we may use the flip operator $|b\rangle\langle a|$ to write the density matrix as the quantum statistical average of this operator

$$\rho_{ab} = \text{tr}\left\{\hat{W}|b\rangle\langle a|\right\} \, . \tag{3.84}$$

Since the density operator $\hat{W}$ is Hermitian the density matrix fulfills the relation

$$\rho_{ab} = \rho^*_{ba} \, , \tag{3.85}$$

from which one simply deduces

$$\text{Re}\;\rho_{ab} = \text{Re}\;\rho_{ba} \, , \qquad \text{Im}\;\rho_{ab} = -\text{Im}\;\rho_{ba} \, . \tag{3.86}$$

In particular, it follows from this expression that the diagonal elements of the density matrix are real

$$\rho_{aa} = \text{Re}\;\rho_{aa} \, . \tag{3.87}$$

Alternatively, one can use the definition (3.71) of the density operator to write

$$\begin{aligned}\rho_{aa} &= \langle a|\hat{W}|a\rangle = \sum_{\nu} \langle a|w_\nu|\Psi_\nu\rangle\langle\Psi_\nu|a\rangle \\ &= \sum_{\nu} w_\nu \left|\langle a|\Psi_\nu\rangle\right|^2 \,, \end{aligned} \tag{3.88}$$

which also yields real diagonal elements. Additionally, it shows that $\rho_{aa}$ gives us the probability for the state $|a\rangle$ being contained in the statistical mixture described by $\hat{W}$. And indeed, $\rho_{aa} \geq 0$ since $w_\nu$ and $|\langle a|\Psi_\nu\rangle|^2$ are larger than zero. Taking the off–diagonal matrix elements of the density operator it follows

$$\rho_{ab} = \sum_{\nu} w_\nu c_a(\nu) c_b^*(\nu) \,, \tag{3.89}$$

with the expansion coefficients $c_a(\nu) = \langle a|\Psi_\nu\rangle$. Apparently, the density matrix $\rho_{ab}$ describes an *incoherent* superposition of contributions from different pure states. Depending on the basis set $\{|a\rangle\}$ the different terms on the right–hand side of Eq. (3.89) can cancel each other or give a finite $\rho_{ab}$. The off–diagonal density matrix are also called *coherences*.

Since the definition of the density matrix, Eq. (3.82) represents a quadratic form the Schwarz inequality

$$\rho_{aa}\rho_{bb} \geq |\rho_{ab}|^2 \,, \tag{3.90}$$

holds. Eq. (3.90) is particularly useful for checking the quality of any numerical or analytical approximation to the density matrix.

The representation of the statistical operator Eq. (3.82) via the density matrix introduced in Eq. (3.83) is frequently termed *state representation*. If eigenstates of some Hamiltonian are used is also called *energy representation*. Alternatively, it is possible to use, for example, *eigenstates* of the coordinate operator

$$|s\rangle = \prod_j |s_j\rangle \,, \tag{3.91}$$

or the momentum operator

$$|p\rangle = \prod_j |p_j\rangle \,, \tag{3.92}$$

with coordinate $|s_j\rangle$ and momentum states $|p_j\rangle$ for the $j$th degree of freedom of the system, respectively. Consequently, the *coordinate representation* of the statistical operator (density matrix in the coordinate representation) reads

$$\rho(s,\bar{s}) = \langle s|\hat{W}|\bar{s}\rangle \tag{3.93}$$

In the same way the *momentum representation* can be introduced. This allows us to define the respective probability distributions in coordinate ($\rho(s,s)$) and momentum ($\rho(p,p)$) space. However, both types of density matrices cannot straightforwardly be related to the *classical* distribution function in phase space. This limit is conveniently approached using the so–called *Wigner* representation which is defined as

$$\rho(x,p;t) = \int dr \, e^{-ipr/\hbar} \, \rho(x+r/2, x-r/2;t) \, . \tag{3.94}$$

To simplify the notation we first concentrate to the case of a single coordinate. The arguments of the density matrix in the coordinate representation, $\rho(s,\bar{s})$ have been transformed to a difference, $r = s - \bar{s}$, and sum, $x = (s+\bar{s})/2$, coordinate. The dependence on the momentum $p$ enters via a Fourier transformation with respect to the difference coordinate. Apparently, $\rho(x,p;t)$ is a *phase space* representation of the density operator. (Obviously, its generalization to the case of many coordinates requires the introduction of difference and sum coordinates for every DOF). Given $\rho(x,p;t)$, the probability distribution with respect to the coordinate $x$ and the momentum $p$ can be obtained by integration over $p$ and $x$, respectively. As already mentioned the great advantage of this representation is that in the classical limit ($\hbar \to 0$) the density matrix $\rho(x,p;t)$ can be directly related to the phase space distribution of classical statistical physics (see Section 3.4.4).

### 3.4.3 Equation of Motion for the Density Operator

According to the definition of the density operator $\hat{W}$, Eq. (3.71), the probabilities $w_\nu$ represent our reduced knowledge about the state of the system. Furthermore, we note that the state vectors $|\Psi_\nu\rangle$ of the mixed state evolve in time, of course, according to the time–dependent Schrödinger equation

$$i\hbar\frac{\partial}{\partial t}|\Psi_\nu\rangle = H|\Psi_\nu\rangle \, . \tag{3.95}$$

Although any individual state of the mixture changes in time there is no change whatsoever in the amount of our knowledge about the system. In particular, the probabilities $w_\nu$ weighting the contribution of the different states $|\Psi_\nu\rangle$ to the mixed state are constant ($w_\nu \neq w_\nu(t)$). The only exception is given if a measurement has been done on the system. It is known from the basics of quantum mechanics that the result of a measurement process is a reduction of the state of the system onto an eigenstates of the operator corresponding to the observable which has been measured. This means that the mixed state collapses into a pure state. If the pure state is, for example, $|\Psi_{\nu_0}\rangle$, all $w_\nu$ will be zero except the one related to the final pure state which is equal to unity: $w_\nu = \delta_{\nu,\nu_0}$.

According to this reasoning the time–dependent density operator has the following form

$$\hat{W}(t) = \sum_\nu w_\nu |\Psi_\nu(t)\rangle\langle\Psi_\nu(t)| \, . \tag{3.96}$$

In order to derive an equation of motion we write the solution of the time–dependent Schrödinger equation by means of the time–evolution operator, Eq. (3.11), $|\Psi_\nu(t)\rangle = U(t,t_0)|\Psi_\nu(t_0)\rangle$. Then, we get for the density operator

$$\begin{aligned} \hat{W}(t) &= \sum_\nu w_\nu U(t,t_0)|\Psi_\nu(t_0)\rangle\langle\Psi_\nu(t_0)|U^+(t,t_0) \\ &= U(t,t_0)\hat{W}(t_0)U^+(t,t_0) \, . \end{aligned} \tag{3.97}$$

Taking the time derivative of this expression it follows

$$\frac{\partial}{\partial t}\hat{W}(t) = -\frac{i}{\hbar}\left(H\hat{W}(t) - \hat{W}(t)H\right) \equiv -\frac{i}{\hbar}\left[H, \hat{W}(t)\right]_- . \tag{3.98}$$

This is the equation of motion for the density operator $\hat{W}$. It is called *Liouville–von Neumann* or *Quantum Liouville* equation because of its formal analogy to the equation for the classical statistical distribution function.[9] The advantage of the Liouville–von Neumann equation is its capability to directly propagate mixed states without reference to the underlying time–dependent Schrödinger equations. It is also obvious from Eq. (3.98) that any density operator which is given by a mixture of *eigenstates* of the respective Hamiltonian remains stationary. For a concrete example we refer to the canonical density operator, Eq. (3.75).

Next we give the Liouville–von Neumann equation (3.98) in the state representation, Eq. (3.83). One easily derives:

$$\frac{\partial}{\partial t}\rho_{ab} = -i\frac{H_{aa} - H_{bb}}{\hbar}\rho_{ab} - \frac{i}{\hbar}\sum_c \left(H_{ac}\rho_{cb} - H_{cb}\rho_{ac}\right) . \tag{3.99}$$

The difference of the diagonal matrix elements of the Hamiltonian defines the transition frequency $\omega_{ab} = (H_{aa} - H_{bb})/\hbar$, whereas the off–diagonal matrix elements describe the inter–state coupling.

There exists an alternative notation of the Liouville–von Neumann equation which has its origin in the Liouville space formulation of quantum statistical dynamics. The Liouville space is a linear vector space whose elements are the usual operators of the Hilbert space. An operator acting in Liouville space is called a *superoperator*. We will not make full use of this concept here, but introduce superoperators as a convenient shorthand notation (for more details see, e.g. [Hu89,Oht89]). The most important example for a superoperator is the *Liouville superoperator* defined via the commutator with the Hamiltonian:

$$\mathcal{L}\bullet = \frac{1}{\hbar}\left[H, \bullet\right]_- . \tag{3.100}$$

We see immediately that the Liouville–von Neumann equation can be written as

$$\frac{\partial}{\partial t}\hat{W}(t) = -i\mathcal{L}\hat{W}(t) , \tag{3.101}$$

with the solution

$$\hat{W}(t) = e^{-i\mathcal{L}(t-t_0)}\,\hat{W}(t_0) . \tag{3.102}$$

The exponential function of the superoperator is defined via the respective power expansion. In analogy to Eq. (3.11) one can introduce the time–evolution superoperator as follows:

$$\mathcal{U}(t,t_0) = e^{-i\mathcal{L}(t-t_0)} . \tag{3.103}$$

Comparing Eqs. (3.102) and (3.97) we see that $\mathcal{U}(t,t_0)$ is acting on some operator from the left *and* the right, i.e.

$$\hat{W}(t) = \mathcal{U}(t,t_0)\hat{W}(t_0) = U(t,t_0)\hat{W}(t_0)U^+(t,t_0) . \tag{3.104}$$

This is, of course, a consequence of the definition of $\mathcal{L}$ in terms of a commutator.

[9]The classical distribution function depends on all coordinates and momenta and is defined in the so–called phase space spanned by all coordinates and momenta.

### 3.4.4 Wigner Representation of the Density Operator

It has been discussed in Section 3.4.2 that the matrix elements of the density operator can be considered in the coordinate representation, in the momentum representation, but also in a mixture of both which is the Wigner representation $\rho(x,p)$, Eq. (3.94). Here, we will derive the Liouville–von Neumann equation for the density operator in the Wigner representation. As we will see, for instance, in Section 3.8.8 it is not only of conceptional, but also of great practical interest to carry out the classical limit and to demonstrate in which way $\rho(x,p)$ can be interpreted as a phase space distribution of classical statistical physics.

Inspecting Eq. (3.98) it is clear that one needs to find the Wigner representation of some operator product $\hat{Z} = \hat{X}\hat{Y}$; this could be the product $H_S\rho(t)$, for example. First, we introduce the coordinate representation of $\hat{Z}$ as follows

$$Z(s,\bar{s}) = \int ds'\; X(s,s')\; Y(s',\bar{s}) \;. \tag{3.105}$$

As in Eq. (3.94) we first concentrate to the case of a single coordinate and obtain the Wigner representation for $Z(s,\bar{s})$ as:

$$Z(x,p) = \int dr\; ds'\; e^{-ipr/\hbar} X(x+r/2,s')\; Y(s',x-r/2) \;. \tag{3.106}$$

This expression is not yet satisfactory since it contains the coordinate representation of $\hat{X}$ and $\hat{Y}$ on the right–hand side. We introduce the Wigner representation for these functions by using the inverse of Eq. (3.94) and obtain the expression

$$\begin{aligned} Z(x,p) &= \frac{1}{(2\pi\hbar)^2}\int dr\; ds'\; dp' dp'' \\ &\quad\times \exp\left\{\frac{i}{\hbar}\Big(-rp + (x+r/2-s')p' - (x-r/2-s')p''\Big)\right\} \\ &\quad\times X((x+r/2+s')/2,p')\; Y((x-r/2+s')/2,p'') \;. \end{aligned} \tag{3.107}$$

For the following the quantities $X$ and $Y$ have to be written as functions of the single coordinate argument $x$ only, that is, the $r$ and $s'$ dependence has to be eliminated. We achieve this by changing the coordinate arguments and by using the shift operator introduced in Eq. (2.87). For example, for $X$ it gives

$$\begin{aligned} X((x+r/2+s')/2,p') &= X(x-(x-r/2-s')/2,p') \\ &= \exp\left\{\frac{x-r/2-s'}{2}\frac{\partial}{\partial x}\right\} X(x,p') \;. \end{aligned} \tag{3.108}$$

Inserting this result and the similar one for the function $Y$ into Eq. (3.107) gives

$$\begin{aligned} Z(x,p) &= \frac{1}{(2\pi\hbar)^2}\int dr\; ds'\; dp' dp'' \\ &\times \exp\left\{\frac{i}{\hbar}\Big(-rp+(x+r/2-s')p'-(x-r/2-s')p''\Big)\right\} \\ &\times\Big\{\exp\Big\{-\frac{x-r/2-s'}{2}\frac{\partial}{\partial x'}-\frac{x+r/2-s'}{2}\frac{\partial}{\partial x''}\Big\} \\ &\times X(x',p')Y(x'',p'')\Big\}_{x'=x''=x} . \end{aligned} \tag{3.109}$$

A more compact notation is obtained if we take into account that the prefactors of the coordinate derivatives in the shift operators appear again in the exponent. Therefore, we write the integrand in Eq. (3.109) (shorthand notation $[\bullet]$) as

$$\begin{aligned} [\bullet] &= \Big\{\Big\{\exp\Big(-\frac{i\hbar}{2}\frac{\partial}{\partial \bar{p}''}\frac{\partial}{\partial x'}+\frac{i\hbar}{2}\frac{\partial}{\partial \bar{p}'}\frac{\partial}{\partial x''}\Big) \\ &\times \exp\Big(\frac{i}{\hbar}\Big((x+r/2-s')\bar{p}'-(x-r/2-s')\bar{p}''\Big)\Big)\Big\}_{|\bar{p}''=p''}^{|\bar{p}'=p'} \\ &\times X(x',p')Y(x'',p'')\Big\}_{x'=x''=x} . \end{aligned} \tag{3.110}$$

This notation enables us to carry out all four integrations in Eq. (3.109). To do this we order the terms with respect to $s'$, $x$, and $r$. The integration with respect to $s'$ results in the delta function $\delta(p'-p'')$. At the same time the term proportional to $x$ and the $p''$–integration vanishes. Finally, the integration with respect to $r$ leads to $p=p'$, what removes the $p'$–integration. The final result can be put into a compact notation if one introduces the operator

$$\hat{\Theta} = \frac{\partial}{\partial x}\frac{\partial}{\partial p'} - \frac{\partial}{\partial x'}\frac{\partial}{\partial p} . \tag{3.111}$$

It results the Wigner representation of the operator product $\hat{Z}=\hat{X}\hat{Y}$ as

$$Z(x,p) = \Big\{e^{i\hbar\hat{\Theta}/2}X(x,p)Y(x',p')\Big\}_{|p=p'}^{|x=x'} . \tag{3.112}$$

Although exact, this compact expression can only be handled after expanding the exponential function.

To introduce the Wigner representation of the Liouville–von Neumann equation we consider from now on the case where any coordinate argument $x$ and any momentum argument $p$ has to be understood as a set of coordinates and momenta, $x=\{x_j\}$ , and $p=\{p_j\}$, respectively. This requires to generalize Eq. (3.111) to an expression where a summation with respect to all derivatives has to be taken. Additionally, we take into account that the Wigner representation of an operator exclusively defined via the coordinate operator or the momentum operator is a function depending on the coordinate or the momentum alone. Therefore, one obtains for the potential and kinetic energy operator $U(x)$ and $T(p)$, respectively. Let us start with the following form of the Liouville–von Neumann equation

$$\frac{\partial \rho(x,p;t)}{\partial t} = -\frac{i}{\hbar}\int dr\; e^{-ipr/\hbar}\,\langle x+r/2|\left[H,\hat{W}(t)\right]_-|x-r/2\rangle . \tag{3.113}$$

To obtain the classical limit, i.e. the limit $\hbar \to 0$, the $\hat{\Theta}$–operator, Eq. (3.111), has to be taken into account up to the first order:

$$\frac{\partial \rho(x,p;t)}{\partial t} = -\frac{i}{\hbar}\Big\{[1+\frac{i\hbar}{2}\hat{\Theta}][H(x,p)\rho(x',p';t) - \rho(x,p;t)H(x',p')]\Big\}_{|p=p'}^{|x=x'} . \tag{3.114}$$

The zeroth–order contribution vanishes and the classical limit results as

$$\frac{\partial \rho(x,p;t)}{\partial t} = \sum_j \Big\{\frac{\partial U(x)}{\partial x_j}\frac{\partial}{\partial p_j}\rho(x,p;t) - \frac{\partial T(p)}{\partial p_j}\frac{\partial}{\partial x_j}\rho(x,p;t)\Big\} . \tag{3.115}$$

This relation in known from classical statistical mechanics as the *Liouville equation*. It describes the reversible time evolution of the phase space probability distribution. To determine $\rho(x,p;t)$ one has to fix an initial distribution $\rho_0(x,p)$. Then, one can solve the partial differential equation (3.115). The solution can be written as

$$\rho(x,p;t) = \int d\bar{x}d\bar{p}\,\delta\big(x - x(\bar{x},\bar{p};t)\big)\,\delta\big(p - p(\bar{x},\bar{p};t)\big)\rho_0(\bar{x},\bar{p}) , \tag{3.116}$$

where $x(\bar{x},\bar{p};t)$ and $p(\bar{x},\bar{p};t)$ denote the solution of the classical equations of motion for the coordinates and momenta, respectively, following from the initial values $\bar{x}$ and $\bar{p}$. The $\bar{x},\bar{p}$–integral accounts for all those initial values which have some probability according to the initial distribution $\rho_0(\bar{x},\bar{p})$.

### 3.4.5 Dynamics of Coupled Multi–Level Systems in a Heat Bath

As a first application of the density operator method we consider two coupled multi–level systems as already introduced in Eq. (3.54). It will be not the aim here to derive new results, rather we would like to give an alternative derivation of what has been introduced in Section 3.3.2. In particular, a number of approximations will be introduced which we will meet again later on in Section 3.7.1. Following Section 3.3.2 each multilevel system is described by the energies $E_a$ and $E_\alpha$, respectively and the coupling between them is due to the matrix element $V_{a\alpha}$ of the coupling operator $V$. For both quantum numbers, i.e. $a$ and $\alpha$, we use the running indices $\mu$, $\nu$, etc. in the following. Accordingly the Hamiltonian, Eq. (3.54) can be expressed in the common energies $E_\mu$ and coupling matrices $V_{\mu\nu}$ (of course, $V_{ab} = V_{\alpha\beta} = 0$). The density matrix relevant for this system is $\rho_{\mu\nu}(t) = \langle\mu|\hat{W}(t)|\nu\rangle$ and it obeys an equation of motion of the type given in Eq. (3.99) with the transition frequencies $\omega_{\mu\nu} = (E_\mu - E_\nu)/\hbar$.

As in Section 3.3 the subject of the following consideration is to derive a closed set of equations of motion for the total population of the state manifold $\{|a\rangle\}$ (the initial state, note $P_a = \rho_{aa}$)

$$P_i(t) = \sum_a P_a(t) , \tag{3.117}$$

and of the manifold $\{|\alpha\rangle\}$ (the final state, note $P_\alpha = \rho_{\alpha\alpha}$)

$$P_f(t) = \sum_\alpha P_\alpha(t) . \tag{3.118}$$

The coupling of the two multi–level systems to the heat bath will not be specified any further here (but it will be just that type of interaction we will concentrate on in the following sections). The only assumption we will take is that this coupling is much stronger than the interstate coupling $V_{a\alpha}$ (this is identical to the assumption of Section 3.3.2). Thus the rates for transitions within the two manifolds, $k_{aa'}$ and $k_{\alpha\alpha'}$ are supposed to be much larger than those for interstate probability transfer. As a consequence, the populations of the initial and final state can be assumed to be thermalized within the two manifolds on the time–scale of the inter–manifold transfer. Accordingly, the populations can be written as

$$P_a(t) = P_i(t)\, f(E_a)\,, \qquad P_\alpha(t) = P_f(t)\, f(E_\alpha)\,. \tag{3.119}$$

We remind on the fact that this ansatz corresponds to a coarse graining of the time axis which has already been introduced in Section 3.3.2. Within this framework we search for equations of motion obeyed by the total populations $P_i$ and $P_f$, and which are based on Eq. (3.99). Since the coupling matrix element should be small, a perturbational treatment is appropriate. We start with an equation of motion for the diagonal elements of the density matrix, $\rho_{\mu\mu} = P_\mu$, an get from Eq. (3.99)

$$\frac{\partial}{\partial t} P_\mu = -\frac{i}{\hbar} \sum_\kappa \left( V_{\mu\kappa}\rho_{\kappa\mu} - V_{\kappa\mu}\rho_{\mu\kappa} \right)\,. \tag{3.120}$$

The off–diagonal density matrix elements which appear on the right–hand side have to be determined, too. They obey

$$\begin{aligned}\frac{\partial}{\partial t}\rho_{\kappa\mu} &= -i\omega_{\kappa\mu}\rho_{\kappa\mu} - \frac{i}{\hbar}\sum_\lambda \left( V_{\kappa\lambda}\rho_{\lambda\mu} - V_{\lambda\mu}\rho_{\kappa\lambda}\right) \\ &\approx -i\omega_{\kappa\mu}\rho_{\kappa\mu} - \frac{i}{\hbar}V_{\kappa\mu}\left(\rho_{\mu\mu} - \rho_{\kappa\kappa}\right)\,. \end{aligned}\tag{3.121}$$

Since we are looking for the lowest–order approximation in $V_{\mu\nu}$, off–diagonal density matrix elements have been neglected in the second line. Fixing the initial condition as $\rho_{a\alpha}(0) = 0$ (absence of a superposition state between both subsystems), we can solve Eq. (3.121) by formal integration and obtain

$$\rho_{\kappa\mu}(t) = -\frac{i}{\hbar} V_{\kappa\mu} \int_0^t d\bar{t}\, e^{-i\omega_{\kappa\mu}(t-\bar{t})} \left[ P_\mu(\bar{t}) \; - \; P_\kappa(\bar{t}) \right]. \tag{3.122}$$

Inserting the result into Eq. (3.120) yields (note the replacement of $\bar{t}$ by $t - \tau$)

$$\frac{\partial}{\partial t} P_\mu = -\frac{1}{\hbar^2} \sum_\kappa |V_{\mu\kappa}|^2\, 2\mathrm{Re} \int_0^t d\tau\, e^{-i\omega_{\kappa\mu}\tau} \left[ P_\mu(t-\tau) \; - \; P_\kappa(t-\tau) \right]. \tag{3.123}$$

The total state populations $P_i$ and $P_f$ are obtained by making use of the thermalization condition, Eq. (3.119). If these expressions are introduced into Eq. (3.123) we get

$$\frac{\partial}{\partial t} P_i = -\int_0^t d\tau \left[ K_{if}(\tau)\, P_i(t-\tau) - K_{fi}(\tau)\, P_f(t-\tau) \right], \tag{3.124}$$

with the integral kernel given by

$$K_{if}(\tau) = \frac{2}{\hbar^2} \sum_{a,\alpha} |V_{a\alpha}|^2 \, f(E_a) \, \cos(\omega_{a\alpha}\tau) \; . \tag{3.125}$$

Interchanging $i$ and $f$ leads to the equation for $P_f(t)$. The quantity $G_{if}(\tau)$ is usually named *memory kernel* since it reflects that Eq. (3.124) is not an ordinary rate equation like Eq. (3.2). As a consequence of the time integral the state populations enter the equation at a time $\tau$ earlier than $t$. In other words, the system retains the *memory* of its past dynamics. Master Equations which, like Eq. (3.124), include memory effects are called *Generalized Master Equation* (GME).

The time dependence of the memory kernel is determined by the structure of the energy spectrum related to the initial as well as the final state. If these spectra are dense $K_{if}(\tau)$ would decay in a certain time interval $\tau_{\text{mem}}$ due to destructive interference (c.f. Section 3.2). If $\tau_{\text{mem}}$ is short compared to the characteristic time where the populations $P_i$ and $P_f$ change, the variation of both quantities within the interval $[t-\tau_{\text{mem}}, t]$ can be neglected and we can replace $P_{i/f}(t-\tau)$ by $P_{i/f}(t)$ in the integrand. Note that this corresponds to a further coarse graining of the time axis. According to both coarse graining approximations the populations $P_i$ and $P_f$ are only valid for times much larger than $\tau_{\text{mem}}$. Therefore, the result of the integration does not change if the upper limit is put to infinity, and we arrive at the ordinary rate equation

$$\frac{\partial}{\partial t} P_i = -k_{if} \, P_i(t) + k_{fi} \, P_f(t) \; , \tag{3.126}$$

where the transition rates take the form

$$k_{if} = \int\limits_0^\infty dt \; K_{if}(t) = \frac{2}{\hbar^2} \sum_{a,\alpha} |V_{a\alpha}|^2 \, f(E_a) \, \text{Re} \int\limits_0^\infty d\tau \; \exp(i\omega_{a\alpha}\tau) \; . \tag{3.127}$$

We note that $\text{Re}z = (z + z^*)/2$ (where $z$ is an arbitrary complex number) and replace the integral by one along the total time–axis. Using the Fourier representation of the $\delta$–function

$$\delta(\omega) = \frac{1}{2\pi} \int\limits_{-\infty}^{\infty} dt \; e^{i\omega t} \; . \tag{3.128}$$

we get

$$k_{if} = \frac{2\pi}{\hbar} \sum_{a,\alpha} f(E_a) \, |V_{a\alpha}|^2 \, \delta(E_a - E_\alpha) \; . \tag{3.129}$$

The derived rate formula is identical to the *Golden Rule* expression of transition rates of Eq. (3.60). Of course, this is not surprising since our derivation of Eq. (3.129) followed the same arguments, i.e., a strong coupling to the reservoir is assumed to give fast thermalization and a quasi–continuous final–state energy spectrum is required to prevent probability revivals from the final to the initial states. Note that the demand for a quasi–continuous energy spectrum was found to correspond to a short memory time of the kernel entering the GME.

It is instructive to view the transition rates from a different perspective. Let us go back to Eq. (3.125) and write

$$\begin{aligned} K_{if}(t) &= \frac{2}{\hbar^2}\,\mathrm{Re}\sum_{a,\alpha} |V_{a\alpha}|^2\, f(E_a)\, e^{i(E_a-E_\alpha)t/\hbar} \\ &= \frac{2}{\hbar^2}\,\mathrm{Re}\sum_{a,\alpha} f(E_a)\langle a|e^{iE_a t/\hbar}Ve^{-iE_\alpha t/\hbar}|\alpha\rangle\langle\alpha|V|a\rangle\;. \end{aligned} \tag{3.130}$$

Introducing the part $H_0 = \sum_a E_a|a\rangle\langle a| + \sum_\alpha E_\alpha|\alpha\rangle\langle\alpha|$ of the total Hamiltonian, Eq. (3.54), we can replace the energies $E_a$ and $E_\alpha$ by $H_0$. Using the completeness relation with respect to the state manifold $|\alpha\rangle$ gives

$$K_{if}(t) = \frac{2}{\hbar^2}\mathrm{Re}\sum_a \langle a|\hat{W}_{\mathrm{eq}}e^{iH_0 t/\hbar}Ve^{-iH_0 t/\hbar}V|a\rangle = 2\mathrm{Re}\;C(t)\;. \tag{3.131}$$

In the last line we introduced the correlation function $C(t)$ which can be written as

$$C(t) = \frac{1}{\hbar^2}\;\mathrm{tr}_a\{\hat{W}_{\mathrm{eq}}V^{(\mathrm{I})}(t)V^{(\mathrm{I})}(0)\}\;. \tag{3.132}$$

It represents an *autocorrelation function* of the interstate coupling $V^{(\mathrm{I})}(t)$ written in the interaction representation and taken with respect to thermal equilibrium. The distribution $f(E_a)$ has been replaced by the equilibrium density operator $\hat{W}_{\mathrm{eq}}$, Eq. (3.75) and we used $\mathrm{tr}_a\{\bullet\} = \sum_a \langle a|\bullet|a\rangle$. Thus, the memory kernel turns out to be proportional to the auto-correlation function of the inter–state coupling. A short memory time thus implies a rapid decay of this correlation function. The fact that rate expressions like (3.129) in general can be written in terms of correlation functions of the perturbational part of the Hamiltonian is of great importance for the understanding as well as the numerical modelling of condensed phase processes. We will frequently return to this point in the following considerations.

# 3.5 The Reduced Density Operator and the Reduced Density Matrix

## 3.5.1 The Reduced Density Operator

Having discussed the concept of the density operator we are ready to put the idea of the reduced density matrix introduced in Section 3.1 into a more rigorous frame. The starting point will be a Hamiltonian $H$ which is separable into a system part $H_\mathrm{S}$, a reservoir part $H_\mathrm{R}$, and the system–reservoir interaction $H_{\mathrm{S-R}}$ (cf. Eq. (3.3))

$$H = H_\mathrm{S} + H_{\mathrm{S-R}} + H_\mathrm{R}\;. \tag{3.133}$$

First, as in Section 3.1 we introduce the density matrix in the coordinate representation, Eq. (3.93) by using the states Eq. (3.91) separated now into

$$|s\rangle = \prod_j |s_j\rangle\;, \tag{3.134}$$

defined in the state space of the relevant system, and into the states

$$|Z\rangle = \prod_\xi |Z_\xi\rangle \,, \tag{3.135}$$

defined in the state space of the reservoir. According to the general form of the time–dependent density operator $\hat{W}(t)$, Eq. (3.96) the density matrix in the coordinate representation reads

$$\rho(s, Z; s', Z'; t) = \langle s| \langle Z| \hat{W}(t) |Z'\rangle |s'\rangle = \sum_v w_\nu \Psi_\nu(s, Z; t) \; \Psi^*_\nu(s', Z'; t) \,, \tag{3.136}$$

with $\Psi_\nu(s, Z; t) = \langle s|\langle Z|\Psi_\nu(t)\rangle$. Following the reasoning of Section 3.1 we introduce the reduced density matrix defined in the state space of the relevant system only. This quantity is obtained by carrying out an integration with respect to the set of reservoir coordinates $Z$, that is,

$$\begin{aligned} \rho(s, \bar{s}; t) &= \int dZ \sum_\nu w_\nu \Psi_\nu(s, Z; t) \; \Psi^*_\nu(\bar{s}, Z; t) \\ &= \int dZ \; \langle s|\langle Z| \hat{W}(t) |Z\rangle|\bar{s}\rangle \,. \end{aligned} \tag{3.137}$$

Alternatively, we may write

$$\rho(s, \bar{s}; t) = \langle s| \int dZ \; \langle Z|\hat{W}(t)|Z\rangle \; |s'\rangle = \langle s|\hat{\rho}(t)|s'\rangle \,, \tag{3.138}$$

where the *reduced density operator* (RDO) of the relevant system

$$\hat{\rho}(t) = \int dZ \; \langle Z|\hat{W}(t)|Z\rangle \tag{3.139}$$

has been introduced. It is defined by taking the trace of the total density operator with respect to a particular basis in the reservoir state space. Instead of the coordinate states $|Z\rangle$ any basis $|\alpha\rangle$ in the reservoir space may be chosen,

$$\hat{\rho}(t) = \sum_\alpha \langle\alpha|\hat{W}(t)|\alpha\rangle = \mathrm{tr_R}\left\{\hat{W}(t)\right\} \,, \tag{3.140}$$

i.e., the trace with respect to the reservoir states reduces the total density operator $\hat{W}$ to the RDO $\hat{\rho}$.

Besides the coordinate representation of the density matrix Eq. (3.138), any basis $|a\rangle$ in the state space of the system can be used to define the *reduced* density matrix

$$\rho_{ab}(t) = \langle a|\hat{\rho}(t)|b\rangle \,. \tag{3.141}$$

As in the case of the total density operator we expect the following relation to be fulfilled

$$\mathrm{tr_S}\{\hat{\rho}(t)\} \equiv \sum_a \rho_{aa}(t) = 1 \,. \tag{3.142}$$

The relation is easily confirmed if we note that $\hat{W}(t)$ entering Eq. (3.140) obeys $\mathrm{tr}\{\hat{W}(t)\} = 1$.

### 3.5.2 Equation of Motion for the Reduced Density Operator

An equation of motion for the reduced density matrix is derived by starting from the respective operator equation for the RDO. From the Liouville–von Neumann equation (3.98) we obtain

$$\begin{aligned}\frac{\partial}{\partial t}\hat{\rho}(t) &= \mathrm{tr}_{\mathrm{R}}\left\{\frac{\partial}{\partial t}\hat{W}(t)\right\} = -\frac{i}{\hbar}\,\mathrm{tr}_{\mathrm{R}}\left\{\left[H_{\mathrm{S}}+H_{\mathrm{S-R}}+H_{\mathrm{R}},\hat{W}(t)\right]_{-}\right\} \\ &= -\frac{i}{\hbar}\left[H_{\mathrm{S}},\hat{\rho}(t)\right]_{-} - \frac{i}{\hbar}\,\mathrm{tr}_{\mathrm{R}}\left\{\left[H_{\mathrm{S-R}}+H_{\mathrm{R}},\hat{W}(t)\right]_{-}\right\}. \end{aligned} \quad (3.143)$$

In the first part of this equation we used the fact that the basis which defines the trace in the reservoir space state is time–independent. Then we took into account that the system Hamiltonian $H_{\mathrm{S}}$ is not influenced by the reservoir trace. Therefore, it is possible to introduce the commutator of $H_{\mathrm{S}}$ with respect to the RDO directly. Indeed, we could have anticipated such a contribution since for $H_{\mathrm{S-R}} = 0$ the equation for the RDO should reduce to the Liouville–von Neumann equation (3.98).

The commutator notation for the RDO is not possible for the contributions proportional to $H_{\mathrm{S-R}}$ and $H_{\mathrm{R}}$. To calculate the commutator with $H_{\mathrm{R}}$ we take into account Eq. (3.74). The cyclic interchange of operators can be carried out here since $H_{\mathrm{R}}$ exclusively acts in the state space of the reservoir. As a result the term proportional to $H_{\mathrm{R}}$ vanishes and the equation of motion for the RDO follows as

$$\frac{\partial}{\partial t}\hat{\rho}(t) = -\frac{i}{\hbar}\left[H_{\mathrm{S}},\hat{\rho}(t)\right]_{-} - \frac{i}{\hbar}\,\mathrm{tr}_{\mathrm{R}}\left\{\left[H_{\mathrm{S-R}},\hat{W}(t)\right]_{-}\right\}. \quad (3.144)$$

Before dealing with the case $H_{\mathrm{S-R}} \neq 0$ we note that the type of equation (3.98) is recovered if $H_{\mathrm{S-R}}$ is neglected. But this Liouville–von Neumann equation is defined by $H_{\mathrm{S}}$ instead of the full Hamiltonian $H$ and it describes the isolated time evolution of the relevant quantum system. As already pointed out in Section 3.1 the density matrix description of coherent dynamics contains some redundancy and a wave function formulation is more appropriate in this case. However, if the RDO describes a mixed state of the isolated system a generalization of the ordinary time–dependent Schrödinger equation has been achieved. Changing to the more interesting case of the presence of $H_{\mathrm{S-R}}$ we realize that Eq. (3.144) is not yet a closed equation for the RDO. Because of the appearance of $H_{\mathrm{S-R}}$ in the commutator on the right–hand side it still contains the total density operator. It will be the main task of the following sections to develop approximations which yield the second term in Eq. (3.144) as a functional of the RDO only such that one has a closed equation for the RDO.

### 3.5.3 Mean–Field Approximation

In a first attempt to close Eq. (3.144) we take the most simple way. Since the total density operator appears on the right–hand side of Eq. (3.144) which includes $H_{\mathrm{S-R}}$ in all orders (according to the given time dependence of $\hat{W}(t)$) we expect that a perturbation theory with respect to $H_{\mathrm{S-R}}$ can be developed. Let us start with the first–order approximation which is obtained if we replace the total density operator by its $H_{\mathrm{S-R}} \to 0$ limit. In this limit there are no interactions between the two subsystems. $\hat{W}(t)$ factorizes into $\hat{\rho}(t)$ and an operator $\hat{R}(t)$ which is defined only in the Hilbert space of the reservoir and which obeys $\mathrm{tr}_{\mathrm{R}}\{\hat{R}\} = 1$.

According to our assumptions the approximated equation of motion for the RDO becomes

$$\frac{\partial}{\partial t}\hat{\rho}(t) = -\frac{i}{\hbar}\left[H_{\rm S} + {\rm tr}_{\rm R}\{H_{\rm S-R}\hat{R}(t)\}, \hat{\rho}(t)\right]_{-} . \tag{3.145}$$

This equation is of the type of a Liouville–von Neumann equation for the RDO, but with the only difference here that $H_{\rm S}$ has been supplemented by ${\rm tr}_{\rm R}\{H_{\rm S-R}\hat{R}(t)\}$. The additional term is the expectation value of the system–reservoir coupling taken with respect to the actual state of the reservoir. (Note, that $H_{\rm S-R}$ and $\hat{R}(t)$ can be interchanged under the trace giving the compact notation of Eq. (3.145).) Since the bath part of $H_{\rm S-R}$ has been replaced by an expectation value the result is called *mean–field approximation* [10]. The meaning becomes more obvious if we assume that the system–reservoir interaction Hamiltonian can be factorized into system parts $K_u = K_u(s)$, and into reservoir parts, $\Phi_u = \Phi_u(Z)$, that is,

$$H_{\rm S-R} = \sum_u K_u \Phi_u . \tag{3.146}$$

The index $u$ counts the different contributions which may follow from a particular microscopic model for the coupling of the system to the reservoir. Expression (3.146) is called the *multiple factorized form* of the system–reservoir interaction Hamiltonian. Note, that it is not necessary that the single operator $K_u$ or $\Phi_u$ is Hermitian. Only the complete coupling Hamiltonian needs to be Hermitian. Since no further restriction has been introduced with respect to these two functions, Eq. (3.146) is sufficiently general to comprise all cases of practical importance. In the following chapters we will discuss several examples supporting this point of view.

Taking the multiple factorized form of $H_{\rm S-R}$, Eq. (3.145) becomes

$$\frac{\partial}{\partial t}\hat{\rho}(t) = -\frac{i}{\hbar}\left[H_{\rm S} + \sum_u K_u\, {\rm tr}_{\rm R}\{\Phi_u\hat{R}(t)\}, \hat{\rho}(t)\right]_{-} . \tag{3.147}$$

For further use we define the mean–field Hamiltonian

$$H_{\rm mf} = \sum_u K_u\, {\rm tr}_{\rm R}\{\Phi_u\hat{R}(t)\} . \tag{3.148}$$

Because the time–dependence of the reservoir density operator is not known the equation for the RDO is not closed. But taking an equilibrium assumption for the reservoir and replacing $\hat{R}(t)$ by

$$\hat{R}_{\rm eq} = e^{-H_{\rm R}/k_{\rm B}T}/{\rm tr}_{\rm R}\{e^{-H_{\rm R}/k_{\rm B}T}\} , \tag{3.149}$$

Eq. (3.147) defines a closed equation. The effect of ${\rm tr}_{\rm R}\{\Phi_u\hat{R}_{\rm eq}\} \equiv \langle\Phi_u\rangle_{\rm R}$, and thus of the mean–field term is a shift of the energy scale, i.e., it does not give the relaxation behavior discussed in the context of the Golden Rule approach. As we will see below relaxation is

[10]The term mean–field indicates that quantum fluctuations are absent and that the quantum mechanical operators act only via the "mean–field" given by their expectation values. Such a type of approximation has been already considered in Section 2.4 in the framework of the derivation of the Hartree–Fock equations. Therefore, the mean–field approximation is often also called Hartree approximation.

caused by *fluctuations*, $\Phi_u - \langle\Phi_u\rangle_{\rm R}$, around the mean–field energies. In order to take these into account we need to go one step further in our perturbation expansion.

But before doing this we consider the more general case where the mean–field term remains time–dependent. In such a situation we have to set up an additional equation for $\hat{R}(t)$. Understanding it as the RDO of the reservoir and setting

$$\hat{R}(t) = {\rm tr_S}\{\hat{W}(t)\} \ , \tag{3.150}$$

we can repeat the derivation which leads us to Eq. (3.145) (or Eq. (3.147)) and obtain

$$\frac{\partial}{\partial t}\hat{R}(t) = -\frac{i}{\hbar}\Big[H_{\rm R} + \sum_u \Phi_u {\rm tr_S}\{K_u\hat{\rho}(t)\}, \hat{R}(t)\Big]_- \ . \tag{3.151}$$

This equation together with Eq. (3.147) represents a closed set to determine the coupled evolution of the relevant system and the reservoir once respective initial conditions for both types of RDO have been set up. Because the solution of Eq. (3.151) for a macroscopic reservoir becomes impossible the approach is not suited to describe energy dissipation out of the relevant system. An application of the coupled set of Eqs. (3.147) and (3.151) only make sense when both subsystems are sufficiently small.

### 3.5.4 The Interaction Representation of the Reduced Density Operator

In the foregoing section an equation of motion for the RDO has been derived which is of first order in $H_{\rm S-R}$. In the following section we will apply an projection operator technique. It allows to handle separately the projection of the operator equation onto the subspace of the relevant system and the formulation of a perturbation theory with respect to the system–reservoir coupling $H_{\rm S-R}$. The latter is conveniently developed by changing to the interaction representation as explained in the following.

Recall that the formal solution of the Liouville–von Neumann equation can be written as (Eq. (3.97))

$$\hat{W}(t) = U(t-t_0)\hat{W}(t_0)U^+(t-t_0) \ , \tag{3.152}$$

where the time–evolution operator $U(t-t_0)$ is defined with respect to the total Hamiltonian $H$. One can separate this operator into the "free" time–evolution operator

$$\begin{aligned} U_0(t-t_0) &= \exp\Big(-\frac{i}{\hbar}H_{\rm S}(t-t_0)\Big)\exp\Big(-\frac{i}{\hbar}H_{\rm R}(t-t_0)\Big) \\ &\equiv U_{\rm S}(t-t_0)\,U_{\rm R}(t-t_0) \ , \end{aligned} \tag{3.153}$$

(note that $H_{\rm S}$ and $H_{\rm R}$ commute with each other) and the related $S$–operator (cf. Section 3.2.2)

$$S(t,t_0) = \hat{T}\exp\Big(-\frac{i}{\hbar}\int_{t_0}^{t} d\tau H^{\rm (I)}_{\rm S-R}(\tau)\Big) \ . \tag{3.154}$$

This expression contains the system–reservoir coupling Hamiltonian in the interaction representation

$$H^{\rm (I)}_{\rm S-R}(t) = U_0^+(t-t_0)H_{\rm S-R}U_0(t-t_0) \ . \tag{3.155}$$

For the total density operator we can write

$$\hat{W}(t) = U_0(t - t_0)\hat{W}^{(\mathrm{I})}(t)U_0^+(t - t_0) , \tag{3.156}$$

where the density operator in the interaction representation reads

$$\hat{W}^{(\mathrm{I})}(t) = U_0^+(t - t_0)\hat{W}(t)U_0(t - t_0) = S(t, t_0)\hat{W}(t_0)S^+(t, t_0) . \tag{3.157}$$

Using this equation the time derivative of Eq. (3.156) can be written as

$$\frac{\partial}{\partial t}\hat{W}(t) = -\frac{i}{\hbar}[H_0, \hat{W}(t)]_- + U_0(t - t_0)\frac{\partial}{\partial t}\hat{W}^{(\mathrm{I})}(t)U_0^+(t - t_0) . \tag{3.158}$$

If we set this expression equal to the right–hand side of the Liouville–von Neumann equation, $-i[H, \hat{W}(t)]_-/\hbar$, we get after some rearrangement

$$\frac{\partial}{\partial t}\hat{W}^{(\mathrm{I})}(t) = -\frac{i}{\hbar}[H_{\mathrm{S-R}}^{(\mathrm{I})}(t), \hat{W}^{(\mathrm{I})}(t)]_- . \tag{3.159}$$

Notice that this equation can be viewed as the generalization of Eq. (3.28) to the case of a density operator. Next, we transform the RDO into the interaction representation (using Eq. (3.153))

$$\begin{aligned}\hat{\rho}(t) &= \mathrm{tr}_{\mathrm{R}}\{\hat{W}(t)\} = \mathrm{tr}_{\mathrm{R}}\left\{U_0(t - t_0)\hat{W}^{(\mathrm{I})}(t)U_0^+(t - t_0)\right\} \\ &= U_{\mathrm{S}}(t - t_0)\mathrm{tr}_{\mathrm{R}}\left\{U_{\mathrm{R}}(t - t_0)\hat{W}^{(\mathrm{I})}(t)U_{\mathrm{R}}^+(t - t_0)\right\}U_{\mathrm{S}}^+(t - t_0) .\end{aligned} \tag{3.160}$$

Using the cyclic invariance of the trace we can write

$$\hat{\rho}(t) = U_{\mathrm{S}}(t - t_0)\hat{\rho}^{(\mathrm{I})}(t)U_{\mathrm{S}}^+(t - t_0) \tag{3.161}$$

with the RDO in the interaction representation defined as

$$\hat{\rho}^{(\mathrm{I})}(t) = \mathrm{tr}_{\mathrm{R}}\left\{\hat{W}^{(\mathrm{I})}(t)\right\} . \tag{3.162}$$

With these definitions the equation of motion for $\rho^{(\mathrm{I})}(t)$ follows from Eq. (3.159) as

$$\frac{\partial}{\partial t}\hat{\rho}^{(\mathrm{I})}(t) = -\frac{i}{\hbar}\mathrm{tr}_{\mathrm{R}}\left\{[H_{\mathrm{S-R}}^{(\mathrm{I})}(t), \hat{W}^{(\mathrm{I})}(t)]_-\right\} . \tag{3.163}$$

### 3.5.5 The Projection Operator

The generation of equations for the RDO of higher order in the system–reservoir coupling requires the combination of a perturbation theory with a scheme for restricting the operator equations to the state space of the relevant system. Suppose $\hat{O}$ is an operator acting in the space of the system and the reservoir states. Let us consider the operator $\mathcal{P}$ which acts on $\hat{O}$ as follows

$$\mathcal{P}\hat{O} = \hat{R}\,\mathrm{tr}_{\mathrm{R}}\{\hat{O}\} . \tag{3.164}$$

By definition, $\mathcal{P}$ separates $\hat{O}$ defined in the full space into the part $\mathrm{tr}_{\mathrm{R}}\{\hat{O}\}$ acting only in the system space and an operator $\hat{R}$ which by definition exclusively acts in the state space of the reservoir. In other words, the operator $\mathcal{P}$ factorizes any operator into a system part and into a reservoir part. If we apply $\mathcal{P}$ to the full density operator we obtain by definition the RDO $\hat{\rho}$ and some reservoir operator

$$\mathcal{P}\hat{W}(t) = \hat{R}\,\hat{\rho}(t)\,. \tag{3.165}$$

If $\mathrm{tr_R}\{\hat{R}\} = 1$, which we will assume in the following, the operator $\mathcal{P}$ is a projector, i.e., $\mathcal{P}^2 = \mathcal{P}$, as can be easily proved

$$\mathcal{P}^2\hat{O} = \hat{R}\,\mathrm{tr_R}\{\hat{R}\,\mathrm{tr_R}\{\hat{O}\}\} = \hat{R}\,\mathrm{tr_R}\{\hat{R}\}\,\mathrm{tr_R}\{\hat{O}\} = \mathcal{P}\hat{O}\,. \tag{3.166}$$

Since $\hat{R}$ has a trace equal to unity it can be interpreted as a statistical operator restricted to the state space of the reservoir. Although in principle a time dependence is possible we take $\hat{R}$ as the (time–independent) equilibrium density operator of the reservoir, i.e., we define

$$\mathcal{P}\bullet = \hat{R}_{\mathrm{eq}}\,\mathrm{tr_R}\{\bullet\}\,. \tag{3.167}$$

It is useful to introduce in addition to $\mathcal{P}$ its orthogonal complement

$$\mathcal{Q} = 1 - \mathcal{P}\,. \tag{3.168}$$

The operator $\mathcal{Q}$ is a projection operator as well and by construction we have

$$\mathcal{Q}\mathcal{P} = \mathcal{P}\mathcal{Q} = 0\,. \tag{3.169}$$

The action of $\mathcal{Q}$ on the total density operator leads to

$$\mathcal{Q}\hat{W}(t) = \hat{W}(t) - \hat{\rho}(t)\hat{R}_{\mathrm{eq}}\,, \tag{3.170}$$

i.e. $\mathcal{Q}\hat{W}$ is the irrelevant part of the statistical operator.

Both projectors, $\mathcal{P}$ and $\mathcal{Q}$, can be used to systematically develop a perturbation expansion with respect to $H_{\mathrm{S-R}}$ in the equation of motion for the RDO. To achieve this goal we start our considerations in the interaction representation. We have

$$\mathcal{P}\hat{W}^{(\mathrm{I})}(t) = \hat{R}_{\mathrm{eq}}\mathrm{tr_R}\{\hat{W}^{(\mathrm{I})}(t)\} = \hat{R}_{\mathrm{eq}}\hat{\rho}^{(\mathrm{I})}(t)\,. \tag{3.171}$$

Using the identity $\hat{W}^{(\mathrm{I})}(t) = \mathcal{P}\hat{W}^{(\mathrm{I})}(t) + \mathcal{Q}\hat{W}^{(\mathrm{I})}(t)$ the Liouville–von Neumann equation (3.98) can be split into two coupled equations. First we have

$$\mathcal{P}\frac{\partial}{\partial t}\hat{W}^{(\mathrm{I})} = -\frac{i}{\hbar}\mathcal{P}[H^{(\mathrm{I})}_{\mathrm{S-R}}, \mathcal{P}\hat{W}^{(\mathrm{I})} + \mathcal{Q}\hat{W}^{(\mathrm{I})}]_-\,. \tag{3.172}$$

Taking the trace with respect to the reservoir states it follows that

$$\mathrm{tr_R}\{\mathcal{P}\frac{\partial}{\partial t}\hat{W}^{(\mathrm{I})}\} = \frac{\partial}{\partial t}\hat{\rho}^{(\mathrm{I})} = -\frac{i}{\hbar}\mathrm{tr_R}\Big\{\Big[H^{(\mathrm{I})}_{\mathrm{S-R}}, \hat{R}_{\mathrm{eq}}\hat{\rho}^{(\mathrm{I})} + \mathcal{Q}\hat{W}^{(\mathrm{I})}\Big]_-\Big\}\,. \tag{3.173}$$

In a similar manner one obtains the equation of motion for $\mathcal{Q}\hat{W}^{(\mathrm{I})}$ as

$$\frac{\partial}{\partial t}\mathcal{Q}\hat{W}^{(\mathrm{I})} = -\frac{i}{\hbar}\mathcal{Q}\Big[H^{(\mathrm{I})}_{\mathrm{S-R}}, \hat{R}_{\mathrm{eq}}\hat{\rho}^{(\mathrm{I})} + \mathcal{Q}\hat{W}^{(\mathrm{I})}\Big]_- \tag{3.174}$$

By means of these formal manipulations we have been able to reduce the equation of motion for $\hat{W}^{(\mathrm{I})}$ to a coupled set of equations for $\hat{\rho}^{(\mathrm{I})}$ and $\mathcal{Q}\hat{W}^{(\mathrm{I})}$.

Next we will show that an iterative solution of these coupled equations generates a perturbation expansion with respect to $H_{\mathrm{S-R}}$ on the right–hand side of Eq. (3.173). If we neglect $\mathcal{Q}\hat{W}^{(\mathrm{I})}$ altogether we recover the result of the previous section, i.e., we obtain the mean–field correction to the system dynamics which is of first order in $H_{\mathrm{S-R}}$ (see Eq. (3.145)). The second–order contribution is calculated from Eq. (3.173) by inserting the formal solution of Eq. (3.174) which is of first order in $H_{\mathrm{S-R}}$. The commutator structure of the right–hand side of Eq. (3.173) then results in second–order terms. The formal first–order solution of the equation for $\mathcal{Q}\hat{W}^{(\mathrm{I})}$ is obtained by neglecting $\mathcal{Q}\hat{W}^{(\mathrm{I})}$ on the right–hand side of Eq. (3.174). One gets

$$\mathcal{Q}\hat{W}^{(\mathrm{I})}(t) = \mathcal{Q}\hat{W}^{(\mathrm{I})}(t_0) - \frac{i}{\hbar}\int_{t_0}^{t} dt'\ \mathcal{Q}\Big[H^{(\mathrm{I})}_{\mathrm{S-R}}(t'), \hat{R}_{\mathrm{eq}}\hat{\rho}^{(\mathrm{I})}(t')\Big]_- . \tag{3.175}$$

Here, the first part on the right–hand side tells us, whether or not $\hat{W}^{(\mathrm{I})}$ initially factorizes into a system and reservoir part. It is easy to verify that this term vanishes if the total density operator factorizes at $t = t_0$, $\hat{W}(t_0) \to \rho(t_0)\hat{R}_{\mathrm{eq}}$. If such a factorization is not possible so–called *initial correlations* between the relevant system and the reservoir have to be taken into account. The time scale for the decay of these initial correlations depends on the details of the system–reservoir coupling. For simplicity we will not consider this effect in the following, i.e., we assume that $\mathcal{Q}\hat{W}^{(\mathrm{I})}(t_0) = 0$.

The third–order contribution to Eq. (3.173) can be obtained by inserting Eq. (3.175) into the right–hand side of Eq. (3.174). The formal solution of the resulting equation is then used in Eq. (3.173). This iteration procedure can be continued to generate all orders of the perturbation expansion. Needless to say that with increasing order the complexity of the equations increases as well. However, one of the advantages of the projection operator approach is that an exact summation of the perturbation series is possible. The resulting *Nakajima–Zwanzig* equation is discussed in Section 3.12.1. In the following we will focus on the second–order contribution to the equations of motion of the reduced density operator.

### 3.5.6 Second–Order Equation of Motion for the Reduced Statistical Operator

Inserting Eq. (3.175) into Eq. (3.173) we obtain the equation of motion for the RDO which is of second–order with respect to $H_{\mathrm{S-R}}$ as

$$\begin{aligned}\frac{\partial}{\partial t}\hat{\rho}^{(\mathrm{I})}(t) &= -\frac{i}{\hbar}\mathrm{tr_R}\left\{\hat{R}_{\mathrm{eq}}\Big[H^{(\mathrm{I})}_{\mathrm{S-R}}(t), \hat{\rho}^{(\mathrm{I})}(t)\Big]_-\right\} \\ &\quad -\frac{1}{\hbar^2}\int_{t_0}^{t} d\tau\ \mathrm{tr_R}\left\{[H^{(\mathrm{I})}_{\mathrm{S-R}}(t), (1-\mathcal{P})[H^{(\mathrm{I})}_{\mathrm{S-R}}(\tau), \hat{R}_{\mathrm{eq}}\hat{\rho}^{(\mathrm{I})}(\tau)]_-]_-\right\} .\end{aligned} \tag{3.176}$$

In the following we will discuss this equation for the multiple factorized form Eq. (3.146) of the system–reservoir coupling. The first–order term on the right–hand side corresponds to that in Eq. (3.145). In order to show this one has to use the cyclic invariance of the trace, Eq. (3.74) in the space of the reservoir states. The mean–field contribution (in the interaction representation) to the dynamics of the relevant system becomes

$$\mathrm{tr}_{\mathrm{R}}\{\hat{R}_{\mathrm{eq}}[H^{(\mathrm{I})}_{\mathrm{S-R}}(t),\hat{\rho}^{(\mathrm{I})}(t)]_-\} = \sum_v [K^{(\mathrm{I})}_v(t)\langle\Phi_v\rangle_{\mathrm{R}},\hat{\rho}^{(\mathrm{I})}(t)]_- \equiv [H^{(\mathrm{I})}_{\mathrm{mf}}(t),\hat{\rho}^{(\mathrm{I})}(t)]_- \,. \tag{3.177}$$

The general form of the mean–field Hamiltonian $H_{\mathrm{mf}}$ has been introduced in Eq. (3.148). Here, the expectation value has to be taken with the equilibrium reservoir density operator.

Next the second term in Eq. (3.176) is considered in more detail. First let us focus on the integrand which contains a double commutator. Due to the factor $(1-\mathcal{P})$ there are altogether 8 terms where those containing the factor $\mathcal{P}$ include two trace operations. We consider the four terms corresponding to the unit operator "1" of $(1-\mathcal{P})$. It reads, using Eq. (3.146) and the cyclic invariance of the trace,

$$\mathcal{M}_1 = \mathrm{tr}_{\mathrm{R}}\{[H^{(\mathrm{I})}_{\mathrm{S-R}}(t),[H^{(\mathrm{I})}_{\mathrm{S-R}}(\tau),\hat{R}_{\mathrm{eq}}\rho^{(\mathrm{I})}(\tau)]_-]_-\}\,, \tag{3.178}$$

or in more detail

$$\begin{aligned}\mathcal{M}_1 \;=\; \sum_{uv}\Big(&\mathrm{tr}_{\mathrm{R}}\{\Phi^{(\mathrm{I})}_u(t)\Phi^{(\mathrm{I})}_v(\tau)\hat{R}_{\mathrm{eq}}\}\;K^{(\mathrm{I})}_u(t)K^{(\mathrm{I})}_v(\tau)\hat{\rho}^{(\mathrm{I})}(\tau)\\ &-\mathrm{tr}_{\mathrm{R}}\{\Phi^{(\mathrm{I})}_u(t)\hat{R}_{\mathrm{eq}}\Phi^{(\mathrm{I})}_v(\tau)\}K^{(\mathrm{I})}_u(t)\hat{\rho}^{(\mathrm{I})}(\tau)K^{(\mathrm{I})}_v(\tau)\\ &-\mathrm{tr}_{\mathrm{R}}\{\Phi^{(\mathrm{I})}_v(\tau)\hat{R}_{\mathrm{eq}}\Phi^{(\mathrm{I})}_u(t)\}K^{(\mathrm{I})}_v(\tau)\hat{\rho}^{(\mathrm{I})}(\tau)K^{(\mathrm{I})}_u(t)\\ &+\mathrm{tr}_{\mathrm{R}}\{\hat{R}_{\mathrm{eq}}\Phi^{(\mathrm{I})}_v(\tau)\Phi^{(\mathrm{I})}_u(t)\}\hat{\rho}^{(\mathrm{I})}(\tau)K^{(\mathrm{I})}_v(\tau)K^{(\mathrm{I})}_u(t)\Big)\,.\end{aligned} \tag{3.179}$$

For the second term proportional to $\mathcal{P}$ we write using Eq. (3.148)

$$\mathcal{M}_2 = \mathrm{tr}_{\mathrm{R}}\{[H^{(\mathrm{I})}_{\mathrm{S-R}}(t),\hat{R}_{\mathrm{eq}}\mathrm{tr}_{\mathrm{R}}\{[H^{(\mathrm{I})}_{\mathrm{S-R}}(\tau),\hat{R}_{\mathrm{eq}}\hat{\rho}^{(\mathrm{I})}(\tau)]_-\}]_-\}\,, \tag{3.180}$$

what leads to

$$\begin{aligned}\mathcal{M}_2 \;&=\; \sum_{uv}\langle\Phi_u\rangle_{\mathrm{R}}\langle\Phi_v\rangle_{\mathrm{R}}[K^{(\mathrm{I})}_u(t),[K^{(\mathrm{I})}_v(\tau),\hat{\rho}^{(\mathrm{I})}(\tau)]_-]_-\\ &\equiv [H^{(\mathrm{I})}_{\mathrm{mf}}(t),[H^{(\mathrm{I})}_{\mathrm{mf}}(\tau),\hat{\rho}^{(\mathrm{I})}(\tau)]_-]_-\,.\end{aligned} \tag{3.181}$$

Next we will apply the results of Section 3.5.3 to rewrite the expectation values of the reservoir part of $H_{\mathrm{S-R}}$ as follows (first term in Eq. (3.179))

$$\begin{aligned}\langle\Phi^{(\mathrm{I})}_u(t)\Phi^{(\mathrm{I})}_v(\tau)\rangle_{\mathrm{R}} \;&=\; \mathrm{tr}_{\mathrm{R}}\{\Phi^{(\mathrm{I})}_u(t)\Phi^{(\mathrm{I})}_v(\tau)\hat{R}_{\mathrm{eq}}\}\\ &=\; \mathrm{tr}_{\mathrm{R}}\{\hat{R}_{\mathrm{eq}}U^{+}_{\mathrm{R}}(t-\tau)\Phi_u U_{\mathrm{R}}(t-\tau)\Phi_v\}\\ &=\; \langle\Phi^{(\mathrm{I})}_u(t-\tau)\Phi^{(\mathrm{I})}_v(0)\rangle_{\mathrm{R}}\,.\end{aligned} \tag{3.182}$$

Using similar steps we get for the remaining terms in Eq. (3.178)

$$\mathrm{tr}_{\mathrm{R}}\{\Phi^{(\mathrm{I})}_u(t)\hat{R}_{\mathrm{eq}}\Phi^{(\mathrm{I})}_v(\tau)\} = \langle\Phi^{(\mathrm{I})}_v(0)\Phi^{(\mathrm{I})}_u(t-\tau)\rangle_{\mathrm{R}}\,, \tag{3.183}$$

$$\mathrm{tr_R}\{\Phi_v^{(\mathrm{I})}(t)\hat{R}_{\mathrm{eq}}\Phi_u^{(\mathrm{I})}(\tau)\} = \langle\Phi_u^{(\mathrm{I})}(t-\tau)\Phi_v^{(\mathrm{I})}(0)\rangle_{\mathrm{R}} , \tag{3.184}$$

and

$$\mathrm{tr_R}\{\hat{R}_{\mathrm{eq}}\Phi_v^{(\mathrm{I})}(\tau)\Phi_u^{(\mathrm{I})}(t)\} = \langle\Phi_v^{(\mathrm{I})}(0)\Phi_u^{(\mathrm{I})}(t-\tau)\rangle_{\mathrm{R}} . \tag{3.185}$$

Apparently, the integrand of Eq. (3.176) can be cast into a form which has only four terms each containing the following type of function (the superscript (I) on the bath operators will be suppressed in the following).

$$C_{uv}(t) = \frac{1}{\hbar^2}\langle\Delta\Phi_u(t)\Delta\Phi_v(0)\rangle_{\mathrm{R}} = \frac{1}{\hbar^2}\langle\Phi_u(t)\Phi_v(0)\rangle_{\mathrm{R}} - \frac{1}{\hbar^2}\langle\Phi_u\rangle_{\mathrm{R}}\langle\Phi_v\rangle_{\mathrm{R}} . \tag{3.186}$$

Here we combined the reservoir operators with their expectation values to the operator

$$\Delta\Phi_u(t) = \Phi_u(t) - \langle\Phi_u\rangle_{\mathrm{R}} . \tag{3.187}$$

This operator describes the *fluctuations* of the reservoir part of $H_{\mathrm{S-R}}$ with respect to its average value. The function $C_{uv}(t)$ in Eq. (3.186) which is called *reservoir correlation function* therefore establishes a connection between the fluctuations of the operators $\Phi_v(t)$ and $\Phi_u(t)$ at different times (see also Sec. 3.4.5, a detailed discussion of the correlation functions can be found in the following section). For most systems the correlations of the fluctuations decay after a certain *correlation time* $\tau_{\mathrm{c}}$. Note that these fluctuations do *not* change the quantum mechanical state of the reservoir which is still described by the canonical density operator.

If $\Phi_u$ is a Hermitian operator we have

$$\langle\Phi_v(0)\Phi_u(t)\rangle_{\mathrm{R}} = [\langle\Phi_u(t)\Phi_v(0)\rangle_{\mathrm{R}}]^* = \langle\Phi_v(-t)\Phi_u(0)\rangle_{\mathrm{R}} \tag{3.188}$$

from which we get the important property

$$C_{uv}^*(t) = C_{vu}(-t) . \tag{3.189}$$

Using the definition of the correlation function the equation of motion for the RDO finally follows as

$$\begin{aligned}
\frac{\partial}{\partial t}\hat{\rho}^{(\mathrm{I})}(t) &= -\frac{i}{\hbar}\sum_u \langle\Phi_u\rangle[K_u^{(\mathrm{I})},\hat{\rho}^{(\mathrm{I})}(t)]_- \\
&\quad -\sum_{uv}\int_{t_0}^{t} d\tau \Big(C_{uv}(t-\tau)[K_u^{(\mathrm{I})}(t),K_v^{(\mathrm{I})}(\tau)\hat{\rho}^{(\mathrm{I})}(\tau)]_- \\
&\quad -C_{vu}(-t+\tau)[K_u^{(\mathrm{I})}(t),\hat{\rho}^{(\mathrm{I})}(\tau)K_v^{(\mathrm{I})}(\tau)]_-\Big) .
\end{aligned} \tag{3.190}$$

This equation is valid for non–Hermitian operators $K_u$ and $\Phi_u$. If the reservoir operators $\Phi_u$ are Hermitian then $C_{vu}(-t+\tau)$ can be replaced by $C_{uv}^*(t-\tau)$. Since every term on the right–hand side of Eq. (3.190) is given by a commutator it is easy to demonstrate that the RDO equation ensures conservation of total probability, i.e. $\mathrm{tr_S}\{\partial\hat{\rho}(t)/\partial t\} = 0$. Furthermore, by computing the Hermitian conjugated of the right–hand side of Eq. (3.190) one may demonstrate that the Hermiticity of $\hat{\rho}^{(\mathrm{I})}$ is assured for all times (note that in the case of non–Hermitian operators $K_u$ and $\Phi_u$ the whole $u, v$–summation realizes Hermitian operators).

Eq. (3.190) is frequently called *Quantum Master Equation* (QME) since it generalizes ordinary rate equations (Master equations) of the type given in Eq. (3.2) to the quantum case (represented by the reduced density operator). Alternatively, the term *density matrix equation in the second Born approximation* is common. Here, one refers to the second–order perturbation theory applied to the system–reservoir coupling. (The treatment of higher–order contributions can be found in [Jan02,Lai91,Kor97,Tan91].)

The right–hand side of this equation reveals that the change in time of the RDO is not only determined by its actual value but by the history of its own time dependence. Therefore, Eq. (3.190) is specified as the QME with memory effects. This type of memory effect has been already encountered in our introductory example in Section (3.4.5). In the present case the memory time $\tau_{\mathrm{mem}}$ is obviously determined by the reservoir correlation function but is not necessarily identical to the correlation time $\tau_{\mathrm{c}}$. Before we concentrate on the properties of the QME, the reservoir correlation function will be discussed in the following section.

# 3.6 The Reservoir Correlation Function

## 3.6.1 General Properties of $C_{uv}(t)$

The importance of the reservoir correlation function for the dynamics of some relevant system interacting with a reservoir is apparent from the QME (3.190). Before turning to specific models for $C_{uv}(t)$ we will discuss some of the more general properties of this function as well as of its Fourier transform

$$C_{uv}(\omega) = \int dt\; e^{i\omega t} C_{uv}(t) \; . \tag{3.191}$$

If Eq. (3.189) is valid it follows immediately that

$$C_{vu}(-\omega) = \int dt\; e^{i\omega t} C^*_{uv}(t) \; , \tag{3.192}$$

and that $C^*_{uv}(\omega) = C_{vu}(\omega)$. It will further be convenient to introduce symmetric and anti-symmetric correlation functions

$$C^{(+)}_{uv}(t) = C_{uv}(t) + C^*_{uv}(t) \; , \qquad C^{(-)}_{uv}(t) = C_{uv}(t) - C^*_{uv}(t) \; , \tag{3.193}$$

respectively. Note, that $C^{(+)}_{uv}(t)$ is a real function while $C^{(-)}_{uv}(t)$ is imaginary. Moreover, $C^{(+)}_{uv}(-t) = C^{(+)}_{vu}(t)$ as well as $C^{(-)}_{uv}(-t) = -C^{(-)}_{vu}(t)$ holds.

Another fundamental property of $C_{uv}(\omega)$ can be derived if one starts from the definitions (3.182) and (3.186) and introduces eigenstates $|\alpha\rangle$ and eigenvalues $E_\alpha$ of the reservoir Hamiltonian. Using these eigenstates to perform the trace operation we obtain

$$\begin{aligned} C_{uv}(\omega) &= \frac{1}{\hbar^2} \int dt\; e^{i\omega t} \sum_{\alpha\beta} \langle\alpha|\hat{R}_{\mathrm{eq}} e^{iH_{\mathrm{R}}t/\hbar} \Delta\Phi_u e^{-iH_{\mathrm{R}}t/\hbar}|\beta\rangle\langle\beta|\Delta\Phi_v|\alpha\rangle \\ &= \frac{1}{\hbar^2} \sum_{\alpha\beta} \int dt\; e^{i(\omega-\omega_{\beta\alpha})t} f(E_\alpha) \langle\alpha|\Delta\Phi_u|\beta\rangle\langle\beta|\Delta\Phi_v|\alpha\rangle \; . \end{aligned} \tag{3.194}$$

The $\omega_{\beta\alpha} = (E_\beta - E_\alpha)/\hbar$ are the transition frequencies between the reservoir energy levels and

$$f(E_\alpha) \equiv \langle\alpha|\hat{R}_{\text{eq}}|\alpha\rangle = \exp(-E_\alpha/k_{\text{B}}T)/\sum_\beta \exp(-E_\beta/k_{\text{B}}T) \tag{3.195}$$

denotes the thermal distribution function with respect to the reservoir states. The time integration of the exponential function produces the delta function (Eq. (3.128)), i.e. we get

$$C_{uv}(\omega) = \frac{2\pi}{\hbar^2} \sum_{\alpha\beta} f(E_\alpha)\langle\alpha|\Delta\Phi_u|\beta\rangle\langle\beta|\Delta\Phi_v|\alpha\rangle\delta(\omega - \omega_{\beta\alpha}) \,. \tag{3.196}$$

Now we consider the Fourier transform of the correlation function where the indices $u$ and $v$ are interchanged. Interchanging also $\alpha$ and $\beta$ gives

$$C_{vu}(\omega) = \frac{2\pi}{\hbar^2} \sum_{\alpha\beta} f(E_\beta)\langle\alpha|\Delta\Phi_u|\beta\rangle\langle\beta|\Delta\Phi_v|\alpha\rangle\delta(\omega - \omega_{\alpha\beta}) \,. \tag{3.197}$$

According to the identity

$$\exp\left\{-\frac{E_\beta}{k_{\text{B}}T}\right\}\delta(\omega - \omega_{\alpha\beta}) = \exp\left\{-\frac{E_\alpha - \hbar\omega}{k_{\text{B}}T}\right\}\delta(\omega + \omega_{\beta\alpha}) \tag{3.198}$$

we arrive at the important result

$$C_{uv}(\omega) = \exp\left\{\frac{\hbar\omega}{k_{\text{B}}T}\right\} C_{vu}(-\omega) \,, \tag{3.199}$$

which relates the positive frequency part of the correlation function to its negative frequency part. Note, that Eq. (3.199) builds upon the definition of $C_{uv}(\omega)$ with respect to the thermal equilibrium of the reservoir.

Using Eq. (3.193) the Fourier transform of the symmetric and antisymmetric part of the correlation function can be written as

$$C_{uv}^{(\pm)}(\omega) = C_{uv}(\omega) \pm C_{vu}(-\omega) \,. \tag{3.200}$$

If we replace $C_{vu}(-\omega)$ in Eq. (3.200) by the result of Eq. (3.199) it follows that

$$C_{uv}(\omega) = \frac{C_{uv}^{(\pm)}(\omega)}{1 \pm \exp\{-\hbar\omega/k_{\text{B}}T\}} \equiv (1 + n(\omega))\, C_{uv}^{(-)}(\omega) \,. \tag{3.201}$$

Here, the Bose–Einstein distribution function

$$n(\omega) = \frac{1}{\exp\{\hbar\omega/k_{\text{B}}T\} - 1} \tag{3.202}$$

has been used to rewrite the expression for $C_{uv}$. Combining the two parts of Eq. (3.201) we get a relation between the Fourier transforms of the symmetric and antisymmetric parts of the correlation function which reads

$$C_{uv}^{(+)}(\omega) = \coth\left(\frac{\hbar\omega}{2k_{\text{B}}T}\right) C_{uv}^{(-)}(\omega) \,. \tag{3.203}$$

Since a relation between the correlation function and its antisymmetric part $C_{uv}^{(-)}(\omega)$ has been established it is easy to express $C_{uv}(t)$ by $C_{uv}^{(-)}(\omega)$. The inverse Fourier transformation can then be written in terms of the half–sided Fourier integral

$$\begin{aligned} C_{uv}(t) &= \int_{-\infty}^{\infty} \frac{d\omega}{2\pi} e^{-i\omega t}[1+n(\omega)]C_{uv}^{(-)}(\omega) \\ &= \int_{0}^{\infty} \frac{d\omega}{2\pi} \left(e^{-i\omega t}[1+n(\omega)]C_{uv}^{(-)}(\omega) + e^{i\omega t}n(\omega)C_{vu}^{(-)}(\omega)\right). \end{aligned} \tag{3.204}$$

To summarize, it is possible to express the reservoir correlation function either by its symmetric or antisymmetric part. This freedom of choice will be particularly useful in the context of classical simulations of the reservoir as we will see in Section 3.6.6.

### 3.6.2 Harmonic Oscillator Reservoir

The explicit quantum mechanical calculation of $C_{uv}(t)$ is not feasible in practice since there is no way to calculate the quantum mechanical states of a general macroscopic reservoir such as a solvent surrounding some solute molecule. To overcome this difficulty several models for the reservoir and its interaction with the system have been developed.

In the case of a reservoir which is characterized by a stable crystalline structure the correlation function can readily be calculated using the following reasoning: In many of such systems where the atoms (or molecules) form a regular lattice with high symmetry lattice vibrations only appear as small oscillations around the equilibrium positions at sufficiently low temperature. In this case a harmonic approximation is possible, i.e., the force driving the atoms back to their equilibrium position can be taken to be proportional to the deviation from this equilibrium position. In Section 2.6.1 we have seen that a harmonic approximation to some global potential energy surface allows to introduce normal mode vibrations whose quantum counterparts in the case of a crystalline structure are called (lattice–) *phonons*. As the main result of the introduction of normal mode oscillations the individual atom coordinates are mapped on a set of harmonic oscillator coordinates which are independent of each other.

It should be remarked that this situation is not the rule: For example, in low temperature solutions the solvent is essentially frozen into a disordered solid. Here, it is more difficult to calculate $C_{uv}(t)$ because the solute is likely to interact with system specific localized vibrational modes of its immediate surroundings. If the temperature is increased such that the reservoir becomes a liquid the notion of normal modes as small amplitude motions around stable structures loses its meaning at all. In such situations one has to resort to classical simulations of the reservoir. This approach will be discussed in Section 3.11. In fact as we will see in Section 4.3.2 on ultrashort time scales it is often possible to introduce instantaneous normal modes.

Having in mind the important concept of a normal mode bath we will adapt the correlation function to this situation now. In a first step we introduce a more specific structure of the coupling Hamiltonian, $H_{\rm S-R}$. Let us assume that we have performed a Taylor expansion of

$H_{\text{S-R}}$ with respect to the reservoir coordinates. If we focus on the lowest–order contribution only, $H_{\text{S-R}}$ will become linear with respect to the harmonic oscillator reservoir coordinates $Z = \{Z_\xi\}$. Further, $H_{\text{S-R}}$ given in Eq. (3.146) is assumed to contain a single term only. This restriction is made basically to simplify the notation. The extension to more general expressions for the coupling Hamiltonian is straightforward. Dropping the index $u$ we can write

$$H_{\text{S-R}} = K(s) \sum_\xi \hbar\gamma_\xi Z_\xi \, . \tag{3.205}$$

Here, $s$ comprises the coordinates of the system and $\gamma_\xi$ is the system–reservoir coupling constant. The given expression for $H_{\text{S-R}}$, if compared with Eq. (3.146), corresponds to a reservoir part $\Phi = \sum_\xi \hbar\gamma_\xi Z_\xi$. Note that $\langle Z_\xi \rangle_{\text{R}} = 0$, that is, the thermal fluctuations of the reservoir coordinates are taking place symmetrically around $Z_\xi = 0$. Since we are dealing with decoupled normal mode oscillators the reservoir Hamiltonian can be written as $H_{\text{R}} = \sum_\xi H_\xi^{(\text{R})}$. Here the single–mode Hamiltonian is given by $H_\xi^{(\text{R})} = \hbar\omega_\xi (C_\xi^+ C_\xi + 1/2)$, where $C_\xi^+$ and $C_\xi$ denote normal mode oscillator creation and annihilation operators (cf. Section 2.6.2). In terms of the creation and annihilation operators the reservoir coordinates are written as $Z_\xi = \sqrt{\hbar/2\omega_\xi} \times (C_\xi + C_\xi^+)$ (see Eq. (2.81)). Further, $\omega_\xi$ is the normal mode frequency and the harmonic oscillator eigenstates, $|M_\xi\rangle = (C_\xi^+)^{M_\xi}|0_\xi\rangle/\sqrt{M_\xi!}$, will be labelled by the oscillator quantum number $M_\xi$. The respective eigenenergies are given by $E_M = \sum_\xi E_{M_\xi} = \sum_\xi \hbar\omega_\xi(M_\xi + 1/2)$ (cf. Eq. (2.71)). With the help of these assumptions and Eq. (3.196) the correlation function takes the form

$$C(\omega) = 2\pi \sum_{\xi\xi'} \gamma_\xi \gamma_{\xi'} \sum_{NM} f(E_N) \langle N|Z_\xi|M\rangle\langle M|Z_{\xi'}|N\rangle \delta(\omega - (E_M - E_N)/\hbar) \, . \tag{3.206}$$

The matrix elements of the reservoir coordinates simply reduce to (cf. Eq. (2.69))

$$\langle N|Z_\xi|M\rangle = \langle N_\xi|Z_\xi|M_\xi\rangle \prod_{\xi' \neq \xi} \delta_{N_{\xi'},M_{\xi'}} \, . \tag{3.207}$$

Inserting this result into Eq. (3.206) enables us to carry out the complete summation with respect to the set of quantum numbers $M$. Here, care has to be taken when performing the sum with respect to $M_\xi$ and $N_{\xi'}$ if $\xi \neq \xi'$, and the sum with respect to $M_\xi$ alone if $\xi = \xi'$. We distinguish between these two cases in the $\xi$–$\xi'$–summations and get ($\omega_{M_\xi N_\xi} = (E_{M_\xi} - E_{N_\xi})/\hbar$)

$$\begin{aligned} C(\omega) = 2\pi \sum_{\xi,\xi'} \gamma_\xi \gamma_{\xi'} \Big\{ &\delta_{\xi,\xi'} \sum_N f(E_N) \sum_{M_\xi} |\langle N_\xi|Z_\xi|M_\xi\rangle|^2 \, \delta(\omega - \omega_{M_\xi N_\xi}) \\ &+ (1 - \delta_{\xi,\xi'}) \sum_N f(E_N) \sum_{M_\xi} \sum_{M_{\xi'}} \langle N_\xi|Z_\xi|M_\xi\rangle \, \delta_{N_{\xi'},M_{\xi'}} \\ &\times \langle M_{\xi'}|Z_{\xi'}|N_{\xi'}\rangle \, \delta_{M_\xi,N_\xi} \, \delta(\omega - \omega_{M_\xi N_\xi} - \omega_{M_{\xi'} N_{\xi'}}) \Big\} \, . \end{aligned} \tag{3.208}$$

Since $\langle N_\xi | Z_\xi | M_\xi \rangle \neq \delta_{N_\xi, M_\xi}$, all terms with $\xi \neq \xi'$ vanish and it follows that

$$C(\omega) = 2\pi \sum_\xi \sum_{N_\xi, M_\xi} \gamma_\xi^2 f(E_{N_\xi}) |\langle N_\xi | Z_\xi | M_\xi \rangle|^2 \delta\left(\omega - \omega_{M_\xi N_\xi}\right) . \tag{3.209}$$

The matrix elements of the oscillator coordinate can be obtained using Eqs. (2.77) and (2.79). We get for the squared matrix elements entering (3.209):

$$|\langle M_\xi | Z_\xi | N_\xi \rangle|^2 = \frac{\hbar}{2\omega_\xi} (N_\xi \delta_{M_\xi, N_\xi - 1} + (N_\xi + 1)\delta_{M_\xi, N_\xi + 1}) . \tag{3.210}$$

Inserting this into Eq. (3.209) yields

$$C(\omega) = 2\pi \sum_\xi \gamma_\xi^2 \frac{\hbar}{2\omega_\xi} \sum_{M_\xi} f(E_{M_\xi}) \Big( (M_\xi + 1)\delta(\omega - \omega_\xi) + M_\xi \delta(\omega + \omega_\xi) \Big) . \tag{3.211}$$

The summations with respect to the oscillator quantum numbers can be carried out after introducing the mean occupation number of a harmonic oscillator mode (Bose–Einstein distribution, see Eq. (3.202))

$$\sum_{M_\xi} M_\xi f(E_{M_\xi}) = n(\omega_\xi) . \tag{3.212}$$

With the help of this expression we obtain

$$C(\omega) = 2\pi \sum_\xi \gamma_\xi^2 \frac{\hbar}{2\omega_\xi} \left[ (n(\omega_\xi) + 1)\delta(\omega - \omega_\xi) + n(\omega_\xi)\delta(\omega + \omega_\xi) \right] . \tag{3.213}$$

In principle, higher order correlation functions of oscillator coordinates could be obtained along the same lines. However, there is a more elegant way to such multi–time correlation functions which makes use of the concept of generating functions. This will be explained in the next section.

### 3.6.3 Nonlinear Coupling to a Harmonic Oscillator Reservoir

In the preceding section we argued that the type of system–reservoir coupling introduced in Eq. (3.205) is the result of a Taylor expansion of $H_{S-R}$ with respect to the reservoir coordinates. Here we will consider the case where higher contributions have been taken into account. For this reason we introduce a general expansion of the reservoir part $\Phi$ (again only a single coupling term is taken and the indices *u* and *v* will be dropped). In order to simplify the notation we change from $Z_\xi$ to the dimensionless coordinate operators $Q_\xi = \sqrt{2\omega_\xi/\hbar} \times Z_\xi = C_\xi + C_\xi^+$ and write

$$\Phi/\hbar = \sum_{\xi_1} g_{\xi_1}^{(1)} Q_{\xi_1} + \sum_{\xi_1,\xi_2} g_{\xi_1,\xi_2}^{(2)} Q_{\xi_1} Q_{\xi_2} + \sum_{\xi_1,\xi_2,\xi_3} g_{\xi_1,\xi_2,\xi_3}^{(3)} Q_{\xi_1} Q_{\xi_2} Q_{\xi_3} \dots . \tag{3.214}$$

Such type of bath operator we will encounter, for instance, in Section 4.4. According to Eq. (3.186) such an expansion results in a correlation function $C(t)$ which comprises different reservoir coordinate correlation functions. It will be shown in the following that all these different types can be calculated since the reservoir is formed by independent *harmonic* oscillators.

All the expansion contributions result in the following general type of reservoir coordinate correlation function (compare the notation of Eq. (3.186)).

$$C^{(m,n)}_{\xi\bar{\xi}}(t) = \langle Q_{\xi_1}(t)...Q_{\xi_m}(t)\, Q_{\bar{\xi}_1}...Q_{\bar{\xi}_n}\rangle_{\mathrm{R}} \,. \tag{3.215}$$

In our notation the first set of $m$ different time–dependent coordinate operators are positioned left from the set of $n$ different time–independent operators. Although one can try to compute $C^{(m,n)}_{\xi\bar{\xi}}(t)$ directly as presented in Section 3.6.2 the mentioned ordering scheme offers a much more efficient way. It is based on the introduction of the following so–called *generating function*

$$\mathcal{F}(\{\sigma_\xi\},\{\vartheta_{\bar{\xi}}\};t) = \langle \exp\Big[\sum_\xi \sigma_\xi Q_\xi(t)\Big] \exp\Big[\sum_{\bar{\xi}} \vartheta_{\bar{\xi}} Q_{\bar{\xi}}\Big]\rangle_{\mathrm{R}} \,. \tag{3.216}$$

This function contains the set of constants $\sigma_\xi$ referring to the time–dependent coordinate operators and the set $\vartheta_{\bar{\xi}}$ referring to the time–independent operators. Taking multiple derivatives with respect to $\sigma_\xi$ and $\vartheta_{\bar{\xi}}$ the correlation functions $C^{(m,n)}_{\xi\bar{\xi}}(t)$ are recovered provided that we take the limit that all $\sigma_\xi$ and $\vartheta_{\bar{\xi}}$ vanish afterwards. Let us first calculate such multiple derivatives

$$\begin{aligned}
\mathcal{F}^{(m,n)} &= \frac{\partial^m}{\partial\sigma_{\xi_1}...\partial\sigma_{\xi_m}} \frac{\partial^n}{\partial\vartheta_{\bar{\xi}_1}...\partial\vartheta_{\bar{\xi}_n}} \mathcal{F}(\{\sigma_\xi\},\{\vartheta_{\bar{\xi}}\};t) \\
&= \langle \exp\Big[\sum_\xi \sigma_\xi Q_\xi(t)\Big]\, Q_{\xi_1}(t)...Q_{\xi_m}(t)\, Q_{\bar{\xi}_1}...Q_{\bar{\xi}_n} \exp\Big[\sum_{\bar{\xi}} \vartheta_{\bar{\xi}} Q_{\bar{\xi}}\Big]\rangle_{\mathrm{R}} \,.
\end{aligned} \tag{3.217}$$

Obviously, the correlation function $C^{(m,n)}_{\xi\bar{\xi}}(t)$ is reproduced in the limit $\sigma_\xi \to 0$ and $\vartheta_{\bar{\xi}} \to 0$. But the importance of this formulation comes from the possibility to derive a simple analytical expression for the generating function Eq. (3.216). It is explained in detail in the supplementary Section 3.12.2 how such a computation can be carried out. The result is simply

$$\mathcal{F}(\{\sigma_\xi\},\{\vartheta_{\bar{\xi}}\};t) = \exp\Big\{\sum_\xi \big[A_\xi(\sigma_\xi^2 + \vartheta_\xi^2) + B_\xi(t)\sigma_\xi\vartheta_\xi\big]\Big\} \,, \tag{3.218}$$

with

$$A_\xi = \frac{1}{2} + n(\omega_\xi) \,, \qquad B_\xi(t) = (1 + n(\omega_\xi))e^{-i\omega_\xi t} + n(\omega_\xi)e^{i\omega_\xi t} \,. \tag{3.219}$$

Notice that the Fourier transform of the last expression already appeared in Section 3.6.2.

Let us utilize Eq. (3.218) to compute the coordinate correlation functions up to the type $C^{(3,3)}_{\xi\bar{\xi}}(t)$ which is related to the third–order part of the expansion Eq. (3.214) (higher–order functions are calculated in [Oht89]). We take into account that for a harmonic oscillator reservoir all correlation function $C_{\xi\bar{\xi}}^{(m,n)}(t)$ vanish if $m+n$ is an odd number. A systematic way how to carry out the calculation is given in the supplementary part to this chapter, Section 3.12.2. Here we only quote the results. First we consider the correlation function which stems from the linear part of the expansion Eq. (3.214). The result is

$$C_{\xi\bar{\xi}}^{(1,1)}(t) = \langle Q_{\xi_1}(t)\, Q_{\bar{\xi}_1} \rangle_{\mathrm{R}} = \delta_{\xi_1,\bar{\xi}_1} B_{\xi_1}(t) \; . \tag{3.220}$$

The Fourier-transformed expression can be retrieved in Eq. (3.213). Moreover, we obtain

$$C_{\xi\bar{\xi}}^{(2,0)}(t) = \langle Q_{\xi_1}(t) Q_{\xi_2}(t) \rangle_{\mathrm{R}} = \delta_{\xi_1,\xi_2} 2A_{\xi_1} \; , \tag{3.221}$$

For $\langle Q_{\bar{\xi}_1} Q_{\bar{\xi}_2} \rangle_{\mathrm{R}}$ we get $\delta_{\bar{\xi}_1,\bar{\xi}_2} 2A_{\bar{\xi}_1}$. The next type of correlation function is

$$\begin{aligned} C_{\xi\bar{\xi}}^{(2,2)}(t) &= \langle Q_{\xi_1}(t) Q_{\xi_2}(t)\, Q_{\bar{\xi}_1} Q_{\bar{\xi}_2} \rangle_{\mathrm{R}} \\ &= (\delta_{\xi_1,\bar{\xi}_1}\delta_{\xi_2,\bar{\xi}_2} + \delta_{\xi_1,\bar{\xi}_2}\delta_{\xi_2,\bar{\xi}_1}) B_{\xi_1}(t) B_{\xi_2}(t) \\ &\quad + \delta_{\xi_1,\xi_2}\delta_{\bar{\xi}_1,\bar{\xi}_2} 4A_{\xi_1} A_{\bar{\xi}_1} \; . \end{aligned} \tag{3.222}$$

It corresponds to the case of a quadratic dependence of the system–reservoir coupling on the reservoir coordinates. Besides time–dependent contributions there are also time–independent terms. As the next higher type of correlation function we obtain

$$C_{\xi\bar{\xi}}^{(1,3)}(t) = \langle Q_{\xi_1}(t)\, Q_{\bar{\xi}_1} Q_{\bar{\xi}_2} Q_{\bar{\xi}_3} \rangle_{\mathrm{R}} = \delta_{\xi_1,\bar{\xi}_1} B_{\xi_1}(t) \delta_{\bar{\xi}_2,\bar{\xi}_3} A_{\bar{\xi}_2} + (\bar{\xi}_1 \leftrightarrow \bar{\xi}_2) + (\bar{\xi}_1 \leftrightarrow \bar{\xi}_3) \; . \tag{3.223}$$

The first term of the right–hand side written explicitly is followed by those terms where the indicated exchange of arguments has to be carried out ($C^{(3,1)}\xi\bar{\xi}(t)$ has the same structure but different arguments). Finally, we present the correlation function related to a cubic dependence of the system–reservoir coupling on the reservoir coordinates

$$C_{\xi\bar{\xi}}^{(3,3)}(t) = \langle Q_{\xi_1}(t) Q_{\xi_2}(t) Q_{\xi_3}(t)\, Q_{\bar{\xi}_1} Q_{\bar{\xi}_2} Q_{\bar{\xi}_3} \rangle_{\mathrm{R}} \; , \tag{3.224}$$

with

$$\begin{aligned} C_{\xi\bar{\xi}}^{(3,3)}(t) &= \Big\{ \delta_{\xi_1,\bar{\xi}_1}\delta_{\xi_2,\bar{\xi}_2}\delta_{\xi_3,\bar{\xi}_3} + \text{ (all permutations of } \bar{\xi}_1, \bar{\xi}_2, \bar{\xi}_3) \Big\} \\ &\quad \times B_{\xi_1}(t) B_{\xi_2}(t) B_{\xi_3}(t) \\ &+ \Big\{ \delta_{\xi_1,\xi_2} 2A_{\xi_1} \times \big[ \delta_{\xi_3,\bar{\xi}_1}\delta_{\bar{\xi}_2,\bar{\xi}_3} A_{\bar{\xi}_2} + \text{(all permutations of } \bar{\xi}_1, \bar{\xi}_2, \bar{\xi}_3) \big] \\ &\quad + \text{(all permutations of } \xi_1, \xi_2, \xi_3) \Big\} \; . \end{aligned} \tag{3.225}$$

The expression looks quite complicated but simplifies essentially if it is introduced into the overall correlation function $C(t)$ where summations with respect to the different indices are carried out. Some applications can be found in the Sections 3.8.5 and 4.4.

### 3.6.4 The Spectral Density

To simplify the (Fourier transformed) correlation function, Eq. (3.213) or the coordinate correlation functions introduced in the preceeding section one introduces the so–called *spectral density* $J(\omega)$. Here, we will exclusively concentrate on the type of correlation function introduced in Eq. (3.213). Therefore, the spectral density is defined as

$$J(\omega) = \sum_{\xi} g_{\xi}^{2} \delta(\omega - \omega_{\xi}) \ . \qquad (3.226)$$

The dimensionless coupling constant $g_\xi$ is given by $\gamma_\xi \sqrt{\hbar/2\omega_\xi} = \omega_\xi g_\xi$, what can be related to Eq. (3.214) via $g_\xi = g_\xi^{(1)}/\omega_\xi$. With the help of Eq. (3.226) the correlation function (3.213) can be written as

$$C(\omega) = 2\pi\, \omega^2 [1 + n(\omega)]\, [J(\omega) - J(-\omega)] \ . \qquad (3.227)$$

This notation points out the significance of the spectral density which contains the specific information about the reservoir and its interaction with the relevant system. We emphasize that for the case that the reservoir can be modelled as a collection of harmonic oscillators in thermal equilibrium which are linearly coupled to the system degrees of freedom the reservoir correlation function is described by a *single* function $J(\omega)$.

Using the general relation (3.201) we can also write

$$C^{(-)}(\omega) = 2\pi\, \omega^2 \left[J(\omega) - J(-\omega)\right] \ . \qquad (3.228)$$

It is important to note that in the literature the factor $\omega^2$ in (3.227) is often included into the definition of the spectral density. However, the present notation will be more convenient in the following chapters since $g_\xi$ is directly related to the dimensionless shift between PES belonging to different electronic states (cf. Eq. (2.86)).

Although the spectral density, Eq. (3.226) is defined in terms of a sum of delta functions any macroscopic system will in practice have a continuous spectral density. There exist different models for $J(\omega)$ which are adapted to particular system–environment situations. They are often characterized by a frequency dependence showing a power law rise for small frequencies which turns, after reaching a cut–off frequency $\omega_c$, into an exponential decay for large frequencies:

$$\omega^2 J(\omega) = \Theta(\omega)\, j_0\, \omega^p\, e^{-\omega/\omega_c} \ . \qquad (3.229)$$

Here, the unit–step function guarantees that $J = 0$ for $\omega < 0$, and $j_0$ is some normalization factor which depends on the actual definition of the spectral density (see above). For $p = 1$ and a cut–off frequency $\omega_c$ which is much larger than relevant frequencies of the considered system we obtain the *Ohmic* form of the spectral density, $\omega^2 J(\omega) \propto \omega$. This expression has to be used with caution, since a real system cannot have oscillator modes at arbitrarily high frequencies.

A different frequency dependence is given by the so–called Debye spectral density

$$\omega^2 J(\omega) = \Theta(\omega)\, \frac{j_0 \omega}{\omega^2 + \omega_D^2} \ , \qquad (3.230)$$

which is typically used to characterize the coupling between a solute and a polar solvent (cf. the discussion in Section 6.5). The frequency, $\omega_{\rm D}$, appearing in Eq. (3.230) is called the Debye frequency. Note, that this spectral density also reduces to the ohmic case mentioned above if the Debye frequency is assumed to be large.

If there exists an unambiguous relation between the mode index $\xi$ and the mode frequency $\omega_\xi$ the quantity $g_\xi$ can be defined as a frequency–dependent function. Using the abbreviation $\kappa(\omega_\xi) = g_\xi^2$ it is then possible to rewrite the spectral density by introducing the *density of states* of the reservoir

$$\mathcal{N}_{\rm R}(\omega) = \sum_\xi \delta(\omega - \omega_\xi) \ . \tag{3.231}$$

It follows the relation

$$J(\omega) = \kappa(\omega)\mathcal{N}_{\rm R}(\omega) \ , \tag{3.232}$$

which highlights that the spectral density can be viewed as the reservoir density of states which is weighted by the coupling strength between system and reservoir DOF.

Once $J(\omega)$ is fixed the time–dependent correlation function $C(t)$ can be calculated using Eq. (3.204) as well as Eq. (3.228) for $C^{(-)}(\omega)$. The obtained expression can be easily separated into a real and imaginary part [11]

$$C(t) = \int_0^\infty d\omega \left( \cos(\omega t) \coth \frac{\hbar\omega}{2k_{\rm B}T} - i \sin(\omega t) \right) \omega^2 J(\omega) \ . \tag{3.233}$$

Independent on the short–time behavior $C(t)$ decays for larger times with the correlation time $\tau_{\rm c}$ which characterizes the time range for correlated reservoir fluctuations. To give an example we will calculate $C(t)$, Eq. (3.233) for the Debye spectral density introduced in Eq. (3.230). Note that in the case of zero temperature $C(t)$ becomes a purely imaginary function. The real part only enters at finite temperatures. It is easy to calculate $C(t)$ in the high–temperature limit $k_{\rm B}T \gg \hbar\omega_{\rm D}$ where one can approximate $\coth(\hbar\omega/2k_{\rm B}T)$ by $2k_{\rm B}T/\hbar\omega$. If one inserts this approximation into $C(t)$ and replaces $\omega \sin(\omega t)$ by the time derivative of $\cos(\omega t)$ one obtains

$$C(t) = \frac{j_0}{\hbar} \left( 2k_{\rm B}T + i\hbar \frac{\partial}{\partial t} \right) \int_0^\infty d\omega \ \frac{\cos(\omega t)}{\omega^2 + \omega_{\rm D}^2} \ . \tag{3.234}$$

Since the integrand is an even function we can extend the frequency integral up to $-\infty$ and

[11] Here we used $\coth(\hbar\omega/2k_{\rm B}T) = 1 + 2n(\omega)$.

calculate it using the residue theorem

$$
\begin{aligned}
\int d\omega \, \frac{\cos(\omega t)}{\omega^2 + \omega_D^2} &= \frac{i}{4\omega_D} \Bigg( \int_{\mathcal{C}_1} d\omega \, e^{i\omega t} \left( \frac{1}{\omega + i\omega_D} - \frac{1}{\omega - i\omega_D} \right) \\
&\quad - \int_{\mathcal{C}_2} d\omega \, e^{-i\omega t} \left( \frac{1}{\omega + i\omega_D} - \frac{1}{\omega - i\omega_D} \right) \Bigg) \\
&= \frac{\pi}{\omega_D} \, e^{-\omega_D t} \, . \qquad (3.235)
\end{aligned}
$$

Here, $\mathcal{C}_1$ and $\mathcal{C}_2$ are closed integration contours (with mathematically positive orientation) in the upper and lower half of the complex frequency plane, respectively. We obtain for the correlation function

$$
C(t) = \frac{\pi j_0}{2\hbar\omega_D} \left(2k_B T - i\hbar\omega_D\right) e^{-\omega_D t} \, . \qquad (3.236)
$$

The correlation function decays with a time constant $\tau_c$ determined by the inverse of $\omega_D$. Notice that $C(t)$ is defined by bath operators, i.e., the correlation time can be considered as a reservoir property. If the Debye frequency is assumed to be large, the spectral density (3.230) has an ohmic behavior and the correlation time goes to zero, i.e., $C(t) \approx \delta(t)$; this is the *Markovian* limit.[12]

Finally, we point out that in the general case where $H_{S-R}$ has the form (3.146) we arrive at a spectral density which depends on the same indices $u$ and $v$ as the reservoir correlation function, i.e., $J(\omega)$ is replaced by $J_{uv}(\omega)$. The dependence of the coupling on some additional index could occur, for example, if we consider several diabatic electronic states which are characterized by different coupling strengths to the environment. This point will be further discussed in Chapter 6.

### 3.6.5 Linear Response Theory for the Reservoir

In Section 3.5.6 we have seen that the correlation functions $C_{uv}(t)$ automatically enter the QME as a result of the second–order approximation with respect to the system–reservoir coupling. In the following we will demonstrate how these functions, which are exclusively defined by reservoir quantities, can be introduced in an alternative way. For this reason we change the point of view taken so far. We will not ask in which manner the system is influenced by the reservoir but how the reservoir dynamics is modified by the system's motion. To answer this question it will be sufficient to describe the action of the system on the reservoir via classical time–dependent fields $K_u(t)$. Therefore, we replace $H_{S-R}$ by

$$
H_{\text{ext}}(t) = \sum_u K_u(t) \hat{\Phi}_u \, . \qquad (3.237)
$$

[12] There exist other model spectral densities that allow to determine the time dependence of the correlation function analytically (see literature quoted in the Suggested Reading section).

Here $\hat{\Phi}_u$ stands for the various reservoir operators. The bath Hamiltonian becomes time–dependent too, and is denoted by

$$\mathcal{H}(t) = H_{\text{R}} + H_{\text{ext}}(t) \; . \tag{3.238}$$

As a consequence of the action of the fields $K_u(t)$, the reservoir will be driven out of equilibrium. But in the case where the actual nonequilibrium state deviates only slightly from the equilibrium this deviation can be linearized with respect to the external perturbations. We argue that in this limit the expectation value of the reservoir operator $\hat{\Phi}_u$ obeys the relation

$$\langle \hat{\Phi}_u(t) \rangle = \sum_v \int_{t_0}^{t} d\bar{t} \; \chi_{uv}(t, \bar{t}) K_u(\bar{t}) \; . \tag{3.239}$$

The functions $\chi_{uv}(t, \bar{t})$ are called *linear response functions* or *generalized linear susceptibilities*. In order to derive Eq. (3.239) we start with the definition of the expectation value $\langle \hat{\Phi}_u(t) \rangle$ which is given by ($\hat{R}_{\text{eq}}$ is the reservoir equilibrium density operator)

$$\langle \hat{\Phi}_u(t) \rangle \equiv \text{tr}_{\text{R}} \{ \hat{R}_{\text{eq}} U^{+}(t - t_0) \hat{\Phi}_u U(t - t_0) \} \; . \tag{3.240}$$

To linearize this expression with respect to the external fields the time–evolution operator $U(t, t_0)$ (which does not depend on $t - t_0$ since the Hamiltonian is time–dependent ( $\mathcal{H}(t)$, Eq. (3.238))) is first separated into the free part $U_{\text{R}}(t - t_0)$ defined by $H_{\text{R}}$, and the $S$–operator (cf. Section 3.2.2) which reads

$$S(t, t_0) = \hat{T} \exp\Big( - \frac{i}{\hbar} \int_{t_0}^{t} d\tau \; U_{\text{R}}^{+}(\tau - t_0) H_{\text{ext}}(\tau) U_{\text{R}}(\tau - t_0) \Big) \; . \tag{3.241}$$

In a second step the $S$–operator is expanded up to first order in $H_{\text{ext}}(\tau)$. The result is inserted into Eq. (3.240) and we obtain

$$\langle \hat{\Phi}_u(t) \rangle \approx \text{tr}_{\text{R}} \left\{ \hat{R}_{\text{eq}} \hat{\Phi}_u^{(\text{I})}(t) - \frac{i}{\hbar} \int_{t_0}^{t} d\bar{t} \; \text{tr}_{\text{R}} \{ \hat{R}_{\text{eq}} \left[ \hat{\Phi}_u^{(\text{I})}(t), \hat{\Phi}_v^{(\text{I})}(\bar{t}) \right]_{-} \right\} K_v(\bar{t}) \; . \tag{3.242}$$

Here, the time dependence of the reservoir operators $\hat{\Phi}_u^{(\text{I})}(t)$ is given in the interaction representation. Comparing Eq. (3.242) with Eq. (3.239) the linear response function can be identified as[13]

$$\chi_{uv}(t, \bar{t}) = - \frac{i}{\hbar} \text{tr}_{\text{R}} \left\{ \hat{R}_{\text{eq}} \left[ \hat{\Phi}_u^{(\text{I})}(t), \hat{\Phi}_v^{(\text{I})}(\bar{t}) \right]_{-} \right\} \tag{3.243}$$

First, we notice that the right–hand side depends on the time difference $t - \bar{t}$ only (cf. Eq. (3.182)), that is, $\chi_{uv}(t, \bar{t}) = \chi_{uv}(t - \bar{t})$. Second, a comparison with Eq. (3.193) shows that $\chi_{uv}(t) =$

[13]Here we assume that the equilibrium expectation values of $\hat{\Phi}_u$ vanish.

$-iC_{uv}^{(-)}(t)/\hbar$. The important point is that if there exists an experimental setup to measure the various $\langle \hat{\Phi}_u(t) \rangle$ one is able to deduce $\chi_{uv}(t)$ if the $K_v$ can be changed in the measurement. Thus, the response functions $\chi_{uv}(t)$ are quantities which can be experimentally determined at least in principle. In contrast, the correlation functions $C_{uv}(t)$ which are needed to study dissipation into the reservoir are not directly related to an experiment. However, using Eq. (3.204) one can compute $C_{uv}(t)$ if $\chi_{uv}(t)$ is known[14].

Next, we consider how the internal energy of the reservoir changes via the influence of the external fields $K_u(t)$. We obtain the internal energy as

$$E_{\rm R}(t) = {\rm tr_R}\left\{\hat{R}_{\rm eq} U^+(t-t_0)\mathcal{H}(t)U(t-t_0)\right\} . \tag{3.244}$$

The change in time follows as

$$\begin{aligned}\frac{\partial}{\partial t}E_{\rm R} &= {\rm tr_R}\left\{\hat{R}_{\rm eq}\frac{i}{\hbar}\mathcal{H}(t)U^+(t,t_0)\mathcal{H}(t)U(t,t_0)\right\} \\ &-{\rm tr_R}\left\{\hat{R}_{\rm eq}U^+(t,t_0)\mathcal{H}(t)\frac{i}{\hbar}\mathcal{H}(t)U(t,t_0)\right\} \\ &+{\rm tr_R}\left\{\hat{R}_{\rm eq}U^(t,t_0)\left(\frac{\partial}{\partial t}\mathcal{H}(t)\right)U(t,t_0)\right\} .\end{aligned} \tag{3.245}$$

The first two terms compensate each other and one finally gets

$$\frac{\partial}{\partial t}E_{\rm R} = \sum_u {\rm tr_R}\left\{\hat{R}_{\rm eq}U^+(t,t_0)\hat{\Phi}_u U(t,t_0)\right\}\frac{\partial}{\partial t}K_u(t) \equiv \sum_u \langle\hat{\Phi}_u(t)\rangle\frac{\partial}{\partial t}K_u(t) . \tag{3.246}$$

If the disturbance of the reservoir equilibrium state is weak enough we can insert the linear susceptibility, Eq. (3.239), and obtain the change of internal energy expressed by the correlation function $C_{uv}^{(-)}(t)$. The latter describes fluctuations of certain operators of the reservoir whereas the change of internal energy is a measure of energy dissipation. Therefore, the relation is called *fluctuation–dissipation theorem*.

Finally, it should be noted that this discussion is not restricted to the present situation. Whenever some system under the influence of a weak external field is considered, its response can be described in lowest–order using an appropriate linear response function. The latter is completely defined by an equilibrium correlation function of some system operators.

### 3.6.6 Classical description of $C_{uv}(t)$

As long as the reservoir can be described by independent harmonic oscillators it is easy to compute the correlation functions $C_{uv}$ using spectral densities, as has been shown in Section 3.6.2. If this is not possible one can go back to a classical description via molecular dynamics simulations using the Hamilton function $H_{\rm R}(P,Z)$ (which is defined by the sets $P = \{P_\xi\}$ and $Z = \{Z_\xi\}$ of momenta and coordinates, respectively). In such a case one has to clarify

[14] Note that the use of Eq. (3.204) requires in a first step according to Eq. (3.243) the determination of $C_{uv}^{(-)}(\omega)$ from $\chi_{uv}(t)$.

how the quantum statistical correlation functions discussed so far have to be expressed via the results of the classical molecular dynamic simulations. Let us denote the classical correlation functions by

$$\zeta_{uv}(t) = \langle \Phi_u(t) \Phi_v \rangle_{\rm cl} \,. \tag{3.247}$$

Here the $\Phi_u(t)$ are functions $\Phi_u(P(t), Z(t))$ of the canonically conjugated variables, and the classical average is performed with respect to the set $P_0 \equiv \{P_\xi^{(0)}\}$ and $Z_0 \equiv \{Z_\xi^{(0)}\}$ of initial momenta and coordinates corresponding to the thermal equilibrium distribution, $f(P, Z) = \exp(-H_{\rm R}(P, Z)/k_{\rm B}T)/\mathcal{Z}$ ($\mathcal{Z}$ is the partition function). Thus, we have

$$\langle \Phi_u(t) \Phi_v \rangle_{\rm cl} = \int dP_0 dZ_0 \; f(P_0, Z_0) \Phi_u\big(P(t), Z(t)\big) \Phi_v\big(P_0, Z_0\big) \,. \tag{3.248}$$

The classical correlation function is a real quantity which can be determined by a molecular dynamics simulation of the reservoir equilibrium. The problem is that instead of the relation $C_{uv}(t) = C_{vu}^*(-t)$ the classical correlation function fulfills $\zeta_{uv}(t) = \zeta_{uv}(-t)$ as a result of the time–reversal symmetry of the classical equations of motion. This makes upward and downward relaxation equally probable, since the relation $C_{uv}(\omega)/C_{vu}(-\omega) = \exp\{-\hbar\omega/k_{\rm B}T\}$ does not exist (see Eq. (3.199)). In order to solve this problem one identifies $\zeta_{uv}(t)$ with half of the symmetric correlation function $C_{uv}^{(+)}$, Eq. (3.193). For the Fourier transform $\zeta_{uv}(\omega)$ we use Eq. (3.201) and obtain

$$C_{uv}(\omega) = 2\left(1 + \exp\{-\frac{\hbar\omega}{k_{\rm B}T}\}\right)^{-1} \zeta_{uv}(\omega) \,, \tag{3.249}$$

and

$$C_{vu}(-\omega) = 2\left(1 + \exp\{\frac{\hbar\omega}{k_{\rm B}T}\}\right)^{-1} \zeta_{uv}(\omega) \,. \tag{3.250}$$

Due to the temperature dependent prefactor detailed balance is guaranteed by these expressions. However, note that for an arbitrary system the replacement of the symmetrized quantum correlation function by the classical correlation function represents only an approximation (for a systematic study see, for instance, [Ego99]).

## 3.7 Quantum Master Equation

So far we have derived the QME in the interaction representation and studied the properties of the reservoir correlation function which governs energy dissipation out of the relevant system. For a number of applications it may be useful to stay in the interaction representation. Often, however, it is more appropriate to go back to the Schrödinger representation. Following Section 3.5.3 the equation of motion for the RDO can be transformed from the interaction representation into the Schrödinger representation according to

$$\begin{aligned} \frac{\partial}{\partial t}\hat\rho(t) &= \frac{\partial}{\partial t}[U_{\rm S}(t-t_0)\hat\rho^{(\rm I)}U_{\rm S}^+(t-t_0)]_- \\ &= -\frac{i}{\hbar}\Big[H_{\rm S}, \hat\rho(t)\Big]_- + U_{\rm S}(t-t_0)\frac{\partial}{\partial t}\hat\rho^{(\rm I)}(t)U_{\rm S}^+(t-t_0) \,. \end{aligned} \tag{3.251}$$

For Eq. (3.190) this gives

$$\begin{aligned}\frac{\partial}{\partial t}\hat{\rho}(t) &= -\frac{i}{\hbar}\Big[H_{\rm S} + \sum_u \langle\Phi_u\rangle K_u, \hat{\rho}(t)\Big]_- - U_{\rm S}(t-t_0) \\ &\times \sum_{u,v}\int_{t_0}^{t} d\bar{t}\Big\{C_{uv}(t-\bar{t})\Big[U_{\rm S}^{+}(t-t_0)K_u U_{\rm S}(t-t_0), \\ &U_{\rm S}^{+}(\bar{t}-t_0)K_v U_{\rm S}(\bar{t}-t_0)U_{\rm S}^{+}(\bar{t}-t_0)\hat{\rho}(\bar{t})U_{\rm S}(\bar{t}-t_0)\Big]_- \\ &-C_{vu}(-t+\bar{t})\Big[U_{\rm S}^{+}(t-t_0)K_u U_{\rm S}(t-t_0), \\ &U_{\rm S}^{+}(\bar{t}-t_0)\hat{\rho}(\bar{t})U_{\rm S}(\bar{t}-t_0)U_{\rm S}^{+}(\bar{t}-t_0)K_v U_{\rm S}(\bar{t}-t_0)\Big]_-\Big\} \\ &\times U_{\rm S}^{+}(t-t_0)\,. \end{aligned} \tag{3.252}$$

Combining products of time–evolution operators and replacing $t-\bar{t}$ by $\tau$, we obtain the QME in the Schrödinger representation

$$\begin{aligned}\frac{\partial}{\partial t}\hat{\rho} &= -\frac{i}{\hbar}[H_{\rm S} + \sum_u \langle\Phi_u\rangle K_u, \hat{\rho}]_- \\ &-\sum_{u,v}\int\limits_0^{t-t_0} d\tau\Big(C_{uv}(\tau)\Big[K_u, U_{\rm S}(\tau)K_v\hat{\rho}(t-\tau)U_{\rm S}^{+}(\tau)\Big]_- \\ &\qquad - C_{vu}(-\tau)\Big[K_u, U_{\rm S}(\tau)\hat{\rho}(t-\tau)K_v U_{\rm S}^{+}(\tau)\Big]_-\Big)\,. \end{aligned} \tag{3.253}$$

Before discussing details of this equation we will estimate the range of validity for the second–order perturbation theory. Let us assume that the integrand in Eq. (3.190) is constant within the memory time. Then the contribution of the integral to the right–hand side of the QME is of the order of $\tau_{\rm mem}\langle H_{\rm S-R}\rangle^2/\hbar^2$. In order to justify the perturbation expansion, this quantity (which has the dimension of a rate) has to be small compared to the first term on the right–hand side of Eq. (3.190), $\langle H_{\rm S}\rangle/\hbar$.

The first term on the right–hand side is already known from Section (3.5.3). It contains the mean–field contribution to the system dynamics which is of first order in the system–reservoir interaction. The dynamics including this mean–field term is reversible. The second term on the right–hand side which depends on the complex–valued correlation function $C_{uv}(t)$ leads to a quite different behavior. This can be rationalized by neglecting the time integration for a moment and considering only the diagonal elements of the density operator (in an arbitrary representation) which are real. In this case the resulting differential equation is of the type $\partial f(t)/\partial t = -kf(t)$, where $k$ is proportional to the real part of the correlation function. The solution of this type of equation will decay exponentially in time indicating an irreversible flow of probability in the system. It will be shown in more detail below that the second term in Eq. (3.253) is responsible for energy dissipation from the relevant system into the reservoir. Finally, as already discussed at the end of Section 3.5.6, Eq. (3.253) also guarantees Hermiticity of $\rho$ and conservation of total probability.

In the QME (3.253) the RDO $\hat{\rho}$ appears with an retarded time argument, $t-\tau$, in the integrand. This means that the actual change of probabilistic information in time (i.e., the right–hand side of Eq. (3.253)) is not determined by the probabilistic information at the same time $t$ but also by that of earlier times $t - \tau$. This type of equation is known from probabilistic theory as a *non–Markovian* equation. It is encountered whenever time–local equations of motion are reduced to equations which only describe a part of the original set of degrees of freedom. In the present case we changed from the Liouville–von Neumann equation (3.98) for the full density operator which is Markovian to the non–Markovian QME for the RDO. In Section 3.7.1 we will show under what conditions the right–hand side only depends on $\hat{\rho}(t)$ and the dynamics becomes Markovian again.

The characteristic feature of non–Markovian behavior is the appearance of *memory* effects in the determination of the time dependence of the RDO. As already pointed out in Section 3.5.6 the time span for this memory is mostly determined by the reservoir correlation functions $C_{uv}(t)$. The time dependence of $C_{uv}(t)$ can often be characterized by a single or a set of *correlation times*, $\tau_c$ (for more details see, e.g., [Man01,Mei99,Sua92].

The QME (3.253) is a fundamental result of relaxation theory. It has found many applications in different areas of physics, mainly in quantum optics, nuclear magnetic resonance, solid state physics, and in recent years also for the description of ultrafast phenomena in molecular systems. When using a QME, however, one should keep in mind that the perturbative treatment of the system–reservoir coupling restricts its applicability and demands for a careful separation of the full system at hand. In Chapters 5 – 9 we will discuss several examples in this respect.

### 3.7.1 Markov Approximation

In the following we we explain in detail the transition from the non–Markovian Quantum Master Equation (3.253) to a Markovian equation. Let us assume that a characteristic time $\tau_{\rm mem}$ (*memory time*) exists which characterizes the time span of memory effects. Now, if the reduced density operator $\hat{\rho}$ (i.e. any of its matrix elements) does not change substantially on the time scale given by $\tau_{\rm mem}$, memory effects will be negligible. In this case one can invoke the *Markov approximation* which amounts to setting

$$\hat{\rho}(t - \tau) \approx \hat{\rho}(t) \tag{3.254}$$

in the time integral Eq. (3.253).

An alternative view is provided, if we suppose that within the Markov approximation the minimum time step, $\Delta t$, for which information on the reduced density matrix is obtainable is restricted by the memory time, i.e., $\Delta t > \tau_{\rm mem}$. In case that the continuous time axis is *course grained* with a mesh size dictated by $\tau_{\rm mem}$, memory effects do not play any role for the dissipative dynamics of the system. Due to this requirement the upper limit of the integration in Eq. (3.253) exceeds the time interval where the integrand is finite. Thus, we can increase this limit without changing the value of the integral, i.e., we will set $t - t_0 \to \infty$ in the following.

In order to discuss in more detail what precisely it means when we assume that the reduced density operator does not change on the time scale of $\tau_{\rm mem}$ we change to the representation of $\hat{\rho}(t)$ in the eigenstates of $H_{\rm S}$, Eq. (3.141). Without any coupling to the reservoir the solution for $\rho_{ab}$ can be directly deduced from Eq. (3.99) as

$$\rho_{ab}(t) = e^{-i\omega_{ab}(t-t_0)}\rho_{ab}(t_0) \; . \tag{3.255}$$

The diagonal elements are time–independent but the off–diagonal elements may be rapidly oscillating functions. If $1/\omega_{ab} \ll \tau_{\text{mem}}$ the above given reasoning leading to the Markov–approximation breaks down. Thus, it is advisable to split off the oscillatory factor $e^{-i\omega_{ab}t}$ from the reduced density matrix and invoke the Markov approximation for the remaining slowly varying envelope. Therefore, we carry out the following replacement

$$\begin{aligned} \rho_{ab}(t-\tau) &= e^{-i\omega_{ab}(t-\tau-t_0)}\tilde{\rho}_{ab}(t-\tau) \\ &\approx e^{-i\omega_{ab}(t-\tau-t_0)}\tilde{\rho}_{ab}(t) = e^{i\omega_{ab}\tau}\rho_{ab}(t) \; , \end{aligned} \tag{3.256}$$

where the tilde denotes the envelope part of the reduced density matrix. This approximation scheme is equivalent to perform the Markov approximation in the interaction representation since

$$\begin{aligned} \rho_{ab}(t) &= \langle\varphi_a|e^{-iH_{\text{S}}(t-t_0)/\hbar}\hat{\rho}^{(\text{I})}(t)e^{iH_{\text{S}}(t-t_0)/\hbar}|\varphi_b\rangle \\ &= e^{-i\omega_{ab}(t-t_0)}\langle\varphi_a|\hat{\rho}^{(\text{I})}(t)|\varphi_b\rangle \; . \end{aligned} \tag{3.257}$$

Thus, the general prescription is that first we have to change to the interaction representation and only then the Markov approximation is made:

$$\begin{aligned} \hat{\rho}(t-\tau) &= U_{\text{S}}(t-\tau-t_0)\hat{\rho}^{(\text{I})}(t-\tau)U_{\text{S}}^{+}(t-\tau-t_0) \\ &\approx U_{\text{S}}(-\tau)U_{\text{S}}(t-t_0)\hat{\rho}^{(\text{I})}(t)U_{\text{S}}^{+}(t-t_0)U_{\text{S}}^{+}(-\tau) \\ &= U_{\text{S}}^{+}(\tau)\hat{\rho}(t)U_{\text{S}}(\tau) \; . \end{aligned} \tag{3.258}$$

Using this approximation the dissipative part of the QME (3.253) becomes

$$\begin{aligned} \left(\frac{\partial\hat{\rho}}{\partial t}\right)_{\text{diss}} = -\sum_{u,v}\int_0^\infty d\tau \, \Big\{ & C_{uv}(\tau)\left[K_u, K_v^{(\text{I})}(-\tau)\hat{\rho}(t)\right]_- \\ & - C_{vu}(-\tau)\left[K_u, \hat{\rho}(t)K_v^{(\text{I})}(-\tau)\right]_- \Big\} \; , \end{aligned} \tag{3.259}$$

where $K_v^{(\text{I})}(-\tau) = U_{\text{S}}(\tau)K_v U_{\text{S}}^{+}(\tau)$. A more compact form of this equation is obtained after introduction of the operator

$$\Lambda_u = \sum_v \int_0^\infty d\tau \, C_{uv}(\tau)K_v^{(\text{I})}(-\tau) \; . \tag{3.260}$$

and the operator $\Lambda_u^{(+)}$ following from $\Lambda_u$ upon replacing $C_{uv}(\tau)$ by $C_{vu}(-\tau)$ (if any term of $H_{\text{S-R}}$ is Hermitian, then $\Lambda_u^{(+)} = \Lambda_u^{+}$). With this definition Eq. (3.259) can be written as

$$\left(\frac{\partial\hat{\rho}}{\partial t}\right)_{\text{diss}} = -\sum_u \left[K_u, \Lambda_u\hat{\rho}(t) - \hat{\rho}(t)\Lambda_u^{(+)}\right]_- \; . \tag{3.261}$$

Carrying out the commutator the resulting expression suggests to supplement the system Hamiltonian by non–Hermitian contributions which are proportional to $K_u\Lambda_u$. Therefore, we introduce the effective non–Hermitian system Hamiltonian

$$H_{\rm S}^{(\rm eff)} = H_{\rm S} + \sum_u K_u\big[\langle\Phi_u\rangle - i\hbar\Lambda_u\big] \,. \tag{3.262}$$

Note that for convenience we included the first–order mean–field term in the definition of the effective Hamiltonian as well. Using Eq. (3.262) we obtain for the QME in the Markov approximation the final result ($H_{\rm S}^{(\rm eff)+}$ has to be understood as the Hermitian conjugated of $H^{(\rm eff)}$ except that all $\Lambda_u$ have been replaced by $\Lambda_u^+$)

$$\frac{\partial}{\partial t}\hat\rho(t) \;=\; -\frac{i}{\hbar}\left(H_{\rm S}^{(\rm eff)}\hat\rho(t) - \hat\rho(t)H_{\rm S}^{(\rm eff)+}\right) + \sum_u\left(K_u\hat\rho(t)\Lambda_u^{(+)} + \Lambda_u\hat\rho(t)K_u\right) \,. \tag{3.263}$$

This equation can be interpreted as follows. We first note that the part of the dissipative contributions acting exclusively from the left or from the right on the reduced density operator can be comprised to a non–Hermitian Hamiltonian. According to the general structure of the density operator, Eq. (3.71), this action can be understood as changing of the state vector norm. However, the remaining dissipative part acting on the reduced density operator from the left and the right simultaneously compensates for this normalization change. As a result the condition $\mathrm{tr_S}\{\rho\} = 1$ is fulfilled (together, of course, with conservation of total probability).

We conclude the discussion of this section by giving an alternative notation of the QME, Eq. (3.263) based on the superoperator formulation in Liouville space which has already been introduced in connection with the Liouville–von Neumann equation in Section 3.4.3. In the present case a Liouville superoperator can only be introduced for the reversible part of the QME. We set $\mathcal{L}_{\rm S}\bullet = [H_{\rm S},\bullet]_-/\hbar$ and obtain from Eq. (3.261):

$$\frac{\partial}{\partial t}\hat\rho(t) = -i\mathcal{L}_{\rm S}\hat\rho(t) - \mathcal{D}\hat\rho(t) \,. \tag{3.264}$$

In contrast to first term on the right–hand side, the second one cannot be given via a Liouville superoperator abbreviating a simple commutator. Instead, the so–called dissipative (or relaxation) superoperator $\mathcal{D}$ has been introduced. Its concrete action on the RDO can be obtained from the right–hand side of Eq. (3.261). Somctimes it is of useful to introduce the formal solution of Eq. (3.264) as

$$\hat\rho(t) = \mathcal{U}(t-t_0)\rho(t_0) \,, \tag{3.265}$$

with the time–evolution superoperator

$$\mathcal{U}(t-t_0) = \exp\big(-i(\mathcal{L}_{\rm S} - i\mathcal{D})(t-t_0)\big) \,. \tag{3.266}$$

The action of $\mathcal{D}$ can be characterized by considering the change of the internal energy of the relevant system $E_{\rm S} = \mathrm{tr_S}\{\hat\rho(t)H_{\rm S}\}$. Using Eq. (3.264) one immediately obtains

$$\frac{\partial}{\partial t}E_{\rm S} = \mathrm{tr_S}\{H_{\rm S}\mathcal{D}\hat\rho(t)\} = \sum_u \mathrm{tr_S}\{[H_{\rm S},K_u]_-\big(\Lambda_u\hat\rho(t) - \hat\rho(t)\Lambda_u^{(+)}\big)\} \,. \tag{3.267}$$

The second part of the right–hand side follows if $\mathcal{D}$ is introduced according to Eq. (3.261). The resulting expression shows that for cases where the commutator of the system Hamiltonian with every operator $K_u$ vanishes, dissipation does not alter the internal energy. This may be interpreted as an action of the environment reduced to elastic scattering processes which do not change the system energy but probably the phase of the system. Because of this particular property dissipative processes which do not change the system energy are related to what is known as *pure dephasing* (for further details see the following section). Assuming that the $|a\rangle$ are *eigenstates* of $H_S$ the coupling operator $K_u = |a\rangle\langle a|$ represents an example for a system–reservoir coupling which guarantees the conservation of the internal energy $E_S$. This has to be expected since the system part $K_u$ of the system–reservoir coupling does not change the system–state. To be complete we also remark that the internal energy remains constant if the dissipation is of such a type that the second term in the trace expression of Eq. (3.267) vanishes. We will discuss this case in more detail in Section 3.8.2

So far we did not take into account the specific structure Eq. (3.186) of the reservoir correlation function which may contain the factorized and time–independent part $\langle\Phi_u\rangle_R\langle\Phi_v\rangle_R$. However, by carrying out the Markov–approximation it needs a separate treatment which is detailed in the supplementary Section 3.12.4.

# 3.8 Reduced Density Matrix in Energy Representation

## 3.8.1 The Quantum Master Equation in Energy Representation

In the following we will transform the QME (3.253) into the energy (state) representation with respect to the system Hamiltonian. Suppose we have solved the eigenvalue problem for $H_S$,

$$H_S|a\rangle = E_a|a\rangle \ . \tag{3.268}$$

Then the reduced density matrix (RDM) is given by $\rho_{ab}(t) = \langle a|\hat{\rho}(t)|b\rangle$ (cf. Eq. (3.141)). Furthermore, we will introduce the matrix elements of the system part of the system–reservoir coupling according to

$$\langle a|K_u|b\rangle = K_{ab}^{(u)} \ . \tag{3.269}$$

It should be pointed out that even though any other choice of a complete basis set for representing the density matrix is possible, the energy representation offers the advantage that

$$U_S(\tau)|a\rangle = e^{-iE_a\tau/\hbar}|a\rangle \ , \tag{3.270}$$

which simplifies the description of the coherent system dynamics. Taking the respective matrix element of the QME (3.253) we obtain after some rearrangement on the right–hand side the following equation of motion for the RDM ($\omega_{ab} = (E_a - E_b)/\hbar$)

$$\begin{aligned}\frac{\partial}{\partial t}\rho_{ab} &= -i\omega_{ab}\rho_{ab} + \frac{i}{\hbar}\sum_c \sum_u \langle\Phi_u\rangle_{\rm R}(K^{(u)}_{cb}\rho_{ac} - K^{(u)}_{ac}\rho_{cb}) \\ &- \sum_{c,d}\sum_{u,v}\int_0^{t-t_0} d\tau \Big( C_{vu}(-\tau)K^{(u)}_{db}K^{(v)}_{cd}e^{i\omega_{da}\tau}\rho_{ac}(t-\tau) \\ &\qquad + C_{uv}(\tau)K^{(u)}_{ac}K^{(v)}_{cd}e^{i\omega_{bc}\tau}\rho_{db}(t-\tau) \\ &\qquad - \{C_{vu}(-\tau)K^{(u)}_{ac}K^{(v)}_{db}e^{i\omega_{bc}\tau} \\ &\qquad + C_{uv}(\tau)K^{(u)}_{db}K^{(v)}_{ac}e^{i\omega_{da}\tau}\}\rho_{cd}(t-\tau)\Big) . \end{aligned} \tag{3.271}$$

A more compact notation of this equation is achieved by introducing the tetradic matrix

$$M_{ab,cd}(t) = \sum_{u,v} C_{uv}(t)K^{(u)}_{ab}K^{(v)}_{cd} , \tag{3.272}$$

which satisfies the relation (note Eq. (3.189))

$$M^*_{ab,cd}(t) = \sum_{u,v} C_{vu}(-t)K^{(u)}_{ba}K^{(v)}_{dc} = M_{dc,ba}(-t) . \tag{3.273}$$

Apparently, $M_{ab,cd}(t)$ determines the time span for correlations. For this reason, it will be called *memory matrix* or *memory function*. Using this notation we can write the dissipative part of the non–Markovian density matrix equation (3.271) as

$$\begin{aligned}\left(\frac{\partial\rho_{ab}}{\partial t}\right)_{\rm diss.} &= -\sum_{cd}\int_0^{t-t_0} d\tau \Big( M_{cd,db}(-\tau)e^{i\omega_{da}\tau}\rho_{ac}(t-\tau) \\ &\qquad + M_{ac,cd}(\tau)e^{i\omega_{bc}\tau}\rho_{db}(t-\tau) \\ &\qquad - [M_{db,ac}(-\tau)e^{i\omega_{bc}\tau} + M_{db,ac}(\tau)e^{i\omega_{da}\tau}]\rho_{cd}(t-\tau)\Big) . \end{aligned} \tag{3.274}$$

In the following let us discuss two important properties of the solutions of the QME in the energy representation. The first one concerns the normalization condition for the RDM, Eq. (3.142), which expresses the fact that the total occupation probability of the different eigenstates of $H_{\rm S}$ is conserved, i.e. $\sum_a \partial\rho_{aa}/\partial t = 0$. It should be noted here that the basic property of a probability to be positive, i.e., $\rho_{aa} \geq 0$, cannot be proven in the general case (cf. Section 3.12.3). This requires careful analysis when carrying out the numerical calculation.

As a further property we expect that the stationary solution of the equations of motion for $\rho_{ab}$ must correspond to a state which is in equilibrium with the reservoir. Since the reservoir is at temperature $T$ we demand for the density matrix the limiting behavior

$$\lim_{t\to\infty} \rho_{ab}(t) = \delta_{ab}e^{-E_a/k_{\rm B}T} / \sum_c e^{-E_c/k_{\rm B}T} . \tag{3.275}$$

To verify Eq. (3.275) we demonstrate that its right–hand side is an asymptotic solution of the QME (3.271). This means that the right–hand side of the QME should vanish in the stationary limit, $\lim_{t\to\infty} \partial \rho_{aa}/\partial t = 0$. In the first step of the proof we introduce the limit $t \to \infty$ in every integral in Eq. (3.274). Since the reservoir correlation time $\tau_c$ is finite we can replace the time–dependent RDM $\rho_{ab}(t-\tau)$ in the integrand by its asymptotic expression $\rho_{ab}(\infty)$. For $\rho_{ab}(\infty)$ we substitute Eq. (3.275) (omitting the normalization constant) which is supposed to be the correct solution. It follows

$$\begin{aligned} 0 &= -\frac{i}{\hbar}\sum_{c,u}\langle\Phi_u\rangle_{\rm R}(K^{(u)}_{ac}\delta_{ca} - \delta_{ac}K^{(u)}_{ca})e^{-E_a/k_{\rm B}T} \\ &\quad -\sum_{cd}\int_0^\infty d\tau\Big[\{M_{cd,da}(-\tau)e^{i\omega_{da}\tau}\delta_{ac} + M_{ac,cd}(\tau)e^{i\omega_{ac}\tau}\delta_{da}\}e^{-E_a/k_{\rm B}T} \\ &\qquad - \{M_{da,ac}(-\tau)e^{i\omega_{ac}\tau} + M_{da,ac}(\tau)e^{i\omega_{da}\tau}\}\delta_{cd}e^{-E_c/k_{\rm B}T}\Big] . \end{aligned} \tag{3.276}$$

Next we use the properties of the memory matrix and combine various terms of the dissipative part. Afterwards, the Fourier transform of the correlation function $C_{uv}(\omega)$ will be introduced

$$\begin{aligned} 0 &= -\sum_c\int_0^\infty d\tau\Big(\{M_{ac,ca}(-\tau)e^{i\omega_{ca}\tau} + M_{ac,ca}(\tau)e^{i\omega_{ac}\tau}\}e^{-E_a/k_{\rm B}T} \\ &\qquad - \{M_{ca,ac}(-\tau)e^{i\omega_{ac}\tau} + M_{ca,ac}(\tau)e^{i\omega_{ca}\tau}\}e^{-E_c/k_{\rm B}T}\Big) \\ &= -\sum_c\int_{-\infty}^\infty d\tau\Big(M_{ac,ca}(\tau)e^{i\omega_{ac}\tau}e^{-E_a/k_{\rm B}T} - M_{ca,ac}(\tau)e^{i\omega_{ca}\tau}e^{-E_c/k_{\rm B}T}\Big) \\ &= -\sum_{c,u,v}\Big(C_{uv}(\omega_{ac})K^{(u)}_{ac}K^{(v)}_{ca}e^{-E_a/k_{\rm B}T} - C_{uv}(\omega_{ca})K^{(u)}_{ca}K^{(v)}_{ac}e^{-E_c/k_{\rm B}T}\Big) . \end{aligned} \tag{3.277}$$

To see that the last part vanishes we use relation (3.199). Introducing it into Eq. (3.276) gives

$$\sum_c\sum_{u,v}\Big(C_{uv}(\omega_{ac})K^{(u)}_{ac}K^{(v)}_{ca} - C_{vu}(-\omega_{ca})K^{(u)}_{ca}K^{(v)}_{ac}\Big) = 0 . \tag{3.278}$$

The final result is obtained after an interchange of $u$ and $v$ in the second term. Thus, the above given reasoning demonstrates that the asymptotic form of the RDO determined by the QME reads

$$\lim_{t\to\infty}\hat\rho(t) = \frac{1}{\mathcal{Z}}e^{-H_S/k_{\rm B}T} \tag{3.279}$$

The asymptotic form of the RDO which is the equilibrium density operator of the *relevant* system was obtained as a result of the second–order perturbational treatment of the system–reservoir coupling $H_{\rm S-R}$. Including all orders in a nonperturbative treatment the exact asymptotic form of the RDO has to be derived from the equilibrium density operator of the total system with the Hamiltonian $H$. This is achieved by restricting it to the state space of the relevant system according to ${\rm tr_R}\{\exp(-H/k_{\rm B}T)\}$.

### 3.8.2 Multi–Level Redfield Equations

After having introduced the energy (state) representation of the RDO let us discuss the Markov limit. Eq. (3.274) gives the dissipative part of the reduced density matrix equations of motion. Carrying out the Markov approximation, i.e., using Eq. (3.256) and shifting the upper bound of the time integral to infinity, we obtain

$$\begin{aligned}\left(\frac{\partial\rho_{ab}}{\partial t}\right)_{\text{diss.}} &= -\sum_{cd}\int_0^\infty d\tau\, \Big(M_{cd,db}(-\tau)e^{i\omega_{dc}\tau}\rho_{ac}(t) + M_{ac,cd}(\tau)e^{i\omega_{dc}\tau}\rho_{db}(t) \\ &\quad - [M_{db,ac}(-\tau)e^{i\omega_{bd}\tau} + M_{db,ac}(\tau)e^{i\omega_{ca}\tau}]\rho_{cd}(t)\Big) \; . \end{aligned} \tag{3.280}$$

(Note that we could have started from the operator equation Eq. (3.263) as well.) The time integrals can be viewed as half–sided Fourier transforms of the memory functions. These complex quantities define the dissipative part of the QME in the Markov approximation. Their real part describes an irreversible redistribution of the amplitudes contained in the various parts of reduced density matrix. The imaginary part introduces terms which can be interpreted as a modification of the transition frequencies and the respective mean–field matrix elements. These frequency shifts often give no qualitatively new contribution to the reduced density matrix equations. They can in these cases be accounted for by changing the energy scale or adjusting the transition frequencies. Therefore, we restrict ourselves to the discussion of the real part only leading to the following (damping) matrix

$$\Gamma_{ab,cd}(\omega) = \text{Re}\int_0^\infty d\tau\; e^{i\omega\tau} M_{ab,cd}(\tau) = \text{Re}\sum_{u,v} K_{ab}^{(u)} K_{cd}^{(v)} \int_0^\infty d\tau\; e^{i\omega\tau} C_{uv}(\tau) \; . \tag{3.281}$$

In the second part we introduced Eq. (3.272) indicating that the damping matrix is mainly determined by the half–sided Fourier transform of the reservoir correlation functions. To establish the connection to the operator equation (3.263) derived in the previous section, we note that the damping matrix can be written in the alternative form:

$$\Gamma_{ab,cd}(\omega_{dc}) = \text{Re}\sum_u \langle a|K_u|b\rangle\langle c|\Lambda_u|d\rangle \; . \tag{3.282}$$

(Note that the actual frequency argument is fixed by the matrix elements of $\langle c|K_v^{(\text{I})}(-\tau)|d\rangle$ in Eq. (3.260).) Using Eq. (3.281) the dissipative part of the QME in the state representation, Eq. (3.280) becomes

$$\begin{aligned}\left(\frac{\partial\rho_{ab}}{\partial t}\right)_{\text{diss.}} &= -\sum_{cd}\Big(\Gamma_{bd,dc}(\omega_{cd})\rho_{ac}(t) + \Gamma_{ac,cd}(\omega_{dc})\rho_{db}(t) \\ &\quad - [\Gamma_{ca,bd}(\omega_{db}) + \Gamma_{db,ac}(\omega_{ca})]\rho_{cd}(t)\Big) \; , \end{aligned} \tag{3.283}$$

If we further introduce the *relaxation matrix*

$$R_{ab,cd} = \delta_{ac}\sum_e \Gamma_{be,ed}(\omega_{de}) + \delta_{bd}\sum_e \Gamma_{ae,ec}(\omega_{ce}) - \Gamma_{ca,bd}(\omega_{db}) - \Gamma_{db,ac}(\omega_{ca}) \; , \tag{3.284}$$

the dissipative contribution to the reduced density matrix equations of motion can be finally written as

$$\left(\frac{\partial \rho_{ab}}{\partial t}\right)_{\text{diss.}} = -\sum_{cd} R_{ab,cd}\rho_{cd}(t) . \tag{3.285}$$

It should be noted that in the literature the tetradic relaxation matrix, Eq. (3.284), is frequently termed *Redfield* tensor after A. G. Redfield who introduced it in the theory of nuclear magnetic resonance spectroscopy in the early sixties [Red65].

Let us discuss in more detail the Redfield tensor and its effect on the dynamics of the reduced density matrix $\rho_{ab}(t)$. Since the density matrix elements can be distinguished as populations ($a = b$) and the coherences ($a \neq b$) it is reasonable to discuss $R_{ab,cd}$ according to its effect on the dynamics of $\rho_{aa}$ and $\rho_{ab}$.

1. Population transfer: $a = b$, $c = d$

Using Eq. (3.284) the respective matrix elements of the Redfield tensor can be written as

$$R_{aa,cc} = 2\delta_{ac}\sum_{e}\Gamma_{ae,ea}(\omega_{ae}) - 2\Gamma_{ca,ac}(\omega_{ca}) = \delta_{ac}\sum_{e} k_{ae} - k_{ca} . \tag{3.286}$$

Here we introduced the rate $k_{ab}$ ($\equiv k_{a\to b}$) for the transition from state $|a\rangle$ to state $|b\rangle$ according to

$$\begin{aligned} k_{ab} &= 2\Gamma_{ab,ba}(\omega_{ab}) = 2\text{Re}\int_0^\infty d\tau e^{i\omega_{ab}\tau} M_{ab,ba}(\tau) \\ &= \int_0^\infty d\tau e^{i\omega_{ab}\tau} M_{ab,ba}(\tau) + \int_0^\infty d\tau e^{-i\omega_{ab}\tau} M^{*}_{ab,ba}(\tau). \end{aligned} \tag{3.287}$$

The two terms on the last line can be combined to give

$$k_{ab} = \int d\tau\, e^{i\omega_{ab}\tau} M_{ab,ba}(\tau) \equiv M_{ab,ba}(\omega_{ab}) . \tag{3.288}$$

From Eq. (3.286) we see that $R_{aa,cc}$ combines rates for transitions between different system eigenstates. The first term in Eq. (3.286) corresponds to transitions from the state $|a\rangle$ into all other system states $|e\rangle$ thus decreasing the occupation probability of the state $|a\rangle$. Conservation of probability is established then by the second term in Eq. (3.286) which represents transitions from all other states into the state $|a\rangle$.

Eq. (3.288) shows that the transfer rate can also be written in terms of the Fourier–transformed memory matrix at the transition frequency $\omega_{ab}$. Using Eq. (3.272) for the memory matrix gives the following alternative expression for the energy relaxation rates

$$k_{ab} = \sum_{u,v} K_{ab}^{(u)} K_{ba}^{(v)} C_{uv}(\omega_{ab}) . \tag{3.289}$$

The amplitude of the rate for a particular transition is determined by the matrix elements of the operators $K_u$ and by the value of the correlation function taken at the respective transition

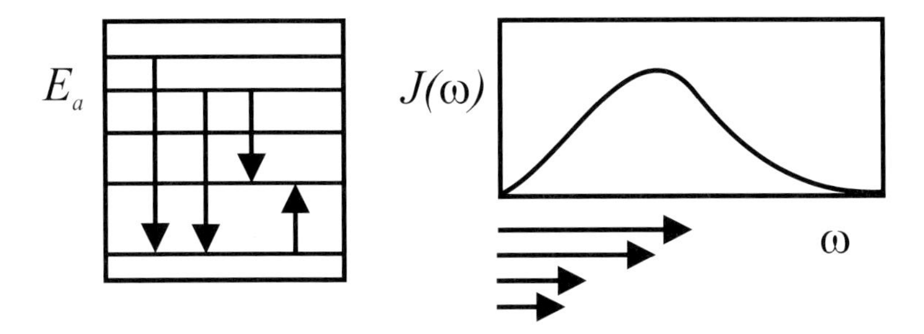


**Figure 3.5:** Transitions among five different quantum states $|a\rangle$ of the relevant system with energies $E_a$ ($a = 1, 2, .., 5$) (left part). The transitions are induced by the interaction with the reservoir which is characterized by the spectral density $J(\omega)$ (right part). The respective transition energies are related to the shape of the spectral density which particular values represent one measure (besides the coupling matrix elements) for the strength of the transition.

frequency, $C_{uv}(\omega = \omega_{ab})$. This last dependence can be viewed as a "probing" of the spectral density at this frequency (cf. Fig. 3.5). In terms of the harmonic reservoir model this implies that there has to be a reservoir oscillator mode which can absorb or emit a reservoir quantum at the transition frequency of the system. Since the transitions between the system states are therefore accompanied by energy dissipation into the reservoir, the rates (3.288) are also called *energy relaxation* rates.

We can use Eq. (3.199) for $C_{uv}(\omega)$ to relate the forward rate for the transition from $|a\rangle$ to $|b\rangle$ to the respective backward rate. Interchanging the summation indices $u$ and $v$ in Eq. (3.289) yields

$$\begin{aligned} k_{ab} &= \sum_{u,v} K^{(v)}_{ab} K^{(u)}_{ba} C_{vu}(\omega_{ab}) = e^{\hbar\omega_{ab}/k_\mathrm{B}T} \sum_{u,v} K^{(u)}_{ba} K^{(v)}_{ab} C_{uv}(\omega_{ba}) \\ &= e^{\hbar\omega_{ab}/k_\mathrm{B}T} k_{ba} \; . \end{aligned} \tag{3.290}$$

This result which is a direct consequence of Eq. (3.199) guarantees the proper relation between excitations and de–excitation of the system's quantum states yielding the equilibrium according to Eq. (3.275). Eq. (3.290) is also known as the *principle of detailed balance*.

2. Coherence dephasing. $a \neq b$, $a = c$, $b = d$

In this case we have according to Eq. (3.284)

$$R_{ab,ab} \equiv \gamma_{ab} = \sum_e \left(\Gamma_{ae,ea}(\omega_{ae}) + \Gamma_{be,eb}(\omega_{be})\right) - \Gamma_{aa,bb}(0) - \Gamma_{bb,aa}(0) \; . \tag{3.291}$$

The expression determines the damping of the off–diagonal elements of the reduced density matrix $\rho_{ab}(t)$. These off–diagonal elements are usually called *coherences* since they represent phase relations between different states (here eigenstates of $H_\mathrm{S}$). Consequently, the decay of coherences is known as the *dephasing* process, and the $\gamma_{ab}$ are called *dephasing rates*. We notice that the first part of the dephasing rate can be written as $\gamma_a + \gamma_b$ where $\gamma_a$ and $\gamma_b$ equals half of the relaxation rates Eq. (3.288) for the transitions out of the state $|a\rangle$ and $|b\rangle$, respectively. Thus, within the present model energy relaxation is a source of coherence dephasing.

The second part of Eq. (3.291) denoted by $\gamma_{ab}^{(\text{pd})}$ is defined by the reservoir correlation function at zero frequency, i.e., it represents an elastic type of collision where no energy is exchanged between system and reservoir. These rates are usually named *pure dephasing* rates and we write

$$\gamma_{ab} = \frac{1}{2}\sum_e k_{ae} + \frac{1}{2}\sum_e k_{be} + \gamma_{ab}^{(\text{pd})} , \tag{3.292}$$

with

$$\gamma_{ab}^{(\text{pd})} = -\sum_{u,v} K_{aa}^{(u)} K_{bb}^{(v)} C_{uv}(\omega = 0) . \tag{3.293}$$

However, the presence of pure dephasing not only requires non–zero correlation functions at zero frequency but also non–vanishing diagonal matrix elements of the operators $K_u$. We already met this requirement at the end of Section 3.7.1 where we discussed types of dissipation which do not change the internal energy.

Traditionally, the relation $1/T_2 = 1/2T_1 + 1/T_2^*$ is used to indicate the different contributions to the dephasing rate. Here, the total dephasing time $T_2$ is called transverse relaxation time. (The term "transverse" is connected with its early use in the field of magnetic resonance experiments where only two–level systems with a single relaxation time have to be considered.) $1/T_2$ has to be identified with $\gamma_{ab}$ for a particular pair of levels and the pure dephasing rate $\gamma_{ab}^{(\text{pd})}$ with $1/T_2^*$. Moreover, $T_1$ is called longitudinal relaxation time and corresponds to the lifetime $(\sum_e k_{ae})^{-1}$ and the lifetime $(\sum_e k_{be})^{-1}$ (provided that both lifetimes are equal). It is important to note that we have related the different relaxation times, which often serve as phenomenological parameters, to a particular microscopic model for the system–reservoir interaction.

3. All elements of $R_{ab,cd}$ which do not correspond to cases (1) and (2):

The remaining elements of the Redfield tensor do not have a simple interpretation in terms of energy relaxation and coherence dephasing rates. However, we can distinguish the following transitions induced by $R_{ab,cd}$. First coherences can be transferred between different pairs of states: $\rho_{ab} \to \rho_{cd}$ $(R_{ab,cd})$. Second, populations can change to coherences: $\rho_{aa} \to \rho_{cd}$ $(R_{aa,cd})$. And finally, the coherences can be transformed into populations: $\rho_{ab} \to \rho_{cc}$ $(R_{ab,cc})$. As a consequence there is a mixing between different types of reduced density matrix elements. The conditions under which this reservoir induced mixing of populations and coherences is negligible will be discussed in the following section.

Before doing this we shortly demonstrate that the multi–level Redfield equations also guarantee that the equilibrium density matrix Eq. (3.275) is a stationary solution. The demand immediately leads to $0 = \sum_c R_{aa,cc} \exp(-E_c/k_\text{B}T)$. Noting Eq. (3.286) and the principle of detailed balance, Eq. (3.290), it becomes obvious that the required relation is fulfilled.

### 3.8.3 The Secular Approximation

The present form of the dissipative contribution to the QME in the state representation, Eq. (3.285) mixes diagonal and off–diagonal elements of the reduced density matrix, as pointed out at the end of the previous section. In order to see under what conditions this mixing between population and coherence type density matrix elements can be neglected consider Eq. (3.285) in the interaction representation with respect to the system Hamiltonian (see also Eq. (3.257)) :

$$\left(\frac{\partial \rho_{ab}^{(\mathrm{I})}}{\partial t}\right)_{\mathrm{diss}} = -\sum_{cd} R_{ab,cd}\, e^{i(\omega_{ab}-\omega_{cd})(t-t_0)} \rho_{cd}^{(\mathrm{I})}(t) \,. \tag{3.294}$$

The right–hand side contains various contributions which oscillate with the combined frequency $\omega_{ab} - \omega_{cd}$. All contributions to the equations of motion where $1/|\omega_{ab} - \omega_{cd}|$ is much smaller than the time increment $\Delta t$ for which the QME is solved will cancel each other upon integration of the equations of motion due to destructive interference. Let us suppose that we can neglect all those contributions to the dissipative part for which the condition $1/|\omega_{ab} - \omega_{cd}| \ll \Delta t$ is fulfilled. There are at first glance two types of contributions which cannot be neglected since $|\omega_{ab} - \omega_{cd}| = 0$ holds. These are related to those elements of $R_{ab,cd}$ which were discussed as cases (1) and (2) in the previous section. However, for systems with degenerate transition frequencies such as a harmonic oscillator $|\omega_{ab} - \omega_{cd}| = 0$ can be fulfilled even if $R_{ab,cd}$ belongs to the category (3) of the previous section. In general the approximation which builds upon the consideration of only those terms in the dissipative part of the QME (3.285) for which $|\omega_{ab} - \omega_{cd}| = 0$ holds is called *secular approximation*.[15]

Note that within the Markov approximation the smallest possible time step, $\Delta t$, is determined by the memory time $\tau_{\mathrm{mem}}$. If however, in systems with nearly degenerate transition frequencies the condition $1/|\omega_{ab} - \omega_{cd}| > \tau_{\mathrm{mem}}$ is realized the secular approximation determines the *coarse graining* of the time axis and therefore imposes a lower limit on the time resolution of the reduced density matrix. On the other hand, even in anharmonic systems the condition $|\omega_{ab} - \omega_{cd}| = 0$ can also be fulfilled accidentally. In other words, in practice one should always carefully examine the system at hand and its time scales before using the secular approximation. All contributions to the QME which are beyond the secular approximation will be called *nonsecular* in the following.

Thus, we have seen that even in the secular approximation there is a chance that populations and coherences are coupled via $R_{ab,cd}$. If we neglect this coupling, *i.e.* if we suppose that $|\omega_{ab} - \omega_{cd}| = 0$ holds only in the cases (1) and (2) of the previous section we are at the level of the so called *Bloch model*. This type of approximation is likely to be good in rather anharmonic systems. Within the Bloch model the right–hand side of Eq. (3.8.3) can be separately written down for the diagonal part of the reduced density matrix, $\rho_{uu}^{(I)} \equiv \rho_{uu} - P_u$, and the off–diagonal part. We obtain for the former using $a = b$ and $c = d$

$$\left(\frac{\partial P_a}{\partial t}\right)_{\mathrm{diss}} = -\sum_{c} R_{aa,cc} P_c(t) \,. \tag{3.295}$$

Next, we consider the off–diagonal part of the Eq. (3.8.3), i.e. , $a \neq b$. Assuming within the Bloch model that all transition frequencies are different we obtain from the secular condition $\omega_{ab} = \omega_{cd}$ the relations $a = c$ and $b = d$, *i.e.* case (2) of the previous section. Changing from

[15] The approximation is often also termed *rotating wave approximation*

the interaction representation of the reduced density matrix to the Schrödinger representation the off–diagonal part becomes

$$\left(\frac{\partial \rho_{ab}}{\partial t}\right)_{\text{diss}} = -(1-\delta_{ab})R_{ab,ab}\rho_{ab} \,. \tag{3.296}$$

Inspecting Eqs. (3.295) and (3.296) we find that these elements of the Redfield tensor do not mix diagonal and off–diagonal elements of the reduced density matrix as desired. This means that we can consider the equations for the populations and the coherences separately. The influence of the reservoir on these two types of reduced density matrix elements is characterized by the energy relaxation and coherence dephasing rates introduced in the foregoing section (Eqs. (3.289) and (3.289), respectively).

### 3.8.4 State Expansion of the System–Reservoir Coupling

To illustrate the formulas presented for the damping matrix in Section 3.8.2 we introduce an expansion of $H_{\text{S-R}}$ in the eigenstates of $H_{\text{S}}$:

$$H_{\text{S-R}} = \sum_{a,b} \langle a|H_{\text{S-R}}|b\rangle\, |a\rangle\langle b| \,. \tag{3.297}$$

This expansion is very fundamental, and we will meet different versions of it in in the following sections. However, Eq. (3.297) is also a special version of the multiple factorized ansatz, Eq. (3.146) for the system–reservoir interaction Hamiltonian. This conclusion is obvious when identifying the index $u$ with $(ab)$, $K_u$ with $|a\rangle\langle b|$ (i.e. $K_{cd}^{(u)} = \delta_{ca}\delta_{db}$), and $\Phi_u$ with $\langle a|H_{\text{S-R}}|b\rangle$. We also stress the fact that the $K_u$–operators do not represent Hermitian operators. In a first step and in accordance with Eq. (3.214) we set $\langle a|H_{\text{S-R}}|b\rangle \equiv \Phi_{ab} = \hbar \times \sum_\xi \omega_\xi g_{ab}(\xi) Q_\xi$ (note the replacement $g_\xi^{(1)}/\omega_\xi$ by the dimensionless coupling constant $g_{ab}(\xi)$ introduced in Section 3.6.4). From Eq. (3.289) the (energy) relaxation rates are obtained as

$$k_{ab} = C_{ab,ba}(\omega_{ab}) \,. \tag{3.298}$$

In accordance with Eq. (3.227) we get for the correlation function

$$C_{ab,cd}(\omega) = 2\pi\omega^2[1+n(\omega)]\,[J_{ab,cd}(\omega) - J_{ab,cd}(-\omega)] \,, \tag{3.299}$$

where we introduced the generalized spectral density

$$J_{ab,cd}(\omega) = \sum_\xi g_{ab}(\xi) g_{cd}(\xi) \delta(\omega - \omega_\xi) \,. \tag{3.300}$$

The dephasing rate $\gamma_{ab}$ can be derived from Eq. (3.292). The pure dephasing contribution may vanish if the correlation function equals zero for $\omega = 0$. A microscopic model for nonvanishing pure dephasing is introduced below (see also Section 4.4).

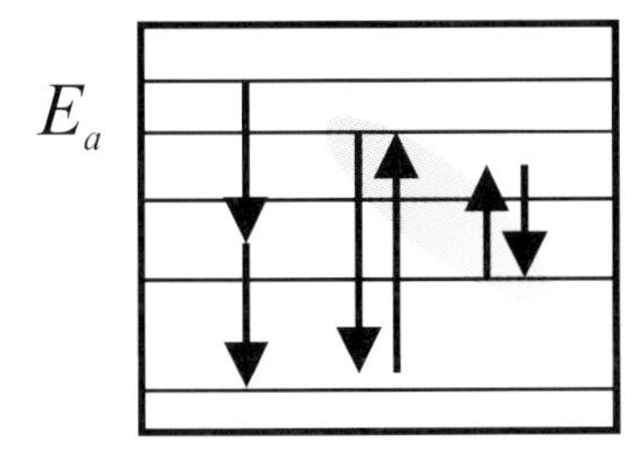


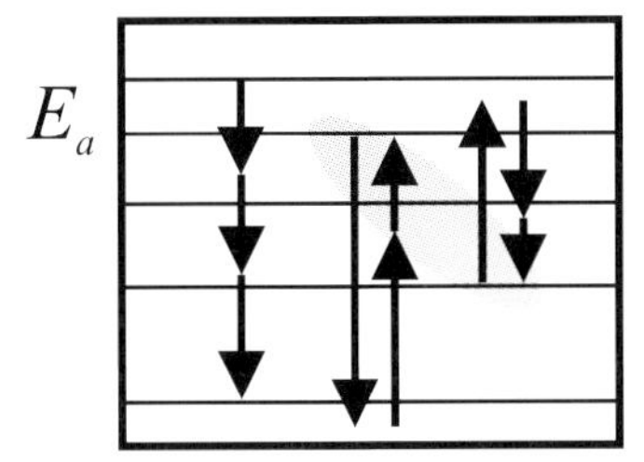


**Figure 3.6:** Multiquantum transitions among five different states $|a\rangle$ of the relevant system with energies $E_a$ $(a = 1, 2, .., 5)$. Left part: Case of quadratic coupling with a transition from level 5 to level 1 accompanied by the emission of two reservoir quanta. Pure dephasing between level 4 and level 2 is also drawn (pair of states for transition indicated by shaded area). It proceeds via an emission and afterwards an absorption of a reservoir quantum (level 4), and via an absorption followed by an emission of a reservoir quantum (level2). (The positioning at level 4 and level 2 is not necessary but reminds on those levels which are concerned.) Right part: Case of cubic coupling with a transition from level 5 to level 1 accompanied by the emission of three reservoir quanta and of pure dephasing between level 4 and level 1 (pair of states for transition indicated by shaded area). (Processes which go upwards in energy do contribute, but are not shown.)

## 3.8.5 Model for Pure Dephasing Processes

We already indicated in Section 3.6.3 that a coupling to the reservoir which is nonlinear in the reservoir coordinates offers a mechanism which results in pure dephasing rates. In the following this statement will be put into more concrete terms. To this end we use the representation of the system–reservoir coupling as introduced in the preceding section but with $\Phi_{ab}$ expanded according to Eq. (3.214). Since the reservoir part $\Phi_{ab}$ of the system–reservoir coupling depends on the system quantum numbers $a$ and $b$ this will be also the case for the expansion coefficients, i.e. we get instead of the quantities $g^{(1)}_{\xi_1}$, $g^{(2)}_{\xi_1,\xi_2}$, and $g^{(3)}_{\xi_1,\xi_2,\xi_3}$ the coefficients $g^{(1)}_{ab}(\xi_1)$, $g^{(2)}_{ab}(\xi_1,\xi_2)$, and $g^{(3)}_{ab}(\xi_1,\xi_2,\xi_3)$, respectively. The transition processes which are caused by these couplings can be classified by the following equations for the related transition frequencies (it will be justified below in considering the respective reservoir correlation functions, cf. Fig 3.6). For the process described by the linear term we get $\omega_{ab} = \omega_{\xi_1}$, indicating as already discussed that transitions in the spectrum of the relevant system are originated by the absorption or emission of a single reservoir quantum. The quadratic coupling can be characterized by $\omega_{ab} = \omega_{\xi_1} \pm \omega_{\xi_2}$ and the cubic one by $\omega_{ab} = \omega_{\xi_1} \pm \omega_{\xi_2} \pm \omega_{\xi_3}$. In both cases transitions in the spectrum of the relevant system may be accompanied by a combined emission and absorption of reservoir quanta.

As already underlined we meet processes related to pure dephasing if the given equations are specified to the case $a = b$. It requires to take the the limit $\omega_{\xi_1} \to 0$ for the linear coupling and $\omega_{\xi_1} = \omega_{\xi_2}$ for the case of quadratic coupling. Here, the absorption and emission of the same energy quantum appears. But in the case of cubic coupling we get $\omega_{\xi_1} \pm \omega_{\xi_2} \pm \omega_{\xi_3} = 0$ what underlines that the absorption of a single quantum may be combined with the emission of two different quanta (case $\omega_{\xi_1} = \omega_{\xi_2} + \omega_{\xi_3}$), and vice versa.

Using Eq. (3.293) we may introduce the pure dephasing rate $\gamma^{(\text{pd},2)}$ caused by the quadratic coupling and the rate $\gamma^{(\mathrm{pd},3)}$ caused by the cubic coupling (the interference term between contributions from the linear and cubic coupling will be not discussed here). We may write

$$\gamma_{ab}^{(\mathrm{pd},2)} = -\int dt \sum_{\xi_1,\xi_2} \sum_{\bar{\xi}_1,\bar{\xi}_2} g_{aa}^{(2)}(\xi_1,\xi_2) g_{bb}^{(2)}(\bar{\xi}_1,\bar{\xi}_2) \Big\{ \langle Q_{\xi_1}(t) Q_{\xi_2}(t)\, Q_{\bar{\xi}_1} Q_{\bar{\xi}_2} \rangle_{\mathrm{R}} - \langle Q_{\xi_1}(t) Q_{\xi_2}(t)\rangle_{\mathrm{R}}\, \langle Q_{\bar{\xi}_1} Q_{\bar{\xi}_2}\rangle_{\mathrm{R}} \Big\} , \quad (3.301)$$

and

$$\gamma_{ab}^{(\mathrm{pd},3)} = -\int dt \sum_{\xi_1,\xi_2,\xi_3} \sum_{\bar{\xi}_1,\bar{\xi}_2,\bar{\xi}_3} g_{aa}^{(3)}(\xi_1,\xi_2,\xi_3) g_{bb}^{(3)}(\bar{\xi}_1,\bar{\xi}_2,\bar{\xi}_3) \times \langle Q_{\xi_1}(t) Q_{\xi_2}(t) Q_{\xi_3}(t)\, Q_{\bar{\xi}_1} Q_{\bar{\xi}_2} Q_{\bar{\xi}_3}\rangle_{\mathrm{R}} . \quad (3.302)$$

The first rate $\gamma_{ab}^{(\mathrm{pd},2)}$ contains a nonvanishing factorized part (compare Eq. (3.221)) whereas such a part equals to zero for the second type of dephasing rate. We continue by inserting the expressions Eqs. (3.222) and (3.224) (together with Eq. (3.221)). Noting Eq. (3.222) for $\langle Q_{\xi_1}(t) Q_{\xi_2}(t)\, Q_{\bar{\xi}_1} Q_{\bar{\xi}_2}\rangle_{\mathrm{R}}$ we see that the time independent part is cancelled by the factorized correlation function. In a next step we account for the fact that the coupling constants $g^{(2)}$ and $g^{(3)}$ are even functions of the mode indices and get (cf. Eq. (3.219) for $B_\xi(t)$)

$$\gamma_{ab}^{(\mathrm{pd},2)} = -2 \sum_{\xi_1,\xi_2} g_{aa}^{(2)}(\xi_1,\xi_2) g_{bb}^{(2)}(\xi_1,\xi_2) \int dt\, B_{\xi_1}(t) B_{\xi_2}(t) . \quad (3.303)$$

The third–order coordinate correlation function determining $\gamma^{(\mathrm{pd},3)}$ is given by a term cubic in $B_\xi(t)$ and a linear term. The latter can be neglected since we consider the limit $\omega \to 0$. Therefore, it follows

$$\gamma_{ab}^{(\mathrm{pd},3)} = -6 \sum_{\xi_1,\xi_2,\xi_3} g_{aa}^{(3)}(\xi_1,\xi_2,\xi_3) g_{bb}^{(3)}(\xi_1,\xi_2,\xi_3) \int dt\, B_{\xi_1}(t) B_{\xi_2}(t) B_{\xi_3}(t) . \quad (3.304)$$

According to Eq. (3.219) for $B_\xi(t)$ we may systematically arrange the different terms which contribute. We obtain

$$\gamma_{ab}^{(\mathrm{pd},2)} = -4\pi \sum_{\xi_1,\xi_2} g_{aa}^{(2)}(\xi_1,\xi_2) g_{bb}^{(2)}(\xi_1,\xi_2) \Big\{ \big[(1+n(\omega_{\xi_1}))(1+n(\omega_{\xi_2})) + n(\omega_{\xi_1}) n(\omega_{\xi_2})\big] \delta(\omega_{\xi_1}+\omega_{\xi_2}) + 2(1+n(\omega_{\xi_1})) n(\omega_{\xi_2}) \delta(\omega_{\xi_1}-\omega_{\xi_2}) \Big\} , \quad (3.305)$$

and

$$\gamma_{ab}^{(\mathrm{pd},3)} = -12\pi \sum_{\xi_1,\xi_2,\xi_3} g_{aa}^{(2)}(\xi_1,\xi_2,\xi_3) g_{bb}^{(2)}(\xi_1,\xi_2,\xi_3) \Big\{ \big[(1+n(\omega_{\xi_1}))(1+n(\omega_{\xi_2}))(1+n(\omega_{\xi_3})) + n(\omega_{\xi_1}) n(\omega_{\xi_2}) n(\omega_{\xi_3})\big] \times \delta(\omega_{\xi_1}+\omega_{\xi_2}+\omega_{\xi_3}) + 3\big[(1+n(\omega_{\xi_1}))(1+n(\omega_{\xi_2})) n(\omega_{\xi_3}) + n(\omega_{\xi_1}) n(\omega_{\xi_2})(1+n(\omega_{\xi_3}))\big] \times \delta(\omega_{\xi_1}+\omega_{\xi_2}-\omega_{\xi_3}) \Big\} . \quad (3.306)$$

In both formulas we used the definition of the $\delta$–function and took advantage of the symmetry of the coupling constants. The rate $\gamma_{ab}^{(\mathrm{pd},2)}$ as well as the rate $\gamma_{ab}^{(\mathrm{pd},3)}$ would only give nonvanishing contributions via the term proportional to $\delta(\omega_{\xi_1} - \omega_{\xi_2})$ and proportional to $\delta(\omega_{\xi_1} + \omega_{\xi_2} - \omega_{\xi_3})$, respectively. However, in the case of the quadratic coupling pure dephasing is realized by the absorption and emission of reservoir quanta which are degenerated in energy. This is different from the cubic coupling. Here, a single quantum may be absorbed by the relevant system and afterwards two quanta with the same total energy may be emitted, or alternatively, two quanta are absorbed and a single quantum is emitted (cf. Fig. 3.6). This difference between the two coupling mechanisms indicates that the cubic coupling seems more important since the variety of frequency combinations is much larger then in the case of quadratic coupling.

Finally, we would like to point out that as for the linear coupling case the summation on the various mode–indices $\xi_1$, $\xi_2$, etc. can be removed by introducing spectral densities (cf. Section 4.4). However, these spectral densities will dependent on two frequency arguments in the case of quadratic coupling and on three in the case of cubic coupling.

### 3.8.6 Some Estimates

After Eq. (3.253), we already discussed the range of validity of the QME. Using the energy representation introduced in this section, a more detailed account is possible. To do this we concentrate on the energy representation of the Markovian version of Eq. (3.253) with the dissipative part given by Eq. (3.285). A necessary criterion for the validity of the QME would be that the absolute value of any transition frequency $\omega_{ab}$ is larger than the respective level broadening determined by the dephasing rates $\gamma_{ab}$, Eq. (3.292). Using the expression for $H_{\mathrm{S-R}}$ introduced in Eq. (3.297) and noting the absence of pure dephasing we have to compare $|\omega_{ab}|$ with the dephasing rates following from the Eqs. (3.289) and (3.299). Since every term stemming from Eq. (3.299) has to be small and assuming zero temperature we get $|\ \omega_{ab}\ | > \omega_{ae}^2\ J_{ae,ea}(\omega_{ae}) + \omega_{be}^2\ J_{be,eb}(\omega_{be})$. If $\omega_{ab} \approx \omega_{ae}, \omega_{be}$ the respective values of the spectral densities have to be small compared to $\omega_{ab}^{-1}$. This restriction can be relaxed whenever the cut–off frequencies of the spectral densities are smaller than $\omega_{ae}, \omega_{be}$. If $\omega_{ab}$ is much larger than $\omega_{ae}, \omega_{be}$, then the spectral densities have to be small compared to $\omega_{ab}/\omega_{ae}^2$. This latter case imposes to the spectral density a much stronger constraint of smallness as the foregoing relations. The discussion indicates that the concrete structure of the spectrum of the relevant system decides on the extend at which the system–reservoir coupling can be increased such that the QME is still valid

Finally, we demonstrate that in case of the Bloch model as introduced in Section 3.8.3 it is possible to change back from the energy representation to an operator notation of the QME. One immediately arrives at

$$\begin{aligned}\left(\frac{\partial \hat{\rho}(t)}{\partial t}\right)_{\mathrm{diss}} &= -\sum_{a,b} \left\{ \frac{1}{2} \Big[ k_{ab} |a\rangle\langle a|, \hat{\rho}(t) \Big]_+ - k_{ab} |b\rangle\langle a|\hat{\rho}(t)|a\rangle\langle b| \right\} \\ &\quad + \sum_{a,b} \gamma_{ab}^{(\mathrm{pd})} |a\rangle\langle a|\hat{\rho}(t)|b\rangle\langle b| \end{aligned} \tag{3.307}$$

The first sum including an anti–commutator is exclusively determined by the energy relaxation rate $k_{ab}$ whereas the second sum incorporates the pure dephasing part $\gamma_{ab}^{(\mathrm{pd})}$, Eq. (3.293).

Once pure dephasing vanishes the whole dissipative part resembles what is often called the *Lindblad form*. It is possible to derive this type of dissipative contribution to the equation of motion of the reduced density operator in a more formal way starting from the assumption that the diagonal elements of the reduced density operator have to be greater or equal to zero in any basis set. This has been shown by Lindblad in the 1970s. The advantage of Eq. (3.307) is that the condition $\rho_{aa}(t) \geq 0$ is guaranteed by construction in contrast to the case of the QME. The Lindblad form of the dissipative contribution to the equations of motion for the reduced density operator has become very popular in the context of the so–called *Monte Carlo wave function* method which is discussed in Section 3.12.6.

Using the Lindblad form of dissipation (or the Bloch model) which guarantees positivity of the density matrix one has to pay attention not to over–interpret the results. In contrast to the multi–level Redfield theory one may increase the system reservoir coupling strength without obtaining results which apparently behave in a wrong way. Nevertheless, one has already left the region of applicability of the whole approach, which is of second–order in the system reservoir coupling, and obtained formally meaningless results.

### 3.8.7 From Coherent to Dissipative Dynamics: A Simple Example

In the following we will discuss the dynamics of a coupled two–level system using the methods developed in Sections 3.2 and 3.8.2. It should be noted that despite its simplicity the model of a two–level system provides an important reference for understanding the dynamics in complicated condensed phase situations. We will start by solving the time–dependent Schrödinger equation for the two–level system. Afterwards the density matrix theory based on the QME in the Markov approximation will be applied.

#### Coherent Dynamics

In Section 2.8.3 we obtained the eigenvalues $\mathcal{E}_{a=\pm}$ and eigenvectors $|a=\pm\rangle$ for a system consisting of two zeroth–order states $|m=1,2\rangle$ with energies $\varepsilon_{m=1,2}$ coupled by some interaction $V$ (cf. Eq. (2.151)). The time–evolution operator for the isolated two–level system $U(t) = e^{-iHt/\hbar}$ is conveniently expressed in terms of the eigenstates $|a=\pm\rangle$. One obtains

$$U(t) = \sum_{a,b=\pm} \langle a|U(t)|b\rangle\, |a\rangle\langle b| = \sum_{a=\pm} e^{-i\mathcal{E}_a t/\hbar} |a\rangle\langle a| \,. \qquad (3.308)$$

This expression can be used to determine, for instance, how the initially prepared zeroth–order state $|1\rangle$ evolves in time. To this end we calculate the probability for transitions between $|1\rangle$ and $|2\rangle$ which is defined as

$$P_{1\to 2}(t) = |\langle 2|U(t)|1\rangle|^2 \,. \qquad (3.309)$$

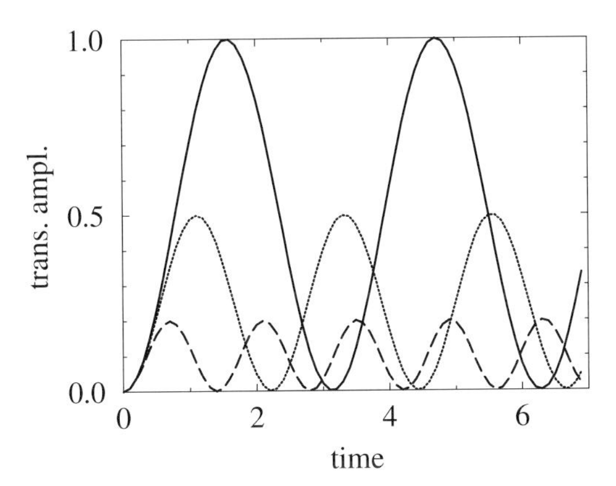


**Figure 3.7:** Transition amplitude $P_{1\to 2}(t)$ following from Eq. (3.313) for different (scaled) detunings $|\varepsilon_1 - \varepsilon_2|/2|V| = 0$ (solid), 1 (dotted), 2 (dashed), plotted versus (scaled) time $|V|t/\hbar$.

Once this quantity is known the survival probability is obtained as $P_{1\to 1}(t) = 1 - P_{1\to 2}(t)$. Using Eqs. (2.163) and (2.165) we get ($\hat{a} = \mp$, if $a = \pm$)

$$\begin{aligned}
\langle 2|U(t)|1\rangle &= \sum_{a=\pm} e^{-i\mathcal{E}_a t/\hbar}\langle 2|a\rangle\langle a|1\rangle = \sum_{a=\pm} e^{-i\mathcal{E}_a t/\hbar}\, c_a(2)\, c_a^*(1) \\
&= \sum_{a=\pm} e^{-i\mathcal{E}_a t/\hbar}\left(\frac{\mathcal{E}_a - \varepsilon_2}{\mathcal{E}_a - \mathcal{E}_{\hat{a}}}\;\frac{\mathcal{E}_a - \varepsilon_1}{\mathcal{E}_a - \mathcal{E}_{\hat{a}}}\right)^{\frac{1}{2}} e^{i(\chi_2(a)-\chi_1(a))} \\
&= e^{-i\arg(V)}\frac{|V|}{\mathcal{E}_+ - \mathcal{E}_-}\left(e^{-i\mathcal{E}_+ t/\hbar} - e^{-i\mathcal{E}_- t/\hbar}\right) .
\end{aligned} \tag{3.310}$$

This gives for the transition probability ($\omega_{+-} = (\mathcal{E}_+ - \mathcal{E}_-)/\hbar$)

$$\begin{aligned}
P_{1\to 2}(t) &= \frac{|V|^2}{(\mathcal{E}_+ - \mathcal{E}_-)^2}\left|e^{-i\mathcal{E}_- t/\hbar}\{e^{-i\omega_{+-}t} - 1\}\right|^2 \\
&= \frac{|V|^2}{(\mathcal{E}_+ - \mathcal{E}_-)^2}\left([\cos(\omega_{+-}t) - 1]^2 + \sin^2(\omega_{+-}t)\right) .
\end{aligned} \tag{3.311}$$

Using

$$[\cos(\omega_{+-}t) - 1]^2 + \sin^2(\omega_{+-}t) = 2\,(1 - \cos(\omega_{+-}t)) = 4\sin^2\frac{\omega_{+-}t}{2} \tag{3.312}$$

we finally get

$$P_{1\to 2}(t) = \frac{4|V|^2}{(\varepsilon_1 - \varepsilon_2)^2 + 4|V|^2}\quad \sin^2\left(\frac{1}{2\hbar}\sqrt{(\varepsilon_1 - \varepsilon_2)^2 + 4|V|^2 t}\right) . \tag{3.313}$$

For the case that the zeroth–order states have the same energy this expression simplifies to

$$P_{1\to 2}(t) = \sin^2\left(|V|t/\hbar\right) . \tag{3.314}$$

The time dependence of the transition probability is shown in Fig. 3.7 for different detunings, $|\varepsilon_1 - \varepsilon_2|/2|V|$, between the zeroth–order states. From (3.313) we realize that $P_{1\to 2}(t)$ will oscillate with a frequency $\Omega$ which depends on the detuning: $\Omega = \sqrt{(\varepsilon_1 - \varepsilon_2)^2 + 4|V|^2}/2\hbar$. Given a constant coupling $V$ the oscillation frequency will increase with increasing detuning. At the same time, due to the prefactor in Eq. (3.313), the transfer will be less complete. A complete population switching occurs only if the two zeroth–order states are degenerate. The oscillation frequency is then $\Omega = V/\hbar$ and according to Eq. (3.314) a complete transfer is realized for the condition $t = (2N + 1)\pi\hbar/2|V|$ where $N$ is an integer.

We would like to point out that this simple result reflects the general statement made earlier namely that time–dependent phenomena in a closed quantum system appear whenever a non–eigenstate, that is, a superposition of eigenstates has been prepared initially. In the present case the initial preparation of state $|1\rangle$ corresponds to a particular superposition of the two eigenstates $|+\rangle$ and $|-\rangle$.

### Dissipative Dynamics using Eigenstates

The dissipative dynamics of the two–level system will be described using the density matrix in the state representation. Here in principle we have two possibilities: In a situation where some zeroth–order initial state has been prepared, one is often interested in the survival amplitude related to this initial state which is given by $\rho_{mm}(t)$ (cf. Chapter 3). On the other hand, one could also use the representation in terms of the eigenstates: $\rho_{ab}$. It seems as if there were no difference between these two representations because we can relate them via

$$\rho_{mn}(t) = \sum_{a,b} c_a(m) c_b^*(n)\, \rho_{ab}(t)\ . \tag{3.315}$$

However, we should recall that in Sections 3.7 and 3.263 the equations of motion for the reduced density matrix have been derived in the *eigenstate* representation. As a consequence all approximations (Markovian dynamics, secular approximation) make reference to the spectrum of the *full* Hamiltonian (2.150) and not only to the zeroth–order states. In the following we will show that simulations using either eigenstates or zeroth–order states can yield different results.

Let us start by specifying the coupling of the two–level system to its environment. We will assume that the latter can be described by uncoupled harmonic oscillators with coordinates $Z = \{Z_\xi\}$. To account for energy dissipation from the two–level system into the reservoir we will consider the simplest version of the coupling Hamiltonian (cf. the final part of Section 3.8.2). Using the general notation, Eq. (3.146), we take the system part to be

$$K_u = |m\rangle\langle n|\ . \tag{3.316}$$

The index $u$ in Eq. (3.146) has to be identified with the pair $(m, n)$ and the reservoir part of Eq. (3.146) is written as a linear expression in the reservoir coordinates:

$$\Phi_u \equiv \Phi_{mn} = \sum_\xi \hbar\omega_\xi g_{mn}(\xi)(C_\xi^+ + C_\xi)\ . \tag{3.317}$$

Here we have introduced the dimensionless coupling constant $g_{mn}(\xi)$ which has already been used in Section 3.6.4. Concentrating on energy exchange with the environment via transitions between both zeroth–order system states we will assume that $g_{mn}(\xi)$ has only off–diagonal elements.

The definition of the system–environment coupling in terms of the zeroth–order states will often have practical reasons. For instance, in Chapter 6 we will discuss the electron transfer between a donor and an acceptor state (i.e., in an electronic two–level system). Since the electronic donor and acceptor states are well–defined, it may be more straightforward to model their interaction with the environment *separately*, i.e., without taking into account their mutual interaction.

The eigenstate representation of the system–reservoir coupling Hamiltonian $H_{\text{S-R}}$ is easily obtained. For the system part we have

$$K_u = |a\rangle\langle b| \tag{3.318}$$

($u \equiv (a,b)$ in Eq. (3.146)). The reservoir part has diagonal and off–diagonal contributions

$$\Phi_u \equiv \Phi_{ab} = \sum_\xi \hbar\omega_\xi g_{ab}(\xi)(C_\xi^+ + C_\xi) , \tag{3.319}$$

where the coupling matrix is now given by

$$g_{ab}(\xi) = \sum_{a,b} c_a^*(m) g_{mn}(\xi) c_b(n) . \tag{3.320}$$

The density matrix equations in the eigenstate representation are directly obtained from the QME (3.283). Restricting ourselves to the secular approximation (cf. Section 3.8.3), we get for the state populations the equation of motion

$$\frac{\partial}{\partial t}\rho_{++} = -k_{+-}\,\rho_{++} + k_{-+}\,\rho_{--} . \tag{3.321}$$

Due to the secular approximation this equation is decoupled from the equation for the coherence which reads with $\omega_{+-} = (\mathcal{E}_+ - \mathcal{E}_-)/\hbar$:

$$\frac{\partial}{\partial t}\rho_{+-} = -i\big[\omega_{+-} - i\gamma_{+-}\big]\rho_{+-} . \tag{3.322}$$

(The other two matrix elements follow from $\rho_{--} = 1 - \rho_{++}$ and $\rho_{-+} = \rho_{+-}^*$.) The transition rates can be directly adapted from Eq. (3.289). We obtain for the rate $k_{a\to b} \equiv k_{ab}$

$$\begin{aligned} k_{ab} &= \text{Re}\, C_{ab,ba}(\omega_{ab}) \\ &= 2\pi\omega_{ab}^2(1 + n(\omega_{ab}))[J_{ab}(\omega_{ab}) - J_{ab}(-\omega_{ab})] , \end{aligned} \tag{3.323}$$

where the spectral density is given by

$$J_{ab}(\omega) = \sum_\xi g_{ab}^2(\xi)\,\delta(\omega - \omega_\xi) . \tag{3.324}$$

The dephasing rates follow from Eq. (3.292) as (note that $\hat{a} = \mp$ if $a = \pm$)

$$\gamma_{a\hat{a}} = \frac{1}{2}(k_{a\hat{a}} + k_{\hat{a}a}) + \gamma_{a\hat{a}}^{(\text{pd})} . \tag{3.325}$$

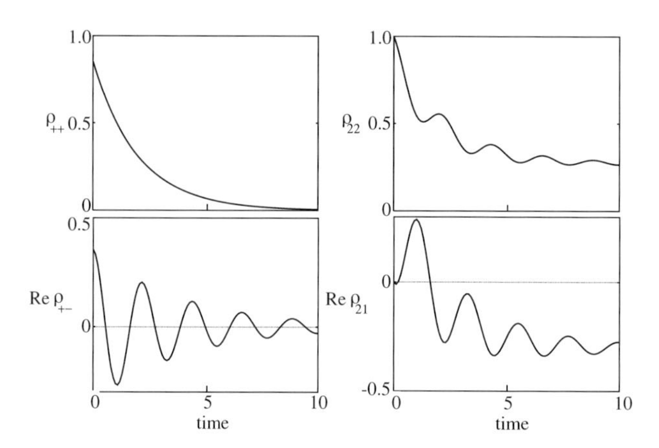


**Figure 3.8:** Dissipative dynamics in a coupled two–level system as obtained from the eigenstate representation (left) and the zeroth–order state representation (right). The parameters are: detuning $|\varepsilon_1 - \varepsilon_2|/2|V| = 1$, relaxation rate $k_{+-} = 0.5|V|/\hbar$, initial state: $|1\rangle$. The time is given in units of $\hbar/V$ (figure courtesy of H. Naundorf).

In the low–temperature limit, $k_\mathrm{B}T \ll \hbar\omega_{+-}$, we can neglect $n(\omega_{+-})$, and the rate for upward transitions $k_{-+}$ vanishes.

The solutions for the equations of motion, (3.321) and (3.322), can be given right away. Considering the low–temperature limit for simplicity we obtain

$$\rho_{++}(t) = \rho_{++}(0)\, e^{-k_{+-}t} , \tag{3.326}$$

and

$$\rho_{+-}(t) = \rho_{+-}(0)\, e^{-i(\omega_{+-}-i\gamma_{+-})t} . \tag{3.327}$$

Thus, nonequilibrium populations of the eigenstates decay exponentially while the coherences will oscillate with the transition frequency $\omega_{+-}$. The amplitude of this oscillation will decrease exponentially, too, i.e., any initial coherence between the two eigenstates is destroyed.

This time dependence can be easily translated into the picture of zeroth–order states using Eq. (3.315). For illustration let us consider the case of an initially prepared zeroth–order state, i.e., $\rho_{mn}(t=0) = \delta_{mn}\delta_{m1}$. Since the transformation (3.315) couples populations and coherences in the different representations, the respective initial density matrix in the eigenstate representation will have non–zero diagonal and off–diagonal elements. In Fig. 3.8 we show the subsequent time evolution of the density matrix in both representations. Notice that the dynamics of the zeroth–order state population reflects the oscillatory behavior obtained for the coherent regime in Fig. 3.7. Further the off–diagonal elements $\rho_{12}$ do not vanish at long times. This reflects the fact that the eigenstate $|-\rangle$, which is populated in the stationary limit, is a superposition state with respect to the zeroth–order states $|1\rangle$ and $|2\rangle$ (cf. Eq. (2.153)).

### Dissipative Dynamics using Zeroth–Order States

Let us compare the eigenstate formulation of the equations of motion with their zeroth–order version. In this case the equations of motion read

$$\frac{\partial}{\partial t}\rho_{11} = \frac{2V}{\hbar}\mathrm{Im}\,[\rho_{21}] - k_{12}\,\rho_{11} + k_{21}\,\rho_{22} \tag{3.328}$$

and

$$\frac{\partial}{\partial t}\rho_{21} = -i[\omega_{21} - i(\gamma_2 + \gamma_1)]\,\rho_{21} - \frac{i}{\hbar}V\,(\rho_{11} - \rho_{22})\,. \tag{3.329}$$

Suppose we would consider a problem which involves many zeroth–order states such that the diagonalization of the Hamiltonian may be rather time consuming. In this case it would be tempting to formulate the rates in the zeroth–order states only, i.e.

$$\begin{aligned} k_{mn} &= \mathrm{Re}\,C_{mn,nm}(\omega_{mn}) \\ &= 2\pi\omega_{mn}^2(1 + n(\omega_{mn}))[J_{mn}(\omega_{mn}) - J_{mn}(-\omega_{mn})]\,, \end{aligned} \tag{3.330}$$

with the spectral density now given by

$$J_{mn}(\omega) = \sum_{\xi} g_{mn}^2(\xi)\,\delta(\omega - \omega_\xi)\,. \tag{3.331}$$

At first glance there appears to be nothing wrong with this expression. However, writing down the detailed balance condition which follows from Eq. (3.330) (cf. Eq. (3.290))

$$k_{21} = e^{\hbar\omega_{21}/k_{\mathrm{B}}T}k_{12} \tag{3.332}$$

we realize that this will guide the system to an equilibrium distribution with respect to the *zeroth–order* states, i.e., the coupling $V$ is not accounted for. In order to understand the reason for this behavior we have to go back to Section 3.7.1. There we had introduced the operators $\Lambda_u$ in Eq. (3.260) which contain the information about the system–bath interaction. In particular they include the operators $K_u$ defined in the interaction representation with respect to $H_{\mathrm{S}}$. Let us inspect the matrix elements of the operator $\Lambda_u$ with respect to the zeroth–order basis ($u \equiv (mn)$)

$$\langle\bar{m}|\Lambda_{mn}|\bar{n}\rangle = \sum_{k,l}\int_0^\infty d\tau\,C_{mn,kl}(\tau)\langle\bar{m}|K_{kl}^{(\mathrm{I})}(-\tau)|\bar{n}\rangle\,, \tag{3.333}$$

Strictly speaking the calculation of $\langle\bar{m}|K_{kl}^{(\mathrm{I})}(-\tau)|\bar{n}\rangle$ would require to use the eigenstates. This would result in the expression

$$\langle\bar{m}|K_{kl}^{(\mathrm{I})}(-\tau)|\bar{n}\rangle = \sum_{a,b} e^{-i(\mathcal{E}_a - \mathcal{E}_b)\tau/\hbar}\,c_a(\bar{m})c_a^*(k)c_b(l)c_b^*(\bar{n})\,. \tag{3.334}$$

Inserting this into Eq (3.333) gives

$$\langle\bar{m}|\Lambda_{mn}|\bar{n}\rangle = \sum_{kl}\sum_{a,b} c_a(\bar{m})c_a^*(k)c_b(l)c_b^*(\bar{n}) \int_0^\infty d\tau\, e^{-i(\mathcal{E}_a-\mathcal{E}_b)\tau/\hbar} C_{mn,kl}(\tau) . \tag{3.335}$$

On the other hand neglecting the coupling $V$ in Eq. (3.334) one gets

$$\langle\bar{m}|K_{kl}^{(\mathrm{I})}(-\tau)|\bar{n}\rangle = \delta_{\bar{m}k}\delta_{l\bar{n}}e^{-i\omega_{cd}\tau} . \tag{3.336}$$

Whether the half–sided Fourier transform of the correlation function $C_{mn,cd}(t)$ is taken with respect to $(\mathcal{E}_a - \mathcal{E}_b)/\hbar$ or $\omega_{cd}$ determines the frequency argument in the Bose–Einstein distribution function. This in turn fixes the detailed balance condition to the respective spectrum.

Finally, we point out that the quality of the approximation which neglects the detailed structure of the spectrum and only takes into account some zeroth–order states depends of course on the strength of the coupling. Moreover, if one is only interested in the short–time behavior and not in the stationary solutions of the equations of motion, a formulation of the relaxation rates in terms of zeroth–order states may be acceptable. We will return to this point in the context of electron (Section 6.8) and exciton (Section 8.6) transfer.

### 3.8.8 Coordinate and Wigner Representation of the Reduced Density Matrix

In the preceding parts of this section we concentrated on the energy representation of the density matrix. There may be situations where the eigenstate of the Hamiltonian are not easily available, for example, for problems involving dissociation. In this case the coordinate representation may offer a convenient alternative. In the following we will derive the coordinate representation of the QME in the Markov approximation, Eq. (3.263).

As in Section 3.1 we assume that the total system has been separated into a relevant part and a reservoir. The relevant system will be described by the set of coordinates $s \equiv \{s_j\}$. Then, according to Eq. (3.138) the density matrix in the coordinate representation follows as $\rho(s,\bar{s};t) = \langle s|\hat{\rho}(t)|\bar{s}\rangle$, i.e., the matrix elements of the RSO are taken with the eigenstates $|s\rangle$ of the coordinate operator. In contrast to the energy representation the RDM introduced here is a *continuous* function of the coordinates $s_j$.

The equation of motion for $\rho(s,\bar{s};t)$ is obtained by taking the respective matrix elements of the Markovian QME, Eq. (3.263). First, we have to calculate matrix elements of the system Hamiltonian, $\langle s|H_\mathrm{S}|\bar{s}\rangle$. It is well–known from quantum mechanics that these matrix elements follow as $H_\mathrm{S}(s,p)\delta(s-\bar{s})$ (here and in the following the $\delta$–function abbreviates a product of the single coordinate expressions $\delta(s_j - \bar{s}_j)$). The momentum operators in $H_\mathrm{S}$ are given by $p_j = -i\hbar\partial/\partial s_j$. (The notation $H_\mathrm{S}(s,p)$ used in the following indicates the coordinate representation of the system Hamiltonian.)

The QME in the Markov approximation follows as

$$\frac{\partial}{\partial t}\rho(s,\bar{s};t) = -\frac{i}{\hbar}\Big(H_\mathrm{S}(s,p) - H_\mathrm{S}(\bar{s},\bar{p})\Big)\rho(s,\bar{s};t) + \langle s|\Big(\frac{\partial\hat{\rho}}{\partial t}\Big)_\mathrm{diss}|\bar{s}\rangle . \tag{3.337}$$

Here, the mean–field contribution is not considered explicitly; it is supposed to be included into the definition of $H_S$ (see Eq. (3.262)). The dissipative part can be rewritten as a nonlocal (integral) operator. For the present purpose it is sufficient to assume that the operators $K_u$ in Eq. (3.146) only depend on the coordinates $s$. Thus, we have $\langle s|K_u|\bar{s}\rangle = \delta(s-\bar{s})K_u(s)$ and the dissipative part reads

$$\begin{aligned}\langle s|\left(\frac{\partial\hat{\rho}}{\partial t}\right)_{\text{diss}}|\bar{s}\rangle &= -\sum_u \Big(K_u(s) - K_u(\bar{s})\Big) \\ &\times \int ds' \Big(\langle s|\Lambda_u|s'\rangle\rho(s',\bar{s};t) - \langle\bar{s}|\Lambda_u|s'\rangle^*\rho(s,s';t)\Big)\end{aligned} \tag{3.338}$$

Note, that $\int ds'$ abbreviates the multi–dimensional integration with respect to all coordinates $\{s'_j\}$. In order to elucidate the influence of the reservoir further let us combine the dissipative part and the reversible part of the equation of motion. This can be done by introducing a complex potential as follows

$$\Omega(s,s';\bar{s}) = \delta(s-s')U(s) - i\hbar\sum_u \Big(K_u(s) - K_u(\bar{s})\Big)\langle s|\Lambda_u|s'\rangle\,. \tag{3.339}$$

It is a combination of the real and local part $U(s)$ stemming from $H_S$ and a nonlocal and complex contribution which results from the coupling to the environment. Using this potential the QME can be written as ($T$ denotes the kinetic energy operator)

$$\begin{aligned}\frac{\partial}{\partial t}\rho(s,\bar{s};t) &= -\frac{i}{\hbar}\Big(T(p) - T(\bar{p})\Big)\rho(s,\bar{s};t) \\ &- \frac{i}{\hbar}\int ds' \Big(\Omega(s,s';\bar{s})\rho(s',\bar{s};t) - \rho(s,s';t)\Omega^*(\bar{s},s';s)\Big)\,.\end{aligned} \tag{3.340}$$

This equation shows that the action of the environment has the same effect as a complex and nonlocal potential added to the system potential $U(s)$. We notice that using such a potential within the formalism of the time–dependent Schrödinger equation, the related Hamiltonian would no longer be Hermitian and the conservation of the wave function norm would not be guaranteed (cf. the discussion in Section 3.7.1). For the reduced density matrix the appearance of a complex potential reflects phenomena known from the energy representation as finite state lifetime and coherence dephasing.

To establish the relation to the approximations discussed in Section 3.8.2 we compute the coordinate matrix elements of the $\Lambda$–operator Eq. (3.260). In doing so it is necessary to determine $\langle s|K_v^{(\mathrm{I})}(-\tau)|\bar{s}\rangle$ which will have nonvanishing off–diagonal elements for $\tau > 0$. All elements are easily calculated if one uses the eigenstates $\varphi_a(s)$ of $H_S$. Inserting the result into the matrix elements of the $\Lambda$–operator gives (for the matrix elements of the operators $K_v$ see Eq. (3.269))

$$\langle s|\Lambda_u|\bar{s}\rangle = \sum_{a,b}\varphi_a(s)\varphi_b^*(\bar{s})\sum_v\int_0^\infty d\tau\; C_{uv}(\tau)e^{-i\omega_{ab}\tau}K_{ab}^{(v)}\,. \tag{3.341}$$

As in Section 3.8.2 we would like to relate the given description for the complex potential Eq. (3.339) to the concept of the spectral density Eq. (3.226) of a harmonic oscillator environment. Therefore the system–reservoir coupling of Eq. (3.205) is used, resulting in a single $K$–operator and a single correlation function $C(t)$. The approximation made in Section 3.8.2 which takes into account only the real expression $\Gamma_{ab,cd}(\omega)$, Eq. (3.281) ,is in the present context equivalent to the replacement of the half–sided Fourier transform of $C(t)$ by half of the fully transformed expression $C(\omega)$[16]. Therefore, Eq. (3.341) is expressed by $C(-\omega_{ab})$. Noting Eq. (3.227) which relates $C(\omega)$ to $J(\omega)$ we finally obtain the following specific expression for the complex potential Eq. (3.339)

$$\begin{aligned}\Omega(s,s';\bar{s}) &= \delta(s-s')U(s) - i\pi\hbar\big(K(s)-K(\bar{s})\big) \\ &\quad\times\sum_{a,b}\varphi_a(s)\varphi_b^*(s')K_{ab}\,\omega_{ba}^2\big(1+n(\omega_{ba})\big)\big(J(\omega_{ba})-J(-\omega_{ba})\big)\,.\end{aligned} \tag{3.342}$$

Let us discuss a case where the potential reduces to a local one. First, we concentrate on the high–temperature limit where $n(\omega)\approx k_\mathrm{B}T/\hbar\omega$ holds. If one takes an Ohmic spectral density, $\omega^2 J(\omega) = j_0\omega$, the $\omega_{ba}$ stemming from $n(\omega_{ba})$ and those coming from the spectral density cancel each other. If we assume $K_{aa} = 0$ there is no need to care about the case $a = b$ and the $a, b$–summation gives $\langle s|K(s)|s'\rangle = \delta(s-s')K(s)$. According to Eq. (3.338) we obtain the dissipative part of the QME in the coordinate representation as

$$\left(\frac{\partial\rho(s,\bar{s};t)}{\partial t}\right)_\mathrm{diss} = -\frac{2\pi k_\mathrm{B}Tj_0}{\hbar^2}\big(K(s)-K(\bar{s})\big)^2\rho(s,\bar{s};t)\,. \tag{3.343}$$

The $s,\bar{s}$–dependence of the right–hand side nicely reflects the destruction of coherences contained in the off–diagonal elements of the reduced density matrix.

Next we use the coordinate representation of the reduced density operator to introduce the respective Wigner representation. For simplicity we consider the case of a single coordinate and a coupling function to the reservoir $K(s) = s$ (i.e. the so–called bilinear system–reservoir coupling is used). In Section 3.4.4 the change to the Wigner representation has been demonstrated for the total density operator, putting emphasis on the relation to classical statistical mechanics. This will be repeated here for the RDO but including the dissipative part. Following Section 3.4.4 we can directly adopt Eq. (3.115) to transform the reversible part of the QME. One obtains

$$\left(\frac{\partial\rho(x,p;t)}{\partial t}\right)_\mathrm{rev} = \frac{\partial U(x)}{\partial x}\frac{\partial}{\partial p}\rho(x,p;t) - \frac{\partial T(p)}{\partial p}\frac{\partial}{\partial x}\rho(x,p;t)\,. \tag{3.344}$$

Next we determine the dissipative part of the QME in the Wigner representation using expression (3.343). To compute the respective Wigner representation we introduce sum and

[16] The half–sided Fourier transform if expressed by the complete Fourier–transformed correlation function reads $\hat{C}(\omega) = \int_0^\infty dt\exp(i\omega t)\int d\bar{\omega}/2\pi\times\exp(-i\bar{\omega}t)C(\bar{\omega})$. A rearrangement of the integrations leads to $\int_0^\infty dt\exp(i\Delta\omega t)$ where we introduced $\Delta\omega = \omega-\bar{\omega}$. The integral gives $i/(\Delta\omega+i\varepsilon)$ ( with $\varepsilon\to+0$). As a result we obtain $\hat{C}(\omega) = -\int d\bar{\omega}/2\pi i\times C(\bar{\omega})/(\Delta\omega+i\varepsilon)$. Since $C(\bar{\omega})$ is a real function (c. f. Section 3.6.1) the separation of $\hat{C}(\omega)$ results in a principle–value integral and a $\delta$–function from which the relation $\mathrm{Re}\hat{C}(\omega) = C(\omega)$ can be verified.

difference coordinates and take into account that

$$\int dr\, e^{-ipr/\hbar}\, r^2 \rho(x,r;t) = -\hbar^2 \frac{\partial^2}{\partial p^2}\rho(x,p;t) \; . \tag{3.345}$$

This directly gives

$$\left(\frac{\partial \rho(s,p;t)}{\partial t}\right)_{\text{diss}} = 2\pi k_{\text{B}} T j_0 \frac{\partial^2}{\partial p^2}\rho(x,p;t) \; . \tag{3.346}$$

Combining this expression with Eq. (3.344) we obtain the Markovian Quantum Master Equation in the Wigner representation as follows

$$\left(\frac{\partial}{\partial t} - \frac{\partial U(x)}{\partial x}\frac{\partial}{\partial p} + \frac{\partial T(p)}{\partial p}\frac{\partial}{\partial x} - 2\pi k_{\text{B}} T j_0 \frac{\partial^2}{\partial p^2}\right)\rho(x,p;t) = 0 \; . \tag{3.347}$$

As required for a classical limit, Eq. (3.347) is of zeroth–order in $\hbar$. Eq. (3.347) is also known as the *Fokker–Planck equation* (note that it is common to replace $2\pi j_0$ by the friction constant $\eta$) [17].

# 3.9 Generalized Rate Equations: The Liouville Space Approach

In the previous sections we have focused on a second–order perturbational treatment of the system–reservoir coupling. This approach is particularly useful if it can be combined with a normal mode description of the reservoir. Of course, second–order perturbation theory may not always be appropriate, even if we did our best to separate the total system into active and spectator degrees of freedom. Including higher–order perturbation terms is of course a way for improvement, but the resulting expressions become very soon rather cumbersome (see, for example [Jan02,Kor97,Lai91]). In the next sections we will deal with approaches which may overcome the second–order treatment of the system–reservoir coupling. In Section 3.10 we introduce the path integral representation of the density operator, and in the last Section 3.11 we present a classical description of the reservoir coordinates. The present section is devoted to a derivation of generalized rate equations which are also going beyond a perturbational treatment of the system–reservoir coupling.

The approach focuses on the derivation of Generalized Rate Equations (Generalized Master Equations) for the populations $P_a(t)$ of the system eigenstates (an elementary version of what will follow here we already encountered in Section 3.4.5). Once such equations have been established one can easily extract the transition rates which are valid in any order of perturbation theory. To this end we will use the projection operator technique. Since the projection operator $\mathcal{P}$ is a superoperator acting in the Liouville space formed by the usual operators, we will refer to the following treatment as the *Liouville space approach* [Hu89,Spa88]. But

[17] A term proportional to $\partial\rho(x,p;t)/\partial p$ may also appear for a slightly different derivation. Moreover, if necessary, a term proportional $x^2$ (the counter term) has to be added to $U(x)$ to avoid divergencies caused by the spectral density (for details see [Cal83,Tan91]).

before introducing the projection operator $\mathcal{P}$ we will separate the total Hamiltonian Eq. (3.3) into a zeroth–order and coupling term. This separation starts from the expansion of $H_{\rm S-R}$ with respect to the system eigenstates (cf. Section 3.8.6). Here we assume that the diagonal elements of $\Phi_{ab} = \langle a|H_{\rm S-R}|b\rangle$ are much larger than the off–diagonal ones. Therefore, a different treatment of the two types of couplings is reasonable. In particular a perturbational description of the off–diagonal elements might be possible. But we will assume in the following that the diagonal elements are so large that they cannot be handled in a perturbation theory. (Such a situation, for example, is typical for nonadiabatic electron transfer and will be discussed in greater detail in the Sections 6.4, 6.6, and 6.7.)

We write the system–reservoir Hamiltonian as follows

$$H = H_0 + \hat{V} \ , \tag{3.348}$$

where the "zeroth–order" part is given by

$$H_0 = H_{\rm S} + \sum_a \Phi_{aa}(Z)|a\rangle\langle a| + H_{\rm R} \equiv \sum_a \Big(E_a + H_{\rm R} + \Phi_{aa}(Z)\Big)|a\rangle\langle a| \ . \tag{3.349}$$

The second part suggests that $E_a + H_{\rm R} + \Phi_{aa}(Z)$ can be understood as the reservoir Hamiltonian $H_a$ defined with respect to the reservoir coordinates and valid for the system eigenstate $|a\rangle$. The perturbation $\hat{V}$ accounts for the off–diagonal elements of $\Phi_{ab}(Z)$ and reads

$$\hat{V} = \sum_{a,b}(1-\delta_{ab})\Phi_{ab}(Z)|a\rangle\langle b| \ . \tag{3.350}$$

Once the diagonal matrix elements $\Phi_{aa}$ can be accounted for exactly, a nonperturbative description of the system–reservoir coupling has been achieved.

### 3.9.1 Projection Operator Technique

In order to establish a nonperturbative description of the system–reservoir coupling let us introduce an appropriate projection operator. Since a simultaneous description of various states $|a\rangle$ is necessary, we generalize the projection operator, Eq. (3.164). The new quantity acting on an arbitrary operator $\hat{O}$ follows as

$$\mathcal{P}\hat{O} = \sum_a \hat{R}_a \, {\rm tr}_{\rm R}\{\langle a|\hat{O}|a\rangle\}|a\rangle\langle a| \ . \tag{3.351}$$

Instead of including the full state space related to the system Hamiltonian (as it would be the case for the projection operator according to Eq. (3.5.5)) the quantity $\mathcal{P}$ projects on the diagonal matrix element of some operator. The reservoir coordinates are fixed by the equilibrium statistical operators $\hat{R}_a$ defined by $H_a = E_a + H_{\rm R} + \Phi_{aa}$ (cf. (3.75)). If $\mathcal{P}$ acts on the complete statistical operator we obtain

$$\mathcal{P}\hat{W}(t) = \sum_a \hat{R}_a P_a(t)|a\rangle\langle a| \ . \tag{3.352}$$

The state populations are extracted if we take the trace with respect to the complete system–reservoir state space. This procedure is equivalent to choosing the respective diagonal matrix element of (3.352) and taking the trace with respect to the reservoir states

$$P_a(t) = \mathrm{tr}_{\mathrm{R}} \left\{ \langle a|\mathcal{P}\hat{W}(t)|a\rangle \right\} . \tag{3.353}$$

Starting with the Liouville–von Neumann equation[18]

$$\frac{\partial}{\partial t}\hat{W}(t) = -i\mathcal{L}\hat{W}(t) , \tag{3.354}$$

and introducing the orthogonal complement, $\mathcal{Q} = 1 - \mathcal{P}$, a separation into two orthogonal parts yields

$$\frac{\partial}{\partial t}\mathcal{P}\hat{W}(t) = -i\mathcal{P}\mathcal{L}\left(\mathcal{P}\hat{W}(t) + \mathcal{Q}\hat{W}(t)\right) , \tag{3.355}$$

and

$$\frac{\partial}{\partial t}\mathcal{Q}\hat{W}(t) = -i\mathcal{Q}\mathcal{L}\left(\mathcal{P}\hat{W}(t) + \mathcal{Q}\hat{W}(t)\right) . \tag{3.356}$$

The solution of the equation for $\mathcal{Q}\hat{W}$ including the assumption $\mathcal{Q}\hat{W}(t_0) = 0$ can be written as follows

$$\mathcal{Q}\hat{W}(t) = -i\int\limits_{t_0}^{t} d\bar{t}\, \mathcal{U}_Q(t-\bar{t})\mathcal{Q}\mathcal{L}\mathcal{P}\hat{W}(\bar{t}) , \tag{3.357}$$

where the time–propagation superoperator

$$\mathcal{U}_Q(t) = \exp\left\{-i\mathcal{Q}\mathcal{L}t\right\} \tag{3.358}$$

has been introduced. The resulting equation for $\mathcal{P}\hat{W}$ is closed and reads

$$\frac{\partial}{\partial t}\mathcal{P}\hat{W}(t) = -i\mathcal{P}\mathcal{L}\mathcal{P}\hat{W}(t) - \int\limits_{t_0}^{t} d\bar{t}\, \mathcal{P}\mathcal{L}\,\mathcal{U}_Q(t-\bar{t})\mathcal{Q}\mathcal{L}\,\mathcal{P}\hat{W}(\bar{t}) . \tag{3.359}$$

Using Eq. (3.353) it is possible to derive the related equations of motion for the state populations. This derivation will be explained step by step in the following. First, we determine $\langle a|\mathrm{tr}_{\mathrm{R}}\{\mathcal{P}\mathcal{L}\hat{O}\}|a\rangle$, where $\hat{O}$ is a operator acting in the electron reservoir state space. We obtain

$$\mathrm{tr}_{\mathrm{R}}\left\{\langle a|\mathcal{P}\mathcal{L}\hat{O}|a\rangle\right\} = \mathrm{tr}_{\mathrm{R}}\left\{\langle a|\mathcal{L}\hat{O}|a\rangle\right\} . \tag{3.360}$$

[18]In the following we do not change to the interaction representation as it had been done, for example, in Section 3.5.6. Instead we stay in the Schrödinger representation which has the technical advantage that we can avoid the introduction of a time–ordered $S$–superoperator (compare Eq. (3.433)). However, the basic idea to arrive at a closed equation for $\mathcal{P}\hat{W}$ is the same.

This type of matrix element results from both terms on the right–hand side of Eq. (3.359) ($\hat{O} = \mathcal{P}\hat{W}(t)$ as well as $\hat{O} = \mathcal{U}_Q(t-\bar{t})\mathcal{QLP}\hat{W}(\bar{t})$). If we replace $\hat{O}$ by $\mathcal{P}\hat{W}$, we easily verify that this term vanishes. The second term leads to the memory kernel of a Generalized Master Equation for the populations of the system eigenstates (cf. Section 3.4.5 and in particular Eq. (3.124)). After introducing the unit–step function $\Theta(t-\bar{t})$ it reads

$$\frac{\partial}{\partial t}P_a(t) = \sum_b \int_{t_0}^{\infty} d\bar{t}\; K_{ab}(t-\bar{t})P_b(\bar{t}) \;, \tag{3.361}$$

where we have defined the memory kernel

$$K_{ab}(t) = -\Theta(t)\text{tr}_\text{R}\left\{\langle a|\left(\mathcal{L}\mathcal{U}_Q(t)\mathcal{QL}\hat{R}_b|b\rangle\langle b|\right)|a\rangle\right\} \;. \tag{3.362}$$

It will be the aim of the following considerations to derive a more comprehensible expression for the memory kernel.

### 3.9.2 Rate Equations

Before deriving an expression for the $K_{ab}(t)$ we briefly explain their relation to transition rates. Suppose we can neglect memory effects and change to the rate equation limit (cf. Section 3.4.5). Then Eq. (3.361) becomes (note the change from $\bar{t}$ to $\tau = t - \bar{t}$)

$$\frac{\partial}{\partial t}P_a(t) = \sum_b K_{ab}P_b(t) \;, \tag{3.363}$$

with

$$K_{ab}(\omega = 0) = \int d\tau\; K_{ab}(\tau) \;. \tag{3.364}$$

For further application we interpreted the total time integral with respect to the memory kernel as the zero–frequency part of the respective Fourier–transformed quantity. The Eqs. (3.363) have to fulfill the conservation of total probability,

$$\sum_a \frac{\partial}{\partial t}P_a(t) = 0 = \sum_{ab} K_{ab}P_b(t) \;. \tag{3.365}$$

Since this expression should be valid for any time $t$ one can deduce $0 = \sum_a K_{ab}$, which yields $K_{bb} = -\sum_{a\neq b} K_{ab}$. Therefore, we obtain the standard rate equation

$$\frac{\partial}{\partial t}P_a = -\sum_b (k_{a\to b}P_a - k_{b\to a}P_b) \;, \tag{3.366}$$

with transition rates from state $a$ to state $b$ given by $k_{a\to b} = K_{ba}(\omega = 0)$. To carry out a perturbational expansion of the memory kernel it is advantageous to introduce the projector

$$\hat{\Pi}_a = |a\rangle\langle a| \;. \tag{3.367}$$

Taking the trace with respect to the complete system–reservoir state space we obtain

$$K_{ba}(t) = -\Theta(t)\mathrm{tr}\left\{\hat{\Pi}_b \mathcal{L} \mathcal{U}_Q(t) \mathcal{Q} \mathcal{L} \hat{R}_a \hat{\Pi}_a\right\} . \tag{3.368}$$

This is still a rather complicated expression. A more appealing form of the memory kernel can be obtained if a number of formal manipulations are carried out.

### 3.9.3 Perturbational Expansion of the Rate Expressions

The memory kernel will be rewritten using the superoperator notation and finally by introducing a power expansion with respect to the coupling. For this reason we write

$$\mathcal{L} = \mathcal{L}_0 + \mathcal{L}_V , \tag{3.369}$$

where $\mathcal{L}_0$ corresponds to the commutator with the zeroth–order Hamiltonian $H_0$ (divided by $\hbar$). The commutator with the coupling operator $\hat{V}/\hbar$ is denoted by $\mathcal{L}_V$. By construction we have the important properties

$$\mathcal{P}\mathcal{L}_0 = \mathcal{L}_0\mathcal{P} = 0 , \tag{3.370}$$

and

$$\mathcal{P}\mathcal{L}_V\mathcal{P} = 0 . \tag{3.371}$$

Both relations are simply verified using the definition of the projector and taking into account that $H_0$ is diagonal with respect to the states $|a\rangle$, whereas $\hat{V}$ has only off–diagonal contributions. Eq. (3.370) leads to

$$\mathcal{Q}\mathcal{L} = \mathcal{L}_0 + \mathcal{Q}\mathcal{L}_V . \tag{3.372}$$

This results in the following notation of the time–propagation superoperator $\mathcal{U}_Q$, Eq. (3.358),

$$\mathcal{U}_Q(t) = \exp\{-i(\mathcal{L}_0 + \mathcal{Q}\mathcal{L}_V)t\} . \tag{3.373}$$

Next let us have a closer look at the Fourier transformed memory function, Eq. (3.368). The one–sided Fourier transform of the time–propagation superoperator leads to

$$\int_0^\infty dt\, e^{i\omega t}\, \mathcal{U}_Q(t) = \frac{i}{\omega - \mathcal{L}_0 - \mathcal{Q}\mathcal{L}_V + i\epsilon} \equiv i\mathcal{G}_Q(\omega) . \tag{3.374}$$

If introduced for ordinary operators the quantity $\mathcal{G}_Q(\omega)$ is known as the resolvent operator or *Green's operator*. The present formula gives an extension of this concept to superoperators. Inserting $\mathcal{G}_Q(\omega)$ into the Fourier transformed version of Eq. (3.368) yields

$$K_{ba}(\omega) = -i\ \mathrm{tr}\left\{\hat{\Pi}_b\ \mathcal{L}\mathcal{Q}\mathcal{G}_Q(\omega)\mathcal{Q}\mathcal{L}\hat{R}_a\hat{\Pi}_a\right\} . \tag{3.375}$$

A detailed inspection of the trace expression shows that the operators $\mathcal{L}$ appearing on the left– and right–hand side can be replaced by $\mathcal{L}_V$. ($\mathcal{L}_0$ on the left–hand side vanishes because the trace of the commutator equals zero. On the right–hand side $\mathcal{L}_0$ gives no contribution because of the property (3.370).)

The desired perturbation expansion with respect to the coupling follows if an equation of motion for $\mathcal{G}_Q(t)$ is established. The Fourier transformed version reads

$$(\omega - \mathcal{L}_0 - \mathcal{Q}\mathcal{L}_V)\, \mathcal{G}_Q(\omega) = 1\,. \tag{3.376}$$

We introduce a zeroth–order Green's operator according to

$$(\omega - \mathcal{L}_0)\, \mathcal{G}_0(\omega) = 1\,, \tag{3.377}$$

and write

$$(\mathcal{G}_0^{-1}(\omega) - \mathcal{Q}\mathcal{L}_V)\mathcal{G}_Q(\omega) = 1\,. \tag{3.378}$$

A multiplication by $\mathcal{G}_0(\omega)$ leads to

$$\mathcal{G}_Q(\omega) = \mathcal{G}_0(\omega) + \mathcal{G}_0(\omega)\mathcal{Q}\mathcal{L}_V\mathcal{G}_Q(\omega)\,. \tag{3.379}$$

This equation is a version of the ubiquitous *Dyson equation* . If rearranged it gives a solution for $\mathcal{G}_Q(\omega)$ in terms of $\mathcal{G}_0(\omega)$ and $\mathcal{Q}\mathcal{L}_V$.

Let us inspect the Dyson equation in more detail using a special property of the memory function, Eq. (3.375). Apparently, the memory function is defined as a diagonal matrix element by virtue of the trace operation. We expect that a power expansion with respect to the coupling part $\hat{V}$ will contain only non–vanishing terms with even powers of $\hat{V}$. Therefore, it is advantageous to choose the formal solution for $\mathcal{G}_Q$ in such a manner that it also contains even powers of the coupling only. We get from Eq. (3.379)

$$\begin{aligned}
\mathcal{G}_Q(\omega) &= (1 - \mathcal{G}_0(\omega)\mathcal{Q}\mathcal{L}_V)^{-1}\, \mathcal{G}_0(\omega) = \sum_{n=0}^{\infty} (\mathcal{G}_0(\omega)\mathcal{Q}\mathcal{L}_V)^n\, \mathcal{G}_0(\omega) \\
&\approx \sum_{n=0}^{\infty} (\mathcal{G}_0(\omega)\mathcal{Q}\mathcal{L}_V\mathcal{G}_0(\omega)\mathcal{Q}\mathcal{L}_V)^n\, \mathcal{G}_0(\omega) \\
&= (1 - \mathcal{G}_0(\omega)\mathcal{Q}\mathcal{L}_V\mathcal{G}_0(\omega)\mathcal{Q}\mathcal{L}_V)^{-1}\, \mathcal{G}_0(\omega)\,.
\end{aligned} \tag{3.380}$$

his expression is inserted into Eq. (3.375). Let us concentrate on the pure superoperator part first. We write

$$\mathcal{L}_V\mathcal{Q}\mathcal{G}_Q(\omega)\mathcal{Q}\mathcal{L}_V = \sum_{n=0}^{\infty} \mathcal{L}_V\mathcal{Q}\Big(\mathcal{G}_0(\omega)\mathcal{Q}\mathcal{L}_V\mathcal{G}_0(\omega)\mathcal{Q}\mathcal{L}_V\Big)^n \mathcal{G}_0(\omega)\mathcal{Q}\mathcal{L}_V\,. \tag{3.381}$$

The expression can be arranged in a more symmetric manner by using a particular property of $\mathcal{G}_0(\omega)$. We consider

$$\mathcal{U}_0(t) = e^{-i\mathcal{L}_0 t}\,, \tag{3.382}$$

and use Eqs. (3.370) and (3.371). It follows that

$$\mathcal{P}\mathcal{U}_0(t) = \mathcal{U}_0(t)\mathcal{P} = \mathcal{P} \; . \qquad (3.383)$$

This gives

$$\mathcal{Q}\mathcal{U}_0(t) = \mathcal{U}_0(t)\mathcal{Q} = \mathcal{U}_0(t) - \mathcal{P} \equiv \mathcal{Q}\mathcal{U}_0(t)\mathcal{Q} \; . \qquad (3.384)$$

If changed to the Fourier–transformed Green's operator $\mathcal{G}_0(\omega)$ it reads (note that the Fourier transform of the unit–step function appears here)

$$\mathcal{Q}\mathcal{G}_0(\omega) = \mathcal{G}_0(\omega)\mathcal{Q} = \mathcal{Q}\mathcal{G}_0(\omega)\mathcal{Q} = \mathcal{G}_0(\omega) - \frac{1}{\omega + i\epsilon}\mathcal{P} \equiv \tilde{\mathcal{G}}_0(\omega) \; . \qquad (3.385)$$

In our derivation of the more symmetric version of Eq. (3.381) we obtain

$$\begin{aligned} \mathcal{L}_V \mathcal{Q}\mathcal{G}_Q(\omega)\mathcal{Q}\mathcal{L}_V &= \sum_{n=0}^{\infty} \mathcal{L}_V \left(\tilde{\mathcal{G}}_0(\omega)\mathcal{L}_V\tilde{\mathcal{G}}_0(\omega)\mathcal{L}_V\right)^n \tilde{\mathcal{G}}_0(\omega)\mathcal{L}_V \\ &\equiv \sum_{m=0}^{\infty} \mathcal{L}_V \left(\tilde{\mathcal{G}}_0(\omega)\mathcal{L}_V\right)^{2m+1} \; . \end{aligned} \qquad (3.386)$$

Combined with the two projection operators $\hat{\Pi}_a$ and $\hat{\Pi}_b$ (Eq. (3.367)) the resulting frequency–dependent memory kernel $K_{ba}$ which will be identified with the frequency–dependent transition rate $k_{a\to b}$ reads

$$k_{a\to b}(\omega) = \sum_{m=0}^{\infty} k_{a\to b}^{(2m)}(\omega) \; . \qquad (3.387)$$

It is given as an infinity summation of rates $k_{a\to b}^{(2m)}$ which are of $m$th order in the square of the coupling $\hat{V}$:

$$k_{a\to b}^{(2m)}(\omega) = -i \,\mathrm{tr}\left\{\hat{\Pi}_b\mathcal{L}_V\left(\tilde{\mathcal{G}}_0(\omega)\mathcal{L}_V\right)^{2m+1}\hat{R}_a\hat{\Pi}_a\right\} \qquad (3.388)$$

This expression may serve as a starting point for different perturbation expansions. Considering a system of two coupled levels, for instance, one is able to go beyond simple second–order Golden Rule rate expression like those derived in Section 3.4.5. If more than two states are involved already the lowest–order contribution is higher than second order. Applications of this kind will be discussed in more detail in Chapters 5 – 8.

In order to connect this formal results to the discussion in Section 3.4.5 we concentrate on the lowest–order two–state version of Eq. (3.388). The transition rate is obtained as (note the replacement of $\mathcal{P}\hat{\Pi}_1$ by $\hat{R}_1\hat{\Pi}_1$ and of $\tilde{\mathcal{G}}_0(\omega)$ by $\mathcal{G}_0(\omega)$)

$$k_{1\to 2}^{(2)}(\omega = 0) = -i \,\mathrm{tr}\left\{\hat{\Pi}_2\mathcal{L}_V\mathcal{G}_0(\omega = 0)\mathcal{L}_V\hat{R}_1\hat{\Pi}_1\right\} \; . \qquad (3.389)$$

In a next step we replace the Green's operator (in the frequency–domain) by the time–evolution superoperator and introduce the explicit form of $\mathcal{L}_V$

$$
\begin{aligned}
k_{1\to 2} &= -\frac{1}{\hbar^2}\int_0^\infty dt\, \mathrm{tr}_\mathrm{R}\Big\{\langle 2|\big[(\Phi_{12}|1\rangle\langle 2| + \text{h.c.}),\\
&\qquad \times U_0(t)\big[(\Phi_{12}|1\rangle\langle 2| + \text{h.c.}),\ \hat{R}_1\hat{\Pi}_1\big]_-\ U_0^+(t)\ \big]_-|2\rangle\Big\}\ .
\end{aligned}
\tag{3.390}
$$

Performing the commutation operations we finally get

$$
k_{1\to 2} = \frac{1}{\hbar^2}\int_0^\infty dt\ \mathrm{tr}_\mathrm{vib}\{\hat{R}_1 U_0^+(t)\Phi_{12}U_0(t)\Phi_{21}\} + \text{c.c.}\ . \tag{3.391}
$$

If the Hermitian conjugate part is replaced by the time integral from $-\infty$ to 0 the rate $k_{1\to 2}$ follows as the Fourier–transform (with zero frequency argument) of the coupling potential correlation function (cf. Eq. (3.132)). The perturbation theory with respect to this small quantity can be extended as will be shown in different applications in the following sections. However, in any case the approach is beyond a perturbation theory with respect to the diagonal coupling functions $\Phi_{aa}$ since those are incorporated in the Hamiltonian $H_1$ and $H_2$ defining the time–evolution operators in Eq. (3.391). This result looks very promising, and indeed the whole Liouville Space Approach enables one to derive Master Equations with rate expressions beyond simple Golden Rule formulas. However, the approach is less suited when transitions among the states $|a\rangle$ are induced by external fields (for example via optical excitations). For these cases off–diagonal density matrix elements may become important which do not appear in the present technique.

## 3.10 The Path Integral Representation of the Density Matrix

The second–order perturbational treatment of the system–reservoir coupling and the Markov approximation are restrictions inherent to the density matrix theory presented particularly in Section 3.8.2. If we focus on harmonic oscillator reservoirs it is possible to derive an exact, i.e., nonperturbative and non–Markovian, expression for the reduced density matrix within the framework of Feynman's path integral approach to quantum dynamics. This will be demonstrated in the present section.

In order to illustrate the basic idea we go back to Section 3.4.3 where the time evolution of the total density operator was given in Eq. (3.97). Let us suppose that we slice the time interval $[t_0, t = t_N]$ into $N$ pieces of length $\Delta t = (t_N - t_0)/N$, i.e., $t_i = t_0 + i\Delta t$ $(i = 0, \ldots, N)$. If we use the decomposition property of the time–evolution operator, Eq. (3.13), the matrix elements of this operator with respect to the coordinate representation become

$$
\langle x_N|U(t_N,t_0)|x_0\rangle = \langle x_N|U(t_N,t_{N-1})U(t_{N-1},t_{N-2})\ldots U(t_2,t_1)U(t_1,t_0)|x_0\rangle\ . \tag{3.392}
$$

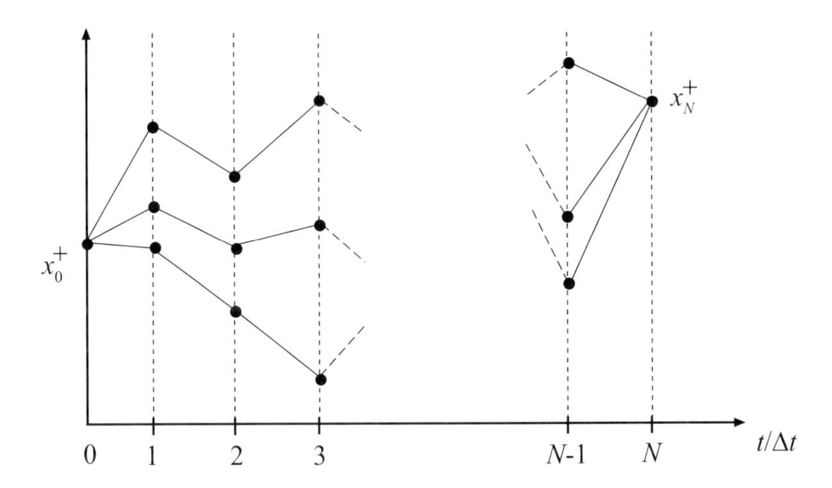


**Figure 3.9:** Visualization of different time–sliced paths leading from $x_0^+$ to $x_N^+$ in the time interval $[t_0, t_N]$.

In a next step we insert the identity $1 = \int dx_i |x_i\rangle\langle x_i|$ between all operator products in Eq. (3.392). This gives

$$\langle x_N|U(t_N,t_0)|x_0\rangle = \prod_{j=1}^{N-1}\left[\int dx_j\right]\prod_{j=1}^{N}\langle x_j|U(t_j,t_{j-1})|x_{j-1}\rangle \ . \tag{3.393}$$

Within this representation the matrix elements of the time–evolution operator, i.e., the transition amplitudes for the particle for going from point $x_0$ to point $x_N$ in the time interval $[t_0, t_N]$, have a simple interpretation which is illustrated in Fig. 3.9. The vertical axis in this figure represents the coordinate, the horizontal one is the discretized time. Starting from a particular $x_0$ the system explores *all* possible paths which lead to $x_N$ in the interval $[t_0, t_N]$, because at each intermediate time step $t_i$, Eq. (3.393) demands for an integration with respect to the coordinate $x_i$. Within this intuitive picture the differential equations for the time–evolution operator presented in the previous sections are replaced by a high–dimensional integration in coordinate space.

A fundamental property of the representation (3.393) can be derived starting for simplicity from the single particle Hamiltonian $H = T(\hat{p}) + V(\hat{x})$. We further suppose that the time step $\Delta t$ is small enough to justify the decomposition

$$e^{-iH\Delta t/\hbar} \approx e^{-iV(\hat{x})\Delta t/\hbar}e^{-iT(\hat{p})\Delta t/\hbar} \ . \tag{3.394}$$

According to Eq. (2.95) the error here will be of the order $\Delta t^2[V(\hat{x}), T(\hat{p})]_-$. For the matrix elements of the time–evolution operator in Eq. (3.393) we can then write

$$\begin{aligned}\langle x_j|U(t_j,t_{j-1})|x_{j-1}\rangle &\approx \langle x_j|e^{-iV(\hat{x})\Delta t/\hbar}e^{-iT(\hat{p})\Delta t/\hbar}|x_{j-1}\rangle\\ &= \int dx \int \frac{dp}{2\pi\hbar}\int\frac{dp_j}{2\pi\hbar}\langle x_j|e^{-iV(\hat{x})\Delta t/\hbar}|x\rangle\\ &\quad\times\langle x|p_j\rangle\langle p_j|e^{-iT(\hat{p})\Delta t/\hbar}|p\rangle\langle p|x_{j-1}\rangle\\ &= \int\frac{dp_j}{2\pi\hbar}e^{-iV(x_j)\Delta t/\hbar}e^{ip_j(x_j-x_{j-1})/\hbar}e^{-iT(p_j)\Delta t/\hbar}\,,\end{aligned} \tag{3.395}$$

where we used $\langle x_j|V(\hat{x})|x\rangle = V(x_j)\delta(x-x_j)$, $\langle x|p_j\rangle = \exp\{ip_jx/\hbar\}$, and $\langle p_j|T(\hat{p})|p\rangle = 2\pi\hbar T(p_j)\delta(p-p_j)$. Inserting this expression into Eq. (3.393) we have

$$\begin{aligned}\langle x_N|U(t_N,t_0)|x_0\rangle &= \prod_{j=1}^{N-1}\left[\int dx_j\right]\prod_{j=1}^{N}\left[\int\frac{dp_j}{2\pi\hbar}\right]\\ &\quad\times\exp\left\{\frac{i}{\hbar}\sum_{j=1}^{N}\left[p_j(x_j-x_{j-1})-\Delta t[T(p_j)+V(x_i)]\right]\right\}.\end{aligned} \tag{3.396}$$

The momentum integrals can be performed analytically if the kinetic energy operator has the form $T = p^2/2m$. One obtains

$$\begin{aligned}\langle x_N|U(t_N,t_0)|x_0\rangle &= \frac{1}{\sqrt{2\pi\hbar i\Delta t/m}}\prod_{j=1}^{N-1}\left[\int\frac{dx_j}{\sqrt{2\pi\hbar i\Delta t/m}}\right]\\ &\quad\times\exp\left\{\frac{i}{\hbar}\Delta t\left[\sum_{j=1}^{N}\frac{m}{2}\left(\frac{x_j-x_{j-1}}{\Delta t}\right)^2-V(x_j)\right]\right\}.\end{aligned} \tag{3.397}$$

If we now take the continuum limit, $\Delta t\to 0$, the sum in the exponent becomes an integral over time in the interval $[t_0,t_N]$. Since $(x_j-x_{j-1})/\Delta t\to\dot{x}(t)$, the integrand turns into the Lagrange function known from classical mechanics. Introducing the symbol $\mathcal{D}x$ for the $\Delta t\to 0$ limit of the integration over the different intermediate points along the time sliced path the matrix elements of the time–evolution operator can be written as

$$\langle x_N|U(t_N,t_0)|x_0\rangle = \int\mathcal{D}x\exp\left\{\frac{i}{\hbar}\int_{t_0}^{t_N}dt\;L(x,\dot{x},t)\right\}. \tag{3.398}$$

The exponent is just the classical action corresponding to the Lagrangian $L(x,\dot{x},t) = m\dot{x}^2/2 - V(x,t)$. (Note that we have tacitly assumed a general time–dependent potential here for which the derivation proceeds along the same lines.) Since $\int\mathcal{D}x$ denotes the integration over all possible paths leading from $x_0$ to $x_N$ the interpretation of (3.398) is as follows: The transition amplitude for going from $x_0$ to $x_N$ during the time interval $[t_0,t_N]$ is obtained by summing all different paths and assigning a *phase* to each path which corresponds to $i/\hbar$ times the classical action. The exponent in Eq. (3.398) is of course a rapidly oscillating function on a scale given by Planck's constant. Thus, most of the terms in this sum will interfere in such a way that their net contributions are small. The main contribution is likely to come, however, from the classical path which makes the action stationary. This offers the intriguing possibility to understand quantum mechanical transition amplitudes in terms of classical paths supplemented by fluctuations around these paths which are responsible for quantum effects.

We now turn to the density matrix. Here we have to account for the time evolution of the bra and ket vectors. Going back to Eq. (3.152) we can write for the total density operator using Eq. (3.398)

$$\begin{aligned} W(x_N^+, x_N^-; t_N) &= \int dx_0^\pm \langle x_N^+|U(t_N,t_0)|x_0^+\rangle\langle x_0^+|W(t_0)|x_0^-\rangle\langle x_0^-|U^+(t_N,t_0)|x_N^-\rangle \\ &= \int dx_0^\pm \int \mathcal{D}x^+\mathcal{D}x^- \exp\left\{\frac{i}{\hbar}\int_{t_0}^{t_N} dt L(x^+,\dot{x}^+,t)\right\} \\ &\quad\times\langle x_0^+|W(t_0)|x_0^-\rangle \exp\left\{-\frac{i}{\hbar}\int_{t_0}^{t_N} dt L(x^-,\dot{x}^-,t)\right\} . \end{aligned} \tag{3.399}$$

This path integral expression for the total system's density matrix shows that the bra and the ket part of the density operator have to be propagated forward and backward, respectively. Thereby all paths $x^\pm(t)$ connecting $x_0^+$ with $x_N^+$ and $x_0^-$ with $x_N^-$, respectively, in the interval $[t_0, t_N]$ have to be explored. The initial points are subject to an additional integration.

In order to develop the expression for the reduced density matrix for a one–dimensional system ($s$) embedded in a harmonic bath ($q = \{q_\xi\}$) we start with the Lagrangian corresponding to the generic bilinear system–bath Hamiltonian (cf. Section 2.6.3)

$$L = \frac{m_s\dot{s}^2}{2} - V(s) + \sum_\xi \left(\frac{\dot{q}_\xi^2}{2} - \frac{\omega_\xi^2}{2}\left(q_\xi - \frac{c_\xi s}{\omega_\xi^2}\right)^2\right) \tag{3.400}$$

Apparently this Lagrangian is of the form $L = L_\mathrm{S} + L_\mathrm{R} + L_\mathrm{S-R}$. Using Eq. (3.400) and (3.399) with $x = (s, q)$ we obtain for the reduced density matrix

$$\rho(s_N^+, s_N^-; t_N) = \int dq_N W(s_N^+, q_N, s_N^-, q_N; t_N) \tag{3.401}$$

the expression

$$\begin{aligned} \rho(s_N^+, s_N^-; t_N) &= \int ds_0^\pm dq_0^\pm dq_N \int \mathcal{D}s^+\mathcal{D}s^-\mathcal{D}q^+\mathcal{D}q^- W(s_0^+, q_0^+, s_0^-, q_0^-; t_0) \\ &\quad\times\exp\left\{\frac{i}{\hbar}\int_{t_0}^{t_N} dt\left[L_\mathrm{S}(s^+,\dot{s}^+,t) - L_\mathrm{S}(s^-,\dot{s}^-,t)\right]\right\} \\ &\quad\times\exp\left\{\frac{i}{\hbar}\int_{t_0}^{t_N} dt\left[L_\mathrm{R}(q^+,\dot{q}^+,t) - L_\mathrm{R}(q^-,\dot{q}^-,t)\right]\right\} \\ &\quad\times\exp\left\{\frac{i}{\hbar}\int_{t_0}^{t_N} dt\left[L_\mathrm{S-R}(s^+,\dot{s}^+,q^+,\dot{q}^+,t)\right.\right. \\ &\qquad\left.\left. - L_\mathrm{S-R}(s^-,\dot{s}^-,q^-,\dot{q}^-,t)\right]\right\} . \end{aligned} \tag{3.402}$$

Note that due to the trace operation with respect to the reservoir degrees of freedom the end-points for the forward and backward reservoir paths are identical ($q_N$). For simplicity let us assume that the density matrix factorizes initially according to

$$W(s_0^+, q_0^+, s_0^-, q_0^-; t_0) = \rho(s_0^+, s_0^-; t_0)\hat{R}_{\text{eq}}(q_0^+, q_0^-; t_0) \, . \tag{3.403}$$

Then Eq. (3.402) can be written as

$$\begin{aligned} \rho(s_N^+, s_N^-; t_N) &= \int ds_0^\pm \int \mathcal{D}s^+ \mathcal{D}s^- \rho(s_0^+ s_0^-; t_0) \mathcal{F}(s^+, s^-) \\ &\times \exp\left\{ \frac{i}{\hbar} \int_{t_0}^{t_N} dt \, (L_{\text{S}}(s^+, \dot{s}^+, t) - L_{\text{S}}(s^-, \dot{s}^-, t) \right\} \, . \end{aligned} \tag{3.404}$$

The structure of the equation is such that all the influence of the environment on the system dynamics is contained in the so–called *Feynman–Vernon influence functional* which is defined as

$$\begin{aligned} \mathcal{F}(s^+, s^-) &= \int dq_0^\pm dq_N \int \mathcal{D}q^+ \mathcal{D}q^- \\ &\times \exp\left\{ \frac{i}{\hbar} \int_{t_0}^{t_N} dt \, (L_{\text{R}}(q^+, \dot{q}^+, t) - L_{\text{R}}(q^-, \dot{q}^-, t) \right\} \hat{R}_{\text{eq}}(q_0^+, q_0^-; t_0) \\ &\times \exp\left\{ \frac{i}{\hbar} \int_{t_0}^{t_N} dt \, (L_{\text{S-R}}(s^+, \dot{s}^+, q^+, \dot{q}^+, t) \right. \\ &\qquad \left. - L_{\text{S-R}}(s^-, \dot{s}^-, q^-, \dot{q}^-, t) \right\} \, . \end{aligned} \tag{3.405}$$

Even though we arrived at Eqs. (3.404) and (3.405) in a rather straightforward manner, the physical content of these expressions is remarkable. The system's density matrix evolution is given as a path integral over all paths $s^+(t)$, connecting $s_0^+$ with $s_N^+$ (forward time evolution), and all paths $s^-(t)$, connecting $s_0^-$ with $s_N^-$ (backward time evolution). The free system dynamics is modified by the interaction with the environment introduced by the influence functional $\mathcal{F}(s^+, s^-)$. Inspecting Eq. (3.405) we realize immediately that $\mathcal{F}(s^+, s^-)$ contains interactions which are *nonlocal* in time and span, in principle, the whole time interval of the density matrix evolution. This is just another way of saying that Eq. (3.404) contains all memory effects. Further we notice that $L_{\text{S-R}}$ can be interpreted as an extra potential for the environmental oscillators; they experience a force which changes along the system paths. For the Lagrangian (3.400) it is possible to obtain an analytical expression for the influence functional. It reads

$$\begin{aligned} \mathcal{F}(s^+, s^-) &= \exp\left\{ \frac{1}{\hbar} \int_{t_0}^{t_N} dt' \int_{t_0}^{t'} dt'' \, [s^+(t') - s^-(t')] \right. \\ &\times \left. [\alpha(t' - t'')s^+(t'') - \alpha^*(t' - t'')s^-(t'')] \right\} \\ &\times \exp\left\{ -\frac{i}{\hbar} \int_{t_0}^{t_N} dt' \sum_\xi \frac{c_\xi^2}{2\omega_\xi^2} [s^+(t')^2 - s^-(t')^2] \right\} \, . \end{aligned} \tag{3.406}$$

Here, $\alpha(t)$ is the correlation function for the harmonic oscillator reservoir coordinates which was discussed in Section 3.6.2.

Eqs. (3.404) and (3.405) represent an *exact* analytical solution to the problem of the time evolution of the reduced density matrix for a one–dimensional system in a harmonic oscillator bath. However, the bottleneck for its straightforward numerical implementation is the multi–dimensional integration of a highly oscillatory function. Since the time step and the number of integrations are directly related (cf. Eqs. (3.392) to (3.394)) until recently only the short–time dynamics was accessible by this method. However, in a typical condensed phase situation the memory time of the bath degrees of freedom does not extend over arbitrarily large time intervals. Stated in another way, the spectral density of the environment is often a rather smooth function in the spectral range of interest. Thus, the correlation function $C(t)$ can be expected to decay rather rapidly. Note, that the extreme limit $C(t) \propto \delta(t)$ leads directly to the Markov approximation. Here the interactions introduced by the influence functional become local in time and the time evolution of the density matrix can be solved in an iterative fashion. Needless to say that this does not yet correspond to the level of theory adopted in Section 3.7.1; the system–bath interaction is still treated nonperturbatively. Along these lines it is possible now, to develop a systematic procedure for incorporation of finite memory effects. A limitation of the approach outlined so far is its restriction to harmonic oscillator reservoirs. A generalization to arbitrary reservoirs is possible, however, only at the expense of additional assumptions.

Finally, we refer the reader to Chapter 6 dealing with electron transfer reactions. There, an application for the path integral propagation of the reduced density matrix will be given in Fig. 6.33.

# 3.11 Quantum–Classical Hybrid Methods

## 3.11.1 The Mean–Field Approach

The theoretical description of condensed phase dynamics given so far has been focused on situations where the reservoir degrees of freedom can be modelled by a collection of harmonic oscillators. Of course, in many cases, for example, in a liquid environment, this will be an oversimplification. Here, one can often resort to a description of the reservoir (the solvent) in terms of classical mechanics. For the solute, however, quantum effects may be important and a quantum approach is necessary. In the following we will introduce some basic concepts behind such a *quantum–classical hybrid* description . Specific realizations will be discussed in Sections 4.3.2, 7.3, and 7.5.2.

In a first step let us consider an approach based on the time–dependent Schrödinger equation. We start with the Hamiltonian (3.133). The classical reservoir coordinates and momenta are denoted $Z = \{Z_\xi\}$ and $P = \{P_\xi\}$, respectively. As before we assume that $H_{\text{S-R}} = H_{\text{S-R}}(s, Z)$, where $s$ is the coordinate (set of coordinates) of the relevant quantum system. $H_{\text{S-R}}$ can be considered as an external potential acting on the relevant quantum system and the Schrödinger equation for the quantum system reads

$$i\hbar\frac{\partial}{\partial t}\Psi(s,t) = (H_{\text{S}} + H_{\text{S-R}}(s, Z(t)))\,\Psi(s,t)\,. \qquad (3.407)$$

It describes the quantum part of the coupled quantum mechanical and classical dynamics of the full system. $H_{\mathrm{S-R}}(s, Z(t))$ is now time–dependent, where the time dependence is induced by the classical reservoir coordinates.

Let us take the perspective of the classical reservoir now. If the coupling among both subsystems vanishes, i.e., if we have $H_{\mathrm{S-R}} = 0$, the reservoir coordinates $Z_\xi$ and momenta $P_\xi$ satisfy the canonical equations of motion

$$\frac{\partial Z_\xi}{\partial t} = \frac{\partial H_{\mathrm{R}}}{\partial P_\xi} , \qquad \frac{\partial P_\xi}{\partial t} = -\frac{\partial H_{\mathrm{R}}}{\partial Z_\xi} . \tag{3.408}$$

If the coupling is taken into account we would expect to have an additional potential on the right–hand side of (3.408). In order to link classical and quantum mechanics this potential is taken as the quantum mechanical expectation value of the system–reservoir coupling according the instantaneous values of the classical coordinates and the wave function of the quantum degrees of freedom

$$\langle H_{\mathrm{S-R}}(Z,t)\rangle_{\mathrm{S}} = \int ds\, \Psi^*(s,t) H_{\mathrm{S-R}}(s,Z)\Psi(s,t) . \tag{3.409}$$

The canonical equations are then given by Eqs. (3.408), where $H_{\mathrm{R}}$ has to be replaced by $H_{\mathrm{R}}+\langle H_{\mathrm{S-R}}(Z,t)\rangle_{\mathrm{S}}$. This leads to an additional force

$$F_\xi(t) = -\frac{\partial}{\partial Z_\xi} \int ds\; \Psi^*(s,t)\, H_{\mathrm{S-R}}(s,Z)\, \Psi(s,t) \tag{3.410}$$

on the classical coordinates $Z$.

The coupled equations of motion (3.407) and (3.408) are solved simultaneously by imposing appropriate initial conditions such as

$$\begin{aligned} \Psi(s,t=0) &= \Psi_0(s) , \\ Z_\xi(t=0) &= Z_\xi^{(0)} , \qquad P_\xi(t=0) = P_\xi^{(0)} . \end{aligned} \tag{3.411}$$

The set of Eqs. (3.407) and (3.408) (extended by the system–reservoir coupling) determines the quantum dynamics of the relevant quantum system coupled to the reservoir which is described by classical mechanics. While the presence of the classical degrees of freedom results in an additional potential for the motion of the quantum system, the classical reservoir feels the quantum mechanically averaged force of the relevant quantum system. Therefore, this is just another example for a *mean–field* approach.[19]

Within this description it is possible to determine how the small quantum system is influenced by the macroscopic reservoir. In many applications the reservoir is in thermal equilibrium characterized by a temperature $T$. Then the quantum system is influenced by external forces which change in time according to a thermal equilibrium state of the reservoir. We also note that besides the action of the reservoir on the system dynamics, it is also possible that the system dynamics drives the reservoir out of the equilibrium. In this case a nonequilibrium

[19]Note that this approach is also called *Time–Dependent Self–Consistent Field* method which expresses the fact that the simultaneous time–evolution of both subsystems is self–consistently coupled by the Hamiltonian $H_{\mathrm{ext}}(t)$.

state of the reservoir influences the system dynamics. This can also be viewed as a back reaction of the system dynamics on itself by means of the nonequilibrium state of the reservoir. Such a behavior may be observed if the system–reservoir coupling is strong enough. In the case of a sufficiently weak coupling it is appropriate to study the system dynamics under the influence of a reservoir staying in thermal equilibrium.

To describe the coupling between the thermal reservoir and the quantum system within the hybrid approach, classical molecular dynamics simulations for the reservoir have to be performed. Although the concept of a thermal reservoir is not applied directly, a temperature can be introduced by choosing the initial conditions for the reservoir according to a thermal distribution. To this end we introduce a set of different initial conditions $(Z^{(0,i)}, P^{(0,i)})$ with $i = 1 \ldots N$ for which the coupled equations (3.407) and (3.408) have to be solved. The initial conditions are taken from the related thermal distribution, i.e. they have to fulfill

$$f(Z^{(0,i)}, P^{(0,i)}) = \frac{1}{\mathcal{Z}_{\rm R}} e^{-E_{\rm R}/k_{\rm B}T} , \tag{3.412}$$

where $E_{\rm R}$ denotes the internal energy of the reservoir and $\mathcal{Z}_{\rm R}$ is the related partition function. The solutions of the time–dependent Schrödinger equation can then be labelled by the specific initial condition,

$$\Psi_i(s,t) = \Psi(s,t; Z^{(0,i)}, P^{(0,i)}) . \tag{3.413}$$

Calculating the expectation value of a system operator $\hat{O}$ we have to introduce an additional averaging with respect to the distribution of the initial values of the reservoir coordinates

$$\langle\langle \hat{O} \rangle_{\rm S} \rangle_{\rm R} = \sum_{i=1}^{N} f(Z^{(0,i)}, P^{(0,i)}) \int ds\ \Psi_i^*(s,t)\ \hat{O}\ \Psi_i(s,t) . \tag{3.414}$$

The success of a numerical implementation of this approach depends on the proper choice of representative initial values $(Z^{(0,i)}, P^{(0,i)})$ such that $N$ does not become too large. The thermal distribution function can be used in this respect since it gives a weight to the different $(Z^{(0,i)}, P^{(0,i)})$.

There are some characteristics of the hybrid approach which we would like to point out: First, the dynamics is obtained starting with a specific initial state of the reservoir, i.e., no thermal averaged equation of motion is introduced. Any averaging with respect to the reservoir can be carried out after solving the coupled equations of motion (see Eq. (3.414)). Second, the approach does not contain any approximation with respect to the coupling between $S$ and $R$, and any type of complex reservoir dynamics can be introduced. Third, it is necessary to note that the simulation of a thermal reservoir acting on a quantum system assumes that there are no *initial correlations* between system and reservoir. Any coupling between $S$ and $R$ in the initial state of the correlated dynamics has been removed by choosing the initial reservoir state from a distribution defined by the reservoir Hamiltonian function $H_{\rm R}(P, Z)$ alone. Finally, a clear disadvantage of the described method is the fact that the reservoir coordinates move only under the influence of the quantum mechanically averaged system–reservoir coupling. This is very different from the quantum dynamics where the reservoir wave function would "feel" a concrete system–reservoir coupling Hamiltonian (as it is the case for a dependence on the electronic states of the system). The method outlined in the following section is designed such as to compensate for this disadvantage.

### 3.11.2 The Surface Hopping Method

In the foregoing section we did not elaborate on the solution of the time–dependent Schrödinger equation (3.407) which has to be determined simultaneously with the classical propagation of the coordinates $Z$. For condensed phase systems this is a tremendous task. Therefore, most applications focus on the special case of a light particle (electron, proton) in a heavy atom environment. Of course, this is a situation where one would also expect that a quantum–classical hybrid treatment is sensible. In this case one can, in the spirit of the Born–Oppenheimer ansatz introduced in Section 2.3, solve the electron–nuclear Schrödinger equation, define *adiabatic* basis functions according to the stationary Schrödinger equation for fixed classical coordinates:

$$\big(H_{\rm S}(s) + H_{\rm S-R}(s,Z)\big)\phi_a(s,Z) = E_a(Z)\phi_a(s,Z) \ . \tag{3.415}$$

It includes the influence of the classical coordinates via $H_{\rm S-R}$ and, in this way, it may account for the influence of the reservoir in a quantum mechanically correct way. In analogy to Eq. (2.12) the eigenstates as well as the eigenvalues of Eq. (3.415) depend parametrically on the reservoir coordinates $Z$.

The solution of the time–dependent Schrödinger equation (3.407) can then be obtained by expanding $\Psi(s,t)$ with respect to $\phi_a(s,Z)$:

$$\Psi(s,t) = \sum_a c_a(t)\phi_a(s,Z) \ . \tag{3.416}$$

Inserting the expansion into Eq. (3.407) one may derive an equation of motion for the $c_a(t)$. However, we have to be aware that the basis functions become time–dependent according to the time–dependence of the reservoir coordinates. It follows

$$i\hbar\frac{\partial}{\partial t}c_a(t) = E_a\big(Z(t)\big) - i\hbar\sum_b \langle\phi_a\big(Z(t)\big)|\,\frac{\partial}{\partial t}\,|\phi_b\big(Z(t)\big)\rangle c_b(t) \ . \tag{3.417}$$

The matrix elements which couple different adiabatic basis states read

$$\begin{aligned}\langle\phi_a\big(Z(t)\big)|\,\frac{\partial}{\partial t}\,|\phi_b\big(Z(t)\big)\rangle &= \int ds\ \phi_a^*(s,Z(t))\,\frac{\partial}{\partial t}\,\phi_b(s,Z(t)) \\ &= \sum_\xi \langle\phi_a\big(Z(t)\big)|\,\frac{\partial}{\partial Z_\xi}\,|\phi_b\big(Z(t)\big)\rangle\frac{\partial Z_\xi}{\partial t} \ . \end{aligned} \tag{3.418}$$

This coupling is a consequence of the fact that the reservoir coordinates $Z$ change with time so that there will be an overlap between state $\phi_a$ and the time–derivative of state $\phi_b$. Consequently, the coupling is of the *nonadiabatic* type. Note that according to the correspondence principle we can replace the classical velocity by the respective momentum operator, i.e., $-i\hbar\partial Z_\xi/\partial t \to \hbar^2/M_\xi^2\nabla_\xi$. Thus the coupling between different adiabatic states in Eq. (3.418) corresponds to the second term in the nonadiabaticity operator in Eq. (2.17). The first term in Eq. (2.17), that is, the matrix elements of the kinetic energy operator are neglected here.

If one computes the expectation value, Eq. (3.409),

$$\langle H_{\rm S-R}(Z,t)\rangle_{\rm S} = \sum_{a,b} c_a^*(t)c_b(t)\int ds\ \phi_a^*(s,Z(t))H_{\rm S-R}(s,Z(t))\phi_b(s,Z(t)) \ , \tag{3.419}$$

it appears to be a sum of diagonal as well as off–diagonal contributions with respect to the basis states. Thus, in principle, Eq. (3.419) contains the information on the forces which are specific for a given adiabatic state of the quantum system. Moreover, there is a contribution from the nonadiabatic transition between these states. However, this information has to be disentangled in order to go beyond the mean–field treatment of the previous section. A recipe for turning this concept into a practical scheme is the surface hopping approach which is sketched in the following.

Let us assume that there is some initial set of classical reservoir coordinates $Z^{(0)}$ for which the Schrödinger equation (3.416) has been solved. Suppose that the quantum system is initially in the state $\phi_i(s, Z^{(0)})$ such that $c_a(0) = \delta_{ai}$ in Eq. (3.417). Thus only the initial state contributes the expectation value $\langle\phi_i|H_{\mathrm{S-R}}|\phi_i\rangle$ to Eq. (3.419). The forces (3.410) are calculated and the classical degrees of freedom are propagated for one time step $\Delta t$. For the new configuration Eq. (3.416) is solved and the new expansion coefficients are obtained from Eq. (3.417). Due to the nonadiabatic couplings entering Eq. (3.417) other states $|\phi_{a\neq i}\rangle$ may become populated at time $\Delta t$.

The information about these population changes can now be used to decide whether the adiabatic state for calculating the expectation value of $H_{\mathrm{S-R}}$, i.e., the forces in Eq. (3.410), is changed during the propagation of the classical trajectory.

Let us consider two adiabatic states for simplicity. Suppose we have calculated the coefficients as $|c_i(t + \Delta t)|^2 < |c_i(t)|^2$ and $|c_a(t + \Delta t)|^2 > |c_a(t)|^2$, i.e., a transition from $|\phi_i\rangle$ to $|\phi_a\rangle$ occurred. The probability $p_{i\to a}(t, t + \Delta t)$ for switching from $|\phi_i\rangle$ to $|\phi_a\rangle$ within the time step $\Delta t$ is therefore equal to

$$p_{i\to a}(t, t + \Delta t) = \frac{|c_i(t)|^2 - |c_i(t + \Delta t)|^2}{|c_i(t)|^2} . \tag{3.420}$$

Whether the potential for the actual trajectory is really changed is decided by drawing a random number $\delta$ in between 0 and 1; if $p_{i\to a}(t, t + \Delta t) < \delta$, the potential for the trajectory is changed, otherwise it stays the same during the next propagation step. If the state is changed the trajectory is interrupted and continues to move according to the potential $\langle\phi_a|H_{\mathrm{S-R}}|\phi_a\rangle$. This procedure is repeated for every time–step and it introduces a discontinuous classical dynamics which mimics the population changes of the quantum states. Notice, however, that these interruptions are influencing the time–evolution of the expansion coefficients as well since Eq. (3.417) also depends on $Z(t)$. Moreover, it should be clear that along a trajectory many jumps between different quantum states are possible which introduces some averaging with respect to the random choice of the number $\delta$ at every time step. Finally, the whole procedure has to be repeated for an ensemble of classical trajectories starting from different initial conditions.

Since the method introduces jumps from one potential $\langle\phi_a|H_{\mathrm{S-R}}|\phi_a\rangle$ to another potential (potential energy surface) $\langle\phi_b|H_{\mathrm{S-R}}|\phi_b\rangle$ it is frequently called *surface hopping method* [Tul90]. It is important to emphasize that the quantum jumps are introduced in an *ad hoc* fashion, i.e., the method has no strict theoretical foundation. There are numerous studies of the performance of surface hopping which enable one to appreciate its advantages and shortcomings; an example is given in Fig. 3.10. After all it often presents the only choice to treat complex problems such as, e.g., condensed phase proton transfer (cf. Section 7.5).

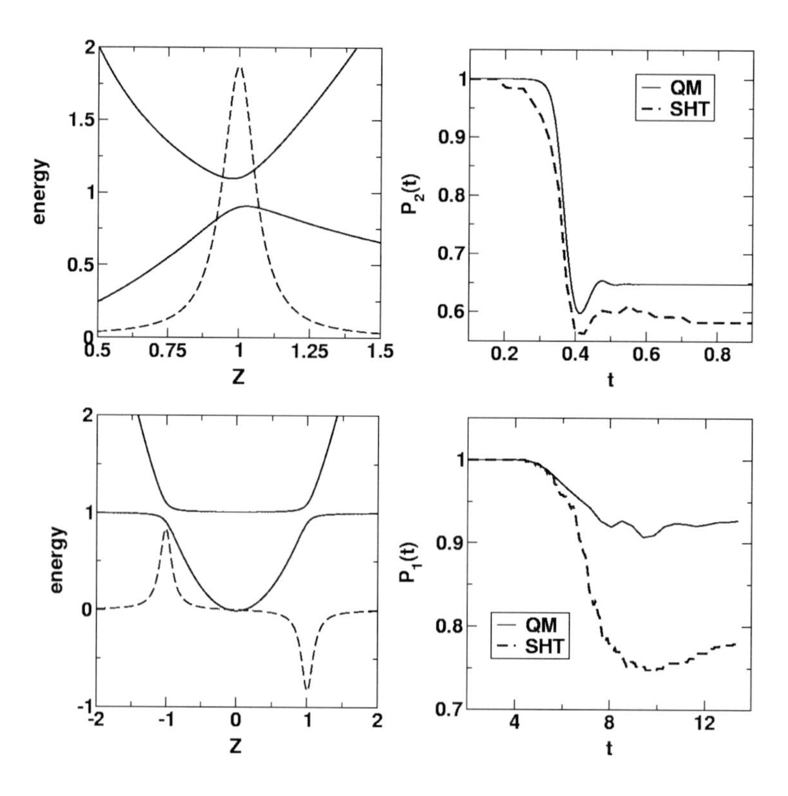


**Figure 3.10:** Performance of the surface hopping quantum–classical hybrid method for two examples with a single (upper panels) and a double (lower panels) avoided crossing between two adiabatic potential curves along a single reservoir coordinate $Z$. The two states (solid lines) correspond to adiabatic eigenstates of Eq. (3.415) which, by virtue of the interaction $H_{\text{S-R}}$, depend parametrically on the reservoir coordinate $Z$. The respective nonadiabatic couplings which are also due to $H_{\text{S-R}}$ are shown as dashed lines. In the right panels exact quantum mechanical (QM) and surface–hopping (SHT) results are shown for the population of the upper (upper panel) and lower (lower panel) adiabatic state. The initial condition has been given by a Gaussian wave packet centered at $Z = 0.4$ on the upper adiabatic state in the single crossing case, whereas it was centered at $Z = -4$ on the lower adiabatic potential in the double crossing case. For the single crossing setup surface hopping trajectories perform rather well, but the performance deteriorates dramatically for the double crossing case. This is a consequence of the incorrect treatment of quantum mechanical interference in the surface hopping approach. Notice that the actual calculations have been performed in the context of the partial Wigner representation method (cf. Section 3.11.3). (figure courtesy of B. Schmidt, for more details see also [Hor02])

### 3.11.3 Partial Wigner Representation as a Quantum–Classical Hybrid Method

In the following the quantum–classical hybrid method is formulated in a more general way as compared to the preceding sections by using the method of the Wigner representation of the statistical operator (cf. Section 3.4.4). This will give us a *systematic* technique which allows for a classical description of the reservoir dynamics and which enables us, at least in principle, to compute quantum corrections to this classical dynamics. (This is not possible when using the wave function approach of the previous sections.)

Since the classical description will be restricted to the reservoir DOF only, the Wigner representation of the density operator $\hat{W}(t)$ is exclusively introduced in the state space of the reservoir. The action of $\hat{W}(t)$ in the state space of the relevant system will be not specified. First, we introduce the representation of the total density operator with respect to the coordinates $Z$ of the reservoir (cf. Eq. (3.136)). We obtain the expression

$$\hat{\rho}(Z, \bar{Z}; t) = \langle Z| \hat{W}(t) |\bar{Z}\rangle , \tag{3.421}$$

which remains an operator in the state space of the relevant system. To establish the relation to a classical description of the reservoir we introduce the *partial* Wigner representation of $\hat{\rho}(Z, \bar{Z}; t)$ with respect to the reservoir coordinates, that is, $\hat{\rho}(X, P; t)$. Here, $X$ denotes the set of coordinates and $P$ the set of respective momenta. Taking the *eigenstates* $|a\rangle$ of the system–Hamiltonian to form matrix elements of $\hat{\rho}(X, P; t)$ we end up with $\langle a|\hat{\rho}(X, P; t)|b\rangle = \rho_{ab}(Z, \bar{Z}; t)$. The derived expression has to be understood as an ordinary density matrix defined by the relevant–system states. Additionally, the dependence on the reservoir coordinates and momenta as explained in Section 3.4.4 allows for the classical description of the reservoir dynamics. Note that this description is different from the one taken in the foregoing Sections 3.11 and 3.11.2, because $\rho_{ab}(Z, \bar{Z}; t)$ is a continuous function of the reservoir coordinates and momenta. Forming a density matrix from the wave function expansion Eq. (3.416), there would be no continuous dependence on the set of coordinates $Z$ and $\bar{Z}$; instead all quantities become well defined functions of the time. This property of the present approach does not rely on any assumptions for the classical reservoir dynamics like the use of a mean coupling potential or the surface hopping approach.

In order to determine $\rho_{ab}(Z, \bar{Z}; t)$ we have to introduce the mixed representation of the whole Liouville–von Neumann equation (3.98). However, in contrast to Eq. (3.98) the total Hamiltonian has to be separated according to Eq. (3.3). The matrix elements with respect to the reservoir coordinates are obtained as

$$\langle Z|H|\bar{Z}\rangle = \Big\{H_{\rm S} + H_{S-R}(Z) + H_{\rm R}(Z)\Big\}\delta(Z - \bar{Z}) . \tag{3.422}$$

The equation contains $H_{\rm S}$ and $H_{\rm S-R}(Z)$ which are both operators in the state space of the relevant system. The additional dependence of $H_{\rm S-R}(Z)$ on the reservoir coordinates $Z$ accounts for the system–reservoir coupling. The coordinate representation $H_{\rm R}(Z)$ of the reservoir Hamiltonian acts via the kinetic energy part $T_{\rm R}$ as a differential operator on the $Z$–dependence of the $\delta$–function. The potential energy of the reservoir is denoted by $U_{\rm R}(Z)$.

According to Eq. (3.422) the density operator, Eq. (3.421) obeys the following equation of motion (which is still exact)

$$\begin{aligned}\frac{\partial}{\partial t}\hat{\rho}(Z, \bar{Z}; t) \quad &= \quad -\frac{i}{\hbar}\left(H_{\rm S} + H_{\rm S-R}(Z) + H_{\rm R}(Z)\right)\hat{\rho}(Z, \bar{Z}; t) \\ &\quad -\hat{\rho}(Z, \bar{Z}; t)\left(H_{\rm S} + H_{\rm S-R}(Z) + H_{\rm R}(\bar{Z})\right. .\end{aligned} \tag{3.423}$$

In order to derive the partial Wigner representation of this equation of motion we follow the procedure given in Section 3.4.4. The result is written in the following form

$$\frac{\partial}{\partial t}\hat{\rho}(X,P;t) = \left(\frac{\partial\hat{\rho}(X,P;t)}{\partial t}\right)_{\text{quantum}} + \left(\frac{\partial\hat{\rho}(X,P;t)}{\partial t}\right)_{\text{qc-coupling}} + \left(\frac{\partial\hat{\rho}(X,P;t)}{\partial t}\right)_{\text{classical}} . \tag{3.424}$$

The first term of the right–hand side describes the quantum part of the dynamics determined by $H_S$ and $H_{S-R}$

$$\left(\frac{\partial\hat{\rho}(X,P;t)}{\partial t}\right)_{\text{quantum}} = -\frac{i}{\hbar}\left[H_S + H_{S-R}(X), \hat{\rho}(X,P;t)\right]_- . \tag{3.425}$$

This is a standard commutator expression we are already familiar with. But the coupling to the reservoir enters via a dependence on the classically described reservoir coordinates $X$. However, $H_{S-R}$ is also responsible for the coupling between the quantum and the classical dynamics (qc-coupling, second term on the right–hand side of Eq. (3.424))

$$\left(\frac{\partial\hat{\rho}(X,P;t)}{\partial t}\right)_{\text{qc-coupling}} = \frac{1}{2}\sum_{\xi}\Big[\frac{\partial}{\partial X_\xi}H_{S-R}(X)\frac{\partial}{\partial P_\xi}\hat{\rho}(X,P;t) + \frac{\partial}{\partial P_\xi}\hat{\rho}(X,P;t)\frac{\partial}{\partial X_\xi}H_{S-R}(X)\Big] . \tag{3.426}$$

This expression would become more involved in the case of a momentum dependence of $H_{S-R}$. Finally, the classical part of Eq. (3.424) reads as

$$\left(\frac{\partial\hat{\rho}(X,P;t)}{\partial t}\right)_{\text{classical}} = \sum_{\xi}\Big[\frac{\partial U_R}{\partial X_\xi}\frac{\partial}{\partial P_\xi}\hat{\rho}(X,P;t) - \frac{\partial T_R}{\partial P_\xi}\frac{\partial}{\partial X_\xi}\hat{\rho}(X,P;t)\Big] , \tag{3.427}$$

what is identical to Eq. (3.115).

In a next step we will expand $\hat{\rho}(X,P;t)$ and Eq. (3.423) in the basis set used in the foregoing section (cf. Eq. (3.415)). The density matrix follows as $\langle\phi_a(X)|\hat{\rho}(X,P;t)|\phi_b(X)\rangle = \rho_{ab}(X,P;t)$ (note that expansion functions have been taken which depend on the same set of reservoir coordinates) and the different parts of the equation of motion (3.424) read

$$\left(\frac{\partial\rho_{ab}(X,P;t)}{\partial t}\right)_{\text{quantum}} = -i\omega_{ab}(X)\rho_{ab}(X,P;t) , \tag{3.428}$$

and

$$\left(\frac{\partial\rho_{ab}(X,P;t)}{\partial t}\right)_{\text{qc-coupling}} = \frac{1}{2}\sum_c\sum_\xi\Big[\langle\phi_a(X)|\frac{\partial H_{S-R}(X)}{\partial X_\xi}|\phi_c(X)\rangle\frac{\partial}{\partial P_\xi}\rho_{cb}(X,P;t) + \frac{\partial}{\partial P_\xi}\rho_{ac}(X,P;t)\langle\phi_c(X)|\frac{\partial H_{S-R}(X)}{\partial X_\xi}|\phi_b(X)\rangle\Big] , \tag{3.429}$$

as well as

$$\left(\frac{\partial \rho_{ab}(X,P;t)}{\partial t}\right)_{\text{classical}} = \sum_{\xi}\left[\frac{\partial U_{\mathrm{R}}}{\partial X_{\xi}}\frac{\partial}{\partial P_{\xi}}\rho_{ab}(X,P;t) - \frac{\partial T_{\mathrm{R}}}{\partial P_{\xi}}\langle\phi_a(X)|\frac{\partial}{\partial X_{\xi}}\hat{\rho}(X,P;t)|\phi_b(X)\rangle\right]. \quad (3.430)$$

Here, we introduced the coordinate dependent transition frequencies $\omega_{ab}(X)$ between the energy levels $E_a(X)$ and $E_b(X)$, which completely determine the quantum part of the total equation of motion. The coupling term between quantum and classical degrees of freedom includes derivatives of $H_{\mathrm{S-R}}$ with respect to the reservoir coordinates. When comparing the obtained equation of motion (3.424) with that derived in the foregoing Section 3.11, we note the absence of such derivatives in the time–dependent Schrödinger equation. The systematic construction of mixed quantum–classical dynamics by means of the partial Wigner representation introduces such derivatives leading to a nonlocal system–reservoir coupling. If necessary the matrix elements of coordinate derivatives of operators can be rewritten as coordinate derivatives of matrix elements (using standard rules of partial differentiation).

The practical application of the given quantum–classical hybrid description is connected with a number of difficulties (cf. discussion in [Kap99]). First of all, one has to solve partial differential equations for a potentially large number of coordinates and momenta. In practice this is impossible and can be circumvented in a way already used in Sections 3.11. To this end one replaces all $X$ and $P$ by time–dependent functions determined by the classical canonical equations (cf. Eq. (3.408)). But there are further problems related, for example, to the conservation of energy (for a time–independent total Hamiltonian) and the positivity of the density matrix. Despite of that, the present approach can be used, for instance, to justify the surface hopping method discussed in Section 3.11.2. The latter turns out to be based on neglecting certain coherences described by the off–diagonal density matrix elements (for more details see also [Kap99]).

The partial Wigner representation method will be used in the Sections 5.3.5 and 6.7.

# 3.12 Supplement

## 3.12.1 Different Equations of Motion for the Reduced Density Operator

### The Nakajima–Zwanzig Equation

In Section 3.5.6 we derived the Quantum Master Equation as a closed equation of motion for the reduced density operator by restricting ourselves to the second order in the coupling, $H_{\mathrm{S-R}}$, between the active system and the reservoir. In the following we will show that, in principle, one can sum up the complete perturbation series. To this end we start again with Eqs. (3.173) and (3.174). To have a more compact notation the commutator with the interaction Hamiltonian $H^{(\mathrm{I})}_{\mathrm{S-R}}$ is replaced by the interaction Liouville superoperator $\mathcal{L}^{(\mathrm{I})}_{\mathrm{S-R}}$ defined as

$$\mathcal{L}^{(\mathrm{I})}_{\mathrm{S-R}}\bullet = \frac{1}{\hbar}[H^{(\mathrm{I})}_{\mathrm{S-R}},\bullet]_{-}\,. \quad (3.431)$$

We introduce this notation into Eq. (3.174) and obtain after integration

$$\mathcal{Q}\hat{W}^{(\mathrm{I})}(t) = \mathcal{S}_{\mathcal{Q}}(t,t_0)\mathcal{Q}\hat{W}^{(\mathrm{I})}(t_0) - i\int_{t_0}^{t} d\bar{t}\; \mathcal{S}_{\mathcal{Q}}(t,\bar{t})\mathcal{Q}\mathcal{L}^{(\mathrm{I})}_{\mathrm{S-R}}(\bar{t})\hat{R}_{\mathrm{eq}}\hat{\rho}^{(\mathrm{I})}(\bar{t})\;. \tag{3.432}$$

Here, the time–ordered superoperator

$$\mathcal{S}_{\mathcal{Q}}(t,\bar{t}) = \mathcal{T}\exp\left\{-i\int_{\bar{t}}^{t} d\tau\; \mathcal{Q}\mathcal{L}^{(\mathrm{I})}_{\mathrm{S-R}}(\tau)\right\} \tag{3.433}$$

has been introduced. As all other types of $S$–operators Eq. (3.433) is defined via the Taylor expansion of the exponential function.

Since we are not interested in the problem of initial correlations the first term on the right–hand side of Eq. (3.432) will be neglected by assuming that at time $t_0$ the density operator of the total system factorizes into the density operator of the relevant system and the reservoir, $W(t_0) = \hat{\rho}(t_0)\hat{R}_{\mathrm{eq}}$.

Inserting Eq. (3.432) into Eq. (3.173) we get an equation of motion which allows an *exact* determination of the reduced statistical operator of the relevant system. This so–called *Nakajima–Zwanzig equation* reads

$$\begin{aligned}\frac{\partial}{\partial t}\hat{\rho}^{(\mathrm{I})}(t) &= -i\,\mathrm{tr}_{\mathrm{R}}\left\{\mathcal{L}^{(\mathrm{I})}_{\mathrm{S-R}}(t)\hat{R}_{\mathrm{eq}}\right\}\hat{\rho}^{(\mathrm{I})}(t)\\ &\quad -\int_{t_0}^{t} d\bar{t}\;\mathrm{tr}_{\mathrm{R}}\left\{\mathcal{L}^{(\mathrm{I})}_{\mathrm{S-R}}(t)\mathcal{S}_{\mathcal{Q}}(t,\bar{t})\mathcal{Q}\mathcal{L}^{(\mathrm{I})}_{\mathrm{S-R}}(\bar{t})\hat{R}_{\mathrm{eq}}\right\}\hat{\rho}^{(\mathrm{I})}(\bar{t})\;.\end{aligned} \tag{3.434}$$

Since according to Eq. (3.433), $\mathcal{L}_{\mathrm{S-R}}$ is contained in the time–ordered exponential operator, the system–reservoir interaction enters the right–hand side of Eq. (3.434) in *infinite* order. In other words, the whole perturbation series with respect to $H_{\mathrm{S-R}}$ is summed up in Eq. (3.434). A more compact notation of the Nakajima–Zwanzig equation is given by

$$\frac{\partial}{\partial t}\hat{\rho}^{(\mathrm{I})}(t) = -i\,\langle\mathcal{L}^{(\mathrm{I})}_{\mathrm{S-R}}(t)\rangle_{\mathrm{R}}\,\hat{\rho}^{(\mathrm{I})}(t) - \int_{t_0}^{t} d\bar{t}\;\mathcal{M}^{(\mathrm{I})}(t,\bar{t})\hat{\rho}^{(\mathrm{I})}(\bar{t})\;. \tag{3.435}$$

Here we used the short–hand notation for the averaging with respect to the reservoir equilibrium density operator: $\langle ...\rangle_{\mathrm{R}} = \mathrm{tr}_{\mathrm{R}}\{...\hat{R}_{\mathrm{eq}}\}$. The first term in (3.435) can easily be identified as as the mean–field contribution (cf. Section 3.5.3). In the second term of Eq. (3.435) we have introduced the memory kernel superoperator (in the interaction representation)

$$\mathcal{M}^{(\mathrm{I})}(t,\bar{t}) = \langle\mathcal{L}^{(\mathrm{I})}_{\mathrm{S-R}}(t)\mathcal{S}_{\mathcal{Q}}(t,\bar{t})\mathcal{Q}\mathcal{L}^{(\mathrm{I})}_{\mathrm{S-R}}(\bar{t})\rangle_{\mathrm{R}}\;. \tag{3.436}$$

According to the definition of the kernel we have $t > \tau$, and, additionally, any expansion of the $S$–superoperator guarantees time–ordered expressions. Therefore, the approach leading to the Nakajima–Zwanzig equation is often named *chronological time ordering prescription* (COP). In most practical cases it is impossible to derive a closed expression for this memory kernel even if we restrict ourselves to a partial summation only. However, the formal exact Nakajima–Zwanzig equation is well suited for the development of approximation schemes to the propagation for the reduced density operator. For example, the QME (3.176) is easily recovered if the $\mathcal{S}_Q$–operator is treated in zeroth–order with respect to the system–reservoir coupling.

Finally, it is important to note here that the special choice Eq. (3.167) for the projection operator does not imply that the reservoir stays in equilibrium in the course of the time evolution of the relevant system. Instead, a time dependence of the reduced density operator of the reservoir can be expected. This nonequilibrium behavior is induced by the coupling to the relevant system. Let us define the reduced density operator for the *reservoir* as

$$\hat{R}^{(\mathrm{I})}(t) = \mathrm{tr}_{\mathrm{S}}\{\hat{W}^{(\mathrm{I})}(t)\} \equiv \mathrm{tr}_{\mathrm{S}}\{\mathcal{P}\hat{W}^{(\mathrm{I})}(t) + \mathcal{Q}\hat{W}^{(\mathrm{I})}(t)\} \, . \tag{3.437}$$

Using the definition of the projection operator $\mathcal{P}$ and Eq. (3.432) gives

$$\hat{R}^{(\mathrm{I})}(t) = \hat{R}_{\mathrm{eq}} - i \int_{t_0}^{t} d\bar{t} \; \mathrm{tr}_{\mathrm{S}} \Big\{ \mathcal{S}_{\mathcal{Q}}(t,\bar{t}) \mathcal{Q} \mathcal{L}^{(\mathrm{I})}_{\mathrm{S-R}}(\bar{t}) \hat{R}_{\mathrm{eq}} \hat{\rho}^{(\mathrm{I})}(\bar{t}) \Big\} \, . \tag{3.438}$$

Since it is only the equilibrium density operator $\hat{R}_{\mathrm{eq}}$ which enters the Nakajima–Zwanzig equation (3.434), the time dependence of $\hat{R}^{(\mathrm{I})}(t)$ does not affect $\hat{\rho}^{(\mathrm{I})}(t)$ directly. It is only indirectly accounted for via $\mathcal{L}_{\mathrm{S-R}}$ which is contained in the $S$–superoperator Eq. (3.433).

### Convolution–Less Equations of Motion

So far we have discussed equations of motion for the RDO which are originally of the non–Markovian type, that is, earlier dynamics of the RDO influence the actual time–behavior. In the following it shall be demonstrated that such a type of RDO equation is not necessarily obtained and that equations of motion which do not refer to the RDO at earlier times can be derived. Since a time–convolution between the memory kernel superoperator and the RDO is absent, such types of equations are often termed *convolution–less* equations of motions.

Let us start from the Liouville–von Neumann equation (3.159), but now written with the interaction Liouville superoperator $\mathcal{L}^{(\mathrm{I})}_{\mathrm{S-R}}$, Eq. (3.431). A formal integration yields

$$\hat{W}^{(I)}(t) = \mathcal{S}(t,t_0)\hat{W}^{(I)}(t_0) \, , \tag{3.439}$$

with the the $S$–superoperator defined similar to Eq. (3.433) except that $\mathcal{Q}\mathcal{L}^{(\mathrm{I})}_{\mathrm{S-R}}$ has to be replaced by $\mathcal{L}^{(\mathrm{I})}_{\mathrm{S-R}}$. The time–propagation of the density operator can also be written as a transformation from the actual time $t$ back to the initial time $t_0$:

$$\hat{W}^{(I)}(t_0) = \mathcal{S}^{+}(t,t_0)\Big(\mathcal{P}\hat{W}^{(I)}(t) + \mathcal{Q}\hat{W}^{(I)}(t)\Big) \tag{3.440}$$

Note that we introduced the projection superoperator $\mathcal{P}$ and its orthogonal complement $\mathcal{Q}$ on the right–hand side. In a next step we use this relation to rewrite Eq. (3.432) which determines

$\mathcal{Q}\hat{W}^{(I)}(t)$. As in the foregoing section we assume the absence of initial correlations and introduce the time–evolution superoperator

$$\mathcal{A}(t,t_0) = \int_{t_0}^{t} d\bar{t}\; \mathcal{S}_{\mathcal{Q}}(t,\bar{t})\mathcal{Q}\mathcal{L}^{(I)}_{\text{S-R}}(\bar{t})\mathcal{P}\mathcal{S}^{+}(t,\bar{t})\;. \tag{3.441}$$

Then, Eq. (3.432) can be cast into the following form:

$$\Big\{1 + i\mathcal{A}(t,t_0)\Big\}\mathcal{Q}\hat{W}^{(I)}(t) = -i\mathcal{A}(t,t_0)\mathcal{P}\hat{W}^{(I)}(t)\;. \tag{3.442}$$

Next we assume the existence of the inverse of $(1{+}i\mathcal{A})$ and derive an expression for $\mathcal{Q}\hat{W}^{(I)}(t)$. Afterwards we insert the result into Eq. (3.173) and obtain a closed equation for the RDO. In particular, the time integral with respect to the memory kernel superoperator appearing in Eq. (3.435) is replaced by the time–local expression

$$-\mathcal{D}^{(I)}(t)\hat{\rho}^{(I)}(t) = -\langle\mathcal{L}^{(I)}_{S-R}(t)[1 + i\mathcal{A}(t,t_0)]^{-1}\mathcal{A}(t,t_0)\rangle_{\text{R}}\hat{\rho}^{(I)}(t)\;. \tag{3.443}$$

The time–local character of this *time–dependent* dissipative superoperator has been obtained by introducing a complete propagation backward in time into Eq. (3.441). Because $\mathcal{A}$ contains two different types of $S$–superoperators ($\mathcal{S}$ and $\mathcal{S}_{\mathcal{Q}}$) and determines $\mathcal{D}^{(I)}$ via an infinite power expansion a complete time–ordering is not achieved. This property is different as compared to the case of the memory kernel superoperator Eq. (3.436). Therefore, the approach is called *partial time ordering prescription* (POP).

Of course, it is impossible to further simplify the exact expression of the dissipative superoperator. But we can carry out a second–order approximation of $\mathcal{D}^{(I)}(t)$ with respect to the system–reservoir coupling. This procedure is similar to that leading to the QME in Section 3.5.6. In doing so the inverse of $(1 + i\mathcal{A})$ has to be replaced by the unity operator and the operator $\mathcal{A}$, Eq. (3.441) has to be simplified by removing $\mathcal{S}_{\mathcal{Q}}$ and $\mathcal{P}\mathcal{S}^{+}(t,\bar{t})$. We obtain the second–order approximation

$$-\mathcal{D}^{(I)}(t) \approx -\int_{t_0}^{t} d\bar{t}\; \langle\mathcal{L}^{(I)}_{\text{S-R}}(t)\mathcal{Q}\mathcal{L}^{(I)}_{\text{S-R}}(\bar{t})\rangle_{\text{R}}\;. \tag{3.444}$$

It is identical to the second–order approximation of the memory kernel superoperator, Eq. (3.436) but results in an equation for the RDO which is obtained from Eq. (3.435) when replacing $\hat{\rho}^{(I)}(\tau)$ by $\hat{\rho}^{(I)}(t)$ (Markov approximation).

The derivation demonstrates that there does not exist a unique equation of motion for the RDO, either in an exact form nor in an approximate version. Which of the types given so far is most appropriate can only be decided when comparing with an exact solution of the dissipative dynamics, for example, by a path integral formulation (cf. Section 3.5.6).

At the end of this section we will present some general properties of the dissipative superoperator introduced in Eq. (3.443). We change to the Schrödinger representation and note that the action of $\mathcal{D}(t)$ on the density operator (as well as any other type, for example, that of Eq. (3.5.6)) can be written in the following general form

$$-\mathcal{D}(t)\hat{\rho}(t) = \hat{A}(t)\hat{\rho}(t) + \hat{\rho}(t)\hat{B}(t) + \sum_j \hat{C}_j(t)\hat{\rho}(t)\hat{D}_j(t)\;. \tag{3.445}$$

The action of $\mathcal{D}$ has been expressed by a set of ordinary time–dependent operators, the single operator $\hat{A}(t)$ acting from the left, the operator $\hat{B}(t)$ acting from the right, and the operators $\hat{C}_j$ and $\hat{D}_j$ acting from the left and the right, respectively. However, all these operators are not independent one from another. We first use the Hermiticity of $\hat{\rho}(t)$ to get the identity $\mathcal{D}\hat{\rho} = (\mathcal{D}\hat{\rho})^+$. Provided that all terms in Eq. (3.445) are linear independent we may deduce $\hat{A} = \hat{B}^+$ and $\hat{C}_j = \hat{D}_j^+$. Next we apply the basic property of the RDO equation to conserve probability, i.e. $\mathrm{tr}_S\{\partial\hat{\rho}/\partial t\} = 0$. It follows $\mathrm{tr}\{\mathcal{D}\hat{\rho}\} = 0$ what leads to (after using the cyclic invariance of the trace)

$$\mathrm{tr}\{(\hat{A}(t) + \hat{A}^+(t))\hat{\rho}(t) + \sum_j \hat{C}_j^+(t)\hat{C}_j(t)\hat{\rho}(t)\} = 0 \,. \qquad (3.446)$$

Since this has to be fulfilled for every set of basis functions used for the trace operation we get

$$\hat{A}(t) + \hat{A}^+(t) = -\sum_j \hat{C}_j^+(t)\hat{C}_j(t) \,. \qquad (3.447)$$

This equation may be solved by

$$\hat{A}(t) = -\frac{1}{2}\sum_j \hat{C}_j^+(t)\hat{C}_j(t) + i\hat{h}(t) \,, \qquad (3.448)$$

where $\hat{h}$ is a Hermitian operator. As a result, the so–called *Lindblad form* (cf. Section 3.12.3) is derived, but generalized here to time–dependent operators

$$\begin{aligned} -\mathcal{D}(t)\hat{\rho}(t) &= -\frac{1}{2}\sum_j \left[\hat{C}_j^+(t)\hat{C}_j(t), \hat{\rho}(t)\right]_+ - i\left[\hat{h}(t), \hat{\rho}(t)\right]_- \\ &\quad + \sum_j \hat{C}_j(t)\hat{\rho}(t)\hat{C}_j^+(t) \,. \end{aligned} \qquad (3.449)$$

There exists an alternative solution of $\mathcal{D}\hat{\rho} = (\mathcal{D}\hat{\rho})^+$. This becomes obvious if one groups together pairs of terms $\hat{C}_j(t)\hat{\rho}(t)\hat{D}_j(t)$ and $\hat{C}_{j'}(t)\hat{\rho}(t)\hat{D}_{j'}(t)$ such that one gets a Hermitian expression like $\hat{C}_j(t)\hat{\rho}(t)\hat{D}_j(t) + \hat{D}_j^+(t)\hat{\rho}(t)\hat{C}_j^+(t)$. Here, we identified $\hat{D}_j(t) = \hat{C}_{j'}^+(t)$ and $\hat{C}_j(t) = \hat{D}_{j'}^+(t)$. Then, instead of Eq. (3.447), $\mathrm{tr}\{\mathcal{D}\hat{\rho}\} = 0$ results in

$$A(t) + A^+(t) = -\sum_j \left(\hat{D}_j(t)\hat{C}_j(t) + \hat{C}_j^+(t)\hat{D}_j^+(t)\right) \qquad (3.450)$$

with a possible solution

$$A(t) = -\sum_j \hat{D}_j(t)\hat{C}_j(t) \qquad (3.451)$$

The dissipative part of the density matrix equation reads

$$\begin{aligned} -\mathcal{D}(t)\hat{\rho}(t) &= -\frac{1}{2}\sum_j \Big(\hat{D}_j(t)\hat{C}_j(t)\hat{\rho}(t) + \hat{\rho}(t)\hat{C}_j^+(t)\hat{D}_j^+(t) \\ &\quad + \hat{D}_j(t)\hat{\rho}(t)\hat{C}_j(t) + \hat{C}_j^+(t)\hat{\rho}(t)\hat{D}_j^+(t)\Big)_- \,. \end{aligned} \qquad (3.452)$$

If we neglect the time dependence of all operators this is just the structure of the Markovian QME derived in Section 3.7.1 and we may identify the various operators $K_u$ and $\Lambda_u^+$ with $\hat{D}_j/\sqrt{2}$ and $\hat{C}_j/\sqrt{2}$.

### 3.12.2 Correlation Function for Nonlinear Couplings to the Reservoir

In the first part of this section we will derive the formula for the generating function $\mathcal{F}$ of the reservoir correlation function as given in Eq. (3.218). Afterwards, this generating function is used to determine different types of reservoir coordinate correlation functions $C_{\xi,\bar{\xi}}^{(m,n)}(T)$. In order to calculate a compact expression for $\mathcal{F}$ we note that it can be written as a product with respect to the various coordinates $\mathcal{F} = \prod_\xi \mathcal{F}_\xi$. The expression for $\mathcal{F}_\xi$ looks similar to that of Eq. (3.216) but is restricted to a single coordinate. To simplify the notation the mode index $\xi$ will be neglected as long as the single mode part of $\mathcal{F}$ is considered. Furthermore, we replace the trace by a summation with respect to the oscillator eigenstates $|N\rangle$. This yields (the prefactor is the single–mode partition function)

$$\mathcal{F}_\xi = (1 - e^{-\hbar\omega_{\rm vib}/k_{\rm B}T}) \sum_N \langle N|\, e^{-\hbar\omega_{\rm vib}N/k_{\rm B}T}\, e^{\sigma Q(t)}\, e^{\vartheta Q}|N\rangle \,. \tag{3.453}$$

Before computing the sum with respect to the vibrational quantum number $N$ we rearrange the oscillator annihilation and creation operators $C$ and $C^+$, respectively. This will be done in such a manner that the annihilation operators are positioned to the right from the creation operators. This arrangement is known as the *normal ordering* of boson operators. For the matrix elements of Eq. (3.453) this is achieved using the following identity valid for boson operators (cf. Eq. (2.95))

$$e^{aC\,+\,bC^+} = e^{-ab/2}\, e^{aC}\, e^{bC^+} = e^{ab/2}\, e^{bC^+}\, e^{aC} \,. \tag{3.454}$$

For a further treatment the time–dependence of the coordinate operator in Eq. (3.453) is absorbed in $\sigma(t) = \sigma\exp(-i\omega_{\rm vib}t)$ and we obtain $\sigma Q(t) = \sigma(t)C + \sigma^*(t)C^+$. Consequently, we may write for the oscillator state matrix element in Eq. (3.453)

$$\langle N|e^{\sigma Q(t)}\, e^{\vartheta Q}|N\rangle = \langle N|e^{\sigma(t)C+\sigma^*(t)C^+}\, e^{\vartheta C+\vartheta C^+}|N\rangle \,. \tag{3.455}$$

This expression is rewritten by using Eq. (3.454) twice

$$\begin{aligned}\langle N|e^{\sigma Q(t)}\, e^{\vartheta Q}|N\rangle &= \langle N|e^{|\sigma(t)|^2/2}\, e^{\sigma^*(t)C^+}\, e^{\sigma(t)C}\, e^{\vartheta^2/2}\, e^{\vartheta C^+}\, e^{\vartheta C}|N\rangle \\ &= e^{\frac{1}{2}(|\sigma(t)|^2\,+\,\vartheta^2)}\,\langle N|e^{\sigma^*(t)C^+} e^{\sigma(t)\vartheta} e^{\vartheta C^+} e^{\sigma(t)C} e^{\vartheta C}\,|N\rangle \\ &= e^{\frac{1}{2}(|\sigma(t)|^2\,+\,\vartheta^2+2\sigma(t)\vartheta)}\,\langle N|e^{(\sigma^*(t)+\vartheta)C^+}\, e^{-(\sigma(t)+\vartheta)C}\,|N\rangle \,.\end{aligned} \tag{3.456}$$

Next we introduce the abbreviation

$$\tilde{\sigma}(t) = \sigma(t) + \vartheta = \sigma e^{-\omega_{\rm vib}t} + \vartheta \,, \tag{3.457}$$

and take into account that

$$|\sigma(t)|^2 + \vartheta^2 + 2\sigma(t)\,\vartheta = |\tilde{\sigma}(t)|^2 + (\sigma(t) - \sigma^*(t))\vartheta \,. \tag{3.458}$$

Then we obtain the normal ordering of the original matrix elements in the trace formula as

$$\begin{aligned}\langle N|e^{\sigma Q(t)}\, e^{\vartheta Q}|N\rangle &= \exp\frac{1}{2}\left\{|\tilde{\sigma}|^2 + [\sigma(t) - \sigma^*(t)]\vartheta\right\} \\ &\times \langle N|e^{\tilde{\sigma}^* C^+}\, e^{-\tilde{\sigma} C}|N\rangle\,. \end{aligned} \tag{3.459}$$

To determine the oscillator matrix elements we use Eqs. (2.97)–(2.99) and obtain for Eq. (3.453)

$$\begin{aligned}\mathcal{F}_\xi &= \left(1 - e^{-\hbar\omega_{\rm vib}/k_{\rm B}T}\right) e^{(|\tilde{\sigma}(t)|^2 + [\sigma(t) - \sigma^*(t)]\vartheta)/2} \\ &\times \sum_{N=0}^{\infty} e^{-\hbar\omega_{\rm vib}N/k_{\rm B}T}\, L_N(|\tilde{\sigma}(t)|^2)\,. \end{aligned} \tag{3.460}$$

Here, we introduced the Laguerre polynomials $L_N$ of order $N$ (cf. Eq. (2.99)). The relation between the Laguerre polynomials and their generating function is given by

$$\sum_{N=0}^{\infty} \lambda^N\, L_N(|\tilde{\sigma}(t)|^2) = \frac{1}{1-\lambda}\, e^{-\lambda|\tilde{\sigma}(t)|^2/(1-\lambda)} \qquad (|\lambda| < 1)\,. \tag{3.461}$$

We identify $\lambda = \exp\{-\hbar\omega_{\rm vib}/k_{\rm B}T\}$ and obtain

$$\mathcal{F}_\xi = \exp\left\{\frac{1}{2}\left(|\tilde{\sigma}(t)|^2 + [\sigma(t) - \sigma^*(t)]\vartheta\right) + \frac{e^{-\hbar\omega_{\rm vib}/k_{\rm B}T}}{1 - e^{-\hbar\omega_{\rm vib}/k_{\rm B}T}}|\tilde{\sigma}(t)|^2\right\}\,. \tag{3.462}$$

The last term of the exponent can be rewritten if we introduce the Bose–Einstein distribution $n(\omega_{\rm vib})$ (cf. Section 3.6) which determines the mean number of vibrational quanta excited in the mode with frequency $\omega_{\rm vib}$ at temperature $T$. The final result for $\mathcal{F}_\xi$ will be obtained if the exponent denoted by $\mathcal{G}_\xi$ is rearranged according to

$$\begin{aligned}\mathcal{G}_\xi &= \frac{1}{2}\Big\{\big(1 + 2n(\omega_{\rm vib})\big)\left(\sigma^2 + \vartheta^2 + [\sigma(t) + \sigma^*(t)]\vartheta\right) + [\sigma(t) - \sigma^*(t)]\vartheta\Big\} \\ &= \frac{1}{2}\big(1 + 2n(\omega_{\rm vib})\big)(\sigma^2 + \vartheta^2) + \big((1 + n(\omega_{\rm vib}))e^{-i\omega_{\rm vib}t} + n(\omega_{\rm vib})e^{i\omega_{\rm vib}t}\big)\sigma\vartheta\,. \end{aligned} \tag{3.463}$$

This expression can be easily used to reproduce the multimode generating function, Eq. (3.218). For further use will take here the more compact notation

$$\mathcal{F}(t) = e^{\mathcal{G}(t)}\,, \tag{3.464}$$

with

$$\mathcal{G}(t) \equiv \sum_\xi \mathcal{G}_\xi = \sum_\xi \left[A_\xi(\sigma_\xi^2 + \vartheta_\xi^2) + B_\xi(t)\sigma_\xi\vartheta_\xi\right]\,, \tag{3.465}$$

(for $A_\xi$ and $B_\xi$ and see Eq. (3.219)). Taking into account Eq. (3.217), the various types of reservoir coordinate correlation functions can be obtained from the derivatives of the generating function. It is obvious that multiple derivatives of $\mathcal{F}$ result in derivatives of $\mathcal{G}$. To simplify the notation the derivative in Eq. (3.217) will be abbreviated by $\mathcal{F}^{(\xi_1,\ldots,\xi_M|\bar{\xi}_1,\ldots,\bar{\xi}_N)}$ The indices in the left part correspond to derivatives with respect to $\sigma$ and those in the right part to derivatives with respect to $\vartheta$.

Being interested in correlation functions up to the third order we expect derivatives of the type $\mathcal{F}^{(\xi_1,\xi_2,\xi_3|\bar{\xi}_1,\bar{\xi}_2,\bar{\xi}_3)}$ which in the highest order may lead to $\mathcal{G}^{(\xi_1,\xi_2,\xi_3|\bar{\xi}_1,\bar{\xi}_2,\bar{\xi}_3)}$. Therefore we start to calculate the various derivatives of $\mathcal{G}$. According to Eq. (3.465) we obtain for the first derivatives

$$\mathcal{G}^{(\xi|)} = 2A_\xi \sigma_\xi + B_\xi(t)\vartheta_\xi \ , \tag{3.466}$$

and

$$\mathcal{G}^{(|\bar{\xi})} = 2A_{\bar{\xi}} \vartheta_{\bar{\xi}} + B_{\bar{\xi}}(t)\sigma_{\bar{\xi}} \ . \tag{3.467}$$

Both expressions vanish in the limit $\sigma, \vartheta \to 0$. The second derivatives follow as

$$\mathcal{G}^{(\xi_1,\xi_2|)} = \delta_{\xi_1,\xi_2} 2A_{\xi_1} \ , \tag{3.468}$$

and

$$\mathcal{G}^{(|\bar{\xi}_1,\bar{\xi}_2)} = \delta_{\bar{\xi}_1,\bar{\xi}_2} 2A_{\bar{\xi}_1} \ , \tag{3.469}$$

as well as

$$\mathcal{G}^{(\xi|\bar{\xi})} = \delta_{\xi,\bar{\xi}} B_\xi(t) \ . \tag{3.470}$$

Since there is no further dependence on $\sigma$ and $\vartheta$ any higher derivative vanishes. However, all the expression Eqs. (3.468), (3.469), and (3.470) exist in the limit $\sigma, \vartheta \to 0$. Next we compute the derivatives of $\mathcal{F}$. We obtain for those with respect to $\vartheta$

$$\mathcal{F}^{(|\bar{\xi})} = \mathcal{G}^{(|\bar{\xi})} \mathcal{F} \ , \tag{3.471}$$

and

$$\mathcal{F}^{(|\bar{\xi}_1,\bar{\xi}_2)} = \left\{ \mathcal{G}^{(|\bar{\xi}_1,\bar{\xi}_2)} + \mathcal{G}^{(|\bar{\xi}_1)} \mathcal{G}^{(|\bar{\xi}_2)} \right\} \mathcal{F} \ , \tag{3.472}$$

and finally

$$\begin{aligned} \mathcal{F}^{(|\bar{\xi}_1,\bar{\xi}_2,\bar{\xi}_3)} &= \Big\{ \mathcal{G}^{(|\bar{\xi}_1,\bar{\xi}_3)} \mathcal{G}^{(|\bar{\xi}_2)} + \mathcal{G}^{(|\bar{\xi}_1)} \mathcal{G}^{(|\bar{\xi}_2,\bar{\xi}_3)} + \mathcal{G}^{(|\bar{\xi}_1,\bar{\xi}_2)} \mathcal{G}^{(|\bar{\xi}_3)} \\ &\quad + \mathcal{G}^{(|\bar{\xi}_1)} \mathcal{G}^{(|\bar{\xi}_2)} \mathcal{G}^{(|\bar{\xi}_3)} \Big\} \mathcal{F} \ , \end{aligned} \tag{3.473}$$

i.e. only first and second–order derivatives of $\mathcal{G}$ enter. Now we consider additional derivatives with respect to $\sigma$. We have

$$\mathcal{F}^{(\xi|\bar{\xi})} = \left\{ \mathcal{G}^{(\xi|\bar{\xi})} + \mathcal{G}^{(\xi|)} \mathcal{G}^{(|\bar{\xi})} \right\} \mathcal{F} \ , \tag{3.474}$$

and obtain

$$\mathcal{F}^{(\xi|\bar{\xi})}_{|\sigma,\vartheta=0} \equiv \mathrm{tr}_{\mathrm{vib}}\{\hat{R}_{\mathrm{eq}} Q_\xi(t) Q_{\bar{\xi}}\} = \delta_{\xi,\bar{\xi}} B_\xi(t) \ . \tag{3.475}$$

To get higher–order correlation functions we start to calculate

$$\mathcal{F}^{(\xi|\bar{\xi}_1,\bar{\xi}_2)} = \Big\{\mathcal{G}^{(\xi|\bar{\xi}_1)}\mathcal{G}^{(|\bar{\xi}_2)} + \mathcal{G}^{(|\bar{\xi}_1)}\mathcal{G}^{(\xi|\bar{\xi}_2)} + [\mathcal{G}^{(|\bar{\xi}_1,\bar{\xi}_2)} + \mathcal{G}^{(|\bar{\xi}_1)}\mathcal{G}^{(|\bar{\xi}_2)}]\mathcal{G}^{(\xi|)}\Big\} \times\mathcal{F}\,, \tag{3.476}$$

and

$$\begin{aligned}\mathcal{F}^{(\xi_1,\xi_2|\bar{\xi}_1,\bar{\xi}_2)} = {} & \Big\{\mathcal{G}^{(\xi_1|\bar{\xi}_1)}\mathcal{G}^{(\xi_2|\bar{\xi}_2)} + \mathcal{G}^{(\xi_2|\bar{\xi}_1)}\mathcal{G}^{(\xi_1|\bar{\xi}_2)} \\ & + \mathcal{G}^{(|\bar{\xi}_1,\bar{\xi}_2)}\mathcal{G}^{(\xi_1,\xi_2|)} + \mathcal{G}^{(\xi_2|\bar{\xi}_1)}\mathcal{G}^{(|\bar{\xi}_2)}\mathcal{G}^{(\xi_1|)} \\ & + \mathcal{G}^{(|\bar{\xi}_1)}\mathcal{G}^{(\xi_2|\bar{\xi}_2)}\mathcal{G}^{(\xi_1|)} + \mathcal{G}^{(|\bar{\xi}_1)}\mathcal{G}^{(|\bar{\xi}_2)}\mathcal{G}^{(\xi_1\xi_2|)}\Big\}\mathcal{F} \\ & + \mathcal{G}^{(\xi_2|)}\mathcal{F}^{(\xi_1|\bar{\xi}_1,\bar{\xi}_2)}\,,\end{aligned} \tag{3.477}$$

From this we may deduce

$$\begin{aligned}\mathcal{F}^{(\xi_1\xi_2|\bar{\xi}_1\bar{\xi}_2)}_{|\sigma,\vartheta=0} & \equiv \langle Q_{\xi_1}(t)Q_{\xi_2}(t)Q_{\bar{\xi}_1}Q_{\bar{\xi}_2}\rangle_{\mathrm{R}} \\ & = [\delta_{\xi_1,\bar{\xi}_1}\delta_{\xi_2,\bar{\xi}_2} + \delta_{\xi_1,\bar{\xi}_2}\delta_{\xi_2,\bar{\xi}_1}]B_{\xi_1}(t)B_{\xi_2}(t) + \delta_{\xi_1,\xi_2}\delta_{\bar{\xi}_1,\bar{\xi}_2}4A_{\xi_1}A_{\bar{\xi}_1}\,.\end{aligned} \tag{3.478}$$

Finally, we determine the correlation function of third order. In a first step we compute (note the use of Eq. (3.473))

$$\begin{aligned}\mathcal{F}^{(\xi_1|\bar{\xi}_1,\bar{\xi}_2,\bar{\xi}_3)} = {} & \Big\{\mathcal{G}^{(|\bar{\xi}_1,\bar{\xi}_3)}\mathcal{G}^{(\xi_1|\bar{\xi}_2)} + \mathcal{G}^{(\xi_1|\bar{\xi}_1)}\mathcal{G}^{(|\bar{\xi}_2,\bar{\xi}_3)} + \mathcal{G}^{(|\bar{\xi}_1,\bar{\xi}_2)}\mathcal{G}^{(\xi_1|\bar{\xi}_3)} \\ & + \mathcal{G}^{(\xi_1|\bar{\xi}_1)}\mathcal{G}^{(|\bar{\xi}_2)}\mathcal{G}^{(|\bar{\xi}_3)} + \mathcal{G}^{(|\bar{\xi}_1)}\mathcal{G}^{(\xi_1|\bar{\xi}_2)}\mathcal{G}^{(|\bar{\xi}_3)} + \mathcal{G}^{(|\bar{\xi}_1)}\mathcal{G}^{(|\bar{\xi}_2)}\mathcal{G}^{(\xi_1|\bar{\xi}_3)}\Big\}\mathcal{F} \\ & + \mathcal{G}^{(\xi_1|)}\mathcal{F}^{(|\bar{\xi}_1,\bar{\xi}_2,\bar{\xi}_3)}\,.\end{aligned} \tag{3.479}$$

It follows

$$\begin{aligned}\mathcal{F}^{(\xi_1,\xi_2|\bar{\xi}_1,\bar{\xi}_2,\bar{\xi}_3)} = {} & \Big\{\mathcal{G}^{(\xi_1|\bar{\xi}_1)}\mathcal{G}^{(\xi_2|\bar{\xi}_2)}\mathcal{G}^{(|\bar{\xi}_3)} + \mathcal{G}^{(\xi_1|\bar{\xi}_1)}\mathcal{G}^{(|\bar{\xi}_2)}\mathcal{G}^{(\xi_2|\bar{\xi}_3)} \\ & + \mathcal{G}^{(\xi_2|\bar{\xi}_1)}\mathcal{G}^{(\xi_1|\bar{\xi}_2)}\mathcal{G}^{(|\bar{\xi}_3)} + \mathcal{G}^{(|\bar{\xi}_1)}\mathcal{G}^{(\xi_1|\bar{\xi}_2)}\mathcal{G}^{(\xi_2|\bar{\xi}_3)} \\ & + \mathcal{G}^{(\xi_2|\bar{\xi}_1)}\mathcal{G}^{(|\bar{\xi}_2)}\mathcal{G}^{(\xi_1|\bar{\xi}_3)} + \mathcal{G}^{(|\bar{\xi}_1)}\mathcal{G}^{(\xi_2|\bar{\xi}_2)}\mathcal{G}^{(\xi_1|\bar{\xi}_3)}\Big\}\mathcal{F} \\ & + \mathcal{G}^{(\xi_2|)}[\mathcal{F}^{(\xi_1|\bar{\xi}_1,\bar{\xi}_2,\bar{\xi}_3)} - \mathcal{G}^{(\xi_1|)}\mathcal{F}^{(|\bar{\xi}_1,\bar{\xi}_2,\bar{\xi}_3)}] \\ & + \mathcal{G}^{(\xi_1\xi_2|)}\mathcal{F}^{(|\bar{\xi}_1,\bar{\xi}_2,\bar{\xi}_3)} + \mathcal{G}^{(\xi_1|)}\mathcal{F}^{(\xi_2|\bar{\xi}_1,\bar{\xi}_2,\bar{\xi}_3)}\,.\end{aligned} \tag{3.480}$$

In calculating the final expression $\mathcal{F}^{(\xi_1,\xi_2,\xi_3|\bar{\xi}_1,\bar{\xi}_2,\bar{\xi}_3)}$ we just can consider the limit $\sigma,\vartheta\to 0$. It follows

$$\begin{aligned}\mathcal{F}^{(\xi_1,\xi_2,\xi_3|\bar{\xi}_1,\bar{\xi}_2,\bar{\xi}_3)}_{|\sigma,\vartheta=0} = {} & \Big\{\mathcal{G}^{(\xi_1|\bar{\xi}_1)}\mathcal{G}^{(\xi_2|\bar{\xi}_2)}\mathcal{G}^{(\xi_3|\bar{\xi}_3)} + \mathcal{G}^{(\xi_1|\bar{\xi}_1)}\mathcal{G}^{(\xi_3|\bar{\xi}_2)}\mathcal{G}^{(\xi_2|\bar{\xi}_3)} \\ & + \mathcal{G}^{(\xi_2|\bar{\xi}_1)}\mathcal{G}^{(\xi_1|\bar{\xi}_2)}\mathcal{G}^{(\xi_3|\bar{\xi}_3)} + \mathcal{G}^{(\xi_3|\bar{\xi}_1)}\mathcal{G}^{(\xi_1|\bar{\xi}_2)}\mathcal{G}^{(\xi_2|\bar{\xi}_3)} \\ & + \mathcal{G}^{(\xi_2|\bar{\xi}_1)}\mathcal{G}^{(\xi_3|\bar{\xi}_2)}\mathcal{G}^{(\xi_1|\bar{\xi}_3)} + \mathcal{G}^{(\xi_3|\bar{\xi}_1)}\mathcal{G}^{(\xi_2|\bar{\xi}_2)}\mathcal{G}^{(\xi_1|\bar{\xi}_3)}\Big\} \\ & + \mathcal{G}^{(\xi_2\xi_3|)}\mathcal{F}^{(\xi_1|\bar{\xi}_1,\bar{\xi}_2,\bar{\xi}_3)}_{|\sigma,\vartheta=0} + \mathcal{G}^{(\xi_1\xi_2|)}\mathcal{F}^{(\xi_3|\bar{\xi}_1,\bar{\xi}_2,\bar{\xi}_3)}_{|\sigma,\vartheta=0} \\ & + \mathcal{G}^{(\xi_1\xi_3|)}\mathcal{F}^{(\xi_2|\bar{\xi}_1,\bar{\xi}_2,\bar{\xi}_3)}_{|\sigma,\vartheta=0}\,.\end{aligned} \tag{3.481}$$

From Eq. (3.479) we obtain

$$\begin{aligned}\mathcal{F}^{(\xi_1|\bar{\xi}_1,\bar{\xi}_2,\bar{\xi}_3)}_{|\sigma,\vartheta=0} &= \mathcal{G}^{(|\bar{\xi}_1,\bar{\xi}_3)}\mathcal{G}^{(\xi_1|\bar{\xi}_2)} + \mathcal{G}^{(\xi_1|\bar{\xi}_1)}\mathcal{G}^{(|\bar{\xi}_2,\bar{\xi}_3)} + \mathcal{G}^{(|\bar{\xi}_1,\bar{\xi}_2)}\mathcal{G}^{(\xi_1|\bar{\xi}_3)} \\ &\equiv \Big\{\delta_{\bar{\xi}_1,\bar{\xi}_3}\delta_{\xi_1,\bar{\xi}_2}A_{\bar{\xi}_1} + \delta_{\xi_1,\bar{\xi}_1}\delta_{\bar{\xi}_2,\bar{\xi}_3}A_{\bar{\xi}_2} + \delta_{\bar{\xi}_1,\bar{\xi}_2}\delta_{\xi_1,\bar{\xi}_3}A_{\bar{\xi}_1}\Big\}2B_{\xi_1}(t)\ .\end{aligned} \tag{3.482}$$

Then it follows

$$\begin{aligned}\mathcal{F}^{(\xi_1,\xi_2,\xi_3|\bar{\xi}_1,\bar{\xi}_2,\bar{\xi}_3)}_{|\sigma,\vartheta=0} &\equiv \langle Q_{\xi_1}(t)Q_{\xi_2}(t)Q_{\xi_3}(t)Q_{\bar{\xi}_1}Q_{\bar{\xi}_2}Q_{\bar{\xi}_3}\rangle_{\mathrm{R}} \\ &= \Big\{\delta_{\xi_1,\bar{\xi}_1}\delta_{\xi_2,\bar{\xi}_2}\delta_{\xi_3,\bar{\xi}_3} + (\text{all permutations of } \bar{\xi}_1,\bar{\xi}_2,\bar{\xi}_3\Big\} \\ &\quad \times B_{\xi}(t)B_{\xi_2}(t)B_{\xi_3}(t) \\ &\quad +\Big\{\delta_{\xi_2,\xi_3}2A_{\xi_2}[\delta_{\bar{\xi}_1,\bar{\xi}_3}\delta_{\xi_1,\bar{\xi}_2}A_{\bar{\xi}_1} + \delta_{\xi_1,\bar{\xi}_1}\delta_{\bar{\xi}_2,\bar{\xi}_3}A_{\bar{\xi}_2} \\ &\quad +\delta_{\bar{\xi}_1,\bar{\xi}_2}\delta_{\xi_1,\bar{\xi}_3}A_{\bar{\xi}_1}]2B_{\xi_1}(t) \\ &\quad +(\xi_3\leftrightarrow\xi_1) + (\xi_2\leftrightarrow\xi_1)\Big\}\ .\end{aligned} \tag{3.483}$$

In particular this latter type of correlation function will be used to discuss pure dephasing in the Sections 3.7.1 and 3.8.2.

### 3.12.3 Limit of Ultrashort Reservoir Correlation Time

It has already been pointed out several times that the reservoir correlation functions $C_{uv}(t)$ determine to what extent memory effects influence the dynamics of the reduced density operator. In view of the complications one faces with the practical solution of the QME for multi–level systems there is a natural desire to ask under what conditions these memory effects are negligible. As mentioned in Section 3.5.6 the correlation functions are usually characterized by a correlation time, $\tau_c$, giving the time scale during which $C_{uv}(t)$ decays to zero. In our attempt to simplify the QME we start with the assumption that this correlation time is short compared to any other characteristic time scale of the considered system. This allows us to write

$$C_{uv}(t) \approx \delta(t)\ c(u,v)\ . \tag{3.484}$$

In terms of the spectral density this assumption implies that the latter is a continuous featureless functions which depends only weakly on frequency. Replacing $\omega$ by a representative constant value $\omega_0$ in $C_{uv}(\omega)$ one can identify $c(u,v) = C_{uv}(\omega_0)$.

Inserting Eq. (3.484) into the non–Markovian QME (3.253) it becomes Markovian and the dissipative part reads

$$\left(\frac{\partial\hat{\rho}}{\partial t}\right)_{\text{diss.}} = -\sum_{uv}\{c(u,v)\,[K_u, K_v\hat{\rho}]_- - c^*(u,v)\,[K_u,\hat{\rho}K_v]_-\}\ . \tag{3.485}$$

If the $c(uv)$ are real and diagonal this expression reduces to [20] (note the anticommutator on the right–hand side)

$$\left(\frac{\partial\hat{\rho}}{\partial t}\right)_{\text{diss.}} = -\sum_{u} c(u,u)\left\{\left[K_u^2,\hat{\rho}\right]_+ - 2K_u\hat{\rho}K_u\right\}\ . \tag{3.486}$$

[20] If the $c(u,v)$ are not diagonal we can diagonalize the complete matrix to get a similar result as in Eq. (3.486).

This is another version of the Lindblad form of dissipation already discussed in Section 3.8.6. Quite often in practical calculations one starts from the rather general expression (3.486) without making any particular model for the system–reservoir interaction operator. In this case the choice of the operators $K_u$ as well as of the prefactor $c(u, u)$ has to be guided by intuition.

### 3.12.4 Markov–Approximation and the Factorized Part of the Reservoir Correlation Function

If the reservoir correlation function contains a nonvanishing factorized part $\langle\Phi_u\rangle_{\rm R}\langle\Phi_v\rangle_{\rm R}$ the Markov–approximation needs a separate treatment, what becomes obvious if one takes a closer look at Eq. (3.260). Once the correlation functions $C_{uv}$ have been replaced by their factorized parts only the latter can be taken out of the time–integral. Moreover, expanding $\Lambda_u$ with respect to the *eigenstates* $|a\rangle$ and $|b\rangle$ of $H_{\rm S}$, the time integration yields $\delta(\omega_{ab})$. The appearance of this singular term indicates the inadequacy of the Markov approximation. We shortly demonstrate how to proceed in this case. Let us start from Eq. (3.253) concentrating on the factorized part of $C_{uv}$ only. We obtain

$$\left(\frac{\partial\hat{\rho}}{\partial t}\right)_{\rm fac} = -\frac{1}{\hbar^2}\int\limits_0^{t-t_0} d\tau \Big[H_{\rm mf}, U_{\rm S}(\tau)\big[H_{\rm mf}, \hat{\rho}(t-\tau)\big]_- U_{\rm S}^+(\tau)\Big]_- , \tag{3.487}$$

where we introduced the mean–field Hamiltonian, Eq. (3.148). Next we replace $\tau$ by $t - \bar{t}$ and take the $t$–depended terms out of the time integral. This yields

$$\left(\frac{\partial\hat{\rho}}{\partial t}\right)_{\rm fac} = -\frac{i}{\hbar}\Big[H_{\rm mf}, \hat{\sigma}(t)\Big]_- , \tag{3.488}$$

with the newly introduced operator

$$\hat{\sigma}(t) = -\frac{i}{\hbar}U_{\rm S}(t)\int\limits_t^{t_0} d\bar{t}\, U_{\rm S}^+(\bar{t})\Big[H_{\rm mf}, \hat{\rho}(t-\tau)\Big]_- U_{\rm S}(\bar{t})U_{\rm S}^+(t) \,. \tag{3.489}$$

The advantage of this expression is that it can be easily translated into an equation of motion for $\hat{\sigma}(t)$ which completes the equation for the RDO $\hat{\rho}$ with the part Eq. (3.488). It follows

$$\frac{\partial\hat{\sigma}}{\partial t} = -\frac{i}{\hbar}\Big[H_{\rm S}, \hat{\sigma}(t)\Big]_- - \Big[H_{\rm mf}, \hat{\rho}(t)\Big] \quad . \tag{3.490}$$

Now the original equation for the RDO has to be solved simultaneously with that for $\hat{\sigma}$.

### 3.12.5 Numerical Propagation Methods

In the following we present some numerical techniques for solving the time–dependent Schrödinger equation and the Liouville–von Neumann equation. The list is of course far from being complete and primarily serves as an illustration of some of the general ideas in this respect. We start with the simplest situation, where the eigenfunctions of some zeroth–order part of the Hamiltonian are known. Then we pay attention to the more general case where the Hamiltonian is defined on a grid in coordinate space. The split–operator method for time propagation of the wave function and the density matrix are described in some detail.

### Basis Set Methods

The numerical solution of the time–dependent Schrödinger equation for a Hamiltonian of the form $H = H_0 + V$ is most easily performed if the eigenfunctions of the zeroth–order part $H_0$ are known, $H_0|\alpha\rangle = E_\alpha|\alpha\rangle$. Expanding the wave function in terms of this basis, $|\Psi(t)\rangle = \sum_\alpha c_\alpha(t)|\alpha\rangle$, inserting it into the Schrödinger equation, and multiplying the result with $\langle\alpha'|$ from the left yields

$$i\hbar\frac{d}{dt}c_\alpha(t) = E_\alpha c_\alpha(t) + \sum_{\alpha'} V_{\alpha\alpha'}c_{\alpha'}(t) \ . \tag{3.491}$$

This is a set of coupled ordinary, linear, first–order differential equation which can be solved by standard methods such as the Runge–Kutta procedure.

The density matrix equations of motion in the representation of some zeroth–order basis set has already been given in Eq. (3.99). From the numerical point of view the only difference to Eq. (3.491) is that one deals with $N^2$ instead of $N$ equations if $N$ is the number of basis states which need to be included. Using the density matrix property $\rho_{\alpha\alpha'} = \rho^*_{\alpha'\alpha}$, the number of equations can be reduced to $N(N+1)/2$.

### Grid Representation of Operators

If the eigenfunctions of the Hamiltonian $H = T + V$ are not known but $H$ is defined on $N$ specific points $x_n = x_0 + n\Delta x$ ($n = 0, \ldots, N-1$) of some finite interval $[x_0, x_{N-1}]$, a different strategy is necessary (For simplicity we discuss the one dimensional case only.). First, the wave function is discretized on the same grid, i.e. $\Psi(x) \to \Psi(x_n)$. Next the action of the Hamilton operator on $\Psi(x_n)$ has to be specified. The potential energy operator is diagonal in the coordinate representation and its action on $\Psi(x_n)$ is a simple multiplication: $V(x_n)\Psi(x_n)$. The kinetic energy operator, however, is nonlocal in this representation. Its action on the wave function can be written using a three–point finite difference formula for the derivative operator

$$T\Psi(x_n) = -\frac{\hbar^2}{3m}\frac{d^2}{dx^2}\Psi(x_n) = -\frac{\hbar^2}{2m}\frac{\Psi(x_{n+1}) - 2\Psi(x_n) + \Psi(x_{n-1})}{\Delta x^2} \ . \tag{3.492}$$

Apparently, the error will depend on the choice of the grid spacing $\Delta x$. If one switches to the momentum representation of the kinetic energy operator defined through (3.492) one gets ($p = \hbar k$): $T(k) = \hbar^2/2m[2\sin(k\Delta x/2)/\Delta x]^2$. This approaches the exact expression $\hbar^2 k^2/2m$ only in the limit $\Delta x \to 0$. The correct momentum space representation is, however, trivially obtained, if instead of (3.492) the action of $T$ on $\Psi(x_n)$ is calculated in momentum space. This leads to the so–called Fast Fourier Transform (FFT) method which involves three steps to get $T\Psi(x_n)$. First, a discrete Fourier transform of the wave function is performed,

$$\Psi(k_k) = \frac{1}{\sqrt{2\pi N}}\sum_{n=0}^{N-1}\Psi(x_n)e^{-ik_k x_n} \ . \tag{3.493}$$

Due to the finite grid size, $\Delta x$, the momentum space is also discretized, $k_k = 2\pi j/(x_{N-1} - x_0)$, ($j = -(N/2-1), \ldots, N/2$). This implies that the largest momentum which can be represented is $p_{\max} = \hbar\pi/\Delta x$. Thus, the maximum possible energy is $E_{\max} = \hbar^2\pi^2/2m\Delta x^2 +$

$V_{\max}$. The action of the kinetic energy operator in momentum space is a simple multiplication, i.e.

$$T(k_k)\Psi(k_k) = -\hbar k_k^2/2m\Psi(k_k) = \tilde{\Psi}(k_k) . \tag{3.494}$$

The final step of the FFT method is the back transformation of $\tilde{\Psi}(k_k)$ into coordinate space. Making use of the Fourier transformation the FFT method requires the wave function to be bounded in momentum space. In practice the size and the spacing of the coordinate grid is chosen such that the wave function in coordinate and momentum space is localized well within the grid boundaries. The great popularity the FFT method enjoys is due to the efficiency of the numerical calculation of the Fourier transform. The computational effort scales only like $N\log N$ with the grid size.

In order to propagate the density matrix on a grid one has to specify the action of coordinate and momentum space operators on the density matrix, $\rho(x_n, x_m)$, in the commutator form $\langle x|[T+V,\rho]|x'\rangle$. Assuming that $\rho$ is defined on a grid the commutator with the potential energy operator is simply:

$$\langle x_n|[V,\rho]|x_m\rangle = (V(x_n) - V(x_m))\rho(x_n, x_m) . \tag{3.495}$$

The commutator with the kinetic energy operator is calculated after Fourier transforming the density matrix into momentum space,

$$\rho(k_i, k_k) = \frac{1}{2\pi N} \sum_{m,n=0}^{N-1} \rho(x_n, x_m) e^{-i(k_i x_n - k_k x_m)} . \tag{3.496}$$

The commutation relation in momentum space are local,

$$\langle k_i|[T,\rho]_-|k_j\rangle = \hbar^2/2m(k_i^2 - k_j^2)\rho(k_i, k_j) . \tag{3.497}$$

The final step is again the back-transformation into the coordinate representation.

### Split–Operator Method

Having specified how $H\Psi(x)$ or $\langle x|[H,\rho]|x'\rangle$ can be efficiently calculated on a grid we turn to the time evolution next. Let us first consider the Schrödinger equation. Usually it is more convenient to use the time–evolution operator $U(t', t)$ (Eq. (3.11)) and calculate its action on some initial wave function. Assuming a discretized time axis with spacing $\Delta t$ we are looking for

$$|\Psi(t + \Delta t)\rangle = U(t + \Delta t, t)|\Psi(t)\rangle . \tag{3.498}$$

Since $T$ and $V$ do not commute we have $U(t + \Delta t, t) \neq \exp(-iT\Delta t/\hbar)\exp(-iV\Delta t/\hbar)$. However, the error made in this decomposition is of the order $[V,T]\Delta t^2$, as can be shown using a Taylor expansion of the exponential operators (see also Eq.(2.95)). With the symmetric splitting

$$e^{-iH\Delta t/\hbar} = e^{-iT\Delta t/2\hbar} e^{-iV\Delta t/\hbar} e^{-iT\Delta t/2\hbar} \tag{3.499}$$

the error even goes as $\Delta t^3$ only. We conclude that a splitting of the time evolution operator, for instance, according to Eq. (3.499) is possible in a numerical propagation provided the time step $\Delta t$ is sufficiently small .

The basic propagation step then involves calculating the action of $\exp(-iT\Delta t/2\hbar)$ and $\exp(-iV\Delta t/\hbar)$ on the discretized wave function $\Psi(x_n, t)$. The former is done using the FFT method while the latter is just a multiplication on the coordinate grid. We note in passing that using the FFT method gives itself an upper limit for the time increment $\Delta t$. Since the grid spacing determines a maximum energy $E_{\max}$, the uncertainty principle dictates that $\Delta t < \hbar/E_{\max}$.

Next we turn to the time evolution of the density matrix according to the Quantum Master Equation. It is convenient for the following to use the superoperator notation introduced in Eq. (3.264). The formal solution of Eq. (3.264) can be written as (cf. Eq. (3.103))

$$\begin{aligned} \rho(t) &= \exp\{-(i\mathcal{L}_0 + \mathcal{D})t\}\rho(0) \\ &= \mathcal{U}(t)\rho(0) . \end{aligned} \tag{3.500}$$

Here $\mathcal{U}(t)$ is the time–evolution operator propagating the density operator. Note that due to the commutator structure it acts from the left and from the right on $\rho(0)$. In analogy with the splitting in Eq. (3.499) we can write

$$\mathcal{U}(t + \Delta t) \approx \mathcal{U}_{\text{kin}}(\Delta t/2)\,\mathcal{U}_{\text{pot}}(\Delta t)\,\mathcal{U}_{\text{kin}}(\Delta t/2) , \tag{3.501}$$

where $\mathcal{U}_{\text{kin}}(t)$ and $\mathcal{U}_{\text{pot}}(t)$ are defined with respect to the commutator with the kinetic and potential energy operator, respectively. As in the case of the Schrödinger equation the error of this splitting will be of the order $\Delta t^3$.

### 3.12.6 The Monte Carlo Wave Function Method

In Sections 3.8.2 and 3.8.8 we derived equations of motion for the reduced density matrix in the energy and coordinate representation, respectively. The question whether this type of approach can be used to treat the dynamics of multi–dimensional relevant systems has to be answered negatively. The bottleneck is certainly that the representation of the density matrix requires to keep track of $N^2$ matrix elements, if $N$ is, for example, the number of basis functions needed for the state expansion (see also Section 3.12.5).

In the following we will outline an approximate method which can be used to solve this problem. To this end let us return to the definition Eq. (3.71) of the full (relevant system plus reservoir) density operator $\hat{W}(t)$. At first glance it would appear as if one could calculate this quantity by separate propagation of every pure state contained in $\hat{W}(t)$. And indeed this was the method used in Section 3.4.3 to derive the Liouville–von Neumann equation. However, for the *reduced* density operator a representation like Eq. (3.71) does not exist. Nevertheless, let us postulate the following expansion

$$\hat{\rho}(t) = \sum_j w_j |\psi_j(t)\rangle\langle\psi_j(t)| . \tag{3.502}$$

Since this formula cannot be derived from Eq. (3.71) for the full density operator, we have to accept that the state vectors $|\psi_j(t)\rangle$ are propagated according to an *effective* time–dependent

Schrödinger equation. This effective equation is based on a non–Hermitian system Hamiltonian $H_S^{\text{eff}}$ given in Eq. (3.262). Since a propagation by means of a non–Hermitian Hamiltonian does not conserve the norm of the state vector one has to introduce a particular correction procedure. In the *Monte Carlo wave function method* this correction is combined with a prescription for the introduction of quantum transitions. These transitions account in some way for those terms on the right–hand side of Eq. (3.263) which could not be comprised into the effective non–Hermitian Hamiltonian.

The Monte Carlo wave function approach is based on the *Lindblad* form of the dissipative contribution to the QME as given in Section 3.12.3. Let us assume that we have obtained the state vector at some time $t$ as $|\psi_j(t)\rangle$ and the loss in the norm is equal to $\Delta = 1 - \langle\psi_j(t)|\psi_j(t)\rangle$. At this point we select a random number $\xi$ between 0 and 1. If $\xi < \Delta$ we apply the operator $K_m$ to this state vector to obtain the new state vector $K_m\ |\psi_j(t)\rangle$ which has to be normalized. The change of the state vector is called a *quantum jump*. Otherwise ($\xi > \Delta$) the state vector is just normalized. This procedure is repeated at each time step during the propagation of the state vector.

As a consequence of the random quantum jumps the wave function propagation becomes *stochastic*. In order to obtain a reasonable result one has to carry out the full propagation many times. However, instead of a matrix ($\rho_{ab}$) we only propagate a *vector* ($\psi_j$). Therefore, the numerical effort is reduced considerably and an application to multi–dimensional relevant systems becomes feasible. Moreover, the justification of the ansatz (3.502) can be given for the special case of the Lindblad form of the dissipative part of the QME. One can show that in the limit of an infinite number of stochastic propagations of $\psi_j$ one reproduces the time dependence of the reduced density operator, that is, we have

$$\hat{\rho}(t) = \lim_{M\to\infty} \frac{1}{M} \sum_{j=1}^{M} |\psi_j(t)\rangle\langle\psi_j(t)| \ . \tag{3.503}$$

Comparing this expression with the one given in Eq. (3.502) one notices that the method determines the probabilities $w_j$ to be equal to $1/M$.

---

# Notes


[Cal83] A. O. Caldeira and A. J. Legget, Physica **121A**, 587 (1983).

[Ego99] S. A. Egorov, K. F. Everitt, and J. L. Skinner, J. Phys. Chem. A **103**, 9494 (1999).

[Hu89] Y. Hu and S. Mukamel, J. Chem. Phys. **91**, 6973 (1989).

[Hor02] I. Horenko, Ch. Salzmann, B. Schmidt, and Ch. Schütte, J. Chem. Phys. **117**, 11075 (2002).

[Jan02] S. Jang, J. Cao, and R. J. Silbey, J. Chem. Phys. **116**, 2705 (2002).

[Kap99] R. Kapral and G. Ciccotti, J. Chem. Phys. **110**, 8919 (1999).

[Kor97] M. V. Korolkov and G. K. Paramonov, Phys. Rev. A **55**, 589 (1997).

[Lai91] B. B. Laird, J. Budimir, and J. L. Skinner, J. Chem. Phys. **94**, 4391 (1991).

[Man01] T. Mancal and V. May, Chem. Phys. **268**, 201 (2001).

[Mei99] Ch. Meier and D. Tannor, J. Chem. Phys. **101**, 3365 (1999).

[Oht89] Y. Ohtsuki and Y. Fujimura, J. Chem. Phys. **91**, 3903 (1989).

[Red65] A. G. Redfield, Adv. Magn. Reson. **1**, 1 (1965).

[Spa88] M. Sparpaglione and S. Mukamel, J. Chem. Phys. **88**, 3263 (1988).

[Sua92] A. Suarez, R. Silbey and I. Oppenheim, J. Chem. Phys. **97**, 5101 (1992).

[Tan91] Y. Tanimura and P. G. Wolynes, Phys. Rev. A **43**, 4131 (1991).

[Tul90] J. C. Tully, J. Chem. Phys. **93**, 1061 (1990).

# 4 Vibrational Energy Redistribution and Relaxation

*In Chapter 3 we have introduced some fundamental concepts for the description of quantum dynamics ranging from the coherent (Schrödinger equation) to the incoherent (Pauli Master Equation) regime. Density matrix theory was shown to provide a versatile tool for all types of dynamics thus establishing the link between these two limits. The type of dynamics which is realized in an actual system, of course, depends on the Hamiltonian describing the way active and reservoir degrees of freedom interact with each other. The definition of the active coordinates is closely related to the type of preparation of the initial state, for instance, in laser spectroscopy. A general approach for constructing a Hamiltonian for molecular systems based on the Born–Oppenheimer separation of electronic and nuclear motions has been introduced in Chapter 2. This type of Hamiltonian sets up the stage for inter– and intramolecular vibrational energy flow within a given adiabatic electronic state.*
*In the present chapter we will discuss different scenarios ranging from simple diatomic molecules in solution to large polyatomic molecules in the gas or the condensed phase. The key to the derivation of vibrational energy transfer rates is provided by a low–order expansion of the respective interaction Hamiltonian. In this respect we will introduce the model of instantaneous normal modes for solvents. This approach allows to define the parameters entering the system–reservoir Hamiltonian on a microscopic level.*
*As an example we give a detailed account on the transition between coherent and dissipative dynamics in a single harmonic potential energy surface.*

# 4.1 Introduction

The investigation of vibrational energy flow in polyatomic molecules is of pivotal importance for the understanding of chemical reaction dynamics. In Section 2.6.3 it has been pointed out that the definition of reaction coordinates is closely related to the incorporation of a large number of environmental degrees of freedom. If the latter are only weakly coupled they still might cause energy dissipation out of the reaction coordinate. If there is a stronger coupling to specific environmental motions, this might provide a means, for instance, to accelerate the reaction dynamics by helping to overcome a potential barrier. These two examples already indicate that the type of energy transfer may cover the whole range from incoherent to coherent processes.

The starting point for the description of vibrational energy flow will be a situation in which energy is contained in specific vibrational modes of a molecule. Such a state could have been prepared, for example, by an external laser field (cf. Section 5.5). Taking a time–dependent point of view this initial state often can be thought of as a *superposition* of eigenstates of the vibrational Hamiltonian (cf. Eq. (5.124)). Then the subsequent dynamics of the system as characterized, for example, by the survival amplitude (3.22) will show a behavior similar to the ones shown in Fig. 3.2; an experimental example is given in Fig. 4.1. On the other hand, high resolution spectroscopy of polyatomics in the frequency domain can provide direct insight into the spectrum of the molecular eigenstates itself.

For large polyatomics knowledge of these eigenstates is, however, hardly available. Therefore, the interpretation of both types of experiments conveniently starts with some *zeroth–order states* which are chosen according to the preparation conditions. For an excitation with a laser field, for instance, this leads to a definition of zeroth–order states in close relation to the classification into optically (or infrared) *bright* and *dark* states; depending on whether there is some oscillator strength for the transition to the considered state. Typical choices for zeroth–order states are, for example, normal modes or localized vibrational modes along particular internal coordinates. In terms of the vibrational Hamiltonian this approach implies a Taylor

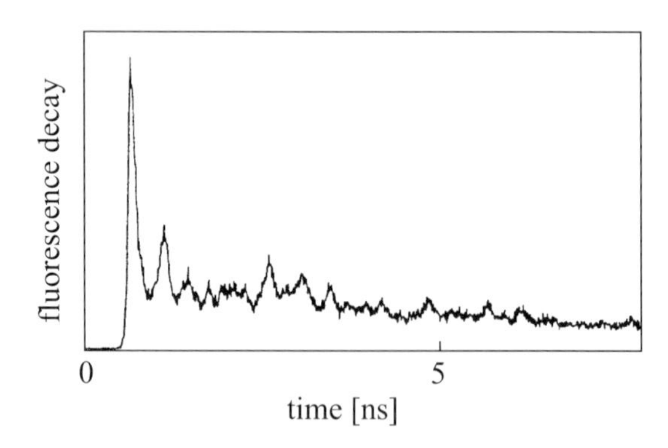


**Figure 4.1:** Fluorescence decay from the $S_1$ excited state of anthracene. The signal is proportional to the survival amplitude $P_{\text{surv}}$. (reprinted with permission from [Fel85]; copyright (1985) American Institute of Physics)

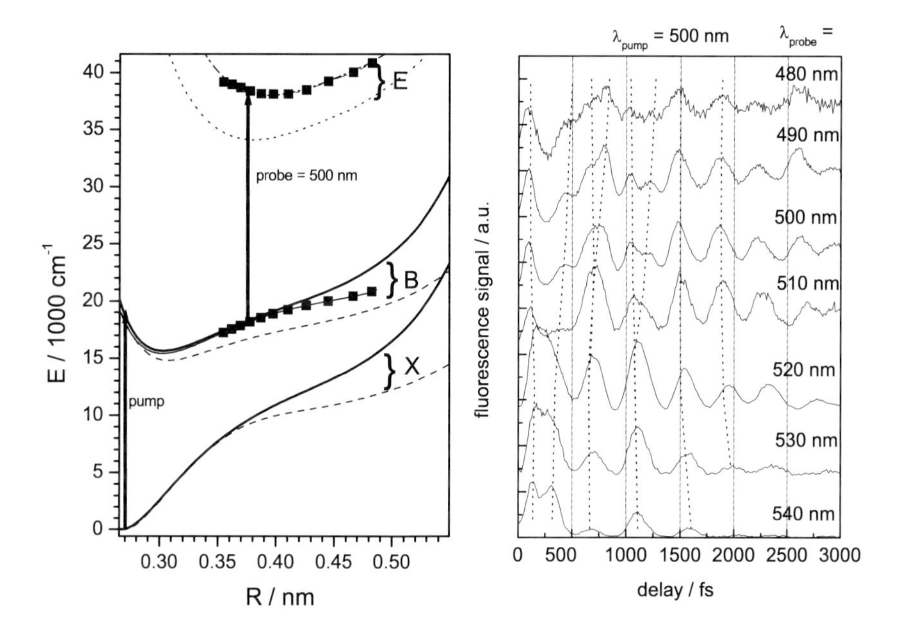


**Figure 4.2:** Pump–probe spectroscopy of the vibrational dynamics of a two–atomic molecule ($I_2$) in a rare gas lattice (Kr). Left panel: Potential energy curves of the ground state $X$, the valence $B$ state, and an ionic $E$ state of $I_2$ if embedded in the Kr lattice. The solid lines correspond to calculations where the Kr lattice is relaxed to its minimum configuration for a given $I_2$ distance. In contrast, the dashed lines display results for the Kr lattice frozen at its equilibrium configuration (in the absence of the guest molecule). The dotted ionic state curve corresponds to a fixed lattice as well. Right panel: Pump–probe spectra (the signal is measured in terms of the fluorescence from the ionic states) for pumping at 500 nm (transition from the $X$ to the $B$ state) and probing at different wavelength (from the $B$ to the $E$ state, cf. the arrows in the left panel). Shown are oscillating spectra of the probe beam absorption versus the delay time between the pump and the probe pulse and for different probe wavelengths (numbers at the curves in right panel). Upon increasing the probe wavelength the wave packet dynamics is tested at different energies and thus bond lengths in the $B$ state. The phase and frequency of the oscillations due to coherent $I_2$ bond vibration changes as does the decay time of the oscillatory signal. This information from different pump–probe measurements can be used to construct an effective potential which is shown with solid squares in the left panel. ([Bar02] – Reproduced by permission of the PCCP Owner Societies)

expansion of the adiabatic potential energy surface allowing the specification of zeroth–order states and the couplings between them. At least in principle, this concept is straightforwardly applied to polyatomics in the gas phase. However, as we will see below under certain conditions it can be transferred to the condensed phase as well where optically bright and dark zeroth–order states constitute the system and the reservoir Hamiltonian, respectively.

The relaxation of energy from an initially excited state into other vibrational degrees of freedom of a polyatomic molecule is called *intramolecular vibrational redistribution* (IVR). IVR will be on focus in Section 4.2 where we consider collision–free polyatomic molecules as being observable in a molecular beam, for instance. Neglecting also radiative couplings leading to emission it is clear from the outset that the total energy of the molecule is conserved. Depending on the number of degrees of freedom and the couplings between them, the vibrational dynamics as observed in time–domain spectroscopy can cover a broad range of regimes. It may extend from a damped oscillatory to an exponential decay at early times, and to a power law decay at intermediate times before a stationary value is reached. In Section 4.2 it will be discussed how these different regimes can be rationalized. Golden Rule type approaches to IVR apparently cannot account for such a rich dynamics. However, anything beyond the simple Golden Rule requires knowledge of the molecule's Hamiltonian which is not easily obtained for larger systems. The useful concepts of the tier model and the state space approach will be introduced in Section 4.2.

The simplest condensed phase situation one can study is the vibrational dynamics of a single diatomic solute molecule in an atomic solvent (infinite dilution limit). Here, the interaction between both subsystems will lead to energy dissipation into the solvent, where it is stored as translational motion. The irreversible energy transfer between solute and solvent is termed *vibrational energy relaxation* (VER) and will be on focus in Section 4.3.1.[1] An example for the case that the solvent is a rare gas matrix is shown in Fig. 4.2.

The dynamics will of course become more complex if we go to molecular solvents. However, as will be shown in Section 4.3.2 the theoretical framework for modelling VER dynamics is rather generic, provided that the above mentioned low–order expansion of the total potential energy surface can be performed. This is most straightforward if the solvent is, for instance, a low temperature solid state matrix. Here, the atoms perform only small amplitude motions around their equilibrium positions (Section 4.3.1). In the case of a solvent at liquid state temperatures, a low–order expansion of the interaction potential leads to relaxation rates which are determined by the equilibrium correlation function of the force which is exerted by the solvent on the static solute (Section 4.3.2). Often the decay of these types of correlation functions takes place on a rather short time scale. This allows for a modelling of the solvent in terms of collective harmonic oscillators in the vicinity of each instantaneous configuration, even though the liquid's dynamics is inherently anharmonic on longer time scales (Section 4.3.2).

The most general situation of a polyatomic molecule in solution will be treated in Section 4.4. Here it is shown how the system–bath concept can be applied resulting in relaxation rates which enter the density matrix equations of motion discussed in Section 3.7. Since energy conservation constrains relaxation dynamics, different types of multiple quantum transitions will become important, in particular for the relaxation of high frequency modes.

There are various motivations for studying vibrational energy flow besides its general importance for reaction dynamics mentioned in the beginning of this section. On a fundamental level the understanding of such processes requires knowledge about the molecular Hamiltonian, and in particular of the potential energy hypersurfaces for nuclear motions, as well as a sophisticated treatment of the dynamics. Thus, combining experimental observations with theoretical predictions provides an excellent testing ground for theoretical models in this respect. On the more practical side one can imagine, for instance, that vibrational energy has been deposited into a particular mode in order to trigger a chemical reaction (cf. Chapter 9).

---

[1]Note, that according to our classification this type of energy transfer would be of the dissipative type. Using the term "vibrational energy relaxation" in this chapter we follow the convention which is widely adopted.

In such a case it would be desirable to know to what extent this energy is lost, for example, due to the release of heat into the surrounding solvent.

## 4.2 Intramolecular Energy Redistribution

### 4.2.1 Zeroth–Order Basis

In this section we consider the case of isolated polyatomic molecules in the electronic ground state. We are interested in the redistribution of vibrational energy after it has been deposited into some state, e.g., by means of an infrared excitation.[2] The central question concerns the microscopic origins and mechanism of IVR? Let us recall that in Section 2.6.1 we learned about the normal mode expansion of the vibrational Hamiltonian obtained from the Born–Oppenheimer separation of electronic and nuclear degrees of freedom. This description was based on the harmonic approximation to the potential energy surface which is supposed to be reasonable close to stationary points. Away from these stationary points, with increasing vibrational energy, the harmonic approximation breaks down and anharmonic effects have to be included. An empirical anharmonic potential is given by the Morse potential shown in Fig. 2.6. In more general terms the PES can be expanded with respect to the (mass–weighted) normal mode coordinates, $\{q_\xi\} = (q_{\xi_1}, q_{\xi_2} \dots q_{\xi_N})$, to obtain the N–dimensional vibrational Hamiltonian

$$\begin{aligned} H_{\text{vib}} &= \frac{1}{2}\sum_{\xi}\left(p_\xi^2 + \omega_\xi^2 q_\xi^2\right) + \sum_{\xi_i\xi_j\xi_k} K_{\xi_i\xi_j\xi_k}\, q_{\xi_i} q_{\xi_j} q_{\xi_k} \\ &+ \sum_{\xi_i\xi_j\xi_k\xi_l} K_{\xi_i\xi_j\xi_k\xi_l}\, q_{\xi_i} q_{\xi_j} q_{\xi_k} q_{\xi_l} + \dots \end{aligned} \tag{4.1}$$

Here, $K_{\xi_i\dots}$ are the anharmonic coupling constants, i.e., the derivatives of the PES with respect to the normal mode coordinates. It should be emphasized that the usually collective normal mode coordinates are not the only choice for a representation of the vibrational Hamiltonian. Alternatively, we could have used, for instance, local modes pertaining to individual bonds. Then the Hamiltonian contains couplings between these local modes.

An eigenstate of the harmonic part of $H_{\text{vib}}$ in Eq. (4.1) can be classified according to the quanta contained in the different normal modes, i.e.$|M_{\xi_1}, M_{\xi_2} \dots M_{\xi_N}\rangle$. The possible types of anharmonic mode couplings can easily be rationalized by inspection of matrix elements with respect to such state vectors. For example, consider two DOF and the anharmonic coupling term $\propto K_{122} q_1 q_2^2$ which gives non–vanishing matrix elements of the type

$$\langle M_{\xi_1}, M_{\xi_2}|q_1 q_2^2|N_{\xi_1}, N_{\xi_2}\rangle \propto \delta_{N_{\xi_1},M_{\xi_1}-1}\delta_{M_{\xi_2},N_{\xi_1}-2}\sqrt{(N_{\xi_1}+1)N_{\xi_2}(N_{\xi_2}-1)}\,. \tag{4.2}$$

This so–called Fermi resonance interaction is typical for the coupling between a hydrogenic vibrational stretching ($q_1$) fundamental and a bending overtone transition ($q_2$) as observed in infrared spectroscopy. The relevant excited states are sketched in Fig. 4.3. In the harmonic

[2] We do not take into account the effect of rotations in the following discussion. Whereas the overall rotation of large molecules can be safely neglected on the time scale of interest here ($< 1$ ns), internal rotations will have an effect on the density of states.

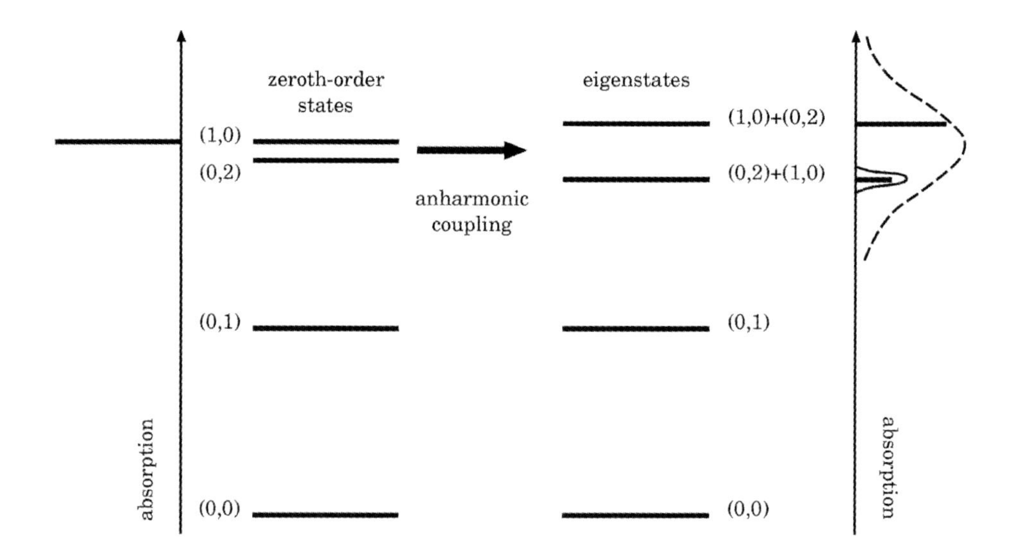


**Figure 4.3:** Vibrational eigenstate formation due to anharmonic coupling. Left part: Zeroth–order harmonic oscillator states with quantum numbers $(M_{\xi_1}, M_{\xi_2})$ for a two–mode model where the second excited state of mode $\xi_2$, $(0,2)$, is close to resonance with the first excited state of mode $\xi_1$, $(1,0)$. In harmonic approximation only the transition $(0,0) \rightarrow (1,0)$ shall be allowed. The respective (infrared) absorption spectrum shows a single peak (as indicated in the outermost left part). Right part: As a consequence of the anharmonic coupling $\propto K_{122}q_1q_2^2$ the states $(1,0)$ and $(0,2)$ are mixed, giving rise to an increase of their energetic difference as well as to a sharing of the oscillator strength among the two eigenstates. Depending on the frequency distribution in the exciting radiation field (indicated by the envelopes in the right spectra), either an eigenstate (solid line) or a superposition state (dashed line) is prepared (right part). The situation is typical, for instance, for hydrogenic stretching ($q_1$ and bending ($q_2$) vibrations in infrared spectroscopy.

limit and assuming a dipole which depends linearly on the coordinates, only the fundamental transition of the $q_1$ mode will carry oscillator strength. However, due to the anharmonic coupling ($K_{122}$), according to Eq. (4.2) the states $|1,0\rangle$ and $|0,2\rangle$ are mixed and two peaks appear in the spectrum, i.e., an eigenstate with overtone transition character ($q_2$) "borrows" oscillator strength. The details can be easily appreciated by using the exact results available for a coupled two–level system (see Section 2.8.3). Here, we want to focus on another issue which concerns the most suitable description of the vibrational dynamics.

In principle, there are different possibilities for the so–called *zeroth–order* basis and the best choice will depend on the experimental situation at hand. For example, using the normal mode Hamiltonian (4.1) the presence of, for instance, a strong Fermi resonance between two DOFs leaves the question whether or not to diagonalize the Hamiltonian in the subspace of these two modes and work with the respective eigenstates whose residual coupling to the remaining modes, of course, would be modified. At this point the type of preparation of the initial state plays an important role as shown in Fig. 4.3. Let us assume that the excitation is due to an infrared laser having a narrow spectral bandwidth. In this case the prepared state will be close to an eigenstate of the full Hamiltonian. Therefore, a pre–diagonalization of the Fermi resonance interaction might be suitable, i.e., the zeroth–order basis set would include the strongest anharmonic coupling. On the other hand, if the excitation is due to an ultrafast broad–band laser infrared pulse, a coherent superposition state is prepared which will be of more local character (cf. Section 5.5), i.e., in our example it would be closer to the fundamental stretching transition. In this case the use of the normal mode basis would be appropriate.

Let us assume that we would be able to obtain the eigenstates of the *full* Hamiltonian, Eq. (4.1), although this is, of course, impossible for larger systems. In terms of high–resolution frequency–domain spectra this give the means to reproduce the numerous spectral lines which upon increasing the excitation energy merge into a quasi–continuum. However, unless these eigenstates are expressed as superposition states of, for instance, harmonic oscillator states there would be little insight gained into the nature of the eigenstates and the resulting transitions. In time–domain experiments, the initially prepared wave packet will be a complicated superposition of many eigenstates. However, as in our Fermi–resonance example, only the choice of a suitable basis which captures the nature of the initially excited state enables a straightforward analysis of the vibrational dynamics .

This discussion reflects the influence of selection rules and the initial state preparation time scale on the choice of the zeroth–order basis for the description of the IVR process. Using another terminology one has, depending on the situation, different choices for the identification of *bright* and *dark* states. In any case there will be a residual coupling between the initially excited bright states and the usually large number of dark states. This situation which comprises the essential theme of IVR dynamics is illustrated in Fig. 4.4.

### 4.2.2 Golden Rule and Beyond

The zeroth–order bright state basis which is shown in Fig. 4.4 reminds of the discussion of the Golden Rule in the previous chapter (see Section 3.3). In the context of IVR the Golden Rule description has been given by Bixon and Jortner in the 1960s [Bix68]. In order to emphasize the basic idea let us assume that there is initially only a single bright state appreciably populated. For simplicity we suppose that the vibrational Hamiltonian has been *pre–diagonalized* with the single bright state $|0\rangle$ with energy $E_0$ projected out. Let $E_\alpha$ denote the energy of the pre–diagonalized bath (dark) states and $V_{0\alpha}$ the coupling of the bright state to the pre–diagonalized bath (cf. Fig. 4.4). If we further assume that the bath has a quasi–continuous spectrum, we recovered the Hamiltonian Eq. (3.40), and the Golden Rule rate for IVR is according to Eq. (3.53) given by

$$k_{\text{IVR}} = \frac{2\pi}{\hbar} \sum_\alpha |V_{0\alpha}|^2 \delta(E_0 - E_\alpha) \, . \tag{4.3}$$

First, we should recall that Eq. (4.3) is a result which is valid in second–order perturbation theory, i.e., $|V_{0\alpha}|^2$ has to be sufficiently small. This assumption is reasonable since upon pre–diagonalization the bath the anharmonic coupling strengths are distributed over many bath eigenstates. This means that on average they are likely to be small, provided that there is a quasi–continuous manifold of eigenstates. It is important to remember, however, that the

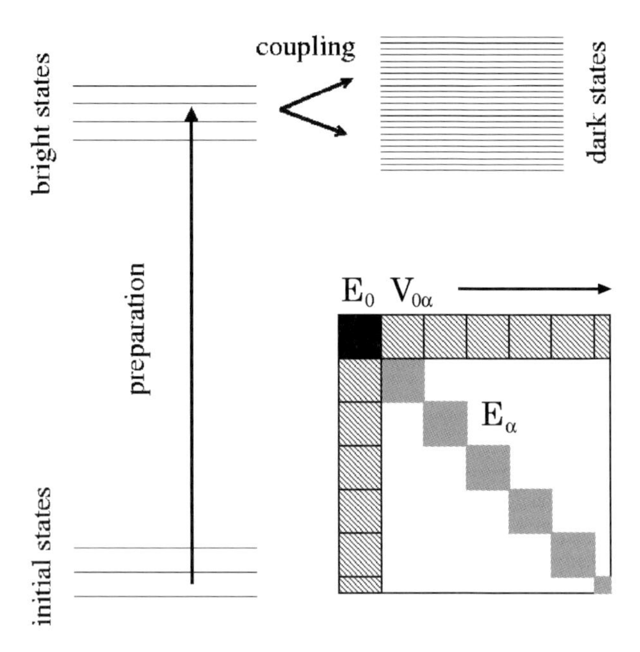


**Figure 4.4:** Preparation of bright (zeroth–order) states (for example, by laser light absorption, cf. Chapter 5) starting from some set of initial states. The subsequent dynamics will be influenced by the coupling between bright and dark states. In the lower right part the structure of the Hamiltonian is sketched for the case of a single bright state $|0\rangle$ ($E_0$, black) and a bath of dark states $|\alpha\rangle$ ($E_\alpha$, grey). The residual couplings $V_{0\alpha}$ ($V_{\beta 0}$) are shown as hatched areas.

coupling has to be much larger than the mean level spacing of the bath in order to validate the Golden Rule description. In a next step we replace the coupling matrix by its root mean square value across the bath spectrum, $V^2_{\rm rms}$. In other words we throw away all details of this coupling such as its energy dependence or correlations between different couplings. Replacing the sum over the delta functions by the *global* density of states,

$$\mathcal{N}(E) = \sum_{\alpha=1}^{N_\alpha} \delta(E - E_\alpha) \tag{4.4}$$

we obtain the Bixon–Jortner rate for IVR

$$k_{\rm IVR} = \frac{2\pi}{\hbar} V^2_{\rm rms} \mathcal{N}(E_0) \ . \tag{4.5}$$

As a matter Eq. (4.5) expresses the IVR rate in terms of the quantities, $V^2_{\rm rms}$ and $\mathcal{N}(E)$, which are *experimentally* available from an analysis of the positions and relative intensities of spectral lines in high–resolution frequency domain vibrational spectroscopy. Thus, provided that we have obtained these values and the conditions leading to the Golden Rule expression are fulfilled the IVR rate will by construction match the experimentally observed rate. But

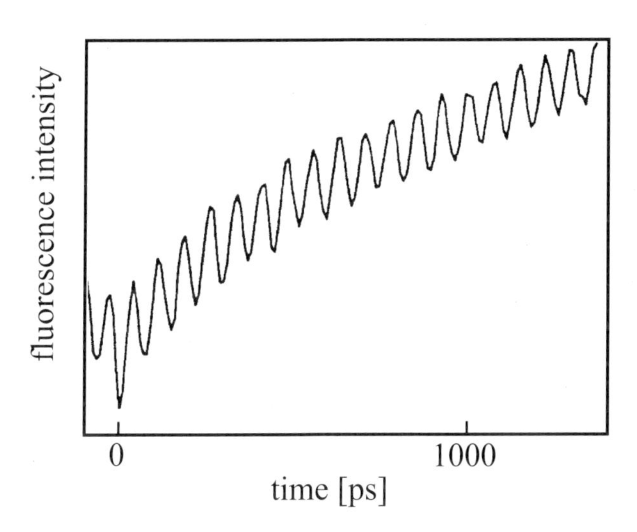


**Figure 4.5:** Fluorescence depletion decay signal from the $S_1$ excited state of $p$–cyclohexylaniline. In this pump–probe experiment a known electron–vibrational transition is first excited by the pump pulse and then stimulated back onto the electronic ground state by the probe pulse. Increasing the delay between pump and probe pulse the probability for stimulated emission decreases due to the decay of the initially prepared state. This gives rise to an enhanced fluorescence signal which is proportional to $1 - P_{\rm surv}$. (reprinted with permission from [Smi90]; copyright (1990) American Institute of Physics)

one should be aware at this point that there is no direct link between $V_{\rm rms}$ and the anharmonic coupling constants entering Eq. (4.1).

So far we have assumed that there is an infinite number of accessible bath states. However, at not too high energies the number of accessible states, $N_{\rm IVR}$, in polyatomic molecules may be rather large but still finite. This fact is characterized by the so–called dilution factor $\sigma = \bar{P}_{\rm surv}(t \to \infty)$, i.e., the time–averaged long–time survival probability. Assuming that all states are equally populated the latter is proportional to $N_{\rm IVR}^{-1}$ and experimentally accessible (see Fig. 4.7 below). Taking this into account the decay of the survival probability should be characterized by the function:

$$P_{\rm surv}(t) \equiv |\langle 0(t)|0\rangle|^2 = (1-\sigma)e^{-k_{\rm IVR}t} + \sigma \,. \tag{4.6}$$

There are many examples where this Golden Rule approach had been successfully applied. In the following we will concentrate, however, on situations where the dynamics cannot be described by a single exponential decay. In Figs. 4.1 and 4.5 we show experimental data indicating a quantum beat type behavior and a nonexponential decay of the survival amplitude as revealed by fluorescence spectroscopy. The possibility of these types of dynamics has already been discussed in more general terms in Section 3.3. In principle the existence of a finite dilution factor already indicates that there is the possibility for a revival of the survival amplitude after a finite time. In the context of IVR there are several more points which need to be addressed:

First, given an initial bright state one can imagine that it is not equally strongly coupled to all energetically possible dark states, in contrast to the assumption which led to Eq. (4.5).

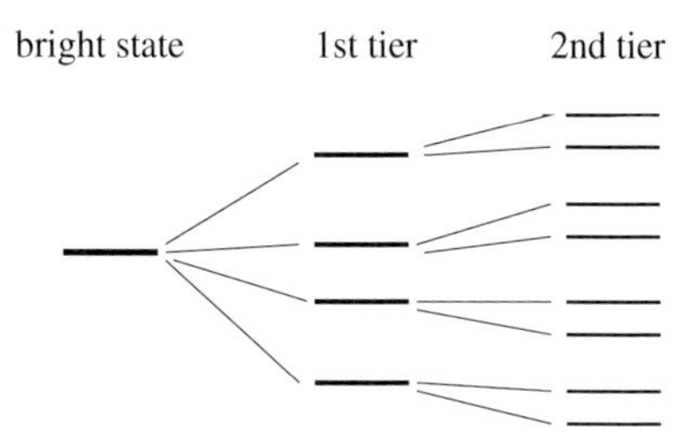


**Figure 4.6:** Hierarchical structure of IVR as described by the tier model. A single bright state is strongly coupled only to a finite set of dark states constituting the first tier. The first tier is then coupled to another set of dark states in the second tier and so on.

Thus there will be a certain finite set of dark states coupled to the initial state $|0\rangle$. This can be seen as a direct consequence of the local character of chemical bonding. This reasoning can be extended to the first set of strongly coupled dark states, i.e., they in turn will be strongly coupled to a finite set of different dark states only. This leads to the so–called *tier* model of IVR which is illustrated in Fig. 4.6. Given a local normal mode expansion of the vibrational Hamiltonian, the partitioning of the dark states into different tiers can proceed, for instance, by using a classification with respect to the order of the anharmonic coupling. However, one has to keep in mind that resonance conditions have to be considered as well when judging efficient IVR pathways. For instance, it is possible that high–order couplings dominate when the respective dark states are in better resonance than those coupled via low–order anharmonicities.

The tier model is in fact a two–dimensional projection of the *vibrational state space* model. Here the energy flow is considered in the space spanned, e.g., by the N–dimensional harmonic oscillator vectors $|M_{\xi_1}, M_{\xi_2} \dots M_{\xi_N}\rangle$. This is illustrated in Fig. 4.7 where we also see another important feature of IVR, namely its energy dependence. For a given state $|0\rangle$ there is a certain threshold below which this state will be effectively isolated from the rest of state space. Upon increasing the energy of the initial state, the average number of dark states increases and so does the average strength of anharmonic couplings. Panels (A) to (C) schematically show three different cases realized when moving across the *IVR threshold*. They correspond to a situation of small, moderate, and large couplings. Clearly, the number of relevant couplings indicated thick lines increases with energy such that from some initial state $|0\rangle$ an increasing number of states becomes accessible. Notice, however, that the IVR threshold is not a strictly defined quantity but reflects the averaged structure of state space.

In the tier model these possible pathways are projected onto a single axis. In terms of the observed absorption the increasing number of couplings results in a spectrum which becomes more and more structured as seen in the right panels of Fig. 4.7(A–C). In other words, due to the coupling, the zeroth–order states share oscillator strength and appear as pairs (see also Fig. 4.3). Still not all states are equally accessible and one has to state that it is *not* the total but a *local* density of states which is responsible for the energy flow. Furthermore, Fig. 4.7 illustrates that there are likely to exist correlations between different couplings. In other words, the basic assumption made in the derivation of the Golden Rule expression Eq. (4.5) are not

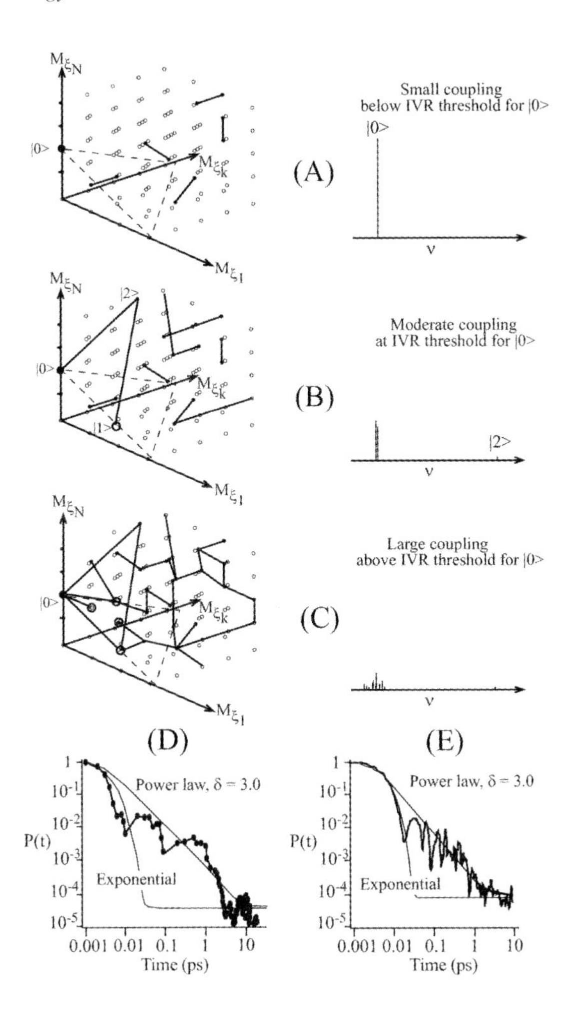


**Figure 4.7:** IVR in a polyatomic system. Panels A – C show the IVR process in a three–dimensional section of the space of quantum numbers $(M_{\xi_1}, M_{\xi_2} \dots M_{\xi_N})$ (shown as open circles, thick solid lines connect strongly coupled states, and the dashed triangle is a surface of constant vibrational energy). The position corresponding to some initial state $|0\rangle$ is indicated by a full circle. The coupling strength of $|0\rangle$ to other states increases from panel A to C (depending on the energetic position relative to the IVR threshold). The related (infrared) absorption lines (right panels) show a an increasing splitting. In panels D and E the decay of the survival probability is shown for $SCCl_2$ as obtained from the experiment (fluorescence and stimulated emission pumping) and quantum dynamics simulation, respectively. Exponential (Eq. (4.6)) and power law (Eq. (4.7)) fits of the decay are shown as thin lines. (figure courtesy of M. Gruebele, for more details see also [Gru03])

fulfilled here.

How is this reflected in the behavior of the survival probability? Once the state is at the IVR threshold there might be a few other states (denoted $|1\rangle$ and $|2\rangle$ in Fig. 4.7B) which dominate the coupling and lead to quantum beats (Fig. 4.5). Going beyond this threshold the number of possible zeroth–order states for energy transfer increases and $P_{\rm surv}$ will decay almost exponentially. During a certain time interval the decay is reasonably reproduced by a single exponential function. After a characteristic time, however, the energy flow will become dominated by the details of the local density of states and the local anharmonic couplings for a given point in state space. It was found that in this intermediate time range, the decay of $P_{\rm surv}$ will no longer be exponential and it is better described by a power law:

$$P_{\rm surv}(t) \sim (1-\sigma)t^{-\delta/2} + \sigma \; . \tag{4.7}$$

Here the exponent $\delta$ is a parameter which reflects the local dimensionality of IVR; usually one finds $\delta \ll N$ (cf. Figs. 4.7D and E). This period of power law decay is considerably longer than the initial exponential decay. Finally, $P_{\rm surv}$ will fluctuate around the value of the dilution factor which is a manifestation of the finite size of the state space.

## 4.3 Intermolecular Vibrational Energy Relaxation

In the previous section we considered the *intra*molecular energy flow in large polyatomic molecules. The latter where assumed to be isolated from any environment such that *inter*molecular processes could be neglected. The complicated IVR dynamics already indicated that large polyatomics may form a reservoir on their own. However, the flow of energy out of an initially prepared state cannot necessarily be described by a single exponentially decaying function at all times. In the following sections we will extend the scope and add an additional external reservoir which is provided by a solid state matrix or a solvent. On general grounds it is to be expected that this will considerably enhance the density of states for vibrational energy flow such that for an appropriate separation of relevant and bath DOFs one may recover rate dynamics. To set the stage we will start with the discussion of diatomic molecules in condensed phase in Sections 4.3.1 and 4.3.2 before focusing on the polyatomic case in Section 4.4.

### 4.3.1 Diatomic Molecule in Solid State Environment

#### The System–Reservoir Hamiltonian

The simplest setup for studying vibrational relaxation is certainly the situation of a single diatomic molecule in a monoatomic solid state environment. Typical examples are dihalogens such as $I_2$ or ClF, for instance, built into a rare gas matrix at low temperatures (cf. Fig. 4.2). A preliminary stage in terms of the system size of such a matrix environment would be a cluster which offer in particular the intriguing possibility to study the transition from small to large bulk–like systems. In this section we will focus, however, on diatomics in low temperature rare gas matrix environments as shown in the left panel of Fig. 4.8. The separation into system and bath degrees of freedom in the sense of Chapter 2 strongly depends on the considered dynamics. If, for instance, the diatomic molecule is photoexcited onto a PES which is dissociative in the gas phase (Fig. 4.8A) the fragments might be able to escape the matrix cage. However, if their energy is not sufficient they will recombine after collision with the

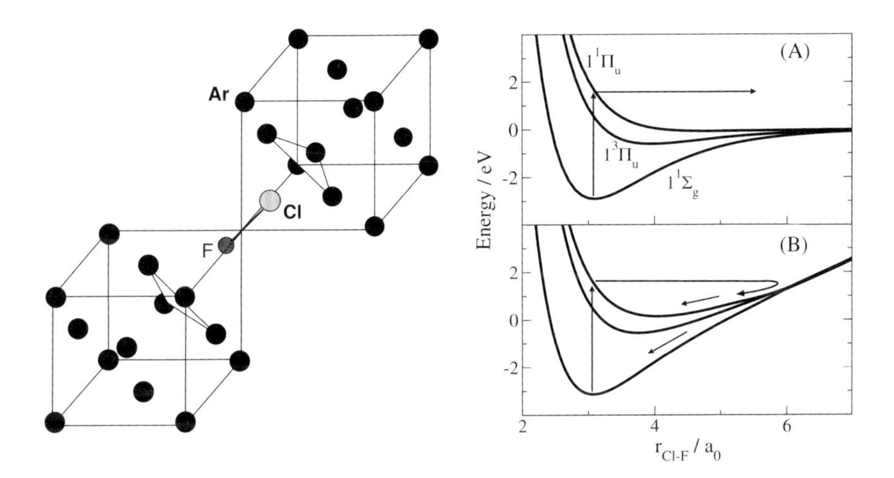


**Figure 4.8:** Photodissociation of a diatomic molecule (ClF) in a rare gas lattice (Ar). Left panel: Ar lattice structure with ClF on a mono–substitutional site. The equilateral triangles mark the window for a possible cage exit after photodissociation. Right panel: Potential energy curves for the lowest singlet and triplet states in the gas phase (A) and in the Ar lattice (B). Upon photoexcitation to the $^1\Pi$ state the Cl–F bond dissociates in the gas phase. In the Ar lattice the fragments recombine after collision with the lattice atoms. (figure courtesy of M. Schröder)

nearest matrix atoms as shown in Fig. 4.8B (cage effect). This process can be accompanied by a nonadiabatic transition, e.g., back onto the electronic ground state. Here, the vibrationally hot diatomic bond will relax into thermal equilibrium as a consequence of the interaction with the matrix environment. Clearly, the theoretical modelling of the collision process requires to take into account the solvent cage explicitly, e.g., in the spirit of a supermolecule approach (cf. Chapter 3). On the other hand, the relaxation in the ground state the matrix acts merely as a thermal heat bath. It is the latter process which we will discuss in the following.

The bond length, $s$, is taken as the relevant system degree of freedom and the remaining coordinates, $Z \equiv \{Z_k\}$, form the bath. The Hamiltonian can therefore be written in the standard form

$$H = H_{\rm S} + H_{\rm R} + H_{\rm S-R}\,, \tag{4.8}$$

where

$$H_{\rm S} = \frac{p^2}{2\mu_{\rm s}} + V_{\rm S}(s)\,, \tag{4.9}$$

$$H_{\rm R} = \sum_k \frac{P_k^2}{2M_k} + V_{\rm R}(Z)\,, \tag{4.10}$$

and

$$H_{\rm S-R} = V(s, Z)\,. \tag{4.11}$$

To simplify our considerations we will restrict ourselves to the case where the diatomic guest molecule does not appreciably disturb the host's lattice of rare gas atoms. Further we will allow for only small vibrations of all atoms with respect to their equilibrium positions (low temperature limit). Thus, we can simplify the system potential by assuming a harmonic approximation to be valid:[3]

$$V_{\rm S}(s) = \frac{\mu_{\rm s}\Omega_{\rm s}^2}{2} s^2 \,, \tag{4.12}$$

with $\mu_{\rm s}$ being the reduced mass of the diatomic and $\Omega_{\rm s}$ its vibrational frequency. In this case the eigenfunctions and eigenenergies of $H_{\rm S}$ are given by Eq. (2.70) and (2.71), respectively.

Concerning the reservoir (matrix) degrees of freedom it is customary to map the small amplitude vibrations of the coordinates $Z$ with respect to their equilibrium values to collective normal modes $q = \{q_\xi\}$ (lattice phonons, see Section 2.6.1). Thus, the reservoir Hamiltonian becomes a sum of decoupled harmonic oscillators (cf. Eq. (2.67))

$$H_{\rm R} = \frac{1}{2}\sum_\xi \left(p_\xi^2 + \omega_\xi^2 q_\xi^2\right) \,. \tag{4.13}$$

The interaction potential in Eq. (4.11) is written in terms of the introduced normal mode coordinates, $V(s, \{q_\xi\})$. Suppose we expand this potential with respect to the equilibrium configuration ($s = 0, \{q_\xi\} = 0$) the lowest–order nontrivial term is of first order with respect to both types of coordinates

$$\begin{aligned} V(s,\{q_\xi\}) &= \frac{1}{2}\sum_\xi \frac{\partial^2 V(s,\{q_\xi\})}{\partial s \partial q_\xi}\Big|_{s=0,\{q_\xi\}=0} s\, q_\xi \\ &= s\sum_\xi c_\xi q_\xi \,. \end{aligned} \tag{4.14}$$

Note that we will frequently assume that the trivial zeroth–order term $V(s = 0, \{q_\xi\} = 0)$ has been included in the definition of either the system or the bath Hamiltonian. Further we assume that the Taylor expansion is performed around a minimum of the total PES where first order derivatives vanish. In the last line of Eq. (4.14) we introduced the coupling constant $c_\xi$, thus transforming the interaction Hamiltonian into the form (3.205) which has been discussed in Section 3.6.2. Note that in Section 3.6.2 the system part $K(s)$ of the interaction Hamiltonian was chosen to be dimensionless. For the present case this means that $K(s) = s\sqrt{2\mu_{\rm s}\Omega_{\rm s}/\hbar}$. Accordingly, the reservoir part of the interaction Hamiltonian is given by $\Phi(q) = \sqrt{\hbar/2\mu_{\rm s}\Omega_{\rm s}}\sum_\xi c_\xi q_\xi$.

In the following we will investigate the influence of the system–reservoir interaction on the system dynamics. First, we will discuss the case of *no* interaction, that is, the regime of *coherent* dynamics. In a second step this is confronted with the *dissipative* dynamics observed in the condensed phase.

---

[3]Note that the extension to anharmonic potentials such as shown in Fig. 4.8B is not a principal problem but might require to calculate matrix elements numerically (see below).

## Coherent Dynamics

If we neglect the system–reservoir interaction Hamiltonian, $H_{\mathrm{S-R}}$, the problem is reduced to the solution of the time–dependent Schrödinger equation for the harmonic oscillator Hamiltonian, $H_{\mathrm{S}}$, given in Eqs. (4.9) and (4.12). A convenient representation of the Hamiltonian is in terms of creation ($C^+$) and annihilation ($C$) operators as introduced in Section 2.6.2 (cf. Eqs. (2.76), (2.81), and (2.82)). Since we are considering only a single mode, the Hamiltonian in this representation is given by $H_{\mathrm{S}} = \hbar\Omega_{\mathrm{s}}\,(C^+C + 1/2)$; the eigenvalues and eigenstates are $E_N = \hbar\Omega_{\mathrm{s}}\,(N + 1/2)$ and $|N\rangle = (C^+)^N/\sqrt{N}\;\;|0\rangle$, respectively. In the following we will use the dimensionless coordinate $S = \sqrt{2\mu_{\mathrm{s}}\Omega_{\mathrm{s}}/\hbar}\;s$.

Let us consider the dynamics of an initially displaced oscillator ground state:

$$\langle S|\chi(t=0)\rangle = \langle S|\chi_0\rangle = \chi_0(S - S^{(0)})\,. \tag{4.15}$$

Here the non–displaced ground state wave function is given by (cf. Eq. (2.70))

$$\chi_0(S) = \langle S|0\rangle = \left(\frac{\mu_{\mathrm{s}}\Omega_{\mathrm{s}}}{\pi\hbar}\right)^{\frac{1}{4}} \exp\left(-\frac{S^2}{4}\right), \tag{4.16}$$

and the shift is $S^{(0)}$. In Chapter 5 we will discuss how this type of initial state can be prepared during an ultrafast optical transition between two electronic potential energy surfaces, for example. It is important to note that $\chi_0(S - S^{(0)})$ is no eigenstate but a superposition state of the oscillator Hamiltonian. This superposition state is called *vibrational wave packet*. The calculation of the subsequent dynamics of this wave packet can be based on an expansion of the time–dependent state vector with respect to the oscillator eigenstates :

$$|\chi(t)\rangle = \sum_N \langle N|\chi_0\rangle\, e^{-i\Omega_{\mathrm{s}}(N+\frac{1}{2})t}\; |N\rangle\,. \tag{4.17}$$

In this case we would have to calculate the Franck–Condon factors (cf. Eq. (2.98))

$$\langle N|\chi_0\rangle = \int dS\; \chi_N^*(S)\,\chi_0(S - S^{(0)})\,. \tag{4.18}$$

In the supplementary Section 4.5.1 a more intuitive expression for the time–dependent state vector is derived using the displacement operator introduced in Section 2.6.2. The resulting time dependence of the state vector $\chi(S,t)$ follows as:

$$\begin{aligned}
\chi(S,t) &= \exp\left\{-(i/2)\left(\Omega_{\mathrm{s}}t - (S^{(0)\,2}/4)\sin(2\Omega_{\mathrm{s}}t)\right)\right\} \\
&\quad \times \langle S|\; \exp\left\{-i(S^{(0)}/2)\sin(\Omega_{\mathrm{s}}t)\;\hat{S}\right\}\; D^+(-(S^{(0)}/2)\,\cos(\Omega_{\mathrm{s}}t))\,|0\rangle \\
&= \exp\left\{-(i/2)\left(\Omega_{\mathrm{s}}t - (S^{(0)\,2}/4)\sin(2\Omega_{\mathrm{s}}t) - S^{(0)}S\;\sin(\Omega_{\mathrm{s}}t)\right)\right\} \\
&\quad \times \chi_0(S - S^{(0)}\cos(\Omega_{\mathrm{s}}t))\,.
\end{aligned} \tag{4.19}$$

First we note that the oscillating phase factor cancels if we consider the time evolution of the probability distribution $P(S,t) = |\chi(S,t)|^2$. This probability distribution will retain

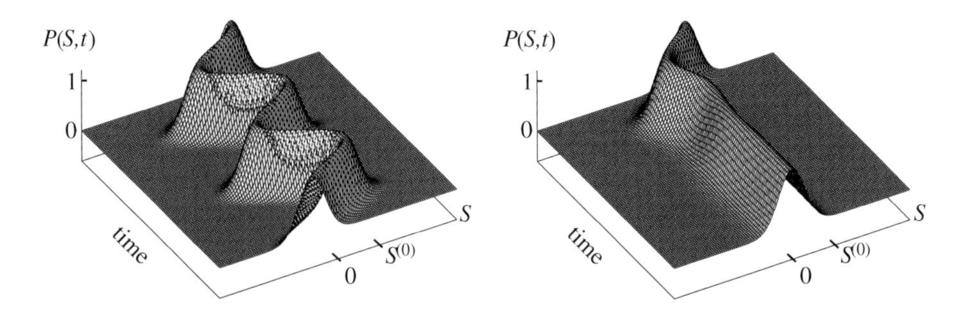


**Figure 4.9:** Illustration of the time–dependent probability distribution for an initially displaced harmonic oscillator ($\chi_0(S - S^{(0)})$) in the coherent (left) and dissipative (right) regime. (figure courtesy of H. Naundorf)

the Gaussian shape of the ground state wave function, but its maximum moves according to $S(t) = S^{(0)} \cos\Omega_s t$ between $\pm S^{(0)}$. Therefore, the time evolution of the expectation value of the coordinate operator follows the classical trajectory if the initial value of the coordinate is given by $S^{(0)}$ and the initial momentum is zero. The coherent wave packet dynamics is illustrated in Fig. 4.9 (left panel).

### Dissipative Dynamics

Next we will incorporate the interaction between relevant system and reservoir, $H_{\text{S-R}}$. This will be done by using the Quantum Master Equation in the representation of the relevant system's eigenstates $\{|N\rangle\}$. Further we invoke the Markov approximation and for simplicity also the secular approximation. Thus, the equations of motion given in Section 3.8.3 can be straightforwardly adapted to the present situation. We obtain the expression

$$\frac{\partial}{\partial t}\rho_{MN} = -\delta_{MN}\sum_K \left(k_{MK}\rho_{MM} - k_{KM}\rho_{KK}\right) - (1-\delta_{MN})\,(i\Omega_s + \gamma_M + \gamma_N)\rho_{MN}\,. \tag{4.20}$$

According to Eq. (3.289) the energy relaxation rates, $k_{MN} = k_{M\to N}$, are given by

$$k_{MN} = |\langle M|K(s)|N\rangle|^2\; C(\omega_{MN})\,. \tag{4.21}$$

Assuming that $C(0) = 0$ holds, we have for the dephasing rates $\gamma_M = \sum_N k_{MN}/2$. The correlation function $C(\omega)$ has been given in Eq. (3.227). The transition rates can be calculated straightforwardly for the interaction Hamiltonian given in Eq. (4.14). We obtain for the matrix elements of the system part (cf. Section 3.6.2)

$$\langle M|K(s)|N\rangle = \left(\sqrt{N}\,\delta_{M,N-1} + \sqrt{N+1}\,\delta_{M,N+1}\right)\,. \tag{4.22}$$

using this result the rate for the vibrational transition between the states $|M\rangle$ and $|N\rangle$,

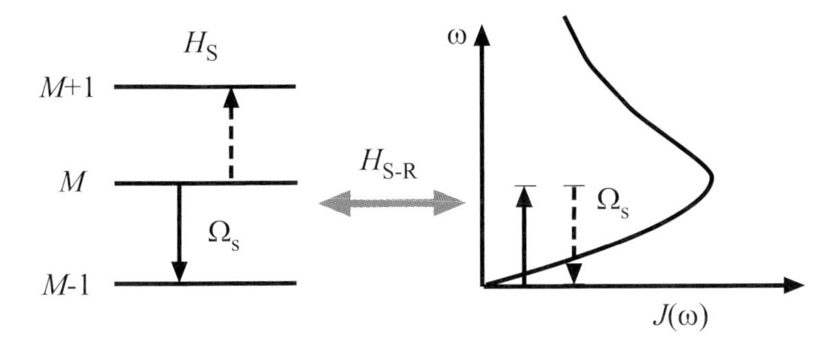


**Figure 4.10:** Schematic view of vibrational relaxation out of the harmonic oscillator state with quantum number $M$ due to the coupling to some environment, characterized by the spectral density $J(\omega)$.

Eq. (3.289), becomes

$$k_{MN} = \Big(\delta_{M,N-1}(M+1)C(-\Omega_{\rm s}) + \delta_{M,N+1}MC(\Omega_{\rm s})\Big) \,. \tag{4.23}$$

For the oscillator reservoir the Fourier transform of the correlation function was given in Eq. (3.213). Inspecting this rate one can draw a number of conclusions as visualized in Fig. 4.10:

First, only transitions between neighboring system oscillator states are possible if $H_{\rm S-R}$ is linear in the system coordinate $s$. Second, the relaxation rates grow linearly with the quantum number of the excited state. If we define the inverse lifetime of the state $|M\rangle$, $\tau_M^{-1}$, in terms of all possibilities to make transitions out of this state we obtain

$$\tau_M^{-1} = \sum_N k_{MN} = \big((M+1)C(-\Omega_{\rm s}) + MC(\Omega_{\rm s})\big) \,. \tag{4.24}$$

Thus, with increasing quantum number the lifetime is decreasing as $1/M$.

Third, according to Eq. (3.227) the correlation function $C(\omega)$ is proportional to the spectral density of the matrix environment. From Eq. (4.23) we see that the spectral density is "probed" only at the system's oscillator frequency $\Omega_{\rm s}$ (cf. Section 3.8.2).

Finally, for the bilinear coupling, upward transitions from $|M\rangle$ to $|M+1\rangle$ require that there is a bath mode at frequency $\Omega_{\rm s}$ which is also thermally accessible according to the distribution function $n(\Omega_{\rm s})$ (see Eq. (3.227)). Downward transitions release energy into the bath and occur due to the factor $1 + n(\Omega_{\rm s})$ even if the temperature goes to zero. They also require a bath mode having the same frequency. Thus, in both cases energy conservation is fulfilled.

A comparison between the coherent and the dissipative dynamics is most conveniently performed by inspection of the time evolution of the probability distribution for the oscillator coordinate. This quantity is readily obtained from the knowledge of the time evolution of the reduced density matrix $\rho_{MN}(t)$:

$$P(S,t) \equiv \rho(S,S;t) = \sum_{M,N} \chi_M(X)\chi_N^*(S)\rho_{MN}(t) \,. \tag{4.25}$$

This result generalizes the information contained in $|\chi(S,t)|^2$ to the case of an open quantum system. In Fig. 4.9 (right panel) we have plotted the resulting *dissipative wave packet motion* of the harmonic oscillator. As in the case of coherent dynamics, the shape of the moving wave packet is not modified. However, stressing the classical analogy we notice that the dynamics resembles the behavior of a classical damped oscillator. This is also what one expects for the motion of a wave packet representing the bond vibration of a diatomic in a solid state matrix as illustrated in Figs. 4.2 and 4.8.

## 4.3.2 Diatomic Molecules in Polyatomic Solution

### Classical Force-Force Correlation Functions

The introduction of normal modes for the bath in Eq. (4.13) presupposes an environment which performs only small oscillations around its equilibrium configuration. The situation in solution at liquid state temperatures is of course different even though normal modes can be introduced under certain assumptions, as we will see below. We will start, however, with the more general case. In the simulation of solvents it is customary to use a classical description since quantum effects are not expected to play any appreciable role for most systems. In recent years classical molecular dynamics based on the numerical solution of Newton's equations has been developed into a versatile tool for studying systems ranging from atomic solvents to complex proteins. Most of the interaction potentials between different atoms, groups of atoms within a molecule, or molecules can ultimately be traced back to the Coulomb interaction between electronic and nuclear charges (see Section 2.2). Especially for the long–range part of the interaction potential this implies that there are contributions which can be identified as purely classical interactions between point charges. However, the quantum mechanical character of the electronic degrees of freedom modifies the classical energy particularly for short distances where the wave functions overlap.

Instead of taking into account the Coulomb interactions on an ab initio level it is customary to use parameterized *empirical* potentials whose parameters are chosen in a way to obtain agreement with experimental results, for example, for thermodynamic properties of the solvent. These empirical interaction potentials are usually partitioned into parts involving only a single atom (for instance, potentials describing external fields or container walls), pairs of atoms (bonding or repulsive interaction), three atoms (for instance, bending motions), four atoms (for instance, dihedral motions), etc. Thus, we can write for a potential depending on the nuclear coordinates $R$ of some system

$$V(R) = \sum_{n} v_1(\mathbf{R}_n) + \sum_{m>n} v_2(\mathbf{R}_m, \mathbf{R}_n) + \sum_{k>m>n} v_3(\mathbf{R}_k, \mathbf{R}_m, \mathbf{R}_n) + \dots \, . \tag{4.26}$$

The pair potential most successfully applied is the so–called *Lennard–Jones potential*

$$V_{\rm LJ}(|\mathbf{R}_m - \mathbf{R}_n|) = 4\epsilon\left(\left(\frac{\sigma}{|\mathbf{R}_m - \mathbf{R}_n|}\right)^{12} - \left(\frac{\sigma}{|\mathbf{R}_m - \mathbf{R}_n|}\right)^{6}\right) . \tag{4.27}$$

It has a steeply rising repulsive wall for inter–particle separations less than $\sigma$ (effective particle diameter due to non–bonding interactions in the region of wave function overlap), a negative well of depth $\epsilon$, and a long–range attractive $r^{-6}$ tail (van der Waals interaction, for instance, due to so–called dispersion interactions originating from the correlated electronic motion in different molecules). We would like to stress that this effective pair potential may contain the effect of complicated many body interactions in an averaged way. Further it should be noted that for situations where long–range electrostatic interactions are important (for example, if the system contains ions), the Lennard–Jones potential is not sufficient and the classical Coulomb interaction has to be taken into account explicitly.

For the following we will assume that the potentials for the interaction between the solvent molecules and between solvent and solute, $V(s, Z)$, are known. Here, $Z = \{Z_k\}$ is the respective set of the nuclear solvent molecule coordinates. In a first step we suppose that this potential can be expanded in terms of the internal coordinate $s$ of the diatomic molecule as follows

$$V(s,Z) = V(s=0,Z) + \left.\frac{\partial V(s,Z)}{\partial s}\right|_{s=0} s + \ldots . \tag{4.28}$$

The first term is a potential for the bath coordinates if the system coordinate is kept fixed at its equilibrium value, $s = 0$. It can be incorporated into the reservoir Hamiltonian $H_{\rm R}$. The second part is the force of the solvent acting on the diatomic (if it is kept fixed at its equilibrium position). This term is the system–reservoir coupling we are looking for, and we write

$$H_{\rm S-R} = s\left.\frac{\partial V(s,Z)}{\partial s}\right|_{s=0} = -sF . \tag{4.29}$$

At this point $F$ is still an operator. Its autocorrelation function $\langle F(t)F(0)\rangle_{\rm R}$ is the reservoir correlation function whose Fourier transform according to Eq. (3.289) enters the desired relaxation rates. Since we are aiming at a classical description of the reservoir the quantum correlation function has to be replaced by its classical counterpart given in Eq. (3.247) ($\Phi_u = \Phi_v = F$). As explained in Section 3.6.6 detailed balance can be fulfilled despite the time reversal symmetry of the classical correlation function, if we identify $\langle F(t)F(0)\rangle_{\rm R}$ with the symmetrized quantum correlation function $C^{(+)}(t)$. Suppose the eigenvalue problem, $H_{\rm S}|M\rangle = E_M|M\rangle$, for the system part of the Hamiltonian has been solved, the relaxation rates are obtained from (3.289) as ($\hbar\Omega_{MN} = E_M - E_N$)

$$\begin{aligned} k_{MN} &= \frac{|s_{MN}|^2}{1+\exp(-\hbar\Omega_{MN}/k_{\rm B}T)} \int_{-\infty}^{\infty} dt\, e^{i\Omega_{MN}t}\, C^{(+)}(t) \\ &= \frac{2|s_{MN}|^2}{1+\exp(-\hbar\Omega_{MN}/k_{\rm B}T)} \int_0^{\infty} dt\, \cos(\Omega_{MN}t)\, \langle F(t)F(0)\rangle_{\rm R} \\ &= \frac{2|s_{MN}|^2\, c(\Omega_{MN})}{1+\exp(-\hbar\Omega_{MN}/k_{\rm B}T)} . \end{aligned} \tag{4.30}$$

Here $c(\Omega_{MN})$ is the Fourier cosine transform of the classical force–force autocorrelation function which is obtained within the present model using classical *equilibrium* molecular dynamics simulations. In the limit that the system is harmonic we can calculate the matrix elements $s_{MN}$ and recover Eq. (3.289) with $C(\Omega_{\rm s}) \rightarrow 2c(\Omega_{\rm s})/(1+\exp(-\hbar\Omega_{\rm s}/k_{\rm B}T))$ (to establish this relation completely, $s$ has to be transformed into a dimensionless form, see above).

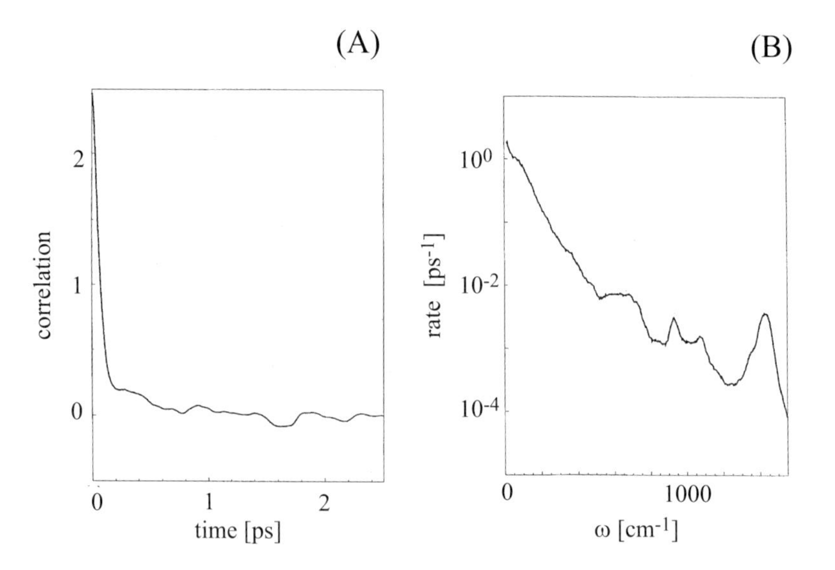


**Figure 4.11:** Molecular dynamics simulations of HgI in ethanol solution. (A) Classical autocorrelation function of the force acting along the HgI bond, and (B) the respective relaxation rate. In the actual experiment the diatomic molecule is produced as a fragment in the photodissociation of $HgI_2$. (reprinted with permission from [Gna96]; copyright (1996) American Institute of Physics)

In Fig. 4.11A we plot as an example the correlation function of the force an ethanol solvent exerts on the bond of HgI which is obtained as a fragment in the photodissociation of $HgI_2$. There are two regimes easily discernible in this figure. First there is a rapid decay on the time scale of about 60 . . . 100 fs which is rationalized as follows: Given some initial configuration chosen from an equilibrium ensemble (see Eq. (3.248)) the molecules perform inertial motions up to a point where collisions with neighboring molecules set in. After such collisions the forces get randomized. Thus, the decay time of the correlation function represents the average time between two collision events, i.e., it reflects mostly the short–range part of the interaction potential. After this initial phase the correlation function does not decay to zero in the time window shown in Fig. 4.11. Instead the decay rate slows done appreciably. In this regime the long–range forces give the major contributions to the correlation function. There appears to be also some oscillatory behavior which can be traced back to the internal vibrations of the solvent. The vibrational relaxation rate (which is proportional to the Fourier transform of this correlation function, see Eq. (4.23)) is shown in Fig. 4.11B. In order to see what types of motions of the solvent are responsible for the relaxation of the HgI bond vibration in the harmonic approximation one has to analyze the spectrum at the vibrational frequency, which is 130 $cm^{-1}$. This leads to the conclusion that mostly the Lennard–Jones type collisions are responsible for the relaxation in this system. The high–frequency peaks ($>$900 $cm^{-1}$) which are due to internal modes of the ethanol have no influence on the relaxation of HgI.

Finally, we would like to point out that there is a limitation involved in the calculation of the relaxation rates we have outlined so far. Remember that the system coordinate has been fixed at its equilibrium value. For the symmetric oscillator potential this can be viewed as if the environment only sees the time–averaged position of the system oscillator. Any type of back reaction, where the reservoir notices the actual position of the system coordinate whose dynamics it influences would be beyond the QME treatment. In the spirit of an adiabatic separation of system and bath motions the rigid bond approximation will be particularly good for high frequency system oscillators.

### Instantaneous Normal Modes

In Fig. 4.11 we have seen that the initial stage of the solvent's force autocorrelation function is characterized by a rapid decay. This behavior is typical also for other solvent correlation functions, for example, of the velocity or the position. Focusing on the short time dynamics of the solvent the *instantaneous normal mode* approach maps the motions of the solvent molecules onto the dynamics of independent collective oscillators which interact with the solute. This results in a microscopically defined system–bath Hamiltonian.

Suppose we have chosen some initial configuration for the solute and solvent coordinates, $(s(0), Z(0))$, from the classical equilibrium distribution function as given in Eq. (3.248).[4] For short enough times it is possible to expand the potential energy in terms of the deviation from this initial configuration, $\Delta s(t) = s(t) - s(0)$ and $\Delta Z(t) = Z(t) - Z(0)$. Considering, for example, the solvent potential in Eq. (4.10) one can write

$$\begin{aligned} V_{\mathrm{R}}(Z(t)) &= V_{\mathrm{R}}(Z(0)) + \sum_k \frac{\partial V_{\mathrm{R}}(Z)}{\partial Z_k}\Big|_{Z=Z(0)} \Delta Z_k(t) \\ &+ \frac{1}{2}\sum_{kl} \frac{\partial^2 V_{\mathrm{R}}(Z)}{\partial Z_k \partial Z_l}\Big|_{Z=Z(0)} \Delta Z_k(t) \Delta Z_l(t) \; . \end{aligned} \tag{4.31}$$

It goes without saying that the time range where this expansion applies is intimately connected to the specific situation, that is, to the form of the potential, the masses of the particles and so on. With this type of potential the Hamiltonian (4.10) is readily diagonalized by a linear normal mode transformation (see Eq. (2.65)), $\Delta Z_k(t) = \sum_\xi M_k^{-1/2} A_{k\xi}(Z(0))\, q_\xi(t)$ leading to the time–dependent collective (mass–weighted) reservoir coordinates $q_\xi(t)$. The reservoir potential then becomes

$$V_{\mathrm{R}}(q_\xi(t)) = V_{\mathrm{R}}(Z(0)) - \sum_\xi \left( F_\xi(Z(0))\, q_\xi(t) + \frac{1}{2}\omega_\xi^2(Z(0))\, q_\xi^2(t) \right) . \tag{4.32}$$

Here, the frequencies are resulting from the diagonalization of the (mass–weighted) Hessian matrix (see Section 2.6.1) at the initial configuration and the forces for this configuration are

$$F_\xi(Z(0)) = -\sum_k M_k^{-1/2} \frac{\partial V_{\mathrm{R}}(Z)}{\partial Z_k}\Big|_{Z=Z(0)} A_{k\xi}(Z(0)) \; . \tag{4.33}$$

[4] In practice it is customary to solve Newton's equations of motion starting from some guessed initial configuration. After the configuration has equilibrated initial conditions for configuration averaging are picked along the trajectories of the equilibrium simulation.

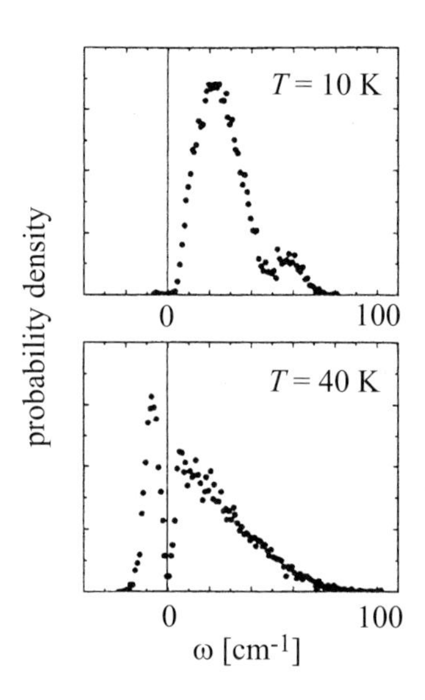


**Figure 4.12:** Instantaneous normal mode density of states for an $Ar_{13}$ cluster at 10 K (upper panel) and 40 K (lower panel). (reprinted with permission from [Str95]; copyright (1995) American Chemical Society)

Let us neglect the presence of the solute for a moment. In this case the spectral distribution of the instantaneous eigenvalues of the Hessian gives important information about the solvent. To discuss the instantaneous density of states for the liquid normal modes it is appropriate to use the following quantity

$$\mathcal{N}_{\mathrm{INM}}(\omega) = \left\langle \sum_{\xi} \delta(\omega - \omega_{\xi}(Z(0)) \right\rangle_{Z(0)} , \tag{4.34}$$

which contains the average with respect to different initial configurations. In Fig. 4.12 we show the instantaneous normal mode density of states for a small Argon cluster at different temperatures. Within the instantaneous normal mode approach it is typical that the Hessian also has negative eigenvalues, $\omega_{\xi}^2(Z(0)) < 0$, implying "imaginary" frequencies. This indicates that the chosen initial configuration is not a minimum in all directions of the potential energy surface as it is reasonable for liquids. Notice, however, that upon decreasing the temperature the cluster is frozen and the negative frequency contributions disappear in the example of Fig. 4.12.

We now return to the solute–solvent system and consider the interaction Hamiltonian (4.11). Assuming that the same type of short time expansion is valid we can write after introducing the solvent normal modes

$$V_{\mathrm{S-R}}(s, Z) = V_{\mathrm{S-R}}(s(0), Z(0)) + \Delta s(t) \sum_{\xi} c_{\xi}(s(0), Z(0))\, q_{\xi}(t) \; . \tag{4.35}$$

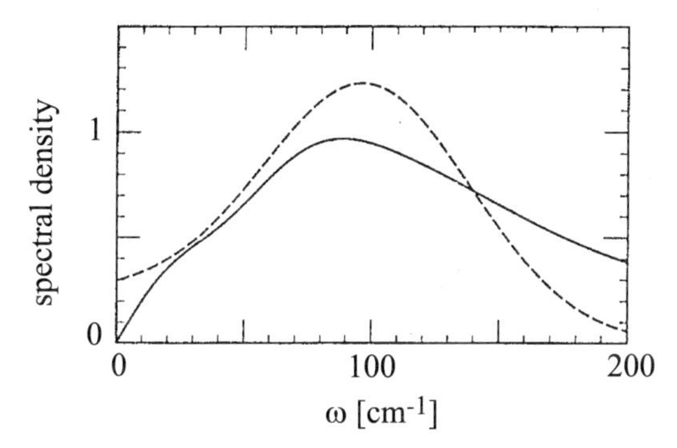


**Figure 4.13:** The instantaneous normal mode spectral density according to Eq. (4.37) (dashed line) is compared with the exact result (solid line) for a model homonuclear diatomic in a rare gas environment. (reprinted with permission from [Goo96]; copyright (1996) American Institute of Physics)

Here, we have retained the bilinear term only (cf. Eq. (4.14)) and the solvent–solute coupling constant is given by

$$c_\xi(s(0), Z(0)) = \sum_k A_{k\xi}(s(0), Z(0)) M_k^{-1/2} \frac{\partial V_{\text{S-R}}(s, Z)}{\partial Z_k \partial s}\Big|_{s=s(0), Z=Z(0)} . \tag{4.36}$$

Comparing Eq. (4.35) with (3.205) we realize that we have obtained the *classical* system–bath Hamiltonian function in the generic form with all parameters microscopically well defined. Therefore, the results of Section 3.6 can readily be adapted to the present case. In particular one can define the spectral density within the instantaneous normal mode approach as

$$J_{\text{INM}}(\omega) = \left\langle \sum_\xi c_\xi^2(s(0), Z(0))\, \delta(\omega - \omega_\xi(Z(0)) \,. \right\rangle_{Z(0)} \tag{4.37}$$

In Fig. 4.13 we show results of a calculation of $J_{\text{INM}}(\omega)$ for a model homonuclear diatomic molecule in an atomic solvent. For comparison also the exact result is shown. The most pronounced differences are for low and for high frequencies. The deviation in the low–frequency part is due to motions which take place on a time scale well beyond the short–time approximation made in the present approach. The failure at high frequencies can be attributed to the low–order expansion of the interaction potential, which neglects transitions involving more than one vibrational quantum. Such so–called multi–phonon transitions would contribute in particular to the high frequency part of the spectral density (see Section 4.4).

Despite the fact that a liquid is inherently unstable as compared with a solid state environment the idea of extracting information on the solute–solvent dynamics from the short time behavior of classical trajectories has proven to be quite powerful. Using simple test systems, results of exact molecular dynamics simulations have been compared with predictions by the instantaneous normal mode approach, thus elucidating the range of validity for the latter (see Fig. 4.13) In addition, from the conceptually desirable link between the formal system–bath Hamiltonian and actual microscopic dynamics, analysis of instantaneous normal modes provides the key to the understanding of energy relaxation processes in terms of *collective* solvent motions. In contrast, within standard molecular dynamics simulations the collective character of the solvent response is usually hidden in the correlation functions.

## 4.4 Polyatomic Molecules in Solution

### 4.4.1 System–Bath Hamiltonian

After having discussed aspects of intramolecular vibrational dynamics of isolated polyatomics as well as vibrational energy relaxation of diatomics in solution we are now in the position to address the general situation of a polyatomic solute in a polyatomic solvent. On the one hand we know that the intramolecular motions can be adequately described in terms of zeroth–order states such as given by the normal mode expansion. On the other hand, we have seen that even the complex dynamics of a polyatomic solvent can be mapped onto a set of collective harmonic oscillators provided one is interested in the short time behavior of solvent correlation functions, for instance. Given this information there appears to be still some freedom in the choice of the zeroth–order states for the solute plus the solvent. The most frequently used approach to this problem is as follows: Assume that we have identified a bright state of the solute corresponding, for example, to a local vibrational mode. In this case the system Hamiltonian is given by

$$H_{\mathrm{S}} = \frac{p^2}{2\mu_{\mathrm{s}}} + U(s) \;, \tag{4.38}$$

with $U(s)$ being the system potential along the considered mode which in the general case may be anharmonic. The remaining degrees of freedom, i.e., the intramolecular as well as the solvent degrees of freedom, are then treated within the harmonic approximation. If the corresponding normal mode coordinates, $\{q_\xi\}$, are introduced, $H_{\mathrm{R}}$ is given by Eq. (4.10). For the solvent this may imply, for instance, the introduction of instantaneous normal modes. Since it is assumed that the bath has been diagonalized without the system degrees of freedom, there will be a coupling between both subsystems

$$H^{\mathrm{I}}_{\mathrm{S-R}} = V_{\mathrm{S-R}}(s, \{q_\xi\}) \equiv V_{\mathrm{S-R}}(s, q) \;. \tag{4.39}$$

In the next step we introduce a Taylor expansion of this interaction potential with respect to both types of coordinates. We write

$$\begin{aligned} V_{\mathrm{S-R}}(s,q) \;&=\; V_{\mathrm{S-R}}(s=0,q=0) + \frac{\partial V_{\mathrm{S-R}}(s,0)}{\partial s}\Big|_{s=0} s + \sum_\xi \frac{\partial V_{\mathrm{S-R}}(0,q)}{\partial q_\xi}\Big|_{q=0} q_\xi \\ &+ \frac{1}{2}\frac{\partial^2 V_{\mathrm{S-R}}(s,q)}{\partial s^2}\Big|_{s=0} s^2 + \frac{1}{2}\sum_{\xi_1\xi_2} \frac{\partial^2 V_{\mathrm{S-R}}(s,q)}{\partial q_{\xi_1}\partial q_{\xi_2}}\Big|_{q=0} q_{\xi_1} q_{\xi_2} \\ &+ \sum_\xi \frac{\partial^2 V_{\mathrm{S-R}}(s,q)}{\partial s \partial q_\xi}\Big|_{s=0,q=0} s q_\xi + \dots \;. \end{aligned} \tag{4.40}$$

Note that this expansion illustrates the appropriateness of the *ansatz* we made for $H_{\mathrm{S-R}}$ in Eq. (3.146) since it is a sum of terms which can be factorized into a system part ($K(s)$)

and a reservoir part ($\Phi(q)$). This is in accord with another statement made in Section 3.5.3, that is, the factorized system–bath interaction Hamiltonian provides sufficient flexibility to model dissipative quantum dynamics. Before continuing it may be useful to compare this approach with an alternative one which assumes that the normal modes of *all* degrees of freedom, denoted here as $\{\tilde{q}_\xi\}$ to distinguish them from the set $\{q_\xi\}$, are used as a zeroth–order basis. For simplicity suppose we are able to identify the bright mode mentioned above as $\tilde{q}_{\xi_1}$. Let us further assume that the Hamiltonian has been expanded in these normal modes according to Eq. (4.1). Then we can partition the Hamiltonian to fourth–order as follows

$$H_{\rm S} = \frac{1}{2}\left(\tilde{p}_{\xi_1}^2 + \omega_{\xi_1}^2 \tilde{q}_{\xi_1}^2\right) + K_{\xi_1\xi_1\xi_1}\tilde{q}_{\xi_1}^3 + K_{\xi_1\xi_1\xi_1\xi_1}\tilde{q}_{\xi_1}^4 \,, \tag{4.41}$$

$$\begin{aligned} H_{\rm R} &= \frac{1}{2}\sum_{\xi_2\neq\xi_1}\left(\tilde{p}_{\xi_2}^2 + \omega_{\xi_2}^2\tilde{q}_{\xi_2}^2\right) + \sum_{\xi_2\xi_3\xi_4\neq\xi_1} K_{\xi_2\xi_3\xi_4}\tilde{q}_{\xi_2}\tilde{q}_{\xi_3}\tilde{q}_{\xi_4} \\ &+ \sum_{\xi_2\xi_3\xi_4\xi_5\neq\xi_1} K_{\xi_2\xi_3\xi_3\xi_4}\tilde{q}_{\xi_2}\tilde{q}_{\xi_3}\tilde{q}_{\xi_4}\tilde{q}_{\xi_5} \,, \end{aligned} \tag{4.42}$$

and

$$\begin{aligned} H_{\rm S-R}^{\rm II} &= \tilde{q}_{\xi_1}\left(\sum_{\xi_2\xi_3\neq\xi_1} K_{\xi_1\xi_2\xi_3}\tilde{q}_{\xi_2}\tilde{q}_{\xi_3} + \sum_{\xi_2\xi_3\xi_4\neq\xi_1} K_{\xi_1\xi_2\xi_3\xi_4}\tilde{q}_{\xi_2}\tilde{q}_{\xi_3}\tilde{q}_{\xi_4}\right) \\ &+\tilde{q}_{\xi_1}^2\left(\sum_{\xi_2\neq\xi_1} K_{\xi_1\xi_1\xi_2}\tilde{q}_{\xi_2} + \sum_{\xi_2\xi_3\neq\xi_1} K_{\xi_1\xi_1\xi_2\xi_3}\tilde{q}_{\xi_2}\tilde{q}_{\xi_3}\right) \\ &+\tilde{q}_{\xi_1}^3\sum_{\xi_2\neq\xi_1} K_{\xi_1\xi_1\xi_1\xi_2}\tilde{q}_{\xi_2} \,. \end{aligned} \tag{4.43}$$

For a large polyatomic molecule in a complex solvent exact microscopic knowledge about the individual anharmonic couplings is out of reach. So one may as well assume that the $K_{\xi_1\xi_2\xi_3}$, $K_{\xi_1\xi_1\xi_2}$, $K_{\xi_1\xi_1\xi_1\xi_2}$ etc. factorize into a system and a bath part. If we treat the bath in harmonic approximation we are back to the level of approximation of Eq. (4.40). Note that the coupling constants are defined differently because of the different basis sets used. Comparing $H_{\rm S-R}^{\rm I}$ and $H_{\rm S-R}^{\rm II}$, however, we notice that there is no bilinear term in $H_{\rm S-R}^{\rm II}$ since we have assumed a normal mode expansion with respect to *all* degrees of freedom around a minimum of the *total* potential surface.

This consideration can be viewed as a variation on the theme of which basis should be used for actual calculations. Depending on the choice, i.e., here zeroth–order system and bath Hamiltonian plus coupling versus expansion of the total Hamiltonian and subsequent partitioning into system and bath, different types of transitions between the considered system and bath states due to the coupling may occur. It goes without saying that for a comparable level of approximation both representations should give the same results for experimental observable quantities such as decay rates.

### 4.4.2 Higher–Order Multi–Quantum Relaxation

In the following we will focus on the effect of higher–order terms in the normal mode expansion Eq. (4.40), i.e., we use the interaction Hamiltonian $H^{\rm I}_{\rm S-R}$ (the superscript will be omitted). For simplicity we will take a single contribution $K(s)\Phi(q)$ of the complete expansion of $V_{\rm S-R}(s,q)$. Which part of the expansion we take, i.e., the concrete structure of $K(s)$ and $\Phi(q)$, will be specified below. In Section 3.8.2 we have seen that the transition rates between vibrational state $|M\rangle$ and $|N\rangle$ of the relevant system ($H_{\rm S}|N\rangle = E_N|N\rangle$) can be written as

$$\begin{aligned} k_{MN} &= \frac{1}{\hbar^2}|\langle M|K(s)|N\rangle|^2 \int\limits_{-\infty}^{+\infty} dt\, e^{i\omega_{MN}t}\langle \Delta\Phi^{(\rm I)}(q,t)\Delta\Phi^{(\rm I)}(q,0)\rangle_{\rm R} \\ &= |\langle M|K(s)|N\rangle|^2 C(\omega_{MN})\,. \end{aligned} \tag{4.44}$$

It has also been shown that this is just a particular element of the damping matrix $\Gamma_{KL,MN}$ which can be calculated by the same procedure we will follow now. The rates for the bilinear form of the interaction Hamiltonian have already been discussed after Eq. (4.23) (we just have to keep in mind that the coupling constant entering the spectral density is defined differently in the present case).

In connection with Eq. (4.23) we have highlighted the role of energy conservation in the relaxation process. A particularly interesting case in this respect is the relaxation of an intramolecular high–frequency mode. If we assume that intramolecular and solvent modes do not mix appreciably there are two possibilities: Either there is another intramolecular mode in this frequency range or the solvent normal mode spectrum supports such a high frequency. Usually the collective solvent modes are of rather low frequency (see Fig. 4.13). Thus, vibrational energy acceptors in the solvent can only be specific intramolecular modes of the solvent molecules in the surrounding of the solute. From these restricting conditions it should become clear that relaxation of high–frequency vibrations due to the bilinear coupling term is not always possible since this implies a one quantum transition.

As an example we mention the studies of the CO asymmetric stretch relaxation of tungsten hexacarbonyl ($W(CO)_6$) in chloroform ($CHCl_3$). Here the 1976 cm$^{-1}$ CO mode cannot relax via a mechanism involving a single transition in the bath only: The next $W(CO)_6$ vibrational normal mode is at 580 cm$^{-1}$ and $CHCl_3$ supports a mode at 1250 cm$^{-1}$, both out of reach for the CO stretch. Therefore, a quartic process is most likely to be responsible for relaxation behavior. Changing the solvent to carbon tetrachloride (highest frequency mode 780 cm$^{-1}$) even led to the conclusion that a quintic order process might be involved (for details, see [Tok94]).

This is where the higher–order terms in the expansion of $H_{\rm S-R}$ in Eq. (4.40) come into play. Here the relaxation proceeds with participation of different bath modes, that is, multi–quantum transitions occur in the bath. For a given order in the system coordinate $s$, we have to identify the bath part of $H_{\rm S-R}$ in Eq. (4.40) with $\Phi(Q_\xi = q_\xi\sqrt{\hbar/2\omega_\xi})$ which will contain a product of harmonic bath coordinates. The calculation of the rates in Eq. (4.44) requires to determine the multi–time correlation function of these bath coordinate. Such expressions have been derived in Sections 3.6.3 and 3.12.2, where a general $m+n$–time correlation function $C^{(m,n)}_{\xi,\bar\xi}(t)$ had been defined in Eq. (3.215). In the present case the respective correlation

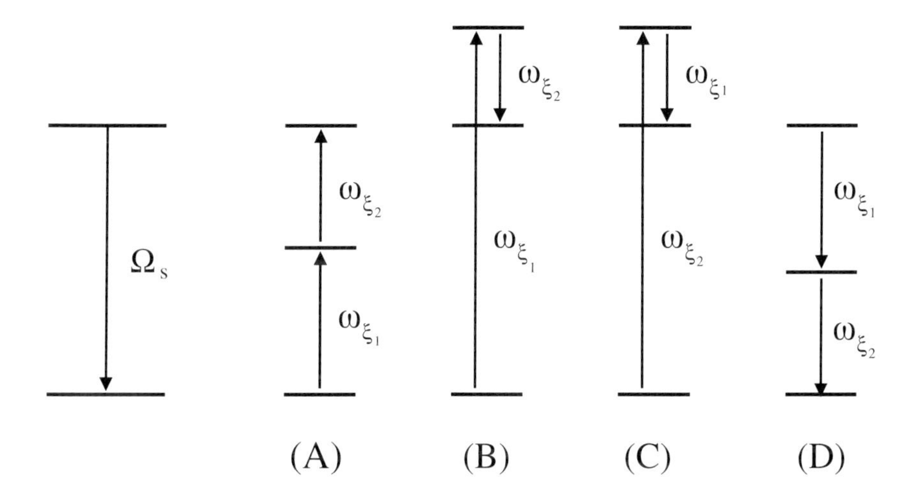


**Figure 4.14:** Multi–quantum relaxation processes in a two–level system. Downward relaxation of the system mode (left) can be accompanied by different transitions in the reservoir if the system–reservoir coupling is quadratic in the reservoir coordinates. Diagrams (A)–(D) correspond to the four terms of the right–hand side of Eq. (4.47).

function takes the form

$$\langle \Delta\Phi^{(\mathrm{I})}(q,t)\Delta\Phi^{(\mathrm{I})}(q,0)\rangle_{\mathrm{R}} = \sum_{m,n}\sum_{\xi,\bar{\xi}} g_{\xi}^{(m)} g_{\bar{\xi}}^{(n)} C_{\xi,\bar{\xi}}^{(m,n)}(t)\,, \tag{4.45}$$

where $\xi = (\xi_1, \xi_2, \ldots, \xi_m)$ and $\bar{\xi} = (\bar{\xi}_1, \bar{\xi}_2, \ldots, \bar{\xi}_n)$. The coupling constant $g_{\xi}^{(m)}$ comprises the $m$th derivative of the coupling with respect to the bath coordinates as well as the factor $\prod_{i=1}^{m}\sqrt{\hbar/2\omega_{\xi_i}}$.[5]

In order to illustrate the principal effect of higher–order system–bath interactions let us consider the term linear in the system coordinate but quadratic in the bath normal mode coordinates. This requires the calculation of the bath correlation function $C_{\xi,\bar{\xi}}^{(2,2)}(t)$ which was given in Eq. (3.222). Notice that the correlation function of $\Delta\Phi$ enters Eq. (4.44), i.e. we need $C_{\xi,\bar{\xi}}^{(0,2)}(t)$ as well which in fact cancels the last term in Eq. (3.222). Collecting the different contributions we obtain

$$\begin{aligned}\langle \Delta\Phi^{(\mathrm{I})}(q,t)\Delta\Phi^{(\mathrm{I})}(q,0)\rangle_{\mathrm{R}} \;&=\; 2\sum_{\xi_1,\xi_2} [g_{\xi_1,\xi_2}^{(2)}]^2 [(1+n(\omega_{\xi_1}))e^{-i\omega_{\xi_1}t} + n(\omega_{\xi_1})e^{i\omega_{\xi_1}t}] \\ &\times [(1+n(\omega_{\xi_2}))e^{-i\omega_{\xi_2}t} + n(\omega_{\xi_2})e^{i\omega_{\xi_2}t}]\,. \end{aligned} \tag{4.46}$$

The frequency–domain correlation function entering Eq. (4.44) can be written as

[5]Notice that in contrast to Section 3.8.5 in the present model the coupling constants do not depend on the system coordinate. This is a consequence of the fact that in Eq. (4.40) the derivatives are taken at the minimum of the PES.

$$\begin{aligned} C(\Omega) &= 4\pi \sum_{\xi_1,\xi_2} [g^{(2)}_{\xi_1,\xi_2}]^2 \Big[\delta(\Omega - \omega_{\xi_1} - \omega_{\xi_2})(1 + n(\omega_{\xi_1}))(1 + n(\omega_{\xi_2})) \\ &+\delta(\Omega - \omega_{\xi_1} + \omega_{\xi_2})(1 + n(\omega_{\xi_1}))n(\omega_{\xi_2}) \\ &+\delta(\Omega + \omega_{\xi_1} - \omega_{\xi_2})n(\omega_{\xi_1})(1 + n(\omega_{\xi_2})) \\ &+\delta(\Omega + \omega_{\xi_1} + \omega_{\xi_2})n(\omega_{\xi_1})n(\omega_{\xi_2})\Big] . \end{aligned} \tag{4.47}$$

If the system is harmonic with frequency $\Omega_s$ we have upward and downward transitions with rates proportional to $C(\Omega_s)$ (see Eq. (4.23)). For illustration of the transitions which are possible according to Eq. (4.47) let us consider *downward* relaxation. Note that all transitions are weighted with the proper thermal equilibrium distribution for the environmental modes. In Fig. 4.14 we have shown diagrams visualizing the different terms in Eq. (4.47). The first term (A) corresponds to an excitation of two vibrational modes of the environment, while the second (B) and the third (C) term incorporate excitation as well as de–excitation. The last term (D) represents the simultaneous de–excitation of the system *and* the environment. This last process is very unlikely and is usually neglected within the so–called rotating wave approximation. However, if we would have considered upward transitions, this process of de–excitation of two bath modes and simultaneous excitation of the system mode in principle could give a contribution.

The *exact* resonance conditions will be hardly met in any realistic solute–solvent system taking into account only high–frequency intramolecular modes. However, we have not yet discussed the low–frequency collective modes of the solvent. Often they will provide the continuum of states which is necessary to ensure energy conservation (see Fig. 4.13). To be specific suppose that the modes with index $\eta$ belong to the low–frequency solvent continuum of states; the intramolecular high–frequency modes are labelled by $\sigma$. Let us further assume that the coupling matrix factorizes with respect to the oscillators having quite different origin such that we can use: $[g^{(2)}_{\eta,\sigma}]^2 \rightarrow c_\eta^2 c_\sigma^2$. If we then introduce the spectral density of the low frequency solvent modes as

$$J_{\rm lf}(\omega) = \sum_\eta c_\eta^2 \delta(\omega - \omega_\eta) , \tag{4.48}$$

we can cast the correlation function into the form

$$\begin{aligned} C(\Omega) &= 4\pi \sum_\sigma c_\sigma^2 \big[(1 + n(\omega_\sigma))(1 + n(\Omega - \omega_\sigma))J_{\rm lf}(\Omega - \omega_\sigma) \\ &- (1 + n(\omega_\sigma))(1 + n(\Omega - \omega_\sigma))J_{\rm lf}(-\Omega + \omega_\sigma) \\ &+ n(\omega_\sigma)(1 + n(\Omega + \omega_\sigma))J_{\rm lf}(\Omega + \omega_\sigma) \\ &-n(\omega_\sigma)(1 + n(\Omega + \omega_\sigma))J_{\rm lf}(-\Omega - \omega_\sigma)\big] . \end{aligned} \tag{4.49}$$

(Here we have taken only those contributions into account which mix low and high–frequency modes.) If we think of the mode $\sigma$ as being a high–frequency mode of the solute itself or of the neighboring solvent molecules whose frequency we know, the transfer rates can be immediately calculated provided we have knowledge about the instantaneous normal mode solvent spectrum, for instance.

Generalizing the results we have obtained for the case of a system–bath interaction quadratic in the bath normal modes, we want to point out that of course the next (cubic) term gives rise to three quantum transitions in the bath and so on. Even though the coupling might at first glance seem to be rather weak compared to the low–order terms in the Taylor expansion, energy conservation can take over such that higher–order quantum transitions provide the only relaxation channel.

Up to this point we have considered the situation typical for the relaxation of a high–frequency solute mode, which is basically a transition between the first excited and the ground vibrational state of that mode. In the following we would like to discuss the relaxation of a mode having some intermediate frequency. The initial condition might be provided by an excitation of an electronic transition. In the electronically excited state we suppose an excitation along a vibrational mode involving high quantum number states (see also Chapter 5). In the bilinear model the relaxation would proceed via relaxation down the vibrational ladder step by step. What happens if we include the next higher order, i.e., the term quadratic in the system coordinate? For simplicity we assume that the environment provides a dense spectrum and all relaxation takes place via single quantum transition in the bath. The system coordinate is assumed to describe a harmonic oscillator with frequency $\Omega_s$. The quadratic contribution to the interaction Hamiltonian can be written in dimensionless form as $K(s) = s^2 2\Omega_s \mu_s/\hbar$. The respective transition rate is given by

$$k_{MN} = |\langle M|K(s)|N\rangle|^2 C(\omega_{MN}) \tag{4.50}$$

with $C(\omega)$ given by Eq. (3.227). The matrix elements can be calculated straightforwardly using, for instance, the second quantization representation for harmonic oscillator states introduced in Section 2.6.2. We obtain

$$\begin{aligned} k_{MN} &= \frac{1}{\hbar^2}\Big[\delta_{MN}(2M+1)^2 + \delta_{M,N+2}M(M-1) \\ &\quad + \delta_{M,N-2}(M+1)(M+2)\Big]\, C(\Omega_{MN}) \,. \end{aligned} \tag{4.51}$$

As to be expected two–quantum transitions become possible at this point. Provided that the spectral density is flat such that $J(\Omega_s) \approx J(2\Omega_s)$, the difference between the one quantum and two–quantum transitions comes from the thermal prefactor. Thus, the rates for downward transitions behave like $(1+n(\Omega_s))/(1+n(2\Omega_s)) = (1+\exp(-\Omega_s/k_B T))$ and for upward transitions we have $n(\Omega_s)/n(2\Omega_s) = (1+\exp(\Omega_s/k_B T))$. Therefore, for moderate temperatures in particular the two–quantum upward transitions will be much less probable than the respective one–quantum transitions. In general however, estimating the relative importance of one– and two–quantum transitions one has to take into account the form of the spectral density. There is another term in Eq. (4.51) which is proportional to $\delta_{MN}$. This term is nothing but energy conserving *pure dephasing*, a process which has already been discussed for a generic multilevel system in Section 3.8.5.

# 4.5 Supplement

## 4.5.1 Coherent Wave Packet Motion in a Harmonic Oscillator

In the following the derivation of Eq. (4.19) for the time–dependent state vector of an initially shifted harmonic oscillator ground state is sketched. In particular we will make use of the displacement operator, $D^+(g) = e^{\,g(C-C^+)}$, which has already been introduced in Eq. (2.89). Here, the quantity $g$ is the dimensionless displacement, $-S^{(0)}/2$.

First, note that the position ket vector $|S\rangle$ is an eigenstate of the coordinate operator $\hat{S} = C + C^+$. Since

$$\hat{S}D|S\rangle = DD^+\hat{S}D|S\rangle = D\,(\hat{S} - S^{(0)})|S\rangle = (S - S^{(0)})D|S\rangle \ ,$$

$D|S\rangle$ is the eigenstate of the coordinate operator with eigenvalue $S - S^{(0)}$. Consequently, $D|S\rangle$ has to be identified with $|S - S^{(0)}\rangle$. With the help of Eq. (2.92), i.e., $D^+ \; C \; D = C + g$, we obtain the following identity

$$\chi_0(S - S^{(0)}) = \langle S|D^+(-S^{(0)}/2)|0\rangle \ . \tag{4.52}$$

In order to calculate the time evolution of the initial state given in Eq. (4.15) we write

$$\chi(S,t) = \langle S|e^{-iH_{\mathrm{S}}t/\hbar}D^+|0\rangle = \langle S|e^{-iH_{\mathrm{S}}t/\hbar}D^+e^{iH_{\mathrm{S}}t/\hbar}e^{-i\Omega_{\mathrm{s}}t/2}|0\rangle \ . \tag{4.53}$$

The Heisenberg representation of the displacement operator contained in the right–hand side of this expression follows as

$$\begin{aligned} e^{-iH_{\mathrm{S}}t/\hbar}D^+e^{iH_{\mathrm{S}}t/\hbar} &= e^{-iH_{\mathrm{S}}t/\hbar}\ \exp\left\{-\frac{S^{(0)}}{2}(C - C^+)\right\}\ e^{iH_{\mathrm{S}}t/\hbar} \\ &= \exp\left\{-\frac{S^{(0)}}{2}\left(e^{-iH_{\mathrm{S}}t/\hbar}(C - C^+)e^{iH_{\mathrm{S}}t/\hbar}\right)\right\} \\ &= \exp\left\{-\frac{S^{(0)}}{2}\left(C\,e^{i\Omega_{\mathrm{s}}t} - C^+\,e^{-i\Omega_{\mathrm{s}}t}\right)\right\} \ . \end{aligned} \tag{4.54}$$

The exponent can be rewritten according to ($\hat{S}$ is the coordinate operator)

$$C\,e^{i\Omega_{\mathrm{s}}t} - C^+\,e^{-i\Omega_{\mathrm{s}}t} = (C - C^+)\cos(\Omega_{\mathrm{s}}t) + i\hat{S}\sin(\Omega_{\mathrm{s}}t) \ . \tag{4.55}$$

With the help of the operator identity Eq. (2.95) we obtain

$$\begin{aligned} \exp\left\{-(S^{(0)}/2)\,(C\,e^{\,i\Omega_{\mathrm{s}}t} - C^+\,e^{-i\Omega_{\mathrm{s}}t})\right\} &= \exp\left\{-i(S^{(0)}/2)\sin(\Omega_{\mathrm{s}}t)\;\hat{S}\right\} \\ &\times\,\exp\left\{-(S^{(0)}/2)\cos(\Omega_{\mathrm{s}}t)\;(C - C^+)\right\} \\ &\times\,\exp\left\{-i(S^{(0)\,2}/8)\cos(\Omega_{\mathrm{s}}t)\sin(\Omega_{\mathrm{s}}t)\;[\hat{S}, C - C^+]_-\right\} \ . \end{aligned} \tag{4.56}$$

Using this result we finally obtain for the time dependence of the state vector $\chi(S,t)$:

$$\begin{aligned} \chi(S,t) &= \exp\left\{-(i/2)\left(\Omega_{\mathrm{s}}t - (S^{(0)\,2}/4)\sin(2\Omega_{\mathrm{s}}t)\right)\right\} \\ &\quad\times\langle S|\ \exp\left\{-i(S^{(0)}/2)\sin(\Omega_{\mathrm{s}}t)\;\hat{S}\right\}\ D^+(-(S^{(0)}/2)\,\cos(\Omega_{\mathrm{s}}t))\,|0\rangle \\ &= \exp\left\{-(i/2)\left(\Omega_{\mathrm{s}}t - (S^{(0)\,2}/4)\sin(2\Omega_{\mathrm{s}}t) - S^{(0)}S\,\sin(\Omega_{\mathrm{s}}t)\right)\right\} \\ &\quad\times\chi_0(S - S^{(0)}\,\cos(\Omega_{\mathrm{s}}t)) \ . \end{aligned} \tag{4.57}$$

which is the desired result quoted in Eq. (4.19).

# Notes

[Bar02] M. Bargheer, M. Gühr, P. Dietrich, and N. Schwentner, Phys. Chem. Chem. Phys. **4**, 75 (2002).

[Bix68] M. Bixon and J. Jortner, J. Chem. Phys. **48**, 715 (1968).

[Fel85] P. M. Felker and A. H. Zewail, J. Chem. Phys. **82**, 2975 (1985).

[Gna96] S. Gnanakaran and R. M. Hochstrasser, J. Chem. Phys. **105**, 3486 (1996).

[Goo96] G. Goodyear and R. M. Stratt, J. Chem. Phys. **105**, 10050 (1996).

[Gru03] M. Gruebele, Theor. Chem. Acc. **109**, 53 (2003).

[Smi90] P. G. Smith and J. D. McDonald, J. Chem. Phys. **92**, 1004 (1990).

[Str95] R. M. Stratt, Acc. Chem. Res. **28**, 201 (1995).

[Tok94] A. Tokmakoff, B. Sauter, and M. D. Fayer, J. Chem. Phys. **100**, 9035 (1994).

# 5 Intramolecular Electronic Transitions

*Photoinduced intramolecular electronic transitions as well as transitions caused by intramolecular state couplings will be discussed in the following. In contrast to more complex transfer reactions, which we will encounter in later chapters, these processes exclusively take place in a single molecule where an electron is promoted from an initial electronic state $|\phi_i\rangle$ to a final electronic state $|\phi_f\rangle$. We will explain in detail that the interplay of the dynamics of the electronic transition and the accompanying vibrational motion is at the heart of such transitions. If the vibrational relaxation within the considered potential energy surface is fast compared to the electronic motion, a simple perturbational treatment with respect to the electronic coupling between the states $|\phi_i\rangle$ and $|\phi_f\rangle$ becomes possible. As an important example for such a transition linear optical absorption is discussed. Here the state coupling is due to an external electromagnetic field. Different ways for calculating the absorption spectrum are introduced which highlight particular aspects of the coupled electronic and nuclear motions during the transition event and which demonstrate different types of approximations. Nonlinear optical processes are also shortly discussed.*
*On the basis of this discussion we are in the position to describe the internal conversion process which is an intramolecular electronic transition that results from the nonadiabatic coupling between different adiabatic electronic states. Here we will focus on the cases of slow and fast vibrational relaxation as compared to the electronic transition rate.*

## 5.1 Introduction

Adiabatic electronic states are approximate solutions of the stationary electronic Schrödinger equation for the molecule (cf. Section 2.3). Since they are *not eigenstates* of the molecular Hamiltonian, there exists a residual interaction between adiabatic states. Different situations can be identified where this nonadiabatic coupling has to be taken into account. For example, if bound electronic states are considered, an overlap of the related PES indicates that nonadiabatic transitions may be important. If a molecule is initially prepared in a particular excited adiabatic state, a spontaneous transition to the electronic ground state will take place. In the general case, this may involve a sequence of transitions via intermediate electronic states. This type of transition is called *internal conversion* (IC). It is characterized by the conservation of the total molecular spin. Hence, starting in an excited singlet state $S_n$, IC proceeds down to the singlet ground state $S_0$. Normally, the transitions down to the first excited singlet state, $S_1$, are rather rapid (so–called Kasha rule). However, the transition $S_1 \to S_0$ is so slow that it competes with possible optical transitions (luminescence). IC processes are also observed within the manifold of triplet states, whereas singlet–triplet transitions (and vice versa) are known as *inter–system crossing* processes.

In Section 2.19 we discussed how the nuclear configuration of a molecule changes upon the change of the electronic state. This is the reason why the dynamics of intramolecular electronic transitions may strongly depend on the coupling between electronic and nuclear degrees of freedom. Let us consider transitions between two adiabatic electronic states, $|\phi_i\rangle$ (initial state) and $|\phi_f\rangle$ (final state). If the relaxation of the nuclear degrees of freedom in these two electronic states is fast compared to the transition time, the process is electronically incoherent. Electronic coherences between the two states can exist if the nuclear motion is comparable or slow in relation to the electron dynamics. Then, a time–dependent wave function is formed which is a superposition of the initial and final state. In general intramolecular electronic transitions are not necessarily induced by internal couplings, such as nonadiabatic or spin–orbit couplings. Scattering processes of the considered molecule with other molecules as well as the interaction with external electromagnetic fields can cause electronic transitions, too. The most common example is the absorption of light energy via an electronic transition from the ground state of the molecule to a particular excited state. This process conserves the total spin; it is a transition from the singlet ground state $S_0$ to an excited singlet state $S_n$ ($n = 1, 2, \ldots$). We expect many similarities between externally induced electronic transitions and those induced by internal perturbations. In particular, this holds when the coupling is weak, which allows for a perturbational treatment. Such an assumption is the main reason to discuss the theoretical description of optical absorption in some detail. Many of the relations we introduce here will be valid also for the other types of transfer processes treated in the following chapters.

### 5.1.1 Optical Transitions

We start with some qualitative considerations of optical absorption in a simple diatomic molecule. Let the ground state be characterized by the PES $U_g(R)$, whereas the excited state PES is given by $U_e(R)$. $R$ denotes the relative distance between the two atoms, and rotational motion will be neglected. Both PES are supposed to have a single minimum at $R_a$ ($a = g, e$).

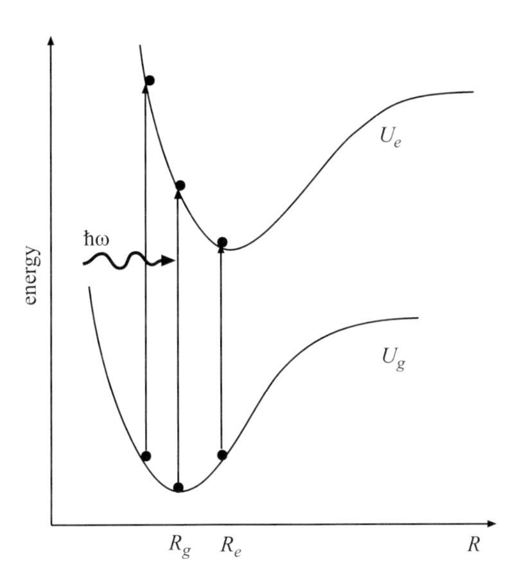


**Figure 5.1:** Ground and excited state PES of a diatomic molecule versus bond distance $R$. Different vertical transition are shown which correspond to different relative positions in the sense of classical physics.

Usually, $R_e > R_g$, since an electronic excitation results in a weakening of the bond. The electronic transition can be considered to take place on a time scale which is fast compared to the relative motion of both nuclei (bond vibration).[1] We can disregard the nuclear motion during the time of the electronic transition (conservation of nuclear momenta). In the picture of PES this means that the electronic transition is *vertical* and the nuclei are frozen during the transition as shown in Fig. 5.1. This scheme for optical transitions in molecules is known as the *Franck–Condon principle*.

Next, we discuss how this principle can affect the details of absorption spectra. We start with a *classical* description of the relative motion of the two nuclei. Although the nuclear motion in molecules is of quantum nature, the classical description is correct if the energy of a characteristic vibrational quantum $\hbar\omega_{\text{vib}}$ is much smaller than a characteristic mean energy of the vibrational motion, for example, smaller than the thermal energy $k_{\text{B}}T$.

According to the classical description the state of lowest energy in the electronic ground state PES $U_g$ corresponds to $R = R_g$, i.e., the bond distance takes its equilibrium value. After the optical excitation the electronic state has been changed to the excited state without any change of $R$. Optical absorption is possible whenever the photon energy $\hbar\omega$ equals to $U_e(R_g) - U_g(R_g)$. It results in a sharp absorption line at this photon energy. Usually we have $R = R_g < R_e$ and the bond is elongated according to the new equilibrium length $R_e$. As a consequence there will be vibrational motion with respect to $R_e$.

[1] Although this picture is useful we remind the reader on the fact that quantum mechanics does not make any statement concerning the actual time at which an electron jumps and how long it takes for the electron to jump. We exclusively get information on the change of the particle wave function with time (cf. Section 5.5).

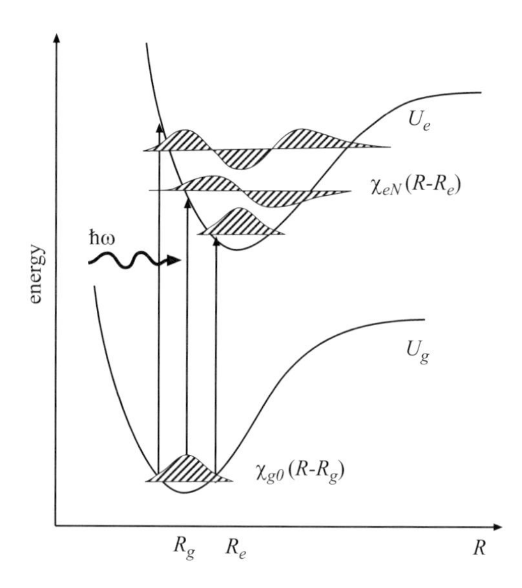


**Figure 5.2:** Ground and excite state PES of a diatomic molecule are shown versus bond distance. Different vibrational states in both electronic states together with the respective wave function are sketched. Transitions to the excited electronic state are possible for different values of the bond length $R$.

If collisions with other molecules take place, for example, in a condensed phase environment (solvent), $R$ may deviate from $R_g$ in the initial state, and absorption becomes possible at photon energies different from $U_e(R_g) - U_g(R_g)$. Since the experiments we have in mind are done with a macroscopically large number of molecules, a multitude of transitions becomes possible at photon energy $U_e(R) - U_g(R)$, with values of $R$ being determined by the type and the strength of the collision processes. If the environment is in thermal equilibrium with temperature $T$, we can use statistical mechanics to determine possible values of $R$ from the thermal distribution

$$f(R) = \frac{1}{\mathcal{Z}} e^{-U_g(R)/k_B T} , \tag{5.1}$$

with the partition function $\mathcal{Z} = \int dR \, e^{-U_g(R)/k_B T}$ . As a consequence of the thermal distribution of bond lengths, the measured absorption spectrum will be broadened having a width which is approximately equal to $k_B T$. Since $R = R_g$ occurs with highest probability, the maximum of the absorption line is located at the photon energy $U_e(R_g) - U_g(R_g)$ (vertical transition).

Next we assume $k_B T < \hbar\omega_{\text{vib}}$ and change to a quantum description as illustrated in Fig. 5.2. In this case the vibrational motion is characterized by discrete vibrational levels $E_{gM}$ and $E_{eN}$ of the electronic ground and excited state, respectively. Consequently, (one–photon) absorption processes of a monochromatic radiation field take place only if the energy $\hbar\omega$ of a photon equals to a possible transition energy $E_{eN} - E_{gM}$. Let us again make use of the Franck–Condon principle to understand details of these transitions. For simplicity we assume that only the vibrational ground state, $\chi_{gM=0}(R)$, is populated before the absorption process.

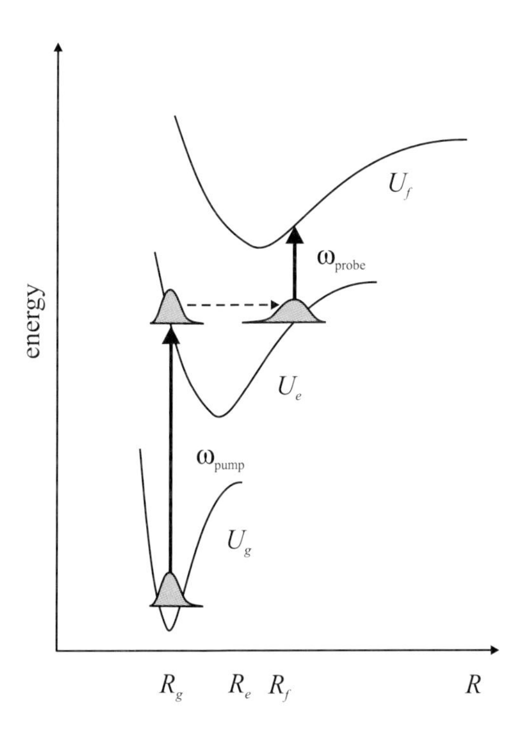


**Figure 5.3:** Scheme of pump–probe spectroscopy at a molecule with three electronic states $|\phi_g\rangle$, $|\phi_e\rangle$ and $|\phi_f\rangle$. One radiation field with frequency $\omega_{\text{pump}}$ drives the transition from the ground state PES $U_g$ to the first excited PES $U_e$. A second radiation field with frequency $\omega_{\text{probe}}$ probes the population of the excited state $|\phi_e\rangle$ by a transition to the higher excited state $|\phi_f\rangle$. (If the pump pulse is shorter than the time scale of vibrational motion, it places the ground state vibrational wave function to the excited state PES around $R_g$ where it moves as a wave packet in the PES $U_e$. For a short enough probe pulse the transition of the wave packet to $U_f$ may occur. This is possible for a probe pulse which carries out the excitation at the right delay time with respect to the pump pulse and with the proper frequency.)

The description in terms of a wave function implies that the coordinate $R$ does not possess the sharp value $R_g$ but is distributed around it. As a consequence vertical transitions to the excited state are also possible for $R \neq R_g$. To what extent the various vibrational states in the PES $U_e$ are excited, is determined by the overlap between the initial wave function $\chi_{g0}(R)$ and the final state wave functions $\chi_{eN}(R)$:

$$\int dR \, \chi_{eN}^*(R)\chi_{g0}(R) \equiv \langle \chi_{eN}|\chi_{g0}\rangle \ . \tag{5.2}$$

The square of these overlap matrix elements is proportional to the respective transition strength (see below); the individual spectral lines will be sharp in the simple one–dimensional model considered here. If excited vibrational states in the electronic ground state are thermally occupied, additional transitions to the excited electronic state are possible. Considering a large number of molecules in thermal equilibrium with a particular environment, every transition is weighted by the probability

$$f(E_{gM}) = \frac{\exp(-E_{gM}/k_{\mathrm{B}}T)}{\sum\limits_{N} \exp(-E_{gN}/k_{\mathrm{B}}T)}, \tag{5.3}$$

by which a molecule of the thermal ensemble is in the initial state $|\chi_{gM}\rangle$.

So far we have assumed that the spectrum can be calculated from the knowledge of the eigenstates of the relevant system. In a condensed phase environment, however, one has to account for the interaction between the relevant system and the reservoir (for example, the solvent). Without adapting a specific model for the reservoir and its interaction with the system we can discuss two principal effects. To this end we assign a certain time scale to the modulation of the system's properties (for example, the transition energies) by the reservoir. If this time scale is long compared with some characteristic time for the experiment (typically given by the optical pulse length which can be nanoseconds for absorption measurements) the effect of the environment is to introduce *static disorder*. This means that there will be a static distribution of transition energies in Eq. (5.33), for example. This induces in addition to a quantum mechanical broadening a further broadening of the molecular absorption caused by the fact that the absorption spectra of individual molecules are not identical. Usually it is referred to as *inhomogeneous broadening* of absorption lines (cf. Section 8.7). In contrast, if the modulation of molecular properties by the environment is fast with respect to the time of the measurement, we have *dynamic disorder*. It results in the so–called *homogeneous broadening*, which can be rationalized in terms of dephasing rates which have to be added to the transition energies in Eq. (5.33). In Section 5.2.5 these two limits will emerge from a particular model for a spectral density which can be introduced in similarity to Section 3.6.4.

Having discussed optical processes where a single photon is absorbed by the molecule we now turn to the case of higher intensities of the incoming light. The increasing number of photons increases the probability to absorb two or even more photons simultaneously. Fig. 5.3 displays the absorption of two photons; one results in the transition from the electronic ground–state to the excited state $|\phi_e\rangle$ and the other from this excited state to a higher excited states $|\phi_f\rangle$. Since in the described process two or even more photons of the incoming radiation field may be absorbed the process is named *nonlinear absorption*. It can also take place as process where one photon comes from one type of radiation field (for example one laser beam) and the other photon from a different field (separate laser beam). In this way one may construct an important scheme of nonlinear optical experiments the so–called *pump–probe spectroscopy* (cf. Fig. 5.3).

In the preceding discussion we concentrated on the absorption of photons, but the reverse process is also possible. Optical recombination, i.e., emission is simply the inversion of the optical absorption process as shown in Fig. 5.3. Since the life–time of the first excited singlet state is in most cases large compared to every characteristic time of the vibrational motion and relaxation, a thermal equilibrium is established of the vibrational motion in the excited electronic state before recombination. Therefore, the initial state of this process has to be characterized by a thermal distribution function like in Eq. (5.3) but related to the first excited state. Emission from this equilibrium distribution in the excited electronic state takes place to different excited vibrational states belonging to the electronic ground state. Finally, we note that emission can take place as a spontaneous as well as stimulated process.

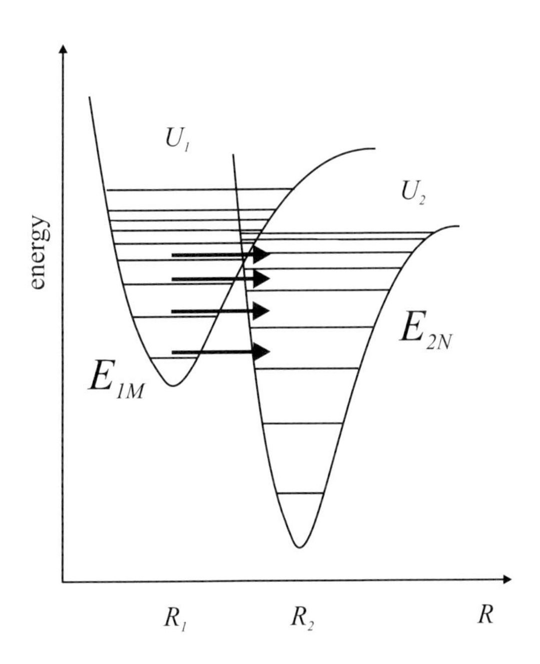


**Figure 5.4:** Internal conversion of the population of electronic levels $E_{1M}$ into a population of electronic levels $E_{2N}$. Every vibrational level loses its population via transitions induced by the nonadiabatic coupling. (If the nonadiabatic transition is fast compared to vibrational relaxation, recurrences of population becomes possible. In the contrary case the transfer is irreversible and energetic degeneracy between the levels of state 1 and of state 2 is a necessary condition for the transition.)

## 5.1.2 Internal Conversion Processes

A possible scheme for IC processes is given in Fig. 5.4 indicating the transition from the adiabatic electron–vibrational states $E_{1M}$ to the states $E_{2N}$. If the nonadiabatic coupling is weak so that the characteristic time for the transition is large compared to the time vibrational relaxation needs to establish thermal equilibrium, the IC process starts from such a thermal distribution among the vibrational levels $E_{1M}$. However, if the transition is also fast, vibrational relaxation is accompanied by transitions to the vibrational levels $E_{2N}$. Of course, the details of this process are not only determined by the strength of the nonadiabatic coupling. There is also a considerable influence of the initial state preparation. For instance, if the preparation time is long compared to the time of the nonadiabatic transition any detail of the IC process may be smeared out (for more details on this problem we refer to Section 5.5).

In the following section different approaches for computing the absorption coefficient as a function of the incoming radiation field frequency will be presented. Nonlinear optical processes will be briefly introduced afterwards. IC is described in Section 5.7. Since the absorption coefficient is determined for a large ensemble of identical molecules, macroscopic electrodynamics for dielectrics has to be used. The essentials of classical electrodynamics needed for the discussion are briefly summarized.

## 5.2 The Optical Absorption Coefficient

### 5.2.1 Basic Theoretical Concepts

In the following we will consider the standard experimental situation in which a macroscopic large number of molecules, for instance, dye molecules in solution, are studied by shining light on them.[2] For simplicity we only consider transitions within the spectrum of electron–vibrational states; any rotational contributions will be neglected. Although light has a quantum nature exemplified in many experiments, often it is sufficient to use classical theory. For example, if the light field is intense, quantum fluctuations can be neglected and the theory can be formulated using a classical description based on Maxwell's equations. In contrast, if the molecular dynamics is mainly characterized by electronic transitions, it must be described by quantum mechanics. This hybrid version of the molecule–light interaction is called *semiclassical.*

Molecules in the condensed phase form a dielectric with electronically polarizable units (cf. Section 2.5). Having in mind the macroscopic nature of the absorption experiments, this justifies the to use *Maxwell's* macroscopic electrodynamics for dielectrics. The term *macroscopic* indicates that the description does not account for those parts of the electromagnetic field varying on a microscopic length scale of some nanometer. These parts are not measurable in standard experiments where the spectrometer is far away from the illuminated sample. In other words, only the so–called *far–field* is measured and any near–field contribution is eliminated from the theory.[3] To include the far–field only, we proceed as in Section 2.5 and carry out an averaging of the field with respect to a volume element $\Delta V$. This volume element should contain a sufficiently large number of molecules. In particular $\Delta V$ has to be large in comparison to the molecular extension. Furthermore, the externally applied field should vary weakly inside $\Delta V$. Then, we can discretize the sample volume by the elements $\Delta V$ and label every element by the spatial vector $\mathbf{x}$. These vectors are discrete on the length scale of the volume elements, but from a macroscopic point of view, they can be considered as continuous quantities.

The key quantity of the electrostatics as well as of the electrodynamics of dielectrics is the *macroscopic* polarization vector $\mathbf{P}(\mathbf{x}, t)$. As long as the description of the response of the dielectric to the applied field can be restricted to the formation of electric molecular dipoles (and not higher multipoles), $\mathbf{P}(\mathbf{x}, t)$ corresponds to the dipole density of the medium. As a macroscopic vector field its definition must contain the averaging with respect to the volume elements $\Delta V$. This has already been done in Eq. (2.47), which can be generalized from the static to the dynamic case as follows

$$\mathbf{P}(\mathbf{x}, t) = \frac{1}{\Delta V(\mathbf{x})} \sum_{m \in \Delta V} \mathbf{d}_m(t) \; . \tag{5.4}$$

It gives the polarization as the sum of molecular dipole moments $\mathbf{d}_m(t)$ contained in the volume element $\Delta V$ in the neighborhood of the point $\mathbf{x}$ versus $\Delta V$ (dipole density). The

[2] It should be noted that in recent years it became possible to measure optical properties of single molecules.

[3] The situation is different in the recently developed *near–field spectroscopy* where the near–field on a sub–wavelength scale is measured.

time–dependent expectation value of the dipole operator can be written in two ways:

$$\mathbf{d}_m(t) = \mathrm{tr}\{\hat{W}_{\mathrm{eq}} U^+(t,t_0)\hat{\mu}_m U(t,t_0)\} \equiv \mathrm{tr}\{\hat{W}(t)\hat{\mu}_m\} . \tag{5.5}$$

In the first expression $\hat{W}_{\mathrm{eq}} = \exp\{-H_{\mathrm{mol}}/k_{\mathrm{B}}T\}/\mathcal{Z}$ is the equilibrium statistical operator in the absence of an external field. The time dependence of the dipole operator is given by the Hamiltonian which includes the electromagnetic field via $H_{\mathrm{field}}(t)$ (see Eq. (5.9)). In the second part of Eq. (5.5) the time evolution has been transferred to the statical operator to give $\hat{W}(t) = U(t,t_0)\hat{W}_{\mathrm{eq}}U^+(t,t_0)$. Both variants to compute $\mathbf{d}_m$ will be used in the following. The operator $\hat{\mu}_m$ is the operator of the (electric) dipole moment of a single molecule labelled by $m$. It contains electronic as well as nuclear contributions (cf. the notation in Eq. (2.46)) $\hat{\mu}_m = -\sum_j e\mathbf{r}_{m,j} + \sum_n ez_{m,n}\mathbf{R}_{m,n}$. Introducing an expansion with respect to the (adiabatic) electronic states of the $m$th molecule $|\phi_{m,a}\rangle$ as

$$\hat{\mu}_m = \sum_{a,b} \langle\phi_{m,a}|\hat{\mu}_m|\phi_{m,b}\rangle|\phi_{m,a}\rangle\langle\phi_{m,b}| , \tag{5.6}$$

it is straightforward to select all those electronic states involved in a particular experiment and to exclude all unimportant states from the summation. For example, it often suffices to use off–diagonal matrix elements only, leading to the *transition dipole moments* (transition matrix elements)

$$\mathbf{d}_{ab}^{(m)} = \langle\phi_{m,a}|\hat{\mu}_m|\phi_{m,b}\rangle . \tag{5.7}$$

To arrive at the Hamiltonian for the coupling between the electromagnetic field and the molecule, we use the expression of medium electrodynamics for the interaction energy $E_{\mathrm{int}} = -\int d^3\mathbf{x}\ \mathbf{E}(\mathbf{x},t)\mathbf{P}(\mathbf{x},t)$. If we replace the various $\hat{\mu}_m$ in Eq (5.4) by the related quantum mechanical operators, one obtains the Hamiltonian $H_{\mathrm{int}}(t)$ of light–matter interaction. Since the electric field, as an external macroscopic field, does not vary within the small volume element $\Delta V(\mathbf{x})$, we can incorporate $\mathbf{E}(\mathbf{x},t)$ with $\mathbf{x} = \mathbf{x}_m$ into the $m$–summation (Eq. (5.4)). After carrying out the volume integration it follows that

$$H_{\mathrm{int}}(t) = \sum_m H_m^{(\mathrm{field})}(t) , \tag{5.8}$$

with

$$H_m^{(\mathrm{field})}(t) = -\mathbf{E}(\mathbf{x}_m,t)\,\hat{\mu}_m \tag{5.9}$$

The various $H_m^{(\mathrm{field})}(t)$ describe the coupling of the $m$th molecule located at $\mathbf{x}_m$ with the external field.

This Hamiltonian of light–matter interaction supplements the Hamiltonian of the molecular system. In contrast to the following chapters we will neglect molecule–molecule interactions responsible for charge and energy transfer here. Therefore, the Hamiltonian of all molecules in the sample can be written as a sum of individual molecular contributions.[4] Neglecting intermolecular interactions the macroscopic optical properties of the material system

[4]This single molecule concept is extended in the so–called *local field approximation* to include the effect of the averaged field of all other molecules on a particular molecule. While this approach is reasonable for weakly to moderately interacting molecules it cannot account for many effects which occur in strongly interacting molecular aggregates (see Chapter 8).

are calculated by determining the interaction of a single molecule with the radiation field first. In a second step all individual molecular contributions are summed to give the macroscopic response.

Before considering the description of the radiation field let us specify Eq. (5.4) for the polarization field to the case where the ensemble of molecular systems does not show the phenomenon if inhomogeneous broadening. In such a situation all molecules within the volume element $\Delta V(\mathbf{x})$ around point $\mathbf{x}$ are identical, and we may replace the summation with respect to the various dipole moments by a representative dipole moment at $\mathbf{x}$ times the volume density $n_{\rm mol}$ of molecules in the sample volume:

$$\mathbf{P}(\mathbf{x};t) = n_{\rm mol}\mathbf{d}(\mathbf{x};t) \; . \tag{5.10}$$

Note that in the present case the spatial variation of $\mathbf{d}$ is exclusively determined by the spatial dependence of the radiation field.

To get this dependence the radiation field and the molecules have to be considered as two coupled dynamic subsystems. Suppose one has solved the dynamic equations for the molecules under the action of the electromagnetic field, i.e., the time–dependent Schrödinger equation or the density operator equation. Then, one knows how the molecules react to the field, and the polarization introduced in Eq. (5.4) is known as a function (functional) $\mathbf{P}[\mathbf{E}]$ of the electric field strength. Inserting this into Maxwell's equations one gets a closed equation for the electric field strength[5]

$$\left(\frac{\partial^2}{\partial t^2} - c^2\Delta\right) \mathbf{E} = -4\pi \frac{\partial^2}{\partial t^2}\mathbf{P}[\mathbf{E}] \; . \tag{5.11}$$

In general the polarization $\mathbf{P}$ will be a nonlinear functional of the electric field $\mathbf{E}$. For the present purpose, however, a linear approximation is sufficient. In the simplest case one has the relation $\mathbf{P} = \chi\mathbf{E}$ (or $\mathbf{D} = \varepsilon\mathbf{E}$) (cf. Section 2.5.1) and the response of the molecular system is completely determined by the electric susceptibility $\chi$ (or dielectric constant $\varepsilon$).

Next, we briefly recall the definition of the linear absorption coefficient $\alpha$. For this reason we consider a platelet of a dielectric medium of thickness $d$ extending into $z$–direction. In the $x$ and $y$–direction (lateral directions) there should be no geometric restriction. The strictly monochromatic light is supposed to propagate in $z$–direction with perpendicular incidence on the platelet. For simplicity we let the platelet thickness $d$ go to infinity, $d \to \infty$ such that there is a single reflecting boundary between the dielectric and the vacuum (dielectric half–space). Thus, the complete electric field strength along the propagation direction can be written as

$$\mathbf{E}(z,t) = \mathbf{e}\, e^{-i\omega t} \left(\Theta(-z)\{E_0 e^{ik_{\rm vac}z} + E_{\rm r} e^{-ik_{\rm vac}z}\} + \Theta(z) E_{\rm t} e^{ik_{\rm med}z}\right) + \text{c.c.} \, . \tag{5.12}$$

The unit–step function $\Theta(z)$ has been used to discriminate between the part in the medium (transmitted part) with amplitude $E_{\rm t}$ and wave number $k_{\rm med} = \sqrt{\varepsilon}\omega/c$ in $z$–direction, and

[5] Actually one has to start from Maxwell's equations for a non–magnetic medium. If changed from their microscopic version containing any details of the fields to the averaged form they read for the curls of the fields $\nabla \times \mathbf{E} = -1/c\; \partial\mathbf{H}/\partial t$ as well as $\nabla \times \mathbf{H} = 1/c\; \partial\mathbf{D}/\partial t$, and for the field sources $\nabla\mathbf{D} = 0$ and $\nabla\mathbf{H} = 0$. Note the absence of external charge and current densities. Taking the curl of the equation for $\nabla \times \mathbf{E}$ one obtains $\Delta\mathbf{E} = 1/c^2\; \partial^2\mathbf{D}/\partial t^2$. Finally we use $\mathbf{D} = \mathbf{E} + 4\pi\mathbf{P}$ and arrive at Eq. (5.11).

the field in the vacuum. The latter contains the incoming part with amplitude $E_0$ and the reflected part with amplitude $E_r$, both with wave number $k_{vac} = \omega/c$. The unit vector $\mathbf{e}$ defines the polarization direction of the field. The different field components are determined via the boundary conditions for the electric and magnetic field. For the field intensity inside the medium one has (Beer's law)

$$I(z) = I(0)\, e^{-\alpha z} \; . \tag{5.13}$$

The exponential decrease is determined by the absorption coefficient $\alpha$, which will depend on the frequency of the light wave travelling through the platelet. The absorption coefficient can be expressed as

$$\alpha(\omega) = \frac{4\pi\omega}{nc} \, \text{Im}\chi(\omega) \; , \tag{5.14}$$

where $n$ denotes the reflection coefficient which is assumed to be constant and $\chi(\omega)$ is the susceptibility.

The actual frequency dependence of the absorption coefficient is determined by the properties of the molecules as will be explained in the following section. There we will calculate the change of light intensity $dI$ versus length segment $dz$ in propagation direction

$$dI(z) = -\alpha\, I(z) dz \; , \tag{5.15}$$

which gives direct access to the absorption coefficient.

### 5.2.2 Golden Rule Formulation

We start our discussion of the absorption coefficient by deriving the expression for the rate $k_{g \to e}$ of transitions between the electronic ground state $|\phi_g\rangle$ and some excited state $|\phi_e\rangle$ due to the absorption of a single photon. During this process energy of the electromagnetic field is converted into molecular excitation energy. For the molecular Hamiltonian the representation in terms of the electron–vibrational states $|\phi_a\rangle|\chi_{aN}\rangle$ will be used (cf. Section 2.7). The interaction with the field is accounted for in the semiclassical approximation discussed in the previous section. Note that describing electronic transitions in a system of two electronic states with the Golden Rule formula means that the coupling between both states is weak and that any coherence between the initial and final state is suppressed.

It will be the main aim of the following to determine the transition rate for a single molecule using a point of view which establishes a relation between optical absorption to other types of electronic transitions induced by *intramolecular* perturbations. Therefore, the absorption process is viewed as the transition from the initial state $|\phi_g\rangle|\chi_{gM}\rangle$ with energy levels $\hbar\omega + E_{gM}$ to the final states $|\phi_e\rangle|\chi_{eN}\rangle$ in the excited electronic state with energy levels $E_{eN}$. In other words, the energy of the absorbed photon $\hbar\omega$ is incorporated by defining the PES for the initial state as $U_g + \hbar\omega$ (cf. Fig. 5.5). We consider situations where this PES overlaps with the final state PES $U_e$. Then the optical absorption can be interpreted as a charge transfer between PES belonging to different adiabatic electronic states. Such a arrangement of two overlapping PES is usually called a *curve–crossing system*.

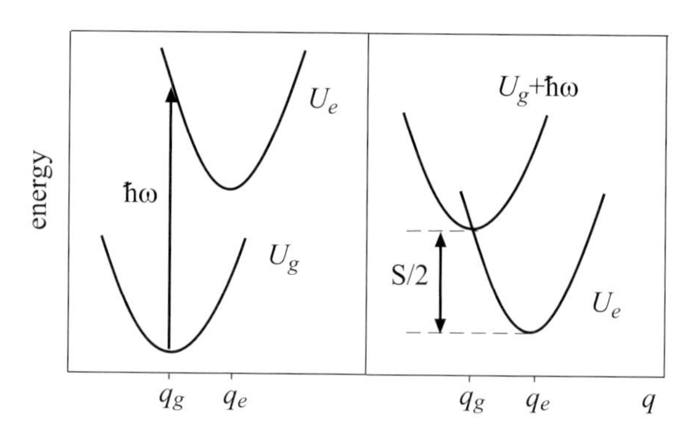


**Figure 5.5:** Description of optical absorption as a curve–crossing problem with initial state PES $\hbar\omega+U_g$ and final state PES $U_e$. The type of curve–crossing determines the degree of initial state and final state vibrational wave function overlap. ($S$ is the Stokes shift parameter.)

First the concept of the energetically shifted electronic ground state PES is introduced into the molecular Hamiltonian (two–level version of Eq. (2.115) with the neglect of nonadiabatic couplings). Including also the external field the Hamiltonian reads (cf. Eq. (5.9))

$$H(t) = \sum_{a=g,e} H_a(q)|\phi_a\rangle\langle\phi_a| - \mathbf{E}(t)\,(\mathbf{d}_{eg}|\phi_e\rangle\langle\phi_g| + \text{h.c.}) \ . \tag{5.16}$$

Here $q$ comprises the relevant nuclear coordinates of the molecule. The external field is taken to be monochromatic: $\mathbf{E}(t) = \mathbf{E}_0 \exp(-i\omega t) + \text{c.c.}$. Let us rearrange the Hamiltonian according to

$$H(t) = \mathcal{H}_0 + \mathcal{H}_1(t) \ , \tag{5.17}$$

with $\mathcal{H}_0 = -\hbar\omega|\phi_g\rangle\langle\phi_g|$. The remaining part gives the Hamiltonian with the shifted electronic ground state spectrum

$$\mathcal{H}_1(t) = (H_g(q) + \hbar\omega)\,|\phi_g\rangle\langle\phi_g| + H_e(q)|\phi_e\rangle\langle\phi_e| - \mathbf{E}(t)\,(\mathbf{d}_{eg}|\phi_e\rangle\langle\phi_g| + \text{h.c.}) \ . \tag{5.18}$$

Next we will use this particular representation to calculate the transition rate $k_{ge} \equiv k_{g\to e}$. To do so, we follow the approach presented in Section 3.3 where the Golden Rule formula has been derived. The manifold of initial states $|a\rangle$ is identified with the adiabatic electron–vibrational states $|\phi_g\rangle|\chi_{gM}\rangle$, whereas the final states $|\alpha\rangle$ are the excited electron–vibrational states $|\phi_e\rangle|\chi_{eN}\rangle$.

To apply the results of Section 3.3 we have to specify the time–evolution operator. This is done here by taking the (zeroth–order) time–evolution operator $U_0(t) = \exp\{-i\mathcal{H}_0 t/\hbar\}$ and changing to the interaction picture with respect to the (zeroth–order) Hamiltonian $\mathcal{H}_0$. According to Section 3.2.2 the total time–evolution operator reads $U(t,t_0) = U_0(t)S(t,t_0)$. The $S$–operator is defined as the time–ordered exponential of $\mathcal{H}_1(t)$ taken in the interaction picture,

$$S(t, t_0 = 0) = \hat{T} \exp\left(-\frac{i}{\hbar}\int_0^t dt'\, \mathcal{H}_1^{(\mathrm{I})}(t')\right) \tag{5.19}$$

with $\mathcal{H}_1^{(I)}(t) = U_0^+(t)\mathcal{H}_1(t)U_0(t)$. One easily verifies that

$$U_0(t)|\phi_g\rangle = e^{i\omega t}|\phi_g\rangle\,, \quad U_0^+(t)|\phi_g\rangle = e^{-i\omega t}|\phi_g\rangle\,, \quad U_0(t)|\phi_e\rangle = |\phi_e\rangle\,. \tag{5.20}$$

Applying these results we get

$$\begin{aligned}\mathcal{H}_1^{(\mathrm{I})}(t) &= (H_g(q) + \hbar\omega)\, e^{-i\omega t}|\phi_g\rangle\langle\phi_g|e^{i\omega t} + H_e(q)|\phi_e\rangle\langle\phi_e| \\ &\quad -\mathbf{E}_0 e^{-i\omega t}\left(\mathbf{d}_{eg}|\phi_e\rangle\langle\phi_g|e^{i\omega t} + \mathbf{d}^*_{eg}e^{-i\omega t}|\phi_g\rangle\langle\phi_e|\right) \\ &\quad -\mathbf{E}^*_0 e^{i\omega t}\left(\mathbf{d}_{eg}|\phi_e\rangle\langle\phi_g|e^{i\omega t} + \mathbf{d}^*_{eg}e^{-i\omega t}|\phi_g\rangle\langle\phi_e|\right)\,.\end{aligned} \tag{5.21}$$

This transformed Hamiltonian contains a *time–independent* part

$$\begin{aligned}\mathcal{H}_{\mathrm{rw}}^{(\mathrm{I})} &= (H_g(q) + \hbar\omega)\,|\phi_g\rangle\langle\phi_g| + H_e(q)|\phi_e\rangle\langle\phi_e| \\ &\quad -\mathbf{E}_0\mathbf{d}_{eg}|\phi_e\rangle\langle\phi_g| - \mathbf{E}^*_0\mathbf{d}^*_{eg}|\phi_g\rangle\langle\phi_e|\,,\end{aligned} \tag{5.22}$$

and a part oscillating at twice the field frequency $\omega$. If one neglects these high–frequency oscillations, the interaction picture introduced in this way leads to a time–independent Hamiltonian $\mathcal{H}_{\mathrm{rw}}^{(\mathrm{I})}$. It is known as the Hamiltonian in the *rotating wave approximation.*[6]

Accepting the time–independent Hamiltonian as a good approximation for $\mathcal{H}_1^{(\mathrm{I})}(t)$, the $S$–operator becomes very simple since no time–ordering is necessary. We obtain the complete time–evolution operator as

$$U(t) = \exp\{i\omega t|\phi_g\rangle\langle\phi_g|\}\ \exp\{i\mathcal{H}_{\mathrm{rw}}^{(\mathrm{I})}t/\hbar\}\,. \tag{5.23}$$

Following the derivation of the Golden Rule formula in Section 3.3 we introduce the transition amplitude, Eq. (3.42), $A_{gM,eN}(t) = \Theta(t)\ \langle\phi_g|\langle\chi_{gM}|U(t)|\phi_e\rangle|\chi_{eN}\rangle$, which simplifies to $A_{gM,eN}(t) = \Theta(t)\ \exp(i\omega t)\langle\phi_g|\langle\chi_{gM}|\exp\{i\mathcal{H}_{\mathrm{rw}}^{(\mathrm{I})}t/\hbar\}|\phi_e\rangle|\chi_{eN}\rangle$. Except for the unimportant time dependent phase factor the transition amplitude is identical with that discussed in Sections 3.3.1 and 3.59. We identify Hamiltonian, Eq. (5.22) with that of Eq. (3.40) and get from Eq. (3.60) the desired transition rate

$$k_{g\to e} = \frac{2\pi}{\hbar}\sum_{M,N} f(E_{gM})|\langle\chi_{gM}|\mathbf{E}^*_0\mathbf{d}^*_{eg}|\chi_{eN}\rangle|^2\,\delta(\hbar\omega + E_{gM} - E_{eN})\,. \tag{5.24}$$

Before discussing this expression in more detail we will give its relation to the absorption coefficient $\alpha(\omega)$. To that end a macroscopic sample volume $V$ is considered, containing $N_{\mathrm{mol}}$ non–interacting molecules absorbing light at frequency $\omega$. The sample volume should have a surface cross section of area $A$ where the light goes through perpendicular. We take a small

[6] Since $\mathcal{H}_{\mathrm{rw}}^{(\mathrm{I})}$ is time–independent, one can imagine that it has been defined in a frame rotating with the frequency of the externally applied light field.

section of length $dz$ and volume $Adz$ and determine the change of radiation field energy $dE$ if absorption takes place. It is given during the time interval $dt$ as

$$dE = -N_{\text{mol}} \frac{Adz}{V} \hbar\omega \, k_{g\to e} \, dt \; . \tag{5.25}$$

Here, $N_{\text{mol}} Adz/V$ gives the fraction of molecules inside the considered segment, and $\hbar\omega \times k_{g\to e} dt$ is the mean energy absorbed by a single molecule in the time interval $dt$. Since the field energy decreases the minus sign has been introduced.

Instead of $dE$ we can calculate the change of field energy density $du = dE/Adz$. Given the volume density $n_{\text{mol}} = N_{\text{mol}}/V$ of absorbing molecules the change of energy density per time follows as

$$\frac{du}{dt} = -n_{\text{mol}} \, \hbar\omega \, k_{g\to e} \; . \tag{5.26}$$

The continuity equation $du/dt = dI/dz$, which is a direct consequence of Maxwell's equations, allows to change from the energy density to the field intensity $I$. We further note that $I = c\,E_0^2/2\pi$ which is valid for a monochromatic field, and get

$$\frac{dI}{dz} = -\frac{2\pi n_{\text{mol}}}{c\,E_0^2} \, \hbar\omega \, k_{g\to e} \, I. \tag{5.27}$$

Comparing this expression with Eq. (5.15) enables us to identify the frequency–dependent absorption coefficient as

$$\alpha(\omega) = \frac{2\pi n_{\text{mol}}}{c\,E_0^2} \, \hbar\omega \, k_{g\to e} \; . \tag{5.28}$$

Using Eq. (5.24) we finally obtain (here and in the following $c$ has to be understood as the medium velocity of light $nc_{\text{vacuum}}$)

$$\alpha(\omega) = \frac{4\pi^2 \omega n_{\text{mol}}}{c} \sum_{MN} f(E_{gM}) |\langle \chi_{gM} | d_{eg} | \chi_{eN} \rangle|^2 \, \delta(\hbar\omega + E_{gM} - E_{eN}) \; . \tag{5.29}$$

This result assumes that the molecular transition matrix elements $\mathbf{d}_{eg}$ of the single molecules are identical, and that all molecules possess the same spatial orientation. Then the scalar product $\mathbf{d}_{eg}\mathbf{E}_0$ can be calculated. The quantity $d_{eg}$ in Eq. (5.29) is the component of the vector $\mathbf{d}_{eg}$ along the direction of the field vector.[7] Often, one simplifies the matrix element $\langle \chi_{gM} | d_{eg} | \chi_{eN} \rangle$ to the expression $d_{eg} \langle \chi_{gM} | \chi_{eN} \rangle$. This approximation is known as the *Condon approximation* which replaces the exact matrix element by the pure electronic matrix element $d_{eg}$ of the dipole operator and the *Franck–Condon factor* $\langle \chi_{gM} | \chi_{eN} \rangle$. The simplification is possible whenever the dependence of $d_{eg}$ on the nuclear degrees of freedom (via the parametric dependence of the electronic wave functions $\phi_a(r; R)$, cf. Section 2.3) is sufficiently weak to be negligible.

[7] In the case of random orientations an additional factor 1/3 appears on the right–hand side of Eq. (5.29) as a consequence of orientational averaging.

### 5.2.3 The Density of States

The obtained result for the absorption coefficient will be transformed into a more compact form by introducing the *lineshape function* $\mathcal{D}_{\text{abs}}$. We assume that the Condon approximation can be used and get for a sample of randomly oriented molecules

$$\alpha(\omega) = \frac{4\pi^2 \omega n_{\text{mol}}}{3c} |d_{eg}|^2 \, \mathcal{D}_{\text{abs}}(\omega - \omega_{eg}) \,, \tag{5.30}$$

with

$$\mathcal{D}_{\text{abs}}(\omega - \omega_{eg}) = \sum_{NM} f(E_{gM}) \, |\langle \chi_{eN} | \chi_{gM} \rangle|^2 \, \delta\left(\hbar\omega - (E_{eN} - E_{gM})\right) \,. \tag{5.31}$$

For convenience the transition frequency

$$\hbar\omega_{eg} = U_e^{(0)} - U_g^{(0)} \,, \tag{5.32}$$

defined via the values of the PES at the respective vibrational equilibrium configuration has been separated.

The lineshape function can be understood as a density of states (DOS) which combines two electronic states, the ground state and the considered excited state. Therefore, it is often called *Franck–Condon weighted and thermally averaged combined density of states.* $\mathcal{D}_{\text{abs}}$ can be written in an alternative way after introduction of the Fourier representation of the delta function. Then it follows from Eq. (5.31) that

$$\mathcal{D}_{\text{abs}}(\omega - \omega_{eg}) = \frac{1}{2\pi\hbar} \sum_{NM} \int dt \, f(E_{gM}) \, |\langle \chi_{eN} | \chi_{gM} \rangle|^2 \, e^{i(\omega - (E_{eN} - E_{gM})/\hbar)t} \,. \tag{5.33}$$

The actual calculation of $\mathcal{D}_{\text{abs}}$ and thus of the absorption coefficient requires the detailed knowledge of the vibrational energy spectrum for both PES. This knowledge may be attained for small systems in the gas phase or whenever only a small number of vibrational degrees of freedom is coupled to an electronic transition. However, Eq. (5.31) is inadequate for more complex systems or for systems in the condensed phase. Therefore, our aim is to formulate $\mathcal{D}_{\text{abs}}$ without making use of any eigenstates of the system. For this purpose Eq. (5.33) is better suited even though at first glance the replacement of the delta function may appear as a formal mathematical trick. However, we should recall that in Section 3.4.5, where the Golden Rule had been derived via the Liouville–von Neumann equation of the statistical operator, this type of time integration emerged in a natural way. In Section 5.3 we will discuss such a *time–dependent* formulation for the absorption coefficient. Here, we only derive some general relations used later in this chapter,

Let us follow Section 3.4.5 and eliminate the vibrational energy spectra by using the vibrational eigenvalue equations $H_a|\chi_{aN}\rangle = E_{aN}|\chi_{aN}\rangle$. We obtain

$$\begin{aligned} |\langle \chi_{eN} | \chi_{gM} \rangle|^2 \, e^{i(E_{eN} - E_{gM})t/\hbar} &= \langle \chi_{gM} | e^{iE_{gM}t/\hbar} e^{-iE_{eN}t/\hbar} | \chi_{eN} \rangle \langle \chi_{eN} | \chi_{gM} \rangle \\ &= \langle \chi_{gM} | e^{iH_g t/\hbar} e^{-iH_e t/\hbar} | \chi_{eN} \rangle \langle \chi_{eN} | \chi_{gM} \rangle \,. \end{aligned} \tag{5.34}$$

If we introduce this result into $\mathcal{D}_{\text{abs}}$ and use the completeness relation for the vibrational states it follows that

$$\begin{aligned}\mathcal{D}_{\text{abs}}(\omega - \omega_{eg}) &= \frac{1}{2\pi\hbar}\int dt\, e^{i\omega t}\sum_M \langle\chi_{gM}|\hat{R}_g\, e^{iH_g t/\hbar}\, e^{-iH_e t/\hbar}|\chi_{gM}\rangle \\ &= \frac{1}{2\pi\hbar}\int dt\, e^{i\omega t}\, \text{tr}_{\text{vib}}\{\hat{R}_g\, e^{iH_g t/\hbar}\, e^{-iH_e t/\hbar}\}\,. \end{aligned} \tag{5.35}$$

Here, we inserted the equilibrium statistical operator for the vibrational motion in the electronic ground state

$$\hat{R}_g = \frac{e^{-H_g/k_B T}}{\text{tr}_{\text{vib}}\{e^{-H_g/k_B T}\}}\,. \tag{5.36}$$

This way the DOS (lineshape function) has been obtained as a Fourier transformed correlation function, which relates the vibrational motion in the electronic ground state PES to the motion in the excited state PES. The averaging has to be taken with respect to the vibrational equilibrium statistical operator for the electronic ground state.

For later use we give an alternative notation of Eq. (5.35) which found many applications (cf. Section 5.3.3). There we introduce $\Delta H_{eg} = H_e - H_g - \hbar\omega_{eg}$, which is used to rewrite the time–evolution operator for $H_e$. This can be achieved in accordance with the identity Eq. (3.31) which is introduced to obtain the $S$–operator. It follows that

$$\begin{aligned} U_e(t) &= e^{-iH_e t/\hbar} = e^{-i\omega_{eg}t} e^{-i(H_g + \Delta H_{eg})t/\hbar} \\ &= e^{-i\omega_{eg}t} U_g(t) S_{eg}(t,0)\,, \end{aligned} \tag{5.37}$$

with $U_g(t) = e^{-iH_g t/\hbar}$. The $S$–operator has the form

$$S_{eg}(t,0) = \hat{T}\exp\left\{-\frac{i}{\hbar}\int\limits_0^t d\bar{t}\,\Delta H_{eg}^{(g)}(\bar{t})\right\}, \tag{5.38}$$

where the abbreviation $\Delta H_{eg}^{(g)}(\bar{t}) = U_g^+(\bar{t})\Delta H_{eg} U_g(\bar{t})$ has been introduced. Thus, the $S$–operator describes that part of the time evolution which is driven by the difference between $U_e(t)$ and $U_g(t)$. Using this notation the lineshape function is obtained in the compact form

$$\mathcal{D}_{\text{abs}}(\omega - \omega_{eg}) = \frac{1}{2\pi\hbar}\int dt\, e^{i(\omega - \omega_{eg})t}\, \text{tr}_{\text{vib}}\{\hat{R}_g S_{eg}(t,0)\}\,. \tag{5.39}$$

If the trace expression in the foregoing formula or in Eq. (5.35) becomes proportional to $\exp(-t/\tau)$ at $t \to \infty$, the time–constant $\tau$ is known as the dephasing time. It describes the decay of the long–time correlations between the electronic ground and excited state vibrational dynamics (cf. Eq. 5.35). In this way it determines the broadening of the transition from the vibrational ground state level $E_{g0}$ in the electronic ground state to the vibrational ground state level $E_{e0}$ in the excited electronic states (so–called zero–phonon transition). The constant $\tau$ is called pure–dephasing time if it is exclusively caused by the vibrational motion (for example by anharmonicities).

Although Eq. (5.35) or Eq. (5.39) do not require any knowledge about the system's eigenstates the determination of the correlation function is still a formidable task. The best strategy depends on the details of the system under consideration. For systems with a few number of vibrational degrees of freedom (3 to 5), the time–dependent Schrödinger equation subject to particular initial conditions can be solved as will be presented in Section 5.3.2. For larger systems a direct solution of the Schrödinger equation (either time–dependent or stationary) is impossible. Instead, approximations have to be introduced. They can be based on an approximate description of the vibrational degrees of freedom as a thermal reservoir in the spirit of the Quantum Master Equation discussed in Section 3.7 (cf. Section 5.3.4 for this approach to the absorption spectrum). A classical description of the vibrational coordinates (or a part of it) is given in Section 5.3.5. In Section 5.3.3 the so–called *cumulant expansion* is introduced which is of particular importance when considering nonlinear optical processes (see, e.g. [Muk95]).

For the case that the vibrational motion proceeds within the harmonic approximation such that it can be mapped onto those of independent harmonic oscillators (via a normal mode analysis, see Section 2.6.1), an analytical computation of the correlation function in Eq. (5.35) becomes possible.

### 5.2.4 Absorption Coefficient for Harmonic Potential Energy Surfaces

In the case that the vibrations are described by independent harmonic oscillators an analytical expression for Eq. (5.35) can be derived starting from the vibrational Hamiltonian introduced in Eq. (2.74). The various vibrational frequencies should be independent on the electronic state but both PES are shifted relative to each other along the different normal mode coordinates. The related dimensionless displacements $g_a(\xi)$ are given in Eq. (2.86). Within this model we obtain the lineshape function as (for details see the supplementary Section 5.8.1)

$$\mathcal{D}_{\text{abs}}(\omega - \omega_{eg}) = \frac{1}{2\pi\hbar} \int dt \, e^{i(\omega-\omega_{eg})t - G(0) + G(t)} \,, \tag{5.40}$$

with the transition frequency introduced in Eq. (5.32). The time–dependent function in the exponent of Eq. (5.40) reads

$$G(t) = \sum_\xi \left(g_e(\xi) - g_g(\xi)\right)^2 \left[(1 + n(\omega_\xi))e^{-i\omega_\xi t} + n(\omega_\xi)e^{i\omega_\xi t}\right] \,. \tag{5.41}$$

This expression includes the dimensionless relative displacement $g_e(\xi) - g_g(\xi)$ between both PES. The $\omega_\xi$ denote the frequencies of the normal mode oscillators, and $n(\omega)$ is the Bose–Einstein distribution which introduces the temperature dependence. Obviously, the function $G(t)$ carries the complete information on the influence of the nuclear degrees of freedom. Neglecting $G(t)$ the absorption profile reduces to a sharp line at $\omega = \omega_{eg}$. The time–independent part

$$G(0) = \sum_\xi (g_e(\xi) - g_g(\xi))^2 (1 + 2n(\omega_\xi)) \tag{5.42}$$

includes the so–called *Huang–Rhys factor* $\sum_\xi (g_e(\xi) - g_g(\xi))^2$. It is related to the expression

$$S = 2\hbar \sum_\xi \omega_\xi \left(g_e(\xi) - g_g(\xi)\right)^2 \,, \tag{5.43}$$

which is known as the *Stokes shift* and experimentally obtainable by comparing the peak shift between absorption and fluorescence spectra (for details see literature list).

Starting with the Golden Rule expression, Eq. (5.31), where the thermal averaging with respect to the initial vibrational eigenstates has to be carried out directly, the final result includes thermally averaged quantities in the exponent. This can be formally rationalized by means of the cumulant expansion as explained in Section 5.3.3. Once the function $G(t)$ is given, a single time integration generates the complete absorption spectrum according to

$$\alpha(\omega) = \frac{2\pi\omega n_{\text{mol}}}{3\hbar c} |d_{eg}|^2 e^{-G(0)} \int dt\; e^{i(\omega-\omega_{eg})t+G(t)} \; . \tag{5.44}$$

The general character of this expression becomes obvious when discussing limiting cases for $G(t)$. We start considering the limit where only a *single* vibrational mode with frequency $\omega_{\text{vib}}$ couples to the electronic transition. From Eq. (5.41) it follows that ($\Delta g = g_e - g_g$)

$$G(t) = \Delta g^2 \left(e^{-i\omega_{\text{vib}}t}(1+n(\omega_{\text{vib}})) + e^{i\omega_{\text{vib}}t} n(\omega_{\text{vib}})\right) \; . \tag{5.45}$$

Expanding the exponential function in Eq. (5.40) yields

$$\begin{aligned} \exp\{G(t)\} \;&=\; \sum_{M=0}^{\infty} \frac{1}{M!} \left[\Delta g^2 (1+n(\omega_{\text{vib}}))\right]^M e^{-iM\omega_{\text{vib}}t} \\ &\times \sum_{N=0}^{\infty} \frac{1}{N!} \left[\Delta g^2\, n(\omega_{\text{vib}})\right]^N e^{iN\omega_{\text{vib}}t} \; . \end{aligned} \tag{5.46}$$

Inserting this result into the expression of the combined DOS allows to carry out the time integration for every contribution in the double sum. It simply gives

$$\begin{aligned} \mathcal{D}_{\text{abs}}(\omega-\omega_{eg}) \;&=\; \frac{1}{\hbar} e^{-\Delta g^2(1+2n(\omega_{\text{vib}}))} \sum_{M,N=0}^{\infty} \frac{1}{M!} \left[\Delta g^2 (1+n(\omega_{\text{vib}}))\right]^M \\ &\times \frac{1}{N!} \left[\Delta g^2 n(\omega_{\text{vib}})\right]^N \delta(\omega - \omega_{eg} - (M-N)\omega_{\text{vib}}) \; . \end{aligned} \tag{5.47}$$

Using this expression, the absorption coefficient, Eq. (5.30), becomes a collection of sharp lines corresponding to transitions at frequencies $\omega_{eg} - (M-N)\,\omega_{\text{vib}}$ (vibrational progression). In contrast to Eq. (5.31) the Franck–Condon factor and the thermal distribution has been replaced by powers of $\Delta g$ and $n(\omega_{\text{vib}})$, respectively.

Before discussing this result further, we consider the zero–temperature case:

$$\mathcal{D}_{\text{abs}}(\omega-\omega_{eg})|_{T=0} = \frac{1}{\hbar} e^{-\Delta g^2} \sum_{M=0}^{\infty} \frac{\Delta g^{2M}}{M!} \delta(\omega - \omega_{eg} - M\omega_{\text{vib}}) \; . \tag{5.48}$$

The absorption spectrum is a sequence of sharp lines (see Fig. 5.6) at frequencies $\omega_{eg} + M\omega_{\text{vib}}$ with weighting factors

$$w_M = e^{-\Delta g^2} \frac{\Delta g^{2M}}{M!} \; . \tag{5.49}$$

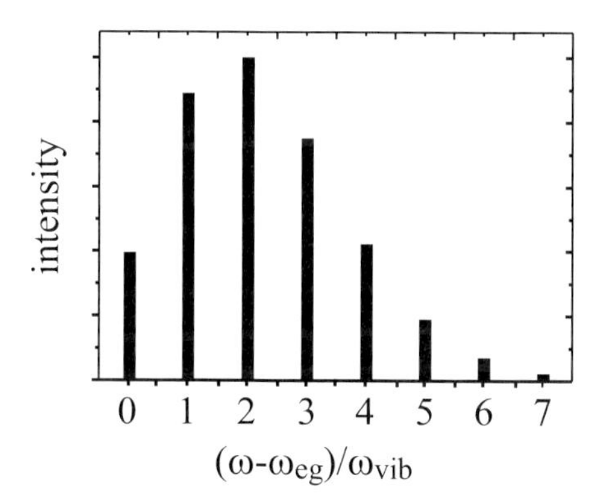


**Figure 5.6:** Stick spectrum of the absorption described by Eq. (5.48). The weighting factors, Eq. (5.49) are plotted versus frequency ($\Delta g = 1.5$).

These weighting factors follow from a so–called *Poisson distribution*. They become maximal at $M \approx \Delta g^2$ or, in terms of energies, at $M\hbar\omega_{\rm vib} \approx \hbar\omega_{\rm vib}\Delta g^2$. Note that the vibrational quantum number at which the absorption reaches its maximum is given by the difference $U_e(q = q_g) - U_e(q = q_e) = \hbar\omega_{\rm vib}\Delta g^2$. This corresponds to a vertical transition which is in accord with the Franck–Condon principle introduced at the beginning of this chapter. The shape of the spectrum following from Eq. (5.48) is illustrated in Fig. 5.5.

At finite temperatures we have to consider the double summation in Eq. (5.47). Nevertheless, a more compact expression can be derived. If $M > N$ we introduce $K = M - N$ and $N = L$, where $L$ and $K$ run from 0 to $\infty$. In case $M < N$ we set $M = L$ which again runs from 0 to $\infty$, but $K$ is in between the interval from 0 to $-\infty$. Rearranging the combined DOS gives

$$\begin{aligned}\mathcal{D}_{\rm abs}(\omega - \omega_{eg}) &= \frac{1}{\hbar}\sum_{K=-\infty}^{\infty}\left(\frac{n(\omega_{\rm vib})}{1 + n(\omega_{\rm vib})}\right)^{|K|/2}\delta(\omega - \omega_{eg} - K\omega_{\rm vib}) \\ &\times\left(\sum_{L=0}^{\infty}\frac{1}{L!(L + |K|!)}(\Delta g^4 n(\omega_{\rm vib})[1 + n(\omega_{\rm vib})]/4)^{L+|K|/2}\right) .\end{aligned} \tag{5.50}$$

Using the definition of the modified Bessel function

$$I_K(z) = \sum_{L=0}^{\infty}\frac{1}{L!(L + |K|!)}(z^2/4)^{L+|K|/2} \tag{5.51}$$

we get

$$\begin{aligned}\mathcal{D}_{\rm abs}(\omega - \omega_{eg}) &= \frac{1}{\hbar}\sum_{K=-\infty}^{\infty} I_K\left(\Delta g^2\sqrt{n(\omega_{\rm vib})[1 + n(\omega_{\rm vib})]}\right)\left(\frac{n(\omega_{\rm vib})}{1 + n(\omega_{\rm vib})}\right)^{|K|/2} \\ &\times\delta(\omega - \omega_{eg} - K\omega_{\rm vib}) .\end{aligned} \tag{5.52}$$

This compact expression contains only a single sum with respect to the difference in vibrational quanta between the electronic ground and the excited state.

### 5.2.5 Absorption Lineshape and Spectral Density

If many vibrational modes couple to the electronic transition, we expect a quasi–continuous spectrum of vibrational frequencies. In such a case it is convenient to introduce the *spectral density*

$$J_{eg}(\omega) = \sum_{\xi} (g_e(\xi) - g_g(\xi))^2 \, \delta(\omega - \omega_\xi) \tag{5.53}$$

into Eq. (5.41) (cf. Section 3.6.4). This function enables us to write

$$G(t) = \int_0^\infty d\omega \, \left[(1 + n(\omega))e^{-i\omega t} + n(\omega)e^{i\omega t}\right] J_{eg}(\omega) \,. \tag{5.54}$$

Furthermore, we can use the spectral density to write the Stokes shift introduced in Eq. (5.43) as

$$S = 2\hbar \int_0^\infty d\omega \, \omega J_{eg}(\omega) \,. \tag{5.55}$$

For further use it is convenient to introduce the real and imaginary parts of the function $G(t)$, Eq. (5.41),

$$G(t) = G_1(t) - iG_2(t) \,, \tag{5.56}$$

where

$$G_1(t) = \int_0^\infty d\omega \, \cos(\omega t)[1 + 2n(\omega)]J_{eg}(\omega) \,, \tag{5.57}$$

and

$$G_2(t) = \int_0^\infty d\omega \, \sin(\omega t) J_{eg}(\omega) \,. \tag{5.58}$$

Note that the imaginary part is temperature independent, whereas the real part includes all temperature effects. Apparently, the real and imaginary part of $G(t)$, Eq. (5.41), have to obey $G_1(t) = G_1(-t)$ and $G_2(t) = -G_2(-t)$, respectively. In particular, these properties have to be fulfilled if both functions are calculated from model spectral density.

According to this separation of $G(t)$ the density of states reads

$$\mathcal{D}_{\text{abs}}(\omega - \omega_{eg}) = \frac{1}{2\pi\hbar} \int dt \, \exp\left\{i\left[(\omega - \omega_{eg})t - G_2(t)\right] + G_1(t) - G_1(0)\right\} \,. \tag{5.59}$$

The imaginary part of $G(t)$ introduces a shift of the electronic transition frequency $\omega_{eg}$, whereas the real part leads to an exponential decay of the integrand in Eq. (5.59). We expect that this exponential decay has something to do with the broadening of the absorption lines.

To be more specific let us discuss the absorption in terms of the spectral density. In real molecular systems the spectrum of possible vibrational frequencies has an upper limit. Thus, according to the definition (5.53) of the spectral density, we can say that $J_{eg}(\omega)$ should go to zero if $\omega$ is larger than a certain cut–off frequency. How $J_{eg}(\omega)$ behaves for $\omega \to 0$ cannot be further specified in general. However, according to Eq. (5.55) the frequency integral of $\omega J_{eg}$ should give the finite quantity $S/2\hbar$. Consequently, the spectral density cannot diverge for $\omega \to 0$. Because of this property, the quantity $\omega J_{eg}$ instead of $J_{eg}$ is often used to characterize a system (cf. discussion in Section 3.6.4).

One example for $J_{eg}$, the Debye spectral density, has already been introduced in Section 3.6.4. Within the present notation it reads

$$\omega J_{eg}(\omega) = \Theta(\omega)\frac{S}{\pi\hbar}\,\frac{\omega_\mathrm{D}}{\omega^2 + \omega_\mathrm{D}^2} \; . \tag{5.60}$$

How this function is related, for example, to the dynamics of a polar solvents will be explained in detail in Section 6.5.

We note that the simple form (5.60) has the advantage that a number of quantities defined via the spectral density (including the absorption coefficient) can be calculated analytically. This has already been demonstrated in Section 3.6.4 for the case of the reservoir correlation function.

Next we calculate the absorption coefficient using the Debye spectral density. It is easy to obtain the imaginary part $G_2(t)$ of $G(t)$. First we extend the frequency integral to $-\infty$. Then, to carry out the Fourier transform, we replace $\sin(\omega t)/\omega$ by a time integral with respect to $\cos(\omega t)$. This gives

$$G_2(t) = \frac{S\omega_\mathrm{D}}{2\pi\hbar}\int_0^t d\tau \int d\omega\, \frac{\cos(\omega\tau)}{\omega^2 + \omega_\mathrm{D}^2} \; . \tag{5.61}$$

The frequency integral has already been calculated in Section 3.6.4 (Eq. (3.235)). The remaining $\tau$–integration results in the final expression

$$G_2(t) = \frac{S}{2\hbar\omega_\mathrm{D}}\left(1 - e^{-\omega_\mathrm{D} t}\right) . \tag{5.62}$$

The calculation of the real part of $G$ is more complicated. Here we only consider the high–temperature limit, $k_\mathrm{B}T \gg \hbar\omega_\mathrm{D}$ and approximate $1 + 2n(\omega) \approx 2k_\mathrm{B}T/\hbar\omega$. If one inserts this approximation into $G_1(t)$, one can proceed as in the case of $G_2(t)$. First, one extends the limits for the frequency integration over the whole axis. Afterwards, $\cos(\omega t)/\omega^2$ is replaced by a double time integral. The resulting frequency integral can be performed as in the case of $G_2(t)$. After carrying out the final double time integration we get

$$G_1(t) = -\frac{k_\mathrm{B}TS}{(\hbar\omega_\mathrm{D})^2}\left(e^{-\omega_\mathrm{D} t} + \omega_\mathrm{D} t - 1\right) . \tag{5.63}$$

The frequency–dependent combined density of states is obtained after the time integration in Eq. (5.59) has been carried out.

In order to discuss this result we introduce two time scales. First we have the time scale for vibrational motion characterized by $T_{\text{vib}} \approx 1/\omega_{\text{D}}$. The second time scale is related to the strength of the coupling between electronic and nuclear motions ($S$). It can be interpreted as the time scale for fluctuations of the electronic energy gap, we have $T_{\text{fluc}} = \hbar/\sqrt{k_{\text{B}}TS}$. We can distinguish two limits:

Slow nuclear motion:

We suppose that $T_{\text{vib}} \gg T_{\text{fluc}}$ and, in particular, that it is possible to perform a short time expansion of $G(t)$ with respect to $\omega_{\text{D}}t$. One obtains

$$\begin{aligned} \mathcal{D}_{\text{abs}}(\omega - \omega_{eg}) &= \frac{1}{2\pi\hbar} \int dt \, \exp\left\{ i(\omega - \omega_{eg} - S/2\hbar)t - \frac{1}{2}\left(\frac{t}{T_{\text{fluc}}}\right)^2 \right\} \\ &= \frac{T_{\text{fluc}}}{\sqrt{2\pi}\,\hbar} \exp\left\{ -\frac{1}{2}\left(T_{\text{fluc}}(\omega - \omega_{eg} - S/2\hbar\right)^2 \right\} . \end{aligned} \tag{5.64}$$

This case is known as the limit of *inhomogeneous broadening*, where the time scale for nuclear motion is such that the nuclei can be considered to be frozen (cf. Fig. 5.1). We have a Gaussian absorption lineshape centered around the vertical Franck–Condon transition $\omega = \omega_{eg} + S/2$.

Fast nuclear motion:

Here, $T_{\text{vib}} \ll T_{\text{fluc}}$ and the exponential factors in $G(t)$ can be neglected. Setting $\omega_{\text{D}}t - 1 \approx \omega_{\text{D}}t$ and neglecting $G_2(t)$ which is of the order of $(T_{\text{vib}}/T_{\text{fluc}})^2\hbar\omega_{\text{D}}/k_{\text{B}}T$, one arrives at $G(t) \approx T_{\text{vib}}|t|/T_{\text{fluc}}^2$. The absorption lineshape follows as a Lorentzian

$$\begin{aligned} \mathcal{D}_{\text{abs}}(\omega - \omega_{eg}) &= \frac{1}{2\pi\hbar} \int dt \, \exp\left\{ (i(\omega - \omega_{eg})t - T_{\text{vib}}|t|/T_{\text{fluc}}^2 \right\} \\ &= \frac{1}{\pi\hbar} \frac{\gamma}{(\omega - \omega_{eg})^2 - \gamma^2} . \end{aligned} \tag{5.65}$$

The linewidth is given by $\gamma = T_{\text{vib}}/T_{\text{fluc}}^2$. This is the limit of *homogeneous broadening*. Note that the absorption is now centered at the electronic transition frequency and the Stokes shift does not appear. This can be rationalized by the fact that the nuclear motion is so fast that only the electronic transition which is averaged with respect to the nuclear dynamics is detected in the experiment.

Finally, we point out that the transition between the limits of inhomogeneous and homogeneous broadening can be observed upon changing the temperature. While at low temperature the nuclear motions are frozen and the lineshape is Gaussian, at higher temperature it becomes Lorentzian. This phenomenon is also known as motional line narrowing .

The model described so far can be generalized to a case which is often met in praxis, a single vibrational coordinate coupled to further vibrations (often named *Brownian oscillator* model). The latter vibrations might be additional intramolecular vibrations or those of a solvent as well as any other type of a condensed–phase environment (bath vibrations). In the most simplest case the single vibrational coordinate (the active vibration) might correspond to a bond vibration of a diatomic molecule. Since it moves like a Brownian particle undergoing random kicks of the environment it is called a Brownian oscillator. However, the active vibration can also be understood as a collective coordinate which is dominantly coupled to

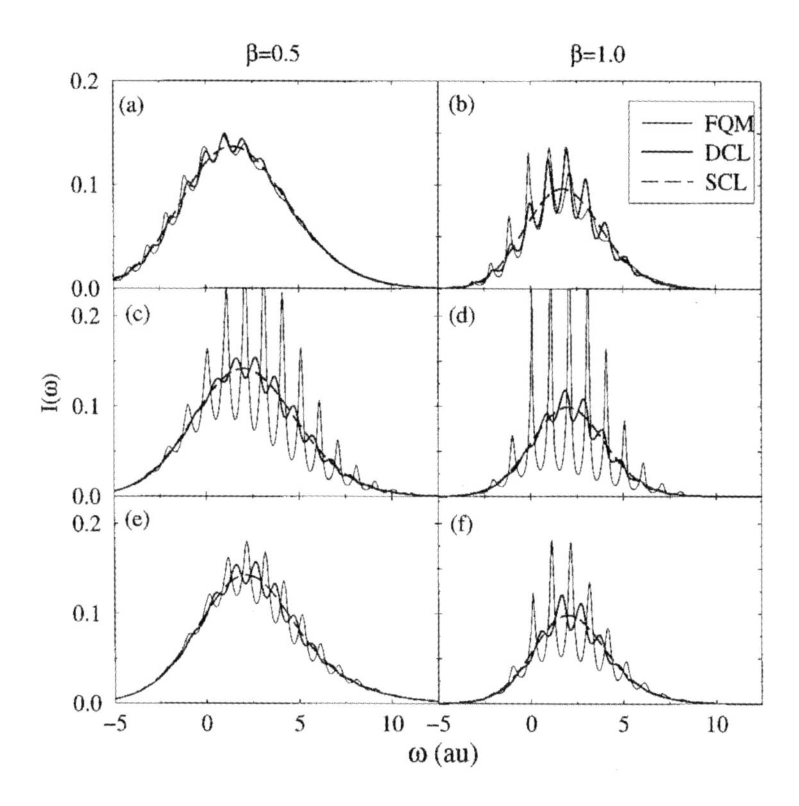


**Figure 5.7:** Linear absorption coefficient (arbitrary units) versus frequency (in units of $\omega_{\rm vib}$ and relative to $\omega_{eg}$) for a diatomic molecule embedded in a condensed–phase environment at two different temperatures (left part: $\beta = k_{\rm B}T/\hbar\omega_{\rm vib} = 1/2$, right part: $\beta = k_{\rm B}T/\hbar\omega_{\rm vib} = 1$, for easy comparison the left curves are multiplied with $1/2$) and different coupling strength to the reservoir vibrations (see text for these numbers and all other parameters). FQM denotes the results of the correct (full) quantum mechanical calculation. DCL and SCL curves are the result of semiclassical descriptions as explained in Section 5.3.5 (reprinted with permission from [Ego98]; copyright (1998) American Institute of Physics).

the electronic transition. In any case it is responsible for a vibrational progression in the linear absorption as displayed in Fig. 5.6. While this progression consists of sharp lines for an isolated oscillator, the present study includes a coupling to additional vibrations which should introduce a broadening of the absorption lines. We expect a dependence of this broadening on the coupling strength to the secondary vibrations and on temperature.

The related Hamiltonian represents a particular example for the generic system–reservoir Hamiltonian, Eq. (3.3). Here, $H_{\rm S}$ describes the active vibration, $H_{\rm R}$ denotes the Hamiltonian of all bath vibrations, and $H_{\rm S-R}$ is responsible for the coupling between both. We assume an expression for $H_{\rm S}$ which is identical with the molecular part of Eq. (5.16) (of course restricted to a single vibration). $H_{\rm R}$, for simplicity, should be independent on the electronic states entering $H_{\rm S}$. However, the system–bath coupling Hamiltonian is written as $H_{\rm S-R} =$

$\sum_a |\phi_a\rangle\langle\phi_a|\ K_a(q)\Phi_a(Z)$, with $K_a$ ($\Phi_a$) depending on the primary (secondary) vibrational coordinates, i.e. on $q$ ($Z = \{Z_\xi\}$).

As already pointed out there exists a case where the linear absorption spectrum of a Brownian oscillator can be described by a combined density of states $\mathcal{D}_{\text{abs}}$ according to Eq. (5.59). This becomes possible if all vibrations are described by harmonic oscillators, and if the coupling between the Brownian oscillator and all secondary vibrations is linear in both types of coordinates. This corresponds to the case $K_a(q) = q - q_a$ (here $q_a$ is the respective equilibrium position) and $\Phi_a = \sum_\xi \hbar k_\xi Z_\xi$. Because of the bilinear form of the coupling Hamiltonian, $H_{\text{S-R}}$, the complete Hamiltonian $H_{\text{S}} + H_{\text{S-R}} + H_{\text{R}}$ can be diagonalized for both electronic states just leading again to a Hamiltonian of the type Eq. (5.16) whose linear absorption spectrum can be calculated according to Eq. (5.59).

Fig. 5.7 displays the absorption spectra for such a particular case of a Brownian oscillator. The exact results are obtained after a discretization of the bath modes followed by the introduction of 100 normal mode vibrations (derived from 99 bath modes and the single Brownian oscillator). Related spectra (FQM results, thin lines) are displayed for two different temperatures each with three different coupling strength between the active vibration and the bath vibrations. Furthermore, $\Delta g = -4$ for the primary oscillator (cf. Eq. (5.45)). The spectral density $J(\omega)$ is of the type given in Eq. (3.229) but differs between the ground and the excited electronic state ($J_a = \eta_a J, a = g, e$). The factors $\eta_a$ have been set equal to $\eta_g = 0.05$, $\eta_e = 0.125$ (part (A) and (B) of Fig. 5.7), $\eta_g = 0.0625$, $\eta_e = 0.025$ (part (C) and (D)), and $\eta_g = 0.125$, $\eta_e = 0.05$ (part (E) and (F)). Upon decreasing the temperature and the coupling to the secondary vibrations. the vibrational progression can be clearly identified.

## 5.3 Time–Dependent Formulation of the Absorption Coefficient

In the following different derivations of the frequency–dependent absorption coefficient will be given, based on the solution of dynamic equations for respective time–dependent wave functions or density matrices. The basic idea behind these approaches has been already encountered in Section 5.2.4, where the absorption coefficient was expressed in terms of a Fourier–transform of a particular correlation function (see, Eq. (5.35)). Having different ways of calculating particular correlation functions at least approximately is of great importance not only to get the linear absorption coefficient but to compute any type of transfer rates in molecular systems.

In order to keep the connection with the previous section we will adopt the same electronic two–level model for our discussion. The time–dependent description of the stationary absorption will also enable us to bridge a gap between fast intramolecular dynamic phenomena, i.e., phenomena in the time domain, and properties observed in the frequency domain. We start our considerations with an alternative derivation of the linear susceptibility and consequently of the absorption coefficient. This will be based on the idea of general *linear response theory* introduced already in Section 3.6.5.

### 5.3.1 Dipole–Dipole Correlation Function

Our objective is to obtain a formula for the absorption coefficient which does not rely on a particular representation of the Hamiltonian, but is based on a general description of the molecular quantum dynamics. Further, we do not want to invoke any approximation for the PES as it was done in the previous section.

For simplicity let us neglect inhomogeneous broadening and therefore let us start with Eq. (5.10) for the macroscopic polarization including Eq. (5.5) for the time–dependent expectation value of the dipole operator. In order to carry out a perturbation expansion with respect to $H_{\text{field}}(t)$, i.e. in powers of the electric field–strength, we separate $U(t,t_0)$ in Eq. (5.5) into the molecular part $U_{\text{mol}}(t-t_0) = \exp(-iH_{\text{mol}}(t-t_0)/\hbar)$ and into the $S$–operator (cf. Section 3.2.2)

$$S(t,t_0) = \hat{T}\exp\left(-\frac{i}{\hbar}\int_{t_0}^{t} dt'\, H^{(\text{I})}_{\text{field}}(t')\right) . \tag{5.66}$$

The coupling Hamiltonian in the interaction representation reads $H^{(I)}_{\text{field}}(t) = U^{+}_{\text{mol}}(t-t_0) \times H_{\text{field}}(t)\, U_{\text{mol}}(t-t_0)$. Accordingly, Eq. (5.5) becomes [8]

$$\mathbf{d}(\mathbf{x};t) = \text{tr}\{\hat{W}_{\text{eq}} S^{+}(t,t_0)\hat{\mu}^{(I)}(t) S(t,t_0)\} . \tag{5.67}$$

The different contributions in powers of the field–strength are obtained by a power expansion of $S(t,t_0)$. Here, we concentrate on the first–order contribution (higher–order terms have to be considered in the case of nonlinear spectroscopy, cf. Section 5.6). The expansion of $\langle\hat{\mu}\rangle$ up to the first order in the field–strength gives (the zeroth–order term does not contribute because of the absence of a macroscopic dipole density in the equilibrium)

$$\begin{aligned}
\mathbf{d}(\mathbf{x};t) &\approx \text{tr}\{\hat{W}_{\text{eq}}[1 + S^{(1)+}(t,t_0)]\hat{\mu}^{(I)}(t)[1 + S^{(1)}(t,t_0)]\} \\
&\approx \frac{i}{\hbar}\int_{t_0}^{t} dt'\ \text{tr}\Big\{\hat{W}_{\text{eq}}[\hat{\mu}^{(\text{I})}(t-t'),\hat{\mu}^{(\text{I})}(0)]_{-}\Big\}\, \mathbf{E}(\mathbf{x};t') .
\end{aligned} \tag{5.68}$$

Here, $S^{(1)}$ denotes the first–order expansion of the $S$–operator, Eq. (5.66) with respect to the electric field–strength (in linear approximation the term containing $S^{(1)+}$ and $S^{(1)}$ is neglected). Inserting the last expression of Eq. (5.68) into Eq. (5.10) for the macroscopic polarization, the *linear dielectric susceptibility* is obtained as (cf. Section 5.2.1)

$$\chi_{jj'}(t-t') = \frac{i}{\hbar}\Theta(t-t')n_{\text{mol}}C^{(\text{d}-\text{d})}_{jj'}(t-t') , \tag{5.69}$$

where we introduced the *dipole–dipole correlation function*

$$C^{(\text{d}-\text{d})}_{jj'}(t) = \text{tr}\left\{\hat{W}_{\text{eq}}\left[\hat{\mu}^{(\text{I})}_{j}(t),\hat{\mu}^{(\text{I})}_{j'}(0)\right]_{-}\right\} . \tag{5.70}$$

---

[8] We remind here on the fact that the operator $\hat{\mu}^{(I)}(t)$ is independent on the spatial position $\mathbf{x}$ since inhomogeneous broadening has been neglected. However, the electric field–strength depends on $\mathbf{x}$, and just this dependence enters the expectation value via the field–dependence of the $S$–operator.

Since the trace has been taken with respect to the equilibrium statistical operator the susceptibility (and thus the correlation function) depends on the time difference only (cf. Section 3.6.5). Furthermore, we note that the linear susceptibility as well as the dipole–dipole correlation function are tensors of second rank which components are given by the vector components of $\hat{\mu}^{(\mathrm{I})}(t-t')$ and $\hat{\mu}^{(\mathrm{I})}(0)$. In the following we will only consider the case of randomly oriented molecules where it is sufficient to use

$$C_{\mathrm{d-d}}(t) = \sum_j C_{jj}^{(\mathrm{d-d})}(t) \, . \tag{5.71}$$

The absorption coefficient follows from Eq. (5.14) as

$$\alpha(\omega) = \frac{4\pi\omega n_{\mathrm{mol}}}{3\hbar c} \mathrm{Re} \int_0^\infty dt \; e^{i\omega t} \; C_{\mathrm{d-d}}(t) \, . \tag{5.72}$$

Expression (5.29) is easily recovered, if the adiabatic electron–vibrational states $|\phi_a\rangle|\chi_{aN}\rangle$ ($a = g, e$) are used to calculate the trace. Moreover, Eq. (5.30) introducing the linear absorption via the combined density of states $\mathcal{D}_{\mathrm{abs}}(\omega)$ shows that we have to identify $\pi|d_{eg}|^2 \, \mathcal{D}_{\mathrm{abs}}(\omega)\hbar$ with the real part of the half–sided Fourier transform of $C_{\mathrm{d-d}}(t)$ (provided that a system with two electronic states is considered).[9]

In the derivation we did not refer so far to any specific form of the molecular Hamiltonian. It can in principle describe any type of PES but can also contain contributions from an environment. Thus, Eq. (5.72) is suitable for computation of the absorption spectrum of molecular systems in the condensed phase. Before studying such a situation we will show in the next section how, in the case of a molecular system with a small number of vibrational degrees of freedom, the dipole–dipole correlation function can be calculated by propagation of a nuclear wave function.

### 5.3.2 Absorption Coefficient and Wave Packet Propagation

Alternatively to Eq. (5.70) one can derive an expression for the dipole–dipole correlation function based on the time evolution of a particular statistical operator. To this end Eq. (5.70) is rewritten as (the Cartesian index on both dipole operators has been omitted since they form a scalar product)

$$C_{\mathrm{d-d}}(t) = \mathrm{tr}\Big\{\hat{\mu} U_{\mathrm{mol}}(t) \, [\hat{\mu}, \hat{W}_{\mathrm{eq}}]_- \, U^+_{\mathrm{mol}}(t)\Big\} \equiv \mathrm{tr}\Big\{\hat{\mu}\hat{\sigma}(t)\Big\} \, . \tag{5.73}$$

This formula results from a simple rearrangement of the various operators under the trace in Eq. (5.70) and we introduced $\hat{\sigma}(t) = U_{\mathrm{mol}}(t) \; [\hat{\mu}, \hat{W}_{\mathrm{eq}}]_- \; U^+_{\mathrm{mol}}(t)$. Now, we may calculate the correlation function via a propagation of the the commutator of the equilibrium statistical operator with the dipole operator (which induces the transitions according to the coupling with the radiation field).

This statement can be put into a more transparent formula, if one changes from the propagation of mixed states to that of pure states. Such a situation is encountered in the gas phase

[9]Note that this result is a particular example for the fluctuation–dissipation theorem introduced in Section 3.6.5.

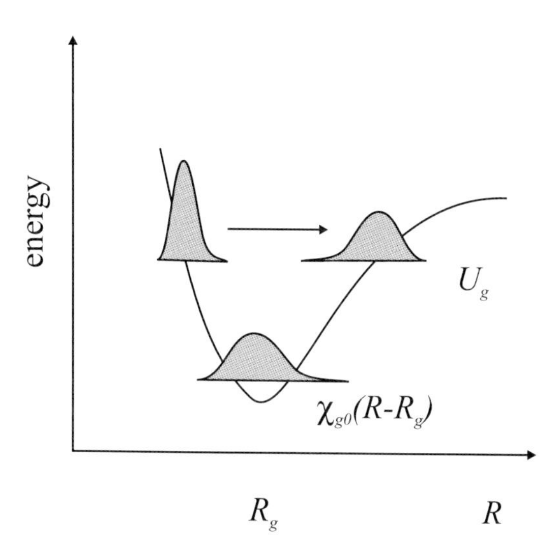


**Figure 5.8:** Wave packet motion used to describe the absorption coefficient according to Eq. (5.75). A single PES versus the vibrational coordinate $R$ is shown (infrared light absorption within a single electronic state). The wave function $\chi_{g0}(R)$ is the unperturbed vibrational ground state. Due to the action of the dipole operator the new wave function $\mu(R)\chi_{g0}(R)$ is created (upper left wave packet, note that this requires that the dipole operator depends on the coordinate). As a non–eigenstate of the vibrational Hamiltonian the wave function $\mu(R)\chi_{g0}(R)$ propagates forth and back in the PES (upper right wave packet). The half–sided Fourier transformation of the time–dependent overlap between the initially displaced and the propagated wave packet gives the absorption spectrum.

in the absence of any environmental influence. Hence, we replace $\hat{W}_{\text{eq}}$ by $|\psi\rangle\langle\psi|$ where $|\psi\rangle$ is an eigenstate of $H_{\text{mol}}$ with energy $\mathcal{E}$. Using $U_{\text{mol}}(t)|\psi\rangle = \exp(-i\mathcal{E}t/\hbar)|\psi\rangle$ we obtain for the dipole–dipole correlation function

$$C_{\text{d-d}}(t) = \text{tr}\left\{\hat{\mu}\left(U_{\text{mol}}(t)\hat{\mu}|\psi\rangle\langle\psi|e^{i\mathcal{E}t/\hbar} - e^{-i\mathcal{E}t/\hbar}|\psi\rangle\langle\psi|\hat{\mu}U^{+}_{\text{mol}}(t)\right)\right\} . \tag{5.74}$$

Next, we rearrange the matrix elements formed by the pure state vector $|\psi\rangle$ and those states used to calculate the trace. We can profit from the completeness relation for the state vectors defining the trace formula and obtain

$$C_{\text{d-d}}(t) = \left(e^{i\mathcal{E}t/\hbar}\langle\psi|\,\hat{\mu}U_{\text{mol}}(t)\,\hat{\mu}|\psi\rangle - \text{c.c.}\right) . \tag{5.75}$$

The interpretation which is illustrated in Fig. 5.8 is straightforward: The dipole operator induces a transition from the initial state $|\psi\rangle$ to the state $\hat{\mu}|\psi\rangle$. This is usually not an eigenstate of $H_{\text{mol}}$. Therefore, one can expect wave packet motion to take place. At time $t$ the propagated state $U_{\text{mol}}(t)\hat{\mu}|\psi\rangle$ is multiplied by $\langle\psi|\hat{\mu}$ to give the *dipole autocorrelation* function $\langle\psi|\hat{\mu}U_{\text{mol}}(t)\hat{\mu}|\psi\rangle$. Its half–sided Fourier transform determines the absorption according to Eq. (5.72).

For the specific case of an electronic two–level system, the state $|\psi\rangle$ is replaced by the vibrational ground state in the electronic ground state, $|\phi_g\rangle|\chi_{g0}\rangle$. Provided that the Condon

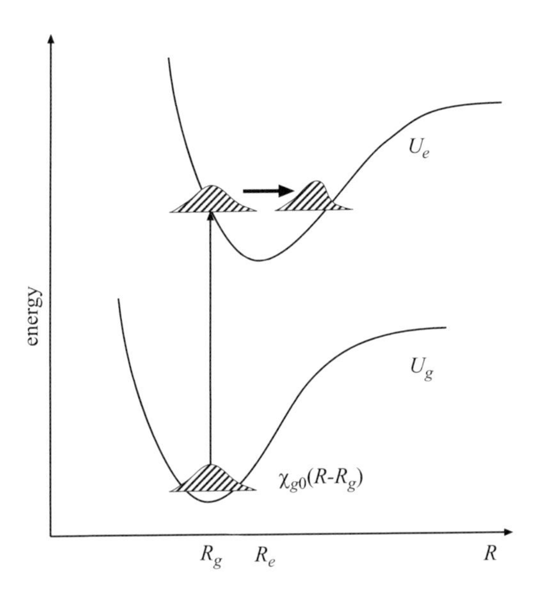


**Figure 5.9:** Wave packet motion used to describe the absorption coefficient according to Eq. (5.77). A system of two PES versus a single vibrational coordinate $R$ is shown. The horizontal displacement of the two PES with respect to each other will give an initial state $d_{eg}|\chi_{g0}\rangle$ which is not an eigenstate of the vibrational Hamiltonian of the electronically excited state. Thus, wave packet dynamics will occur.

approximation is valid, we obtain $\hat{\mu}|\phi_g\rangle|\chi_{g0}\rangle = d_{eg}|\phi_e\rangle|\chi_{g0}\rangle$. Therefore, due to the action of the dipole operator, the vibrational state $|\chi_{g0}\rangle$ of the electronic ground state PES, has been promoted to the excited electronic state $|\phi_e\rangle$. The resulting time dependence reads

$$U_{\text{mol}}(t)\hat{\mu}|\phi_g\rangle|\chi_{g0}\rangle = d_{eg}|\phi_e\rangle e^{-iH_e t/\hbar}|\chi_{g0}\rangle \,. \tag{5.76}$$

The vibrational state $|\chi_{g0}\rangle$ propagates under the action of the vibrational Hamiltonian of the excited electronic state, since $|\chi_{g0}\rangle$ is not an eigenstate of $H_e$. This is illustrated in Fig. 5.9. We neglect the antiresonant contribution and get the absorption coefficient as

$$\alpha(\omega) = \frac{4\pi\omega n_{\text{mol}}}{3\hbar c}\,|\mathbf{d}_{eg}|^2\,\text{Re}\int\limits_0^\infty dt\; e^{i(\omega+E_{g0}/\hbar)t}\,\langle\chi_{g0}|\chi_{g0}^{(e)}(t)\rangle \,. \tag{5.77}$$

Thus, the absorption coefficient is obtained by solving the time–dependent Schrödinger equation for nuclear motion on the electronic excited PES (indicated by the superscript "$e$" at $\chi_{g0}^{(e)}$). The initial condition is given by $|\chi_{g0}^{(e)}(t=0)\rangle = |\chi_{g0}\rangle$. At each time the overlap integral between the propagated and the initial wave function has to be calculated to get the spectrum.

The wave packet description of absorption is particular useful if the excited electronic state $|\phi_e\rangle$ is dissociative. Because no reference is made to eigenstates, there is no need to calculate the continuous energy spectrum. Instead one solves the Schrödinger equation in the coordinate representation as outlined in Section (3.12.5). Additionally, Eq. (5.77) combines

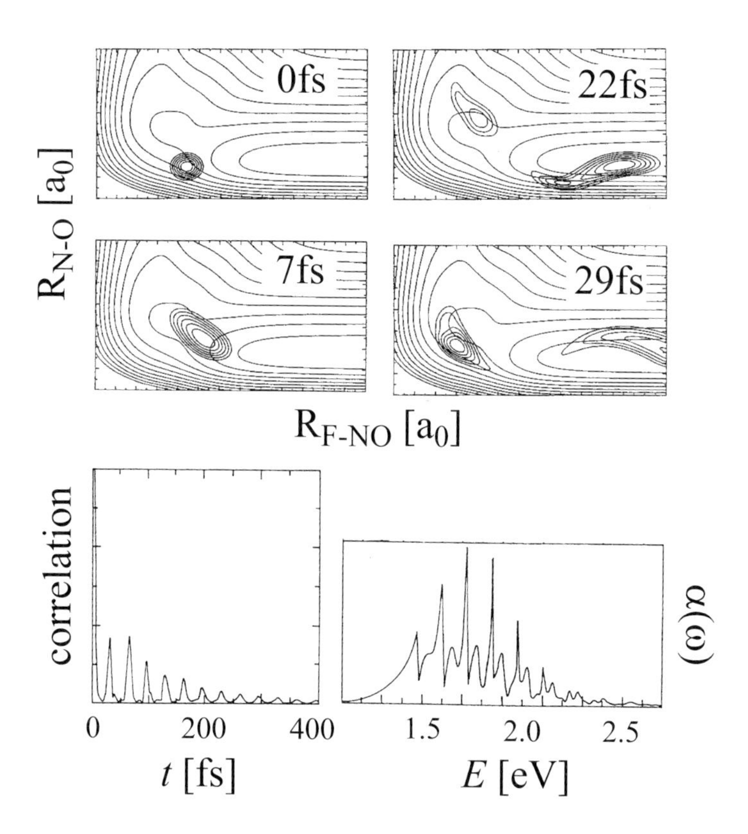


**Figure 5.10:** Numerical results for the $S_0 - S_1$ absorption spectrum of FNO obtained using the wave packet propagation method. The dynamics has been restricted to two dimensions, i.e., the NO bond distance ($R_{\mathrm{N-O}}$ (from 1.8 Å to 3.2 Å)) and the distance between F and the center of mass of the NO fragment ($R_{\mathrm{F-NO}}$ (from 2.5 Å to 5 Å)). In the lower part the correlation function $|\langle\chi_{g0}|\chi_{g0}^{(e)}(t)\rangle|$ (left) as well as the linear absorption spectrum are given (reprinted with permission from [Sut92]; copyright 1992 American Institute of Physics).

a frequency domain quantity, $\alpha(\omega)$, with a time domain quantity, $\chi_{g0}^{(e)}(q,t)$. Therefore, the approach enables one to draw conclusions on the molecular system both in the frequency and in the time domain[10].

In Fig. 5.10 we show the results of a numerical wave packet simulation of the $S_0 - S_1$ absorption spectrum of FNO. The initial wave packet on the excited state is given according to $\chi_{g0}^{(e)}(R_{\mathrm{N-O}}, R_{\mathrm{F-NO}}; t = 0) = \chi_{g0}(R_{\mathrm{N-O}}, R_{\mathrm{F-NO}})$. The subsequent dynamics is characterized by an oscillatory motion in the bound region of the potential and a simultaneous dissociation indicated by those parts of the wave packet which leave along the exit channel $R_{\mathrm{F-NO}} \to \infty$. The interplay between bond vibration and dissociation is reflected in the

[10]As a note in caution we point out that in principle only the *full* time propagation of the wave packet up to $t \to \infty$ gives the absorption coefficient.

damped oscillations of the correlation function shown in the lower part of Fig. 5.10 (left). Consequently, the Fourier transform of this correlation function, which gives the absorption spectrum, is quite structured (right).

It is easy to extend the considerations carried out so far to the case where the absorption process starts from a mixed state (as already assumed in Eq. (5.73)). We may generalize Eq. (5.77) in setting for the equilibrium statistical operator in Eq. (5.73) $\hat{W}_{\text{eq}} = \sum_M f(E_{gM}) |\chi_{gM}\rangle\langle\chi_{gM}|$, i.e. we assume a thermal distribution over the vibrational levels of the electronic ground state. Since every (pure) state in this mixtures enters the formula for the absorption coefficient additively we directly obtain

$$\alpha(\omega) = \frac{4\pi\omega n_{\text{mol}}}{3\hbar c} |\mathbf{d}_{eg}|^2 \sum_M f(E_{gM}) \,\text{Re} \int_0^\infty dt \; e^{i(\omega + E_{gM}/\hbar)t} \, \langle\chi_{gM}|\chi_{gM}^{(e)}(t)\rangle \,. \quad (5.78)$$

Instead of a single wave function overlap as in Eq. (5.77) we obtained a multiple overlap $\langle\chi_{gM}|\chi_{gM}^{(e)}(t)\rangle$ between the vibrational wave functions $\chi_{gM}$ and its time–dependent form propagated at the excited state PES. Moreover, every part is weighted by the thermal distribution function.

A complication appears if several coupled electronic states become accessible after photon absorption. Let us consider the case of two coupled states $|\phi_e\rangle$ and $|\phi_f\rangle$. Here the excited state cannot be described by the single vibrational Hamiltonian $H_e$ but by the Hamiltonians $H_e$ and $H_f$ coupled via the nonadiabaticity operator $\Theta_{fe}$. This is another example for a *curve–crossing* problem. The coupling has to be accounted for in the numerical solution of the Schrödinger equation for the wave function determining the correlation function in Eq. (5.77). As a consequence of the nonadiabatic coupling, the spectrum changes, that is, the positions of the transitions are modified and new transitions appear (cf. Fig. 5.11). We will return to the dynamics within coupled PES in more detail in Section 5.7 as well as in Chapter 6.

### 5.3.3 Cumulant Expansion of the Absorption Coefficient

Assuming that the PES for both electronic states are harmonic it is possible to compute the linear absorption coefficient exactly. This has been demonstrated in Section 5.2.4. For anharmonic PES one has to chose a different approach. Here, we explain the method which is known from probability theory as the *cumulant expansion*. It enables us to put the thermal average appearing in the combined density of states $\mathcal{D}_{\text{abs}}(\omega)$, Eq. (5.35), directly into the exponent (cf. Eq. (5.40)). In practice this corresponds to a partial resummation of the perturbation series, in the present case, with respect to the electron–vibrational coupling and, afterwards, to a computation of certain correlation functions.

The following calculations are based on Eq. (5.39) which defines the dipole–dipole correlation function, Eq. (5.70), by a time–ordered exponential. To determine the trace in Eq. (5.39)

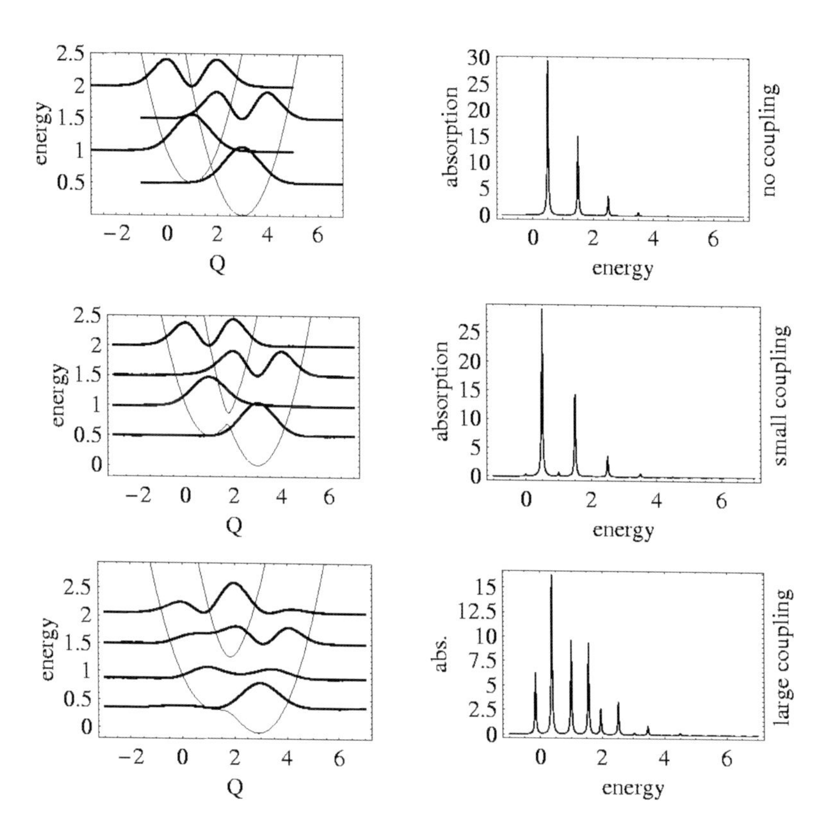


**Figure 5.11:** Linear absorption spectrum for a curve crossing system (states $|\phi_e\rangle$ and $|\phi_f\rangle$) along a one–dimensional reaction coordinate $Q$ (dimensionless oscillator coordinate). The ground state is at $Q = 0$ and the transition dipole moment is nonzero only for the transition to the diabatic state $|\phi_e\rangle$. It is assumed to be constant, that is, only the bare Franck–Condon factors between the ground state wave function and the eigenstates of the coupled excited states determine the spectrum. The left column shows the diagonalized potential curves and the coordinate probability distribution for the lowest eigenstates (energy in units of the oscillator frequency and the detuning between the diabatic potentials is 0.5). The right column shows the absorption spectrum for the case of no, weak ($\Theta_{ef} = 0.1$), and strong ($\Theta_{ef} = 0.5$) inter–state coupling. Upon increasing the inter–state coupling, new peaks appear and the spectrum is shifted.

it is necessary to expand the $S$–operator

$$\begin{aligned}\langle S_{eg}(t,0)\rangle_g &= \text{tr}_{\text{vib}}\{\hat{R}_g S_{eg}(t,0)\} = 1 - \frac{i}{\hbar}\int\limits_0^t d\bar{t}_1 \langle \Delta H_{eg}^{(g)}(\bar{t}_1)\rangle_g \\ &+ \left(\frac{i}{\hbar}\right)^2 \int\limits_0^t d\bar{t}_1 \int\limits_0^{\bar{t}_1} d\bar{t}_2 \langle \Delta H_{eg}^{(g)}(\bar{t}_1)\Delta H_{eg}^{(g)}(\bar{t}_2)\rangle_g + \dots \quad . \end{aligned} \tag{5.79}$$

Translating the exact calculation of Section 5.2.4 to the present notation suggests the identification of $\langle S_{eg}(t,0)\rangle_g$ with $\exp\{G(t)\}$ (Eq. (5.40)). This leads us to the ansatz

$$C_{\text{d-d}}(t) = | d_{eg} e^{-i\omega_{eg}} |^2 \langle S_{eg}(t,0)\rangle_g = | d_{eg} |^2 e^{-i\omega_{eg} - \Gamma(t)} , \tag{5.80}$$

where $\Gamma(t)$ has still to be computed. In order to calculate $\Gamma(t)$ let us introduce a power expansion with respect to $\Delta H_{eg}^{(g)}$ like

$$\Gamma(t) = \Gamma_1(t) + \Gamma_2(t) + \Gamma_3(t) + \dots , \tag{5.81}$$

where $\Gamma_n(t)$ is of $n$th order in $\Delta H_{eg}^{(g)}$. It is reasonable that this expansion exists because it exists for $\langle S_{eg}(t,0)\rangle_g$. Because the exponent of the (time–ordered) $S$–operator is of first order with respect to $\Delta H_{eg}^{(g)}$, there is no zeroth–order contribution in Eq. (5.81). Clearly, even if only some low–order contributions like $\Gamma_1$, or $\Gamma_2$ would be known, we would have $\langle S_{eg}(t,0)\rangle_g$ for any order in $\Delta H_{eg}^{(g)}$. However, in general retaining only low–order terms in $\Gamma(t)$ would not give the exact expression for $\langle S_{eg}(t,0)\rangle_g$. The type of expansion which follows from the introduction of the function $\Gamma(t)$ is known as the *cumulant expansion.*

In order to compare the present approach with the direct expansion of $\langle S_{eg}(t,0)\rangle_g$ in Eq. (5.79) we expand Eq. (5.80) and insert Eq. (5.81). This gives

$$\begin{aligned} e^{-\Gamma(t)} &= 1 - (\Gamma_1(t) + \Gamma_2(t) + \Gamma_3(t) + \dots) \\ &\quad + \frac{1}{2}(\Gamma_1(t) + \Gamma_2(t) + \Gamma_3(t) + \dots)^2 + \dots \quad . \end{aligned} \tag{5.82}$$

Restricting ourselves to terms up to the second order with respect to $\Delta H_{eg}^{(g)}$ yields

$$\begin{aligned} e^{-\Gamma(t)} &\approx 1 - (\Gamma_1(t) + \Gamma_2(t)) + \frac{1}{2}(\Gamma_1(t) + \Gamma_2(t))^2 + \dots \\ &\approx 1 - \Gamma_1(t) - \left(\Gamma_2(t) - \frac{1}{2}\Gamma_1^2(t)\right) . \end{aligned} \tag{5.83}$$

A direct comparison with Eq. (5.79) gives

$$\Gamma_1(t) = \frac{i}{\hbar}\int_0^t dt_1 \langle \Delta H_{eg}^{(g)}(t_1)\rangle_g , \tag{5.84}$$

and

$$\Gamma_2(t) = \frac{1}{\hbar^2}\int_0^t dt_1 \int_0^{t_1} dt_2 \langle \Delta H_{eg}^{(g)}(t_1)\Delta H_{eg}^{(g)}(t_2)\rangle_g + \frac{1}{2}\Gamma_1^2(t) . \tag{5.85}$$

Whether the derived expression for the dipole–dipole correlation function

$$C_{\text{d-d}}(t) = | d_{eg} |^2 e^{-i\omega_{eg} - \Gamma_1(t) - \Gamma_2(t)} , \tag{5.86}$$

is a good approximation or not cannot be decided in general. Clearly, if higher–order cumulants do not exist the result is exact. Otherwise, only a comparison with experiments can give some information about the quality of this approach.

In the supplementary Section 5.8.2 the harmonic oscillator description of the vibrational modes as given in Section 5.2.4 is repeated in the cumulant expansion formalism. In particular, it is demonstrated that the second–order cumulant expansion of $\Gamma(t)$, Eq. (5.81) gives the exact result for the absorption coefficient, Eq. (5.44).

### 5.3.4 Absorption Coefficient and Reduced Density Operator Propagation

In the foregoing section we used the rearrangement Eq. (5.73) of the dipole–dipole correlation function to give an interpretation of the absorption coefficient in terms of a wave packet propagation, which holds if the initial state of the transition is a pure or a mixed state. Here, we will briefly demonstrate how to proceed if the system which undergoes the optical transition is in a condensed phase environment and is characterized by a reduced density operator. To be more specific we will assume that the vibrational degrees of freedom described in the preceding section by the states $\chi_{aM}$, $(a = g, e)$ are coupled to a solvent which acts as a thermal bath in the sense of Section 3.7. As a result every state $\chi_{aM}$ has a finite life–time because of transitions accompanied by the emission or absorption of environmental (bath) quanta.

The most simple example for such a situation would be a diatomic molecule in a solvent where the relative coordinate of the molecule is linearly coupled to generalized solvent coordinates. However, if both are assumed to move in parabolic PES the bilinear coupling allows to introduce common harmonic coordinates and the whole system can be described exactly as outlined in Section 5.2.4. It is the advantage of the following treatment that it is also valid in the case of a general coupling to the condensed phase environment. Nevertheless, the treatment of the active system reservoir coupling is still approximately according to the use of the quantum master equation.

To obtain the dipole–dipole correlation function in the present case we again have to start from Eq. (5.10) for the polarization and have to introduce its linearized version with respect to the radiation field. However, if the radiation field only affects the (active) molecular system and does not induce optical transitions in the environment, Eq. (5.5) for the dipole operator expectation value can be written as

$$\mathbf{d}(t) = \mathrm{tr_S}\{\hat{\mu}\hat{\rho}(t)\} \ . \tag{5.87}$$

Here, the trace is taken with respect to the state space of the active molecular system responsible for the absorption processes. $\hat{\rho}(t)$ denotes the reduced density operator which is propagated under the action of the external field starting with the equilibrium value $\hat{\rho}_{\mathrm{eq}}$[11]. The environment enters the expression via the time–evolution of $\hat{\rho}_{\mathrm{eq}}$. In a next step we linearize Eq. (5.87) with respect to the electric field–strength to find a generalization of Eq. (5.73) for

[11]To be complete we remind the reader on the fact that the assumption of an equilibrium reduced density operator implies a factorization of the total initial statistical operator into a system and a reservoir part (neglect of initial correlations, cf. Section 3.12.1).

the dipole–dipole correlation function:

$$C_{\mathrm{d-d}}(t) = \mathrm{tr_S}\left\{\hat{\mu}\hat{\sigma}(t)\right\} \ . \tag{5.88}$$

The operator $\hat{\sigma}(t)$ has to be propagated according to the respective quantum master equation (without the external field) and with the initial value $[\hat{\mu}, \hat{\rho}_{\mathrm{eq}}]_-$ (at $t = 0$). Therefore, we may write $\hat{\sigma}(t) = \mathcal{U}(t)[\hat{\mu}, \hat{\rho}_{\mathrm{eq}}]_-$, where $\mathcal{U}(t)$ denotes the dissipative time–evolution superoperator (cf. Eq. (3.265)).

To obtain a deeper insight into the expression for the dipole–dipole correlation function we specify it to a system with two electronic states ($\phi_g$ and $\phi_e$) already used in the preceding sections. Introducing an expansion with respect to these electronic states and noting first $\hat{\rho}_{\mathrm{eq}} = \hat{R}_g|\phi_g\rangle\langle\phi_g|$, where $\hat{R}_g$ is the vibrational equilibrium statistical operator of the electronic ground state, we get from Eq. (5.88)

$$C_{\mathrm{d-d}}(t) = \mathrm{tr_{vib}}\left\{\mathbf{d}_{ge}\langle\phi_e|\hat{\sigma}(t)|\phi_g\rangle + \mathbf{d}_{eg}\langle\phi_g|\hat{\sigma}(t)|\phi_e\rangle\right\} \ . \tag{5.89}$$

This expression generalizes Eq. (5.75) or Eq. (5.77) to the condensed phase situation. The propagation of electronic ground state vibrational wave functions in the excited state PES (cf. Eq. (5.77)) has been replaced by a propagation of the off–diagonal electronic matrix elements of the density operator $\hat{\sigma}$.

If the equation of motion for the density matrix elements follow, e.g. from the Bloch approximation, Section 3.8.3, an analytical expression for the linear absorption coefficient may be derived. The Bloch approximation has the advantage that the propagation of the off–diagonal density matrix elements is separated from that of the diagonal elements. We introduce the electron–vibrational energy representation $\sigma_{aN,bM}(t) = \langle\chi_{aN}|\langle\phi_a|\hat{\sigma}(t)\ |\phi_g\rangle|\chi_{bM}\rangle$ and obtain for $a \neq b$

$$\frac{\partial}{\partial t}\sigma_{aN,bM}(t) = -i\big(\omega_{aN,bM} - i\gamma_{aN,bM}\big)\sigma_{aN,bM}(t) \ . \tag{5.90}$$

The $\omega_{aN,bM}$ are transition frequencies, and the $\gamma_{aN,bM}$ describe dephasing rates (cf. Eqs. (3.291), (3.292) and (3.296)). There is no need for the following to specify the dephasing rates any further (for concrete expressions compare Section 6.8). After specifying initial conditions, $\sigma_{eN,gM}(0) = \mathbf{d}_{eg}\ \langle\chi_{eN}|\chi_{gM}\rangle f(E_{gM})$, the time–dependence of $\sigma_{aN,bM}$ is simply obtained as a damped oscillation. A Fourier–transformation like in Eq. (5.72) results in the final expression for the absorption coefficient. If $\sigma_{eN,gM}(t)$ is Fourier–transformed there appears the exponent $\omega - \omega_{eN,gM}$ which describes resonant transitions. In contrast $\sigma_{gN,eM}(t)$ results in the exponent $\omega - \omega_{gN,eM}$ which is completely off–resonant. These contributions can be neglected (cf. Eq. (5.77)) and we only consider $\sigma_{eN,gM}(t)$. Then, the trace formula in Eq. (5.89) can be specified as follows:

$$\begin{aligned}\mathrm{tr_{vib}}\{\langle\phi_e|\hat{\sigma}(t)|\phi_g\rangle\} &= \sum_M \langle\chi_{gM}|\langle\phi_e|\hat{\sigma}(t)|\chi_{gM}\rangle|\phi_g\rangle \\ &= \sum_{M,N} \langle\chi_{gM}|\chi_{eN}\rangle\ \sigma_{eN,gM}(t) \ ,\end{aligned} \tag{5.91}$$

where in the last part the complete set of vibrational states belonging to the excited electronic state has been introduced. The final expression for the absorption coefficient reads

$$\alpha(\omega) = \frac{4\pi\omega n_{\rm mol}}{3\hbar c}|d_{ge}|^2 \sum_{M,N} |\langle\chi_{gM}|\chi_{eN}\rangle|^2 f(E_{gM}) \frac{\gamma_{eN,gM}}{(\omega-\omega_{eN,gM})^2+\gamma_{eN,gM}^2} . \tag{5.92}$$

This expression is a direct generalization of Eq. (5.29) since the various sharp transitions (described by a $\delta$–function) are broadened here to a Lorentzian–like lineshape. The amount of broadening is determined by the dephasing rates $\gamma_{eN,gM}$. Pure dephasing which was introduced in Section 5.2.3 is described here by the line–broadening $\gamma_{e0,g0}$ corresponding to a transition between the vibrational ground states of both considered electronic levels. It depends on the concrete model of the system–reservoir coupling whether $\gamma_{e0,g0}$ gives a contribution or not (cf., for example, [Rei96,Ski86]).

If the present model is reduced to the case of a single active coordinate (relative coordinate in a diatomic system) coupled linearly to a harmonic oscillator bath it may result in an absorption spectrum similar (but not identical) to that shown by the full lines in Fig. 5.7.

### 5.3.5 Quasi–Classical Computation of the Absorption Coefficient

We finish our considerations of different ways to compute the linear absorption spectrum of a molecular system by an approach which accounts for the vibrational degrees of freedom (either of an intramolecular nature or of the surrounding solvent) by using classical mechanics. Again, we will demonstrate this for a transition including an electronic excitation. The classical (as well as quasi–classical) description of the vibrational dynamics is invoked by changing to a partial Wigner representation of the statistical operator. This partial Wigner representation has to be taken since the electronic degrees of freedom will be considered quantum mechanically. In similarity to Eq. (5.89) and after specifying the trace with respect to the vibrational states to coordinate operator eigenstates we write

$$C_{\rm d-d}(t) = \int dq \, \{ \mathbf{d}_{ge} \langle\phi_e|\langle q|\hat\sigma(t)|q\rangle|\phi_g\rangle \cdot \mathbf{d}_{eg}\langle\phi_g|\langle q|\hat\sigma(t)|q\rangle|\phi_e\rangle \} . \tag{5.93}$$

Here, $q = \{q_\xi\}$ denotes the $\mathcal{N}$ vibrational coordinates, $|q\rangle$ is the product of respective eigenstates of the coordinate operators, and $\hat\sigma(t)$ is given by $U_{\rm mol}(t)\ [\hat\mu, \hat R_g|\phi_g\rangle\langle\phi_g|]_-\ U^+_{\rm mol}(t)$. The introduction of the partial Wigner representation of $\hat\sigma(t)$ results in the quantity $\hat\sigma(x,p;t)$ which is an operator in the electronic state space. Having introduced the partial Wigner representation we can directly use Eqs. (3.424)– (3.427) to obtain an equation of motion for $\hat\sigma(x,p;t)$ in the semiclassical limit. Here, we introduced a more compact notation (note the appearance of commutators and anti–commutators), that is:

$$\begin{aligned}\frac{\partial}{\partial t}\hat\sigma(x,p;t) &= -\frac{i}{\hbar}[H_{\rm mol},\hat\sigma]_- \\ &+\frac{1}{2}\sum_\xi \left\{ \left[\frac{\partial H_{\rm mol}}{\partial x_\xi}, \frac{\partial\hat\sigma}{\partial p_\xi}\right]_+ - \left[\frac{\partial H_{\rm mol}}{\partial p_\xi}, \frac{\partial\hat\sigma}{\partial x_\xi}\right]_+ \right\} .\end{aligned} \tag{5.94}$$

As it is the case for $\hat\sigma$, the Hamiltonian $H_{\rm mol}$ depends on the classical coordinates and momenta but remains an operator in the electronic state space. Therefore, also Planck's constant appears in Eq. (5.94). Let us neglect for a moment the second term on the right–hand side. Then, the remaining equation is solved by

$$\hat{\sigma}(x,p;t) = U_{\rm mol}(x,p;t)\hat{\sigma}(x,p;0)U^{+}_{\rm mol}(x,p;t) \; . \tag{5.95}$$

The time–evolution operators have been defined by $H_{\rm mol}$ and, therefore, they depend on the classical coordinates and momenta, too. Moreover, we note that

$$\hat{\sigma}(0) = [\hat{\mu}, \hat{R}_g|\phi_g\rangle\langle\phi_g|]_- \; , \tag{5.96}$$

what leads to the Wigner representation as $\hat{\sigma}(x,p;0) = \mathbf{d}_{eg} f_g \; |\phi_e\rangle\langle\phi_g| - \text{h.c.}$ (the quantity $f_g$ denotes the classical distribution function defined by the vibrational Hamilton function $H_g(x,p)$). Then, after taking electronic matrix elements, it follows for $\hat{\sigma}(x,p;t)$ (as in the foregoing Section 5.3.4 we concentrate on the resonant contribution)

$$\begin{aligned} \langle\phi_e|\hat{\sigma}(x,p;t)|\phi_g\rangle &= \langle\phi_e|U_{\rm mol}(x,p;t)\mathbf{d}_{eg} f_g|\phi_e\rangle\langle\phi_g|U^{+}_{\rm mol}(x,p;t)|\phi_g\rangle \\ &= \mathbf{d}_{eg} f_g(x,p) \exp\big(-\frac{i}{\hbar}(U_e(x) - U_g(x))t\big) \; . \end{aligned} \tag{5.97}$$

Since the time–evolution operators are defined in terms of the classical vibrational Hamiltonians $H_g(x,p)$ and $H_e(x,p)$ the exponent in the last part only contains the difference $U_e(x) - U_g(x)$ of the related PES. If we insert Eq. (5.97) into Eq. (5.93) for the dipole–dipole correlation function the absorption coefficient is obtained as

$$\begin{aligned} \alpha(\omega) &= \frac{4\pi\omega n_{\rm mol}}{3\hbar c}|d_{eg}|^2 {\rm Re}\int\limits_0^\infty dt\; e^{i\omega t} \int dx \frac{dp}{(2\pi\hbar)^N} f_g(x,p) e^{-i(U_e(x)-U_g(x))t/\hbar} \\ &= \frac{4\pi^2\omega n_{\rm mol}}{3c}|d_{eg}|^2 \int dx f_g(x)\delta\big(\hbar\omega - (U_e(x) - U_g(x))\big) \; . \end{aligned} \tag{5.98}$$

In the final expression the momentum integration has been carried out leading to the reduced distribution function $f_g(x) \sim \exp(-U_g/k_{\rm B}T)$. The $\delta$–function follows from the time–integral. The result for the linear absorption coefficient reflects the qualitative discussion carried out in Section 5.1 and related to Fig. 5.1. As explained in this first part of the present chapter, absorption of a radiation quantum becomes possible if its energy $\hbar\omega$ equals to the difference, $U_e(x) - U_g(x)$, between the PES of the ground and the excited electronic state (vertical transitions). This is the essence of Eq. (5.98).

Assuming parabolic PES the $x$–integration in Eq. (5.98) can be carried out. This will be demonstrated in detail in Section 6.4.1, where electron transfer between two states is discussed. We refer the reader to all calculation following Eq. (6.64). Here we use only the result (cf. Eq. (6.73)) adopted to the case of the absorption coefficient

$$\alpha(\omega) = \frac{4\pi^2\omega n_{\rm mol}}{3c}|d_{eg}|^2 \sqrt{\frac{1}{2\pi k_{\rm B}TS}} \exp\left\{-\frac{\left[\hbar(\omega-\omega_{eg}) - S/2\right]^2}{2Sk_{\rm B}T}\right\} . \tag{5.99}$$

Note the introduction of the transition frequency $\omega_{eg}$ according to Eq. (5.32) and of the Stokes shift, Eq. (5.43). In Section 5.2.5 this type of expression had been derived for the case of slow nuclear motion (cf. Eq. (5.64)). This can be brought into a different perspective, that is, the neglect of any dynamic corrections to the solution, Eq. (5.97), of the density operator equation in the Wigner representation (Eq. (5.94)) results in a *static approximation*. In other words, during the absorption process there is no classical vibrational motion. Respective numerical results are given in Fig. 5.7 by the curves labeled as SCL (static classical limit). They wrap the more structured spectra showing the progression of a single vibration (for more details cf. Section 5.2.5). Because the static approximation is of fundamental character we will meet it again at various places in later chapters.

In order to go beyond the static approximation one has to include nonvertical transitions but also account for quantum effects. Respective results can also be found in Fig. 5.7. These have been derived for a single vibration coupled to additional vibrations. The absorption shows the progression of this single vibrations broadened to some extent (see Section 5.2.5 for more details). The curves labelled by DCL (dynamical classical limit) reproduce those curves quite well which are obtained by accounting for the derivatives in Eq. (5.94) (cf. also Eq. (3.116)). However, one has to note that the mentioned DCL–approximation is based on the classical counterpart of Eq. (5.39), where the $S$–operator has been obtained by a classical quantity which is not time–ordered and where the trace has been replaced by a multiple integral over initial vibrational coordinates and momenta weighted by the thermal distribution. Although the DCL–approximation cannot reproduce the sharp absorption lines at low temperature (or weaker coupling to the secondary oscillators) it gives a satisfactory reproduction of the spectrum and the correct position of the various lines in the overall spectrum.

## 5.4 The Rate of Spontaneous Emission

If an excited electronic state has been prepared as a result of a photoabsorption process in the molecule, this state has a finite lifetime. This is a consequence of spontaneous transitions to the electronic ground state accompanied by a photoemission process. Alternatively, a radiationless internal conversion type transition to the ground state is also possible. The radiative decay results from the coupling of the molecule to the vacuum state of the electromagnetic field. The appropriate description therefore demands for a quantization of the radiation field. Here, we only give an intuitive picture and present some formulas for further use in Chapter 8.

Since the radiation field can be considered as a reservoir coupled to the molecular electronic states, the spontaneous emission of a photon is described in analogy to the transition processes resulting from system–reservoir coupling discussed in Chapter 3. The specific point here, of course, is the form of the coupling operator $H_{\text{int}}$ between electronic states and quantized radiation field. One usually starts with the so–called minimal coupling Hamiltonian. There, the momenta of the electrons are replaced by $\mathbf{p}_j - \mathbf{A}(\mathbf{r}_j)e/c$, with the vector potential $\mathbf{A}$. The appropriate interaction Hamiltonian takes the form

$$H_{\text{int}} = -\frac{e}{m_{\text{el}}c} \sum_j \mathbf{p}_j \mathbf{A}(\mathbf{r}_j) \ . \tag{5.100}$$

Field quantization can be achieved by expanding the vector potential $\mathbf{A}(\mathbf{r}_j)$ in terms of plane waves with wave vectors $\mathbf{k}$:

$$\mathbf{A}(\mathbf{r}) = \int \frac{d^3\mathbf{k}}{(2\pi)^3} \, \mathbf{A}(\mathbf{k}) \, e^{i\mathbf{k}\mathbf{r}} \, . \tag{5.101}$$

The expansion coefficients $\mathbf{A}(\mathbf{k})$ have the particular property that they are perpendicular to the actual wave vector $\mathbf{k}$, i.e. $\mathbf{A}(\mathbf{k})\,\mathbf{k} = 0$. The vector potential is a transverse field and every partial wave can be characterized by two (linear independent) transverse polarization directions with unity vectors $\mathbf{e}_\lambda$ ($\lambda = 1, 2$). Upon quantization the expansion coefficients become operators describing creation or annihilation of a photon with energy $\hbar\omega_{\mathbf{k}} = \hbar c|\mathbf{k}|$ (photon dispersion relation).

In analogy to the absorption rate, Eq. (5.24), we can introduce the rate $k_{e\to g}$ for transitions from the excited electronic state $|\phi_e\rangle$ to the ground state $|\phi_g\rangle$ accompanied by the spontaneous emission of a photon with energy $\hbar\omega_{\mathbf{k}}$. It reads

$$\begin{aligned} k_{e\to g} &= \frac{2\pi}{\hbar} \int d^3\mathbf{k} \sum_{M,N} f(E_{eM}) |\langle 0|\langle\phi_e|\langle\chi_{eM}|H_{\text{int}}|\chi_{gN}\rangle|\phi_g\rangle|\mathbf{k}\rangle|^2 \\ &\quad \times \delta(E_{eM} - \hbar\omega_{\mathbf{k}} - E_{gN}) \, . \end{aligned} \tag{5.102}$$

The excited electron–vibrational state with energy $E_{eM}$ and with zero photons (vacuum state $|0\rangle$) decays into the state with energy $E_{gN}$ of the electronic ground state releasing a photon in state $|\mathbf{k}\rangle$ and of energy $\hbar\omega_{\mathbf{k}}$. The initial population of the vibrational levels of the excited electronic states has been described by the thermal distribution $f(E_{eM})$, and a summation with respect to all vibrational levels of the final state is performed. Inserting the expression for the interaction Hamiltonian one obtains

$$\begin{aligned} k_{e\to g} &= \frac{e^2}{m_{\text{el}}^2 c^3} \int \frac{d\Omega}{2\pi} \int_0^\infty d\omega_{\mathbf{k}} \, \omega_{\mathbf{k}} \sum_\lambda \sum_{M,N} f(E_{eM}) \\ &\quad \times \sum_j | \, \langle\phi_e|\langle\chi_{eM}|e^{i\mathbf{k}\mathbf{r}_j}\mathbf{e}_\lambda\mathbf{p}_j|\chi_{gN}\rangle|\phi_g\rangle \, |^2 \; \delta(E_{eM} - E_{gN} - \hbar\omega_{\mathbf{k}}) \, . \end{aligned} \tag{5.103}$$

Here, the three–dimensional wave vector integral has been rewritten by introducing spherical coordinates, and, afterwards, by replacing the $| \; \mathbf{k} \; |$–integral by a frequency integral. The integration with respect to the unit sphere in $\mathbf{k}$–space gives $\int d\Omega/2\pi$. As a consequence of the minimal coupling Hamiltonian the coordinates $\mathbf{r}_j$ and momenta $\mathbf{p}_j$ of the electrons appear in the matrix elements.

For light in the visible region the wavelength is much larger than typical molecular sizes. Therefore, one can consider $\mathbf{k}\mathbf{r}_j$ as a small quantity and replace the exponential function in the matrix element by 1. This corresponds to the dipole approximation, since one can replace the electronic matrix element of the momentum operator by the transition dipole moment. To show this we start with the equation of motion for the electronic coordinate operator $\mathbf{r}_j$ given by

$$i\hbar\frac{\partial}{\partial t}\mathbf{r}_j = [\mathbf{r}_j, H_{\text{el}}]_- = \frac{\mathbf{p}_j}{m_{\text{el}}} \, . \tag{5.104}$$

Thus, the matrix elements of the momentum operator can be written as

$$\begin{aligned}\langle\phi_e|\sum_j \mathbf{e}_\lambda \mathbf{p}_j|\phi_g\rangle &= m_{\text{el}}\mathbf{n}_\lambda \sum_j \langle\phi_e\,|(\mathbf{r}_j H_{\text{el}} - H_{\text{el}}\,\mathbf{r}_j)|\,\phi_g\rangle \\ &= m_{\text{el}}\mathbf{n}_\lambda(E_g - E_e)\sum_j \langle\phi_e\,|\mathbf{r}_j|\,\phi_g\rangle \\ &= -\frac{m_{\text{el}}}{e}(E_e - E_g)\mathbf{n}_\lambda \mathbf{d}_{eg}\,. \end{aligned} \tag{5.105}$$

Here, $\mathbf{d}_{eg}$ is the transition dipole matrix element and $E_e$ and $E_g$ are electronic energy levels.

The radiative lifetime $\tau_{\text{rad}}$ of the excited electronic state follows from the inverse of $k_{e\to g}$. We write

$$k_{e\to g} \equiv \frac{1}{\tau_{\text{rad}}} = \int_0^\infty d\omega\; I(\omega)\,, \tag{5.106}$$

where $I(\omega)$ is the emission rate per frequency interval. The quantity $I(\omega)$ can be identified with the emission spectrum of the considered molecule. Performing the orientational averaging in Eq. (5.103) one obtains for the emission spectrum

$$I(\omega) = \frac{4\hbar\omega^3}{3c^3}|\mathbf{d}_{eg}|^2 \sum_{M,N} f(E_{eM})|\langle\chi_{eM}|\chi_{gN}\rangle|^2\delta(E_{eM} - E_{gN} - \hbar\omega)\,. \tag{5.107}$$

This expression will be encountered in Chapter 8 in the context of the Förster theory of resonance energy transfer in molecular aggregates. It is obvious from the general structure that the emission spectrum can be calculated in complete analogy to the absorption coefficient. Following Section 5.2.2 we obtain

$$I(\omega) = \frac{4\hbar\omega^3}{3c^3}|\mathbf{d}_{eg}|^2\mathcal{D}_{\text{em}}(\omega - \omega_{eg})\,, \tag{5.108}$$

where the combined density of states $\mathcal{D}_{\text{em}}$ characterizes the emission process. In complete analogy to Eq. (5.31) we may write

$$\mathcal{D}_{\text{em}}(\omega - \omega_{eg}) = \sum_{M,N} f(E_{eM})|\langle\chi_{eM}|\chi_{gN}\rangle|^2\delta(E_{eM} - E_{gN} - \hbar\omega)\,. \tag{5.109}$$

This expression differs in two respects from the density of states characterizing the absorption process, Eq. (5.35). First, the electronic quantum numbers $g$ and $e$ have been interchanged, and second $\omega$ has been replaced by $-\omega$. Thus we obtain

$$\mathcal{D}_{\text{em}}(\omega - \omega_{eg}) = \frac{1}{2\pi\hbar}\int dt\; e^{-i\omega t}\,\text{tr}_{\text{vib}}\{\hat{R}_e\, e^{iH_e t/\hbar}\, e^{-iH_g t/\hbar}\}\,. \tag{5.110}$$

Reducing this expression to the case of harmonic PES as demonstrated in Section 5.2.4 we have

$$\mathcal{D}_{\text{em}}(\omega - \omega_{eg}) = \frac{1}{2\pi\hbar}\int dt\, e^{-i(\omega-\omega_{eg})t - G(0) + G(t)}\,, \tag{5.111}$$

directly demonstrating the property $\mathcal{D}_{\text{em}}(\omega - \omega_{eg}) = \mathcal{D}_{\text{abs}}(-\omega + \omega_{eg})$. This shows that for harmonic oscillator PES (with the same curvature in the ground as well as excited state) the absorption and emission spectrum lie mirror symmetric to $\omega = \omega_{eg}$.

## 5.5 Optical Preparation of an Excited Electronic State

The following considerations will focus on the preparation of an excited electronic state via an optical transition from the electronic ground state. A detailed understanding of such a transition is of great importance for the study of photoinduced transfer phenomena occurring in excited electronic states. This goes beyond the linear absorption and thus linear optics, since the preparation and possible detection of the excited state dynamics includes the field in higher than first order. Before studying the preparation in the general frame of density matrix theory, a simpler approach will be given which is based on the solution of the time–dependent Schrödinger equation.

### 5.5.1 Wave Function Formulation

Let us consider again the two–state Hamiltonian Eq. (5.16), however, instead of a strictly monochromatic electromagnetic field, a *pulsed field* (laser pulse) will be assumed. It has the form

$$\mathbf{E}(t) = \mathbf{e}E(t)e^{-i\omega t} + \text{c.c.} \,. \tag{5.112}$$

Here, the vector $\mathbf{e}$ defines the polarization of the field and $\omega$ is the carrier–frequency (center frequency of the pulse spectrum). $E(t) \equiv Af(t)$ denotes the pulse envelope with pulse amplitude $A$ and normalized pulse envelope $f(t)$. The pulse duration is fixed by the pulse envelope. Frequently, a Gaussian pulse shape centered at $t = t_p$ is used:

$$f(t) = \frac{1}{\sqrt{2\pi\tau_p^2}} \exp\left\{ -\frac{(t-t_p)^2}{2\tau_p^2} \right\} \,. \tag{5.113}$$

This envelope is normalized to unity and contains $\tau_p$ as the pulse duration.[12] To solve the time–dependent Schrödinger equation

$$\frac{\partial}{\partial t}|\psi(t)\rangle = -\frac{i}{\hbar}H(t)|\psi(t)\rangle \tag{5.114}$$

defined by the time–dependent Hamiltonian, Eq. (5.16) and the initial condition $|\psi(t_0)\rangle$, we change to the interaction representation (compare Section 3.2.2). The unperturbed Hamiltonian is given by the molecular part $H_{\text{mol}} = \sum_{a=g,e} H_a(q)|\phi_a\rangle\langle\phi_a|$ of Eq. (5.16), whereas the perturbation is represented by the external–field contribution $H_{\text{field}}(t)$. Consequently, the state vector in the interaction representation reads $|\psi^{(\text{I})}(t)\rangle = U^+_{\text{mol}}(t-t_0)|\psi(t)\rangle$, with the time–evolution operator $U_{\text{mol}}$ defined via $H_{\text{mol}}$. The determination of $|\psi^{(\text{I})}(t)\rangle$ can be reduced to the solution of an integral equation of type (3.36). Here, we concentrate on the first–order correction with respect to the external field and obtain

$$|\psi^{(\text{I})}(t)\rangle \approx |\psi(t_0)\rangle - \frac{i}{\hbar}\int\limits_{t_0}^{t} d\bar{t}\; H^{(\text{I})}_{\text{field}}(\bar{t})\; |\psi^{(\text{I})}(t_0)\rangle \,. \tag{5.115}$$

[12] Gaussian pulse envelopes are also characterized by the full width at half maximum which is $\sqrt{8\ln 2}\,\tau_p$.

After switching back to the Schrödinger representation the complete state vector including the first–order correction reads[13]

$$
\begin{aligned}
|\psi(t)\rangle &= |\psi^{(0)}(t)\rangle + |\psi^{(1)}(t)\rangle = U_{\rm mol}(t-t_0)|\psi(t_0)\rangle \\
&\quad -\frac{i}{\hbar}\int_{t_0}^{t} d\bar{t}\, U_{\rm mol}(t-t_0)U^{+}_{\rm mol}(\bar{t}-t_0)H_{\rm field}(\bar{t})U_{\rm mol}(\bar{t}-t_0)|\psi(t_0)\rangle\,.
\end{aligned}
\tag{5.116}
$$

The initial condition will be specified to

$$
|\psi(t_0)\rangle = \delta_{ag}|\phi_g\rangle|\chi_{g0}\rangle\,, \tag{5.117}
$$

thus assuming that the system is in the vibrational ground state of the electronic ground state. In a next step we take into account that

$$
U_{\rm mol}(t) = \sum_a e^{-iH_a t/\hbar}|\phi_a\rangle\langle\phi_a|\,, \tag{5.118}
$$

and expand the time–dependent state vector, Eq. (5.116), with respect to the electronic basis. The zeroth–order part $|\psi^{(0)}(t)\rangle$ corresponds to a stationary vibrational state on the electronic ground state PES. The expansion of the first–order contribution with respect to the excited electronic state gives the related state vector for vibrational motion:

$$
\begin{aligned}
|\chi_e(t)\rangle &= \langle\phi_e|\psi^{(1)}(t)\rangle \\
&= -\frac{i}{\hbar}\int_{t_0}^{t} d\bar{t}\, e^{-iH_e(t-\bar{t})/\hbar}\left(-\mathbf{d}_{eg}\mathbf{E}(\bar{t})\right)e^{-iH_g(\bar{t}-t_0)/\hbar}|\chi_{g0}\rangle\,.
\end{aligned}
\tag{5.119}
$$

This expression offers a simple picture of the excitation process. From the initial time $t_0$ up to time $\bar{t}$ the initial vibrational state $|\chi_{g0}\rangle$ is propagated on the electronic ground state PES ($t_0$ has to be taken well before the pulse arrives). Since the initial state is an eigenstate of $H_g$, one obtains the phase factor $\exp\{-iE_{g0}(\bar{t}-t_0)/\hbar\}$. The dynamics on the excited state PES starts at $\bar{t}$ and proceeds up to the actual time $t$. Note that the transition occurs *instantaneously*, which supports the picture of vertical transitions discussed in the introduction to this chapter. But the moment of transition to the excited state is not fixed, instead an integration over all possible times $\bar{t}$ has to be performed. If the time $t$ lies in the region where the pulse is present, Eq. (5.119) describes the preparation process. For larger times it shows how the optically prepared state develops further in the absence of the field.

To determine the $\bar{t}$–dependence of the integrand in (5.119) we change to the representation given by the eigenstates $|\chi_{eM}\rangle$ of the excited state Hamiltonian $H_e$. The expansion of $|\chi_e(t)\rangle$ gives an example for a wave packet (see Eq. (3.17))

$$
|\chi_e(t)\rangle = \sum_M c_{eM}(t)|\chi_{eM}\rangle\,. \tag{5.120}
$$

[13]Note that as long as we concentrate on a strict linearization with respect to the external field, a normalization of the state vector is not necessary since it gives terms which are of higher order in the field.

The expansion coefficients are obtained as

$$\begin{aligned} c_{eM}(t) &= \langle \chi_{eM} | \chi_e(t) \rangle \\ &= i \frac{\mathbf{e}\mathbf{d}_{eg} A}{\hbar} \, e^{-iE_{eM}t/\hbar} \int_{t_0}^{t} d\bar{t} \, e^{i(E_{eM}-E_{g0}-\hbar\omega)\bar{t}/\hbar} \, f(\bar{t}) \\ &\quad \times e^{iE_{g0}t_0/\hbar} \langle \chi_{eM} | \chi_{g0} \rangle \, . \end{aligned} \tag{5.121}$$

The magnitude of the expansion coefficients is determined by the Franck–Condon factors $\langle \chi_{eM} | \chi_{g0} \rangle$ and by a certain time integral. If we let $t_0 \to -\infty$ and take $t$ at a time at which the pulse has already passed through the sample ($t \gg t_p$) the time integral reduces to the Fourier–transformed pulse envelope $f(\Omega)$ with $\Omega = \omega_{eM} - \omega_{g0} - \omega$. Thus, the different vibrational states $|\chi_{eM}\rangle$ are weighted according to the form of the Fourier–transformed pulse envelope. For the Gaussian–shaped envelope, Eq. (5.113), we get

$$f(\Omega) = \int_{-\infty}^{+\infty} dt \, e^{i\Omega t} \frac{1}{\sqrt{2\pi\tau_p^2}} \exp\left\{ -\frac{(t-t_p)^2}{2\tau_p^2} \right\} = e^{-\Omega^2\tau_p^2/2} \, e^{i\Omega t_p} \, . \tag{5.122}$$

This function becomes maximal for $\Omega \approx \omega_{eM} - \omega_{g0}$ (vertical Franck–Condon transition) and goes to zero for values of $\Omega$ which are larger than the inverse pulse duration $1/\tau_p$. Apparently, the shorter the pulse duration the larger its spectral width.

Case of short pulse duration:

If some vibrational levels away from the vertical Franck–Condon transition should be populated, the inverse pulse duration $1/\tau_p$ has to amount to the respective frequency difference at the excited state PES. The situation is sketched in Fig. 5.12. If sufficient vibrational levels are covered, a transfer of the "frozen" ground state wave function $\chi_{g0}(q)$ to the excited electronic state becomes possible. This picture can be idealized in the limit of impulsive excitation ($\tau_p \to 0$), where one replaces the pulse envelope by a delta function. Then the expansion coefficients introduced in Eq. (5.121) read ($t > t_p$)

$$c_{eM}(t) = i \frac{\mathbf{e}\mathbf{d}_{eg} A}{\hbar} \, e^{-i\omega_{eM}(t-t_p)} \, e^{-i\omega t_p} \, e^{-i\omega_{g0}(t_p-t_0)} \, \langle \chi_{eM} | \chi_{g0} \rangle \, . \tag{5.123}$$

A vibrational wave packet is formed on the excited state PES which results from a projection of the initial state $|\chi_{g0}\rangle$ onto the excited state PES at time $t_p$. The related population of the vibrational levels is exclusively determined by the Franck–Condon factors,

$$P_{eM}(t) = |c_{eM}(t)|^2 = \Theta(t-t_p) \frac{1}{\hbar^2} |\mathbf{e}\mathbf{d}_{eg} A|^2 \, |\langle \chi_{eM} | \chi_{g0} \rangle|^2 \, . \tag{5.124}$$

It is also of interest to obtain the electronic occupation probability, $P_e(t)$, transferred to the excited state. It can be calculated from Eq. (5.124) or directly from Eq. (5.119) as $P_e(t) = \langle \psi^{(1)}(t) | \phi_e \rangle \langle \phi_e | \psi^{(1)}(t) \rangle$. According to the case of impulsive excitation one gets

$$P_e(t) = \sum_M |c_{eM}(t)|^2 = \frac{\Theta(t-t_p)}{\hbar^2} |\mathbf{d}_{eg} \mathbf{E}(t_p) \tau_p|^2 \, . \tag{5.125}$$

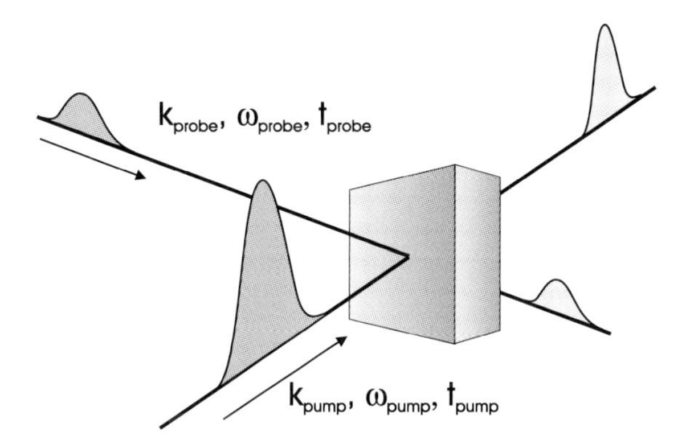


**Figure 5.13:** Scheme of a pump–probe experiment with the pump pulse (probe pulse) characterized by the wave vector $\mathbf{k}_{\rm pump}$ ($\mathbf{k}_{\rm probe}$), the frequency $\omega_{\rm pump}$ ($\omega_{\rm probe}$), and the time $t_{\rm pump}$ ($t_{\rm probe}$) the pulse reaches its maximum in the probe ($t_{\rm probe} - t_{\rm pump}$ defines the so–called delay time). Besides the two incoming pulses only the related transmitted pulses are drawn (any scattered pulse is not shown).

# 5.6 Nonlinear Optical Response

The field of nonlinear optics is of a diversity which goes far beyond the scope of this book. Therefore, we will only outline some basic concepts which provide the background for understanding the methods for detecting elementary charge and energy transfer processes. For any more detailed discussion we refer to the various textbooks existing on the different facets of this rapidly developing field (see Suggested Reading section).

Two basic ways to characterize the optical response of an ensemble of molecular systems can be distinguished. On the one hand side one can solve exactly the time–dependent Schrödinger equation (or the density operator equation) in the presence of the radiation field. As a result the polarization field, Eq. (5.10), given by the time–dependent expectation value of the dipole operator according to Eq. (5.67) is obtained as a function of the electric field strength. In the general case this function of the electric field would be a nonlinear function. However, often such a rigorous treatment is unnecessary since the field is so weak that an expansion of the polarization field in powers of the field strength is sufficient. This expansion represents the second way to analyze the optical response.

Before concentrating on one of the two mentioned ways to study the optical response of molecular system we start with some rather general considerations. In order to obtain some insight into the characteristics of the nonlinear response of a molecular system we go back to Eq. (5.67) which expresses the expectation value of the molecular dipole operator via the field–dependent $S$–operator, Eq. (5.66). Once the $S$–operator has been expanded in powers of $H_{\rm field}^{(\rm I)}$ we have, at the same time, an expansion in powers of the field strength $\mathbf{E}(\mathbf{x},t)$. Let us proceed with the following ansatz for the externally applied field:

$$\mathbf{E}(\mathbf{x},t) = \sum_{p=1}^{N} \mathbf{n}_p E_p(t) \exp\{i(\mathbf{k}_p\mathbf{r} - \omega_p t)\} + \text{c.c.} \,. \tag{5.131}$$

The expression generalizes Eq. (5.112) to the case where $N$ different partial waves (counted by $p$) with polarization unit vector $\mathbf{n}_p$, envelope $E_p$, wave vector $\mathbf{k}_p$, and frequency $\omega_p$ form the total field. These partial waves may interfere or may be separated in time to give different independent pulses (see Fig. 5.13).

Because the expectation value of the molecular dipole operator and thus the total polarization field contains all powers of the electric field strength, every partial wave of the total electric field, Eq. (5.131) should appear with any power. Therefore, we expect the following form for the polarization field:

$$\mathbf{P}(\mathbf{r},t) = \sum_{n_1=-\infty}^{\infty} \dots \sum_{n_N=-\infty}^{\infty} \mathbf{e}(n) P^{(n)}(t) \exp\{i(\mathbf{K}(n)\mathbf{r} - \Omega(n)t)\} \,. \tag{5.132}$$

Note that $n$ abbreviates the whole set $\{n_1, ..., n_N\}$ of numbers counting the power at which the respective partial wave appears in the actual part of the total polarization field. The multiples of the wave vector and the frequency are abbreviated by $\mathbf{K}(n) = \sum_p n_p \mathbf{k}_p$ and by $\Omega(n) = \sum_p n_p \omega_p$, respectively.

In a first step of our analysis of Eq. (5.132) we consider the case that the total electric field is given by a single wave (case $N = 1$). As a result we obtain $\Omega(n) \equiv \Omega(n_1) = \omega_1, 2\omega_1, 3\omega_3, ...$ (negative frequencies appear, too). Here, the $n$th multiple of $\omega_1$ corresponds to the $n$th–order nonlinear response of the molecular system (generation of the $n$th–order harmonic). Apart from the linear response there may appear frequency doubling as the quadratic response and frequency triplication as the third–order nonlinear response (two– and three–photon absorption, respectively). Besides the frequency of the polarization field we have to consider the wave vector $\mathbf{K}(n)$. It contains multiples of $\mathbf{k}_1$ but all having the same direction what indicates that the field corresponding to the nonlinear response propagates in the same direction as the incoming field.

Changing to the case of two partial waves with frequency $\omega_1$ and $\omega_2$ the resulting frequencies of the polarization $\Omega(n) \equiv \Omega(n_1, n_2)$ may take the following values: $\omega_1$, $\omega_1 \pm \omega_2$, $\omega_1 \pm 2\omega_2$, etc. as well as $\omega_2$, $\omega_2 \pm \omega_1$, $\omega_2 \pm 2\omega_1$, etc.. These frequency combinations may be of the first, second and third order in the field–strength, respectively, but may also contain higher orders. This is due to the fact that the combination of the positive and the negative frequency part of every partial wave, i.e. $\exp(-i\omega_p t) \times \exp(i\omega_p t)$, results in a vanishing contribution to the total frequency. For example, we have $\omega_1 = \omega_1 + \omega_2 - \omega_2$ indicating that it may belong to a third–order process. This type of process is of basic importance for the pump–probe spectroscopy which has already been explained in Fig. 5.3. The partial wave with frequency $\omega_2$ introduces a nonequilibrium population (of some electronic and/or vibrational levels) in the molecule (the respective amplitude enters via $|E_2(t)|^2$ as in Eq. (5.130) describing the optical preparation of an excited state). Via the partial wave with frequency $\omega_1$ one measures the linear response of the excited molecule. Thus partial wave 2 corresponds to the pump–beam and partial wave 1 to the probe–beam. Since for this case $\mathbf{K}(n_1, n_2)$ equals $\mathbf{k}_1$ the optical response of the molecular system propagates into the direction of the probe–beam.

Considering, however, the frequency doubling with $\Omega(n_1 = 1, n_2 = 1) = \omega_1 + \omega_2$, in the general case the respective wave vector $\mathbf{k}_1 + \mathbf{k}_2$ neither shows into the direction of partial wave 1 nor in that of partial wave 2. This is also valid for the case with $\Omega(n_1 = 2, n_2 = -1) = 2\omega_1 - \omega_2$ where the wave vector is given by $2\mathbf{k}_1 - \mathbf{k}_2$. If the magnitude of both frequencies is

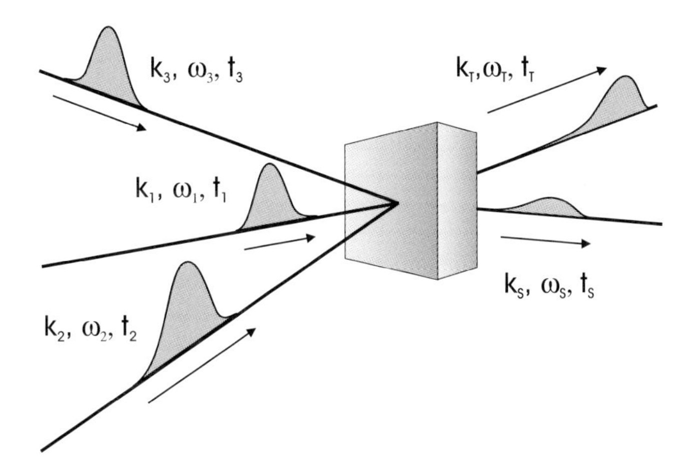


**Figure 5.14:** Scheme of a three pulse experiment with the different pulses $p = 1, 2, 3$ with wave vector $\mathbf{k}_p$, frequency $\omega_p$, and time $t_p$ where the pulse reaches its maximum in the probe. Besides the three incoming pulses two outgoing as representatives of the many scattered pulses are shown, too.

comparable the *mixed* frequency $2\omega_1 - \omega_2$ lies in the same region. Thus this type of nonlinear response does not include frequency multiplication. But it forms a type of response which propagates in the direction $2\mathbf{k}_1 - \mathbf{k}_2$ and which is known as the *photon echo* (if subpulse 1 corresponding to partial wave 1 and subpulse 2 corresponding to partial wave 2 are clearly separated in time and pulse 2 comes first) and the *transient grating* technique (pulse 1 comes first).

When studying the case of three partial waves $N = 3$ the situation becomes even more complex and we will restrict ourselves to some selected examples. First of a all the new frequency $\Omega(n_1 = 1, n_2 = 2, n_3 = 3) = \omega_1 + \omega_2 + \omega_3$ may be generated which has to be considered as a generalization of the third–order harmonic generation. Furthermore, we may introduce $\Omega(n_1 = 1, n_2 = 2, n_3 = -1) = \omega_1 + \omega_2 - \omega_3$. If all three basic frequencies lie in the same spectral region this is also valid for the mixed frequency $\Omega(1, 1, -1)$ as well all other combinations $\Omega(1, -1, 1)$ and $\Omega(-1, 1, 1)$. However, the wave vector $\mathbf{k}_1 + \mathbf{k}_2 - \mathbf{k}_3$ as well as the two other combinations may define directions different from that of $\mathbf{k}_1$, $\mathbf{k}_2$, and $\mathbf{k}_3$ in which all these third–order response signals propagate. If every partial wave forms a separate pulse this type of spectroscopy is known as *three pulse technique* including the *three pulse photon echo* (for details see, e.g. [Muk95]).

Apparently, the extent to which higher order processes become important essentially depends on the considered molecular system. To obtain a more detailed understanding we have to carry out an expansion with respect to the field–strength.

## 5.6.1 Nonlinear Susceptibilities

The expansion of the polarization field in powers of the field–strength reads

$$\mathbf{P}(\mathbf{x};t) = \mathbf{P}^{(1)}(\mathbf{x};t) + \mathbf{P}^{(2)}(\mathbf{x};t) + \mathbf{P}^{(3)}(\mathbf{x};t) + \dots \tag{5.133}$$

Here, the quantity $\mathbf{P}^{(n)}$ $(n = 1, 2, ...)$ should be of the $n$th order in the field–strength (a zeroth–order contribution does not appear since we assumed the absence of a permanent polarization). The function which relates $\mathbf{P}^{(n)}$ to the $n$th power of $\mathbf{E}$ is known as the $n$th–order *nonlinear response function* or, alternatively the *nonlinear susceptibility*:

$$\mathbf{P}^{(n)}(\mathbf{x};t) = \int dt_1...dt_n \chi^{(n)}(t, t_1, ..., t_n)\mathbf{E}(\mathbf{x};t_1) \cdot ... \cdot \mathbf{E}(\mathbf{x};t_n) . \tag{5.134}$$

The expression has to be understood as a $n$–fold multiplication of the electric field–strength with the nonlinear susceptibility which forms a tensor of rank $n+1$. Note that Eq. (5.134) implies the limit $t_0 \to -\infty$. The introduced nonlinear susceptibilities are a direct generalization of the linear susceptibility introduced in Section 5.3.1.

The concrete structure of the various susceptibilities follow from the expansion Eq. (5.134). We do not present details of the respective expressions but refer , for instance, to [Muk95]. The expansion of the $S$–operator up to third order in the field strength reads

$$\begin{aligned} \mathbf{d}(t) &= \mathrm{tr}\{\hat{W}_{\mathrm{eq}}\big[1 + S^{(1)+}(t,t_0) + S^{(2)+}(t,t_0) + S^{(3)+}(t,t_0)\big]\hat{\mu}^{(I)}(t) \\ &\quad \times\big[1 + S^{(1)}(t,t_0) + S^{(2)}(t,t_0) + S^{(3)}(t,t_0)\big]\} . \end{aligned} \tag{5.135}$$

The parts of the $S$–operator which are of different order in the field–strength have the following form

$$S^{(1)}(t,t_0) = \frac{1}{i\hbar}\int_{t_0}^{t} d\tau\ \mathbf{E}(\tau)\hat{\mu}^{(\mathrm{I})}(\tau) , \tag{5.136}$$

$$S^{(2)}(t,t_0) = \frac{1}{(i\hbar)^2}\int_{t_0}^{t} d\tau_1 \int_{t_0}^{\tau_1} d\tau_2\ \mathbf{E}(\tau_1)\hat{\mu}^{(\mathrm{I})}(\tau_1)\ \mathbf{E}(\tau_2)\hat{\mu}^{(\mathrm{I})}(\tau_2) , \tag{5.137}$$

and

$$S^{(3)}(t,t_0) = \frac{1}{(i\hbar)^3}\int_{t_0}^{t} d\tau_1 \int_{t_0}^{\tau_1} d\tau_2 \int_{t_0}^{\tau_2} d\tau_3\ \mathbf{E}(\tau_1)\hat{\mu}^{(\mathrm{I})}(\tau_1)\ \mathbf{E}(\tau_2)\hat{\mu}^{(\mathrm{I})}(\tau_2)\ \mathbf{E}(\tau_3)\hat{\mu}^{(\mathrm{I})}(\tau_3) . \tag{5.138}$$

If inserted into Eq. (5.135) the various nonlinear response functions can be deduced as higher–order dipole–dipole correlation functions. Their computation has been extensively documented in literature and will not be repeated here (see, for instance, [Muk95]).

## 5.7 Internal Conversion Dynamics

In Section 5.2.2 we already stressed the similarity of optical absorption studied so far and intramolecular electronic transitions induced by the nonadiabatic coupling (internal conversion). We will focus on the latter in more detail now. If a higher lying singlet state $S_n$ $(n > 1)$

is excited, it is the internal conversion process which induces transitions to lower electronic states. Within this process the electronic excitation energy is distributed among the different vibrational degrees of freedom. Since the radiation field does not take part in this type transition, it is also called *radiationless*.[14]

Often the internal conversion is slow compared to the time scale of vibrational relaxation within an electronic state and therefore it can be characterized by a transition rate. This is the situation where the Golden Rule formula introduced in Chapter 3 can be applied. The respective rate will be calculated in the following section. If the nonadiabatic coupling becomes stronger, one cannot assume complete vibrational equilibrium for every step of the transition. If vibrational relaxation can be completely neglected a description of ultrafast internal conversion in terms of wave functions becomes possible as shown in Section 5.7.2. The intermediate case needs a more involved description. However, having followed the present discussion and in anticipation of Chapter 6 dealing with electron transfer, the reader will realize a number of similarities. Therefore, the subject of internal conversion dynamics is outlined very briefly only. More involved questions related to charge transfer dynamics will be discussed in the context of electron transfer reactions in Chapter 6.

### 5.7.1 The Internal Conversion Rate

In order to describe internal conversion via a rate expression like $k_{a\to b}^{(\text{IC})}$ the condition has to be fulfilled that the characteristic time for the transition process $1/k_{a\to b}^{(\text{IC})}$ is long compared to any vibrational relaxation (in the initial as well as in the final state of the transition, compare also the similar discussion on electron transfer reactions in Section 6.3). If this condition implies that the nonadiabatic coupling can be considered as a weak perturbation, we can follow the argument of Section 5.2.2. In analogy to Eq. (5.24), the rate of nonadiabatic transitions from state $|\phi_a\rangle$ to state $|\phi_b\rangle$ can be determined by a "Golden rule"–expression. Before doing this we briefly recall the Hamiltonian describing the system which undergoes an internal conversion process. In Chapter 2 we introduced the following notation for the molecular Hamiltonian (compare Eq. (2.114)):

$$H_{\text{mol}} = \sum_{a,b} (\delta_{ab} H_a + (1 - \delta_{ab})\Theta_{ab})|\phi_a\rangle\langle\phi_b| \,. \tag{5.139}$$

Here, $H_a$ denotes the vibrational Hamiltonian for the adiabatic electronic state $|\phi_a\rangle$. The nonadiabatic coupling between different states is described by the nonadiabaticity operator $\Theta_{ab}$ acting on the nuclear coordinates. The Hamiltonian is similar to expression (5.16) but with $\Theta_{ab}$ replacing the interaction term, $-\mathbf{E}(t)\mathbf{d}_{ab}$.

According to the form of the molecular Hamiltonian the internal conversion rate follows as

$$k_{a\to b}^{(\text{IC})} = \frac{2\pi}{\hbar} \sum_{MN} f(E_{aM}) |\langle\chi_{aM}|\Theta_{ab}|\chi_{bN}\rangle|^2 \, \delta(E_{aM} - E_{bN}) \,. \tag{5.140}$$

[14] *Kasha's rule* states that because of strong internal conversion processes any appreciable fluorescence from $S_n$ states ($n > 1$) is absent in polyatomic molecules. Even fluorescence from the $S_1$ state can be reduced by radiationless transitions to the ground state.

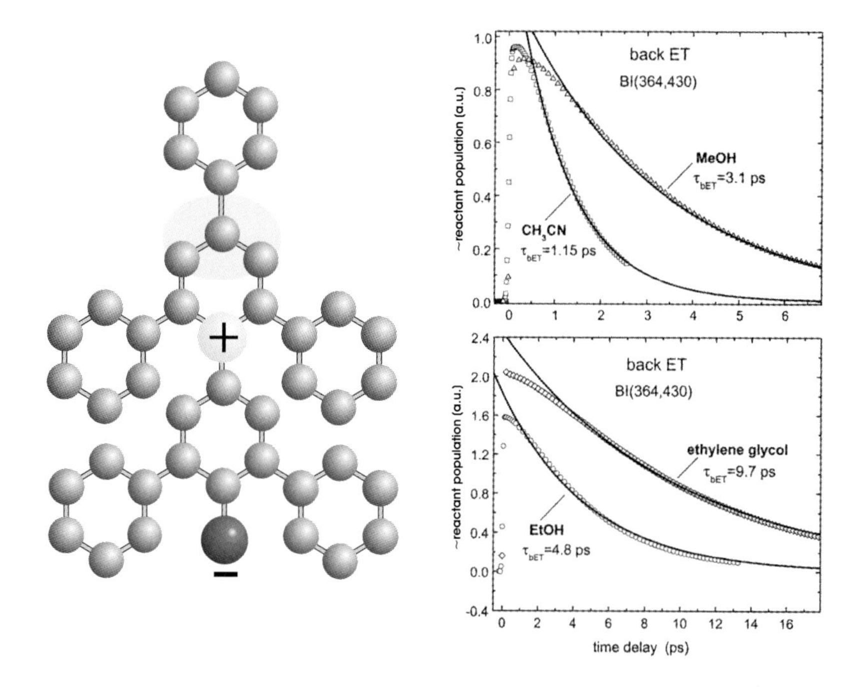


**Figure 5.15:** Ultrafast internal conversion from the $S_1$ to the $S_0$–state of the pyridinium N–phenolate dye betaine–30. (Betaine–30 represents a very sensitive solvatochromic probe often used for polarity measurements.) Left panel: molecular structure (carbon atoms are shown in grey, nitrogen in weak grey, and oxygen in black). The ground state is characterized by a large dipole moment mainly according to the charge separation between nitrogen and oxygen. Upon excitation into the $S_1$–state the dipole moment is reduced since the negative charge moves to a carbon atom (shown by a grey sphere above the nitrogen atom). Right panel: decay of the reactant population (proportional to the transient absorption signal) for different solvents (solid lines are a monoexponential fit of the measured data). (reprinted with permission from [Kov01]; copyright (2001) American Chemical Society).

As indicated in Fig. 5.4 the internal conversion process is a transition from an initial manifold $E_{aM}$ of vibrational levels into the final manifold $E_{bN}$. How to tackle such a problem within lowest–order perturbation theory has already been explained in detail in Section 3.59. Taking the general rate expression Eq. (3.129) for the present case one recovers Eq. (5.140).

Since there is no time–dependent external field involved, the argument in the delta function of Eq. (5.140) contains only the *bare* molecular transition frequencies (cf. Eq. (5.24)). For optical transitions one often neglects the nuclear coordinate dependence of the electronic transition dipole moment (Condon approximation). A similar approximation which replaces the operator $\Theta_{ab}$ by a constant (or by a certain averaged value $\overline{\Theta}_{ab}$ with respect to the nuclear coordinates) results here in the replacement of the full matrix element by the simpler expression $\overline{\Theta}_{ab}\langle\chi_{aM}|\chi_{bN}\rangle$. Introducing the zero–frequency density of states (see Eq. (5.31))

$$\mathcal{D}_{ab}(0) = \sum_{NM} f(E_{aM})\, |\langle\chi_{aN}|\chi_{bM}\rangle|^2\, \delta(E_{bN} - E_{aM}) \, , \tag{5.141}$$

we can write

$$k_{a\to b}^{(\mathrm{IC})} = \frac{2\pi}{\hbar}|\overline{\Theta}_{ab}|^2 \mathcal{D}_{ab}(0) \,. \tag{5.142}$$

As in the case of the absorption coefficient the result for the zero–frequency density of states can be put into a more specific form if the model of parabolic PES is used. Then we obtain (cf. Eq. (5.40))

$$\mathcal{D}_{ab}(0) = \frac{1}{2\pi\hbar}\int dt\, e^{i\omega_{ab}t + G_{ab}(t) - G_{ab}(0)} \,. \tag{5.143}$$

The function $G_{ab}(t)$ has been introduced in Eq. (5.41). The index "$ab$" indicates for which states the displacements enter in Eq. (5.41). This clearly demonstrates the formal analogy between the different transition processes as stressed above.

### 5.7.2 Ultrafast Internal Conversion

In many organic molecules the internal conversion process proceeds on a time scale which is much shorter than any vibrational relaxation time. As already pointed out, it is possible to neglect in this ultrafast limit any vibrational energy dissipation and to describe the internal conversion process by the solution of the respective time–dependent Schrödinger equation. Since different coupled adiabatic electronic states have to be considered, the time–dependent state vector includes different electronic contributions,

$$|\Psi(t)\rangle = \sum_a |\chi_a(t)\rangle|\phi_a\rangle \,. \tag{5.144}$$

Obviously, this ansatz represents the time–dependent version of the solution of the time–independent Schrödinger equation (2.13) describing stationary states of the molecule. The different vibrational state vectors follow from the solution of the equation

$$i\hbar\frac{\partial}{\partial t}|\chi_a(t)\rangle = H_a|\chi_a(t)\rangle + \sum_b \Theta_{ab}|\chi_b(t)\rangle \,. \tag{5.145}$$

This equation is often called *coupled–channel* equation since different electronic levels (channels) are possible for the propagation of vibrational wave packets. In the general case the coupled–channel equations can be solved numerically using, for instance, a discrete representation of the vibrational coordinates on a grid (see Section 3.12.5).

According to the normalization of $|\Psi(t)\rangle$ and of the adiabatic electronic state vectors, the single vibrational state vectors are not normalized. Instead, the $|\chi_a(t)\rangle$ determine the electronic state population as

$$P_a(t) = \langle\chi_a(t)|\chi_a(t)\rangle \,. \tag{5.146}$$

But, the probability density $P(R)$ for the nuclear coordinates $R \equiv \{\mathbf{R}_n\}$ contains contributions from different electronic states:

$$P(R;t) = \sum_a |\chi_a(R,t)|^2 \,. \tag{5.147}$$

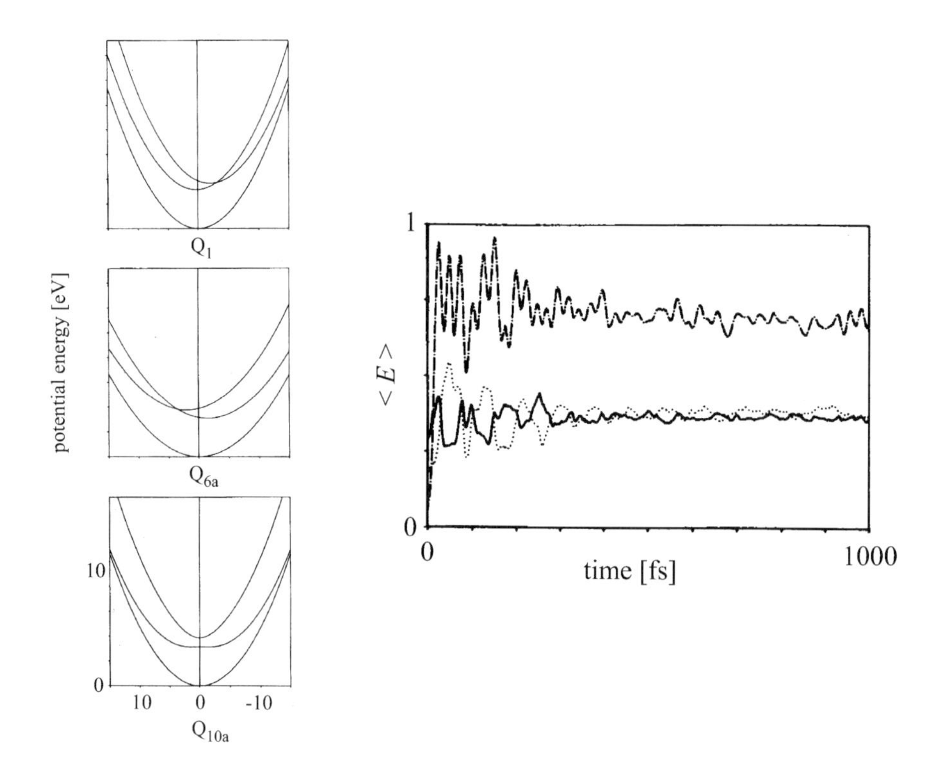


**Figure 5.16:** Left part: PES of the $S_0$, $S_1$, and $S_2$ states of pyrazine for an empirical model of reduced dimensionality with three normal mode coordinates $Q_1$, $Q_{6a}$, and $Q_{10a}$. Note that the $Q_1$ and $Q_{6a}$ modes modulate the energy gap between the $S_1$ and $S_2$ states (tuning modes). The $Q_{10a}$ mode ($B_{1g}$ symmetry), on the other hand, couples both electronic states which leads to a conical intersection modulate the energy gap between the $S_1$ and $S_2$ states (tuning modes). The $Q_{10a}$ mode ($B_{1g}$ symmetry), on the other hand, couples both electronic states which leads to a conical intersection (cf. the two–dimensional quantum chemical calculations for the $S_1$ and $S_2$ states shown in Fig. 2.15). Right part: Vibrational energy redistribution during the $S_2$–$S_1$ internal conversion in pyrazine. The expectation values for the vibrational energy of the pyrazine normal modes $Q_1$ (solid line), $Q_{6a}$ (dotted line), and $Q_{10a}$ (dash–dot line) are plotted. The $S_2$ state has been initially prepared. (reprinted with permission from [Sch90]; copyright (1990) American Institute of Physics)

If only two adiabatic electronic states are involved one usually speaks about the *curve–crossing problem*.

In Section 2.7 it has been discussed that the numerical treatment of electron–vibrational dynamics in coupled electronic states is conveniently done in the *diabatic* representation. Let us suppose that we have obtained a diabatic Hamiltonian where the nuclear motion in the different diabatic PES is harmonic with frequencies $\omega_{\text{vib}}$ (the extension to many vibrational modes with state dependent frequencies is straightforward). In the present case the eigenstates of the displaced harmonic oscillator PES, $|\chi_{aM}\rangle \equiv D_a^+|M\rangle$ can be used. Expanding the time–

dependent state vector in terms of the diabatic states (in analogy to Eq. (5.144)) and setting

$$\langle \chi_{aM} | \chi_a(t) \rangle = C_{aM}(t) \,, \tag{5.148}$$

we obtain from Eq. (5.145) the coupled set of first–order differential equations

$$i\hbar \frac{\partial}{\partial t} C_{aM}(t) = \left( U_a^{(0)} + \hbar\omega_{\text{vib}} M \right) C_{aM}(t) + \sum_{bN} V_{aM,bN} C_{bN}(t) \,. \tag{5.149}$$

The matrix elements of the diabatic coupling operator, $V_{aM,bN}$, were introduced in Section 2.7. In actual calculations the infinite set of equations (5.149) can be truncated to a finite number by simple energetic arguments.

In Fig. 5.16 we show the adiabatic PES and the time evolution of the expectation values of the vibrational energy for different normal modes, respectively, for a three–dimensional model calculation of the $S_2 \to S_1$ internal conversion in pyrazine. The rapid increase of the vibrational energy content in the different modes indicates that, as a result of the electron–vibrational coupling, electronic excitation energy is transformed into vibrational energy.

## 5.8 Supplement

### 5.8.1 Absorption Coefficient for Displaced Harmonic Oscillators

In this section we will show how to simplify expression (5.35) for the lineshape function $\mathcal{D}_{\text{abs}}(\omega)$ of linear absorption, if the two vibrational Hamiltonians $H_g$ and $H_e$ describe independent harmonic oscillators (normal mode vibrations). The derivation will be a particularly illuminating since the model is exactly solvable. For simplicity the normal mode oscillators should not change their vibrational frequencies if the electronic state changes but merely attain a new equilibrium position (see Section 2.6.1 and Fig. 2.11). Using the displacement operator (cf. Eq. (2.89))

$$D_a^+ = \exp\left\{ \sum_\xi g_a(\xi)(C_\xi - C_\xi^+) \right\} \equiv \prod_\xi D_\xi^+(g_a(\xi)) \,, \tag{5.150}$$

the two vibrational Hamiltonians can be generated from the Hamiltonian of a non–shifted oscillator

$$H_a = U_a^{(0)} + D_a^+ H_{\text{vib}} D_a \,. \tag{5.151}$$

Here,

$$H_{\text{vib}} = \sum_\xi \hbar\omega_\xi (C_\xi^+ C_\xi + 1/2) \tag{5.152}$$

denotes the *reference* vibrational Hamiltonian. Accordingly, the trace formula introduced in Eq. (5.35) can be rewritten as (the statistical operator $\hat{R}_g$ is given in Eq. (5.36))

$$\begin{aligned}\mathrm{tr}_g\{\hat{R}_g\, e^{iH_g t/\hbar}\, e^{-iH_e t/\hbar}\} &= e^{-i\omega_{eg}t}\, \mathrm{tr}_{\mathrm{vib}}\{D_g\, D_g^+\, \hat{R}_{\mathrm{vib}}\, D_g\, D_g^+\, e^{iH_{\mathrm{vib}}t/\hbar} \\ &\quad \times D_g\, D_e^+ e^{-iH_{\mathrm{vib}}t/\hbar} D_e\, D_g^+\} \\ &= e^{-i\omega_{eg}t} \\ &\quad \times \mathrm{tr}_{\mathrm{vib}}\{\hat{R}_{\mathrm{vib}}\, e^{iH_{\mathrm{vib}}t/\hbar}\, D_g D_e^+ e^{-iH_{\mathrm{vib}}t/\hbar} D_e D_g^+\} \\ &= e^{-i\omega_{eg}t}\, \mathrm{tr}_{\mathrm{vib}}\{\hat{R}_{\mathrm{vib}}\, e^{iH_{\mathrm{vib}}t/\hbar}\, D_{ge} e^{-iH_{\mathrm{vib}}t/\hbar} D_{ge}^+\}\, .\end{aligned} \tag{5.153}$$

Here we have used the notations $\mathrm{tr}_g$ and $\mathrm{tr}_{\mathrm{vib}}$ to distinguish between the trace taken with respect to the electronic ground state vibrations and the eigenstates $|N\rangle$ of the *non–displaced* reference Hamiltonian $H_{\mathrm{vib}}$, respectively. Additionally, we introduced $\hbar\omega_{eg} = U_e^{(0)} - U_g^{(0)}$,

$$\hat{R}_{\mathrm{vib}} = \frac{\exp\{-H_{\mathrm{vib}}/k_{\mathrm{B}}T\}}{\mathrm{tr}\{\exp -H_{\mathrm{vib}}/k_{\mathrm{B}}T\}} \equiv \frac{1}{\mathcal{Z}} e^{-H_{\mathrm{vib}}/k_{\mathrm{B}}T}\, , \tag{5.154}$$

and the combined displacement operator

$$D_{ge} = D_g D_e^+\, . \tag{5.155}$$

Using the Heisenberg representation of $D_{ge}$ which is given by

$$D_{ge}(t) = e^{iH_{\mathrm{vib}}t/\hbar}\, D_{ge} e^{-iH_{\mathrm{vib}}t/\hbar}\, , \tag{5.156}$$

the trace formula becomes

$$T(t) = \mathrm{tr}_g\{\hat{R}_g\, e^{iH_g t/\hbar}\, e^{-iH_e t/\hbar}\} = e^{-i\omega_{eg}t}\, \mathrm{tr}_{\mathrm{vib}}\{\, \hat{R}_{\mathrm{vib}}\, D_{ge}(t) D_{ge}^+(0)\}\, . \tag{5.157}$$

This is the autocorrelation function of the combined displacement operators taken with respect to the equilibrium of the non–displaced reference oscillators.

Since we are dealing with normal mode oscillators, there is no coupling among the modes. The vibrational Hamiltonian $H_{\mathrm{vib}}$ is additive with respect to the mode index $\xi$ and the vibrational state $|N\rangle$ factorizes into the single oscillator states $|N_\xi\rangle$. As a result the trace in Eq. (5.157) factorizes into single mode traces

$$T(t) = \prod_\xi T_\xi(t)\, . \tag{5.158}$$

Therefore, we can deal in the following with a single–mode contribution $T_\xi(t)$ to the complete trace. To simplify the notation the mode index $\xi$ will be dropped, and $\omega_\xi$ is replaced by $\omega_{\mathrm{vib}}$. First we note that

$$D_{ge} = D(g_g)\, D^+(g_e) = D(g_g - g_e) = D(\Delta g)\, , \tag{5.159}$$

where $\Delta g = g_g - g_e$. The time–dependent displacement operator appearing in Eq. (5.157) (the single–mode contribution to it) can be written as

$$\begin{aligned} D_{ge}(t) = D(\Delta g; t) &= e^{i\omega_{\mathrm{vib}} C^+ C t}\, D(\Delta g)\, e^{-i\omega_{\mathrm{vib}}\, C^+ C t} \\ &= \exp\{-\Delta g(C e^{-i\omega_{\mathrm{vib}}t} - C^+ e^{i\omega_{\mathrm{vib}}t})\}\, .\end{aligned} \tag{5.160}$$

Consequently, the single–mode contribution to the trace reads ($\mathcal{Z}$ is the single–mode partition function)

$$T_\xi(t) = \frac{1}{\mathcal{Z}} \sum_N \langle N| \, e^{-\hbar\omega_{\rm vib} N/k_{\rm B}T} \; D(\Delta g; t) \; D^+(\Delta g; 0)|N\rangle \, . \tag{5.161}$$

For the further treatment of this expression we may profit from the calculations done in Section 3.12.2. (Note that we calculated there a correlation function where the oscillator coordinate operator appeared in the exponent. Here, we have the momentum operator instead of the coordinate operator in the exponent.) Of particular interest will be Eq. (3.454) which allows to rearrange the oscillator operators in the matrix element of Eq. (5.161) (cf. Eq. (2.95)). Using the relation

$$\begin{aligned} \mathcal{M}(N) &= \langle N|D(\Delta g; t) \; D^+(\Delta g; 0) \; |N\rangle \\ &= \langle N|e^{-\alpha(t)C + \alpha^*(t)C^+} \; e^{\alpha(0)C \; - \; \alpha^*(0)C^+} |N\rangle \, , \end{aligned} \tag{5.162}$$

with $\alpha(t) = \Delta g \, \exp(-i\omega_{\rm vib} t)$ we can write

$$\begin{aligned} \mathcal{M}(N) &= \langle N|e^{-|\alpha(t)|^2/2} \; e^{\alpha^*(t) \; C^+} \; e^{-\alpha(t) \; C} \; e^{-|\alpha(0)|^2/2} \; e^{-\alpha^*(0) \; C^+} \; e^{\alpha(0) \; C} |N\rangle \\ &= e^{-\frac{1}{2}(\alpha(t)|^2 \; + \; |\alpha(0)|^2)} \cdot \\ &\quad \times \langle N|e^{\alpha^*(t) \; C^+} \; e^{\alpha(t) \; \alpha^*(0)} \; e^{-\alpha^*(0) \; C^+} \; e^{-\alpha(t) \; C} \; e^{\alpha(0) \; C} \; |N\rangle \\ &= e^{-\frac{1}{2}(|\alpha(t)|^2 \; + \; |\alpha(0)|^2 - 2\alpha(t)\alpha^*(0))} \\ &\quad \times \langle N|e^{(\alpha^*(t) - \alpha^*(0))C^+} \; e^{-(\alpha(t) - \alpha(0))C} \; |N\rangle \, . \end{aligned} \tag{5.163}$$

Next we introduce the abbreviation $\Delta\alpha(t) = \alpha(t) - \alpha(0) = \Delta g(\exp(-i\omega_{\rm vib} t) - 1)$, and take into account that

$$|\alpha(t)|^2 + |\alpha(0)|^2 - 2\alpha(t) \; \alpha^*(0) = |\Delta\alpha(t)|^2 \; - 2 \; i \; {\rm Im}(\alpha(t) \; \alpha^*(0)) \, . \tag{5.164}$$

Then we obtain the normal ordering of the original matrix elements in the trace formula as

$$\begin{aligned} \langle N|D(\Delta g; t) \; D^+(\Delta g; 0)|N\rangle &= \exp\left\{-|\Delta\alpha|^2/2 - i \; {\rm Im}(\alpha^*(t) \; \alpha(0))\right\} \\ &\quad \times \; \langle N|e^{\Delta\alpha^* C^+} \; e^{-\Delta\alpha C}|N\rangle \, . \end{aligned} \tag{5.165}$$

To determine the oscillator matrix elements we use Eqs. (2.97)–(2.99) and obtain for Eq. (5.161)

$$T_\xi(t) \;\; = \;\; (1 - e^{-\hbar\omega_{\rm vib}/k_{\rm B}T}) \; e^{-z/2 - i{\rm Im} \; (\alpha^*(t)\alpha(0))} \sum_{N=0}^{\infty} e^{-\hbar\omega_{\rm vib} N/k_{\rm B}T} \; L_N(z) \, . \tag{5.166}$$

Note the introduction of $z = |\Delta\alpha(t)|^2$, and of the Laguerre polynomial $L_N(z)$ of order $N$ (cf. Eq. (2.99)). The relation between the Laguerre polynomials and their generating function Eq. (3.461) results in $T_\xi(t) = \exp E_\xi(t)$ with

$$E_\xi(t) = -z/2 - i{\rm Im}(\alpha^*(t) \; \alpha(0)) \;\; - \;\; \frac{e^{-\hbar\omega_{\rm vib}/k_{\rm B}T}}{1 - e^{-\hbar\omega_{\rm vib}/k_{\rm B}T}} z \, . \tag{5.167}$$

The Bose–Einstein distribution $n(\omega_{\text{vib}})$ (cf. Section 3.6.1) allows us to rewrite the last term of the exponent. The final result for $T_\xi(t)$ will be obtained if the exponent is rearranged with respect to $\Delta\alpha(t)$ and $\Delta\alpha^*(t)$ according to

$$\begin{aligned} E_\xi(t) &= -z/2 - i\,\text{Im}\,\alpha^*(t)\alpha(0) - n(\omega_{\text{vib}})z \\ &= -\frac{1}{2}(1+2n(\omega_{\text{vib}}))\Delta g^2(2-e^{i\omega_{\text{vib}}t}-e^{-i\omega_{\text{vib}}t}) \\ &\quad -\frac{1}{2}\Delta g^2\,(e^{i\omega_{\text{vib}}t}-e^{-i\omega_{\text{vib}}t}) \\ &= \frac{\Delta g^2}{2}\Big(2(1+n(\omega_{\text{vib}}))(e^{-i\omega_{\text{vib}}t}-1)+2n(\omega_{\text{vib}})(e^{i\omega_{\text{vib}}t}-1)\Big)\,. \end{aligned} \tag{5.168}$$

The result will be denoted by $E_\xi(t) = -G_\xi(0) + G_\xi(t)$ with

$$G_\xi(t) = \Delta g^2(\xi)\ \left[e^{-i\omega_\xi t}(1+n(\omega_\xi))\ +\ e^{i\omega_\xi t}n(\omega_\xi)\right]\,. \tag{5.169}$$

The complete trace is the product with respect to the various single–mode contributions $T_\xi(t)$, hence the total exponent is determined by

$$G(t) = \sum_\xi G_\xi(t)\,, \tag{5.170}$$

This *exact* result is used in the definition of the density of states, Eq. (5.40).

### 5.8.2 Cumulant Expansion for Harmonic Potential Energy Surfaces

In the following we will demonstrate that the second–order cumulant expansion (up to $\Gamma_2(t)$) for the absorption coefficient as introduced Section 5.3.3 is identical with the exact result provided that the harmonic oscillator description of the vibrational modes has been assumed (cf. Section 5.2.4). We will prove this statement by reproducing Eq. (5.40) in the following. Within the model of shifted harmonic potentials the vibrational Hamiltonians $H_e$ and $H_g$ take the form Eq. (2.93) and we obtain $\Delta H_{eg} = D_e^+ H_{\text{vib}} D_e - D_g^+ H_{\text{vib}} D_g$,

$$\begin{aligned} \Delta H_{eg}^{(g)}(t) &= D_g^+ U_{\text{vib}}^+(t) D_g \Delta H_{eg} D_g^+ U_{\text{vib}}(t) D_g \\ &= D_g^+ U_{\text{vib}}^+(t)\left(D_{eg}^+ H_{\text{vib}} D_{eg} - H_{\text{vib}}\right) U_{\text{vib}}(t) D_g\,. \end{aligned} \tag{5.171}$$

Here, $D_{eg}$ abbreviates $D_e D_g^+$. The Hamiltonian $H_{\text{vib}}$ defines a fictitious reference oscillator system which had been introduced in Eq. (2.93). Inserting the explicit expressions for the Hamiltonians yields finally (note the application of the displacement operator and the abbreviation $g_{eg}(\xi) = g_e(\xi) - g_g(\xi)$)

$$\Delta H_{eg}^{(g)}(t) = D_g^+ \sum_\xi \hbar\omega_\xi\,(Q_\xi(t) + g_{eg}(\xi))\,g_{eg}(\xi) D_g\,. \tag{5.172}$$

The time dependence of the dimensionless oscillator coordinate $Q_\xi(t)$ is determined by $H_{\text{vib}}$ and reads $Q_\xi(t) = C_\xi \exp(-i\omega_\xi t) + C_\xi^+ \exp(i\omega_\xi t)$. The trace operation necessary to calculate the different correlation functions of $\Delta H_{eg}^{(g)}(t)$ will be rewritten in terms of to the non–displaced vibrational states of the oscillator Hamiltonian $H_{\text{vib}}$ as $\langle ... \rangle_g \equiv \text{tr}_{\text{vib}}\{\hat{R}_{\text{vib}} D_g \ldots D_g^+\}$, with $\hat{R}_{\text{vib}} = \exp(-H_{\text{vib}}/k_B T)/\mathcal{Z}$.

Now we can determine the correlation functions of interest. The correlation function of first order becomes time independent

$$\begin{aligned}\langle \Delta H_{eg}^{(g)}(t)\rangle_g &= \text{tr}_{\text{vib}}\left\{\hat{R}_{\text{vib}}\sum_\xi \hbar\omega_\xi \left(Q_\xi(t)+g_{eg}(\xi)\right)g_{eg}(\xi)\right\} \\ &= \sum_\xi \hbar\omega_\xi g_{eg}^2(\xi) \equiv \frac{S}{2}\,. \end{aligned} \tag{5.173}$$

Here, we took into account that the equilibrium expectation values of $C_\xi$ and $C_\xi^+$ vanish. Further, we recovered the expression for the *Stokes shift* already introduced in Eq. (5.55). In a similar manner we obtain

$$\begin{aligned}\langle \Delta H_{eg}^{(g)}(t)\Delta H_{eg}^{(g)}(\bar{t})\rangle_g &= \text{tr}_{\text{vib}}\Big\{\hat{R}_{\text{vib}}\sum_{\xi\bar{\xi}} \hbar\omega_\xi\hbar\omega_{\bar{\xi}}\;\left(Q_\xi(t)+g_{eg}(\xi)\right) \\ &\quad\times g_{eg}(\xi)\left(Q_{\bar{\xi}}(\bar{t})+g_{eg}(\bar{\xi})\right)g_{eg}(\bar{\xi})\Big\} \\ &= \sum_\xi \left(\hbar\omega_\xi g_{eg}(\xi)\right)^2 \text{tr}_{\text{vib}}\{\hat{R}_{\text{vib}}Q_\xi(t)Q_{\bar{\xi}}(\bar{t})\} + S^2/4\,. \end{aligned} \tag{5.174}$$

The trace with respect to the dimensionless oscillator coordinate vanishes for $\xi \neq \bar{\xi}$ and reads

$$\text{tr}_{\text{vib}}\{\hat{R}_{\text{vib}}Q_\xi(t)Q_\xi(\bar{t})\} = (1+n(\omega_\xi))\,e^{-i\omega_\xi(t-\bar{t})} + n(\omega_\xi)e^{i\omega_\xi(t-\bar{t})}\,. \tag{5.175}$$

Now it is easy to determine $\Gamma_1$ and $\Gamma_2$. It follows directly from Eqs. (5.84) and (5.173) that $\Gamma_1(t) = iSt/2\hbar$. To obtain $\Gamma_2$ some additional calculations have to be carried out. First, we note that the double time integration necessary to get $\Gamma_2$ gives the contribution $(St/\hbar)^2/2$. This follows directly from the part of Eq. (5.174) containing $S^2$. It can be written as $-\Gamma_1^2(t)/2$. With Eq. (5.85) we obtain

$$\Gamma_2(t) = \sum_\xi (\omega_\xi g_{eg}(\xi))^2 \int_0^t dt_1 \int_0^{t_1} dt_2 \left\{(1+n(\omega_\xi))\,e^{-i\omega_\xi(t_1-t_2)} + n(\omega_\xi)e^{i\omega_\xi(t_1-t_2)}\right\}\,. \tag{5.176}$$

The calculation of the time integrals results in $\Gamma_2(t) = -\Gamma_1(t) - G(t)$ with the function $G(t)$ introduced in Eq. (5.41). Thus, we get as the final result

$$\Gamma_1(t) + \Gamma_2(t) = G(0) - G(t)\,, \tag{5.177}$$

and reproduced the expression Eq. (5.40) for the combined density of states $\mathcal{D}_{\text{abs}}(\omega)$. Since Eq. (5.40) gives the exact result for the combined density of states, the cumulant expansion only contains contributions up to the second–order cumulant $\Gamma_2$. Cumulants of type $\Gamma_3$ and higher are not necessary to calculate $\langle S_{eg}(t,0)\rangle_g$ within the model of shifted harmonic oscillators PES as mentioned above. However, in principle the cumulant expansion provides a means for incorporating also the effect of anharmonic contributions to the PES.

# Notes


[Ego97] S. A. Egorov and B. J. Berne, J. Chem. Phys. **107**, 6050 (1997).

[Ego97a] S. A. Egorov, E. Gallicchio, and B. J. Berne, J. Chem. Phys. **107**, 9312 (1997).

[Ego98] S. A. Egorov, E. Rabani, and B. J. Berne, J. Chem. Phys. **108**, 1407 (1998).

[Kov01] S. A. Kovalenko, N. Eilers–König, T. A. Senyushkina, and N. P. Ernsting, J. Phys. Chem. A **105**, 4834 (2001).

[Muk95] S. Mukamel, *Principles of Nonlinear Optical Spectroscopy*, Oxford University Press, 1995.

[Rei96] D.Reichmann, R. J. Silbey and A. Suarez, J. Chem. Phys. **105**, 10500 (1996).

[Sch90] R. Schneider, W. Domcke, H. Köppel, J. Chem. Phys. **92**, 1045 (1990).

[Ski86] J. L. Skinner and D. Hsu, J. Chem. Phys. **90**, 4931 (1986).

[Sut92] H. U. Suter, J. R. Huber, M. von Dirke, A. Untch, and R. Schinke, J. Chem. Phys. **96**, 6727 (1992).

# 6 Electron Transfer

*Spatial electronic charge redistribution in single molecules as well as in molecular complexes will be described. The charge transfer process occurs as a spontaneous transition from a metastable initial state to a stable final state. The initial state is prepared either by photoabsorption or charge injection from external sources. The electronic transition can be understood as a tunneling process through barriers separating different localization centers of the moving electron. This causes a modification of the electrostatic field in the molecule which leads to a change of the nuclear equilibrium configuration. To develop an understanding for the interplay between electron transfer and the accompanying nuclear rearrangement is the principal aim of electron transfer theories.*
*Different transfer regimes will be discussed which are distinguished by the time scales for electronic and nuclear motion. The limits of classical and quantum mechanical descriptions of the nuclear dynamics are described and appropriate rate expressions will be derived. Finally, ultrafast photoinduced electron transfer is introduced as a phenomenon whose theoretical description is beyond a simple rate equation.*

## 6.1 Introduction

Electron transfer (ET) must be considered as one of the basic types of chemical processes. It represents the initial step of a number of reactions like the making and breaking of chemical bonds or the change of molecular conformations (see Fig. 6.1). In all fields of inorganic, organic and biochemistry ET reactions are common. For example, corrosion is caused by the ET between some metal surface and oxygen. In biological systems ET reactions are a basic step of enzymatic activity in the living cells of bacteria, plants and animals. Electron transfer in proteins or protein complexes plays an important role in the cell metabolism and energy balance. ATP, for instance, is produced in oxidative phosphorylation where NADH releases electrons.[1] These are captured by dioxygen to form water; the total process generates a large amount of excess energy. Another prominent example is given by the electron transferring system of photosynthesis. Here, a transmembrane potential is created which acts as a proton pump to produce ATP (cf. Chapter 7). ET in the reaction center of purple bacteria has been unraveled on an atomic length scale and a time scale down to the femtosecond region (see Fig. 6.2).

To give a working definition of ET we will characterize it as a *spontaneous* charge redistribution between an initially prepared reactant state and a well defined product state. ET reactions proceed in such a manner that the transferred electron remains in a bound state with respect to the particular molecule or molecular system. In other words, the electron is *not* activated above the ionization threshold and in this way transferred to a different region of the molecular complex. This means that ET reactions occur as *tunneling processes*; the reaction barriers which the moving electron experiences are penetrated via tunneling.[2]

In this respect it is important to note that the motion of the considered electron does *not* take place while the configuration of the other electrons is fixed. Instead the electronic wavefunction changes from that describing the reactant state ($\phi_{\rm rea}$) to that of the product state ($\phi_{\rm pro}$). According to the ansatz Eq. (2.29) introduced in Section 2.4 we may say that all molecular orbitals (MO) of the system will be modified during this transition. This change can be characterized by the electronic charge density (a = rea, pro, the whole set of spatial electron coordinates is denoted by $r$)

$$\varrho_a^{(\rm el)}(\mathbf{r}) = eN_{\rm el}\Big(\prod_{\mathbf{r}_j\neq\mathbf{r}}\int d^3\mathbf{r}_j\Big)\,|\phi_a(\{\mathbf{r}_j\})|^2\,, \tag{6.1}$$

which is the probability distribution for the $N_{\rm el}$ electrons reduced to a *single*–particle density. Although the whole electronic wavefunction changes in the course of the ET, in many reactions the change of the electronic charge density corresponds to the change induced by a single electron. Therefore, it is often sufficient to discuss ET as the result of the transition of a *single* electron from an initial MO (*donor* state) to the MO of the final state (*acceptor* state).

Due to the change of the electronic charge distribution during an ET reaction the internal electrostatic field of the molecular complex is modified. This in turn causes new equilibrium

[1] ATP is the abbreviation for adenosine triphosphate, the compound which acts as an energy storage in any living system. NADH stands for nicotinamide adenine dinucleotide which plays an important rule in respiration.

[2] This definition of ET excludes processes in biological systems where special enzymes act as charges carriers transporting electrons over large spatial distances.

$$\text{Na} + \langle\bigcirc\rangle\text{-CH}_2\text{-Br} \longrightarrow \text{Na}^+ + \langle\bigcirc\rangle\text{-CH}_2^\bullet + \text{Br}$$

**Figure 6.1:** ET from sodium to benzyl halide resulting in bond breaking and benzyl halide radical formation.

positions of the nuclei. First, when mentioning nuclei, we have in mind those of the considered molecule. If the environment is polarizable as it is the case for a polar solvent a polarization and rearrangement of the solvent molecules may take place, too. Hence, an ET reaction (as any other change of the electronic state of the molecule) is accompanied by a change of the equilibrium configuration of the nuclei. This process may be viewed as the motion of the electron carrying along a polarization cloud with respect to the surrounding molecular structure.

There is some apparent similarity between intramolecular electronic transitions, induced by the radiation field or by the nonadiabatic coupling, discussed in Chapter 5, and the ET reaction. Therefore, one expects that also in the case of ET there exists a coupling between the reactant and the product state. If this inter–state coupling $V$ is small, one is in the limit of *nonadiabatic* ET. The opposite case is called *adiabatic* transfer. These terms and also the meaning of a small or large coupling will be explained in more detail below. At the moment we only state that in most cases nonadiabatic ET can be understood as a particular type of *spatial* charge redistribution as shown in Fig. 6.3.

In the reactant state the transferred electron is localized at the electron donor part of the molecular systems (it occupies the donor MO). From the donor it moves to the acceptor region, where it is in some spatially localized acceptor MO (product state).

In contrast, the adiabatic ET is *not* connected with a characteristic spatial redistribution of charge. It is usually described in terms of chemical reaction kinetics, for which the double–well potential provides a good model (see Fig. 6.4). In this approach the internal energy of the reaction (or if entropic effects are important, the free energy) is considered in dependence on a reaction coordinate, which is a particularly chosen collective coordinate for the nuclei (cf. Section 2.6.3). The metastable initial (reactant) state and the stable final (product) state are separated by a potential barrier along this reaction coordinate. To overcome this barrier the reaction requires thermal activation. Alternatively, a tunneling transition through the barrier is possible. This is in contrast to the electron motion which exclusively occurs via a tunneling process.

Given the definition of donor (D) and acceptor (A) states of a molecular system the ET reaction is most simply characterized by the following scheme

$$D^-A \rightarrow DA^- . \tag{6.2}$$

$D^-$ means that in the reactant state there is a so–called *excess* electron localized at the donor. After the electron has moved to the acceptor the product state is formed.

This basic event can occur in different variants and numerous generalizations are possible. First, we have to distinguish whether or not the donor and acceptor belong to the same molecule. In the first case the reaction is called *intramolecular* ET or alternatively *unimolec-*

*ular* ET. In contrast, if at least two distinct molecules are involved, the reaction is called *intermolecular* ET or *bimolecular* ET. Independently of this distinction the common structure formed by the donor and acceptor which enables the ET is called donor–acceptor (DA) *complex*.

In Fig. 6.5 the energy level diagram is shown for the ET reaction according to scheme (6.2). We will concentrate on the highest occupied MO (HOMO) (cf. Section 2.4) as well as the lowest unoccupied MO (LUMO) of the donor–acceptor complex (remember that these states have to be computed in a selfconsistent way as explained in Section 2.4). The excess electron initially occupies the LUMO of the donor (donor state) and then moves to the LUMO of the acceptor (acceptor state). The excess electron can be injected into the DA complex, for example, from a metal electrode to which the complex is attached, from a redox compound contained in the solution where DA complex has been dissolved, or via an electron beam. Alternatively, the transferred electron may come from the donor itself. This is the case if the ET reaction involves an excited electron:

$$\mathrm{D^*A} \to \mathrm{D^+A^-} \,. \tag{6.3}$$

The excitation of the donor may be the result of a scattering process with another molecule, or it may be introduced via excitation energy transfer (exciton transfer, see Chapter 8). The excitation can also be achieved via optical absorption as discussed in Chapter 5. After optical excitation an electron of the donor is placed into LUMO, $\mathrm{D} \to \mathrm{D}^*$, where $\mathrm{D}^*$ indicates the excited state of the donor. Then the ET proceeds between donor and acceptor LUMOs

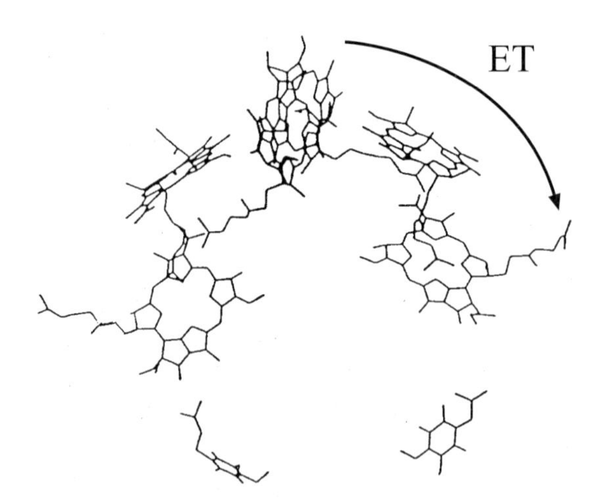


**Figure 6.2:** Chromophores of the photosynthetic bacterial reaction center (the protein the chromophores are embedded in is not shown). In the upper part the special pair of two bacteriochlorophyll molecules is shown acting after an excitation process as the ET donor. The ET takes place along the right branch of the reaction center formed by a further bacteriochlorophyll molecule, a metal–free bacteriochlorophyll molecule, and a quinone as the ET acceptor (from top to bottom ).

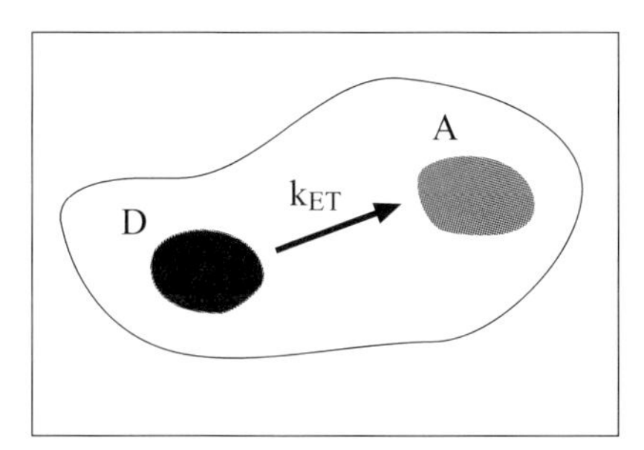


**Figure 6.3:** ET in a schematically drawn DA complex. The initial spatial localization of the transferred electron at the donor is shown by the left hatched area. The right hatched area corresponds to the final localization at the acceptor, and $k_{\mathrm{ET}}$ denotes the transfer rate.

as shown in Fig. 6.6. This type of ET is usually called a *photoinduced* reaction. After the transfer event there is an electron missing at the donor, which becomes positively charged. Accordingly, the acceptor is negatively charged resulting in the formation of a dipole moment in the DA complex.

Fig. 6.6 suggests the possibility of a back reaction where the transferred electron moves directly from the acceptor LUMO into the empty donor HOMO. In most cases this back reaction is much slower than the forward ET, which makes it possible to clearly describe the reaction displayed in Fig. 6.6 as an ET.

Alternatively to the transfer of the excited electron from the donor LUMO to the acceptor LUMO, an unexcited electron may move in the opposite direction from the acceptor HOMO to the donor HOMO:

$$\mathrm{D^*A} \rightarrow \mathrm{D^-A^+} \,. \tag{6.4}$$

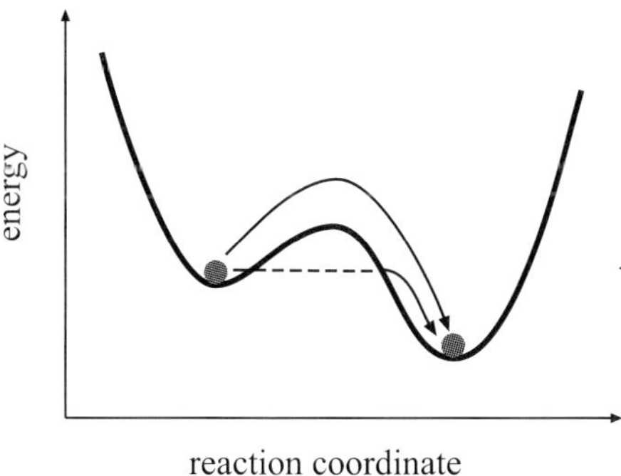


**Figure 6.4:** Double well–potential versus reaction coordinate which can be some collective nuclear coordinate triggering the adiabatic ET. In the initial state of the ET reaction the system is localized in the left metastable well. It can reach the right stable state by crossing the barrier (full line), or by tunneling through the barrier (broken line).

Fig. 6.7 shows this so–called *hole transfer*. The name has been introduced since the reaction can be alternatively understood as the motion of a missing electron (hole) from the donor to the acceptor. (The reverse of this process, i.e. the optical excitation of the acceptor and the motion of the whole from the acceptor to the donor is also possible.)

Let us turn to the discussion of bimolecular ET as it occurs for an intermolecular reaction in solution. If we suppose that initially the donor as well as the acceptor molecules are moving randomly, any reasonable description should include the mechanism which leads to the *formation* of the DA complex. For this purpose, it is also necessary to assess the probability at which the two molecules meet to form the so–called *encounter complex*. Within this encounter complex the donor and the acceptor are close together allowing for the ET to proceed. Afterwards, the encounter complex is destroyed and the donor and acceptor molecule move again independently. The following scheme displays the complete reaction assuming that the side groups R and R′ of the individual molecules are relevant for the formation of the encounter complex:

$$\begin{aligned}\mathrm{D}^- - \mathrm{R} + \mathrm{R}' - \mathrm{A} &\rightarrow \mathrm{D}^- - \mathrm{R}\cdots\mathrm{R}' - \mathrm{A} \rightarrow \mathrm{D} - \mathrm{R}\cdots\mathrm{R}' - \mathrm{A}^- \\ &\rightarrow \mathrm{D} - \mathrm{R} + \mathrm{R}' - \mathrm{A}^- \,. \end{aligned} \tag{6.5}$$

As already indicated those ET reactions are of particular significance which occur in polar solvents. Here, every solvent molecule carries a permanent dipole moment which will be very sensitive to the change of the charge distribution taking place in the DA complex upon ET. If the ET is not too fast the solvent molecules react via the formation of a polarization cloud with an extension large compared to that of the DA complex. Then the macroscopic dielectric properties of the solvent comprised in the dielectric function can be used to characterize the

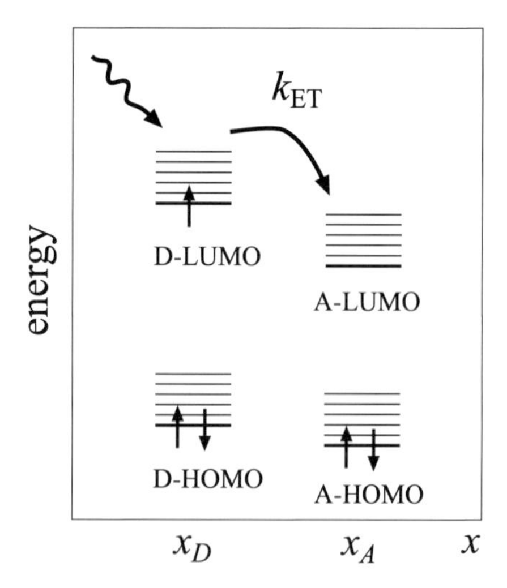


**Figure 6.5:** ET reaction of an excess electron in a DA complex with spatial donor position $x_\mathrm{D}$ and acceptor position $x_\mathrm{A}$ (thick lines – electronic levels, thin lines – vibrational levels). The reactant state electron configuration is shown. The electron spin is represented by an upward or downward arrow, and the curved arrow indicates the pathway the transferred electron takes toward the product state.

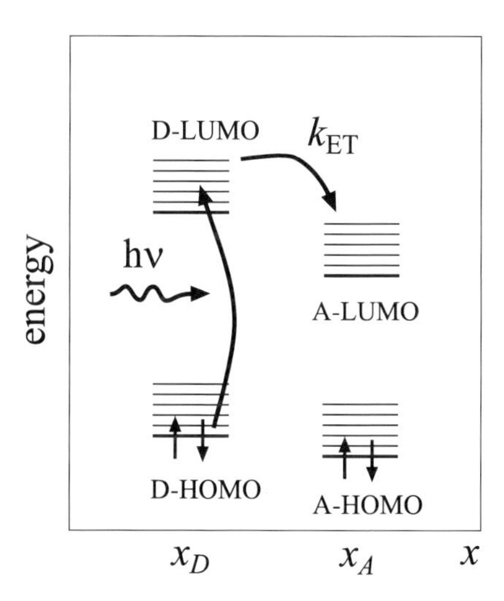


**Figure 6.6:** Photoinduced ET reaction in a DA complex. (thick lines – electronic levels, thin lines – vibrational levels, for further details compare Fig. 6.5)

solvent influence on the ET. If the ET is influenced mainly by solvent molecules it is of *outer–sphere* type. On the other hand, it is of *inner–sphere* type whenever intramolecular nuclear motions are dominant.

The intramolecular ET reactions discussed above for a simple two–state DA complex may

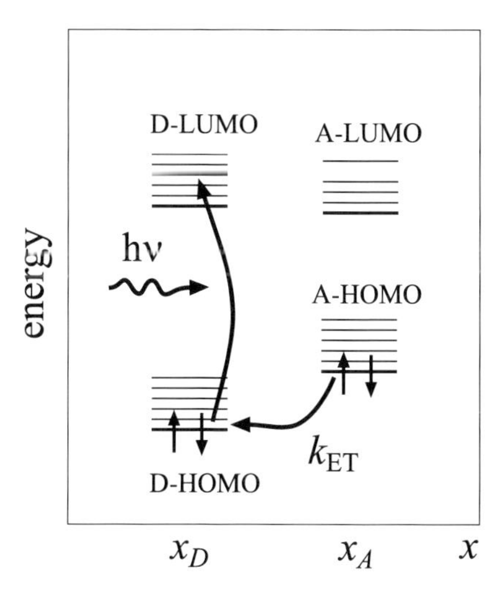


**Figure 6.7:** Photoinduced *hole* transfer reaction in a DA complex (thick lines – electronic levels, thin lines – vibrational levels, for further details compare Fig. 6.5).

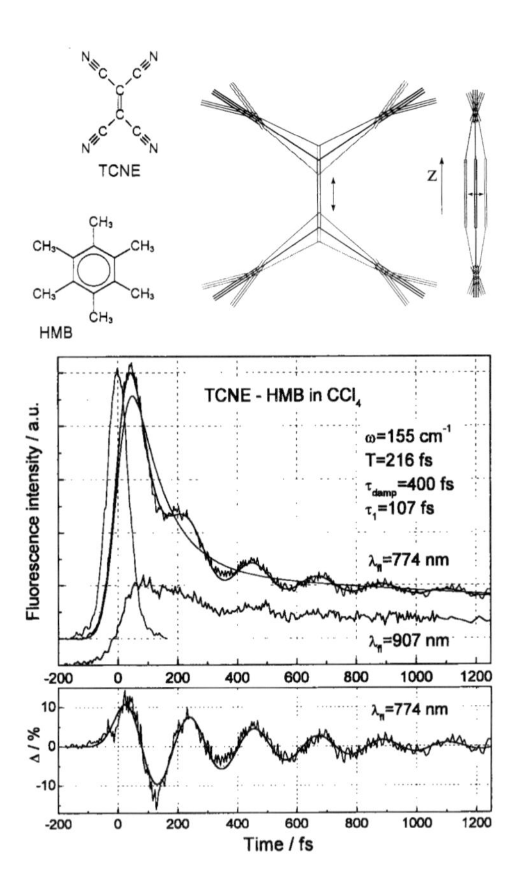


**Figure 6.8:** Ultrafast photoinduced ET reaction from hexamethylbenzene (HMB, donor, see upper part) to tetracyanoethylene (TCNE, acceptor) forming a DA–complex in a $CCl_4$–solvent. The fluorescence decay (at two wavelengths) together with the oscillatory component of the signal at 774 nm are shown in the lower part. An out–off–plane vibrational mode of TCNE could be assigned to these vibrations (top view on TCNE: upper central part part, side view on TCNE: upper right part). (reprinted with permission from [Rub99]; copyright (1999) American Chemical Society)

also take place in this more difficult framework. From scheme (6.5) it is obvious that a particular bimolecular ET reaction can be realized at various geometries and orientations of the donor and acceptor parts of the complex. Therefore, an experimental investigation of an ensemble of encounter complexes will include an averaging with respect to these different realizations. (Note the similarity to inhomogeneous broadening of optical lineshapes introduced in Chapter 5.) An additional complication arises if one takes into account that in the experiment the actual charge transfer act is masked by the random sequence of formation and destruction of the encounter complex. In view of these difficulties it is of great advantage to focus on ET in systems with *fixed* DA distance. Such experiments can be carried out, for example, in frozen solutions or in other types of solid carrier matrices such as polymer layers. But if the ET pro-

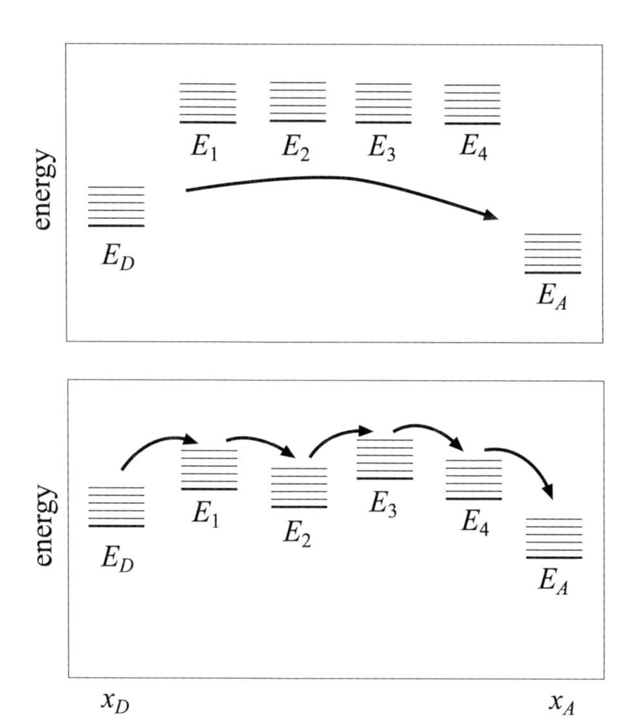


**Figure 6.9:** Bridge–mediated ET between a donor and an acceptor level connected by a linear chain of bridging units (thick lines – electronic levels, thin lines – vibrational levels). In the upper panel the bridge levels are energetically well separated from the donor and acceptor levels. In the lower panel a situation is shown where the energy levels of donor and acceptor are approximately resonant with the bridge levels.

ceeds as a bimolecular reaction the DA distance still enters as a random quantity. In order to avoid all these complications we will focus on the simpler case of intramolecular ET reactions in the following.

Next we discuss ET reactions which are beyond the two–state model used so far for the DA complex. For instance, various types of molecules or molecular building blocks can bridge the donor and the acceptor. If the ET proceeds directly from the donor to the acceptor, although some bridging units are separating them, the process is called *through–space* transfer. If some LUMOs of the bridge participate in the ET, the reaction is called *through–bond* transfer. The through–space transfer is only possible for DA distances less than 20 Å. (We will see below that this value is mainly determined by the overlap of the wave functions of the transferred electron in the reactant and product state.) The ET distance can become larger for the case of the through–bond transfer. This long–range ET is typical for conducting polymers or for the ET in proteins. (The ET distance in the photosynthetic reaction center, Fig. 6.2, amounts to 27 Å.) Through–bond ET is alternative named *bridge–assisted* ET (see Fig. 6.9):

$$D^-BA \rightarrow DB^-A \rightarrow DBA^- \ . \tag{6.6}$$

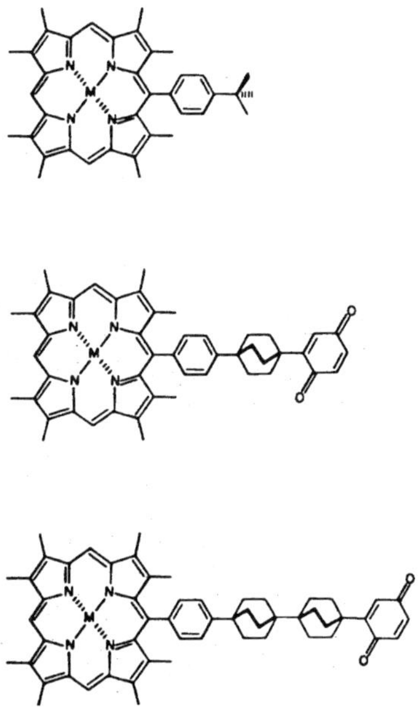

**Figure 6.10:** A Porphyrin quinone complex (M = Zn) with different bridging molecules as a typical example for the through–bond ET reaction is shown. The ET proceeds from the photoexcited zinc porphyrin (left part) to the quinone (right part), and the ET rate is determined according to $k_{\rm ET} = 1/\tau_{\rm obs} - 1/\tau_0$, where $\tau_{\rm obs}$ is the observed fluorescence lifetime and $\tau_0$ denotes the natural fluorescence lifetime of the porphyrin taken from the upper compound. Varying the solvent, $k_{\rm ET}$ equals $5 \times 10^9$ to $1.5 \times 10^{10}$ s$^{-1}$ for the middle compound and is $\leq 10^7$ s $^{-1}$ for the lower compound (reprinted with permission from [Lel85]; copyright (1985) American Chemical Society).

The electron moves from the donor to the acceptor via different bridge molecules (comprised here by the symbol B). Often, these bridging molecules are called *spacers* since they fix the donor and the acceptor at a particular distance one from another (this term is also common if the bridging molecules do not participate in the ET). Note that in contrast to the bimolecular ET, in the encounter complex the properties of the bridge, B, are rather well–defined as compared to $\mathrm{R} \cdots \mathrm{R}'$. There are two distinct mechanisms for bridge–mediated ET as shown in Fig. 6.9. The LUMOs of donor and acceptor may be either resonant or off–resonant with respect to the bridge levels. In the latter case it is reasonable to assume that there will be only a very small probability for population of these levels by the transferred electron.[3] This situation is called *superexchange* ET (upper panel of Fig. 6.9). Here the most important function of the bridge units is to provide a means for delocalization of the donor state wave function across the whole bridge. In the case of ET the charge jumps stepwise from one part to the other of the whole DBA chain. This process is often called *sequential* or *hopping* transfer. Obviously, superexchange as well as sequential ET are through–bond transfer reactions. Fig.

[3]Often one characterizes this very small population of the bridge state as being a virtual one.

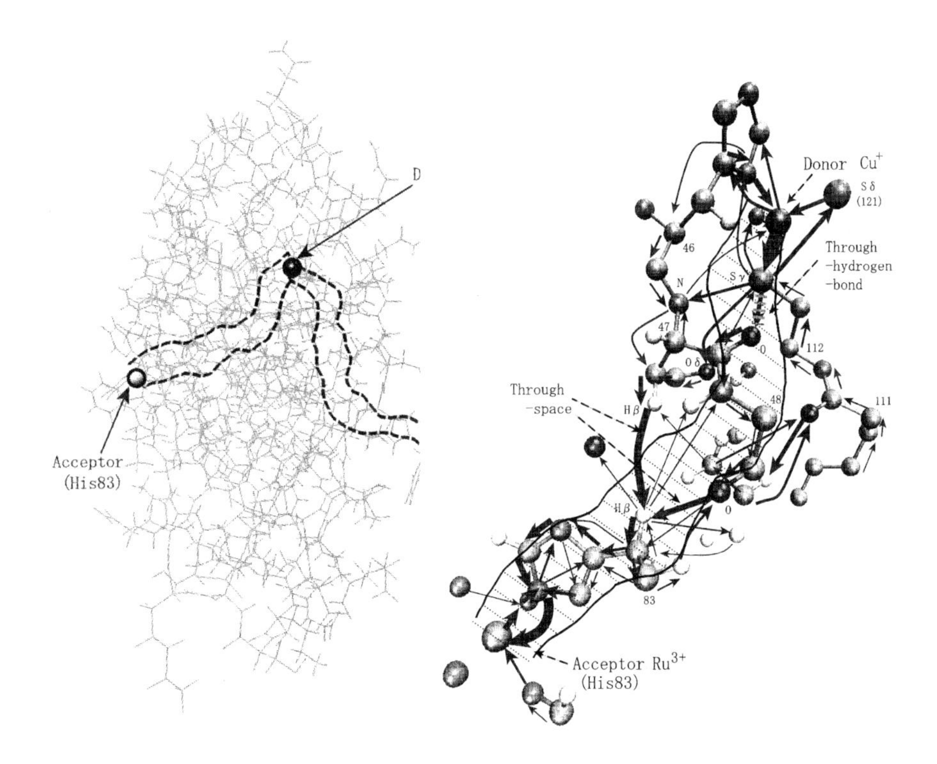


**Figure 6.11:** ET pathways in the blue copper protein azurin. There are six surface positions of azurin where a ruthenium complex can be coordinated. ET reactions are possible from the native copper atom (donor) to the ruthenium atom (acceptor, cf. left part). There exist different pathways for the electron to move from the donor to acceptor (cf. right part, the width of the arrows corresponds to the magnitude of the electronic coupling matrix elements). (reprinted with permission from [Kaw01]; copyright (2001) American Chemical Society)

6.10 displays a concrete example for bridge–mediated ET (for more details we also refer to Section 6.6). If the different bridge molecules are not positioned in a linear arrangement but form a three–dimensional network the electron may move on different *pathways* from the donor to the acceptor. Such a situation is typical for ET reactions in proteins (cf. Fig. 6.11 and [Reg99]).

# 6.2 Theoretical Models for Electron Transfer Systems

The derivation of the ET Hamiltonian proceeds in close analogy to the reasoning which led to the molecular Schrödinger equation in Section 2.3. The problem is split up into an electronic part at frozen nuclear configuration and a nuclear part. However, since there are a number of approximations which are special to the ET problem we will explain in some detail how to arrive at the electron–vibrational Hamiltonian governing the transfer of a single electron through a DA complex. The electronic Hamiltonian of the DA complex including possible bridging units will be denoted as $H_{\mathrm{el}}^{(\mathrm{DBA})}$ whereas the full Hamiltonian including vibrational contributions is written as $H_{\mathrm{DBA}}$.

### 6.2.1 The Electron Transfer Hamiltonian

Although ET comes along with the modification of many molecular orbitals, and thus has to be considered as a process in which different electrons take part, we will proceed here with a simple and intuitive picture. It is based on the notion of a single *excess* electron injected from the outside into the DA complex. The transfer of this excess electron will be described by introducing an effective potential experienced by the excess electron after entering the DA complex

$$V(\mathbf{r}) = \sum_m V_m(\mathbf{r}) \, . \tag{6.7}$$

The individual contributions $V_m(\mathbf{r})$ belong to the donor, the acceptor, or to some bridging molecules. (In the following, the bridging units are counted by $m = 1, \ldots, N_{\mathrm{B}}$, starting at the donor site, whereas the donor and acceptor are labeled by $m = D$ and $A$, respectively.)

The introduction of the effective potential $V(\mathbf{r})$ appears to be reasonable even though there is no unique way of separating it into the various $V_m(\mathbf{r})$. Only in the case of bimolecular ET reactions where independent molecules are involved the separation scheme is obvious. For unimolecular transfer one would relate the $V_m(\mathbf{r})$ to those fragments of the DA complex on which the excess electron is localized for an appreciable time.

Each contribution $V_m(\mathbf{r})$ can be understood as a so–called *pseudo–potential* which mimics the action of the total electronic system of the molecular fragment on the excess electron. Within this picture all exchange and correlation effects among the excess electron and the electrons of the molecule are replaced by a simple single–particle potential which is local in space. The techniques and approximation schemes for establishing these pseudo–potentials are provided by the theory of many–particle systems. Here, we define the various $V_m(\mathbf{r})$ by demanding that their ground state energy level $E_m$ should coincide with the electronic ground state of the isolated molecular unit *plus* the excess electron. The mentioned approach is *not* identical to a single–particle model which neglects any charge relaxation in the course of the excess electron motion. The excess electron does not move through an arrangement of frozen molecular orbitals. It is taken into account that the *full* many–electron wave function adjusts itself during the ET reaction. But this is done by reducing the many–particle dynamics to the action of an *effective* local single–particle potential.

Independent of the specific definition of the localized states, the single level treatment is only a good approximation if the next unoccupied orbital has a much higher energy. Otherwise, the number of unoccupied orbitals per pseudo–potential $V_m(\mathbf{r})$ has to be adjusted. The pseudo–potential $V_m(\mathbf{r})$ enters the single–electron Schrödinger equation which determines the single–particle energies $E_m$ and wave functions $\varphi_m(\mathbf{r})$:

$$(T_{\mathrm{el}} + V_m(\mathbf{r}))\, \varphi_m(\mathbf{r}) = E_m \varphi_m(\mathbf{r}) \, . \tag{6.8}$$

Again, only the lowest eigenvalue $E_m$ is of interest in the following, although higher–energetic solutions may exist. Since the energies $E_m$ correspond to different sites in the complex they

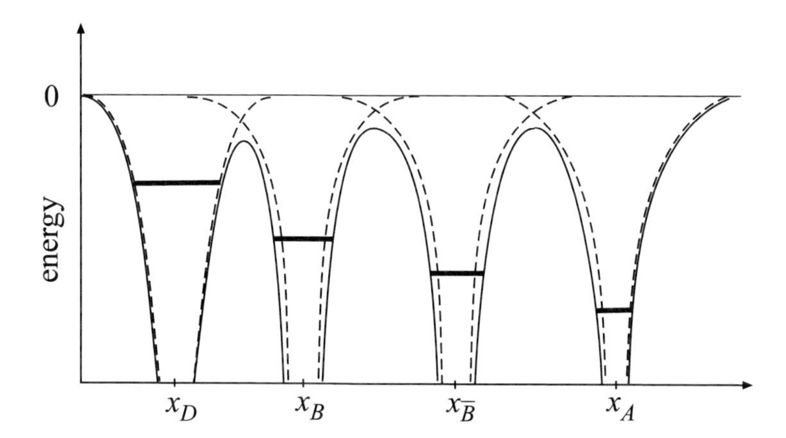


**Figure 6.12:** One–dimensional sketch of the pseudo–potential $V(\mathbf{r})$, introduced in Eq. (6.7) (full line). $x_D$, $x_B$, etc. mark the spatial position of the different units of the ET system. The pseudo–potentials $V_m(\mathbf{r})$ of the individual molecular units (broken lines) and the levels $E_m$ occupied by the excess electron are also shown. The motion of the excess electron among the various energy levels proceeds as a tunneling process through the barriers separating different potential wells.

are usually called *site energies*. The states $\varphi_m$ are reminiscent of the diabatic states introduced in Section 2.7. Hence, using these states as an expansion basis a *diabatic representation* is provided. (Whether these states are diabatic states in the strict sense of Section 2.7, that is, whether they minimize the nonadiabatic coupling, will not be discussed any further.) According to its definition the set of states $\varphi_m$ does not form a normalized and orthogonal basis, i.e., nonvanishing *overlap integrals* exist: $\langle\varphi_m|\varphi_n\rangle \neq \delta_{mn}$. If convenient we will write in the following $|D\rangle$, $|B\rangle$, and $|A\rangle$ instead of $|\varphi_D\rangle$, $|\varphi_m\rangle$, and $|\varphi_A\rangle$, respectively (here, $|B\rangle$ stands for a particular bridge state).

In order to construct the ET Hamiltonian we consider the electronic Schrödinger equation for the total DA complex. Since we assume spin–degeneracy of the considered excess electron states spin quantum numbers do not appear in the following. The total DA Schrödinger equation reads

$$(T_{\rm el} + V)|\phi\rangle = \mathcal{E}|\phi\rangle \ . \tag{6.9}$$

Let us expand this equation with respect to the basis set $|\varphi_m\rangle$:

$$|\phi\rangle = \sum_m c_m|\varphi_m\rangle \ . \tag{6.10}$$

Inserting this into Eq. (6.9) and multiplication by $\langle\varphi_n|$ from the left gives

$$\langle\varphi_n|T_{\rm el} + \sum_k V_k|\phi\rangle = \mathcal{E}\langle\varphi_n|\phi\rangle \ , \tag{6.11}$$

or

$$\sum_m c_m\Big(E_m\langle\varphi_n|\varphi_m\rangle + \sum_{k\neq m}\langle\varphi_n|V_k|\varphi_m\rangle\Big) = \mathcal{E}\sum_m c_m\langle\varphi_n|\varphi_m\rangle \ . \tag{6.12}$$

This set of equations contains the overlap integrals and the *three–center integrals* $\langle\varphi_n|V_k|\varphi_m\rangle$.[4]

Although a more general description is possible, in the following we will introduce two approximations. First, we assume that the two–center overlap integrals can be neglected, i.e., we set $\langle\varphi_n|\varphi_m\rangle \approx \delta_{nm}$. Within this approximation the set of states $\varphi_m$ forms an orthogonal basis.[5] Second, because of their smallness compared to the two–center integrals, all three–center integrals are neglected. We will only take into account one– and two–center integrals. The latter contain terms of the type $\langle\varphi_m|V_k|\varphi_m\rangle$, which introduce a shift of the site energies $E_m$ due to the presence of the pseudo–potential $V_k$ at site $k$. The other two–center integrals are of the type $\langle\varphi_n|V_n|\varphi_m\rangle$. This expression couples the state $|\varphi_m\rangle$ to the state $|\varphi_n\rangle$ via the tail of the potential $V_n$ at site $m$.

An expansion of the electronic part of the DA Hamiltonian gives

$$H_{\text{el}}^{(\text{DBA})} = \sum_{m,n} \langle\varphi_m|H_{\text{el}}^{(\text{DBA})}|\varphi_n\rangle|\varphi_m\rangle\langle\varphi_n| \tag{6.13}$$

with the matrix elements of the Hamiltonian given by

$$\begin{aligned}\langle\varphi_m|H_{\text{el}}^{(\text{DBA})}|\varphi_n\rangle &= \delta_{mn}\Big(E_m + \sum_{k\neq m}\langle\varphi_m|V_k|\varphi_m\rangle\Big)\\ &\quad +(1-\delta_{mn})\langle\varphi_m|T_{\text{el}} + V_m + V_n|\varphi_n\rangle\ .\end{aligned} \tag{6.14}$$

The off–diagonal part can be rewritten in different forms. We use the eigenvalue equation (6.8) and get

$$\begin{aligned}\langle\varphi_m|T_{\text{el}} + V_m + V_n|\varphi_n\rangle &= \frac{1}{2}\langle\varphi_m|(T_{\text{el}}+V_m) + (T_{\text{el}}+V_n) + (V_m+V_n)|\varphi_n\rangle\\ &= \frac{1}{2}\langle\varphi_m|E_m + E_n + (V_m+V_n)|\varphi_n\rangle\\ &= \frac{1}{2}\langle\varphi_m|V_m + V_n|\varphi_n\rangle = V_{mn}\ .\end{aligned} \tag{6.15}$$

The final expression $V_{mn}$ is usually called *transfer integral* or alternatively *inter–state coupling*. Since the motion of the electron through the DA complex proceeds via tunneling processes, the term *tunneling* matrix element is also common. Alternatively to Eq. (6.15) one can also write $V_{mn}$ $(m \neq n)$ in terms of the matrix element of the kinetic energy operator:

$$\begin{aligned}V_{mn} &= \langle\varphi_m|T_{\text{el}} + V_m + T_{\text{el}} + V_n - T_{\text{el}}|\varphi_n\rangle\\ &= \langle\varphi_m|E_m + E_n - T_{\text{el}}|\varphi_n\rangle = -\langle\varphi_m|T_{\text{el}}|\varphi_n\rangle\ .\end{aligned} \tag{6.16}$$

The complete electronic Hamiltonian for the DA complex reads

$$H_{\text{el}}^{(\text{DBA})} = \sum_m E_m|\varphi_m\rangle\langle\varphi_m| + \sum_{m,n} V_{mn}|\varphi_m\rangle\langle\varphi_n|\ . \tag{6.17}$$

Here we included the diagonal matrix elements of the pseudo–potentials into the definition of the site energies $E_m$.[6]

[4]The integrand has contributions from sites $n$, $k$, and $m$, i.e., from three different spatial positions.

[5]Of course, a transformation to a new set of states may remove the overlap integrals. But in this case the intuitive picture of the states $|\varphi_m\rangle$ is lost.

[6]In the solid–state physics literature the electronic part of the full DA complex Hamiltonian (6.17) is often called *tight–binding* Hamiltonian. This emphasizes the strong coupling of the electrons to the various localization sites.

The construction of the Hamiltonian, Eq. (6.17) was mainly based on the concept of a single excess electron moving in a particular spatial arrangement of pseudo–potentials which refer to the donor, the bridge, and the acceptor. But a derivation would also be possible if a many–electron generalization of the wave functions for the excess electron $\varphi_m(\mathbf{r})$ could be achieved. Let $\Phi(r,\sigma)$ be the full many–electron wave function of the neutral DA complex in its ground state (with the set of spatial and spin coordinates $r$ and $\sigma$, respectively). Then, we assume the existence of the wave function $\Phi_m^{(-)}(r,\sigma;\mathbf{r}_{\rm exc})$ referring to the DA complex *plus* the excess electron which is in a diabatic state localized at site $m$. The expansion of some suitable many–electron generalization of the DA Hamiltonian $H_{\rm el}^{(\rm DBA)}$ (whose definition starts from the general expressions of Section 2.2) then gives the respective diabatic energies $E_m$ in the diagonal parts.

Although this many–electron extension seems to be simple, it essentially depends on the proper definition of the states $\Phi_m^{(-)}(r,\sigma;\mathbf{r}_{\rm exc})$. In particular one has to clarify how to define these states as diabatic states of the total DA complex, as well as how to separate the DA complex into isolated units and to define the localized excess electron states for these units. We will not further comment on these more involved issues but refer the reader to the literature listed in the Suggested Reading section below.

In a similar way it becomes possible to construct the Hamiltonian which describes photoinduced ET (cf. Fig. 6.6). Here, we have to assume the existence of the many–electron wave functions $\Phi_m(r,\sigma)$ which correspond to the excited donor state as well as the presence of the transferred electron at the bridge units and the acceptor. One may argue that for those states where the electron already left the donor the Coulomb interaction should be accounted for between the donor without one electron ($D^+$) and the other parts of the DA complex with an additional electron (e.g. $A^-$). But often one interprets the states $\Phi_m$ (and the related energies $E_m$) as defined in such a manner that this Coulomb interaction is already contained in their definition. Of course, all these states are not eigenstates of the electronic part of the molecular Hamiltonian. They describe spatial charge localization in the DA complex and should be coupled weakly one to another. (If this latter restriction is not fulfilled the introduction of the $\Phi_m$ becomes meaningless.) Expanding the electronic part $H_{\rm el}^{(\rm DBA)}$ of the DA Hamiltonian with respect to these states $\Phi_m$ we again arrive at an expression as given by Eq. (6.17). This indicates the universal form of the ET Hamiltonian, Eq. (6.17). Only the actual interpretation of the involved matrix elements and expansion states specifies $H_{\rm el}^{(\rm DBA)}$ to a concrete type of ET reaction.

## 6.2.2 The Electron–Vibrational Hamiltonian of a Donor–Acceptor Complex

The vibrational degrees of freedom may couple in two different ways to the transferred electron as has been already explained for the case of nonadiabatic transitions in Section 5.7. In principle one can distinguish between *accepting modes* which change their equilibrium configuration if the electronic charge density changes, and *promoting modes*. The latter enter the transfer integral, Eq. (6.15) and thus may accelerate the ET. (We note that in general this distinction is not always clear, i.e., accepting modes may act simultaneously as promoting modes and vice versa.)

Including the vibrational degrees of freedom, $\{R_u\} \equiv R$ (note that we use the index $u$ instead of $n$ as in Chapter 2 to avoid confusion with the site indices), the complete Hamiltonian of the DA complex becomes

$$\begin{aligned} H_{\text{DBA}} &= H_{\text{el}}^{(\text{DBA})}(R) + T_{\text{nuc}} + V_{\text{nuc-nuc}}(R) \\ &= \sum_m (T_{\text{nuc}} + E_m(R) + V_{\text{nuc-nuc}}(R)) + \Theta_{mm}|\varphi_m\rangle\langle\varphi_m| \\ &+ \sum_{m\neq n} (V_{mn}(R) + \Theta_{mn})|\varphi_m\rangle\langle\varphi_n| \,. \end{aligned} \tag{6.18}$$

Now, the $E_m$ and $V_{mn}$ which belong to the electronic part of the complete Hamiltonian have been labeled by their dependence on the vibrational coordinates.[7] $T_{\text{nuc}}$ gives the kinetic energy of all the vibrations coupled to the ET reaction and $V_{\text{nuc-nuc}}(R)$ results from the coupling among the vibrational degrees of freedom (electrostatic coupling among the nuclei, see Section 2.2). The off–diagonal part includes the nonadiabaticity operator $\Theta_{mn}$ defined in Eq. (2.17). According to the specific form of $H_{\text{DBA}}$ we can introduce PES which relate to those state with the excess electron localized at site $m$:

$$U_m(R) = E_m(R) + V_{\text{nuc-nuc}}(R) + \Theta_{mm} \,. \tag{6.19}$$

For the following we assume that the nonadiabatic coupling is small and neglect its contribution to the off–diagonal part of Hamiltonian Eq. (6.18). This assumption is motivated by the localization of the wave functions $\varphi_m(\mathbf{r})$ at the various units of the DA complex (its diabatic character). Then the total electron–vibrational Hamiltonian is obtained as

$$H_{\text{DBA}} = \sum_m (T_{\text{nuc}} + U_m(R))\,|\varphi_m\rangle\langle\varphi_m| + \sum_{m\neq n} V_{mn}(R)|\varphi_m\rangle\langle\varphi_n| \,. \tag{6.20}$$

The dependence on the nuclear coordinates can be made more concrete by introducing PES which depend on normal mode coordinates $\{q_\xi\} \equiv q$ (see Section 2.6.1). In this case it is advantageous to choose a particular electronic state to define a reference configuration of the nuclei. We take the state of the neutral DA complex for that purpose, that is, the state where the excess electron is absent. This state is supposed to be characterized by the PES $U_0(R)$ having the equilibrium configuration at $\{R_u^{(0)}\} \equiv R^{(0)}$. Next we carry out an expansion of $U_0(R)$ around $R^{(0)}$ up to the second order with respect to the deviations $\Delta R_u^{(0)} = R_u - R_u^{(0)}$ and obtain

$$U_0(R) = U_0^{(0)} + \frac{1}{2}\sum_{u,u'} \kappa_{uu'}^{(0)} \Delta R_u^{(0)} \Delta R_{u'}^{(0)} \,. \tag{6.21}$$

Here, we used $U_0(R^{(0)}) = U_0^{(0)}$, and the $\kappa_{uu'}^{(0)}$ denote the second derivatives of the PES (Hessian matrix). As discussed in Section 2.6.1 the introduction of (mass-weighted) normal

[7] There is also a dependence of the expansion states $\varphi_m$ on the vibrational coordinates. But this dependence has been already accounted for by the introduction of the nonadiabaticity operator it can be neglected when considering the $\varphi_m$.

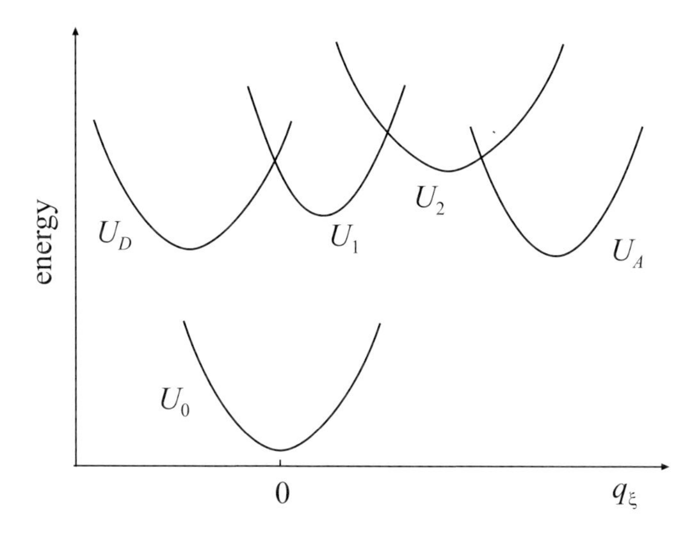


**Figure 6.13:** PES of the DA complex according to Eq. (6.26) versus a single normal mode coordinate $q_\xi$ (all other coordinates with $\xi' \neq \xi$ are fixed at $q_{\xi'}^{(m)}$ for every PES $U_m$). While $U_0$ is the reference PES of the neutral complex, the PES $U_m$ correspond to the situation where one excess electron is present at the donor ($m = D$), the acceptor ($m = A$) or at a bridge unit ($m = 1, 2$, note that the position of the PES along the $q_\xi$–axis has nothing to do with the spatial position of the related electronic wavefunctions $\varphi_m$).

mode coordinates according to the linear transformation

$$\Delta R_u^{(0)} = \sum_\xi M_u^{-1/2} A_{u\xi}^{(0)} q_\xi \ . \tag{6.22}$$

leads to a diagonalization of the vibrational Hamiltonian with the potential energy given by

$$U_0(R) \equiv U_0(q) = U_0^{(0)} + \frac{1}{2} \sum_\xi \omega_{0,\xi}^2 q_\xi^2 \ . \tag{6.23}$$

Next we consider the PES $U_m(R)$ relevant if the excess electron is localized at site $m$. Let us suppose that this electronic state can be described by the same normal mode coordinates (Eq. (6.22)). This means that the quadratic form resulting from the expansion of the PES $U_m(R)$ with respect to the Cartesian nuclear coordinates can be also diagonalized by the same transformation matrix $A_{u\xi}^{(0)}$. The expansion of the PES $U_m(R)$ reads

$$U_m(R) = U_m^{(0)} + \frac{1}{2} \sum_{u,u'} \kappa_{uu'}^{(m)} \Delta R_u^{(m)} \Delta R_{u'}^{(m)} \ , \tag{6.24}$$

with $U_m(R^{(m)}) = U_m^{(0)}$ and $\Delta R_u^{(m)} = R_u - R_u^{(m)}$. Since

$$\begin{aligned} \Delta R_u^{(m)} &= (R_u - R_u^{(0)}) - (R_u^{(m)} - R_u^{(0)}) \\ &= \sum_\xi M_u^{-1/2} (A_{u\xi}^{(0)} q_\xi - A_{u\xi}^{(0)} q_\xi^{(m)}) \ , \end{aligned} \tag{6.25}$$

where $q_\xi^{(m)}$ corresponds to the linear transformation of the difference $R_u^{(m)} - R_u^{(0)}$, we get (see also Section 2.6.1)

$$U_m(q) = U_m^{(0)} + \frac{1}{2}\sum_\xi \omega_{m,\xi}^2 (q_\xi - q_\xi^{(m)})^2 . \tag{6.26}$$

In this manner we have been able to introduce a particular model for the PES of the DA system. The PES are parabolic, their minima are shifted with respect to each other in the space of the normal mode coordinates, and they have different energetic offsets (see Fig. 6.13). If necessary, they can additionally be characterized by vibrational frequencies depending on the site index $m$. The related vibrational Hamiltonian reads

$$H_m(q) = T_{\text{vib}} + U_m(q) = U_m^{(0)} + \frac{1}{2}\sum_\xi \left\{ p_\xi^2 + \omega_{m,\xi}^2 \left(q_\xi - q_\xi^{(m)}\right)^2 \right\} . \tag{6.27}$$

In the general case the inter–site couplings $V_{mn}$ also depend on the nuclear coordinates. Since the magnitude of $V_{mn}$ is mainly determined by the overlap of the exponential tail of the wave functions localized at sites $m$ and $n$ (see Eq. (6.15)), we expect an exponential dependence on the distance $x_{mn}$ between the two sites:

$$V_{mn}(R) = V_{mn}^{(0)} \exp\left\{-\beta_{mn}(x_{mn} - x_{mn}^{(0)})\right\} . \tag{6.28}$$

The reference value $V_{mn}^{(0)}$ of the inter–site couplings is reached for the reference (equilibrium) distance $x_{mn}^{(0)}$ and $\beta_{mn}$ is some characteristic inverse length determined by the wave function overlap. Often the dependence of $V_{mn}$ on the nuclear coordinates is neglected in comparison with the on–site vibrational dynamics. We will use this simplification in the following and set $V_{mn}(R) \approx V_{mn}$ (Condon approximation).

Using this model the total Hamiltonian (6.20) becomes

$$H_{\text{DBA}} = \sum_{m,n} \{\delta_{mn} H_m(q) + (1 - \delta_{mn}) V_{mn}\} |\varphi_m\rangle\langle\varphi_n| . \tag{6.29}$$

The simplest but non–trivial version of the DA Hamiltonian, Eq. (6.29), is obtained if one neglects any bridging unit and if the vibrational frequencies are independent of the actual electronic state. We then have

$$H_{\text{DA}} = H_{\text{D}}|D\rangle\langle D| + H_{\text{A}}|A\rangle\langle A| + V_{DA}|D\rangle\langle A| + V_{AD}|A\rangle\langle D| . \tag{6.30}$$

There exists a widely used alternative notation, which employs the formal similarity of a two–level system with a spin one–half system. In analogy to the quantum mechanical treatment of the spin we define the spin–operator components $\sigma_x$, $\sigma_y$, and $\sigma_z$ as

$$\begin{aligned} \sigma_x &= |D\rangle\langle A| + |A\rangle\langle D| \\ \sigma_y &= i\,(|D\rangle\langle A| - |A\rangle\langle D|) \\ \sigma_z &= |D\rangle\langle D| - |A\rangle\langle A| . \end{aligned} \tag{6.31}$$

Furthermore, we write the Hamiltonian (6.30) in a way similar to Section 2.6.2 where we introduced a reference vibrational Hamiltonian $H_{\rm vib}$ and a linear electron–vibrational coupling with dimensionless vibrational coordinates $q_\xi \sqrt{2\omega_\xi/\hbar} = C_\xi^+ + C_\xi$ and coupling constants $g_m(\xi)$

$$\begin{aligned} H_{\rm DA} &= U_{\rm D}^{(0)}|D\rangle\langle D| + U_{\rm A}^{(0)}|A\rangle\langle A| + H_{\rm vib} \\ &+\sum_\xi \hbar\omega_\xi(C_\xi^+ + C_\xi)\big(g_D(\xi)|D\rangle\langle D| + g_A(\xi)|A\rangle\langle A|\big) \\ &+V_{DA}|D\rangle\langle A| + V_{AD}|A\rangle\langle D| \,. \end{aligned} \tag{6.32}$$

If we introduce

$$U_{\rm D}^{(0)} = \bar{U}^{(0)} + \varepsilon/2 \,, \tag{6.33}$$

and

$$U_{\rm A}^{(0)} = \bar{U}^{(0)} - \varepsilon/2 \,, \tag{6.34}$$

with $\bar{U}^{(0)} = (U_{\rm D}^{(0)} + U_{\rm A}^{(0)})/2$, $\varepsilon = U_{\rm D}^{(0)} - U_{\rm A}^{(0)}$, assume that the shift of the two PES along the $q_\xi$–axis is symmetric ($g_D(\xi) = -g_A(\xi) = g_\xi$), and take a real transfer coupling $V = V_{DA} = V_{AD}$, the Hamiltonian (6.32) becomes

$$H_{\rm sb} = \bar{U}^{(0)} + H_{\rm nuc} + \frac{\varepsilon}{2}\sigma_z + \sum_\xi \hbar\omega_\xi g_\xi(C_\xi^+ + C_\xi)\sigma_z + V\sigma_x \,. \tag{6.35}$$

(Note that $|D\rangle\langle D| + |A\rangle\langle A|$ defines the completeness relation for the electronic states which can be replaced by the unit operator.)

According to this prescription the electronic two–state system (the acceptor and the donor) has been mapped onto the problem of an effective spin one–half particle. Eq. (6.35) gives the so–called *spin–boson* Hamiltonian. The term "boson" indicates that the equilibrium distribution of vibrational energy quanta for the various normal modes follows Bose–Einstein statistics. The spin–boson model represents the archetype to study the interplay of particle (electron) transfer and vibrational motion [Leg87,Wei98].

### 6.2.3 Two Independent Sets of Vibrational Coordinates

In the foregoing considerations we assumed that there exists a common set of normal mode coordinates modulating the donor as well as acceptor electronic states. In the case of a bimolecular ET reaction, however, the electron moves between two independent molecules. Therefore, one needs separate sets of coordinates for the donor ($q_D \equiv \{q_{D\xi}\}$) and the acceptor ($q_A \equiv \{q_{A\xi}\}$) molecule. For simplicity we consider the reaction scheme (6.5) without additional bridging units and introduce four different electronic states for the expansion of the Hamiltonian. They will be denoted by $|\varphi_{D-}\rangle$, $|\varphi_D\rangle$, $|\varphi_A\rangle$, and $|\varphi_{A-}\rangle$, and describe the donor with the excess electron, the neutral donor, the neutral acceptor, and the acceptor plus the excess electron, respectively. Thus, the reactant state is given by $|\varphi_{D-}, \varphi_A\rangle$, whereas the product state is $|\varphi_D, \varphi_{A^-}\rangle$. Generalizing Eq. (6.30) the Hamiltonian for a bimolecular ET reaction follows as

$$H_{\rm DA}^{\rm (bimol)} = [H_{D^-}(q_D) + H_A(q_A)]\,|\varphi_{D^-}, \varphi_A\rangle\langle\varphi_{D^-}, \varphi_A| + \big(H_D(q_D) + H_{A^-}(q_A)\big)|\varphi_D, \varphi_{A^-}\rangle\langle\varphi_D, \varphi_{A^-}| + \big(V_{DA}|\varphi_{D^-}, \varphi_A\rangle\langle\varphi_D, \varphi_{A^-}| + \text{h.c.}\big) \,. \tag{6.36}$$

All PES are of the type Eq. (6.26), but here we have four sets of equilibrium configurations $\{q_{m\xi}^{(0)}\}$ with $m = D^-, D, A, A^-$ and four corresponding values of $U_m^{(0)}$.

### 6.2.4 State Representation of the Hamiltonian

For further applications it is useful to give a representation of the DA Hamiltonian using the complete diabatic electron–vibrational basis defined by the states (cf. Section 2.7)

$$|\mu\rangle \equiv |mM\rangle \equiv |\varphi_m\rangle|\chi_{mM}\rangle \,. \tag{6.37}$$

The vibrational states $|\chi_{mM}\rangle$ which belong to the electronic states $|\varphi_m\rangle$ are the eigenstates of the vibrational Hamiltonian, Eq. (6.27),

$$H_m|\chi_{mM}\rangle = E_{mM}|\chi_{mM}\rangle \,. \tag{6.38}$$

The respective eigenvalues

$$E_{mM} = U_m^{(0)} + \sum_\xi \hbar\omega_{m\xi}(M_\xi + 1/2) \tag{6.39}$$

give the energy spectrum of the normal mode oscillators. According to the introduction of normal mode vibrations the state vector $|\chi_{mM}\rangle$ factorizes into products corresponding to the different normal modes with mode index $\xi$

$$|\chi_{mM}\rangle = \prod_\xi |\chi_{mM_\xi}\rangle \,. \tag{6.40}$$

The Hamiltonian (6.18) can be expanded in the diabatic electron–vibrational basis as follows

$$H_{\rm DBA} = \sum_{\mu\nu} \Big(\delta_{\mu\nu}E_\mu + (1-\delta_{mn})V_{\mu\nu}\Big)|\mu\rangle\langle\nu| \,. \tag{6.41}$$

If the transfer integral $V_{mn}$ is coordinate independent the coupling matrix element follows as $V_{\mu\nu} = V_{mn}\langle\chi_{mM}|\chi_{nN}\rangle$, where $\langle\chi_{mM}|\chi_{nN}\rangle$ is the overlap integral of the vibrational wave functions belonging to different sites (Franck–Condon factor).[8]

The total DA system described by the Hamiltonian (6.41) can be viewed as a set of multi–level systems with energy spectrum $E_\mu$ and mutual level coupling $V_{\mu\nu}$. A similar system has already been considered in Section 3.4.5 from a more formal point of view. There, we calculated the total transition rate for the transfer of occupation probability from site $m$ to site $n$. Here, we expect a similar relation for the description of ET reactions. Under what precise conditions this rate formula is valid will be discussed in detail below.

[8]Obviously, this approximation is identical to the Condon approximation introduced for computing the absorption spectrum in Chapter 5.

## 6.3 Regimes of Electron Transfer

In the forthcoming discussion we will concentrate on a simple model of a DA complex neglecting any further bridging units. The respective Hamiltonian has been introduced in Eq. (6.30) where the diabatic donor and acceptor electronic states have been abbreviated by $|D\rangle \equiv |\varphi_D\rangle$, and $|A\rangle \equiv |\varphi_A\rangle$, respectively (see also Fig. 6.14). For the time being it will be convenient to consider the vibrational degrees of freedom as classical quantities.

It is obvious from the previous section that one of the crucial parameters of ET theory should be the inter–site coupling. For the considered electronic two–state model it is a simple task to consider this coupling exactly, what means to change to the *adiabatic* representation as an alternative to the diabatic representation. Therefore, we start our discussion by introducing this representation. It is obtained after diagonalization of the Hamiltonian (6.30). According to the results for a two–level system (see Section 2.8.3) and the fact that the vibrational kinetic energies are not affected[9], we obtain the two adiabatic PES $U_+$ and $U_-$ as (compare Section 2.7)

$$U_\pm(q) = \frac{1}{2}\left(U_D(q) + U_A(q) \pm \sqrt{(U_D(q) - U_A(q))^2 + 4|V_{DA}|^2}\right) . \tag{6.42}$$

These adiabatic PES together with the diabatic PES are plotted in Fig. 6.14 versus a single coordinate $q$. The crossing point $q^*$ of the two diabatic PES is defined by $U_D(q^*) = U_A(q^*)$. According to Eq. (6.42) there is a splitting of the adiabatic PES by $2|V_{DA}|$ at the crossing point. The splitting becomes smaller if $q$ deviates from $q^*$ and the adiabatic and diabatic curves coincide for $|q - q^*| \gg 0$. Clearly, the shape of the adiabatic PES is much more complicated if two or more vibrational coordinates are involved.

Which type of representation is more appropriate depends on the problem under discussion. To give some guidance using *quantitative* arguments we introduce two different characteristic times. A time typical for electronic quantum motion is

$$t_{\text{el}} = \frac{\hbar}{|V_{AD}|} . \tag{6.43}$$

This quantity is proportional to the time the electronic wave function needs to move from the donor site to the acceptor site if the respective energy levels are degenerate (cf. the discussion of the two–level system dynamics in Section 3.8.7). The degeneracy of the electronic levels occurs if one fixes the vibrational configuration at the crossing point $q^*$. Provided that the vibrational motion is not effected by strong damping the characteristic time of the vibrational motion is given by the vibrational frequency[10]

$$t_{\text{vib}} = \frac{2\pi}{\omega_{\text{vib}}} . \tag{6.44}$$

---

[9]The nuclear kinetic energy operators enter as $T_{\text{nuc}}|B\rangle\langle B| + T_{\text{nuc}}|A\rangle\langle A|$. According to the completeness relation for the electronic states this is identical with $T_{\text{nuc}}$, what demonstrates that the kinetic energy part remains unaffected by the transformation.

[10]Usually molecular systems will have vibrational modes with different vibrational frequencies. Then $\omega_{\text{vib}}$ represents a mean frequency. Clearly, if the various frequencies are quite different, the introduction of a mean frequency is meaningless. Instead, one can use different groups of frequencies (high–frequency and low–frequency vibrations etc.). Of course, particular relations valid for one type of frequencies may be invalid for the other type.

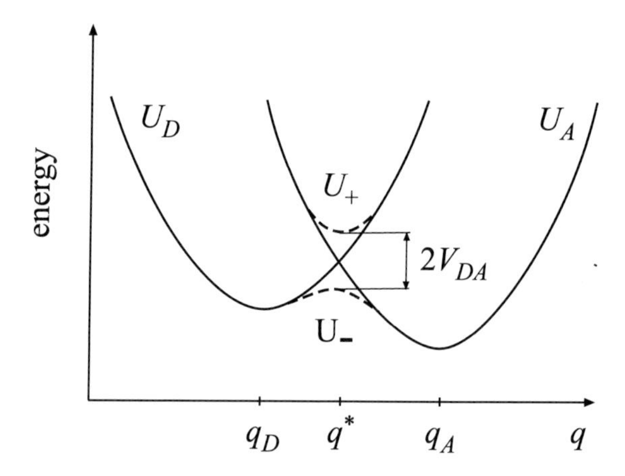


**Figure 6.14:** Donor and acceptor PES versus a single reaction coordinate. The diabatic (full line) as well as adiabatic curves (dashed line) are shown. There is a splitting between the adiabatic curves which has a magnitude of $2|V_{DA}|$ at the crossing point $q^*$.

Let us first assume $t_{\rm el} \ll t_{\rm vib}$. In this case the electron will move many times between the donor and acceptor before any change of the nuclear configuration occurs. This is just the situation we used in Section 2.3 to motivate the introduction of the Born–Oppenheimer (adiabatic) approximation. We expect that the electronic states will be delocalized over the whole DA complex. The electron is in an *adiabatic* state, and if one is interested in a time scale much larger than $t_{\rm el}$ it becomes advantageous to change from the localized diabatic to the delocalized adiabatic representation. In particular, any vibrational motion has to be described within the adiabatic PES. Note that for the case that the vibrational motion triggers electronic transitions, a quantum mechanical treatment including the nonadiabaticity operator may be required (cf. Eq. (6.18)).

If the energetic difference between the lower and the upper adiabatic PES is large enough one has the situation shown in Fig. 6.4 where the motion along the reaction coordinate is subject to a double–well potential. Now, we can specify Fig. 6.4 noting that for adiabatic ET the formal reaction coordinate can be identified with some – possibly collective – vibrational coordinate coupled to the ET. Therefore, adiabatic ET has to be understood as the rearrangement of the vibrational degrees of freedom from their reactant configuration (minimum of diabatic donor PES) to the product configuration (minimum of diabatic acceptor PES). This rearrangement is connected with a barrier crossing, and we expect for the ET rate an expression of the standard Arrhenius type

$$k_{\rm ET} \propto e^{-E_{\rm act}/k_{\rm B}T} , \tag{6.45}$$

with the respective activation energy $E_{\rm act}$.[11]

[11]The use of the term activation *energy* becomes inadequate if there is a macroscopic number of nuclear degrees of freedom involved. For example, in the case of ET in polar solvents entropic effects enter the description and one has to replace the activation energy by the activation *free energy*.

The opposite situation is encountered, if $t_{\rm el} \gg t_{\rm vib}$, i.e., if the vibrational motion is much faster than the electronic one. This reaction type is called *nonadiabatic* ET. (This should not be confused with the "nonadiabatic" coupling.) The initial and final states of the nonadiabatic ET reaction are spatially rather localized, and the motion of the reaction coordinate through the crossing region is so fast that the electronic wave function has not enough time to move completely from the donor to the acceptor. Only a small fraction of electronic probability density will reach the donor state.[12]

Since the coupling $V_{DA}$ is small, it is possible to describe the ET carrying out a perturbation expansion with respect to $V_{DA}$ where the diabatic states represent the zeroth–order states. In the lowest order of perturbation theory (Golden Rule formula) ET occurs, if the donor and acceptor levels are degenerated, that is, in the crossing region of the two PES. The transfer rate becomes proportional to $|V_{DA}|^2$, but it also depends on the probability at which the crossing region on the donor PES $U_D$ is reached by the vibrational coordinates. Accordingly, we expect the following expression for the ET rate

$$k_{\rm ET} \propto |V_{DA}|^2 \, e^{-E_{\rm act}/k_{\rm B}T} \, . \tag{6.46}$$

Following Fig. 6.14, $E_{\rm act}$ denotes the activation energy needed to enter the crossing region starting at the minimum position of the donor PES, hence we have $E_{\rm act} = U_D(q^*) - U_D(q^{(D)})$. Of course, this activation energy is different from the one appearing in the case of the adiabatic ET, since the latter has been introduced with respect to the barrier in the lower adiabatic PES $U_-$, Eq. (6.42).

Although the two types of ET introduced so far are the result of very different values of the two characteristic times $t_{\rm el}$ and $t_{\rm vib}$, the adiabatic as well as the nonadiabatic ET can cover a wide range of time scales up to milliseconds or even slower. On the other hand, if there is a strong DA coupling, ET reactions can proceed ultrafast (in the picosecond or even femtosecond region). This situation is usually encountered in photoinduced ET reactions (see below). Here, experimental observation requires to use ultrafast preparation and detection schemes as well (cf. discussion in Chapter 5).

The foregoing discussion was based on characteristic times for the electronic and vibrational motion. Alternatively, one can introduce characteristic energies. Let us concentrate on the model of an excess electron. If the excess electron is absent the minimum position of the PES $U_0$ is given by $q_\xi = 0$ (compare Eq. (6.23)). If the excess electron is introduced into the complex and its wave function is localized at the state $m = D, A$ the *localization* energy

$$E_{\rm loc} = \frac{1}{2} \sum_\xi \omega_\xi^2 q_\xi^{(m)\,2} \tag{6.47}$$

is gained. On the contrary, if the vibrational coordinates are fixed at the crossing point such that the electronic wave function becomes delocalized, the system may gain the *delocalization*

[12] It should be noted here that we use the term "electronic probability density" instead of "electron" indicating that quantum mechanics only fixes the change of the wave function in the course of the time propagation. It is meaningless to ask how fast or slow the electron itself moves within the ET reaction. The reader should also note the similarity between the nonadiabatic ET and the electronic transition occurring in a linear absorption experiment. There the weak transfer coupling of ET is replaced by the weak external electromagnetic field, both realizing an inter–state coupling (cf. discussion in Chapter 5).

energy (according to the energetic splitting between the D and the A level)

$$E_{\rm del} = |V_{AD}| \,. \tag{6.48}$$

According to the definition of the characteristic energies we can conclude that nonadiabatic ET occurs if $E_{\rm loc} \gg E_{\rm del}$. If it is energetically more favorable for the electron to be in a delocalized state ($E_{\rm loc} \ll E_{\rm del}$) the ET is adiabatic.

Let us return to ET rate formulas whose limiting cases, the adiabatic and the nonadiabatic ET, were estimated above. Both ET rates have been characterized by formulas of type

$$k_{\rm ET} = \nu e^{-E_{\rm act}/k_{\rm B}T} \,. \tag{6.49}$$

This rate includes the activation energy for barrier crossing and the quantity $\nu$. The latter has the dimension of a frequency and is usually called *frequency factor*. In the case of adiabatic ET (which has also been studied in the framework of *Kramers*–theory [Hae90]) the inverse of the frequency factor is simply given by the time it takes to move along the reaction coordinate over the top of the barrier. A more detailed inspection shows that this reasoning is not correct for every type of adiabatic ET. For instance, consider a DA complex dissolved in a polar solvent. Here, the transferred electron may be strongly coupled to the solvent and the motion of the reaction coordinate of the ET is overdamped. The rate will be mainly determined by the way the reaction coordinate reaches the crossing point. In particular, the frequency factor $\nu$ becomes proportional to $1/\tau_{\rm rel}$, the inverse of the solvent relaxation time (more precisely, to the inverse of the longitudinal dielectric relaxation time, for more details see Section 6.5). This type of ET is often called *solvent–controlled.* If the nuclear coordinates (and solvent degrees of freedom) which are coupled to the ET reaction move in a manner such that the reaction coordinate is only weakly perturbed the motion is called *uniform.* If, however, there is a strong perturbation, the motion on the respective PES becomes irregular *diffusion*–like.

This example shows that it is of importance to understand how one can formulate a theory for ET reactions which is not only valid for the two described limiting cases but also in the intermediate regime. To bridge the gap between the nonadiabatic ET and the adiabatic ET with uniform reaction coordinate dynamics *Landau–Zener* theory is appropriate as will be explained in the next section. In passing we note that the transition from the solvent–controlled ET to the uniform adiabatic reaction dynamics as well as to the nonadiabatic ET reaction has also been investigated [Bix99a].

The classical consideration of the vibrational motion assumes for the ET a thermal activation of the vibrational degrees of freedom to reach the crossing region. If the temperature decreases such that $k_{\rm B}T \ll E_{\rm act}$, ET has to proceed via tunneling through the barrier between the donor and acceptor nuclear equilibrium configurations. This so–called *nuclear tunneling* case can be found in nonadiabatic as well as adiabatic ET reactions. It requires a quantum mechanical consideration of the vibrational coordinates.

In the case of nonadiabatic ET one can use the Hamiltonian, Eq. (6.41) with energy levels $E_{mM}$. Considering a DA complex without further intervening bridge units and neglecting the coupling to any environment, the system of the two sets of energy spectra, $E_{DM}$ and $E_{AN}$ represents a closed quantum system and reversible quantum dynamics has to be expected in this multi–level systems (cf. Fig. 6.15). In the presence of an environment, one has an open system and every electron–vibrational state $|mM\rangle$ has a *finite* lifetime $\tau_{mM}$. For simplicity we

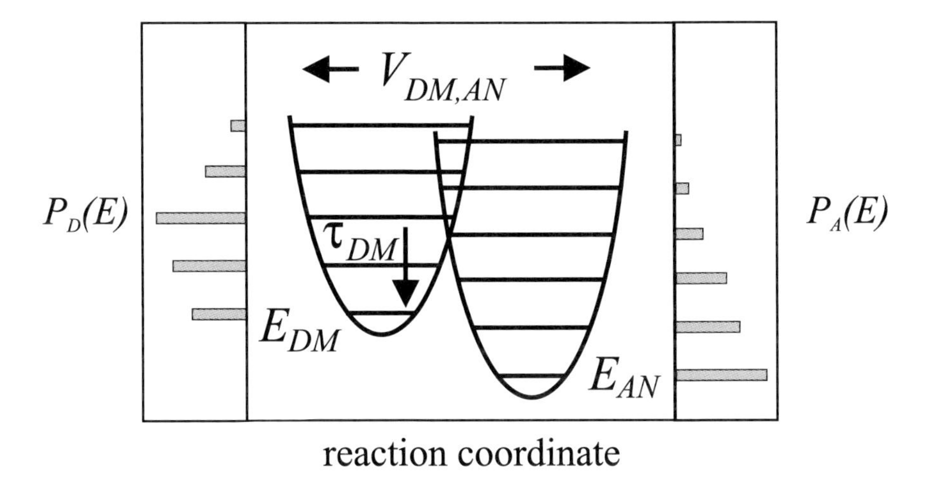


**Figure 6.15:** Ultrafast ET in a system of two coupled PES with donor vibrational levels $E_{DM}$ and acceptor vibrational levels $E_{AN}$ (the coupling matrix elements $V_{DM,AN}$ are also drawn). Left scheme: population $P_D$ of the donor levels after optical excitation (cf. Fig. 5.12), right scheme: population $P_A$ of the acceptor levels after relaxation took place. (If both spectra are degenerated a direct transfer from a selected level $E_{DM}$ to a level $E_{AN}$ becomes possible, probably connected with a back transfer. If degeneracy is absent a set of different levels is coupled simultaneously.)

will assume the existence of a single representative lifetime $\tau_{\text{rel}}$ for the following discussion. If $\tau_{\text{rel}} < t_{\text{vib}}$ and $t_{\text{el}}$, a fast relaxation occurs before any ET takes place,[13] and a description based on transition rates is suitable. This will be the case for the nonadiabatic ET discussed in Section 6.4.

If the lifetime of electron–vibrational states is larger than the characteristic times $t_{\text{vib}}$ and $t_{\text{el}}$, only a weak disturbance of wave–like nuclear motions in the course of the ET appears. This is typical for photoinduced ET where vibrational coherences at the donor and acceptor state can be observed on a sub–picosecond time scale (see Section 6.8).

### 6.3.1 Landau–Zener Theory of Electron Transfer

To characterize the ET in a DA complex we will make the reader familiar with a *classical* treatment which is widely used and which has been developed by Landau and independently by Zener. Originally, Landau considered a scattering of two atoms whereas Zener focused on the electronic levels of a diatomic molecule. In both cases, level coupling has been considered under the condition that the level separation is changed by an external perturbation. This approach is easily mapped on the description of ET in a DA complex. The advantage is that one can derive an analytical formula for the transfer rate which is valid for any value of the coupling $V_{DA}$ spanning the range between adiabatic and nonadiabatic ET.

[13]Note that we have to guarantee that the coupling to the particular environment is not too strong. If $1/\tau_{\text{rel}} \gg \omega_{\text{vib}}$ the energy levels $E_{mM}$ become meaningless since in this case of strong coupling a separate definition of the diabatic energy spectrum $E_{mM}$ cannot be justified.

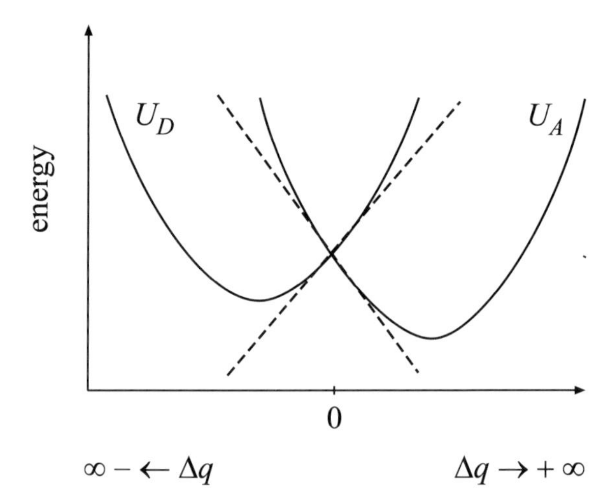


**Figure 6.16:** The coupled PES of a DA complex versus a single reaction coordinate. According to the treatment in the Landau–Zener theory the PES are approximated by straight lines around the crossing point. The asymptotic regions $\Delta q \rightarrow \pm\infty$ are also indicated.

In order to deal with the ET reaction in a DA complex according to Landau and Zener one has to choose a *classical* description for a single vibrational coordinate. To obtain the ET rate we let the vibrational coordinate start to move on the donor PES far away from the crossing point $q^*$ with the acceptor PES. If the coordinate moves through the crossing region we will determine the probability that the electron is transferred to the acceptor level as well as the probability for remaining at the donor level.

The respective Hamiltonian of the DA complex has been introduced in Eq. (6.30). The vibrational Hamiltonian $H_m(q)$ $(m = D, A)$ includes the donor and acceptor PES $U_D(q)$ and $U_A(q)$, respectively. Both depend on the single vibrational coordinate $q$ and may in principle have an arbitrary shape. Since the crossing point $q^*$ between the PES is crucial for the transfer we expand both PES around $q^*$

$$U_m(q) = U^* - F_m(q^*)\Delta q \ . \tag{6.50}$$

Here,

$$F_m(q^*) = -\left.\frac{\partial U_m(q)}{\partial q}\right|_{q=q^*} , \tag{6.51}$$

denotes the force the vibrational coordinate experiences at the crossing point when the electron is in state $m = D, A$. Furthermore, we introduced $\Delta q = q - q^*$, and $U^*$ abbreviates $U_D(q^*) = U_A(q^*)$ (cf. Fig. 6.16).

The time dependence of the coordinate $q$ (or $\Delta q$) is unknown so far. Since we expect the ET reaction to take place at the curve crossing around $\Delta q \approx 0$, we set $\Delta q \approx v^* t$, where $v^*$ is the yet unknown velocity at the crossing point. It represents a parameter of the theory which should be estimated. By virtue of these approximations the Hamiltonian becomes formally time–dependent

$$H_{\rm DA} = T_{\rm vib} + U^* + H_0(t) + \hat{V} \ . \tag{6.52}$$

The classical part $T_{\rm vib} + U^*$ is of less interest for the following; the time–dependent part reads

$$H_0(t) = -F_D \, v^* t \, |D\rangle\langle D| - F_A \, v^* t \, |A\rangle\langle A| \ . \tag{6.53}$$

Further the inter–state coupling is comprised in

$$\hat{V} = V_{DA}|D\rangle\langle A| + \text{h.c.} \ . \tag{6.54}$$

The reactant state of the transfer corresponds to $t = -\infty$ ($\Delta q = -\infty$) whereas the product state is characterized by $t = \infty$ ($\Delta q = \infty$, see Fig. 6.16).

In a first step we will calculate the asymptotic value of the survival probability of the electron for remaining at the donor, $P_D \equiv P_D(t = \infty)$. This quantity follows as the square of the transition amplitude (compare Section 3.3.1)

$$P_D = |\langle D|U(\infty, -\infty)|D\rangle|^2 \ , \tag{6.55}$$

where the time–evolution operator $U(t', t)$ is given by the Hamiltonian $H_0(t) + \hat{V}$, Eq. (6.52). Interestingly, the present model allows to calculate this transition amplitude exactly, as demonstrated in detail in the supplementary Section 6.9.1. Here, we only quote the result for the donor survival probability:

$$P_D = e^{-\Gamma} \ . \tag{6.56}$$

It depends on the so–called Massey parameter which is defined as

$$\Gamma = \frac{2\pi}{\hbar v^*} \frac{|V_{DA}|^2}{|F_D - F_A|} \ . \tag{6.57}$$

Although it is of interest to have an expression for the survival probability $P_D$ we aim to get the ET rate $k_{DA}$ based on our knowledge of $P_D$. The ET rate can be defined by the redistribution of probability between donor and acceptor within the characteristic time interval $t_{\rm vib} = 2\pi/\omega_{\rm vib}$. This time interval is a good estimate for the time the vibrational coordinate needs to go from the region around the minimum of the donor PES to that around the minimum of the acceptor PES and back. Considering the coordinate $\Delta q$ this vibrational motion formally corresponds to the motion between the two asymptotic values $\pm\infty$, that is, to the transition from $\Delta q = -\infty$ to $\Delta q = \infty$ and back to $\Delta q = -\infty$ (cf. Fig. 6.16). Determining the rate for a single transition event from the donor to the acceptor within the time interval $t_{\rm vib}$ one has to account for two alternative ways: Either one goes from $\Delta q = -\infty$ to $\Delta q = \infty$ and a transition to the acceptor takes place. Then, on the way back to $\Delta q = -\infty$ the system has to remain at the acceptor state. Or, the system remains at the donor state during the motion from $\Delta q = -\infty$ to $\Delta q = \infty$, but on the way back, when passing the crossing region, it moves to the acceptor state.

To calculate the ET rate for the first pathway we note that the probability for going to the acceptor as the coordinate moves from $\Delta q = -\infty$ to $\Delta q = \infty$ is $1 - P_D$. Due to the symmetry of the problem the probability to make no transition on the way back to $\Delta q = -\infty$ is identical to $P_D$. In conclusion, we obtain the change of the probability within a single oscillation period as $(1 - P_D)P_D$. In the second case of realizing the ET, the system remains with probability $P_D$ in the donor state as the coordinate moves from $\Delta q = -\infty$ to $\Delta q = \infty$. On the way back a transition to the acceptor state occurs with probability $1 - P_D$. This is indeed the same result as for the first transition pathway. Therefore, we obtain the transition rate as

$$k_{DA} = \frac{\omega_{\text{vib}}}{2\pi} \; 2(1 - P_D)P_D \equiv \frac{\omega_{\text{vib}}}{\pi} \; e^{-\Gamma} \; \left(1 - e^{-\Gamma}\right) . \tag{6.58}$$

The expression is valid for *every* value of $V_{DA}$, thus covering the case of adiabatic as well as nonadiabatic ET. For large $\Gamma$ (and hence large $V_{DA}$) we obtain the rate for adiabatic ET[14]

$$k_{DA}^{(\text{adia})} = \frac{\omega_{\text{vib}}}{\pi} \; e^{-\Gamma} , \tag{6.59}$$

while for small $\Gamma$ the nonadiabatic limit follows

$$k_{DA}^{(\text{nonad})} = \frac{2\omega_{\text{vib}}}{\hbar v^*} \; \frac{|V_{DA}|^2}{|F_D - F_A|} , \tag{6.60}$$

with a rate proportional to the square of the electronic coupling. For the intermediate regime that rate has to be calculated according to (6.58).

Introducing the Landau–Zener length

$$l_{\text{LZ}} = \frac{2\pi |V_{DA}|}{|F_D - F_A|} , \tag{6.61}$$

the Massey parameter becomes

$$\Gamma = l_{\text{LZ}} \; \frac{|V_{DA}|}{\hbar v^*} . \tag{6.62}$$

The Landau–Zener length can be understood as the distance from the crossing point where the difference $U_D - U_A$ in the potential energy reaches the magnitude of the electronic coupling $V_{DA}$.[15] Therefore, it gives an estimate for those $\Delta q$ values up to which the coupling $V_{DA}$ has some influence on the transfer dynamics.

Let us consider a liquid phase situation next. Here ET reactions can be characterized by introducing a mean free path length $l_{\text{f}}$ for the reaction coordinate. It corresponds to the average distance between two collision events of the reaction coordinate with solvent molecules. If $l_{\text{f}} \gg l_{\text{LZ}}$ the reaction coordinate can be considered to carry out a ballistic motion (or uniform

---

[14] Obviously the limit $V_{DA} \to \infty$ is meaningless since for this case $k_{DA}$ vanishes.

[15] To show this let us consider the expression $U_D(q^* + \Delta q) - U_A(q^* + \Delta q)$ which measures the potential energy difference between the case where the electron is at the donor and the case where it is at the acceptor. Expanding this expression with respect to the deviation $\Delta q$ from the crossing point of both PES, we estimate the PES difference as $|F_D - F_A| \, |\Delta q|$. Using now Eq. (6.61) we may write $|(F_D - F_A)| l_{\text{LZ}} = 2\pi |V_{\text{DA}}|$, what justifies the given explanation of the the Landau–Zener length.

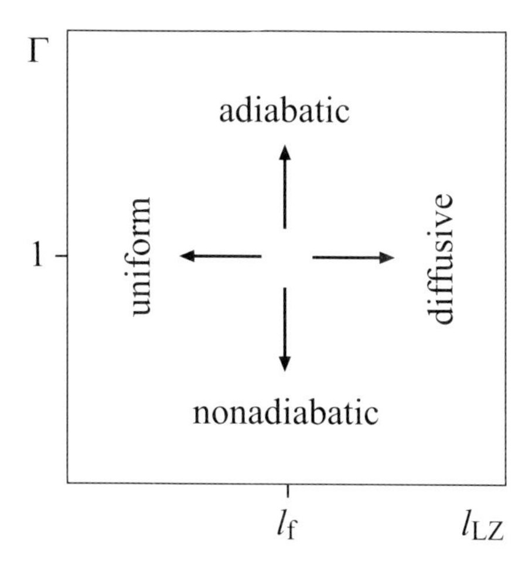


**Figure 6.17:** Schematic representation of the different ET regions. The horizontal axis distinguishes between uniform and diffusive motion of the reaction coordinate by plotting the Landau–Zener length $l_{LZ}$, Eq. (6.61) in relation to a given mean free path length $l_f$. Along the vertical axis adiabatic and nonadiabatic transfer is differentiated using the Massey parameter Γ defined in Eq. (6.62).

motion) on the time–scale of the ET with only minor influence of collisions with the solvent molecules. In the opposite case $l_f \ll l_{LZ}$ the motion is diffusive on the time–scale of the ET. Of course, in both cases the ET can take place either in the adiabatic or nonadiabatic regime, depending on the actual value of Γ. This situation is visualized in Fig. 6.17.

The Landau–Zener theory enjoys great popularity in the description of ET reactions, since it gives a simple but powerful tool for estimating transition rates. In more recent years it has been used to incorporate electronic transitions into the framework of classical simulations of the nuclear degrees of freedom of large molecules.

## 6.4 Nonadiabatic Electron Transfer in a Donor–Acceptor Complex

The concept of nonadiabatic ET has been already introduced in the preceding section as a charge transition process for which the vibrational motion is much faster than the motion of the transferred electron. The type of rate equation we have to expect can be found in Eq. (6.46). In the following we will consider this important type of ET reaction in more detail. The interest in nonadiabatic ET reactions stems from the fact that bridge mediated long–range ET usually proceeds in the this limit. This more complex type of ET will be dealt with in Section 6.6. Here, we concentrate on nonadiabatic ET in a simple DA complex.

Since for nonadiabatic ET we have to account for the transfer coupling between the donor and the acceptor in lowest order of perturbation theory we are in the position to write down the general rate formula (valid at any temperature) using the results of Chapter 5 which deals with transitions between different adiabatic electronic states. There, it has been shown that the computation of the transition rate can be reduced to the determination of a properly defined spectral density. Once this quantity is obtained (which requires the harmonic oscillator approximation for the related PES) the transfer rate can be calculated for all temperatures. In addition vibrational modes with frequencies extending over a broad region can be incorporated. However, to provide the reader with an overview of the various approaches which can be found in the literature, we will discuss different limiting cases of nonadiabatic ET. Thus we approach the most general description of ET dynamics step by step, starting our considerations in the high–temperature limit.

### 6.4.1 High–Temperature Case

The high–temperature limit is applicable if the relation $k_{\rm B}T \gg \hbar\omega_\xi$ holds for all vibrational modes $\xi$. In such a situation it is possible to describe the vibrational dynamics in the framework of classical physics. If the classical description is not possible for all types of vibrations, one has to study the subset of the high–frequency quantum modes by means of quantum mechanics (see the following section).

To derive the rate expression for the ET process we consider the case of an excess electron in the DA complex. Then the appropriate Hamiltonian is given by Eq. (6.30), where the PES $U_m(q)$ ($q \equiv \{q_\xi\}$) for the electron at the donor or acceptor site ($m = D, A$) follows from Eq. (6.26). Since we consider classical vibrational dynamics, the Hamiltonian $H_m$ is replaced by the Hamiltonian functions $H_m(q(t), p(t))$ defined via the time–dependent vibrational momenta and coordinates. As it has been demonstrated in Section 3.11 the incorporation of classical dynamics into the quantum dynamical description of transfer process is conveniently done using the Wigner representation (cf. Section 3.4.4). Its application to reaction rates is outlined in the supplementary Section 6.9.4. In the following we will give a discussion of the rate $k_{\rm ET} \equiv k_{DA}$ for the ET from the donor to the acceptor which is based on more simpler arguments.

Since the vibrational coordinates are described classically the ET system reduces to a simple electronic two–level system (covering the electronic donor and acceptor level). But the two–level system is characterized by a time–dependent modulation of the energetic position of the two levels due to the classical vibrational motion. Furthermore, in standard experimental situations only an average with respect to a large number of identical DA complexes is of interest. This average can be replaced by an ensemble average with respect to the thermal equilibrium distribution function $f(q, p)$ (cf. Section 3.59) which represents the probability distribution for the vibrational degrees of freedom in the reactant state. Applying the reasoning which leads to the Golden Rule of quantum mechanics we arrive at a formula for the rate similar to Eq. (3.59) (however, the summation with respect to the quantum levels has been replaced by an integration over all vibrational coordinates and momenta). Since $H_D(q, p)$ and $H_A(q, p)$ enter the $\delta$–function part of the Golden Rule formula the kinetic energy contributions compensate each other. The average with respect to the momenta can be carried out leading to the coordinate distribution function

$$f(q) = \frac{1}{\mathcal{Z}} e^{-U_D(q)/k_{\rm B}T}\,, \tag{6.63}$$

and the following rate expression is obtained:

$$k_{\rm ET} = \frac{2\pi}{\hbar} \int dq \; f(q) |V_{DA}|^2 \delta \left( U_D(q) - U_A(q) \right) . \tag{6.64}$$

This formula gives the ET rate as the transition rate between the initial electronic state with energy $U_D(q)$ and the final state with energy $U_A(q)$ averaged with respect to all possible configurations of the vibrational coordinates. The averaging is weighted by the thermal distribution, Eq. (6.63). Therefore, Eq. (6.64) implies that there is no change of the vibrational kinetic energy during electron transfer. In the supplementary Section 6.9.4 we will give a justification for the present treatment (as well as a possible extension), and Section 6.4.3 demonstrates that the high–temperature case can be obtained as a certain limit of a rate expression valid for any temperature.

If parabolic PES are used, an analytical expression for the ET rate can be obtained. Let us start with the simple case of a single coordinate $q$ oscillating with frequency $\omega_{\rm vib}$. Note that we also will neglect any dependence of the transfer integral $V_{DA}$ on this coordinate (Condon–like approximation). We obtain for the argument of the delta function in Eq. (6.64)

$$\begin{aligned} U_D(q) - U_A(q) &= U_D^{(0)} - U_A^{(0)} + \frac{\omega_{\rm vib}^2}{2} \left( \left( q - q^{(D)} \right)^2 - \left( q - q^{(A)} \right)^2 \right) \\ &= \Delta E - \omega_{\rm vib}^2 (q^{(D)} - q^{(A)}) q + \frac{\omega_{\rm vib}^2}{2} (q^{(D)2} - q^{(A)2}) . \end{aligned} \tag{6.65}$$

Here, we introduced the energetic difference between the donor and acceptor PES

$$\Delta E = U_D^{(0)} - U_A^{(0)} , \tag{6.66}$$

which is frequently called *driving force* of the ET reaction. The argument of the delta function in (6.64) is linear with respect to $q$ and vanishes at

$$q^* = \frac{\Delta E + \frac{\omega_{\rm vib}^2}{2} \left( q^{(D)2} - q^{(A)2} \right)}{\omega_{\rm vib}^2 (q^{(D)} - q^{(A)})} \tag{6.67}$$

This particular value of $q$ defines the crossing point of both PES (see Fig. 6.18). The thermal distribution introduced in Eq. (6.63) reads

$$f(q) = \sqrt{\frac{\omega_{\rm vib}^2}{2\pi k_{\rm B} T}} \exp\left\{ - \frac{\omega_{\rm vib}^2 (q - q^{(D)})^2}{2 k_{\rm B} T} \right\} , \tag{6.68}$$

and performing the $q$–integration results in

$$k_{\rm ET} = \frac{2\pi}{\hbar} \frac{|V_{DA}|^2}{\sqrt{2\pi k_{\rm B} T \omega_{\rm vib}^2 \left( q^{(D)} - q^{(A)} \right)^2}} \exp\left\{ - \frac{\omega_{\rm vib}^2 \left( q^* - q^{(D)} \right)^2}{2 k_{\rm B} T} \right\} . \tag{6.69}$$

The obtained rate formula is of the type of Eq. (6.46). It represents the activation law for reaching the crossing point $q = q^*$ between the donor and the acceptor PES. The activation energy is given by

$$E_{\text{act}} = \frac{1}{2}\omega_{\text{vib}}^2 \left(q^* - q^{(D)}\right)^2 . \tag{6.70}$$

This expression can be rewritten to give

$$E_{\text{act}} = \frac{(\Delta E - E_\lambda)^2}{4E_\lambda} . \tag{6.71}$$

The quantity

$$E_\lambda = \frac{\omega_{\text{vib}}^2}{2} \left(q^{(A)} - q^{(D)}\right)^2 \tag{6.72}$$

is the potential energy of the vibrational coordinate which corresponds to the following situation: Initially the electron is at the donor and the vibrational coordinate has the value $q = q^{(D)}$. Then a sudden change of the electronic state occurs (see Fig. 6.18). In order to reorganize the vibrational coordinate (nuclear configuration) to the new equilibrium value $q^{(A)}$ the energy $E_\lambda$ has to be removed from the system. Therefore, this energy is usually called *reorganization energy* (cf. the discussion of the reaction path Hamiltonian in Chapter 2). If the ET reaction proceeds in a solvent, the change of the electronic charge density in the DA complex is accompanied by a rearrangement of the solvent polarization field (see the detailed discussion in Section 6.5). Thus, the name *polarization energy* is also common for $E_\lambda$.

The rate expression which follows upon introducing $E_\lambda$ is usually named after R. A. Marcus, who pioneered the theory of ET reactions starting in the 1950s. It reads

$$k_{\text{ET}} = |V_{DA}|^2 \sqrt{\frac{\pi}{\hbar^2 k_{\text{B}} T E_\lambda}} \exp\left\{-\frac{(\Delta E - E_\lambda)^2}{4E_\lambda k_{\text{B}} T}\right\} . \tag{6.73}$$

Before discussing this result in detail we note that the same expression is valid if we consider not a single but a large number of vibrational coordinates for the donor and acceptor PES as introduced in Eq. (6.26) . The only change concerns the reorganization energy, which has to be generalized from Eq. (6.72) to the case of many vibrational degrees of freedom according to (details can be found in the supplementary Section 6.9.2)[16]

$$E_\lambda = \sum_\xi \frac{\omega_\xi^2}{2} \left(q_\xi^{(D)} - q_\xi^{(A)}\right)^2 . \tag{6.74}$$

It is the main advantage of the Marcus formula that it allows to describe the complex vibrational dynamics accompanying the electronic transition by a small number of parameters, namely the transfer coupling $V_{DA}$, the driving force $\Delta E$, and the reorganization energy $E_\lambda$.[17]

[16]If the shapes of the two coupled PES differ, i.e. if the vibrational frequencies become electronic state dependent a generalization of Eq. (6.73) can be derived [Cas02,Tan94].

[17]If the number of vibrational degrees of freedom is macroscopic, the energy difference of the PES minima has to be replaced by the free energy difference.

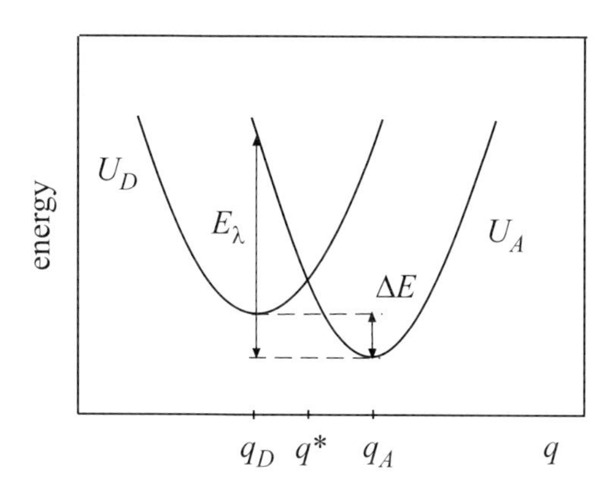


**Figure 6.18:** Potential energy surfaces for a DA complex in harmonic approximation. The definition of the driving force $\Delta E$ and the reorganization energy $E_\lambda$ are indicated.

In particular, the introduction of the reorganization energy reduces the complicated behavior of many intramolecular and intermolecular nuclear degrees of freedom (or the polarization in case of a polar solvent, see below) to a single number.

Since the Marcus formula includes only three unknown quantities a straightforward fit of experimental ET data often becomes possible, particularly if the temperature dependence of the rate is measured. Usually one plots $\log k_{\text{ET}}$ versus $1/T$ in the so–called *Arrhenius plot*. Doing experiments on ET reactions in DA complexes dissolved in a polar solvent, the reorganization energy can be varied using solvents with different polarity (Section 6.5 will provide more details). A controllable change of $\Delta E$ and $V_{DA}$ is also possible altering details of the chemical structure of the complex.

Eq. (6.73) describes the ET reaction proceeding from the donor to the acceptor. The rate for the back transfer from the acceptor to the donor can be easily derived in the used model of donor and acceptor PES with identical parabolic shapes. It is only necessary to interchange the donor and the acceptor index leading to a change of the sign of $\Delta E$. We get

$$k_{AD} = k_{DA}(-\Delta E) = e^{-\Delta E/k_{\text{B}}T}\, k_{DA}(\Delta E)\,. \tag{6.75}$$

The ratio of the forward and backward rate is given by $\exp\{\Delta E/k_{\text{B}}T\}$, that is, the validity of the detailed balance condition is guaranteed.

Let us consider the ET rate in dependence on the driving force $\Delta E$ of the reaction at a given value of $V_{DA}$ and $E_\lambda$. The situation already displayed in Fig. 6.18 is called the *normal region* of ET. Starting in this ET region and increasing $\Delta E$ moves $q^*$ to the left until the activation energy becomes zero for $\Delta E = E_\lambda$ (see Fig. 6.19). Now, the *activationless case* is reached. This regime of ET is observed in the experiment if the rate becomes independent of temperature. Increasing $\Delta E$ further the activation energy increases again. This is the so–called *inverted region*. Looking at the lower panel of Fig. 6.19 one may notice the possible strong overlap of vibrational wave functions corresponding to the presence of the transferred electron at the donor and at the acceptor. Hence, in the inverted region nuclear tunneling may become important instead of the thermally activated transfer studied so far.

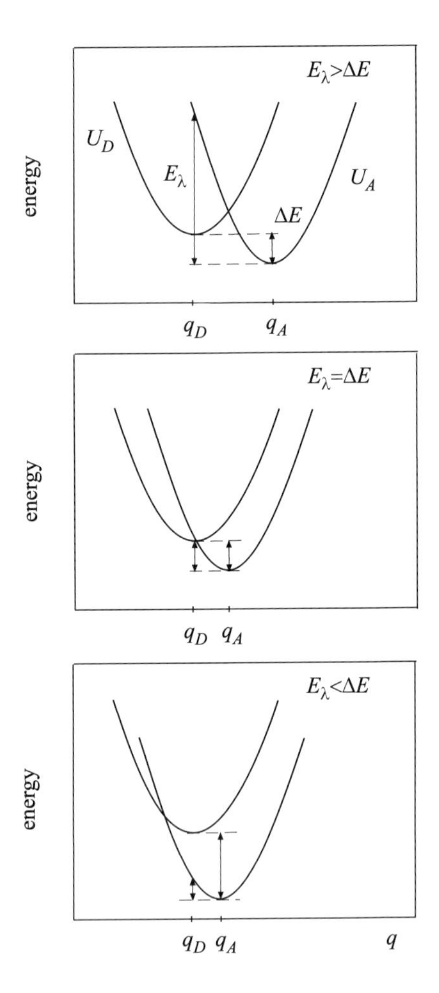


**Figure 6.19:** The normal region (upper panel), the activationless case (middle panel) and the inverted region (lower panel) of ET in a DA complex.

ET in the inverted region has been originally proposed by R. A. Marcus in the 1950s, but it could be verified experimentally only in the late 1980s. Fig. 6.20 shows how one enters the inverted region by changing systematically the acceptor compound such that $\Delta E$ is increased.

### 6.4.2 High–Temperature Case: Two Independent Sets of Vibrational Coordinates

In the foregoing section we applied a model which assumed a common set of vibrational coordinates $q \equiv \{q_\xi\}$ for both the reactant and the product state. This is typical for a unimolecular reaction. Next let us consider a bimolecular ET reaction proceeding in solution where the sep-

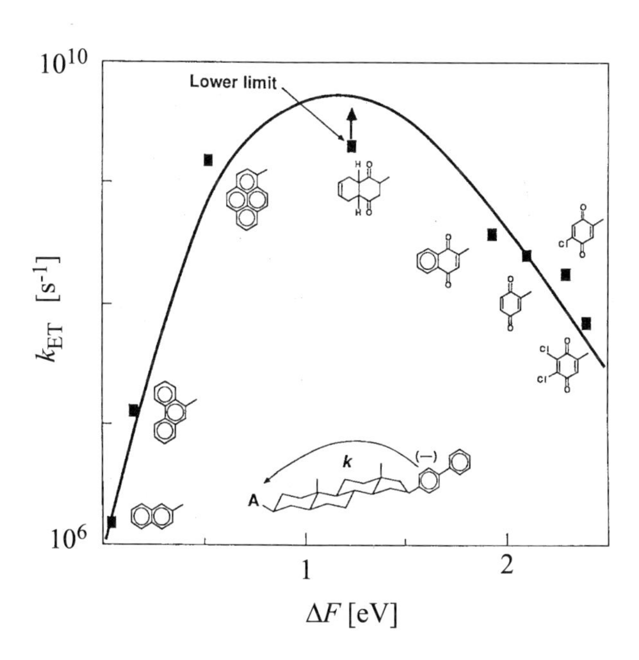


**Figure 6.20:** ET rate versus driving force of the reaction for a DA complex showing transfer in the inverted region. A steroid spacer (androstane) links a 4-biphenylyl donor group with different acceptors (shown below the curve). An excess electron has been attached to the donor by means of a pulsed electron beam. The complex was dissolved in methyltetrahydrofuran. The full curve has been computed from Eq. (6.106) using known parameters for the specific system. (reprinted with permission from [Clo88]; copyright (1988) Science)

arate donor and acceptor molecules form an encounter complex to trigger ET. Here it seems more appropriate to use a separate set of vibrational coordinates $q_D \equiv \{q_{D\xi}\}$ for the donor molecule and $q_A \equiv \{q_{A\xi}\}$ the acceptor molecule as discussed in Section 6.2.3. Of course, if solvent contributions become important the two separate sets of vibrational coordinates have to be supplemented by an additional third set of common coordinates.

Having separate sets of coordinates for the donor and acceptor requires a separate description of the initial and final states of the donor and the acceptor. We consider the scheme (6.5) to be valid, i.e., the reactant state is $|\phi_{D^-}, \phi_A\rangle$ and the product state is $|\phi_D, \phi_{A^-}\rangle$ (cf. Section (6.2.3) for the notation).

The ET rate Eq. (6.64) is easily generalized to the present case and reads (note that $\int dq_D\, dq_A$ is a shorthand notation for the multi–dimensional integration)

$$\begin{aligned} k_{\rm ET}^{\rm (bimol)} &= \frac{2\pi}{\hbar}|V_{DA}|^2 \int dq_D\, dq_A\, f_{D^-}(q_D) f_A(q_A) \\ &\times \delta\left([U_{D^-}(q_D) + U_A(q_A)] - [U_D(q_D) + U_{A^-}(q_A)]\right) . \end{aligned} \tag{6.76}$$

Since there is a separation into vibrational coordinates belonging to the donor and the acceptor, the energy conserving delta function in the rate formula can be split up into a donor part

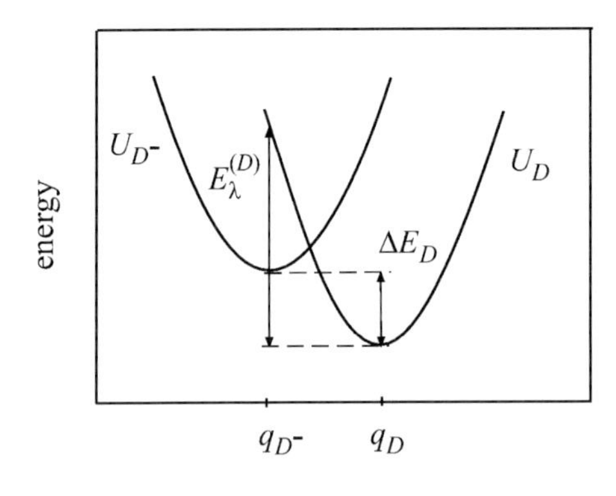

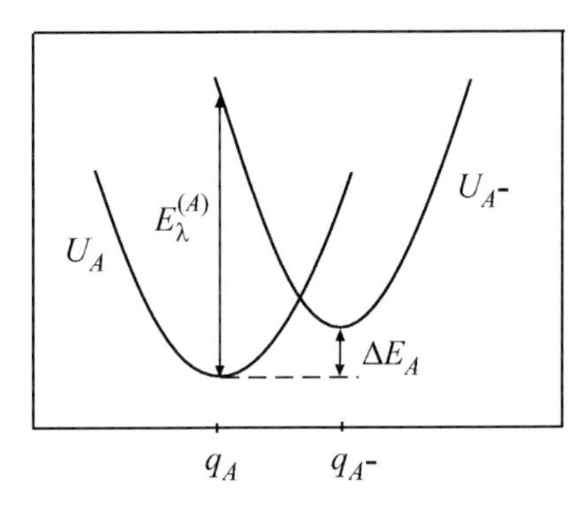


**Figure 6.21:** PES for the case of independent vibrational coordinates of the donor and acceptor part. Left part: PES of the ionized and neutral donor. Right part: PES of the ionized and neutral acceptor. The respective reorganization energies and driving forces are also shown.

and into an acceptor part. This is achieved by introducing an additional frequency integral according to

$$\begin{aligned}\delta(\Delta U) &= \delta\left(U_{D^-}(q_D) + U_A(q_A) - U_D(q_D) - U_{A^-}(q_{A^-})\right)\\ &= \int dE\, \delta\left(U_{D^-}(q_D) - E - U_D(q_D)\right) \delta\left(U_A(q_A) + \hbar\omega - U_{A^-}(q_{A^-})\right) .\end{aligned} \tag{6.77}$$

Let us define the auxiliary functions

$$d(\omega) = \int dq_D\, f_{D^-}(q_D)\, \delta\left(U_{D^-}(q_D) - \hbar\omega - U_D(q_D)\right) , \tag{6.78}$$

and

$$a(\omega) = \int dq_A\, f_A(q_A)\, \delta\left(U_A(q_A) + \hbar\omega - U_A(q_{A^-})\right) . \tag{6.79}$$

Then the rate follows as a frequency overlap of the two auxiliary functions

$$k_{\rm ET}^{\rm (bimol)} = 2\pi |V_{DA}|^2 \int d\omega\, d(\omega) a(\omega) . \tag{6.80}$$

Inspecting the argument of the delta functions in Eqs. (6.78) and (6.79), it is possible to give a physically appealing interpretation. The function $d(\omega)$ can be understood as the frequency–resolved strength of the detachment process where the excess electron is removed from the donor. In the same manner we understand $a(\omega)$ as the spectrum for the attachment of the excess electron at the acceptor.[18]

[18] We will encounter such a type of rate formula again in Chapter 8 when studying the transfer of intramolecular excitation energy.

Using the results of the preceding section we are able to derive explicit expressions for the auxiliary functions. We obtain

$$d(\omega) = \frac{1}{\sqrt{4\pi k_B T E_\lambda^{(D)}}} \exp\left\{ -\frac{\left(\Delta E_D - E_\lambda^{(D)}\right)^2}{4E_\lambda^{(D)} k_B T} \right\} . \tag{6.81}$$

and

$$a(\omega) = \frac{1}{\sqrt{4\pi k_B T E_\lambda^{(A)}}} \exp\left\{ -\frac{\left(\Delta E_A - E_\lambda^{(A)}\right)^2}{4E_\lambda^{(A)} k_B T} \right\} . \tag{6.82}$$

The two newly introduced driving forces read

$$\Delta E_D = U_{D^-}^{(0)} - U_D^{(0)} - \hbar\omega , \tag{6.83}$$

and

$$\Delta E_A = U_A^{(0)} + \hbar\omega - U_{A^-}^{(0)} . \tag{6.84}$$

The reorganization energies are obtained as

$$E_\lambda^{(D)} = \sum_\xi \frac{\omega_\xi^2}{2} \left(q_\xi^{(D^-)} - q_\xi^{(D)}\right)^2 , \tag{6.85}$$

and

$$E_\lambda^{(A)} = \sum_\xi \frac{\omega_\xi^2}{2} \left(q_\xi^{(A)} - q_\xi^{(A^-)}\right)^2 . \tag{6.86}$$

The frequency integration in the rate formula (6.80) can be performed straightforwardly. Again one obtains the Marcus–type expression Eq. (6.73), but with the driving force given by the electronic energy difference of the reactant and product states

$$\Delta E = U_{D^-}^{(0)} - U_D^{(0)} + U_A^{(0)} - U_{A^-}^{(0)} . \tag{6.87}$$

The reorganization energy follows as the sum of the energies of the donor and the acceptor:

$$E_\lambda = E_\lambda^{(D)} + E_\lambda^{(A)} . \tag{6.88}$$

### 6.4.3 Low–Temperature Case: Nuclear Tunneling

We now return to the case of a common set of vibrational coordinates for the donor and acceptor. Additionally we suppose that $k_B T < \hbar\omega_\xi$ holds for all vibrational degrees of freedom participating in the ET reaction. Hence, a quantum mechanical description becomes necessary. In this situation the appropriate DA Hamiltonian is given by Eq. (6.41) (neglecting any bridge units).

As in the case of the classical description of the nuclear motion discussed so far we consider ET reactions which proceed in an ensemble of identical DA complexes. The initial state is characterized by the vibrational energy levels $E_{DM}$, and the levels $E_{AN}$ belong to the final acceptor states. How to obtain the rate for the total probability transfer from the donor state to the acceptor state has been discussed in Section 3.3, and in another context in Section 3.4.5. The similarity between ET reactions and nonadiabatic transition has been already stressed in the introduction to Chapter 5. Again, we remind the reader that the time $\tau_{\text{rel}}$ characterizing vibrational relaxation in the donor or the acceptor state has to be much shorter than all other characteristic times ($t_{\text{el}}$ and $t_{\text{vib}}$, see Eqs. (6.43) and (6.44), cf. discussion at the end of Section 6.3). Adapting these earlier results to the present situation we obtain the rate for nonadiabatic ET as

$$k_{\text{ET}} = \frac{2\pi}{\hbar} \sum_{M,N} f(E_{DM}) |V_{DM,AN}|^2 \delta(E_{DM} - E_{AN}) . \tag{6.89}$$

This Golden Rule formula describes the coupling of the initial manifold of states with that of the final states via a certain interaction matrix element. A thermal averaging of all initial vibrational states is carried out, and the rate contains the sum with respect to all final vibrational states. Neglecting the dependence of the electronic transfer integrals $V_{DA}$ on the nuclear degrees of freedom, $V_{DM,AN}$ splits up into the purely electronic transfer coupling $V_{DA}$ and the Franck–Condon overlap integral $\langle \chi_{DM} | \chi_{AN} \rangle$.

As stated above there is a formal similarity between the nonadiabatic ET from the donor to the acceptor and the optically induced electronic transition from the ground state to a particular excited electronic state of some molecule (cf. Section 5.2). This analogy suggests the introduction of the combined thermally averaged and Franck–Condon weighted density of states, $\mathcal{D}$, such that for the present case the ET rate becomes

$$k_{\text{ET}} = \frac{2\pi}{\hbar} |V_{DA}|^2 \mathcal{D}(\Delta E/\hbar) . \tag{6.90}$$

The density of states reads

$$\mathcal{D}(\omega) = \sum_{M,N} f(E_{DM}) |\langle DM|AN\rangle|^2 \delta\big(\hbar\omega - \sum_{\xi} \hbar\omega_\xi (M_\xi - N_\xi)\big) , \tag{6.91}$$

and has to be taken at the driving force $\Delta E$ introduced in Eq. (6.73) to get $k_{\text{ET}}$. Following the considerations in Section 5.2.4 we can rewrite $\mathcal{D}$ via a time integral as

$$\mathcal{D}(\Delta E/\hbar) = \frac{1}{2\pi\hbar} e^{-G(0)} \int dt\; e^{i\Delta E t/\hbar + G(t)} , \tag{6.92}$$

with[19]

$$G(t) = \sum_{\xi} \big(g_D(\xi) - g_A(\xi)\big)^2 \big(e^{-i\omega_\xi t}(1 + n(\omega_\xi)) + e^{i\omega_\xi t} n(\omega_\xi)\big) . \tag{6.93}$$

[19] Here $g_m(\xi)$, $m = D, A$ are dimensionless displacements of the vibrational coordinates $q_\xi$ and have been defined in Eq. (2.86).

The function $G(t)$ has been already introduced in Section 5.2.4. It is of an universal character and appears whenever transitions among different electronic levels are accompanied by the rearrangement of nuclear coordinates which have been mapped on a set of independent harmonic oscillators (normal mode vibrations).

If the number of different vibrational modes becomes large it is advisable to introduce a special type of *spectral density* responsible for the electron transfer reaction in a DA complex (different types of spectral densities have been already discussed in Sections 3.6.4, 4.3.2 and 5.2.5). For the present application we write in analogy to the case of nonadiabatic transitions (cf. Eq. (5.53))

$$J_{DA}(\omega) = \sum_{\xi} (g_D(\xi) - g_A(\xi))^2 \, \delta(\omega - \omega_\xi) \,. \tag{6.94}$$

If we can write $\kappa_\xi = (g_D(\xi) - g_A(\xi))^2 \equiv \kappa(\omega_\xi)$, we get the more transparent form of the spectral density $J_{DA}(\omega) = \kappa(\omega)\mathcal{N}(\omega)$. Here we used the oscillator density of states defined as $\mathcal{N}(\omega) = \sum_\xi \delta(\omega - \omega_\xi)$. This reminds the reader on the fact that the spectral density can be interpreted as the density of oscillator states weighted by the electron–vibrational coupling constant (cf. Section 3.6.4).

The reorganization energy Eq. (6.72) can be expressed via the spectral density as

$$\int_0^\infty d\omega \, \omega J_{\rm DA}(\omega) = \sum_\xi \omega_\xi \left(g_D(\xi) - g_A(\xi)\right)^2 = \frac{E_\lambda}{\hbar} \,. \tag{6.95}$$

The introduction of the spectral density enables us to write (see Eq. (5.54))

$$G(t) = \int_0^\infty d\omega \left(e^{-i\omega t}(1 + n(\omega)) + e^{i\omega t} n(\omega)\right) J_{\rm DA}(\omega) \,. \tag{6.96}$$

At this point it is useful to clarify what approximations will lead to the rate formula of the high–temperature limit derived in Section 6.4.1. To this end we note that irrespective of the actual frequency dependence, the spectral density rapidly goes to zero beyond a certain cut–off frequency $\omega_{\rm c}$. Hence in the high–temperature limit we have $k_{\rm B}T \gg \hbar\omega_{\rm c}$. This enables us to introduce for all frequencies less than $\omega_{\rm c}$ the approximation $1 + 2n(\omega) \approx 2k_{\rm B}T/\hbar\omega \gg 1$.

To utilize this inequality we separate the function $G(t)$ into its real and imaginary parts as in Eq. (5.56):

$$G(t) = \int_0^\infty d\omega \, \cos\omega t \, (1 + 2n(\omega)) \, J_{DA}(\omega) - i \int_0^\infty d\omega \, \sin\omega t J_{\rm DA}(\omega) \,. \tag{6.97}$$

If $\omega_{\rm c}|t| \ll \pi/2$, the quantity $\exp\{G(t) - G(0)\}$ rapidly approaches zero since the expression $\cos\omega t - 1$, which appears in the exponent, is negative. But for $\omega_{\rm c}|t| > \pi/2$ the different contributions to the time integral may interfere destructively. Consequently, it is possible to approximate $G(t)$ in the exponent by the leading expansion terms of the sine and cosine functions. (This is known as the short–time expansion and identical with the slow fluctuation limit introduced in Section 5.2.5.) Using the definition Eq. (6.94) of the spectral density gives

$$G(t) \approx -\int_0^\infty d\omega \, \frac{(\omega t)^2}{2} 2\frac{k_B T}{\hbar\omega} J_{DA}(\omega) - i\int_0^\infty d\omega \, \omega t J_{DA}(\omega) \,. \tag{6.98}$$

Both frequency integrals define the reorganization energy according to Eq. (6.95), and the combined density of states determining the ET rate follows as

$$\mathcal{D}(\Delta E/\hbar) = \int_{-\infty}^{+\infty} \frac{dt}{2\pi\hbar} \exp\left\{ i\frac{(\Delta E - E_\lambda)t}{\hbar} \right\} \exp\left\{ -\frac{k_B T E_\lambda t^2}{\hbar^2} \right\} . \tag{6.99}$$

The remaining integral is easily calculated as

$$\mathcal{D}(\Delta E/\hbar) = \frac{1}{\sqrt{4\pi k_B T E_\lambda}} \exp\left\{ -\frac{(\Delta E - E_\lambda)^2}{4E_\lambda k_B T} \right\} . \tag{6.100}$$

If inserted into expression (6.90) the classical (high–temperature) limit of the consequent quantum description of nonadiabatic ET reactions reproduces the Marcus formula, Eq. (6.73).

In principle the concept of the spectral density enables us to describe nonadiabatic ET also for vibrational modes differing strongly in their frequencies. For example, it is typical for the ET in dissolved DA complexes that low–frequency solvent modes as well as high–frequency intramolecular vibrations are involved in the ET reaction. Therefore, one has to split up the spectral density into a solvent part $J_{\text{sol}}(\omega)$, and an intramolecular part $J_{\text{intra}}(\omega)$. For simplicity we assume that $J_{\text{sol}}(\omega)$ is different from zero only in the low–frequency region (the cut–off frequency for collective solvent modes typically amounts to values less than 100 cm$^{-1}$). This makes a classical description of the solvent modes possible. Further, it is reasonable to suppose that there is no overlap with $J_{\text{intra}}(\omega)$. The ET rate expressed in terms of these spectral densities will be given in Section 6.4.5. First, however, we will introduce a model for this situation which is more common in the literature in the next section.

### 6.4.4 The Mixed Quantum–Classical Case

Let us consider the case that the ET is coupled to high–frequency intramolecular (quantum) modes and low–frequency (classical) modes, for example those of a solvent. Then the high–frequency modes are conveniently taken into account by using the electron–vibrational representation of the Hamiltonian given in Eq. (6.41). The solvent modes, on the other hand, are described using classical mechanics. Assuming a decoupling of both types of DOF, this leads us to a combination of the Hamiltonian Eq. (6.29) (where the vibrational Hamiltonian $H_m$ is interpreted as a classical Hamiltonian function) with the Hamiltonian, Eq. (6.41) given in the electron–vibrational representation. The respective vibrational energies $E_\mu$ are supplemented by the vibrational Hamiltonian function $H_m(q)$ of low–frequency normal modes $q \equiv \{q_\xi\}$. Accordingly, the complete DA Hamiltonian can be written as

$$H_{\text{DA}} = \sum_{\mu\nu} \Big( \delta_{\mu\nu} (E_\mu + H_m(q)) + (1 - \delta_{mn}) V_{\mu\nu} \Big) |\mu\rangle\langle\nu| \,. \tag{6.101}$$

The PES $U_D(q)$ and $U_A(q)$ related to $H_D$ and $H_A$, respectively, are defined as in Eq. (6.26), but with $U_m^{(0)} = 0$. A more general expression would be obtained if the PES and thus the related Hamilton function differed for different vibrational states $|\chi_{mM}\rangle$. But we will assume that there is no considerable rearrangement of the solvent, if the vibrational state of the intramolecular modes changes.

Using this model one can generalize the ET rate, Eq. (6.64) to the case where transitions from a manifold of donor states $|\phi_D\rangle|\chi_{DM}\rangle$ to many acceptor states $|\phi_A\rangle|\chi_{AN}\rangle$ are included. The ET rate for this mixed case follows as

$$\begin{aligned} k_{\mathrm{ET}} &= \frac{2\pi}{\hbar}\sum_{M,N}\int dq\, f(E_{DM}+U_D(q))|V_{DM,AN}|^2 \\ &\times\delta\left(E_{DM}+U_D(q)-E_{AN}-U_A(q)\right) . \end{aligned} \tag{6.102}$$

The rate can be determined similarly as in Section 6.4.1 leading to the following multi–channel generalization of the Marcus formula (6.73)

$$k_{\mathrm{ET}} = \sum_{M,N} f(E_{DM})\, k_{M\to N} \ , \tag{6.103}$$

with

$$k_{M\to N} = \sqrt{\frac{\pi}{\hbar^2 k_{\mathrm{B}} T E_\lambda}}|V_{DM,AN}|^2 \exp\left\{-\frac{(\Delta E_{DM,AN}-E_\lambda)^2}{4E_\lambda k_{\mathrm{B}}T}\right\} . \tag{6.104}$$

Each transfer channel from the initial vibrational level $E_{DM}$ to the final level $E_{AN}$ contributes its own ET rate $k_{MN}$. The reorganization energy is identical with that in Eq. (6.72), but the driving forces appear in the generalized form

$$\Delta E_{DM,AN} = E_{DM} - E_{AN} \ , \tag{6.105}$$

which accounts for the different initial and final states of the high–frequency mode.

The rate expression simplifies if we note that usually the energy of the high–frequency vibrational quanta exceeds the thermal energy even at room temperature. Therefore, only the vibrational ground state of this mode is occupied in the reactant state. We will concentrate on a single high–frequency normal mode, i.e., $E_{AN} = E_A + \hbar\omega_{\mathrm{intra}}(N+\frac{1}{2})$ (see Fig. 6.22), and get the rate as

$$\begin{aligned} k_{\mathrm{ET}} &= \sqrt{\frac{\pi}{\hbar^2 k_{\mathrm{B}} T E_\lambda}}|V_{DA}|^2 \sum_{N=0}^{\infty} |\langle\chi_{D0}|\chi_{AN}\rangle|^2 \\ &\times\exp\left\{-\frac{(\Delta E-\hbar\omega_{\mathrm{intra}}N-E_\lambda)^2}{4E_\lambda k_{\mathrm{B}}T}\right\} . \end{aligned} \tag{6.106}$$

Here, the reference driving force $\Delta E \equiv E_{D0} - E_{A0}$ has been introduced; its actual value is reduced by $\hbar\omega_{\mathrm{intra}}N$. Often Eq. (6.106) for the ET rate is written using a more explicit expression for the Franck–Condon factor $|\langle\chi_{D0}|\chi_{AN}\rangle|^2$. Making use of the derivations given in Section 2.98 and replacing the shift $g_{\mathrm{intra}}$ of the PES of the intramolecular vibration by

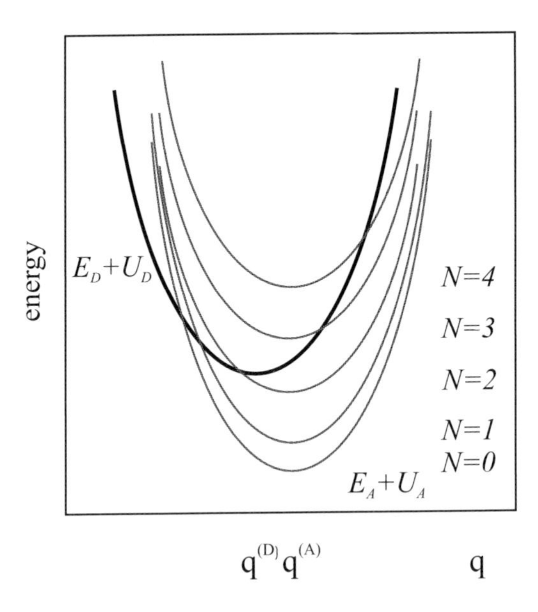


**Figure 6.22:** PES for the ET in the case of a single high–frequency intramolecular vibration and a low–frequency solvent coordinate $q$. According to the assumption $\hbar\omega_{\text{intra}} \gg k_{\text{B}}T$ a single solvent coordinate PES $E_D + U_D(q)$ has been drawn for the reactant state. The various product state solvent coordinate PES $E_A + \hbar\omega_{\text{intra}}N + U_A(q)$ which can be reached in the course of the reaction are also shown. The product state PES with $N = 0$ and $N = 1$ correspond to the ET in the inverted region.

$E_\lambda^{(\text{intra})}/\hbar\omega_{\text{intra}}$ one easily obtains

$$|\langle\chi_{D0}|\chi_{AN}\rangle|^2 = \frac{1}{N!}\left(\frac{E_\lambda^{(\text{intra})}}{\hbar\omega_{\text{intra}}}\right)^N \exp\left\{-\frac{E_\lambda^{(\text{intra})}}{\hbar\omega_{\text{intra}}}\right\}. \quad (6.107)$$

Fig. 6.22 shows the various PES involved in the ET reaction in the mixed quantum–classical case. The product state PES with $N = 0$ corresponds to the ET in the inverted region. But the character of the ET changes to the normal region with increasing vibrational quantum number $N$. The presence of the intramolecular vibrations opens additional channels for the ET reaction. But starting with the $(N = 0)$–PES in the normal region, the activation energy for the solvent coordinate increases for those PES with $N > 0$. Therefore, the rate will be dominated by the transition into the $(N = 0)$–state of the acceptor. If this $(N = 0)$–state refers to the inverted region, the PES with $N > 0$ may result in a reduction of the activation energy for the solvent coordinate, and even the activationless case may contribute to the total rate.

### 6.4.5 Description of the Mixed Quantum–Classical Case by a Spectral Density

Next we will make consequent use of the spectral density introduced in Eq. (6.94) to calculate the ET rate Eq. (6.90). Further, we incorporate into the discussion different types of spectral densities which were given in Section 5.2.5 in the context of optical absorption.

In the following we shortly show how one can calculate the ET rate derived in the foregoing section for the presence of a single high–frequency intramolecular vibration and low–frequency solvent vibrations. We set

$$J_{DA}(\omega) = J_{\text{intra}}(\omega) + J_{\text{sol}}(\omega) \,. \tag{6.108}$$

The high–frequency contribution reads

$$J_{\text{intra}}(\omega) = j_{\text{intra}}\delta(\omega - \omega_{\text{intra}}) \,. \tag{6.109}$$

The prefactor is given by $j_{\text{intra}} = E_\lambda^{(\text{intra})}/\hbar\omega_{\text{intra}}$, i.e., as the ratio of the related reorganization energy and the energy of a vibrational quantum, what can be easily verified using Eq. (6.95). A widely used form of a solvent spectral density is given by the Debye type (with Debye frequency $\omega_{\text{D}}$, a detailed justification will be given in the following Section 6.5, compare also Section 5.2.5)

$$J_{\text{sol}}(\omega) = \Theta(\omega)\; j_{\text{sol}}\; \frac{1}{\omega}\frac{1}{\omega^2 + \omega_{\text{D}}^2} \,. \tag{6.110}$$

Here, we can identify $j_{\text{sol}} = 2E_\lambda^{(\text{sol})}\omega_{\text{D}}/\pi\hbar$.

According to the partitioning of the spectral density we can split up the function $G(t)$, Eq. (6.96) into the solvent and intramolecular contribution $G_{\text{sol}}(t)$ and $G_{\text{intra}}(t)$, respectively. Using for $G_{\text{intra}}(t)$ Eq. (6.96) and assuming that $\hbar\omega_{\text{intra}} \gg k_{\text{B}}T$ one can write $G_{\text{intra}}(t) \approx e^{-i\omega_{\text{intra}}t} j_{\text{intra}}$. We insert $G_{\text{sol}}(t)$ and $G_{\text{intra}}(t)$ into Eq. (6.92) and afterwards expand the expression $\exp\{G_{\text{intra}}(t)\}$ with respect to $\exp\{-i\omega_{\text{intra}}t\}$. It yields

$$\mathcal{D}(\Delta E/\hbar) = e^{-j_{\text{intra}}} \sum_{N=0}^{\infty} \frac{j_{\text{intra}}^N}{N!} \mathcal{D}_{\text{sol}}(\Delta E/\hbar - N\omega_{\text{intra}}) \,, \tag{6.111}$$

The solvent contribution to the density of states is similar to Eq. (6.92) but with $\Delta E/\hbar - N\omega_{\text{intra}}$ instead of $\Delta E/\hbar$, reflecting the presence of the high–frequency vibrational mode.

If we take $\mathcal{D}_{\text{sol}}$ in the slow fluctuation limit, Eq. (5.64) or Eq. (6.100), we can write (note that $T_{\text{fluc}} = \hbar/\sqrt{2k_{\text{B}}TE_\lambda^{(\text{sol})}}$ in Eq. (5.64))

$$\mathcal{D}_{\text{sol}}(\omega) = \frac{1}{\sqrt{4\pi k_{\text{B}}TE_\lambda^{(\text{sol})}}} \exp\left\{-\frac{(\hbar\omega)^2}{4k_{\text{B}}TE_\lambda^{(\text{sol})}}\right\} \,. \tag{6.112}$$

Since $j_{\text{intra}} = E_\lambda^{(\text{intra})}/\hbar\omega_{\text{intra}}$ we have reproduced Eq. (6.106).

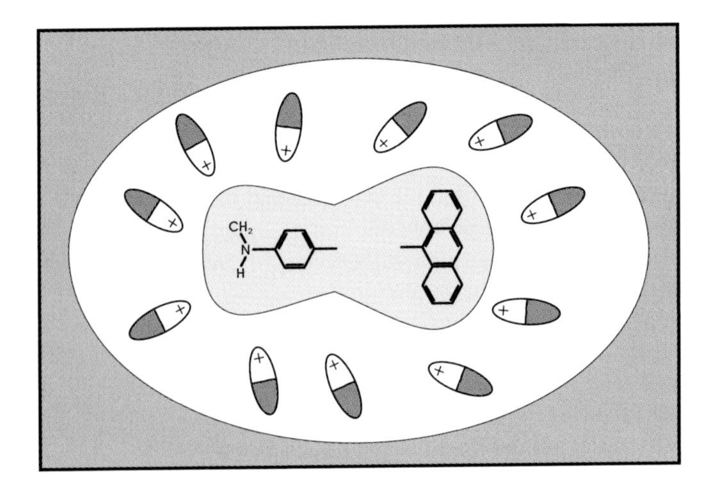


**Figure 6.23:** Dissolved DA complex in a polar solvent. The donor is a methylaniline molecule (left part) and the acceptor anthracene (right part, note that ET only takes place after optical excitation of the acceptor [Hei87]). The first solvation shell has been made obvious.

## 6.5 Nonadiabatic Electron Transfer in Polar Solvents

Most of the relevant ET transfer reactions occur in polar solution. Here, the solvent molecules such as water are characterized by permanent dipoles (see in Fig. 2.4). The ET can take place as a unimolecular reaction or as a transfer between independent donor and acceptor molecules. In either case we expect a pronounced interaction between the charge redistribution in the DA complex and the solvent. By changing the type of solvent it is possible to change the magnitude of the molecular dipoles. In this manner one can control an additional external parameter (besides temperature) influencing the ET characteristics.

Around the dissolved DA complex the so–called *first solvation shell* is formed, which is a layer of solvent molecules (see Fig. 6.23). In nonpolar solvents the mutual interaction is of short range and the first solvation shell will react very sensitively to any change of the charge distribution within the DA complex. However, in polar solvents more distant parts of the solvent will take notice of the ET reaction due to the long–range electrostatic forces. The DA complex interacts with a larger number of solvent molecules, and therefore the dynamics of the first solvation shell does not necessarily dominate the ET reaction. Thus, the application of a macroscopic continuum description neglecting any microscopic details of the solvent becomes possible. This is a main advantage of studying ET in polar solvents. It has been shown in many investigations that this somewhat coarser description of the solvent which utilizes the ideas of the macroscopic theory of dielectrics can yield good results [Cal83,Kes74,Kim90,Rip87].[20] The dielectric continuum model is based on the concept of the dielectric polarization field

[20]The approach fails in the presence of hydrogen bonds in the solvent or between the solvent and the solute. The different states of the ET reaction may alter the characteristics of these bonds and a continuum theory becomes inapplicable. Then the dynamics of the first solvation shell can be included in the description, for example, by supplementing the intramolecular vibrational degrees of freedom by some solvation shell coordinates.

(electric dipole density) $\mathbf{P}$, and the key quantity is given by the frequency dependent dielectric function $\varepsilon(\omega)$ of the solvent (compare Sections 2.5 and 5.2.1). The dielectric function can be determined independently of the ET reaction process and offers a characteristic macroscopic function fixing the dielectric properties of the solvent. The electron transferring DA complex contributes to the macroscopic polarization of the solvent by the electric field the complex generates. This *external* field enters the theory of dielectrics as the dielectric displacement field $\mathbf{D}$ (see Section 2.5).

In the following we concentrate on the nonadiabatic ET in a polar solvent. Therefore, we assume the existence of a diabatic representation with vibrational Hamiltonians $H_D$ and $H_A$ valid for the initial and final electronic state of the ET reaction plus a static coupling $V_{DA}$. For the ET reactions described in the foregoing sections it was typical that the vibrational Hamiltonian depends on intramolecular normal mode coordinates $q \equiv \{q_\xi\}$. Here, these coordinates will be supplemented by the values of the solvent polarization at any spatial point $\mathbf{x}$, that is, by the vector field $\mathbf{P}(\mathbf{x})$. Hence, we may write $H_{m=D,A}(q, \mathbf{P}(\mathbf{x}))$. The dependence on the polarization has to be understood in the sense of a functional dependence, and the set $q$ has to be interpreted as the set of intramolecular coordinates. It will be the task of the next sections to define the polarization field and to find a proper Hamiltonian.

## 6.5.1 The Solvent Polarization Field and the Dielectric Function

Since the dielectric function is of central importance, we briefly recall its definition and physical meaning (cf. Sections 2.5 and 5.2.1). If an (external) field $\mathbf{E}_{\text{ext}}$ is applied to a dielectric material, it becomes polarized. This polarization results from a deformation of the electron cloud of the molecules forming the dielectric. If these molecules possess a permanent dipole moment and if they are not very densely packed they may also reorient in the course of time. This results in the orientational polarization. Usually a deformation of the nuclear geometry is of less importance. All polarization contributions can be comprised in the macroscopic vector field of the dipole density (polarization vector) $\mathbf{P}(\mathbf{x}, t)$ (see Eq. (5.4)). There will be a dependence of the polarization field on the external field, i.e. $\mathbf{P}$ becomes an (in general) nonlinear functional $\mathbf{P}[\mathbf{E}_{\text{ext}}]$ of the electric field. Identifying this external field with the dielectric displacement (compare Eq. (2.51)) one can also write $\mathbf{P}[\mathbf{D}]$.

The mentioned functional can be derived using the general technique to calculate quantum mechanical (or quantum statistical) expectation values in the presence of an external perturbation (see Section 3.6.5). In this case the polarization has to be replaced by the operator $\hat{\mathbf{P}}$ (note the formal similarity to Eq. (5.4)). Moreover, we have to assume the existence of the Hamiltonian $H_{\text{P}}$ which determines the free motion of the solvent polarization (via the evolution operator $U_{\text{P}}(t) = \exp(-iH_{\text{P}}t/\hbar)$), what will be discussed in more detail in the next section. In accordance with the general considerations presented in Section 3.6.5 we have to introduce the perturbation $H_{\text{int}}(t)$ describing the coupling between the polarization field and the external field (similar to the expression in Eq. (5.9)). This coupling is formally accounted for by the field–dependent $S$–operator, and the expectation value of the polarization reads

$$\mathbf{P}(\mathbf{x}, t) = \text{tr}\left\{\hat{W}_{\text{eq}} S^{+}(t, t_0; \mathbf{E}_{\text{ext}}) U_{\text{P}}^{+}(t - t_0) \hat{\mathbf{P}}(\mathbf{x}) U_{\text{P}}(t - t_0) S(t, t_0; \mathbf{E}_{\text{ext}})\right\} , \quad (6.113)$$

where the $S$–operator is defined as (see Section 3.2.2)

$$S(t,t_0) = T\,\exp\left\{-\frac{i}{\hbar}\int\limits_{t_0}^{t} d\tau\; U_{\rm P}^{+}(\tau-t_0)H_{\rm int}(\tau)U_{\rm P}(\tau-t_0)\right\}\,. \tag{6.114}$$

If the externally applied field is weak (in comparison to the strength of the electric field inside and in between the molecules) the functional $\mathbf{P}[\mathbf{D}]$ can be linearized with respect to $\mathbf{D}$, leading to (details of such a linearization have been given in Section 3.6.5)

$$\mathbf{P}(\mathbf{x},t) = \int d^3\bar{\mathbf{x}} \int d\bar{t}\; \alpha(\mathbf{x},t;\bar{\mathbf{x}},\bar{t})\; \mathbf{D}(\bar{\mathbf{x}},\bar{t})\,. \tag{6.115}$$

The function mediating this linear dependence is known as the linear *molecular polarizability*; it is a second–rank tensor. (This symbol should not be confused with the absorption coefficient discussed in Chapter 5.) The linearization of Eq. (6.113) gives an explicit expression for the molecular polarizability

$$\alpha(\mathbf{x},t;\bar{\mathbf{x}},\bar{t}) = -\frac{i}{\hbar}\Theta(t-\bar{t})\mathrm{tr}\left\{\hat{W}_{\rm eq}\left[\hat{\mathbf{P}}(\mathbf{x},t),\hat{\mathbf{P}}(\bar{\mathbf{x}},\bar{t})\right]_-\right\}\,. \tag{6.116}$$

The two–fold dependence of the response function on spatial coordinates – called *spatial dispersion* – expresses spatial correlations between different molecules. The presence of spatial anisotropy in the solvent is accounted for by the tensor character of the molecular polarizability. For the following we assume a homogeneous and isotropic medium and neglect the spatial dispersion and anisotropy. Therefore, the molecular polarizability tensor reduces to a scalar which is independent of the spatial coordinates. [21] As any other type of correlation functions defined with respect to a thermal equilibrium state (cf. Section 3.6.5) the molecular polarizability depends only on the difference of the two time arguments.

Alternatively to the dependence of $\mathbf{P}$ on $\mathbf{D}$, one can define $\mathbf{P}$ as a functional of the total macroscopic electric field $\mathbf{E}$. (In the following we will write $\mathbf{E}$ instead of $\mathbf{E}_{\rm mac}$, cf. Section 2.5.1.) In doing so we already have taken into account that the dielectric medium was influenced by the external field leading to the resulting field $\mathbf{E}$. We use the relation (see Section 2.5.1)

$$\mathbf{D}(\mathbf{E}) = \mathbf{E} + 4\pi\mathbf{P}(\mathbf{E})\,, \tag{6.117}$$

and identify $-4\pi\mathbf{P}(\mathbf{E})$ as the field induced by the dielectric solvent.

If the dependence of the dielectric displacement on the electric field strength remains weak a linear dependence between both fields can be introduced (note that spatial dispersion and anisotropy are again neglected)

$$\mathbf{D}(\mathbf{x},t) = \int d\bar{t}\; \varepsilon(t-\bar{t})\; \mathbf{E}(\mathbf{x},\bar{t})\,. \tag{6.118}$$

Defining the dielectric function according to $\varepsilon(t) = \delta(t) + 4\pi\chi^{(1)}(t)$, the *linear susceptibility* $\chi^{(1)}(t)$ establishes a linear relation between $\mathbf{P}$ and $\mathbf{E}$. The introduction of the susceptibility

[21] More precisely the Cartesian tensor components $\alpha_{ij}$ are reduced to $\delta_{ij}\alpha$ with the single polarizability $\alpha$.

is very appealing because it describes the influence of the total field present in the solvent on the individual solvent molecules.

For the following it is of interest to have a relation between $\mathbf{P}$ and $\mathbf{D}$ expressed via the dielectric tensor. We use Eq. (6.118) to rewrite Eq. (6.117) as $\mathbf{D} = \varepsilon^{-1}\mathbf{D} + 4\pi\mathbf{P}$, which gives

$$\mathbf{P}(\mathbf{x}, t) = \frac{1}{4\pi} \int d\bar{t} \, \left(\delta(t - \bar{t}) - \varepsilon^{-1}(t - \bar{t})\right) \, \mathbf{D}(\mathbf{x}, \bar{t}) \, . \tag{6.119}$$

Here, the inverse of $\varepsilon$ has to be understood as the inverse of the integral operator used in Eq. (6.118). Comparing this with Eq. (6.115) the polarizability follows as

$$\alpha(t) = \frac{1}{4\pi} \, \left(\delta(t) - \varepsilon^{-1}(t)\right) \, . \tag{6.120}$$

Carrying out a Fourier transform of Eq. (6.118) one gets

$$\mathbf{D}(\mathbf{x}, \omega) = \varepsilon(\omega) \, \mathbf{E}(\mathbf{x}, \omega) \, , \tag{6.121}$$

and Eq. (6.119) reads

$$\mathbf{P}(\mathbf{x}, \omega) = \alpha(\omega) \, \mathbf{D}(\mathbf{x}, \omega) \tag{6.122}$$

with

$$\alpha(\omega) = \frac{1}{4\pi} \left(1 - \varepsilon^{-1}(\omega)\right) \, . \tag{6.123}$$

This frequency–domain function as well as the time–domain version, Eq. (6.119), contain the response of the solvent polarization to an externally applied time–dependent field.

For many applications it suffices to consider two distinct contributions to the complete polarization field $\mathbf{P}$. The first contribution $\mathbf{P}_{\mathrm{el}}$ refers to the polarization of the electron cloud of the molecules. Due to the small mass of the electrons compared to that of the nuclei and of the total molecule this polarization may respond to high–frequency external fields. It forms the high–frequency part $\mathbf{P}_\infty$ of the total polarization field, $\mathbf{P}_\infty \equiv \mathbf{P}_{\mathrm{el}}$. The second contribution is related to an orientational polarization $\mathbf{P}_{\mathrm{or}}$ which follows from an reorientation of the molecules carrying a permanent dipole moment. This type of polarization responds much more slowly to an external disturbance than $\mathbf{P}_{\mathrm{el}}$. Together with the electronic contribution it results in the low–frequency part $\mathbf{P}_0$, and we can set $\mathbf{P}_0 \equiv \mathbf{P}_{\mathrm{or}} + \mathbf{P}_{\mathrm{el}}$.

This simple separation of $\mathbf{P}$ into a high and a low–frequency part can be expressed by the following type of dielectric function

$$\varepsilon(\omega) = \varepsilon_\infty + \frac{\varepsilon_0 - \varepsilon_\infty}{1 + i\tau_{\mathrm{D}}\omega}. \tag{6.124}$$

It describes the so–called *Debye dielectric relaxation*, approaching in the high–frequency range ($\omega \to \infty$) the value $\varepsilon_\infty$, whereas for very low frequencies ($\omega \to 0$) it gives $\varepsilon_0$. The intermediate range where the dielectric function becomes frequency dependent is characterized by the relaxation rate $1/\tau_{\mathrm{D}}$.

We briefly comment on the relation between the low– and the high–frequency limits of the dielectric function (6.124) and the respective components of the polarization field. Suppose that the external field $\mathbf{D}$ is monochromatic. Then, in the low–frequency limit, we can write

$$\mathbf{D}(t) = \mathbf{D}_{\text{low}} e^{-i\omega_{\text{low}} t} + \text{c.c.} \ . \tag{6.125}$$

Inserting this into Eq. (6.119) gives

$$\begin{aligned}\mathbf{P}_0 &= \frac{1}{4\pi}\left(\{1-\varepsilon^{-1}(\omega_{\text{low}})\}\mathbf{D}_{\text{low}} e^{-i\omega_{\text{low}} t} + \{1-\varepsilon^{-1}(-\omega_{\text{low}})\}\mathbf{D}^*_{\text{low}} e^{i\omega_{\text{low}} t}\right) \\ &\approx \frac{1}{4\pi}\left(1-\frac{1}{\varepsilon_0}\right)\mathbf{D} \ .\end{aligned} \tag{6.126}$$

If exclusively high–frequency components are contained in the external field, one can deduce in the same manner

$$\mathbf{P}_\infty = \frac{1}{4\pi}\left(1-\frac{1}{\varepsilon_\infty}\right)\mathbf{D} \ . \tag{6.127}$$

In contrast to this expression, Eq. (6.126) for the low–frequency part does not give the orientational contribution alone. $\mathbf{P}_0$ also contains an electronic contribution which has to be eliminated. We get

$$\mathbf{P}_{\text{or}} = \mathbf{P}_0 - \mathbf{P}_\infty = \frac{1}{4\pi}\left(\frac{1}{\varepsilon_\infty}-\frac{1}{\varepsilon_0}\right)\mathbf{D} \ . \tag{6.128}$$

The combination

$$c_{\text{Pek}} = \frac{1}{\varepsilon_\infty} - \frac{1}{\varepsilon_0} \tag{6.129}$$

of the inverse dielectric constants is known as the *Pekar* factor.

## 6.5.2 The Free Energy of the Solvent

In order to complete the total ET Hamiltonian, Eq. (6.29) by solvent contributions we will use the model of two contributions $\mathbf{P}_{\text{el}}$ and $\mathbf{P}_{\text{or}}$ to the solvent polarizations introduced in the preceding section. For simplicity we will not take into account quantum effects, and thus we will not obtain a Hamiltonian but a free energy functional. This treatment implies the interpretation of the polarization field as a special type of generalized coordinate. The essential point for the following considerations is that the solvent will be exclusively described by its dielectric function. Thus, we assume the validity of *linear response* theory for the description of the solvent polarization.

The supplementary Section 6.9.3 is devoted to a detailed introduction of such a free energy functional $F_{\text{P}}[\mathbf{P}_{\text{el}}, \mathbf{P}_{\text{or}}, \mathbf{D}]$, which depends on both types of solvent polarizations as well as on the external field $\mathbf{D}$ (which is due to the electron transferring DA complex). As a central supposition for the following we assume that the rate of ET is much larger than the frequency of the orientational polarization but at the same time much smaller than the frequency of the electronic polarization. This enables us to introduce a reduced free energy functional which is only determined by nonequilibrium values of the orientational polarization. The fast electronic polarization is described by its equilibrium values corresponding to the actual state of the ET as well as the form of the orientational polarization. The derivation of this reduced functional, Eq. (6.230) is also presented in the supplementary Section 6.9.3. It is simplified if one reduces it to a two–state expression with the transferred electron being either at the donor or the acceptor (any intermediate state is neglected). Moreover, we neglect all other vibrational degrees of freedom of the solvent. Then the expectation value of $H_{\mathrm{DBA}}$ entering the reduced functional, Eq. (6.230) is approximated by the electronic energies $E_D$ or $E_A$. The displacement field $\mathbf{D}_m$ is fixed for the case where the electron is at the donor or at the acceptor (cf. Eq. (6.224)). In this manner we obtain two diabatic free energy surfaces ($m = D, A$)

$$\begin{aligned} F_m[\tilde{\mathbf{P}}, \mathbf{D}_m] &= E_m - \frac{1}{2c_{\mathrm{el}}} \int d^3\mathbf{x}\, \mathbf{D}_m^2(\mathbf{x}) \\ &+ \frac{2\pi}{c_{\mathrm{Pek}}} \int d^3\mathbf{x} \tilde{\mathbf{P}}^2(\mathbf{x},t) - \int d^3\mathbf{x} \mathbf{D}_m(\mathbf{x}) \tilde{\mathbf{P}}(\mathbf{x},t) \; . \end{aligned} \tag{6.130}$$

Besides the electronic levels renormalized with respect to the solvation energy (second term on the right–hand side), the free energy contains a freely fluctuating polarization contribution. It is the renormalized orientational polarization $\tilde{\mathbf{P}} = \mathbf{P}_{\mathrm{or}}/\varepsilon_\infty$. In addition we have, as a coupling to an external field, the interaction between the solvent polarization and the DA complex (last term). The new constant $c_{\mathrm{el}}$ is defined in supplementary Section 6.9.3 (cf. Eq. (6.229)).

We can use Eq. (6.130) to introduce the diabatic vibrational Hamiltonian with diabatic PES including the orientational polarization of a polar solvent. Additionally, we are in a position to calculate the reorganization energy of the solvent.

The diabatic PES of the DA complex can be directly deduced from Eq. (6.130) if we identify $F_m[\tilde{\mathbf{P}}, \mathbf{D}]$ with the PES $U_m[\tilde{\mathbf{P}}]$ of the solvent polarization. This identification of a potential energy with a free energy is possible, since the free energy functional has been introduced as a quantity depending on quasi–static fields without any kinetic energy contribution. According to the considered state, the related PES leads to a distinct equilibrium value of $\tilde{\mathbf{P}}$. It is obtained from $\delta U_m/\delta\tilde{\mathbf{P}} = 0$ and given by

$$\tilde{\mathbf{P}}_{\mathrm{or},m}^{(\mathrm{eq})} = \frac{c_{\mathrm{Pek}}}{4\pi} \mathbf{D}_m \; . \tag{6.131}$$

To illustrate the obtained result we calculate the complete equilibrium free energy functional by inserting the equilibrium polarization into Eq. (6.130). One obtains:

$$F_m^{(\mathrm{eq})}[\mathbf{D}_m] = E_m - \frac{1}{2c_{\mathrm{el}}} \int d^3\mathbf{x}\, \mathbf{D}_m^2(\mathbf{x}) - \frac{c_{\mathrm{Pek}}}{8\pi} \int d^3\mathbf{x}\, \mathbf{D}_m^2(\mathbf{x}) \; . \tag{6.132}$$

The last term gives the orientational contribution to the so–called *Born solvation energy* $1/2c_{\mathrm{el}} \times \int d^3\mathbf{x} \mathbf{D}^2(\mathbf{x},t)$ (see also Eq. (6.231)). Therefore, the total free energy expression describes complete solvation at thermal equilibrium conditions including electronic and orientational contributions. It reads

$$F_m^{(\mathrm{eq})}[\mathbf{D}_m] = E_m - \frac{1}{8\pi} \left(1 - \frac{1}{\varepsilon_0}\right) \int d^3\mathbf{x}\, \mathbf{D}_m^2(\mathbf{x}) \; . \tag{6.133}$$

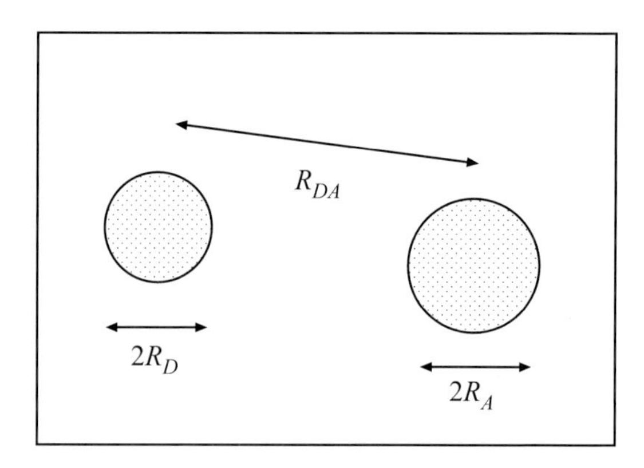


**Figure 6.24:** Simple scheme of a DA complex in a polar solvent. The solvent is approximated as a dielectric continuum. The donor as well as the acceptor are described as conducting spheres of radius $R_D$ and $R_A$, respectively, carrying the charge of the transferred electron. The intra–complex distance is denoted by $R_{\text{AD}}$.

Now, we are in the position to compute the reorganization energy as the energy of the product state (electron at the acceptor) if the polarization has the equilibrium value valid for the reactant state (electron at the donor):

$$E_\lambda^{(\text{sol})} = U_A[\tilde{\mathbf{P}}^{(\text{eq})}_{\text{or},D}] - U_A[\tilde{\mathbf{P}}^{(\text{eq})}_{\text{or},A}] = \frac{c_{\text{Pek}}}{8\pi} \int d\mathbf{x}^3 \; (\mathbf{D}_A(\mathbf{x}) - \mathbf{D}_D(\mathbf{x}))^2 \; . \qquad (6.134)$$

Usually one approximates this expression further in assuming a spherical charge distribution of the transferred electron at the donor as well as the acceptor (i.e., $\mathbf{D}(\mathbf{x}) = e\mathbf{x}/|\mathbf{x}|^3$, see Fig. 6.24):

$$E_\lambda^{(\text{sol})} = \frac{c_{\text{Pek}}}{8\pi}\Big(\frac{e^2}{2R_D} + \frac{e^2}{2R_A} - \frac{e^2}{R_{AD}}\Big) \; . \qquad (6.135)$$

Here, the $R_m$ are the radii of the spheres, and $R_{DA}$ denotes the distance between the donor and acceptor.

Since our treatment of the solvent polarization turns out to be identical with the description of vibrational modes coupled to the ET reaction, we can argue that the reorganization energy $E_\lambda^{(\text{sol})}$, Eq. (6.134), represents the solvent contribution to the complete reorganization energy $E_\lambda$. Together with the vibrational contribution $E_\lambda^{(\text{vib})}$, Eq. (6.74), we obtain $E_\lambda = E_\lambda^{(\text{vib})} + E_\lambda^{(\text{sol})}$. This expression enters the Marcus rate formula Eq. (6.73). An independent justification of interpreting $E_\lambda^{(\text{sol})}$ as the solvent reorganization energy will be given in the following section. This is achieved by embedding the solvent influence on the ET reaction into a general computational scheme for the respective transfer rate.

### 6.5.3 The Rate of Nonadiabatic Electron Transfer in Polar Solvents

In Section 6.4.3 we obtained a general formula for the nonadiabatic ET rate. Since we will concentrate here on the nonadiabatic case, too, we will take the transfer rate according to Eq. (6.89) as $k_{\rm ET} \equiv k_{DA} = 2\pi|V_{\rm DA}|^2\mathcal{D}(\Delta E)/\hbar$. It will be the aim of the following considerations to express the density of states $\mathcal{D}$ by the dielectric function of the polar solvent. Any reference to further intramolecular vibrations will be neglected, but how to include them should be obvious from Section 6.4.5.

As in the foregoing section we will assume a clear separation of the three characteristic time constants involved, the ET time $1/k_{\rm ET}$, and the time constants $1/\omega_{\rm high}$ and $1/\omega_{\rm low}$ for the electronic and orientational polarization of the solvent, respectively. This enables us to use the free energy expression, Eq. (6.130) to introduce the diabatic vibrational Hamiltonians $H_D$ and $H_A$. In principle this would require two additional steps: First, we have to define a proper kinetic energy expression for the polarization field, and second the quantization of the polarization has to be carried out. However, it will become clear in the following that there is no need to have the kinetic energy operator at hand. Furthermore, it is not necessary to carry out the polarization field quantization explicitly. We only suppose that it is possible to replace the nonequilibrium polarization field $\tilde{\mathbf{P}}$ by a quantum mechanical operator $\hat{\mathbf{P}}$. Since $\hat{\mathbf{P}}$ is only needed to define a quantum statistical correlation function related to the dielectric function we do not further comment on the replacement of $\tilde{\mathbf{P}}$ by $\hat{\mathbf{P}}$ (see, e.g. [Kim90]).

A general expression for the density of states valid for nonadiabatic transitions has been already derived in Section 5.3.3. We can directly use the result to obtain a general formula for nonadiabatic ET reactions considered here. Following Eq. (5.39) we write

$$\mathcal{D}(\Delta E/\hbar) = \frac{1}{2\pi\hbar}\int dt\; e^{i\Delta Et/\hbar}\; \mathrm{tr}\{\hat{W}_{\rm eq}^{(D)} S_{AD}(t,0)\}\;, \tag{6.136}$$

where $\Delta E$ denotes the operator–free part of the difference Hamiltonian $H_A - H_D$. The statistical operator $\hat{W}_{\rm eq}^{(D)}$ used to determine the expectation value of the $S$–operator describes the thermal equilibrium of the solvent if the donor state is occupied. The kinetic energy contributions cancel in $H_A - H_D$. This result makes it clear that it is not necessary to specify the kinetic energy part. We identify the difference $H_A - H_D$ by the difference of the free energy expressions Eq. (6.130), replace the orientational polarization field by the related quantum mechanical operator, and obtain

$$H_A - H_D = -\Delta E - \int d^3\mathbf{x}\; \Delta\mathbf{D}_{AD}(\mathbf{x})\; \hat{\mathbf{P}}(\mathbf{x})\;. \tag{6.137}$$

The driving force of the ET, $\Delta E$, includes the reorganization of the electronic levels by the Born solvation energy

$$\Delta E = E_D - E_A - \frac{1}{2c_{\rm el}}\int d^3\mathbf{x}\; \left(\mathbf{D}_D^2(\mathbf{x}) - \mathbf{D}_A^2(\mathbf{x})\right)\;, \tag{6.138}$$

and we used $\Delta\mathbf{D}_{AD}(\mathbf{x}) = \mathbf{D}_A(\mathbf{x}) - \mathbf{D}_D(\mathbf{x})$. The $S$–operator in Eq. (6.136) reads

$$S_{AD}(t,0) = \hat{T}\exp\left\{-\frac{i}{\hbar}\int\limits_0^t d\bar{t}\; \Delta H_{AD}^{(D)}(\bar{t})\right\}\;. \tag{6.139}$$

As in Section 5.3.3 we write

$$\Delta H_{AD}^{(D)}(\bar{t}) = U_D^+(\bar{t})\,(H_A - H_D + \Delta E)U_D(\bar{t}) = -\int d^3\mathbf{x}\,\Delta\mathbf{D}_{\mathrm{AD}}(\mathbf{x})\hat{\mathbf{P}}^{(D)}(\mathbf{x},\bar{t})\;. \tag{6.140}$$

The superscript $D$ indicates that the Heisenberg representation defined via $H_D$ has to be taken. Since the above expressions have been defined by the dielectric displacement fields, we expect to obtain correlation functions which are of the type of molecular polarizabilities $\alpha$, Eq. (6.123) (instead of susceptibilities $\chi$).

The density of states Eq. (6.136) can be calculated using the cumulant expansion discussed in Section 5.3.3. According to Eq. (5.80) we obtain the second–order cumulant approximation as $\mathrm{tr}_{\mathrm{vib}}\{\hat{R}_D S_{AD}(t,0)\} \equiv \langle S_{AD}(t,0)\rangle_D \approx \exp(-\Gamma_1(t) - \Gamma_2(t))$. The two quantities in the exponent are given by (see Eq. (5.84) and Eq. (5.85), respectively)

$$\Gamma_1(t) = \frac{i}{\hbar}\int_0^t dt_1\;\langle \Delta H_{AD}^{(D)}(t_1)\rangle_D\;, \tag{6.141}$$

and

$$\Gamma_2(t) = \frac{1}{\hbar^2}\int_0^t dt_1 \int_0^{t_1} dt_2\;\langle \Delta H_{AD}^{(D)}(t_1)\;\Delta H_{AD}^{(D)}(t_2)\rangle_D + \frac{1}{2}\,\Gamma_1^2(t)\;. \tag{6.142}$$

For the first quantity $\Gamma_1$ we obtain

$$\begin{aligned}\Gamma_1(t) &= -\frac{i}{\hbar}\int_0^t dt_1 \int d^3\mathbf{x}\;\Delta\mathbf{D}_{AD}(\mathbf{x})\;\langle \hat{\mathbf{P}}^{(D)}(\mathbf{x},t_1)\rangle_D \\ &= -\frac{i}{\hbar}\,t\int d^3\mathbf{x}\;\Delta\mathbf{D}_{AD}(\mathbf{x})\;\tilde{\mathbf{P}}_{\mathrm{or},D}^{(\mathrm{eq})}(\mathbf{x}) \\ &\equiv -\frac{i}{\hbar}\,t\,\frac{c_{\mathrm{Pek}}}{4\pi}\int d^3\mathbf{x}\;\Delta\mathbf{D}_{AD}(\mathbf{x})\;\mathbf{D}_D(\mathbf{x})\;.\end{aligned} \tag{6.143}$$

The expectation value of the polarization operator has been replaced by the equilibrium polarization $\tilde{\mathbf{P}}_{\mathrm{or},D}^{(\mathrm{eq})}$ present if the transferred electron is fixed at the donor (see Eq. (6.131)). The quantity $\Gamma_2$ includes a polarization correlation function[22]

$$\begin{aligned}\Gamma_2(t) &= \frac{1}{2}\Gamma_1^2(t) + \frac{1}{\hbar^2}\int_0^t dt_1 \int_0^{t_1} dt_2 \int d^3\mathbf{x}_1 d^3\mathbf{x}_2 \\ &\quad\times \Delta\mathbf{D}_{AD}(\mathbf{x}_1)\;\langle \hat{\mathbf{P}}^{(D)}(\mathbf{x}_1,t_1)\hat{\mathbf{P}}^{(D)}(\mathbf{x}_2,t_2)\rangle_D\;\Delta\mathbf{D}_{AD}(\mathbf{x}_2)\;.\end{aligned} \tag{6.144}$$

[22]This expression has to be understood as formed by two scalar products of the type $\Delta\mathbf{D}_{AD}\hat{\mathbf{P}}^{(D)}$.

This correlation function can be related to the molecular polarizability if one splits up the complete polarization operator according to

$$\hat{\mathbf{P}}^{(D)}(\mathbf{x},t) = \tilde{\mathbf{P}}^{(\text{eq})}_{\text{or},D} + \Delta\hat{\mathbf{P}}^{(D)}(\mathbf{x},t) \ . \tag{6.145}$$

The operator $\Delta\hat{\mathbf{P}}^{(D)}$ describes the deviation from the equilibrium polarization $\tilde{\mathbf{P}}^{(\text{eq})}_{\text{or},D}$ and can be easily related to the dielectric function. This is possible since the fluctuation operator does not dependent on the electronic charge at the donor.

Inserting Eq. (6.145) into Eq. (6.144) the contribution of $\tilde{\mathbf{P}}^{(\text{eq})}_{\text{or},D}$ to the polarization correlation function is equal to $-\Gamma_1^2(t)/2$, and we obtain

$$\Gamma_2(t) = \frac{1}{\hbar^2}\int_0^t dt_1 \int_0^{t_1} dt_2 \int d^3\mathbf{x}_1 d^3\mathbf{x}_2 \ \Delta\mathbf{D}_{AD}(\mathbf{x}_1)\ C_{\text{P-P}}(\mathbf{x}_1,t_1;\mathbf{x}_2,t_2)\ \Delta\mathbf{D}_{AD}(\mathbf{x}_2) \ , \tag{6.146}$$

where the correlation function (tensor) of polarization fluctuations

$$C_{\text{P-P}}(\mathbf{x}_1,t_1;\mathbf{x}_2,t_2) = \langle \Delta\hat{\mathbf{P}}^{(D)}(\mathbf{x}_1,t_1)\Delta\hat{\mathbf{P}}^{(D)}(\mathbf{x}_2,t_2)\rangle \ , \tag{6.147}$$

has been introduced. Note that the index D at the expectation value could have been omitted since the respective statistical operator (cf. Eq. (6.136)) is independent of $\tilde{\mathbf{P}}^{(\text{eq})}_{\text{or},D}$. The correlation function includes freely fluctuating polarization fields, and it can be linked to the dielectric function. To this end, we neglect spatial dispersion and anisotropy to get a scalar and coordinate independent correlation function

$$C_{\text{P-P}}(\mathbf{x}_1,t_1;\mathbf{x}_2,t_2) \rightarrow \delta(\mathbf{x}_1 - \mathbf{x}_2)C_{\text{P-P}}(t_1 - t_2) \ . \tag{6.148}$$

The result is a single spatial integral with respect to the square of $\Delta\mathbf{D}_{AD}$. Using Eq. (6.134) for the solvent reorganization energy one can write

$$\Gamma_2(t) = \frac{8\pi E_\lambda^{(\text{sol})}}{\hbar^2 c_{\text{Pek}}}\int_0^t dt_1 \int_0^{t_1} dt_2 \ C_{\text{P-P}}(t_1 - t_2) \ . \tag{6.149}$$

Next we show how the correlation function can be expressed by the frequency dependent dielectric function. For this purpose we note that $C_{\text{P-P}}(t)$ is defined with respect to the complete time axis, whereas the dielectric function vanishes for $t < 0$. Therefore, we introduce the antisymmetrized correlation function (compare Section 3.6), $C^{(-)}_{\text{P-P}}(t) = C_{\text{P-P}}(t) - C^{*}_{\text{P-P}}(t)$, which directly gives the linear polarizability

$$\alpha(t) = -\frac{i}{\hbar}\Theta(t)C^{(-)}_{\text{P-P}}(t) \ . \tag{6.150}$$

Performing a Fourier transform results in[23]

$$\alpha(\omega) = \int \frac{d\bar{\omega}}{2\pi\hbar}\,\frac{C^{(-)}_{\text{P-P}}(\bar{\omega})}{\omega - \bar{\omega} + i\epsilon} \ . \tag{6.151}$$

[23] Recall that the step function $\Theta(t)$ can be expressed by the Fourier integral $-\int \frac{d\omega}{2\pi i}\exp(-i\omega t)/(\omega + i\epsilon)$.

We further note that $C^{(-)}_{\text{P-P}}(\omega) \equiv \text{Re}\, C^{(-)}_{\text{P-P}}(\omega)$. Therefore, taking the imaginary part of Eq. (6.151) we obtain $\text{Im}\,\alpha(\omega) = -C^{(-)}_{\text{P-P}}(\omega)/2\hbar$. In a next step we relate $\alpha(\omega)$ to $\varepsilon(\omega)$. According to Eq. (6.123) it follows that $\text{Im}\,\alpha = -\text{Im}\,\varepsilon^{-1}/4\pi \equiv \text{Im}\varepsilon/4\pi|\varepsilon|^2$. Finally, we use Eq. (3.204), which establishes a relation between the time–dependent correlation function and its Fourier–transformed antisymmetrized version, to get

$$C_{\text{P-P}}(t) = -\frac{\hbar}{4\pi^2} \int_0^\infty d\omega \, \left(e^{-i\omega t}\,[1+n(\omega)] + e^{i\omega t}\, n(\omega)\right) \frac{\text{Im}\,\varepsilon(\omega)}{|\varepsilon(\omega)|^2} \,. \tag{6.152}$$

Now we have a representation of the correlation function $C_{\text{P-P}}(t)$ which allows us to carry out the double time integration in Eq. (6.149). We obtain

$$\begin{aligned} \Gamma_2(t) &= -\frac{2E_\lambda^{(\text{sol})}}{\pi\hbar c_{\text{Pek}}} \int_0^\infty \frac{d\omega}{\omega^2} \frac{\text{Im}\,\varepsilon(\omega)}{|\varepsilon(\omega)|^2} \\ &\times \left([e^{-i\omega t}-1]\,[1+n(\omega)] + [e^{i\omega t}-1]\, n(\omega) + i\omega t\right) \,. \end{aligned} \tag{6.153}$$

A comparison with the ET rate expression, Eqs. (6.90) and (6.92), and in particular with the function $G(t)$, Eq. (6.93) and (6.96) enables us to introduce the spectral density of the polar solvent:

$$J^{(\text{sol})}(\omega) = \frac{2E_\lambda^{(\text{sol})}}{\pi\hbar c_{\text{Pek}}} \frac{1}{\omega^2} \frac{\text{Im}\,\varepsilon(\omega)}{|\varepsilon(\omega)|^2} \,. \tag{6.154}$$

Inserting the simple dielectric function, Eq. (6.124) one obtains

$$J^{(\text{sol})}(\omega) = \frac{2E_\lambda^{(\text{sol})}}{\pi\hbar} \frac{\omega_{\text{long}}}{\omega} \frac{1}{\omega^2 + \omega_{\text{long}}^2} \,. \tag{6.155}$$

This is just the spectral density of the Debye type already introduced in Eq. (3.230), where the frequency $\omega_{\text{long}}$ is given by $\varepsilon_0/\varepsilon_\infty\,\omega_{\text{D}}$. The inverse of $\omega_{\text{long}}$ is known as the *longitudinal* relaxation time of the solvent.

It can easily be shown that the relation (6.95) is fulfilled, and we have

$$E_\lambda^{(\text{sol})} = \int_0^\infty d\omega \, \hbar\omega J^{(\text{sol})}(\omega) \,. \tag{6.156}$$

In particular, this yields

$$\Gamma_2(t) = G(0) - G(t) - i\frac{E_\lambda^{(\text{sol})}}{\hbar} t \,, \tag{6.157}$$

with the function $G(t)$ from Eq. (6.96) where $J_{DA}$ is replaced by $J^{(\text{sol})}$.

To get the final expression for the ET rate according to Eqs. (6.90) and (6.136), we collect all terms contributing to the exponent under the time integral in the definition of the density of states. All terms except $G(t)$ can be combined to get the free energy difference

$$\Delta F^{(\text{eq})} = F_{\text{D}}^{(\text{eq})}(\mathbf{D}_{\text{D}}) - F_{\text{A}}^{(\text{eq})}(\mathbf{D}_{\text{A}}) \,, \tag{6.158}$$

where $F_m^{(\text{eq})}(\mathbf{D}_m)$, $(m = D, A)$ has been introduced in Eq. (6.132) as the complete equilibrium Born solvation free energy (including electronic and orientational contributions). Accordingly, the final result for the ET rate in polar solvents reads

$$k_{\text{ET}} = \frac{1}{\hbar^2}|V_{DA}|^2 \int dt \, \exp\left\{\frac{i}{\hbar}\Delta F^{(\text{eq})}t + G(t) - G(0)\right\} . \tag{6.159}$$

The expression is similar to that used in Section 6.4.3. But the driving force $\Delta E$ given as a simple energy difference is replaced here by the equilibrium free energy difference (for the polarity dependence of the solvent surrounding a charge transfer reaction we refer to Fig. 5.15).

## 6.6 Bridge–Assisted Electron Transfer

In many cases the simple picture of a direct transfer of an electron from the donor site to the acceptor site does not apply. The reaction may proceed across bridging units between the donor and the acceptor (cf. Section 6.1). In cases where the donor and acceptor are connected by a rather rigid polymer strand, the bridging units can be considered as a linear arrangement of identical sites; an example is given in Fig. 6.25 (see also Fig. 6.10). A less homogeneous bridge structure is encountered in the ET system of the bacterial photosynthetic reaction center shown in Fig. 6.2. Finally, considering ET in proteins the bridge becomes a three–dimensional network of LUMOs (of the amino acid residues) connecting donor and acceptor.

Bridge–mediated ET may take place via two different mechanisms: the superexchange ET or the sequential (hopping) transfer (cf. Fig. 6.9). In the first case the bridge units support a delocalization of the donor state wave function (upper panel of Fig. 6.26). This delocalization will essentially modify the (electronic) coupling between the donor and the acceptor which can be expressed by introducing an *effective* DA transfer integral. Since an extended electronic wave function is formed, a definite phase relation between the electronic states of the different bridge units as well as the bridge and the donor exists. According to this picture superexchange ET is intimately connected to the presence of electronic *coherences*. Due to the off–resonance conditions between the donor level and the bridge levels, small energetic fluctuations of the levels due to a weak coupling to vibrational modes should have a minor effect. However, a *strong* vibrational modulation of the energy levels of the bridge units or of the transfer coupling between them, might prevent the formation of a delocalized electronic wave function.

A delocalized wave function and thus coherences in the ET can be also found if the bridge states are in near resonance to the donor and acceptor levels as is the case in the lower panel of Fig. 6.9. But this requires that the time scale for electronic motion is comparable to or even faster than characteristic vibrational relaxation times. As mentioned above, the vibrational modulation of the electronic states and their mutual coupling can become predominant such that an extended wave function cannot be formed. The electron jumps from bridge level to bridge level and one has a *sequential* ET as shown in the lower panel of Fig. 6.26. We expect that in the case of fast vibrational relaxation, $\tau_{\text{rel}} \ll t_{\text{el}}$, this type of ET can be described by a set of rate equations (cf. Section 3.4.5) which includes various ET rates connecting different bridging sites (see Section 6.4).

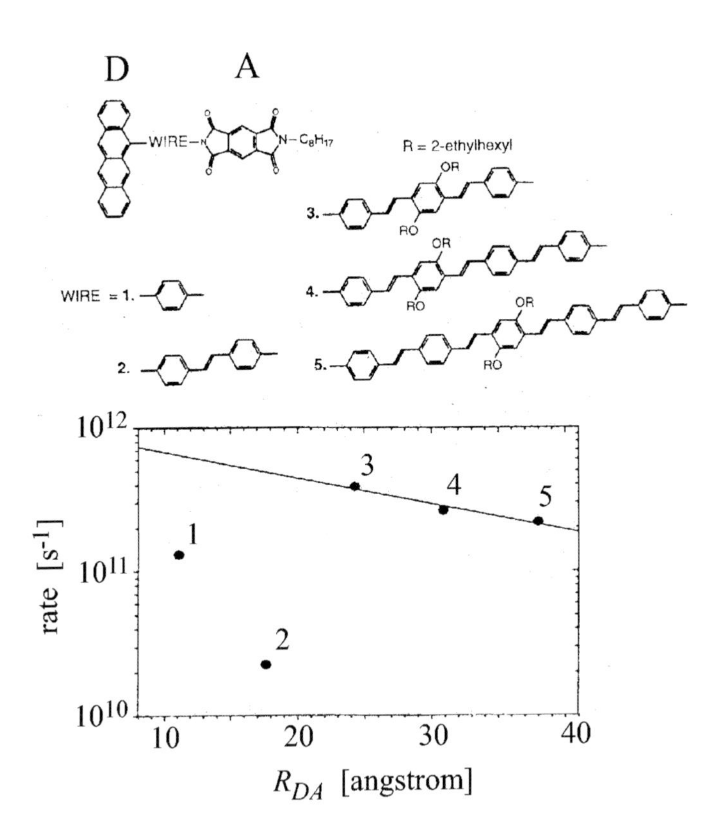


**Figure 6.25:** Bridge–mediated ET using a molecular wire of p–phenylenevinylene oligomers. The donor is given by tetracene and the acceptor by pyroelectricity. The five different types of wires together with the donor and acceptor are shown in the upper panel (DA distances $R_{DA}$ for wire **1** up to **5** are 11.1, 17.7, 24.3, 30.9, and 38.0 Å). The distance dependence of the transfer rate is shown in the lower panel. (reprinted with permission from [Dav98]; copyright (1998) Nature)

In the following discussion of bridge–mediated ET we will concentrate on the superexchange mechanism. Usually, the transfer coupling between the various units is not so strong and the ET takes place in the nonadiabatic regime. Since superexchange ET is of the through–bond type (cf. Section 6.1) the incorporation of intermediate units increases the rate compared to the case where no bridging units are present. In this latter through–space type of reactions, the ET rate is proportional to the square of the transfer integral and therefore determined by the tails of the overlapping donor and acceptor wave functions (cf. Eq. (6.28)). It will be of interest in the following to understand how the intermediate bridge molecules influence the ET rate.

The appropriate Hamiltonian for the present case has already been introduced in Eq. (6.29). In order to have a clear identification of the donor and acceptor levels we write (remember

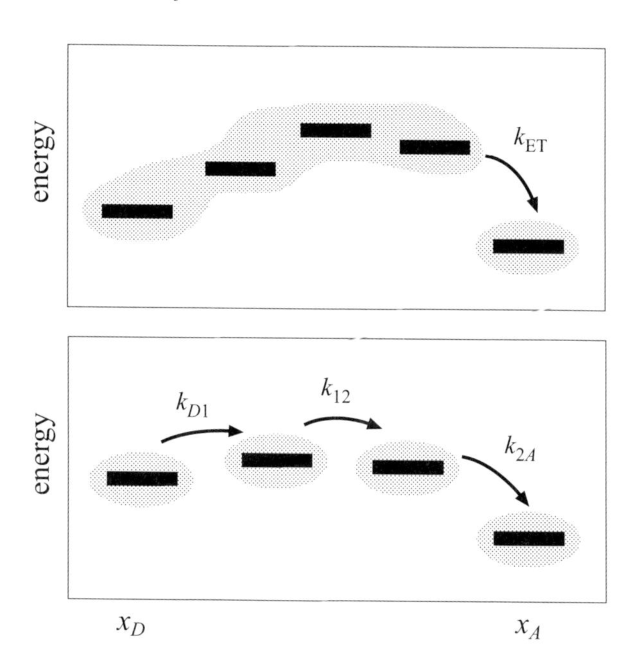


**Figure 6.26:** Bridge–mediated ET between a donor and an acceptor level. The upper part gives a scheme of the superexchange ET where the initial state wave function (shaded area) extends over the whole bridge. For the sequential ET (lower part) the electronic wave function is localized on the various sites during the transfer.

$|\varphi_D\rangle = |D\rangle$ and $|\varphi_A\rangle = |A\rangle$)

$$\begin{aligned} H_{\text{DBA}} &= H_D|D\rangle\langle D| + \sum_m (V_{Dm}|D\rangle\langle\varphi_m| + \text{h.c.}) \\ &+ H_A|A\rangle\langle A| + \sum_m (V_{Am}|A\rangle\langle\varphi_m| + \text{h.c.}) + H_{\text{bridge}} \,. \end{aligned} \tag{6.160}$$

The bridge Hamiltonian $H_{\text{bridge}}$ is identical to expression (6.29) with the summation restricted to the bridge sites, $m = 1, \ldots, N_{\text{B}}$. Note that in the most general way the model should include that donor and acceptor levels may couple to every level of the bridge via the transfer integrals $V_{Dm}$ and $V_{Am}$.

### 6.6.1 The Superexchange Mechanism

To discuss the way a molecular bridge mediates the ET from the donor to the acceptor we first consider the case of a single bridge unit. For such a situation the bridge Hamiltonian is written as: $H_{\text{bridge}} = H_B|B\rangle\langle B|$ (note $|B\rangle \equiv |\varphi_1\rangle$, the related vibrational Hamiltonian is denoted as $H_\text{B}$). Furthermore, two transfer integrals $V_{DB}$ and $V_{AB}$ appear which couple the bridge to the donor and the acceptor, respectively.

Let us first derive the bridge–mediated effective transfer integral without the consideration of vibrational contributions. The delocalization of the donor wave function induced by the bridge can be estimated by perturbation theory. The lowest–order correction to the donor state $|D\rangle$ following from the coupling to the bridge is given by

$$\delta|D\rangle = \frac{V_{DB}^*}{E_D - E_B}|B\rangle \, . \tag{6.161}$$

In the nonadiabatic scheme of ET (cf. the qualitative discussion in Section 6.3) the rate is calculated via the Golden Rule formula. According to Section 3.3 we need the square of the effective coupling matrix element $V_{DA}^{(\text{eff})}$ between the modified donor state [24] $|D\rangle + \delta|D\rangle$ and the acceptor state $|A\rangle$. The coupling is obtained as

$$V_{DA}^{(\text{eff})} = \Big(\langle D| + \frac{V_{DB}}{E_D - E_B}\langle B|\Big) \times V_{BA}|B\rangle\langle A| \times |A\rangle = \frac{V_{DB}V_{BA}}{E_D - E_B} \, . \tag{6.162}$$

The formula holds as long as $|E_D - E_B|$ is nonzero and larger than $\sqrt{|V_{DB}V_{BA}|}$. This simple calculation can easily be extended by incorporating the vibrational levels of the DBA–system. Now, the correction of the electron vibrational donor state $|D\rangle|\chi_{DM}\rangle$ follows as

$$\delta|D\rangle|\chi_{DM}\rangle = \sum_K \frac{V_{DB}^*\langle\chi_{DM}|\chi_{BK}\rangle^*}{E_{DM} - E_{BK}}|B\rangle|\chi_{BK}\rangle \, . \tag{6.163}$$

As in Section 6.2.4 we assumed the independence of the coupling matrix elements on the vibrational coordinates. Additionally, we introduced the electron–vibrational energies, Eq. (6.39). Then, the effective DA–coupling Eq. (6.162) is generalized to the following expression

$$\begin{aligned} V_{DM,AN}^{(\text{eff})} &= \Big(\langle D|\langle\chi_{DM}| + \sum_K \frac{V_{DB}\langle\chi_{DM}|\chi_{BK}\rangle}{E_{DM} - E_{BK}}\langle B|\langle\chi_{BK}|\Big) \\ &\quad \times V_{BA}|B\rangle\langle A| \times |A\rangle|\chi_{AN}\rangle \\ &= \sum_K \frac{V_{DB}V_{BA}\langle\chi_{DM}|\chi_{BK}\rangle\langle\chi_{BK}|\chi_{AN}\rangle}{E_{DM} - E_{BK}} \, . \end{aligned} \tag{6.164}$$

Since this expression directly connects the manifold of vibrational states of the donor with that of the acceptor we can introduce it into formula (6.89) to get the superexchange ET rate as

$$k_\text{ET}^{(\text{sx})} = \frac{2\pi}{\hbar}\sum_{M,N} f(E_{DM}) \mid V_{DM,AN}^{(\text{eff})} \mid^2 \, \delta(E_{DM} - E_{AN}) \, . \tag{6.165}$$

The rate expression itself has been discussed at length in Section 6.4.3. What is of main interest here is the structure of the effective coupling matrix element, Eq. (6.164). Again the

[24]Note, that the proper normalization of the state $|D\rangle + \delta|D\rangle$ can be neglected since it is of higher order in the respective transfer integrals.

energy denominator should not become equal to zero and should be larger then the square root of the numerator to justify perturbation theory. However, the inclusion of vibrational levels may lead to the case $E_{DM} = E_{BK}$. But if the electronic levels $E_D$ and $E_A$ strongly deviate from the energetic degeneracy, this case may be connected with very small vibrational overlap integrals $\langle\chi_{DM}|\chi_{BK}\rangle$ and $\langle\chi_{BK}|\chi_{AN}\rangle$ and the smallness of $V^{(\text{eff})}_{DM,AN}$ is guaranteed. Following this reasoning we may conclude that only terms with $E_{BK} \gg E_{DM}$ contribute to $V^{(\text{eff})}_{DM,AN}$. Hence, it often suffices to replace the denominator by the pure electronic energy difference $E_D = E_B$. The completeness relation for the bridge vibrational states finally results in the effective coupling Eq. (6.162), and we may set in Eq. (6.165) $V^{(\text{eff})}_{DM,AN} \approx V^{(\text{eff})}_{DA}\langle\chi_{DM}|\chi_{AN}\rangle$ to get

$$k^{(\text{sx})}_{\text{ET}} = \frac{2\pi}{\hbar} \mid V^{(\text{eff})}_{DA} \mid^2 \mathcal{D}(\Delta E_{DA}/\hbar)\ . \tag{6.166}$$

The combined density of states $\mathcal{D}$ depends on the driving force $\Delta E_{DA}$ of the donor–acceptor transition (as well as the vibrational overlap integrals $\langle\chi_{DM}|\chi_{AN}\rangle$) and has been introduced in Eq. (6.91). The expressions for the superexchange mediated effective donor acceptor coupling are widely used in literature. But the derivation given so far reveals the shortcomings of the approach. First, it is only valid if the bridge levels are energetically well separated from the donor as well as acceptor levels. And second, any vibrational relaxation of the transferred electron in the bridge has been neglected. Therefore, it is instructive to embed the description of superexchange ET in a more general treatment. It is based on a consequent perturbation expansion with respect to the transfer integral and will be outlined in Section 6.7. Next we will derive expressions for the superexchange ET rate if the bridge is of a more complex structure as discussed so far.

### 6.6.2 Electron Transfer Through Long Bridges

Having discussed bridge mediated ET for the simple case where the whole bridge is given by a single electronic level the more general case of a larger number of bridge units will be described now. There are two possibilities to deal with this case. First, one can extend the perturbational scheme of the foregoing section. This would be possible if the transfer couplings $V_{mn}$ among the bridge levels are sufficiently small. However, one may also be confronted with the situation that all these couplings are large (although the coupling of the bridge levels to the donor and the acceptor remains small to justify the description of the ET as a nonadiabatic process). In this latter case one may change from the description of the bridge levels by localized states (diabatic states) to a description by delocalized adiabatic states.

Case of weak intra–bridge transfer integrals:

For the sake of clarity let us consider a linear arrangement of bridge molecules, which is a realistic model of a molecular bridge realized, for example, if the donor and acceptor are connected by a polymer strand (see Figs. 6.25 and 6.28). Moreover, we will not take into account vibrational levels when determining the effective transfer coupling, i.e. we follow the arguments of the foregoing section which lead us to a description in terms of electronic levels and electronic transfer integrals only. Then the superexchange mechanisms of ET through the bridge is described as follows. We first assume that the state $|D...N_\mathrm{B}-1\rangle$ of the DBA system is known, where the electron is delocalized across all bridge units except the last one. Then, an effective donor–acceptor coupling $V_{DA}^{(\mathrm{eff})}$ is obtained in a way demonstrated in the forgoing section since the last bridge level remains as the single intermediate level:

$$V_{DA}^{(\mathrm{eff})} = \frac{V_{D,N_\mathrm{B}}^{(\mathrm{eff})} V_{N_\mathrm{B},A}}{E_D - E_{N_\mathrm{B}}} . \tag{6.167}$$

The formula contains the effective coupling $V_{D,N_\mathrm{B}}^{(\mathrm{eff})}$ between the state $|D...N_\mathrm{B}-1\rangle$, where the electron is delocalized up to the bridge unit $N_\mathrm{B}-1$, and the last unit $N_\mathrm{B}$ of the bridge. To determine $V_{D,N_\mathrm{B}}^{(\mathrm{eff})}$ we introduce the similar effective coupling $V_{D,N_\mathrm{B}-1}^{(\mathrm{eff})}$ which now describes the interaction between the state $|D...N_\mathrm{B}-2\rangle$, where the electron is delocalized up to the bridge unit $N_\mathrm{B}-2$, and the bridge unit $N_\mathrm{B}-1$. We obtain

$$V_{D,N_\mathrm{B}}^{(\mathrm{eff})} = \frac{V_{D,N_\mathrm{B}-1}^{(\mathrm{eff})} V_{N_\mathrm{B}-1,N_\mathrm{B}}}{E_D - E_{N_\mathrm{B}-1}} . \tag{6.168}$$

In the same way we may compute $V_{D,N_\mathrm{B}-1}^{(\mathrm{eff})}$. If this procedure is repeated until the donor level is reached the effective donor acceptor coupling follows as

$$V_{DA}^{(\mathrm{eff})} = \frac{V_{D1}}{E_D - E_1} \frac{V_{12}}{E_D - E_2} \cdots \frac{V_{N_\mathrm{B}-1,N_\mathrm{B}}}{E_D - E_{N_\mathrm{B}}} V_{N_\mathrm{B},A} . \tag{6.169}$$

Introducing this expression into Eq. (6.166), we obtain the superexchange ET rate for cases where the transfer coupling within the bridge is weak enough to be handled by perturbation theory.

To further characterize this special situation we compute how $k_\mathrm{ET}^{(\mathrm{sx})}$ depends on the number $N_\mathrm{B}$ of bridge units, assuming identical bridge units with energy $E_\mathrm{B}$ and only include nearest–neighbor couplings $V_\mathrm{B}$ (cf. Fig. 6.9). The effective coupling, Eq. (6.168) follows as $V_{DA}^{(\mathrm{eff})}(N_\mathrm{B}) = V_{DA}^{(\mathrm{eff})}(1)\,\zeta^{N_\mathrm{B}-1}$. Here, we introduced $V_{DA}^{(\mathrm{eff})}(1) = V_{D1}V_{N_\mathrm{B}A}/(E_D - E_B)$, what can be interpreted as the effective superexchange coupling for the case of a single bridge unit. The parameter $\zeta = V_B/(E_D - E_B)$ describes the decrease of the coupling with increasing number of bridge units. The decrease of the total rate follows an exponential law: $k_\mathrm{ET}^{(\mathrm{sx})}(N_\mathrm{B}) \propto \zeta^{2(N_\mathrm{B}-1)}$.

Case of strong intra–bridge transfer integrals:

If the intra–bridge transfer integrals are large the situation is best described by introducing the eigenstates of the bridge Hamiltonian. Again we assume that there is a large energetic distance of all bridge levels to the donor as well as to acceptor level and neglect any vibrational level of the bridge. Then, the bridge eigenstates $\phi_a$ and eigenenergies $E_a$ (cf. Fig. 6.27) can be obtained by diagonalization of the electronic part of the bridge Hamiltonian. The eigenstates are written as an expansion with respect to the localized bridge states: $|\phi_a\rangle = \sum_m c_a(m)|\varphi_m\rangle$. In general the coefficients $c_a(m)$ have to be determined numerically; for certain model bridge systems an analytical solution might exist as well (see below). Expressed in the basis of its

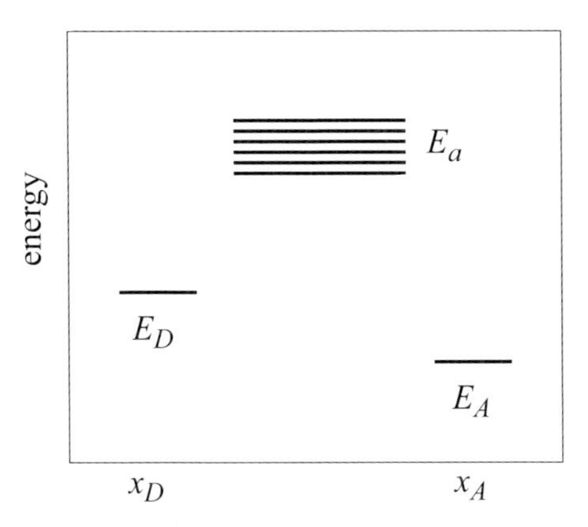


**Figure 6.27:** Bridge–mediated ET between a donor and an acceptor level. The individual LUMOs of the bridge units are replaced by the band of bridge eigenstates $E_a$.

eigenstates the electronic part of the bridge Hamiltonian becomes

$$H_{\text{el}}^{(\text{bridge})} = \sum_a E_a |\phi_a\rangle\langle\phi_a| \,. \tag{6.170}$$

Since the inter–site couplings are transformed to $(X = D, A)$

$$V_{Xa} = \sum_m V_{Xm} c_a(m) \,, \tag{6.171}$$

Eq. (6.29) can be written as

$$H_{\text{el}}^{(\text{DBA})} = \sum_{X=D,A} \Big\{ H_X |X\rangle\langle X| + \sum_a \left( V_{Xa} |X\rangle\langle\phi_a| + \text{h.c.} \right) \Big\} + H_{\text{el}}^{(\text{bridge})} \,. \tag{6.172}$$

Now we are in the position to derive the bridge–mediated effective transfer integral. Although there is not a single intermediate bridge level as discussed in Section 6.6.1, but a whole set of levels labelled by $a$ we can follow the reasoning of this section since all bridge levels couple independently to the donor or acceptor. In generalizing Eq. (6.162) the effective DA–coupling follows as

$$V_{DA}^{(\text{eff})} = \sum_a \frac{V_{Da} V_{aA}}{E_D - E_a} \,. \tag{6.173}$$

If inserted into Eq. (6.166) the superexchange ET rate for the case of a strong transfer coupling in the bridge is obtained.

To compute the bridge–length dependence we consider the model of a regular bridge with common energy levels $E_B$ and nearest–neighbor couplings $V_B$. The bridge energies $E_a$ read $E_0 + 2V_B \cos(a)$ where the quantum number is $a = \pi j/(N_\text{B} + 1)$, with $j = 1, \ldots, N_\text{B}$

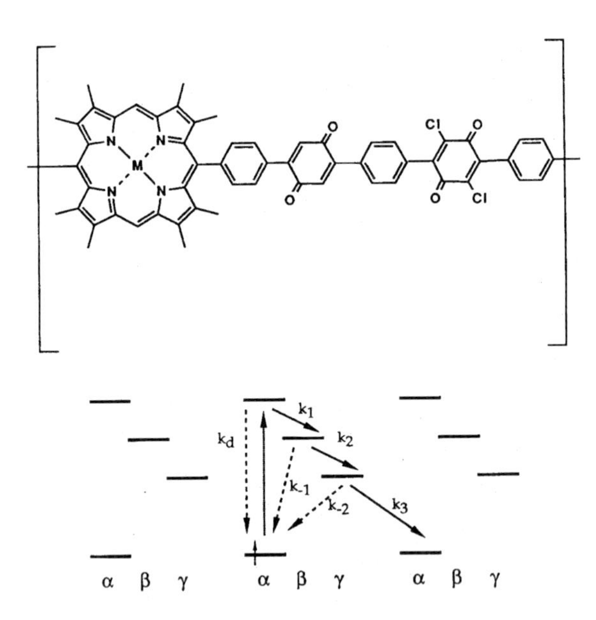


**Figure 6.28:** The porphyrin quinone complex in the upper panel which is similar to one shown in Fig. 6.10 has been proposed to work as the basic unit of a molecular shift register. This requires that the single complexes are polymerized and placed in between microelectrodes. The lower panel of the figure displays the energy level scheme responsible for bridge–mediated ET in the polymeric chain. The energetically lowest reference state is thought to be obtained by an electrochemical oxidation of the porphyrins. If a reduction of the porphyrin at the left terminal part occurs the electron can be shifted to the right. This is triggered by switching a light source on and off to excite the porphyrin. After photoexcitation the electron moves to the right via bridge–mediated ET. (reprinted with permission from [Hop89]; copyright (1989) American Chemical Society)

(cf. Section 2.8.4), and the expansion coefficients $c_a(m)$ follow as $\sqrt{2/(N_{\rm B}+1)}\sin(am)$. Hence, we obtain the coupling matrix elements, Eq. (6.171) between the donor and the various bridge levels as $V_{Da} = V_{D1}\sqrt{2/(N_{\rm B}+1)} \times \sin(\pi j/[N_{\rm B}+1])$ and between the bridge levels and the acceptor as $V_{Aa} = V_{AN_{\rm B}}\sqrt{2/(N_{\rm B}+1)} \times \sin(\pi j N_{\rm B}/[N_{\rm B}+1])$. Then, the bridge–mediated effective DA coupling, Eq. (6.173) can be calculated:

$$V_{DA}^{(\text{eff})}(N_{\rm B}) = V_{DA}^{(\text{eff})}(1)\left(\frac{1-\sqrt{1-4\zeta^2}}{2\zeta}\right)^{N_{\rm B}-1}, \tag{6.174}$$

with the effective superexchange coupling for the case of a single bridge unit $V_{DA}^{(\text{eff})}(1) = -V_{D1}V_{N_{\rm B}A}/(E_0 - E_D)$ and with $\zeta = V_B/(E_0 - E_D)$. (Note, that in the present case the latter quantity may become 0.1 and larger whereas in the case of weak intra–bridge transfer integrals $\zeta$ remains a small quantity.)

To compare bridge–mediated ET with the direct through–space transfer let us recall that the through–space ET rate would become proportional to $|V_{DA}^{(0)}|^2 \exp\{-2\beta x_{DA}\}$ (Eq. (6.28),

$x_{DA}$ denotes the DA distance, and $V_{DA}^{(0)}$ is a reference value of $V_{DA}$ taken at a reference distance). In the same way we may write $V_{DA}^{(\text{eff})}(1) = V_{DA}^{(0,\text{eff})}(1)\exp(-2\beta x_{D1})\exp(-2\beta x_{N_\text{B}A})$. The distance between the donor and the left bridge terminal is given by $x_{\text{D}1}$, and $x_{N_\text{B}A}$ denotes the distance of the right bridge terminal to the acceptor (see also Fig. 6.9). For simplicity we assumed all transfer integrals to vary with the same constant $\beta$. The expression $V_{DA}^{(0,\text{eff})}(1)$ is the reference value of the effective coupling. The superexchange mechanism can increase the ET rate drastically. Compared with the through–space ET rate the small factor $\exp\{-2\beta x_B\}$ ($x_B = x_{DA} - x_{D1} - x_{N_\text{B}A}$ is the bridge length) has been replaced by $|\ V_{DA}^{(0,\text{eff})}(1)\ |^2$ multiplied by $\zeta^{N_\text{B}-1}$ (case of weak intra–bridge transfer integrals) or multiplied by the square of the second factor on the right–hand side of Eq. (6.174). In both cases values larger than $\exp\{-2\beta x_B\} \approx 10^{-9}$ are possible ($\beta \approx 1\text{Å}^{-1}$ and $x_B = 20\text{Å}$).

## 6.7 Nonequilibrium Quantum Statistical Description of Electron Transfer

In this section we will generalize the treatment of ET reactions presented so far. In particular, the generalized approach will enable us to fully include the vibrational degrees of freedom into the bridge–mediated ET as well as to go beyond the nonadiabatic limit for the ET in a DA complex. To achieve this goal the apparatus of nonequilibrium quantum statistics will be utilized as introduced in Section 3.9. There, the ubiquitous system–reservoir Hamiltonian has been rearranged in a manner which is most suitable for the following considerations. First, it has been expanded with respect to the eigenstates of the system part. Following from this, the resulting matrix elements of the system–reservoir coupling separate into diagonal and off–diagonal elements. The former enter the zeroth–order Hamiltonian $H_0$ whereas the latter form the perturbation $\hat{V}$. Then, applying a particular projection operator approach one could derive rate equations for the populations of the system eigenstates containing rate expressions which are given as a complete perturbational expansion with respect to $\hat{V}$.

It is already obvious from this short explanation that such an approach shall be capable of providing a unified description of ET reactions. If we identify the system states of the general approach of Section 3.9 with the diabatic electronic states $\varphi_m$ and the vibrational degrees of freedom of the ET system with the reservoir coordinates of Section 3.9 we may derive general expressions for the ET rates $k_{m\to n}$. These describe all transitions in the system including nonadiabatic processes as well as processes which are of higher order in the inter–state couplings $V_{mn}$, among them the superexchange ET rates.

To establish the relation with the approach of Section 3.9 we separate the ET Hamiltonian Eq. (6.29) according to $H_{\text{DBA}} = H_0 + \hat{V}$ with

$$H_0 = \sum_m H_m(q)|\varphi_m\rangle\langle\varphi_m| \,, \tag{6.175}$$

and

$$\hat{V} = \sum_{m,n}(1-\delta_{mn})V_{mn}|\varphi_m\rangle\langle\varphi_n| \,. \tag{6.176}$$

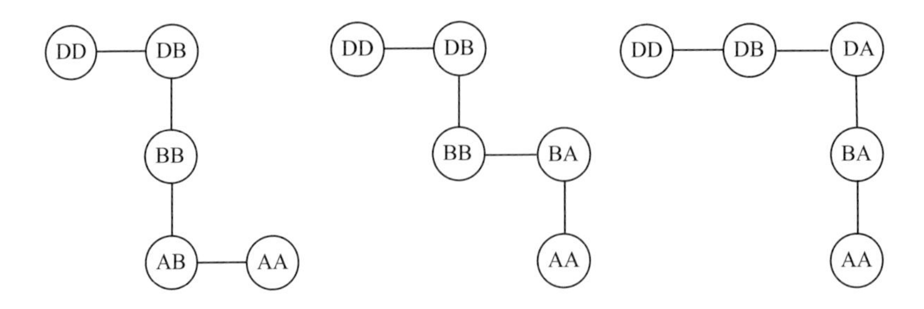


**Figure 6.29:** Graphical representation of the (nonfactorized) fourth–order rate, Eq. (6.178), by three distinct Liouville space pathways [Hu89]. To indicate the sequence of applying the operators in the trace expression of Eq. (6.178) respective electronic quantum numbers are drawn inside the circles. Those correspond to the actual value of the density operator. Its initial value describes population of the donor state and a respective vibrational equilibrium. The first and the second pathways correspond to a process where the density operator becomes diagonal with respect to the bridge state electronic quantum number and vibrational relaxation might be possible in this state (sequential transfer). In contrast, the third pathway describes a superexchange–type of transition without diagonal electronic density operator matrix elements (for more details see the supplementary Section 6.9.5 and Fig. 6.37).

As shown in Section 3.9 the approach has to be based on the general projection superoperator $\mathcal{P}$, Eq. (3.351), here, however, defined by the diabatic states $|\varphi_m\rangle$ instead of the states $|a\rangle$.[25] The formalism leads to rate equations for the diabatic state populations $P_m$ and, simultaneously, to transition rates $k_{m\to n}$ which, according to Eq. (3.388), represent a power expansion with respect to the inter–state couplings $V_{mn}$. The expression which is of the $2N$th order with respect to $V_{mn}$ reads for the present case of ET reactions (note the general frequency dependence of the rates which, however, is reduced here to the case $\omega = 0$)

$$k^{(2N)}_{m\to n}(\omega = 0) = -i \text{ tr}\left\{\hat{\Pi}_n \mathcal{L}_V \left(\tilde{\mathcal{G}}_0(\omega = 0)\mathcal{L}_V\right)^{2N+1} \hat{R}_m \hat{\Pi}_m\right\} . \tag{6.177}$$

The $\hat{\Pi}_m$ and $\hat{\Pi}_n$ project on the diabatic states $|\varphi_m\rangle$ and $|\varphi_n\rangle$, respectively. The Liouville superoperator $\mathcal{L}_V$ is defined via the commutator with the inter–state coupling $\hat{V}/\hbar$, Eq. (6.176). The zeroth–order part $\mathcal{L}_0$ specifies the Green's superoperator $\tilde{\mathcal{G}}_0(\omega)$ (cf. Eqs. (3.385)). It is given as $\mathcal{G}_0(\omega)$ $-(\omega + i\epsilon)^{-1} \cdot \mathcal{P}$ with $\mathcal{G}_0(\omega) = (\omega + i\epsilon - \mathcal{L}_0)^{-1}$.[26] The expression under the trace can be understood as the frequency–domain form of the multiple action of the inter–state coupling $\hat{V}$ (via $\mathcal{L}_V$) and the subsequent action of time–evolution operators (via $\mathcal{G}_0(\omega)$). Therefore, provided that the different Green's superoperators are transformed into the time–domain, one arrives at the multi–time correlation function expression for the transfer rate. The following section should give the reader an impression on the usefulness if this technique.

[25] Noting this definition the action of $\mathcal{P}$ on an arbitrary operator $\hat{O}$ can be written as $\mathcal{P}\hat{O} = \sum_m \text{tr}\{\hat{P}_m\hat{O}\}\hat{P}_m\hat{R}_m$ with the vibrational equilibrium density operator $\hat{R}_m$ of the diabatic state $\varphi_m$.

[26] The combination of $\mathcal{G}_0$ with the part proportional to $\mathcal{P}$ ensures that every rate of the order $2N$ with respect to $\hat{V}$ does not depend on those contributions which are already contained in lower–order rates. Of course, the divergency at $\omega = 0$ in the second term of $\tilde{\mathcal{G}}_0(\omega)$ will be compensated by the first term.

### 6.7.1 Unified Description of Electron Transfer in a Donor–Bridge–Acceptor System

As explained in Section 6.6 the determination of rates for bridge–mediated ET requires the consideration of higher–order contributions with respect to the inter–state coupling. Here we may account for the simultaneous influence of the ordinary nonadiabatic transition rates between neighboring states and, for example, the superexchange rates describing a direct transition from the donor to the acceptor (or vice versa). We remind the reader that both cases correspond to weak intra–bridge transfer integrals as discussed first in Section 6.6.2. The superexchange rates would follow from Eq. (6.177) as $k^{(2N_{\rm B}+2)}_{D\to A}(\omega = 0)$, whereas the nonadiabatic rates are second–order rates $k^{(2)}_{m\to m\pm 1}(\omega = 0)$. Of course, all other types of rates have to be examined concerning their relevance for the whole ET reaction, too.

To keep the matter simple we will deal in the following with a three–site system of a donor, a single–bridge unit, and an acceptor state ($m = D, B, A$). As in Section 6.6.2, we consider the transfer integrals $V_{DB}$, and $V_{BA}$, but neglect the direct coupling $V_{DA}$. In this three–site system we expect the rates $k_{D\to B}$, $k_{B\to A}$, and $k_{D\to A}$ as well as the reverse ones. If expanded with respect to the transfer coupling the first two start with second–order rates (nonadiabatic rates $k^{(2)}_{D\to B}$ and $k^{(2)}_{B\to A}$, cf. Section 6.4). The lowest–order contribution to the rate $k_{D\to A}$ would be of fourth order in $V_{mn}$. Since $\tilde{\mathcal{G}}_0$ (cf. Eq. (6.177)) contains a part proportional to the projector $\mathcal{P}$, there will be a separation of $k^{(4)}_{D\to A}$ into different terms. These can be evaluated using the fact that terms which are of odd order in the inter–state coupling vanish. We obtain (at $\omega \neq 0$)

$$
\begin{aligned}
k^{(4)}_{D\to A}(\omega) = & -\frac{i}{\omega + i\epsilon}\, k^{(2)}_{D\to B}(\omega) k^{(2)}_{B\to A}(\omega) \\
& - i\,\mathrm{tr}\left\{\hat{\Pi}_A \mathcal{L}_V \mathcal{G}_0(\omega) \mathcal{L}_V \mathcal{G}_0(\omega) \mathcal{L}_V \mathcal{G}_0(\omega) \mathcal{L}_V \hat{R}_D \hat{\Pi}_D\right\}. \qquad (6.178)
\end{aligned}
$$

The complete fourth–order rate can be decomposed into two second–order rate expressions and a (non–factorizable) fourth–order part. (The case $\omega \neq 0$ has to be considered because of the divergency in the first term at $\omega = 0$ which, however, will be compensated by the second term.) For the given transfer coupling matrix elements there are no other combinations of second–order rates. These second–order rates are the standard nonadiabatic transition rates which read in the present notation

$$
k^{(2)}_{m\to n}(\omega) = -i\,\mathrm{tr}\left\{\hat{\Pi}_n \mathcal{L}_V \mathcal{G}_0(\omega) \mathcal{L}_V \hat{R}_m \hat{\Pi}_m\right\}. \qquad (6.179)
$$

Both types of rates will be used to solve the respective set of rate equations, i.e. second–order rates for the nearest neighbor transitions ($k_{D\to B}$, $k_{B\to D}$, $k_{B\to A}$, and $k_{A\to B}$) and on fourth–order rates for the donor–acceptor transition ($k_{D\to A}$, and $k_{A\to D}$). Details of the derivation and specification of the second and fourth–order rates can be found in the supplementary Sections 6.9.4 and 6.9.5, respectively. Fig. 6.29 gives a graphical representation of the three terms contributing to the fourth–order rate.

Having discussed the different approximations for the rate expressions we present the rate

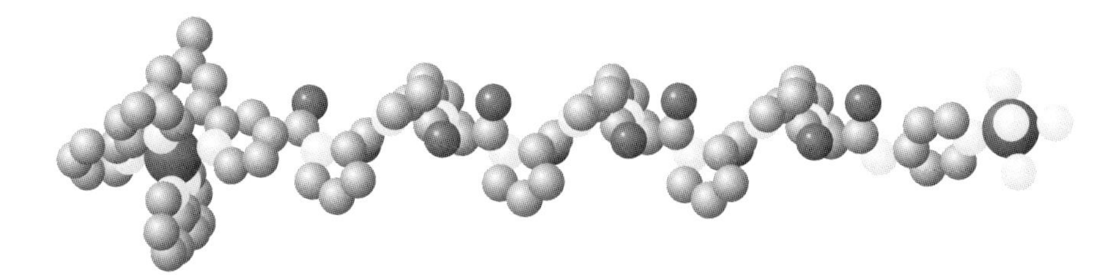

**Figure 6.30:** Bridge mediated ET in a system where the donor given by a metal complex of Ruthenium (left terminal site) and the acceptor by a Cobalt complex (right terminal site). The bridge is formed by an oligomer of the amino acid proline (the chemical structure is $[(bpy)_2Ru(II)L^{\cdot}(Pro)_nCo(III)(NH_3)_5]^{3+}$, carbon atoms are shown in grey, nitrogen in weak grey, oxygen as well as the Ruthenium and Cobalt atoms in black).

equations referring to the simple DBA system:

$$\begin{aligned}\frac{\partial}{\partial t}P_D(t) &= -(k_{D\to B} + k_{D\to A})P_D(t) + k_{B\to D}P_B(t) + k_{A\to D}P_A(t) \, , \\ \frac{\partial}{\partial t}P_B(t) &= -(k_{B\to D} + k_{B\to A})P_B(t) + k_{D\to B}P_D(t) + k_{A\to B}P_A(t) \, , \\ \frac{\partial}{\partial t}P_A(t) &= -(k_{A\to B} + k_{A\to D})P_A(t) + k_{B\to A}P_B(t) + k_{D\to A}P_D(t) \, . \end{aligned} \quad (6.180)$$

As the initial condition we set up $P_m(0) = \delta_{mD}$. A standard way to solve such differential equations is to make the ansatz $P_m(t) = \exp(-Kt)$. In the present case one obtains two rates $K$ which are non–zero, and one which is equal to zero. The first two are simply computed using the conservation of total probability (for example, $P_B$ can be replaced by $1 - P_D - P_A$). Once the resulting two inhomogeneous rate equations have been solved the rates K read:

$$K_\pm = \frac{1}{2}\left(a + b \pm \sqrt{(a-b)^2 + c}\right) \, , \quad (6.181)$$

with $a = k_{D\to B} + k_{B\to D} + k_{B\to A}$, $b = k_{A\to B} + k_{B\to A} + k_{A\to D}$, and $c = 4(k_{D\to A} - k_{B\to A})$ $(k_{A\to D} - k_{B\to D})$. It is apparent from these formulas that the rates characterizing the basic nearest neighbor hopping transitions and the superexchange transitions are strongly mixed. They enter the total rate as independent addend only in a special case. It is characterized by rates $k_{B\to D}$ and $k_{B\to A}$ describing the outflow of charge from the bridge, which are much larger than all other rates. This would be the case for a DBA system with the bridge level being positioned highly above the donor and the acceptor level ($E_B - E_D, E_B - E_A, \gg E_D - E_A$). For such a situation the thermally activated transfer into the bridge level is much smaller than the transfer out of the bridge. Introducing an expansion of the rates $K_\pm$ around the leading contribution $k_{B\to D} + k_{B\to A}$ one obtains $K_+ \approx k_{B\to D} + k_{B\to A}$ and

$$K_- \equiv K_{\rm ET} = k_{D\to A} + k_{A\to D} + \frac{k_{D\to B}k_{B\to A} + k_{A\to B}k_{B\to D}}{k_{B\to D} + k_{B\to A}} \, . \quad (6.182)$$

The rate $K_+$ is responsible for a fast transfer process but at the same time it only causes a small deviation from the initial charge distribution. The actual but slower ET is characterized

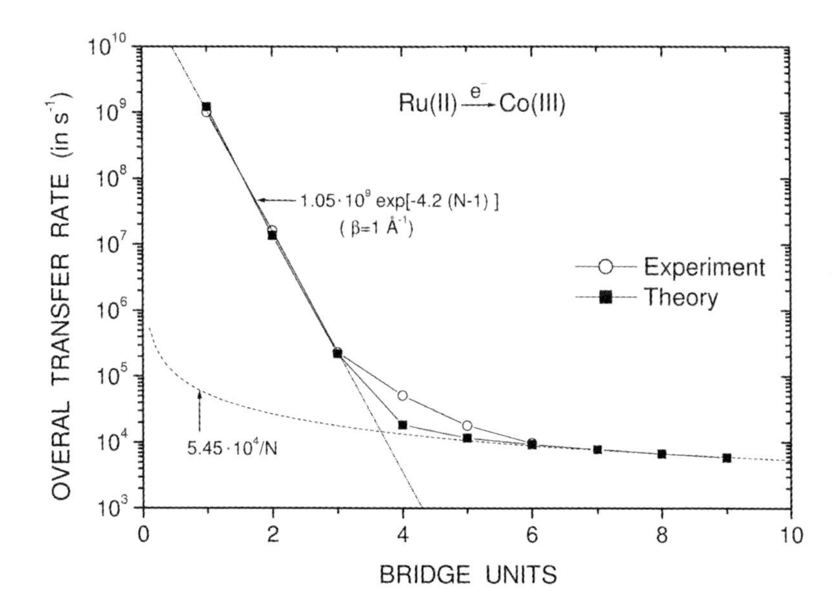


**Figure 6.31:** Length–dependence of the overall ET rate at room temperature for the donor bridge acceptor complex shown in Fig. 6.30). Comparison of experimental data (open circles, after [Isi92] and theoretical computations [Pet01] (full squares). The thin lines show an estimate of the bridge number dependence of the rate if it is dominated by the superexchange mechanisms or the sequential one. (reprinted with permission from [Pet01]; copyright (2001) American Chemical Society)

by the rate $K_{\mathrm{ET}}$ [Pet01]. This overall donor–acceptor transfer rate contains the superexchange forward and backward rate in its first and second term, respectively. The third term comprises the sequential transfer from the donor to the bridge unit and afterwards to the acceptor as well as the reverse part of this transition. In this way $K_{\mathrm{ET}}$ accounts for the superexchange and the sequential mechanism of ET by two independent addend.

Although Eq. (6.182) has been derived for a single bridge unit only, it is well–suited to describe the change of the measured ET rate when increasing the number of bridge units. In particular this is possible when the total bridge is given by a polymer strands whose length can be easily varied. Prominent examples are polymer strands of amino acids [Isi92] (cf. Fig. 6.30) as well DNA fragments [Gie99,Kel99,Lew00]. We briefly justify the use of Eq. (6.182) to determine the ET rate for a case where all bridge units are identical (use of a so called homo–polymer as a bridge). Provided that the bridge–internal hopping transitions are much faster than the transitions into and out of the bridge (and the use of diabatic bridge states is justified) one can assume a bridge–internal equilibrium distribution with $P_m(t) = P_B(t)/N_\mathrm{B}$, where $P_B(t)$ is the total bridge population $\sum_m P_m(t)$ and $m$ runs over all bridge units (from 1 to $N_\mathrm{B}$). It follows a reduction of the multitude of rate equations for the bridge populations to a single one governing the total bridge population. Such an equation is similar to the second one of the Eqs. (6.180), but with the rates $k_{D\to B}$, $k_{B\to A}$ (as well as the reverse ones) replaced by rates divided by the number of bridge units $N_\mathrm{B}$. Of course, the rates $k_{D\to A}$ and $k_{A\to D}$ have to be computed with effective transfer couplings $V_{DA}^{(\mathrm{eff})}$ like that of Eq. (6.169).

If we again assume that the rates $k_{B\to D}$ and $k_{B\to A}$ are much larger than all other rates [Pet01] we can describe the bridge length dependence of the rate by Eq. (6.182). In this case the two basic mechanisms of bridge–mediated ET enter the total rate as separate addend, and we expect that one of both might dominate the other for a given number of bridge units. Fig. 6.31 displays such a behavior via the length–dependence of the overall ET rate in the DBA complex $[(\mathrm{bpy})_2\mathrm{Ru(II)L}^{\cdot}(\mathrm{Pro})_n\mathrm{Co(III)(NH_3)_5}]^{3+}$ (cf. Fig. 6.30). The donor is given by a Ruthenium and the acceptor by a Cobalt complex, whereas a oligopeptide of the amino acid proline connects the donor and the acceptor. The oligopeptide forms a linear bridge which has the advantage to be relatively stiff (when compared with other oligopeptides). As demonstrated by Fig. 6.31 the ET is mainly determined by the superexchange mechanisms if the bridge is short. Followed by a small transition region the sequential mechanism of bridge mediated ET characterizes the length–dependence of the rate for longer bridges. This behavior has also been described for other systems (cf. [Ber02,Bix00]).

### 6.7.2 Transition to the Adiabatic Electron Transfer

Finally, we would like to return to the two–site system, i.e. a simple DA complex and show how to go beyond the second–order approximation with respect to the DA transfer integral, i.e. to leave the regime of nonadiabatic ET. A number of higher–order approximations have been derived in the literature (see, for example, [Rip87,Spa88]). Considering the high–temperature limit a typical expression for the rate reads

$$k_\mathrm{ET}^{(\mathrm{adia})} = \frac{k_\mathrm{ET}^{(\mathrm{nonad})}}{1+\delta_\mathrm{adia}}\,, \tag{6.183}$$

where $k_\mathrm{ET}^{(\mathrm{nonad})}$ is the rate of nonadiabatic transfer given in Eq. (6.73) (Marcus–type formula). The adiabatic correction obtained for example for the ET in polar solvents is given by

$$\delta_\mathrm{adia} = 4\pi\frac{|V_\mathrm{DA}|^2\tau_\mathrm{long}}{\hbar E_\lambda}\,. \tag{6.184}$$

The longitudinal relaxation time $\tau_\mathrm{long}$ of the solvent has been introduced in Eq. (6.155) and $E_\lambda$ denotes the reorganization energy. (More involved expression with a generalization of the definition of $\tau_\mathrm{long}$ have also been derived.) For a very small transfer coupling the nonadiabatic rate expression is recovered, whereas for a large $|V_\mathrm{DA}|$ the rate becomes independent of the coupling. In this manner expression (6.183) interpolates between the two limiting cases of adiabatic and nonadiabatic ET.

An interpolation formula of this kind can also be generated using the so–called *Pade* approximation. We note that Eq. (6.177) gives an expansion of the ET rate like $k_{DA}$ =

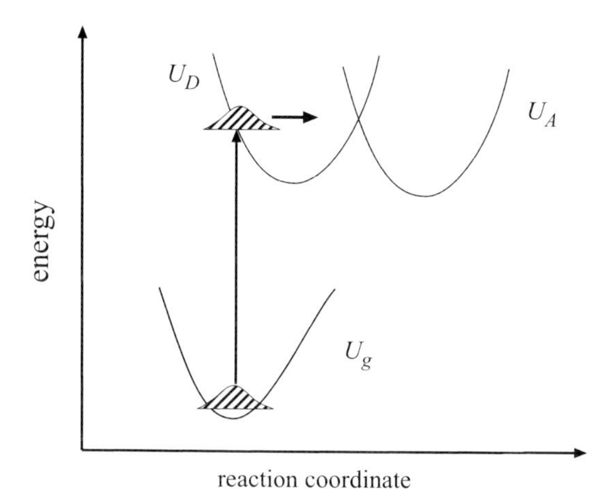


**Figure 6.32:** Ground state PES as well as donor and acceptor diabatic PES appropriate for photoinduced ET. If the initial state preparation becomes ultrafast, the limit of impulsive excitation can be applied resulting in an instantaneous shift of the electronic ground state vibrational wave function to the donor PES.

$|V_{DA}|^2 C^{(2)}(\omega = 0) - |V_{DA}|^4 C^{(4)}(\omega = 0) + \ldots$. Here, the $C^{(N)}$ are Fourier–transformed $N$–time correlation functions of the type encountered in Eq. (6.177). The Pade approximation leads to a rate expression of infinite order in $V_{DA}$, but restricted to the two types, $C^{(2)}$ and $C^{(4)}$, of correlation functions. Such a resummation of the rate reads $k_{DA} = |V_{DA}|^2 C^{(2)}(\omega = 0)/(1 + |V_{DA}|^2 C^{(4)}(\omega = 0)/C^{(2)}(\omega = 0))$. The expression produces a reasonable approximation for the adiabatic case where it becomes independent of the transfer integral. Interpolation formulas inspired by the Landau–Zener ET rate formula have also been proposed.

## 6.8 Photoinduced Ultrafast Electron Transfer

Photoinduced ET reactions have already been introduced in Section 6.1 (cf. Fig. 6.6). In the following we will concentrate on ET processes which are so fast that in the course of the electron motion from the donor to the acceptor no complete vibrational relaxation is possible, that is, $\tau_{\rm rel} > t_{\rm ET}$. Since vibrational relaxation usually occurs on a picosecond ($10^{-12}$ s) or sub–picosecond time scale, *ultrafast* ET reactions have to proceed in the same time region. Of course, this is also valid for the time resolution of the optical pulses used to initiate (to prepare the excited donor state D*, see scheme (6.3)) and to observe the ET. With the advent of laser technology ultrafast laser pulses became available which have a duration comparable to or even shorter than the relevant system time scales $t_{\rm ET}$, and $\tau_{\rm rel}$. The vibrational motion is partly *coherent* and the related wave packet dynamics can be observed using nonlinear optical spectroscopy. In this way it became possible to take snapshots of the electron–vibrational dynamics.

If the inequality $\tau_{\rm rel} > t_{\rm ET}$ is valid the ET is no longer of the nonadiabatic type. In the

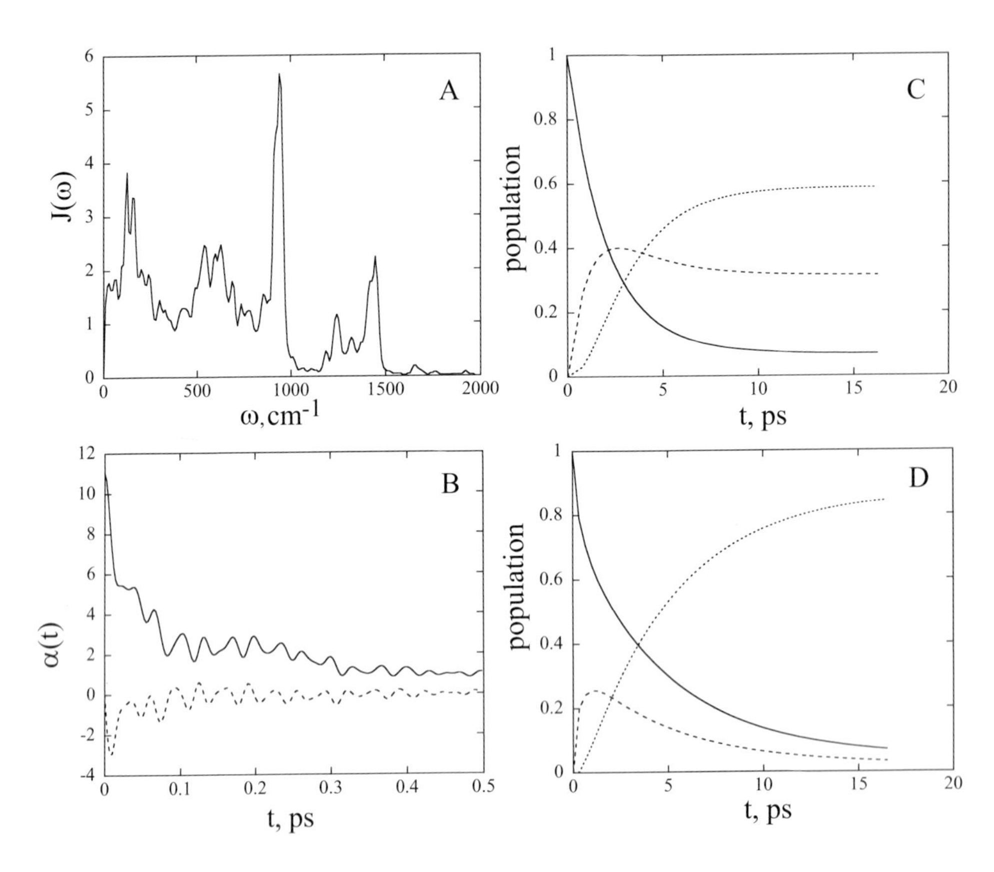


**Figure 6.33:** Room temperature ET in the photosynthetic bacterial reaction center, Fig. 6.2, described within the path integral method of Section 3.10. The approach accounts for the photoexcited special pair (state 1), the reduced accessory bacteriochlorophyll (state 2), and the reduced pheophytin (state 3); the latter was used instead of the wild–type bacteriopheophytin. Panel (A): Spectral density following from classical molecular dynamics simulations including crystallization water molecules. Panel (B): Real (solid) and imaginary (dashed) part of the correlation function corresponding to this spectral density (cf. Eq. (3.406)) . Panel (C): Populations of the three electronic states ($P_1$: solid line, $P_2$: dashed line, $P_3$: dotted line, $E_1 = E_3 = 0$, $E_2 = -400$ cm$^{-1}$, and $V_{12} = V_{23} = 22$ cm$^{-1}$). Panel (D): Populations of a modified setup with $E_1 = E_3 = 0$, $E_2 = 200$ cm$^{-1}$, and $V_{12} = V_{23} = 30$ cm$^{-1}$. Comparing panels (C) and (D) we notice that the intermediate population of the bridge (state 2) depends strongly on its energetic position relative to the donor and acceptor states. (figure courtesy of N. Makri, for more details see [Mak96]; the data for panel (a) have been published in [Mar93])

foregoing section (and in the related supplementary parts) we demonstrated how to go beyond the limit of nonadiabatic ET by improving the lowest–order perturbation theory with respect to the transfer integral $V_{DA}$. An alternative would have been to use the path integral approach outlined in Section 3.10, which can account for arbitrary state couplings, while providing at the same time an exact treatment of the coupling to harmonic oscillator reservoirs. An application of this method is given in Fig. 6.33. Here, we will present yet another alternative way which is numerically less demanding as compared to a path integral study. The drawback, however, is that only a few vibrational degrees of freedom can be incorporated nonperturbatively; the majority forms a heat bath which is treated by the QME approach.

Instead of writing down the perturbation expansion as in the previous section, we derive equations of motion whose solutions describe the ET in a nonperturbative manner with respect to the interaction strength $V_{DA}$. To do this we will introduce a density matrix approach capable of describing the time evolution of an initially defined density matrix. Thus the limits of adiabatic and nonadiabatic ET are naturally included. And, all types of electronic and vibrational coherences are accounted for. Furthermore, it is a particular advantage of this approach that one can easily include the radiation field and compute ultrafast optical spectra. In Section 6.8.2 different prescriptions for defining ET rates starting with the time–dependent density matrix will be also discussed.

According to scheme (6.3) the photoinduced ET considered in the following requires the excitation of an electron from the ground state to the donor state. Therefore, the manifold of electronic states considered so far has to be supplemented by the electronic ground state of the donor part of the complex (compare Fig. 6.6).

Often, only a small number of active vibrational coordinates (at least one coordinate $s$) couple to the external field and to the ET reaction. All remaining inter– and intramolecular vibrational degrees of freedom (denoted $q = \{q_\xi\}$) are only incorporated in an indirect manner in the optical preparation process and the ET reaction. They are assumed to form a heat bath (uncoupled normal mode oscillators) for the active coordinates. This assumption directly results in a separation of the complete Hamiltonian into the (active) system part $H_{\rm S}$ and into the reservoir contribution $H_{\rm R}$. (This has been discussed in Chapter 3 as the basic idea behind the concept of the reduced density operator). We write

$$H = H_{\rm S}(t) + H_{\rm S-R} + H_{\rm R} \; . \tag{6.185}$$

The system part is time dependent since it includes the optical preparation of the initial state. It splits up into the molecular part $H_{\rm mol}$ and the coupling to the external field $H_{\rm F}(t)$ (described in a semiclassical approach and within the electric dipole approximation, cf. Eq. (5.9))

$$H_{\rm S}(t) = H_{\rm mol} + H_{\rm F}(t) \; . \tag{6.186}$$

The molecular part comprises the vibrational Hamiltonian $H_g$ of the electronic ground state $|\varphi_g\rangle$ and the part describing the ET reaction. The latter coincides with the DA Hamiltonian, Eq. (6.32), and we have

$$H_{\rm mol} = H_g|\varphi_g\rangle\langle\varphi_g| + H_{\rm DA} \; . \tag{6.187}$$

This Hamiltonian allows us to describe the excitation process of the donor and thus the preparation process of the ET reactant state. Note that this extends our previous considerations where the preparation process of the initial state of the excess electron had not been considered.

The coupling to the reservoir degrees of freedom can be written as in Eq. (3.146):

$$H_{\rm S-R} = \sum_m K_m(s)\Phi_m(q) \; . \tag{6.188}$$

The structure of the coupling to reservoir degrees of freedom, $H_{S-R}$ can be motivated as follows. First, we note that the original form of $H_{mol}$, Eq. (6.187), determined by the vibrational Hamiltonian of the various diabatic states as well as transfer integrals depends on the active as well as the reservoir coordinates. If we neglect the respective dependencies of the transfer integrals there remains for every diabatic state a coupling potential $U_m(s,Z)$. It depends on both types of coordinates and becomes responsible for energy dissipation from the ET system into the reservoir. Although $U_m(s,Z)$ may depend on the coordinates in a more complicated way we will restrict ourselves in the following on a bilinear coupling, i.e. a potential $U_m(s,Z)$ which depends linearly on all $s$ as well as $Z$. To establish the connection to Eq. (3.146) which gives a multiple factorized expression of the system–reservoir coupling the reservoir part $\Phi_u$ follows (after an identification of the summation index $u$ with the diabatic electronic quantum number $m$) as

$$\Phi_u(Z) \equiv \Phi_m(Z) = \sum_{\xi} k_{\xi}(m) Z_{\xi} \,. \tag{6.189}$$

Concerning the system contribution to $H_{S-R}$ we will restriction ourselves to a single harmonic reaction coordinate (vibrational frequency $\Omega_s$, reduced mass $\mu_s$) and obtain

$$K_m(s) = \sqrt{\frac{2\mu_s\Omega_s}{\hbar}}(s - s^{(m)})|\varphi_m\rangle\langle\varphi_m| \,. \tag{6.190}$$

The used system–reservoir coupling depends on the electronic state and increases with the distance of the system coordinate $s$ from its equilibrium position $s^{(m)}$. (Note that we introduced $K_m(s)$ as a dimensionless quantity (cf. the discussion in Section 4.3.1).) The spectral density $J_{mn}(\omega)$ referring to the given system–reservoir coupling can be obtained in analogy to Eq. (3.226) (but here with an additional dependence on the electronic quantum numbers $m$ and $n$).

Before introducing the density matrix, we specify the preparation process of the reactant state. A general description of the optical preparation of an excited electronic state has been already given in Section 5.5. We follow this scheme here and assume an exclusive population of the donor state (see Fig. 6.32). Obviously, for large values of $V_{DA}$ a certain combination of donor and acceptor levels will be populated and the description in terms of adiabatic states becomes more convenient (see also the discussion below). Additionally, it is assumed that the pulse duration $\tau$ is short compared to the time scales of vibrational motion as well as of ET (limit of impulsive excitation, cf. Section 5.5). As shown in Section 5.5 this enables us to eliminate the nuclear dynamics on the electronic ground state PES from the considerations. The optically prepared initial state for the ET reaction is obtained as (cf. Eq. (5.130))

$$\hat{\rho}(t_0) = \frac{1}{\hbar^2}|\mathbf{d}_{Dg}\mathbf{E}(t_p)\tau|^2\,\hat{R}_m|D\rangle\langle D| \,. \tag{6.191}$$

Here, $\mathbf{E}(t_p)$ is the electric field strength at pulse maximum $t_p$, and $\tau$ denotes the respective pulse duration. We emphasize that, as a result of the impulsive excitation, the vibrational state in the electronic ground state represented by the ground state vibrational equilibrium density operator $\hat{R}_m$ has been instantaneously transferred onto the donor diabatic electronic state. At low temperatures this corresponds to the projection of the ground state vibrational wave function onto the donor state as shown in Fig. 6.32. Again, although Eq. (6.191) contains the vibrational statistical operator of the electronic ground state, the dynamics within the electronic ground state could be eliminated. Eq. (6.191) gives the initial value for the density matrix which is exclusively defined with respect to the electronic DA levels. If the underlying time scale separation is not possible, it is necessary to incorporate the coupled dynamics of ground and excited states into the equations of motion.

Having specified the total Hamiltonian (6.185) we are in a position to use density matrix theory introduced in Section 3.8. However, our molecular Hamiltonian (the $H_{\rm DA}$ part) is not in diagonal form. This might become a problem insofar as a direct use of the state representation of Section 3.8.2 requires the knowledge of the respective eigenstates (see also the similar discussion in Section 3.8.7 dealing with the dynamics of a two–levels system).

Suppose these eigenstates have been calculated from

$$H_{\rm DA}|\psi_\alpha\rangle = \mathcal{E}_\alpha|\psi_\alpha\rangle \, . \tag{6.192}$$

This equation defines the *adiabatic* electron–vibrational states of the DA Hamiltonian. (Although it would be possible to classify the states according to their relation to the upper and lower adiabatic PES, Eq. (6.42), we do not introduce such a specification.) In general terms, diabatic $|\varphi_m\rangle|\chi_{mM}\rangle$ and adiabatic states are related via a linear transformation (see also Section 2.7)

$$|\psi_\alpha\rangle = \sum_{mM} c_\alpha(mM)\, |\varphi_m\rangle|\chi_{mM}\rangle \, . \tag{6.193}$$

Once the adiabatic states and energies are known it is possible to introduce the density matrix in the adiabatic state representation

$$\rho_{\alpha\beta}(t) = \langle\psi_\alpha|\hat{\rho}(t)|\psi_\beta\rangle \, , \tag{6.194}$$

and to derive the respective equations of motion (see Section (3.8.2)). Such an approach is most appropriate for the case of strong inter–state coupling $V_{DA}$. In particular, $V_{DA}$ is nonperturbatively incorporated into the description of dissipative processes via the Redfield tensor, Eq. (3.284).

Alternatively one can define the density matrix in the diabatic state representation

$$\rho_{\mu\nu}(t) \equiv \rho_{mM,nN}(t) = \langle\chi_{mM}|\langle\varphi_m|\hat{\rho}(t)|\varphi_n\rangle|\chi_{nN}\rangle \, , \tag{6.195}$$

where it is straightforward to compute the diabatic electronic state populations

$$P_m(t) = {\rm tr}_{\rm vib}\{\langle\varphi_m|\hat{\rho}(t)|\varphi_m\rangle\} = \sum_M \rho_{mM,mM}(t) \, , \tag{6.196}$$

which can be directly related to the ET rate (for details see Section 6.8.2). To characterize the dynamics of the vibrational mode accompanying the ET reaction one can use the probability distribution

$$P(s,t) = \langle s|{\rm tr}_{\rm el}\{\hat{\rho}(t)\}|s\rangle = \sum_{m,MN} \chi_{mM}(s)\rho_{mM,mN}(t)\chi^*_{mN}(s) \, , \tag{6.197}$$

which replaces the square of the vibrational wave function in the case of dissipative dynamics. For further use we also give the internal energy of the DA electron–vibrational system (for the notation compare also Section 6.2.4):

$$E_{\text{int}}(t) = \sum_{\mu\nu} (\delta_{\mu\nu} E_\mu + (1 - \delta_{mn}) V_{\mu\nu}) \, \rho_{\mu\nu}(t) \, . \tag{6.198}$$

Eqs. (6.196)–(6.198) demonstrate that observables of interest can be determined using the density matrix in the diabatic state representation. Finally, we would like to point out that the diabatic state representation can be used in the case of strong inter–state coupling, too. However, one has to make sure that the Redfield tensor is calculated using the eigenstates (adiabatic states) of $H_{\text{DA}}$. The last point is of less importance if $V_{DA}$ is small. In this case dissipation can often be simulated using the diabatic states. This issue will be addressed in more detail in the next section.

## 6.8.1 Quantum Master Equation for Electron Transfer Reactions

In the following we will consider photoinduced ultrafast ET in the limit where the dissipation of electron–vibrational energy can be described within the Markov approximation. The applicability of this approximation is not straightforward and deserves some comments (cf. the discussion in Section 3.7.1). Let us assume that the Markov approximation is valid in the absence of an external field. It is obvious that the situation would not change if the system interacts with an optical pulse which is long compared to $\tau_{\text{mem}}$ (the characteristic time during which the memory function, describing dissipation, decays, cf. Section 3.7.1). Also in the limit of impulsive excitation we would expect that the Markov approximation is still valid (the initial state preparation is short compared to any other characteristic time of the system). However, if the pulse duration is comparable to $\tau_{\text{mem}}$, the external driving introduces a new characteristic time and no time scale separation is possible. As a consequence the theoretical description based on the QME in the Markov approximation (as presented below) becomes invalid.

In the following we will focus on the limit of impulsive excitation and assume the validity of the Markov approximation. The QME in the adiabatic state representation follows as

$$\frac{\partial}{\partial t} \rho_{\alpha\beta}(t) = -i\omega_{\alpha\beta} \rho_{\alpha\beta}(t) + \left( \frac{\partial \rho_{\alpha\beta}(t)}{\partial t} \right)_{\text{diss}} \, . \tag{6.199}$$

The $\omega_{\alpha\beta}$ are the transition frequencies between the adiabatic energy levels, and the dissipative part can be calculated following the derivation given in Section 3.8.2. We skip further details here and consider the representation of the QME in the diabatic basis. Using $\omega_{\mu\nu} = (E_\mu - E_\nu)/\hbar$ we have

$$\frac{\partial}{\partial t} \rho_{\mu\nu}(t) = -i\omega_{\mu\nu} \rho_{\mu\nu}(t) - \frac{i}{\hbar} \sum_\kappa \Big( V_{\mu\kappa} \rho_{\kappa\nu}(t) - V_{\kappa\nu} \rho_{\mu\kappa}(t) \Big) + \left( \frac{\partial \rho_{\mu\nu}(t)}{\partial t} \right)_{\text{diss}} \, . \tag{6.200}$$

To obtain the dissipative part of the density matrix equation we cannot directly use the general formulas given in Eq. (3.284), since the diabatic representation is no energy representation.

Therefore, we start with the QME in the notation of Eq. (3.263), which does not refer to any specific representation. The indices $u$ and $v$ used in Eq. (3.263) can be directly identified with the electronic state index. Taking the system part of the interaction operator, $K_m(s)$, from Eq. (6.190) we obtain for the respective diabatic state matrix elements

$$\langle\chi_{kK}|\langle\varphi_k|K_m(s)|\varphi_l\rangle|\chi_{lL}\rangle = \delta_{km}\delta_{lm}\left(\delta_{K,L-1}\sqrt{L}+\delta_{K,L+1}\sqrt{L+1}\right). \tag{6.201}$$

The operator $\Lambda_u \equiv \Lambda_m$ (cf. Eq. (3.260))

$$\Lambda_m = \sum_n \int_0^\infty d\tau\, C_{mn}(\tau) U_{\rm DA}(\tau) K_m U^+_{\rm DA}(\tau) \tag{6.202}$$

contains time–evolution operators defined by the complete DA complex Hamiltonian $H_{\rm DA}$. To calculate the diabatic matrix elements of $U_{\rm DA}(\tau)K_m U^+_{\rm DA}(\tau)$ we first use the adiabatic states $|\psi_\alpha\rangle$ introduced in Eq. (6.192) and obtain

$$\langle\psi_\alpha|U_{\rm DA}(\tau)K_m U^+_{\rm DA}(\tau)|\psi_\beta\rangle = e^{-i\omega_{\alpha\beta}\tau}\langle\psi_\alpha|K_m|\psi_\beta\rangle. \tag{6.203}$$

The adiabatic matrix element can be expressed by the diabatic elements, Eq. (6.201), using Eq. (6.193) as

$$\langle\psi_\alpha|K_m|\psi_\beta\rangle = \sum_{kK}\sum_{lL} c^*_\alpha(kK)c_\beta(lL)\langle\chi_{kK}|\langle\varphi_k|K_m|\varphi_l\rangle|\chi_{lL}\rangle. \tag{6.204}$$

As in the general treatment in Section 3.8.2 we omit the imaginary contribution to the Redfield tensor. This is achieved in the present notation by replacing the half–sided Fourier transform of the correlation function $C_{mn}(\tau)$ appearing in Eq. (6.202) by $C_{mn}(\omega)/2$. Then the diabatic matrix elements of the $\Lambda$–operator, Eq. (6.202) read

$$\begin{aligned}\langle kK|\Lambda_m|lL\rangle &= \frac{1}{2}\sum_n\sum_{\alpha,\beta} C_{mn}(-\omega_{\alpha\beta})c_\alpha(kK)c^*_\beta(lL)\langle\psi_\alpha|K_m|\psi_\beta\rangle \\ &= \frac{1}{2}\sum_n\sum_{\alpha,\beta}(1+n(\omega_{\beta\alpha}))\left(J_{mn}(\omega_{\beta\alpha})-J_{mn}(\omega_{\alpha\beta})\right) \\ &\quad\times c_\alpha(kK)c^*_\beta(lL)\langle\psi_\alpha|K_m|\psi_\beta\rangle.\end{aligned} \tag{6.205}$$

Introducing this expression together with the matrix elements of $K_m(s)$ into the dissipative part of Eq. (6.200) gives the full density matrix equation in the diabatic representation. This equation is exact with respect to the transfer integral $V_{DA}$. Consequently, ET for any value of $V_{DA}$ can be studied covering the range from the adiabatic ET to the nonadiabatic ET. In contrast, possible values of the coupling strength to the reservoir are limited by the second–order perturbational treatment. However, this restriction is not severe as long as the total system is properly separated into a relevant system and a reservoir.

The formulation of the relaxation in terms of the eigenstates of the system Hamiltonian guarantees in particular the correct long–time behavior of the system dynamics. The density matrix equations of motion, if expanded with respect to the adiabatic states, will give the canonical equilibrium distribution as a stationary solution (cf. Section 3.8.2). Therefore, the density matrix in the diabatic representation reads

$$\rho_{mM,nN}(t \to \infty) = \sum_{\alpha} c_{\alpha}(mM) c_{\alpha}^{*}(nN) \frac{e^{-\mathcal{E}_{\alpha}/k_{\mathrm{B}}T}}{\sum_{\beta} e^{-\mathcal{E}_{\beta}/k_{\mathrm{B}}T}} . \tag{6.206}$$

Note that although in thermal equilibrium the system is in a mixed state, off–diagonal elements of the density matrix are present in the diabatic representation. For a weak coupling $V_{DA}$ it is possible to consider the relaxation within the diabatic states, thus choosing the dissipative part in zeroth–order approximation with respect to the transfer integral. This has the advantage that there is no need for diagonalizing the system Hamiltonian.

We do not give the complete Redfield tensor, but consider the vibrational energy relaxation rates which can be derived from the general formula (3.289) in the limit of $V_{DA} = 0$:

$$\begin{aligned} k_{mM \to nN} &= \frac{1}{\hbar^2} \delta_{mn} \frac{2\mu_{\mathrm{s}}\Omega_{\mathrm{s}}}{\hbar} |\langle \chi_{mM}|(s - s^{(m)})|\chi_{mN}\rangle|^2 C_{mm}(\omega_{mM,mN}) \\ &= \delta_{mn} \Big[ \delta_{M+1,N}(M+1) n(\Omega_{\mathrm{s}}) + \delta_{M-1,N} M \big(1 + n(\Omega_{\mathrm{s}})\big) \Big] \\ &\quad \times \Omega_{\mathrm{s}}^2 \, J_{mm}(\Omega_{\mathrm{s}}) . \end{aligned} \tag{6.207}$$

Note, that all constants have been included in the definition of the spectral density. In contrast to Eq. (6.205) only electronic diagonal contributions of the spectral density given at a single frequency $\Omega_{\mathrm{s}}$ enter the rate formula. The inverse lifetime of the state $|\mu\rangle$ follows as [27].

$$\frac{1}{\tau_{mM}} = \sum_{N} k_{mM \to mN} = \Big[ (M+1) n(\Omega_{\mathrm{s}}) + M \big(1 + n(\Omega_{\mathrm{s}})\big) \Big] \, \Omega_{\mathrm{s}}^2 \, J_{mm}(\Omega_{\mathrm{s}}) . \tag{6.208}$$

This expression leads to the dephasing rate $\gamma_{mM,nN} = 1/2\tau_{mM} + 1/2\tau_{nN}$ (cf. Eq. (3.292)). Figure 6.34 gives some examples for ultrafast ET reactions by showing the probability distribution Eq. (6.197) of the vibrational coordinate at different instants of the time evolution. The vertical position indicates the actual value of the internal energy, Eq. (6.198). As a consequence of energy dissipation the vibrational wave packet performs a damped motion within the coupled potential energy surfaces. As an initial state the vibrational ground state probability distribution displaced into the donor PES has been taken. Moving in part (A) of Figure 6.34 from the left to the crossing point of both PES, the wave packet splits up into two parts. This results from a partial reflection in the region around the crossing point. A destructive as well as constructive interference of both parts of the wave packet follows. In panel B (activationless case of ET, c.f. Section 6.4.1) and C (inverted case) this behavior is not so clear. The relaxation down to the vibrational ground state of the acceptor PES is most pronounced in part (C).

The derived formulas (with or without the inclusion of $V_{DA}$ in the dissipative part) give a solid basis for simulation of ultrafast photoinduced ET reactions. And if compared with more

[27] The inverse lifetime of the vibrational ground $1/\tau_{m0}$ becomes proportional to $n(\Omega_{\mathrm{s}})$. Because of this fact the bilinear system–reservoir coupling as applied here results in a long vibrational ground–state life time. The inverse lifetime leads to dephasing rates $\gamma_{mM,nN} = 1/2\tau_{mM} + 1/2\tau_{nN} + \gamma_{mM,nN}^{(\mathrm{pd})}$ which may also small if the vibrational ground–state is involved on no pure dephasing is present (cf. Eq. (3.292)).

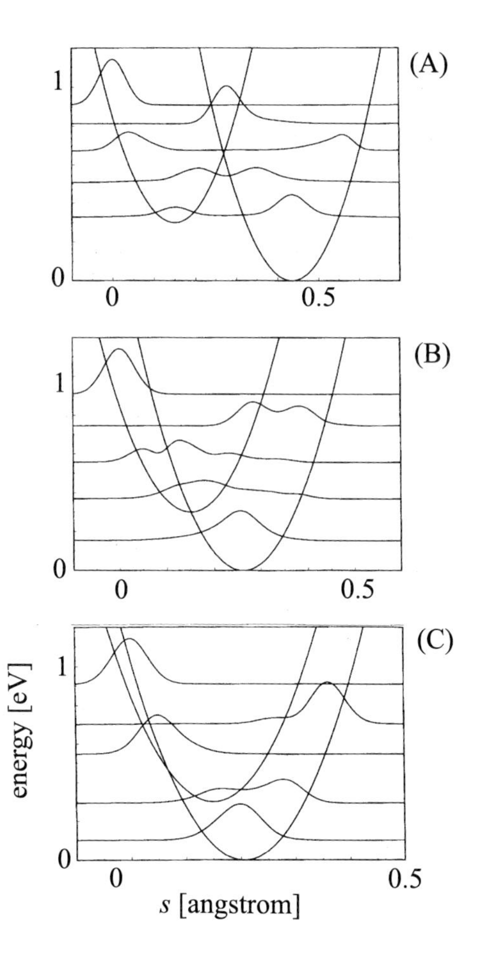


**Figure 6.34:** Probability distribution of the vibrational coordinate, Eq. (6.197) in the system of a coupled donor and acceptor PES ($\hbar\Omega_s = |V_{DA}|^2 = 100\text{meV}$, $\hbar J_{mm}(\Omega_s) = 10\text{meV}$, $k_B T \ll \hbar\Omega_s$). The position of the probability distribution with respect to the energy axis corresponds to the actual internal energy, Eq. (6.198) at different time steps of the propagation ( (A) $t = 0, 20, 40, 100, 500$ fs, (B) $t = 0, 20, 50, 100, 500$ fs, and (C) $t = 0, 20, 45, 115, 500$ fs, (from top)). The chosen configurations of both PES correspond to the different types of ET reactions discussed in Section 6.4.1 ((A) normal region of ET reactions, (B) activationless case, (C) ET in the inverted region).

involved approaches they seem to be sufficiently accurate (cf. Fig. 6.35). Complemented by the study of the response to additional radiation fields, they allow to describe different nonlinear optical experiments. Within the presented approach one can directly include the external fields into the density matrix equations. An alternative description is given by a perturbation expansion with respect to external fields. This leads to nonlinear response functions characterizing the molecular system (cf. Chapter 5). However, as a consequence of the expansion with respect to the electric field strength, one is practically limited to low–order processes.

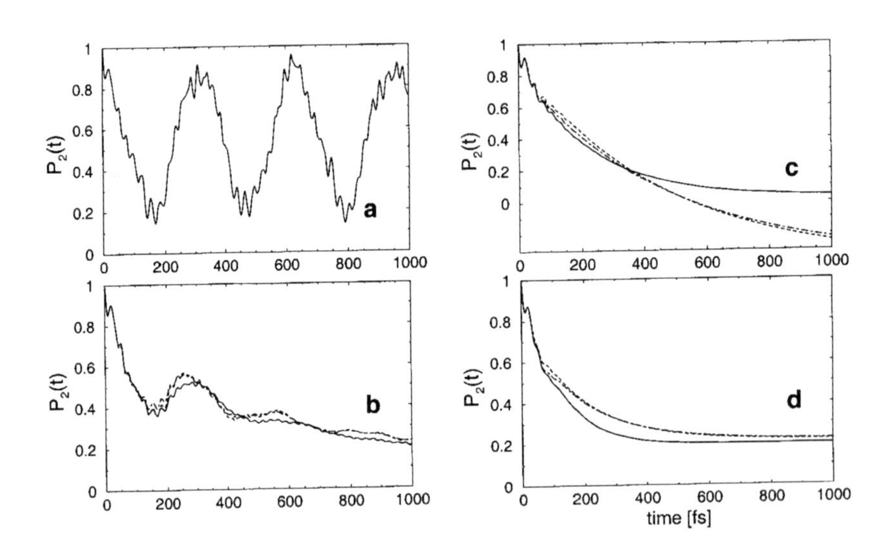


**Figure 6.35:** Population $P_2$ of the acceptor in a system of a coupled donor and acceptor PES similar to case A of Fig. 6.34 ($T = 0$: panel a to c, $T = 300K$: panel d, spectral density according to Eq. (3.229), $j_0 = 0$: panel a, $j_0 = 0.2/\pi$: panel b, $j_0 = 1/\pi$: panel c and d). Dashed lines: solution of Eq. (6.200), dotted–dashed line: use of a time–dependent Redfield–tensor as introduced in (3.444), full line: results of computations based on the description of the reservoir modes by a multi–configuration version of the approach discussed in Section 7.5.1. (reprinted with permission from [Ego03]; copyright (2003) American Institute of Physics)

The outlined density matrix approach, however, provides the tools to study effects depending on the field intensity as well, such as the dynamical Stark effect.

## 6.8.2 Rate Expressions

In the previous section photoinduced ET has been described via the complete time–dependent density matrix. In order to establish the connection with the considerations of Sections 6.6 – 6.7, we will concentrate on the question of how to introduce transfer rates within the present approach.

According to the general type of rate equations (3.2) we expect an exponential decay, for example, for the donor state population, $P_D(t) \propto \exp\{-k_{\text{ET}}t\}$. Although such a behavior is unlikely at early times (here it could be oscillatory and multi–exponential), it is reasonable to expect it after all coherences have decayed. Therefore, one can define the ET rate $k_{\text{ET}}$ as

$$k_{\text{ET}} = -\lim_{t\to\infty} \frac{1}{t} \ln P_D(t) , \tag{6.209}$$

where the donor state population is obtained from the solution of the general density matrix equations (6.200). The deviation from an exponential decay at early times reflects an initial time dependence of $k_{\text{ET}}$.

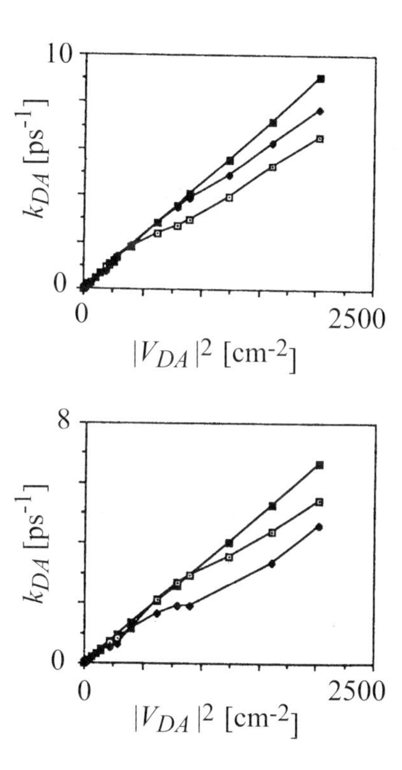


**Figure 6.36:** ET rate $k_{\rm ET}$ as a function of the square of the transfer integral $V_{DA}$ for a donor–acceptor complex including a single relevant coordinate (upper panel: $\Delta E$ = 100 cm$^{-1}$; lower panel: $\Delta E$ = 125 cm$^{-1}$). Full squares: $k_{\rm ET}$ according to the Golden Rule like formula (6.211), open squares: $k_{\rm ET}$ according to the requirement $P_D(1/k_{\rm ET}) = 1/e$ (note that this definition is not completely the same as Eq. (6.209)), full diamonds: $k_{\rm ET}$ according to the averaged expression (6.210). ($\hbar\omega_{\rm vib}$ = 100 cm$^{-1}$), $g_D - g_A = 0.5$, $C_{mm}(\omega_{\rm vib})/\hbar$ = 16 cm$^{-1}$), $T$ = 200K. The excitation energy to prepare the initial state has been placed 25 cm$^{-1}$ above the bottom of the donor PES. (reprinted with permission from [Jea92]; copyright 1992 American Institute of Physics)

Alternatively, one can introduce a transfer rate via the inverse of the mean lifetime of the electron at the donor state as

$$k_{\rm ET} = \left( \int\limits_0^\infty dt\; P_D(t) \right)^{-1} . \tag{6.210}$$

Finally, in the nonadiabatic limit it is also possible to compute an explicit expression for $k_{\rm ET}$. The derivation is similar to that of Section 3.4.5 which resulted in the Golden Rule formula. In correspondence to Eq. (3.120) we start with an equation of motion for the diagonal density matrix elements $\rho_{\mu\mu}$. We get the same type of equation as in Section 3.4.5, but supplemented by relaxation contributions $-\sum_\nu (k_{\mu\nu}\rho_{\mu\mu} - k_{\nu\mu}\rho_{\nu\nu})$. In Eq. (3.121) for the off–diagonal part, the transition frequency $\omega_{\mu\nu}$ have to include dephasing rates $\gamma_\mu$ resulting in the replacement of $\omega_{\mu\nu}$ by the complex transition frequency $\tilde{\omega}_{\mu\nu} = \omega_{\mu\nu} - i(\gamma_\mu + \gamma_\nu)$. Then, one can follow the reasoning in Section 3.4.5 up to Eq. (3.129) taking into account, however, the coupling to a thermal reservoir.

If the rates $k_{\mu\nu}$ are large, the theory describes fast relaxation within the two diabatic states. This point had to be introduced as an additional assumption in the derivation of Section 3.4.5. Here, the approach automatically gives the thermalization introduced via the assumptions in Eqs. (3.119). Taking Eq. (3.129) the only difference is the appearance of complex transition frequencies leading to a "broadened" delta function. Thus we have the final rate

$$\begin{aligned} k_{\text{ET}} &\equiv k_{DA} \\ &= \frac{2\pi|V_{DA}|^2}{\hbar^2} \sum_{M,N} f(E_{DM})|\langle\chi_{DM}|\chi_{AN}\rangle|^2 \frac{(\gamma_{DM} + \gamma_{AN})/\pi}{\omega_{DM,AN}^2 + (\gamma_{DM} + \gamma_{AN})^2} \,. \end{aligned} \tag{6.211}$$

The expression describes nonadiabatic ET from the different vibrational donor levels $E_{DM}$ to the final acceptor levels, both broadened by $\hbar\gamma_{DM}$ and $\hbar\gamma_{AN}$, respectively. A thermal averaging and weighting by the respective Franck–Condon factors is also incorporated.

Figure 6.36 compares this nonadiabatic rate with those following from Eq. (6.209) and Eq. (6.210). Obviously, the Golden Rule expression, Eq. (6.211) increases linearly with $|V_{DA}|^2$. The two other rates obtained from a nonperturbative consideration of the transfer coupling via the solution of the density matrix equations show a sublinear behavior for larger values of $|V_{DA}|^2$. This is to be expected since the rate should become independent of the transfer integral if the ET reaction changes to the adiabatic type (cf. Section 6.3).

## 6.9 Supplement

### 6.9.1 Landau–Zener Transition Amplitude

The description of the ET reaction in a DA complex according to Landau and Zener can be reduced to the calculation of a particular transition amplitude as explained in Section 6.3.1. Here we show how to compute the transition amplitude, Eq. (6.55). In a first step the time–evolution operator is split up into a part $U_0$ defined by $H_0$ and a remaining $S$–operator related to the transfer coupling $\hat{V}$. Reversing the general treatment given in Section 3.2.2, the Hamiltonian defining $U_0$ is time–dependent. Nevertheless, it can be calculated analytically. Let us consider this operator in more detail:

$$U_0(t,\bar{t}) = \hat{T} \exp\left\{ -\frac{i}{\hbar} \int_{\bar{t}}^{t} d\tau \; H_0(\tau) \right\} \,. \tag{6.212}$$

The two parts forming $H_0$ and being proportional to $|D\rangle\langle D|$ and to $|A\rangle\langle A|$, however, commute with each other and the time–evolution operator can easily be calculated as

$$U_0(t,\bar{t}) = \exp\left(i(v^* F_D)(t^2 - \bar{t}^2)/2\hbar\right)|D\rangle\langle D| + \exp\left(i(v^* F_A)(t^2 - \bar{t}^2)/2\hbar\right)|A\rangle\langle A| \tag{6.213}$$

If we want to calculate the $S$–operator, the interaction representation $\hat{V}^{(\mathrm{I})}$ of the inter–state coupling has to be determined. It reads

$$\hat{V}^{(\mathrm{I})}(t,\bar{t}) = \exp\big(iv^*(F_A - F_D)(t^2 - \bar{t}^2)/2\hbar\big)V_{DA}|D\rangle\langle A| + \text{c.c.} \quad . \tag{6.214}$$

Therefore, the transition amplitude can be written as

$$A_{DD} = \langle D|U_0(\infty,-\infty)S(\infty,-\infty)|D\rangle = \langle D|S(\infty,-\infty)|D\rangle \, . \tag{6.215}$$

The part related to $U_0$ can be eliminated from the matrix element, and it reduces to 1 in the limit $t \to \infty$ and $\bar{t} \to -\infty$. It remains to calculate the $S$–operator matrix element. This will be done by expanding the $S$–operator with respect to the inter–state coupling. It gives (cf. Section 3.2.2)

$$\begin{aligned}\langle D|S(\infty,-\infty)|D\rangle &= \sum_{n=0}^{\infty}\left(\frac{i}{\hbar}\right)^n \int_{-\infty}^{+\infty} dt_n \int_{-\infty}^{t_n} dt_{n-1} \ldots \int_{-\infty}^{t_2} dt_1 \\ &\times \langle D|\hat{V}^{(\mathrm{I})}(t_n,-\infty)\,\hat{V}^{(\mathrm{I})}(t_{n-1},-\infty)\,\ldots\,\hat{V}^{(\mathrm{I})}(t_1,-\infty)|D\rangle \, .\end{aligned} \tag{6.216}$$

The matrix element corresponds to $n$ jumps of the electron between the donor and the acceptor level starting at the donor level but also and ending there. Consequently, the number of jumps must be even. Taking into account the concrete structure of the transfer coupling, Eq. (6.214) we can write

$$\begin{aligned}\langle D|S(\infty,-\infty)|D\rangle &= \sum_{n=0}^{\infty}\left(\frac{|V_{DA}|}{i\hbar}\right)^{2n} \int_{-\infty}^{+\infty} dt_{2n} \int_{-\infty}^{t_{2n}} dt_{2n-1} \ldots \int_{-\infty}^{t_2} dt_1 \\ &\times \lim_{\tau\to-\infty} \exp\left\{\frac{i}{\hbar}\frac{v^*(F_A - F_D)}{2}\left((t_{2n}^2 - \tau^2) - (t_{2n-1}^2 - \tau^2)\right.\right. \\ &\left.\left.\pm \ldots -(t_1^2 - \tau^2)\right)\right\} \, .\end{aligned} \tag{6.217}$$

The contributions proportional to $\tau^2$ cancel, and we introduce new time variables $(\tau_1, \ldots, \tau_n)$ and $(T_1, \ldots, T_n)$ which replace the set $(t_1, \ldots, t_{2n})$. This is done such that the Jacobian of this transformation remains unchanged. We set $\tau_1 = t_1$, $\tau_m = t_1 + \sum_{j=1}^{m-1}(t_{2j+1} - t_{2j})$ for $2 \leq m \leq n$, and $T_m = t_{2m} - t_{2m-1}$ for $1 \leq m \leq n$. For the transition amplitude we obtain

$$\begin{aligned}\langle D|S(\infty,-\infty)|D\rangle &= \sum_{n=0}^{\infty}(-1)^n\left(\frac{V_{DA}|}{\hbar}\right)^{2n} \int_{-\infty}^{+\infty} d\tau_1 \int_{\tau_1}^{\infty} d\tau_2 \ldots \\ &\times \ldots \int_{\tau_{n-1}}^{\infty} d\tau_n \int_0^{\infty} dT_1 \ldots dT_n \\ &\times \exp\left\{\frac{i}{\hbar}\frac{v^*(F_A - F_D)}{2}\left(2\sum_{m=1}^{n}\tau_m T_m + \left[\sum_{m=1}^{n} T_m\right]^2\right)\right\} .\end{aligned} \tag{6.218}$$

Any interchange of the variables $\tau_m$ does not alter the total integral. Therefore, we can extend all $\tau_m$–integrations to $-\infty$. Doing so it is necessary to introduce the prefactor $1/n!$. The remaining integrals can be calculated and we have

$$\begin{aligned}\langle D|S(\infty,-\infty)|D\rangle &= \sum_{n=0}^{\infty}(-1)^n|V_{DA}/\hbar|^{2n}\frac{1}{n!}\left(\frac{\pi\hbar}{v^*|F_A-F_D|}\right)^n \\ &= e^{-\Gamma/2}\,, \end{aligned} \tag{6.219}$$

with

$$\Gamma = \frac{2\pi}{\hbar v^*}\frac{|V_{DA}|^2}{|F_D-F_A|}\,. \tag{6.220}$$

This result has been used in Section 6.3.1.

### 6.9.2 The Multi–Mode Marcus Formula

In the following we explain in some detail how to derive the Marcus formula, Eq. (6.73) for the case of many vibrational degrees of freedom leading to the multi–mode reorganization energy Eq. (6.74). To achieve this goal we start again with formula (6.64) for the ET rate. However, to tackle the rate calculation it is advantageous to replace the delta function by its respective Fourier integral. We obtain (notice that $q=\{q_\xi\}$ and the introduction of the partition function $\mathcal{Z}$)

$$k_{\rm ET} = \frac{|V_{DA}|^2}{\hbar^2\mathcal{Z}}\int_{-\infty}^{+\infty}dt\int dq\,\exp\left\{-\frac{U_D(q)-U_D^{(0)}}{k_{\rm B}T}+\frac{i}{\hbar}t\,[U_D(q)-U_A(q)]\right\}\,. \tag{6.221}$$

In a first step one calculates the integrals with respect to the vibrational coordinates. Due to the replacement of the delta function by a Fourier integral the multiple coordinate integral factorizes into a product of simple integrals. These can be reduced to integrals with respect to Gaussian functions. (A detailed inspection shows that one has to shift formally the various $q_\xi$–integrals into the complex $q_\xi$–plane.) We obtain

$$\begin{aligned}&\int dq\,\exp\left\{-\frac{U_D(q)-U_D^{(0)}}{k_{\rm B}T}+\frac{i}{\hbar}t[U_D(q)-U_A(q)]\right\} \\ &= \left(\prod_\xi\sqrt{\frac{2\pi k_{\rm B}T}{\omega_\xi^2}}\right)\exp\left\{-\frac{k_{\rm B}T}{\hbar^2}[U_A(q_D)-U_A^{(0)}]\,t^2\right\} \\ &\times\exp\left\{\frac{i}{\hbar}\left[U_D(0)-U_A(0)-\sum_\xi\omega_\xi^2(q_\xi^{(\rm D)}-q_\xi^{(\rm A)})q_\xi^{(\rm D)}\right]t\right\}\,. \end{aligned} \tag{6.222}$$

The expression $U_A(q_D)-U_A^{(0)}$ can be identified as the reorganization energy, Eq. (6.74). The remaining time integral again is of Gaussian type and can easily be performed. A proper collection of all constants finally yields the multi–mode Marcus–type formula, Eq. (6.73).

### 6.9.3 The Free Energy Functional of the Solvent Polarization

The aim of the present section is the construction of a free energy functional $F_{\rm P}[\mathbf{P}_{\rm el}, \mathbf{P}_{\rm or}, \mathbf{D}]$, which should be determined by the two types of solvent polarizations and the external field $\mathbf{D}$ (cf., for example, [Cal83,Fel77,Kim90]). It is important for the following that all fields involved are quasi–static in the sense that they are time–dependent but do not emit energy via a radiation field. These types of fields are called longitudinal fields.[28]

We will concentrate on such longitudinal fields and, additionally, we will neglect image charge effects at the interface between the dielectric and the DA complex. It is appropriate to start with the external field contribution to the equilibrium free energy of a dielectric[29]

$$F_{\rm eq} = \frac{1}{8\pi}\int d^3\mathbf{x}\, \mathbf{D}(\mathbf{x})\left(\mathbf{E}(\mathbf{x}) - \mathbf{D}(\mathbf{x})\right) \equiv \frac{1}{2}\int d^3\mathbf{x}\, \mathbf{D}(\mathbf{x})\mathbf{P}(\mathbf{x})\ . \tag{6.223}$$

The term *equilibrium* indicates that the polarization field of the dielectric is in equilibrium with the (external and stationary) charge distribution. For the present application this external charge distribution is given by $\varrho_{\rm DA}$, the charge density of the DA complex (at the actual state of the ET reaction). The related dielectric displacement field $\mathbf{D}$ follows as (see also Section 2.5.1)

$$\mathbf{D}(\mathbf{x}) = -\nabla\varphi_{\rm DA}(\mathbf{x}) \equiv -\nabla\int d^3\mathbf{x}'\ \frac{\varrho_{\rm DA}(\mathbf{x}')}{|\mathbf{x}-\mathbf{x}'|}\ . \tag{6.224}$$

The simplest approximation for $\mathbf{D}$ is obtained assuming a spherical charge distribution of the transferred electron located either at the donor or the acceptor site $\mathbf{x}_m$ $(m = D, A)$.

We will take $F_{\rm eq}$, Eq. (6.223) as the reference expression, which should be generalized to the case of an arbitrary polarization field which is not in equilibrium with the external charge distribution. Such a nonequilibrium situation would arise in the course of the ET reaction. The motion of the transferred electron may be faster or slower than the two different contributions to the polarization field. To obtain the *nonequilibrium* generalization of the free energy functional, Eq. (6.223), we make the following guess which will be motivated below [Cal83,Fel77,Kim90] (the nonequilibrium character of the polarization field is characterized by its additional time dependence)

$$\begin{aligned} F_{\rm P}[\mathbf{P}] &= \frac{1}{2}\int d^3\mathbf{x}\, \mathbf{P}(\mathbf{x},t)\int d\bar{t}\ \chi^{-1}(t-\bar{t})\ \mathbf{P}(\mathbf{x},t) \\ &+\frac{1}{2}\int d^3\mathbf{x}\, d^3\mathbf{x}'\ \frac{\nabla\mathbf{P}(\mathbf{x},t)\nabla'\mathbf{P}(\mathbf{x}',t)}{|\mathbf{x}'-\mathbf{x}'|} - \int d^3\mathbf{x}\ \varphi_{\rm DA}(\mathbf{x},t)\nabla\mathbf{P}(\mathbf{x},t)\ . \end{aligned} \tag{6.225}$$

First we note that according to our assumption this expression does not account for any surface charge effects, which could be present at the interface of the cavity (including the DA

[28] It is known from Maxwell's theory that an electromagnetic field which has been emitted from an antenna moves independently from its source as a travelling wave. In addition, it is polarized perpendicular to its propagation direction. These fields are called *transverse* fields. All other types of fields which do not lead to the radiation of energy are longitudinal fields.

[29] This expression is obtained by calculating the work necessary to move the solvent from infinity into the external field $\mathbf{D}$.

complex) and the dielectric continuum (cf. Fig. 6.24) (for their incorporation see the literature mentioned below). The first contribution to the free energy expression can be interpreted as a self–energy term. The second part is nonlocal and describes the Coulomb interaction among different polarization charges (recall the definition of the charge density referring to the dipole density, Eq. (2.49)). The interaction of the polarization charges with the external field related to the DA complex is given by the last contribution to Eq. (6.225).[30]

To show that the guess for $F[\mathbf{P}]$ is correct, in a first step one determines the extremum of $F_{\rm P}[\mathbf{P}]$ with respect to a variation of the polarization, i.e., one computes the functional derivative $\delta F_{\rm P}[\mathbf{P}]/\delta \mathbf{P}(\mathbf{x},t) = 0$.[31] The solution of this equation determines the equilibrium polarization. We obtain the relation

$$\int d\bar{t}\, \chi^{-1}(t-\bar{t})\mathbf{P}(\mathbf{x},\bar{t}) = -\nabla\varphi_{\rm DA}(\mathbf{x},t) + \nabla \int d^3\mathbf{x}' \,\frac{\nabla'\mathbf{P}(\mathbf{x}',t)}{|\mathbf{x}-\mathbf{x}'|}\,, \tag{6.226}$$

which determines the total electric field $\mathbf{E}(\mathbf{x},t)$ as $\chi^{-1}\mathbf{P}$ and splits up into an external contribution $-\nabla\varphi_{\rm DA}$ and an internal contribution. (Again, we stress that the occurrence of the linear susceptibility is a consequence of linear response theory.) Replacing $\mathbf{P}$ by its value at this minimum (at the equilibrium), it is possible to reproduce the equilibrium functional Eq. (6.223).

In a next step we will see how this functional can be rewritten, if the separation of the polarization field into a high– and a low–frequency part is introduced. Therefore, we take the self–energy part $F_{\rm P}^{\rm (se)}$ of the free energy functional, Eq. (6.225) and introduce the two components of the polarization oscillating with a high and a low frequency. Since both frequency ranges differ by orders of magnitude, we introduce a representative high frequency $\omega_{\rm high}$ and a representative low frequency $\omega_{\rm low}$. It follows that

$$\begin{aligned} F_{\rm P}^{\rm (se)} &= \frac{1}{2}\sum_{u,v=\text{high,low}} \int d^3\mathbf{x}\,\left(\mathbf{P}_u e^{-i\omega_u t} + \text{c.c.}\right) \\ &\quad\times \int d\tau\,\chi^{-1}(\tau)\,\left(\mathbf{P}_v e^{-i\omega_v(t-\tau)} + \text{c.c.}\right) \\ &\approx q\frac{1}{2}\int d^3\mathbf{x}\,\left(\frac{\mathbf{P}_{\rm el}^2}{\chi_\infty} + \frac{\mathbf{P}_{\rm or}^2}{\chi_0}\right)\,. \end{aligned} \tag{6.227}$$

In the second part of this expression we took into account that the mixed terms coming from the high and low–frequency polarization oscillate with the difference frequency $\omega_{\rm high} - \omega_{\rm low}$ and can be neglected in the sense of a time averaged free energy functional.

[30] The term is identical with the single contribution to the equilibrium free energy expression Eq. (6.223) if surface charges are neglected. In this case using Gauss' theorem the integrand $\varphi_{\rm DA}\nabla\mathbf{P}$ can be converted to $-\nabla\varphi_{\rm DA}\times\mathbf{P}$ yielding the expression in Eq. (6.223).

[31] Let us consider a functional $\phi[y](x)$ determined by the function $y(\bar{x})$. The functional derivative $\delta\phi[y](x)/\delta y(\bar{x})$ is defined via the special limit $\lim_{\Delta\to 0}\lim_{\Delta y\to 0}\{\phi[y+\eta\Delta y](x) - \phi[y](x)\}/\Delta\cdot\Delta y(\bar{x})$. Here, $\eta = \eta(\bar{x},\Delta)$ is a function which is equal to 1 in the interval $\Delta$ around $\bar{x}$ and zero otherwise, and $\Delta y$ denotes a deviation from the function $y(\bar{x})$. The limit in the definition of the functional derivative has to be taken in such a way that $\delta y(x)/\delta y(\bar{x}) = \delta(x-\bar{x})$. Combining this relation with the chain rule of differential calculus one can determine all functional derivatives which are of interest in this book.

We obtain the desired general energy functional as

$$\begin{aligned} F_{\mathrm{P}}[\mathbf{P}_{\mathrm{el}}, \mathbf{P}_{\mathrm{or}}, \mathbf{D}] &= \frac{1}{2}\int d^3\mathbf{x}\left(\frac{1}{\chi_\infty}\mathbf{P}_{\mathrm{el}}^2(\mathbf{x},t) + \frac{1}{\chi_0}\mathbf{P}_{\mathrm{or}}^2(\mathbf{x},t)\right) \\ &+2\pi\int d^3\mathbf{x}\ (\mathbf{P}_{\mathrm{el}}(\mathbf{x},t) + \mathbf{P}_{\mathrm{or}}(\mathbf{x},t))^2 \\ &-\int d^3\mathbf{x}\ (\mathbf{P}_{\mathrm{el}}(\mathbf{x},t) + \mathbf{P}_{\mathrm{or}}(\mathbf{x},t))\,\mathbf{D}(\mathbf{x},t)\ . \end{aligned} \tag{6.228}$$

Now, the self–energy term contains two types of contributions (with polarizations proportional to $\chi_{0/\infty} = (\varepsilon_{0/\infty} - 1)/4\pi$). The second and third term remain unchanged in comparison to Eq. (6.225). However, since surface effects are supposed to play no role we could change to a representation using the fields instead of the respective charge distributions, that is, the derivatives $\nabla\mathbf{P}$ could be eliminated.

The free energy functional of the complete system including the energy contribution of the DA complex follows if the expectation value $\langle H_{\mathrm{DBA}}\rangle$ of the ET Hamiltonian, Eq. (6.29) is added. The easiest way to compute this expectation value is to use a suitable molecular wave function (a quantum statistical calculation would also be possible). Rearranging the resulting expression yields

$$\begin{aligned} F[\mathbf{P}_{\mathrm{el}}, \mathbf{P}_{\mathrm{or}}, \mathbf{D}] &= \langle H_{\mathrm{DBA}}\rangle + \frac{c_{\mathrm{el}}}{2}\int d^3\mathbf{x}\ \mathbf{P}_{\mathrm{el}}^2(\mathbf{x},t) \\ &+4\pi\int d^3\mathbf{x}\ \mathbf{P}_{\mathrm{el}}(\mathbf{x},t)\mathbf{P}_{\mathrm{or}}(\mathbf{x},t) + \frac{c_{\mathrm{or}}}{2}\int d^3\mathbf{x}\ \mathbf{P}_{\mathrm{or}}^2(\mathbf{x},t) \\ &-\int d^3\mathbf{x}\ \mathbf{D}(\mathbf{x},t)\,(\mathbf{P}_{\mathrm{el}}(\mathbf{x},t) + \mathbf{P}_{\mathrm{or}}(\mathbf{x},t))\ , \end{aligned} \tag{6.229}$$

where we introduced $c_{\mathrm{el}} = 4\pi/(1 - 1/\varepsilon_\infty)$ and $c_{\mathrm{or}} = 4\pi(1 + 1/(\varepsilon_0 - \varepsilon_\infty))$.

For the following we assume that the low–frequency range is characterized by frequencies which are much smaller than the ET rate, i.e., $\omega_{\mathrm{low}} \ll k_{\mathrm{ET}}$. For the high–frequency range the inequality $\omega_{\mathrm{high}} \gg k_{\mathrm{ET}}$ is taken. (Clearly, such a separation into three distinct frequency regions will not always be valid.) Since the solvent electronic polarization should be fast compared to the ET reaction, we can assume that $\mathbf{P}_{\mathrm{el}}$ instantaneously adapts itself to the actual electronic configurations during the ET reaction. In other words, $\mathbf{P}_{\mathrm{el}}$ is equilibrated with respect to $\mathbf{D}$ and $\mathbf{P}_{\mathrm{or}}$. In this case it would be useful to restrict the free energy functional, Eq. (6.229) to the case of an equilibrated high–frequency polarization. It can be determined from the extremum of the functional $F[\mathbf{P}_{\mathrm{el}}, \mathbf{P}_{\mathrm{or}}, \mathbf{D}]$, Eq. (6.229) with respect to $\mathbf{P}_{\mathrm{el}}$. Hence, we have to calculate $\delta F_{\mathrm{P}}/\delta\mathbf{P}_{\mathrm{el}} = 0$, which enables us to express $\mathbf{P}_{\mathrm{el}}$ by means of the two other fields. The calculation gives $\mathbf{P}_{\mathrm{el}} = 1/c_{\mathrm{el}} \times (\mathbf{D} - 4\pi\mathbf{P}_{\mathrm{or}})$. Substituting this relation into the functional Eq. (6.229) leads to the effective functional

$$F_{\mathrm{eff}}[\tilde{\mathbf{P}}_{\mathrm{or}}, \mathbf{D}] = \langle H_{\mathrm{DBA}}^{(\mathrm{eff})}\rangle + \frac{2\pi}{c_{\mathrm{Pek}}}\int d^3\mathbf{x}\ \tilde{\mathbf{P}}^2(\mathbf{x},t) - \int d^3\mathbf{x}\ \mathbf{D}(\mathbf{x},t)\tilde{\mathbf{P}}(\mathbf{x},t)\ . \tag{6.230}$$

The functional includes the effective Hamiltonian

$$H_{\mathrm{DBA}}^{(\mathrm{eff})} = H_{\mathrm{DBA}} - \frac{1}{2c_{\mathrm{el}}}\int d^3\mathbf{x}\,\mathbf{D}^2(\mathbf{x},t)\ , \tag{6.231}$$

where the correction to the DA Hamiltonian is known as the *Born solvation energy*. Furthermore, we defined the scaled polarization $\tilde{\mathbf{P}} = \frac{\mathbf{P}_{\rm or}}{\varepsilon_\infty}$. The effective functional Eq. (6.230) is valid for any type of nonequilibrium *orientational* polarization $\tilde{\mathbf{P}}(\mathbf{x},t)$ and any charge distribution related to the ET reaction. It has been originally introduced by R.A. Marcus in the 1950s and found a widespread application (see Suggested Reading section). Eq. (6.230) forms the basis for the analysis of ET reactions in polar solvents as given in Section 6.5.3.

### 6.9.4 Second–Order Electron Transfer Rate

The rate which is of second–order with respect to the transfer coupling describes nonadiabatic ET and has been already computed in Sections 6.4. Here, we will give an alternative view on the high–temperature version of this rate. It is based on a quasi–classical approximation of the vibrational dynamics accompanying the ET, and in this way, the treatment will justify the ansatz we used in Section 6.4.1 (cf. also Section 5.3.5). Furthermore, the approach will be applied when calculating higher–order transfer rates in the supplementary part 6.9.5.

A quasi–classical approximation can be easily introduced if the vibrational part of the trace expression in Eq. (6.179)

$$\begin{aligned}\mathrm{tr}\left\{\hat{\Pi}_n \mathcal{L}_V \mathcal{G}_0(\omega)\mathcal{L}_V \hat{R}_m \hat{\Pi}_m\right\} &= -i\int dt\; e^{i\omega t}\;\langle\varphi_n|\mathcal{L}_V \mathrm{tr}_{\rm vib}\{\hat{\sigma}^{(m)}(t)\}|\varphi_n\rangle \\ &= -\frac{i}{\hbar}\int dt\; e^{i\omega t}\sum_k\Big(V_{nk}\langle\varphi_k|\mathrm{tr}_{\rm vib}\{\hat{\sigma}^{(m)}(t)\}|\varphi_n\rangle \\ &\quad +(k\leftrightarrow n)\Big)\,. \end{aligned} \tag{6.232}$$

is transformed into the Wigner–representation (cf. Section 3.4.4 and Section 3.11.3). Therefore, we replace the frequency–dependent Green's superoperator by the respective time–evolution operators. The latter quantities appear in the newly introduced operator

$$\hat{\sigma}^{(m)}(t) = U_0(t)\mathcal{L}_V \hat{R}_m \hat{\Pi}_m U_0^+(t)\,. \tag{6.233}$$

The introduction of this quantity is advantageously since a related equation of motion can be simply derived, and, it can be transformed into the Wigner–representation without further difficulties. We note that

$$\mathrm{tr}_{\rm vib}\{\hat{\sigma}^{(m)}(t)\} = \int dq\;\langle q|\hat{\sigma}^{(m)}(t)|q\rangle \equiv \int dx\frac{dp}{(2\pi\hbar)^{\mathcal{N}}}\;\hat{\sigma}^{(m)}(x,p;t)\,. \tag{6.234}$$

In the first part of this equation we specialized the trace by using the complete set of vibrational coordinate operator eigenstates $|q\rangle \equiv |\{q_\xi\}\rangle$. Then, in the second part, this particular choice of the trace has been transformed from the coordinate representation of $\hat{\sigma}^{(m)}$ to the Wigner–representation. This transformation of $\langle q|\hat{\sigma}^{(m)}(t)|q\rangle$ corresponds to a partial Wigner–transformation as introduced in Section 3.11.3. $\hat{\sigma}^{(m)}(x,p;t)$ which depends on the set of $\mathcal{N}$ classical vibrational coordinates $x = \{x_\xi\}$ and vibrational momenta $p = \{p_\xi\}$ remains an operator in the electronic state space (the integrations in Eq. (6.234) are $\mathcal{N}$–fold integrals with respect to the vibrational coordinates and momenta). According to Eq. (6.233) we obtain the initial value

$$\hat{\sigma}^{(m)}(x,p;t=0) = \frac{1}{\hbar} f_m(x,p) \sum_k \left( V_{km} |\varphi_k\rangle\langle\varphi_m| + (k \leftrightarrow m) \right) . \tag{6.235}$$

As a result of the Wigner–representation the equilibrium statistical operator $\hat{R}_m$ has been replaced by the equilibrium distribution $f_m(x,p)$ of the vibrational coordinates and momenta corresponding to the electronic state $\varphi_m$.

Having introduced the (partial) Wigner–representation of $\hat{\sigma}^{(m)}(t)$ we can directly use the Eqs. (3.424), (3.425), (3.426), and (3.427) to obtain an equation of motion for $\hat{\sigma}^{(m)}(x,p;t)$ in the semiclassical limit. We use a more compact notation and obtain (note the appearance of commutators and anti–commutators)

$$\begin{aligned} i\hbar\frac{\partial}{\partial t}\hat{\sigma}^{(m)}(x,p;t) &= [H_0, \hat{\sigma}^{(m)}]_- \\ &+ \frac{i\hbar}{2}\sum_\xi \left\{ \left[\frac{\partial H_0}{\partial x_\xi}, \frac{\partial \hat{\sigma}^{(m)}}{\partial p_\xi}\right]_+ - \left[\frac{\partial H_0}{\partial p_\xi}, \frac{\partial \hat{\sigma}^{(m)}}{\partial x_\xi}\right]_+ \right\} . \end{aligned} \tag{6.236}$$

As it is the case for $\hat{\sigma}^{(m)}$ the Hamiltonian $H_0$ depends on the classical coordinates and momenta but remains an operator in the electronic state space. Let us neglect for a moment the second term on the right–hand side. Then, it is easy to solve the remaining equation. After taking electronic matrix elements it follows

$$\begin{aligned} \langle\varphi_k|\hat{\sigma}^{(m)}(x,p;t)|\varphi_l\rangle &= \langle\varphi_k|U_0(t)\hat{\sigma}^{(m)}(x,p;t=0)U_0^+(t)|\varphi_l\rangle \\ &= \frac{1}{\hbar}(\delta_{ml}V_{km} - \delta_{mk}V_{ml})f_m(x,p)\ \exp\left(-\frac{i}{\hbar}(U_k - U_l)t\right) . \end{aligned} \tag{6.237}$$

Since the vibrational kinetic energy is independent on the actual electronic state the action of the time evolution operators $U_0$ and $U_0^+$ present in the first part of this equation reduces to $\exp(-i[U_k - U_l]t/\hbar)$. The diabatic difference potential carries the only time–dependence of the whole expression. There is no direct time–dependency of the coordinates and momenta. Their distribution remains fixed at the initial equilibrium distribution $f_m(x,p)$. Because of this special property of the solution Eq. (6.237) it is named *static* approximation.

To finally get the (frequency–independent) nonadiabatic rate according to Eq. (6.179) we have to insert Eq. (6.237) into Eq. (6.232) (at $\omega = 0$). It follows

$$k^{(2)}_{m\to n} = \frac{|V_{mn}|^2}{\hbar^2}\int dt \int dx \frac{dp}{(2\pi\hbar)^{\mathcal{N}}}\ f_m(x,p)\ \exp\left(-\frac{i}{\hbar}(U_m - U_n)t\right) . \tag{6.238}$$

The two types of integrations contained in this formula can be carried out. The momentum integration simply reduces $f_m(x,p)$ to the coordinate distribution function $f_m(x)$, Eq. (6.63), and the time–integral can be computed resulting in $\delta(U_m - U_n)$. Thus we arrive at a type of ET rate as given in Eq. (6.64). With this result we justify the ansatz taken to start with Eq. (6.64) and to compute the ET rate in the high–temperature limit. As shown, the static approximation of Eq. (6.236) results in the $\delta$–function which guarantees ET at the crossing of the PES. If one goes beyond the static approximation the restriction on ET at the PES crossing points only is abandoned.

This derivation puts the obtained rate expression into an alternative frame indicating in particular, how MD simulations for the vibrational coordinates may be used to calculate transition rates. Furthermore, the adopted static approximation offers an easy way to compute higher–order rates as it is demonstrated in the subsequent section.

### 6.9.5 Fourth–Order Donor–Acceptor Transition Rate

In the following, details are given for the calculation of the fourth–order rate $k^{(4)}_{D\to A}$, Eq. (6.178), which describes a direct transition from the donor via a single bridge molecule to the acceptor (fourth–order rates which describe the transfer of neighboring sites will be briefly mentioned at the end of this section). In calculating this rate we will follow the approach given in the preceding section which takes advantage of changing to the Winger–representation and applying a static approximation for the classical vibrational dynamics. We start with a consideration of the non–factorized part of the fourth–order rate. The rate is built up by a threefold action of the transfer coupling and the (Fourier–transformed) time–evolution. The expression $\hat{R}_D\hat{P}_D$ on the right–hand of the trace formula can be understood as the initial equilibrium statistical operator $\hat{W}^{(\text{eq})}_{DD}|D\rangle\langle D|$. Then the action of $\mathcal{L}_V$ on this operator results in

$$\begin{aligned} \mathcal{L}_V\hat{W}^{(\text{eq})}_{DD}|D\rangle\langle D| &= \frac{1}{\hbar}V_{BD}\hat{R}_D|B\rangle\langle D| - \frac{1}{\hbar}\hat{R}_D V_{DB}|D\rangle\langle B| \\ &\equiv \hat{W}_{BD}|B\rangle\langle D| - \hat{W}_{DB}|D\rangle\langle B| \,. \end{aligned} \tag{6.239}$$

Thus the electronic matrix element of the initial statistical operator $\hat{W}^{(\text{eq})}_{DD} \equiv \hat{R}_D$ has been transferred to the two new quantities $\hat{W}_{BD} = \hat{R}_D V_{DB}/\hbar$ and $\hat{W}_{DB} = \hat{R}_D V_{DB}/\hbar$. If in a next step $\mathcal{G}_0$ contained in Eq. (6.178) is transformed into the time–domain a propagation of, for example, $\hat{W}_{BD}$ follows with the time–evolution operator $U_B = \exp(-iH_B t_1/\hbar)$ from the left and with $U_D = \exp(iH_D t_1/\hbar)$ from the right. It results the quantity $\hat{W}^{(1)}_{BD}$. The complete time propagation with respect to the time–axis $t_1$ has been indicated by the index 1. Taking a further step in unravelling the matrix elements in Eq. (6.178), $\mathcal{G}_0\mathcal{L}_V$ acts on the two statistical operators $\hat{W}^{(1)}_{mn}$ leading to four new quantities $\hat{W}^{(2)}_{kl}$. The whole procedure starting with $\hat{W}^{(\text{eq})}_{DD}$ and ending with a certain $\hat{W}^{(3)}_{AA}$ is visualized in Fig. 6.37. It contains all six so–called *Liouville space pathways* from $\hat{W}^{(\text{eq})}_{DD}$ to $\hat{W}^{(3)}_{AA}$. But only three pathways have to be considered (see Fig. 6.29), since for each pathway there exists the complex conjugate by construction. Pathways 1 and 2 of Fig. 6.29 contain the statistical operator $\hat{W}^{(2)}_{BB}$, which is diagonal in the electronic quantum number of the intermediate bridge state. Thus it describes the population of this state which may undergo vibrational relaxations (these pathways will be named sequential ones). In contrast, the pathway 3 of Fig. 6.29 does not contain any intermediate diagonal density operator. It will result in an ET rate of the superexchange type.

To obtain a concrete expression for the fourth–order rate we will demonstrate its computation in the most simple case. It is characterized by a classical description of the vibrational dynamics and has been already used in Section 6.9.4 to calculate the nonadiabatic ET rate. In

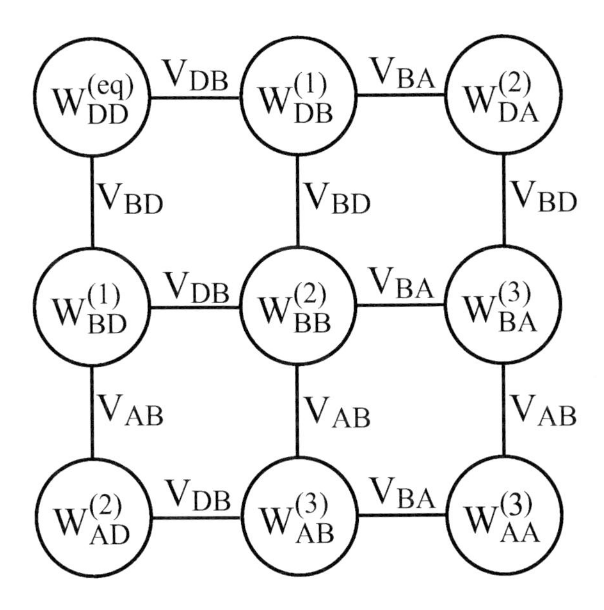


**Figure 6.37:** Liouville space pathways for the bridge–mediated ET rate $k_{DA}$ of fourth–order starting from the donor state and ending at the acceptor state. Every circle represents a particular electronic matrix element of the statistical operator. Lines connecting the circles show the action of a certain transfer integral $V_{mn}$. The different steps of the time evolution are indicated by the subscripts $1, \ldots, 3$. The way from the initial to the final state proceeds from the upper–left to the lower–right circle.

a first step we write the nonfactorized part of the fourth–order rate as

$$\tilde{k}^{(4)}_{D\to A}(\omega) = \int dt_3 dt_2 dt_1 e^{i\omega(t_3+t_2+t_1)} \times \text{tr}\left\{\hat{\Pi}_A \mathcal{L}_V \mathcal{U}_0(t_3)\mathcal{L}_V \mathcal{U}_0(t_2)\mathcal{L}_V \mathcal{U}_0(t_1)\mathcal{L}_V \hat{R}_D \hat{\Pi}_D\right\} . \quad (6.240)$$

To change to the Wigner–representation of the operator expressions under the trace we introduce in analogy to Section 6.9.4, Eq. (6.233) the operator

$$\hat{\sigma}^{(D,3)}(t_3) = \mathcal{U}_0(t)\mathcal{L}_V \hat{\sigma}^{(D,2)}(t_2) . \quad (6.241)$$

The newly introduced operator $\hat{\sigma}^{(D,2)}(t_2)$ is defined in the same way but with $\hat{\sigma}^{(D,1)}(t_1)$ on the right–hand side. Finally, the initial value of $\hat{\sigma}^{(D,1)}(t_1)$ is given by $\mathcal{L}_V \hat{R}_D \hat{\Pi}_D$. Since one operator determines the initial value for the following one a sequential computation of the total rate becomes possible. The related electronic matrix elements will be selected according to the three pathways of Fig. 6.29.

Let us start with the Wigner–representation $\hat{\sigma}^{(D,1)}(x, p; t_1)$ of $\hat{\sigma}^{(D,1)}(t_1)$ (cf. Eq. (6.234)). This is a partial representation taken with respect to the vibrational degrees of freedom. The initial value is given similar to Eq. (6.235) (with $m = D$). The concrete value depends on the chosen pathway of Fig. 6.29. But in any case the coordinate and momentum dependence is exclusively given by the equilibrium distribution $f_D(x, p)$. The determination of $\hat{\sigma}^{(D,1)}(x, p; t_1)$

in the static approximation (cf. Section 6.9.4) follows similar to Eq. (6.237). The particular electronic matrix element $\langle\varphi_m|\hat{\sigma}^{(D,1)}(x,p;t_1)|\varphi_n\rangle$ reads as $\exp(-i(U_m - U_n)t_1/\hbar)$ $\langle\varphi_m|\hat{\sigma}^{(D,1)}(x,p;t_1 = 0)|\varphi_n\rangle$. If we carry out the partial Fourier–transformation (necessary to obtain the $\omega$–dependent rate) we end up with $\hbar\langle\varphi_m|\hat{\sigma}^{(D,1)}(x,p;t_1 = 0)|\varphi_n\rangle/\ \Delta E_{mn}$. The energy denominator is determined by the PES $U_m$ and $U_n$ and has the form $\Delta E_{mn} = \hbar\omega + i\epsilon - (U_m - U_n)$.

To get the contribution of $\hat{\sigma}^{(D,2)}(t_2)$ to the total rate we proceed similar as in the case of $\hat{\sigma}^{(D,1)}(t_1)$. The initial value for $\hat{\sigma}^{(D,2)}(t_2)$ is given by the solution constructed for $\hat{\sigma}^{(D,1)}(t_1)$ and the action of $\mathcal{L}_V$. The electronic matrix elements of the static solution for the Wigner–representation $\hat{\sigma}^{(D,2)}(x,p;t_2)$ read $\hbar\langle\varphi_m|\hat{\sigma}^{(D,2)}(x,p;t_2 = 0)|\varphi_n\rangle/\Delta E_{mn}$. In the same way also the contribution of $\hat{\sigma}^{(D,2)}(t_2)$ to the rate can be calculated. Then, all three single contributions have to be collected to obtain the rate expression. Since we ordered the rate according to the different pathways given in Fig. 6.29 we will start to compute the pathway three. It has already explained how to do this in general. Now, we have concrete expressions for the different intermediate statistical operators at hand. Therefore, pathway three gives

$$\begin{aligned}\tilde{k}^{(4,\mathrm{III})}_{D\to A}(\omega) &= -\frac{i}{\hbar}\int dx\Big\{V_{AB}\frac{V_{BD}}{\Delta\tilde{E}_{BA}}\frac{f_D V_{DB}}{\Delta\tilde{E}_{DB}}\frac{V_{BA}}{\Delta\tilde{E}_{DA}} \\ &+\frac{V_{AB}}{\Delta\tilde{E}_{AD}}\frac{V_{BD}f_D}{\Delta\tilde{E}_{BD}}\frac{V_{DB}}{\Delta\tilde{E}_{AB}}V_{BA}\Big\}\cdot q\end{aligned} \tag{6.242}$$

The pathway III of Fig. 6.29 can be directly identified by the different factors in the first term of the right–hand side (the second term of the right–hand corresponds to the conjugated complex pathway). The third factor originates from $\hat{\sigma}^{(D,1)}$, and the fourth factor from $\hat{\sigma}^{(D,2)}$. The second factor follows from $\hat{\sigma}^{(D,3)}$ and the first factor is the result of the final action of $\mathcal{L}_V$. Since only the thermal distribution function depends on the vibrational momenta the whole expression taken in the Wigner–representation could be reduced to a multiple coordinate integral together with the coordinate distribution function $f_D(x)$.

According to the described scheme one can also calculate the contributions of pathway I and II. If combined with the factorized part of the total rate Eq. (6.178) one notices that these contributions compensate each other in the $\omega = 0$–limit. Therefore, the whole rate reads in this limit

$$k^{(4)}_{D\to A}(\omega = 0) = \tilde{k}^{(4,\mathrm{III})}_{D\to A}(\omega = 0) \tag{6.243}$$

This expression will be specified to the case $\Delta E_{BD}, \Delta E_{BA} \gg \Delta E_{DA} > 0$ (note $\Delta E_{mn} = E_m - E_n$). Hence, we can ignore the vibrational coordinate dependence of the transition energies to the bridge state and obtain

$$\begin{aligned}k^{(4)}_{D\to A} &= \frac{2\pi}{\hbar}\frac{|\,V_{DB}V_{BA}\,|^2}{\Delta E_{BD}\Delta E_{BA}}\int dx\ f_D(x)\delta(U_D(x) - U_A(x)) \\ &= \frac{2\pi}{\hbar}|V^{(\mathrm{eff})}_{DA}|^2\mathcal{D}_{DA}(\Delta E_{DA})\,,\end{aligned} \tag{6.244}$$

with

$$|V^{(\mathrm{eff})}_{DA}|^2 = \frac{|\,V_{DB}V_{BA}\,|^2}{\Delta E_{BD}\Delta E_{BA}}\,. \tag{6.245}$$

and the combined density of states used in Section 6.4.3. The effective bridge molecule mediated coupling $V_{DA}^{(\text{eff})}$ we already met in Eq. (6.166); it describes the superexchange mechanism of ET. However, here we obtained a symmetric expressions which contains $\Delta E_{BD}$ as well as $\Delta E_{BA}$. Nevertheless the structure of the effective coupling, Eq. (6.245) justifies to assign the rate $\tilde{k}_{D\to A}^{(4,\text{III})}$ (and the related pathway of Fig. 6.29) to originate from the superexchange mechanism.

However, $k_{D\to A}^{(4)}$ is determined by the superexchange mechanism alone, only in the limit of the static approximation. Fourth–order rates like $k_{D\to B}^{(4)}$ and $k_{B\to A}^{(4)}$ (which give a higher–order correction to the nonadiabatic transition rates) also vanish for the static approximation. If we allow for internal vibrational relaxation of bridge molecules we are beyond the static approximation and pathway 1 and 2 of Fig. 6.29 as well as $k_{D\to B}^{(4)}$ and $k_{B\to A}^{(4)}$ contribute, too.

---

# Notes


[Ber02] Yu. A. Berlin, A. L. Burin, M. Ratner, Chem. Phys. **275**, 61 (2002).

[Jor99] J. Jortner and M. Bixon (eds.), Adv. Chem. Phys. **106**, **107** (1999), (series eds. I. Prigogine and S. A. Rice).

[Bix99] M. Bixon and J. Jortner, in [Jor99], part one, p. 35

[Bix00] M. Bixon, J. Jortner, J. Phys. Chem. B **104**, 3906 (2000); A. A. Voityuk, J. Jortner, M. Bixon, N. Rösch, J. Chem. Phys. **114**, 5614 (2001); M. Bixon, J. Jortner, Chem. Phys. **281**, 393 (2002).

[Cal83] D. F. Calef and P. G. Wolynes, J. Phys. Chem. **87**, 3387 (1983).

[Cas02] J. Casado-Pascual, I. Goychuk, M. Morillo, and P. Hänggi, Chem. Phys. Lett. **360**, 333 (2002).

[Clo88] G. L. Closs and J. R. Miller, Science **240**, 440 (1988).

[Dav98] W. B. Davis, W. A. Svec, M. A. Ratner, and M. R. Wasielewski, Nature **396**, 60 (1998).

[Ego03] D. Egorova, M. Thoss, W. Domcke, and H. Wang, J. Chem. Phys. **119**, 2761 (2003).

[Fel77] B. U. Felderhof, J. Chem. Phys. **67**, 493 (1977).

[Gie99] B. Giese, S. Wessely, M. Spormann, U. Lindemann, E. Meggers, M. E. Michel–Beyerle, Angew. Chem. Int. Engl. **38**, 996 (1999); B. Giese, J. Amaudrut, A.–K. Köhler, M. Spormann, S. Wessely, Nature **412**, 318 (2001).

[Hae90] P. Hänggi, P. Talkner, abd M. Borkovec, Rev. Mod. Phys. **62**, 251 (1990).

[Hei87] H. Heitele, M. E. Michel–Beyerle, and P. Finkch, Chem. Phys. Lett. **138**, 237 (1987).

[Hop89] J. J. Hopfield, J. N. Onuchic, and D. N. Beratan, J. Phys. Chem. **93**, 6350 (1989).

[Hu89] Y. Hu and S. Mukamel, J. Chem. Phys. **91**, 6973 (1989).

[Isi92] S. S. Isied, M.Y. Ogawa, J. F. Wishart, Chem. Rev., **92**, 381 (1992).

[Jea92] J. M. Jean, R. A. Friesner, G. R. Fleming, J. Chem. Phys. **96**, 5827 (1992).

[Kaw01] T. Kawatsu, T. Kakitani, and T. Yamato, J. Phys. Chem. B, **105**, 4424 (2001).

[Kel99] S. O. Kelley, J. K. Barton, Science **283**, 375 (1999); T.T. Willams, D. T. Odom, J. K. Barton, A. H. Zewail, Proc. Natl. Acad. Sci. U.S.A. **97**, 9048 (2000).

[Kes74] N. R. Kestner, J. Logan, and J. Jortner, J. Phys. Chem. **78**, 2148 (1974).

[Kim90] H. J. Kim and J. T. Hynes, J. Phys. Chem. **94**, 2736 (1990), J. Chem. Phys. **93**, 5194 (1990), **93**, 5211 (1990), **96**, 5088 (1992).

[Lew00] F. D. Lewis, X. Y. Liu, J. Q. Liu, S. E. Miller, R. T. Hayes, M. R. Wasielewski, Nature **406**, 51 (2000).

[Leg87] A. J. Legget, S. Chakravarty, A. Dorsey, M. P. A. Fisher, A. Garg, W. Zwerger, Rev. Mod. Phys. **59**, 1 (1987).

[Lel85] B. A. Leland, A. D. Joran, P. M. Felker, J. J. Hopfield, A. H. Zewail, and P. B. Dervan, J. Phys. Chem. **89**, 5571 (1985).

[Mak96] N. Makri, E. Sim, D. E. Makarov, and M. Topaler, Proc. Natls. Acad. Sci. USA **93**, 3926 (1996).

[Mar93] M. Marchi, J. N. Gehlen, D. Chandler, and M. Newton, J. Am. Chem. Soc. **115**, 4178 (1993).

[Pet01] E. G. Petrov and V. May, J. Phys. Chem. A **105**, 10176 (2001).

[Reg99] J. J. Regan and J. N. Onuchic, in [Jo399], part two, p. 497.

[Rip87] I. Rips and J. Jortner, J. Chem. Phys., **87**, 2090 (1987).

[Rub99] I. V. Rubtsov and K. Yoshihara, J. Phys. Chem. A, **103**, 10202 (1999).

[Spa88] M. Sparpaglione and S. Mukamel, J. Chem. Phys., **88**, 3263 (1988).

[Tan94] J. Tang, Chem. Phys. **188**, 143, (1994).

[Wei98] U. Weiss, *Quantum Dissipative Systems*, World Scientific, Singapore, Sec. Ed., 1998.

# 7 Proton Transfer

*We discuss fundamental aspects of the theory of proton transfer across inter– or intramolecular hydrogen–bonded systems immersed in a solvent or protein environment. Since the strength of the hydrogen bond depends on the distance between proton donor and acceptor entities, vibrational motions modifying the latter are strongly coupled to the proton transfer. We give a classification of such vibrational modes and elaborate on their effect on quantum mechanical proton tunneling.*
*A central observation is that often the proton dynamics can be adiabatically separated from the slow motions of the environmental degrees of freedom. This suggests a close analogy to the treatment of coupled electronic–nuclear dynamics presented in Chapters 2, 5, and 6. Similar to the case of electron transfer, proton transfer can occur in the adiabatic as well as in the nonadiabatic limit. The former requires the proton wave function to adjust instantaneously to any change in the environmental configuration, whereas the latter assumes that the proton dynamics is slow compared to typical relaxation times for the environment.*
*Since proton transfer reactions usually take place in the condensed phase, we discuss the application of approximate quantum and quantum–classical hybrid methods to the solution of the nuclear Schrödinger equation in some detail. A powerful tool in this respect is provided by the surface–hopping method which allows to treat nonadiabatic transitions between the adiabatic protonic states while retaining the classical nature of the environment. This is indispensable in the nonadiabatic limit where proton transfer takes place via tunneling between different diabatic states. In the limit of weak coupling the introduction of diabatic protonic states allows to express transfer rates in close analogy to the case of electron transfer.*

## 7.1 Introduction

As a second type of charge transfer we will consider the proton transfer (PT) in intra– and intermolecular hydrogen bonds as shown in Fig. 7.1. At first glance the reader might wonder why dealing with the transfer of a positive charge if that of a negative charge (electron transfer) has been discussed in quite some detail before. And, indeed we will find many similarities between electron and proton transfer. However, there are also some features which are unique to PT. After all, protons are much heavier than electrons and therefore their wave function will be much more localized in space. On the other hand, the proton is still a quantum particle. This means that its motion has to be treated quantum mechanically and PT is influenced not only by zero–point energy effects but also by quantum tunneling even at room temperature. Further, the simultaneous motion of several protons may be subject to strong correlation effects. As an example we have shown the intermolecular double proton transfer in carboxylic acid dimers in Fig. 7.1B. An important question is related to the fact that the double proton transfer can proceed either step–wise or concerted (as shown in the figure). For larger systems with many hydrogen bonds such as, e.g., water clusters the correlated motion of several protons may lead to interesting collective phenomena.

Many intriguing possibilities in PT studies are opened by the fact that there are four isotopes of hydrogen with mass ratios which are higher than for any other element of the periodic table. This gives rise to the so–called *kinetic isotope effect*, from which, for instance, the relative importance of tunneling can be inferred.

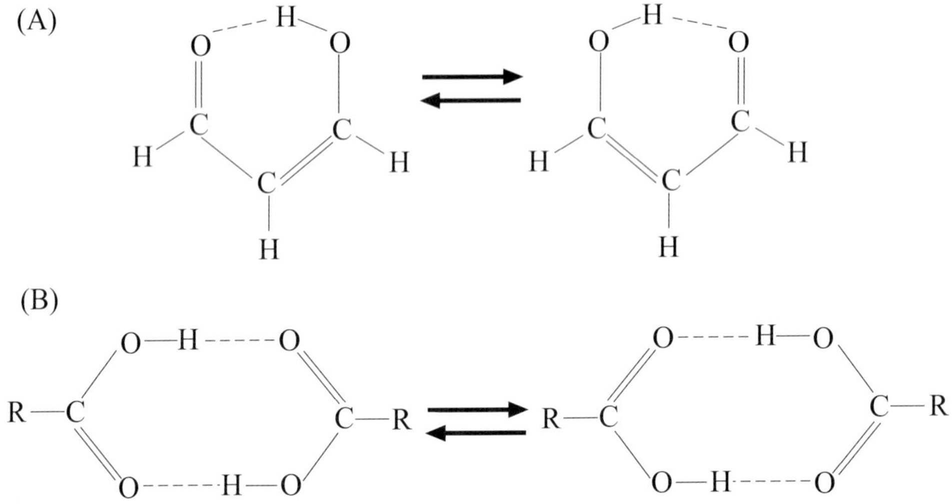

**Figure 7.1:** (A) Single PT in malonaldehyde which is one of the standard examples for an intramolecular proton transfer system with strong coupling between the proton motion and heavy atom vibration. In particular the O–O wagging vibration modulates the reaction barrier for isomerization. (B) Double PT across the intermolecular hydrogen bonds in carboxylic acid dimers (typical choice for R are R=H or R=$CH_3$).

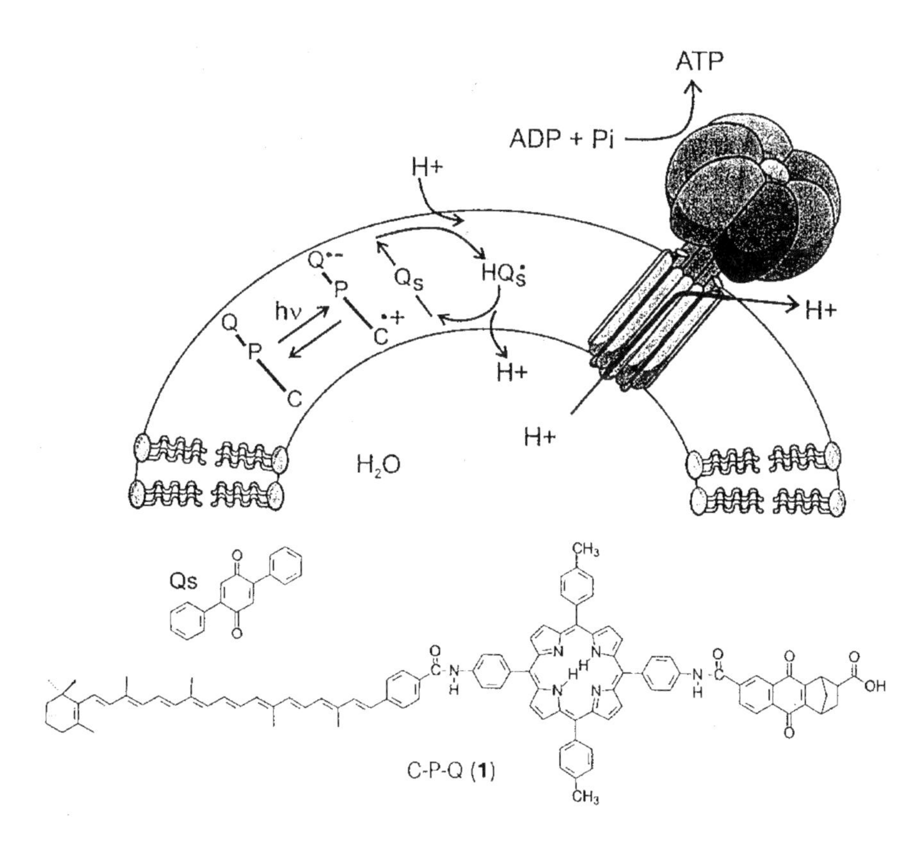


**Figure 7.2:** Artificial proton pump realizing the conversion of light energy into an electrochemical potential which is used to synthesize ATP (adenosine triphosphate). The "reaction center", which is inside a bilayer of a liposome, is a donor–acceptor complex linked by a porphyrin group (bottom). After photoexcitation electronic charge separation occurs establishing a redox potential gradient across the bilayer. A freely diffusing quinone ($Q_S$) experiencing this potential will shuffle protons across the membrane. The established proton concentration gradient then drives ATP production. (reprinted with permission from [Ste98]; copyright (1998) Nature)

Finally, PT transfer is often strongly coupled to low–frequency (heavy atom) modes of its immediate surroundings. A standard example in this respect is malonaldehyde shown in Fig. 7.1A. Here, the intramolecular O–O wagging vibration has a strong influence on the reaction barrier and therefore on the isomerization reaction shown in Fig. 7.1A. The interplay between quantum mechanical tunneling and the strong coupling to low–frequency skeleton modes in PT reactions has some interesting consequences for the tunneling splittings or the related tunneling transfer rates. Some ideas in this respect will be discussed in Section 7.2.3.

PT has an enormous importance for many processes in biology and chemistry. We have already discussed the initial electron transfer steps of photosynthesis in Chapter 6. Subsequently to the electron transfer, a PT across the membrane occurs and the concerted action of electron and proton transfer establishes the storage of solar energy in terms of a transmembrane electrochemical potential. This is one example for a *proton pump*. A second one is given, for instance, by the transmembrane protein bacteriorhodopsin which encapsulates a chain of water molecules (water wire) through which a proton can be transferred. There are also first attempts to mimic photosynthesis by creating artificial proton pumps as shown in Fig. 7.2. The transfer of excess protons in water networks as well as PT processes taking place in ice has also attracted a lot of attention. In particular PT on ice surfaces is believed to have some importance for the ozone depletion in the stratosphere. PT is often a key event in enzyme catalysis where it leads to activation of the proton donor after PT has been triggered, for instance, by polar residues in the protein surroundings. In more general terms one can say that PT is at the heart of acid–base reactions.

Traditionally, hydrogen bonds are characterized by means of their stationary infrared (IR) spectra. Whereas this allows a general characterization, for instance, in terms of the strength of the hydrogen bond (see Section 7.2.1), it was shown to be ultrafast IR spectroscopy which allows to uncover the details of such spectra in the condensed phase. An example is given in Fig. 7.3: In panels C and D IR pump–probe signals are shown for different laser frequencies across the broad absorption band of phthalic acid monomethylester in the OH–stretching region. The analysis of the oscillations in the signal provided evidence for a mechanism where the laser excites a superposition of states involving a low–frequency mode which modulates the O–H–O distance. In the linear absorption spectrum (panel A) these transitions are hidden under a broad band.

Whereas the dynamics observed with IR spectroscopy occurs in the electronic ground state, photochemical reactions involving PT in excited electronic states have also been studied extensively. The sudden change of the electronic state leads to a strong modification of the charge distribution within the hydrogen bond (acidity/basicity), thus giving rise to a large driving force for PT which occurs on a time scale below 100 femtoseconds. This transfer can be strongly coupled to intramolecular vibrational modes of the molecular skeleton and indeed signatures of multidimensional coherent nuclear wave packet motion have been observed for a number of excited state PT reactions (see, Fig. 7.4).

The theoretical description of PT often rests on the large mass difference between the proton and the heavier atoms being involved in the reaction. This makes it possible to introduce a *second Born–Oppenheimer separation* after the electronic problem has been split off as shown in Section 2.3. As with electron transfer, PT can then be characterized as being in the *adiabatic* or in the *nonadiabatic* limit (or in between). If the proton motion is fast and the proton is able to adjust *instantaneously* to the actual configuration of the environmental degrees of freedom a description in terms of adiabatic proton states and a corresponding delocalized wave function is appropriate (Section 7.3). The PT rate may become proportional to some frequency factor characterizing the shape of the adiabatic reaction barrier. On the other hand, in the nonadiabatic limit the proton motion is much slower than characteristic time scales for the environment (Section 7.4). The potential energy surface is conveniently described in terms of weakly interacting diabatic proton states, and the transfer occurs via tunneling. As for the case of electron transfer the related transfer rates will be proportional to the square of the coupling matrix elements (cf. Section 6.46). In both cases the surrounding solvent has an important influence, for it may stabilize reactants and products but also provide the fluctuating force which triggers the transfer event.

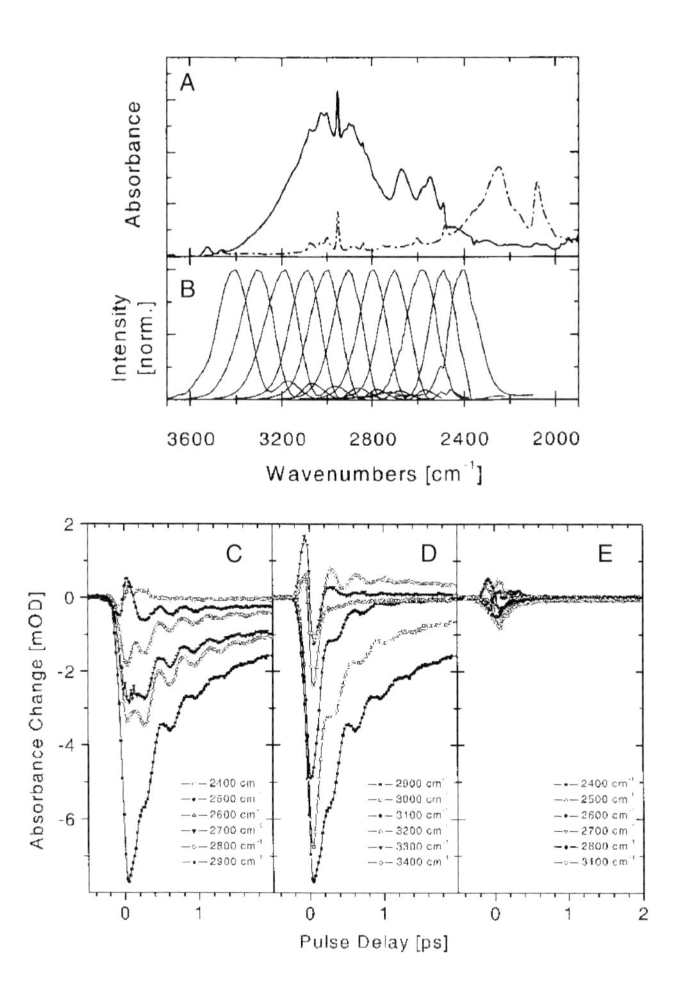


**Figure 7.3:** Coherent oscillations in a hydrogen bond after ultrashort IR pulse excitation. (A) Linear absorption spectrum of phthalic acid monomethylester (solid line, c.f. also the lower part of Fig. 7.6) and its deuterated form (dashed line) in solution ($C_2Cl_4$). (B) IR pump pulse intensity profiles for the different excitation conditions leading to the pump–probe signals which are shown as a function of the delay time between the pulses in panels C and D. The oscillatory component of the signal can be attributed to the excitation of a wave packet with respect to a low–frequency mode (100 $cm^{-1}$) which couples strongly to the OH stretching vibration. The decay of the signal is due to relaxation and dephasing processes introduced by the interaction with the solvent. Panel E shows the signal which comes solely from the solvent. (reprinted with permission from [Mad02]; copyright (2002) Chemical Society of Japan)

Quantum effects are of considerable importance for the proton motion. However, only if the dynamics can be reduced to a reasonably small number of degrees of freedom PT will be amenable to wave packet propagation methods as outlined in Section 7.5.1. On the other

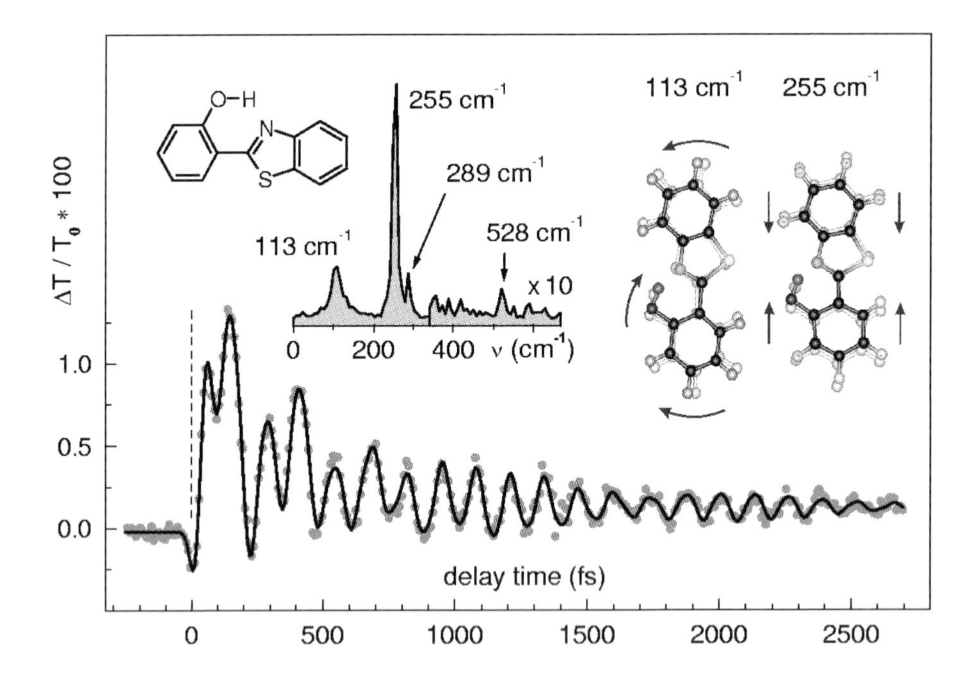


**Figure 7.4:** Infrared transient transmission change due to stimulated emission of 2-(2-hydroxyphenyl)benzothiazole (upper left) detected at 500 nm after ultrafast excitation at 340 nm. PT takes place as a wave packet motion from the enol (shown here) to the keto (-O$\cdots$H–N-) form in about 50 fs. The reaction coordinate is dominated by a low–frequency bending type mode at 113 $\text{cm}^{-1}$ which modulates the hydrogen bond such that the donor–acceptor (O$\cdots$ N) distance is reduced (right part). In the keto–form several modes are coherently excited as seen from the Fourier transform of the oscillatory signal. Most notably is a symmetric mode at 255 $\text{cm}^{-1}$. The normal mode displacements shown in the right correspond to the enol configuration of the electronic ground state. Their character is assumed to change not appreciably in the excited electronic state. (figure courtesy of S. Lochbrunner, for more details see also [Loc00])

hand, for real condensed phase environments solvent and low frequency intramolecular modes have to be treated classically within a quantum–classical hybrid approach (cf. Section 3.11). Here, a unified description of the different regimes of PT is provided by the *surface hopping* method which combines classical trajectories with quantum transitions (see Sections 3.11.2 and 7.5.2).

In the following section we will elaborate on the discussion of the properties of hydrogen bonds. Further, we will introduce the Hamiltonian for a PT complex which sets the stage for the subsequent discussions of the different dynamics regimes.

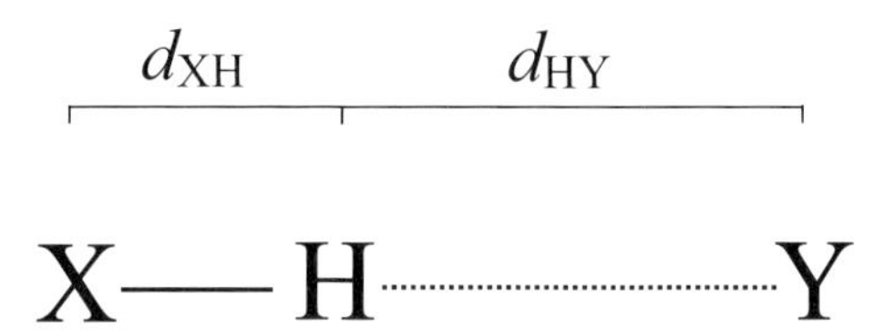


**Figure 7.5:** Linear hydrogen–bonded complex which can be characterized by the distance between the donor (X) and acceptor (Y) fragments, $d_{XH} + d_{HY}$, as well as the relative position of the hydrogen, $s = d_{XH} - d_{HY}$.

## 7.2 Proton Transfer Hamiltonian

### 7.2.1 Hydrogen Bonds

In Section 2.3 the coupled motion of electronic and nuclear degrees of freedom has been treated by making use of their adiabatic separability. This resulted in potential energy surfaces for the nuclear motions corresponding to the various adiabatic electronic states. As a consequence of nonadiabatic couplings electronic transitions between different adiabatic states are possible, especially in the vicinity of avoided crossings. For the following discussion we will assume that the electronic problem has been solved and the adiabatic potential energy surface is known. We will restrict our considerations to the electronic ground state only, although the concepts in principle apply to any other electronic state as long as nonadiabatic couplings can be neglected.

Considering the motion of the proton within the hydrogen–bonded complex, we note that in general it is not a bare proton which is transferred, but part of the electronic charge is dragged with the proton. This makes the distinction between PT and "hydrogen atom" transfer sometimes a little ambiguous. There is also quite some electronic charge flow in the donor–acceptor groups which goes in the opposite direction to that of the PT. This leads to a large variation of the molecule's dipole moment which is a characteristic feature of PT. Consequently the hydrogen bond is highly polarizable and a polar solvent or charged residues in a protein environment can be expected to have a large influence on PT.

The potential energy surface $U(R)$ will be a function of some PT coordinate(s), *all* other nuclear coordinates of the PT complex, plus the environmental coordinates. Of course, such a high–dimensional potential energy surface cannot be obtained on an ab initio level of quantum chemistry. This level of theory is usually reserved for a small subset of *relevant* coordinates only, while the majority of degrees of freedom is treated approximately (see below). The most important coordinates are, of course, those which are directly related to the proton motion. For simplicity let us consider a linear hydrogen–bonded complex as shown in Fig. 7.5. The hydrogen bond is formed between a proton donor, X–H, and a proton acceptor, Y. Here X and Y may represent parts of the same molecule (*intramolecular hydrogen bond*) or of different molecules (*intermolecular hydrogen bond*). In case of Fig. 7.5 the PT (reaction) coordinate $s$ can be chosen as the difference between the X–H distance, $d_{XH}$, and the H–Y distance, $d_{HY}$, i.e., $s = d_{XH} - d_{HY}$.

The minimum requirement for a potential energy surface of the simple system shown in Fig. 7.5 would include besides the PT coordinate $s$, the information about the distance $d_{XH} + d_{HY}$ between donor and acceptor fragments. Depending on the complexity of X and Y, an ab initio calculation of a two–dimensional potential energy surface $U(s, d_{XH} + d_{HY})$, may be possible, for instance, by using the methods introduced in Section 2.6.3.

Before incorporating possible environmental degrees of freedom of a solvent or a protein, let us briefly discuss some features of the hydrogen–bonded complex shown in Fig. 7.5. First of all, hydrogen bonds may be characterized by the fact that proton donor X−H and acceptor Y retain their integrity in the complex. While the X−H bond is covalent, the hydrogen bond H$\cdots$Y is of noncovalent character. A widely accepted point of view is that the hydrogen bond has the characteristics of a strong van der Waals interaction. At long distances $d_{HY}$ this comprises electrostatic, dispersion, and induction energies. At short distances repulsive exchange interactions between the overlapping electron densities of X−H and Y dominate. The process of hydrogen bond formation comes along with a decrease of the donor–acceptor distance $d_{XH} + d_{HY}$ and an increase of the X−H bond length $d_{XH}$. The latter effect weakens the X−H bond and therefore reduces the vibrational frequency of the X−H stretching mode. In addition the infrared X−H absorption band is broadened. This broadening is a consequence of the larger anharmonicity of the potential energy surface in the region of the X−H vibration, which may come along with a pronounced coupling to low–frequency modes of the complex. A comparison between the spectra of a free OH–stretching vibration and of an OH vibrations in inter– and intramolecular hydrogen bonds is shown in Fig. 7.6.

The strength of hydrogen bonding depends on the properties of the donor and acceptor entities, i.e., in particular on their electronegativity. However it also is a function of the separation, $d_{XH} + d_{HY}$, between donor and acceptor which may be imposed by external means. We can distinguish between *weak* and *strong* hydrogen bonds.[1] For weak hydrogen bonds the donor–acceptor distance is relatively large ($> 3$Å) and the potential energy profile for moving the proton along the reaction coordinate $s$ between X and Y shows the typical double–minimum behavior plotted in Fig. 7.7 (top). The barrier will be high enough to allow for several protonic states to be energetically below its top.

This situation might be characteristic for intramolecular hydrogen bonds, where the donor–acceptor distance is more or less fixed by the rigid molecular frame. On the other hand, intermolecular hydrogen bonds are often much stronger. Here the larger structural flexibility allows for relatively short distances between donor and acceptor. Thus the barrier along the PT coordinate $s$ is rather low if existent at all. This is sketched in Fig. 7.7 (bottom). In the middle panel of Fig. 7.7 we have shown schematically the dependence of the potential energy curve for PT on the distance, $d_{XH} + d_{HY}$, between donor and acceptor.

Weak and strong hydrogen bonds may also be distinguished by the extent at which they modify the infrared absorption spectra of the X−H stretching vibration. Upon forming a weak hydrogen bond, the frequency of the X−H stretching vibration moves by about 100–300 cm$^{-1}$ to the red due to the bond lengthening. A strong hydrogen bond, on the other hand, is characterized by a much larger red shift and a considerable broadening of the absorption line (cf. Fig. 7.6).

[1]Note that this classification scheme is not rigorously defined in literature and it may vary in dependence on the properties which are used for characterizing the strength of the hydrogen bond. Often one also distinguishes the range between these two extrema as belonging to medium strong hydrogen bonds.

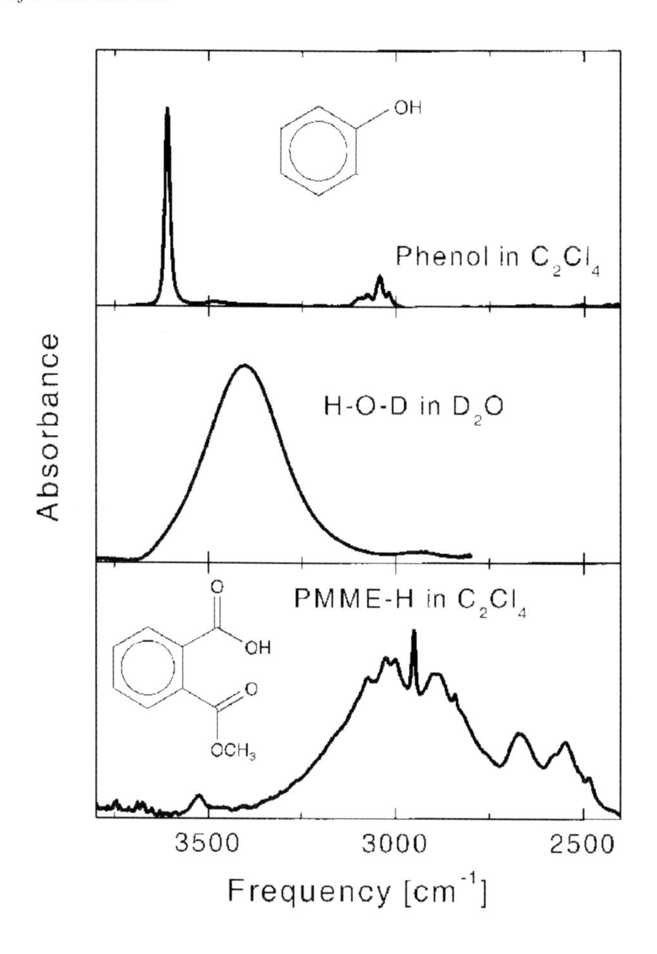


**Figure 7.6:** Infrared absorption spectra show clear signatures of hydrogen bond formation. Compared to the narrow line of a free OH–stretching vibration (upper part), the absorption band shifts to lower frequencies and broadens considerably if a condensed phase situation is considered (middle part: absorption of the OH–vibration in deuterated water). The absorption may also develop a peculiar substructure as shown for an intramolecular hydrogen bond (lower part). (reprinted with permission from [Mad02]; copyright (2002) Chemical Society of Japan)

In the following, we will discuss the effect of the coupling between the PT coordinate and intramolecular modes first. The interaction with environmental degrees of freedom is included in Section 7.2.4. The separate consideration of intramolecular modes is motivated by the distinct influence of strongly coupled intramolecular modes, for instance, on the hydrogen bond geometry. The effect of the environment can often be characterized as leading to phase and energy relaxation or, in the case of polar environments, to a stabilization of a specific configuration of the hydrogen bond. Of course, such a separation is not always obvious, for instance, for intermolecular PT in a protein environment. Here, the motion of the protein in principle may influence the PT distance as well.

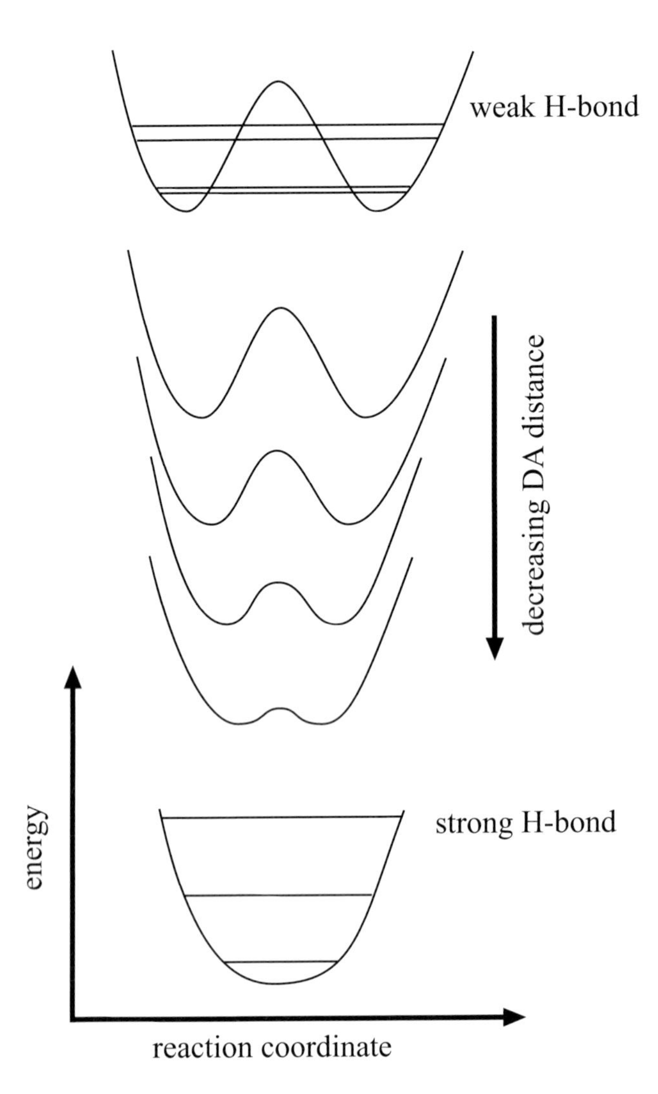


**Figure 7.7:** Potential energy profile along a PT reaction coordinate, for example, $s$ in Fig. 7.5, in dependence on the donor–acceptor (DA) distance $d_{\rm XH} + d_{\rm HY}$. Compounds characterized by a large distance form weak hydrogen bonds (top), while strong hydrogen bonds typically involve a small DA distance. The strength of the hydrogen bond and the exact shape of the potential, of course, depend on the donor and acceptor entities. The symmetric situation plotted here may correspond to the case X = Y.

### 7.2.2 Reaction Surface Hamiltonian for Intramolecular Proton Transfer

We will discuss the potential energy surface for the intramolecular degrees of freedom of a system like that in Fig. 7.5. Note, that malonaldehyde shown in Fig. 7.1 would be particular

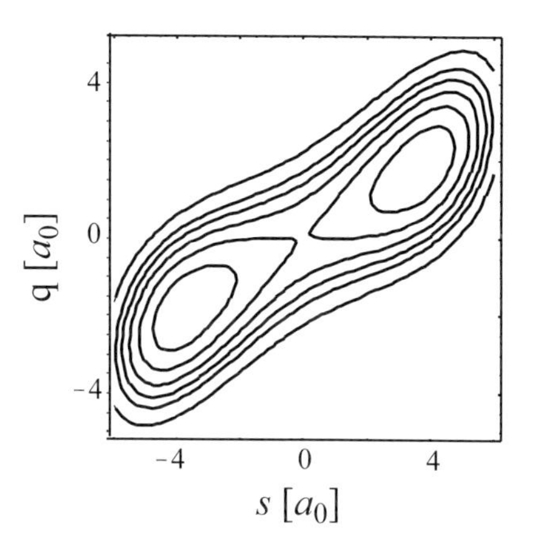


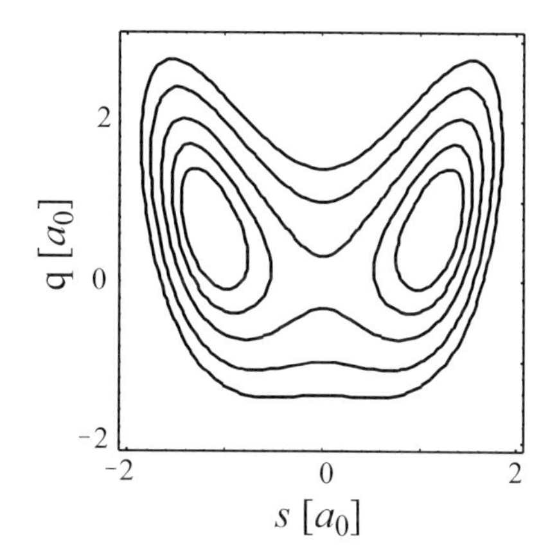


**Figure 7.8:** Schematic view of two–dimensional potential energy surface for linear (left) and quadratic (right) coupling between the PT coordinate $s$ and a harmonic heavy atom mode $q$. The minima for zero coupling are at $s = \pm 1\ a_0$ in both cases. For a specific example see Fig. 7.10.

example for such a system. Suppose that $q = \{q_\xi\}$ comprises all so–called heavy atom vibrational coordinates of the total X–H$\cdots$Y complex which have a strong influence on the PT insofar as they modulate, for instance, the distance $d_{\rm XH} + d_{\rm HY}$. Further, we assume that these modes can be treated in harmonic approximation. This scenario has already been discussed in Section 2.6.3, where we derived a suitable reaction surface Hamiltonian, Eq. (2.105); an example for a PT reaction was shown in Fig. 2.13. If we neglect the dependence of the force constant matrix on the proton coordinate, the reaction surface Hamiltonian can be written as

$$H = T_s + U(s) + \sum_\xi \left[ \frac{p_\xi^2}{2} + \frac{\omega_\xi^2}{2} q_\xi^2 - F(s) q_\xi \right] . \tag{7.1}$$

Here, the $T_s$ is the kinetic energy operator for the proton motion and $U(s)$ is the respective potential as obtained, for example, from a quantum chemistry calculation of the adiabatic electronic ground state energy in dependence on the proton position (cf. Eq. (2.19)). The last term in Eq. (7.1) describes the coupling between the PT coordinate and the heavy atom modes. Note that for a coordinate independent force constant matrix, these modes are not coupled by the motion of the proton (cf. Eq. (2.105)).

The principal effect of the coupling term on the PT dynamics can be highlighted by considering two typical cases, i.e., a linear coupling, $F(s) = c_1 s$ and a quadratic coupling $F(s) = c_2 s^2$. In Fig. 7.8 we show some schematic potential energy surfaces for both situations in the case of a single heavy atom mode. A linear coupling apparently is not favorable to the PT since it effectively increases the distance between donor and acceptor, and therefore according to Fig. 7.7 the barrier for PT will be increased. A mode which is quadratically coupled, however, can reduce the barrier for PT dramatically. In fact if we follow the minimum energy path on the two–dimensional potential energy surface in Fig. 7.8 (right) we find that at the saddle point (transition state) the heavy atom mode is compressed. This type of mode is frequently also called *promoting* or *gating* mode. In fact gating modes will often be

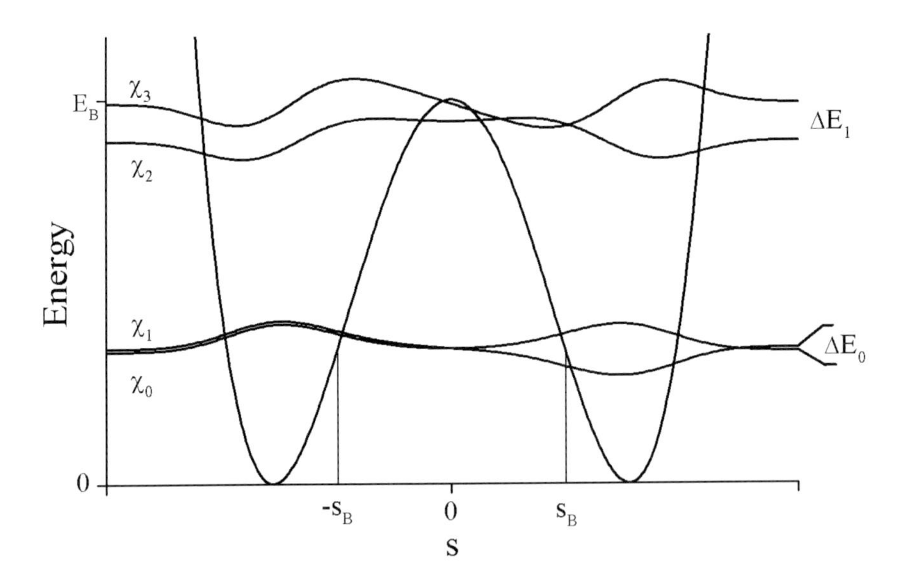


**Figure 7.9:** One–dimensional potential energy curve along a PT coordinate $s$ with the lowest eigenfunctions. There are two tunneling doublets below the barrier with the splitting given by $\Delta E_0$ and $\Delta E_1$. The entrance points for barrier penetration for the ground state doublet are labelled $\pm s_B$.

of donor–acceptor stretching type.[2] A prominent example for a promoting mode is the O–O wagging vibration in malonaldehyde (cf. Fig. 7.1). We note in passing that the principal behavior discussed in Fig. 7.7 can be viewed as representing cuts through the two–dimensional potential energy surface of Fig. 7.8. In the next Section we will elaborate on the influence of intramolecular modes on the quantum tunneling of the proton which is expressed in terms of the spectroscopically accessible tunneling splitting.

### 7.2.3 Tunneling Splittings

Quantum tunneling of the proton through the reaction barrier is one of the most characteristic features in particular for PT in symmetric potentials. Proton tunneling can be viewed in time– and energy domain. Consider, for example, the case of a one–dimensional reaction coordinate shown in Fig. 7.9 and focus on the two lowest eigenstates. If we neglect the higher excited states for the moment, we have essentially recovered the two–level system discussed in Section 2.8.3 (cf. Fig. 2.16). There the coupling between two localized states was shown to give rise to a splitting of the respective eigenstates. In the present case the appearance of a splitting $\Delta E_0$ can be viewed as a consequence of the coupling between two almost localized states (in the left and right well) due to the wave function overlap in the barrier region. The eigenfunctions

[2]There is a third kind of coupling mode which is called *squeezing* type. Here, only the frequency changes upon PT. Such modes are often related to out–of–plane motions of planar molecules.

in Fig. 7.9 are then the symmetric and antisymmetric combinations of these local states.

An alternative view is provided by a time–domain approach. Let us take a state which is localized in one of the minima (i.e. a superposition of the two lowest eigenstates shown in Fig. 7.9) as an initial wave packet. This wave packet will oscillate between the two wells, that is, it will tunnel through the potential barrier. Adopting the results of Section 3.8.7 the oscillation period is given by $2\pi\hbar/\Delta E_0$ (cf. Eq. (3.313)).

In the following we will focus on the (energy–domain) tunneling splitting, which is experimentally accessible, e.g., by high–resolution vibration–rotation spectroscopy. In particular we will address the question how this tunneling splitting is influenced by the coupling to intramolecular modes. However, let us start with the one–dimensional case shown in Fig. 7.9. An expression for the splitting can be obtained from standard quasi–classical Wentzel–Kramers–Brillouin theory, which gives

$$\Delta E_0 = \frac{\hbar\omega}{\pi} \exp\left\{ -\frac{1}{\hbar} \int_{-s_{\mathrm{B}}}^{s_{\mathrm{B}}} ds \sqrt{2m_{\mathrm{proton}}(E - U(s))} \right\} . \tag{7.2}$$

Here $\omega$ is a characteristic frequency in the left/right well and $E$ is the energy of the localized left/right states. From Eq. (7.2) it is obvious that the tunneling splitting is rather sensitive to the details of the potential energy surface and in particular to the energetic separation between the considered state and the top of the barrier as well as to the tunneling distance $2s_{\mathrm{B}}$. Thus the splitting increases for excited states as shown in Fig. 7.9. From the dynamics perspective this implies that, for instance, an initially prepared localized wave packet on the left side of the barrier will be transferred faster with increasing energy.

So far we have considered a one–dimensional situation. However, from Fig. 7.8 it is clear the PT in principle is a multidimensional process and an accurate treatment has to take into account the coupling, e.g., to the heavy atom vibrations of the immediate surrounding. Due to the exponential dependence of the tunneling splitting on the details of the overlapping wave functions in the classically forbidden region, the calculation of tunneling splittings can be considered as a critical test of the accuracy of theoretical methods. This holds in particular as tunneling splittings can be rather accurately measured, e.g., with gas phase high resolution spectroscopy.

Let us discuss the effect of linear coupling (antisymmetric) and promoting (symmetric) modes on the tunneling splitting. In Fig. 7.10 we give an example of a four–dimensional reaction surface calculation (see, Eq. (2.105)) for the PT in a derivative of tropolone (for the reaction scheme see panel (A)). The potential includes the two coordinates for the motion of the proton in the plane of the molecule as well as a symmetrically and an antisymmetrically coupled skeleton normal mode (for the normal mode displacement vectors see panel (B)). We have also plotted two–dimensional projections of the full four–dimensional potential as well as of selected eigenfunctions in Fig. 7.10C.

Let us first consider the effect of a promoting type (symmetric coupling) mode. Already from Fig. 7.8 it is clear that a symmetric coupling leads to an effective reduction of the barrier. In the left panel of Fig. 7.10 it is seen that the overall bending of the two–dimensional potential is reflected in the ground state wave functions. Thus the overlap in the barrier region will be increased and the tunneling splitting is larger as compared to the case of no coupling to this mode. Upon excitation of the symmetric mode only (not to be confused with the excited

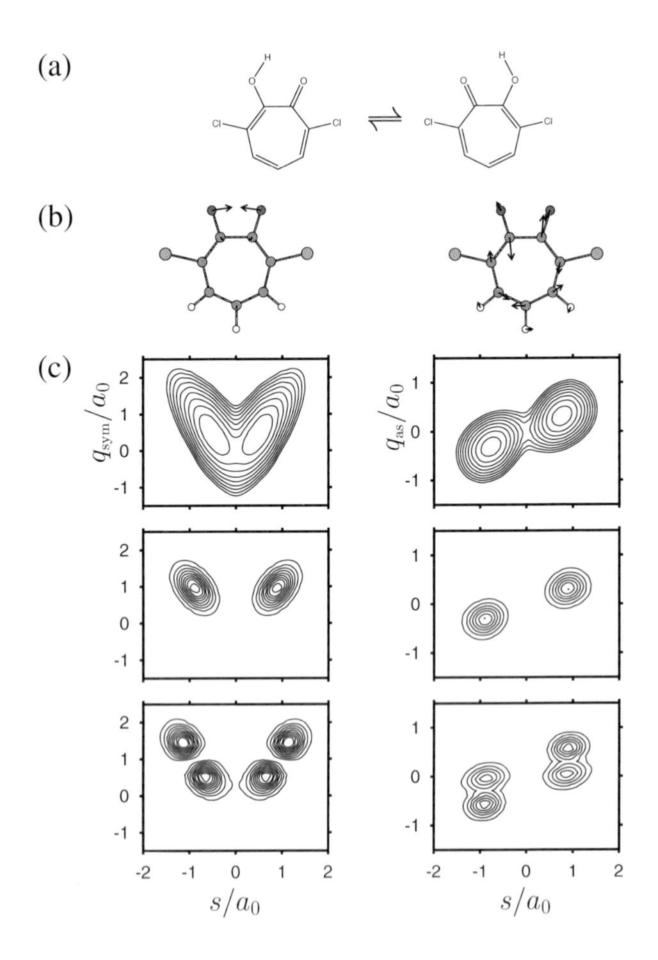


**Figure 7.10:** PES and eigenfunctions of (in plane) PT in 3,7–dichlorotropolone (panel A). The two–dimensional projections of the potential energy surface (first row of panel C) and the related probability densities (second and third row of panel C) correspond to some eigenfunctions of a four–dimensional ab initio quantum chemical Cartesian reaction surface Hamiltonian, Eq. (2.105). The influence of a symmetric (left column) and an antisymmetric (right column) normal mode (displacement vectors in panel B) is shown. The results have been obtained for the case that the proton moves on a straight line ($s$) orthogonal to $C_2$ symmetry axis going through the transition state. The ground state tunneling splitting is 3 $cm^{-1}$ (upper eigenfunctions). For the excitation of the symmetric/antisymmetric mode (lower left/lower right) the splitting amounts to 17 $cm^{-1}$/4 $cm^{-1}$. (figure courtesy of K. Giese)

doublet in Fig. 7.9) the wave function overlap increases further and so does the splitting in the excited doublet.

The situation is more complicated for the linear (antisymmetric) coupling mode shown in the right panel of Fig. 7.10. From the projection of the ground state wave function on the PT coordinate $s$ and the antisymmetric coordinate $q_{as}$ it is seen that the presence of an antisymmetric mode may reduce the tunneling splitting since left and right parts of the ground state wave function are shifted in opposite directions. In principle one would expect such a behavior also for the excited states with respect to this mode (see lower right panel in Fig. 7.10). However, for the excited state wave functions a comparison with the ground state already indicates that the details of the overlap in the barrier region will strongly depend on the position of the nodes along the oscillator coordinate in the left and right well. Therefore, in principle it is possible that the magnitude of the tunnel splitting even may oscillate when going to higher excited states due to interference between the localized wave functions which overlap in the barrier region and give rise to the tunneling splitting.

### 7.2.4 The Proton Transfer Hamiltonian in the Condensed Phase

After having discussed the influence of intramolecular modes which are immediately coupled to the PT coordinate let us next include the interaction with some environmental degrees of freedom such as a solvent. In principle one should distinguish between intramolecular and environmental coordinates in the following discussion. This would be particularly important, if some intramolecular modes have a distinct effect on the PT coordinate such that they cannot be treated on the same level of approximation as the remaining environment (see below). For simplicity, however, we do not make this distinction and comprise all degrees of freedom (intramolecular and environment) into the coordinate $Z = \{Z_k\}$. The total Hamiltonian can then be written as follows

$$H = H_{\text{proton}}(s) + H_{\text{R}}(Z) + V(s, Z) \ . \tag{7.3}$$

Here, the Hamiltonian of the PT coordinate $H_{\text{proton}}(s)$ is given by the first term in Eq. (7.1) (notice that in general $s$ can be 3 three–dimensional vector), $H_{\text{R}}(Z)$ is the Hamiltonian for the environment (solvent or protein plus intramolecular modes), and $V(s, Z)$ comprises the interaction between the PT coordinate and the environment. Notice that Eq. (7.3) has the form of a system–bath Hamiltonian (cf. Eq. (3.3)); in the spirit of Chapter 3 the proton coordinate can be considered as being the relevant system while the remaining coordinates $Z$ form the reservoir.

The interaction potential $V(s, Z)$ can be partitioned into a short–range and a long–range part. Quite often it is reasonable to assume that the short–range part will be dominated by the interaction of the solvent with the intramolecular modes, since the respective donor and acceptor groups will shield the proton from direct collisions with the solvent. The long–range Coulomb interaction, however, influences the PT directly, since the latter is often accompanied by a large change of the dipole moment. In fact a polar solvent is very likely to stabilize one of the two configurations found in the gas phase double–well potential. This is typical, for instance, for hydrogen–bonded acid–base complexes, where the ionic form may be stabilized in polar solution.

In practical condensed phase calculations the environmental degrees of freedom are normally treated by classical mechanics. On the other hand, it is often necessary to describe the proton quantum mechanically. The Hamiltonian $H_{\text{proton}}(s)$ may be obtained, for instance, by performing gas phase quantum chemical calculations for an appropriately chosen reference system which contains the PT coordinate. The interaction $V(s; Z)$ then may enter via effective pair (e.g., Lennard–Jones) and Coulomb potentials. One of the essential ingredients here

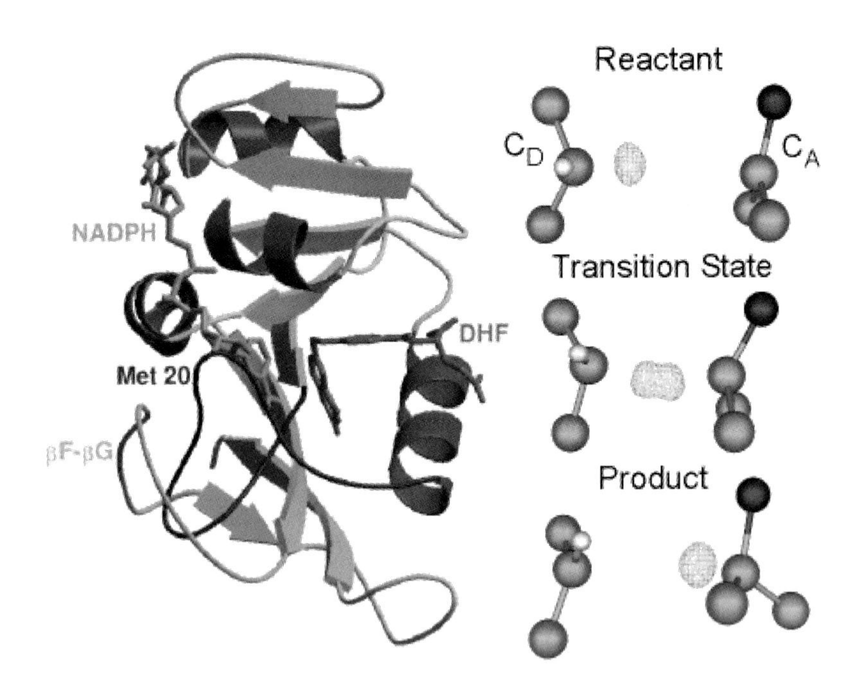


**Figure 7.11:** Quantum–classical hybrid treatment of the hydride ($H^-$) transfer reaction catalyzed by the enzyme dihydrofolate reductase. In the three–dimensional structure (for *Escherichia coli*, left) the nicotinamide adenine dinucleotide phosphate (NADPH) cofactor to the (protonated) 7,8–dihydrofolate (DHF) are labelled. The hydride transfer takes place from the donor carbon ($C_D$) of the NADPH to the acceptor carbon ($C_A$) of DHF. In the right panel adiabatic wave functions are plotted for the hydride at three representative configurations of the environmental (DHF substrate, NADPH cofactor, protein, solvating water molecules) coordinates along the reaction path. The immediate surrounding of the donor and acceptor sites are also shown. (figure courtesy of S. Hammes–Schiffer, for more details see also [Aga02])

is a detailed model for the charge distribution along the PT coordinate. Besides this atomistic view one can also introduce the solvent by means of a continuum model in close analogy to the treatment of electron transfer (cf. Section 6.5).

In the following we will consider two different ways of rewriting the Hamiltonian (7.3) such that it becomes suitable for treating PT in the adiabatic and nonadiabatic limits.

### Adiabatic Representation

The Born–Oppenheimer separation of electronic and nuclear motions provided the key to electronic and vibrational spectra and dynamics (cf. Chapter 2). In fact the small mass of the proton makes it tempting to separate its motion from the slow dynamics of its environment (for example, intramolecular heavy atom modes or collective protein modes etc.). Assuming that the set $\{Z_k\}$ of coordinates and the proton coordinate $s$ are adiabatically separable, it is reasonable to define an *adiabatic* proton wave function as the solution of the following Schrödinger equation for fixed values of the environmental coordinates $Z$:

$$(H_{\text{proton}}(s) + V(s, Z))\,\chi_A(s, Z) = E_A(Z)\chi_A(s, Z) \ . \tag{7.4}$$

Here, the eigenenergies along the proton coordinate $E_A(Z)$ $(A = 0, 1, 2, \ldots)$ and the wave function $\chi_A(s, Z)$ depend parametrically on the coordinates $Z$ in analogy to the parametric dependence of the electronic energies on the nuclear coordinates in Chapter 2. Given the adiabatic basis functions $|\chi_A\rangle$ the total nuclear wave function can be expanded as follows

$$\phi(s, Z) = \sum_A \Xi_A(Z)\,\chi_A(s, Z) \ . \tag{7.5}$$

Stressing the analogy with the electronic–nuclear situation of Section 2.3, the $\Xi_A(Z)$ can be considered as the wave functions for the motion of the slow (environmental) degrees of freedom in the protonic adiabatic state $|\chi_A\rangle$. The respective equations for their determination follow in analogy to Eq. (2.18) and will not be repeated here.

It should be pointed out, that for a condensed phase environment a classical treatment of the reservoir coordinates $Z$ will be necessary using, e.g., the quantum–classical hybrid methods discussed in Section 3.11. In Fig. 7.11 we show an example for an adiabatic protonic wave function in a classical environment. Three snapshots are plotted along the reaction path of a hydride ($H^-$) transfer reaction catalyzed by an enzyme.

### Diabatic Representation

The diabatic representation is convenient if the proton wave function is rather localized at the donor or acceptor site of the hydrogen bond. This will be the case for systems with a rather high barrier (weak hydrogen bonds). Following the strategy of Section 2.7 we define diabatic proton states for the reactant and the product configuration according to some properly chosen Hamiltonian $H_R(s, Z)$ and $H_P(s, Z)$, respectively. This means that we have solved the eigenvalue problem

$$H_{R/P}(s, Z)\chi_{j_R/j_P}(s, Z) = E_{j_R/j_P}(Z)\chi_{j_R/j_P}(s, Z)\,, \quad (j_R/j_P) = 0, 1, 2, \ldots . \tag{7.6}$$

Here, the $E_{j_R/j_P}(Z)$ define the diabatic potential energy surfaces for the motion of the environmental degrees of freedom in the reactant/product state. The total PT Hamiltonian in the diabatic representation can then be written as

$$H = \sum_{j=(j_R,j_P)} \sum_{j'=(j'_R,j'_P)} \Big[\delta_{jj'}\,(E_j(Z) + H_{\text{R}}(Z)) + (1 - \delta_{jj'})V_{jj'}(Z)\Big]|\chi_{j'}\rangle\langle\chi_j| \ . \tag{7.7}$$

Here, the diabatic state coupling $V_{jj'}(Z)$ is given by (cf. Eq. (7.4))

$$V_{jj'}(Z) = \int ds\, \chi_j^*(s, Z)[H_{\text{proton}}(s) + V(s, Z) - H_R(s, Z) - H_P(s, Z)]\chi_{j'}(s, Z) \ . \tag{7.8}$$

The diabatic basis can be used for expansion of the total wave function:

$$\phi(s,Z) = \sum_{j=(j_R,j_P)} \Xi_j(Z)\,\chi_j(s,Z)\,. \tag{7.9}$$

The analogy between the present treatment and that of the electron–vibrational problem discussed in Chapters 2,5, and 6 is apparent. In the spirit of the diabatic representation introduced in Section 2.7, the diabatic Hamiltonians $H_{R/P}(s,Z)$ will be conveniently chosen such that the coupling is only in the potential energy operator (static coupling).

## 7.3 Adiabatic Proton Transfer

The regime of adiabatic PT is characteristic for strong hydrogen bonds. In this situation the potential energy curve often has only a single minimum or a rather low barrier. The heavy atom coordinates will move so slowly that the proton can respond "instantaneously" to any change in $Z$. Thus its wave functions $\chi_A(s,Z)$ as a solution of the Schrödinger equation (7.4) will always correspond to the potential which follows from the actual configuration of $Z$ (see Figs. 7.11 and 7.12).

In order to explore some general features of the potential energy curve for adiabatic PT let us consider the situation of a reactant state with equilibrated heavy atom coordinates as shown in left panel of Fig. 7.12. The potential obtained by varying the PT coordinate but keeping the heavy atom coordinates in $V(s,Z)$ *fixed* will be asymmetric. On the other hand, any displacement of the heavy atom coordinates will influence the potential for PT. Suppose we have moved the heavy atom configuration such that it corresponds to some symmetric transition state. Then the potential for PT will be symmetric (upper panel of Fig. 7.12) with the lowest eigenstate along the proton coordinate being possibly above the top of the barrier. If we promote the heavy atom coordinates to their equilibrated product configuration, the PT potential will become asymmetric again but with the more stable configuration being on the product side (right panel of Fig. 7.12). For the asymmetric reactant and product states it is reasonable to assume that the protonic wave function will be rather localized in these states. In the symmetric case, however, it may be delocalized with respect to the PT coordinate $s$.

Suppose the system was initially in the lowest proton eigenstate $\chi_0(s,Z)$ corresponding to the reactant configuration of $Z$. From the discussion above it is clear that it requires some *fluctuations* of the heavy atom coordinates in order to move the system from the reactant to the product state. In practice it can be either the fluctuation of the dipole moments of the solvent or the fluctuation of some strongly coupled mode. Looking at Fig. 7.12 we notice that adiabatic PT corresponds to the situation where the proton remains in its lowest eigenstate when the heavy atom coordinates move towards the product configuration.

In principle we have separated our total system into a relevant and an environmental part (cf. Eq. (7.3)). This would suggest to use the methods of quantum statistical dynamics which have been introduced in Chapter 3. In particular one could straightforwardly write down a Quantum Master Equation for the time evolution of the reduced proton density matrix. This would require to make some assumptions concerning the spectral density of the environment, or to do some classical simulation of the spectral density as outlined in Section 4.3. In fact there might be cases where such a treatment is justified. However, in general the interaction

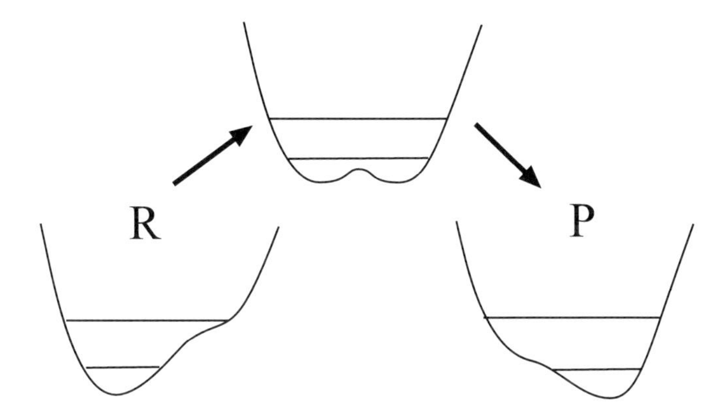


**Figure 7.12:** Schematic view of the potential energy curve for PT in the adiabatic regime. Here the proton wave function adjusts instantaneously to the actual configuration of its environment. The three different panels correspond to environmental degrees of freedom "frozen" at their reactant (R), transition, and product (P) configuration (from left to right). The proton is always in its lowest eigenstate (for an application, see Fig. 7.11).

with the surroundings *cannot* be treated using perturbation theory. This already becomes obvious by inspecting the schematic potential energy surface shown in Fig. 7.12.

Therefore, a realistic modelling of PT in solution can only be achieved by resorting to quantum–classical hybrid approach; the proton coordinate is treated quantum mechanically and the environment classically. We note in passing that there may be situations where some of the strongly coupled modes must be treated quantum mechanically as well. This can occur especially for coupled intramolecular modes whose frequency may exceed $k_B T$ at room temperature.

According to Section 3.11 the hybrid approach requires to solve the coupled set of equations (3.407). In the present case of adiabatic dynamics the simultaneous solution of the time–dependent Schrödinger equation is not necessary. Since the classical particles are assumed to move very slowly it suffices to solve the time–independent Schrödinger equation for fixed positions of the heavy atoms. Thus, the hybrid approach can be cast into the following scheme: Given some configuration of the environment, $Z(t)$, the *stationary* Schrödinger equation (7.4) is solved numerically. This defines, for instance, the "instantaneous" adiabatic ground state proton wave function $\chi_0(s, Z(t))$ (for an example see Fig. 7.11).

This wave function is used to calculate the mean–field force $F_k$ on the environmental degrees of freedom which is given by (cf. 3.410)

$$\begin{aligned} F_k &= -\frac{\partial}{\partial Z_k} \int ds\, \chi_0^*(s, Z(t))\, V(s, Z(t))\, \chi_0(s, Z(t)) \\ &\approx -\left\langle \chi_0 \left| \frac{\partial V}{\partial Z_k} \right| \chi_0 \right\rangle . \end{aligned} \tag{7.10}$$

Here, we used the fact that $|\chi_0\rangle$ depends only adiabatically on the coordinates $Z_k$ such that the respective derivatives are negligibly small. The expression (7.10) is called the *Hellmann–Feynman* force. This force is now used to propagate the classical degrees of freedom by one

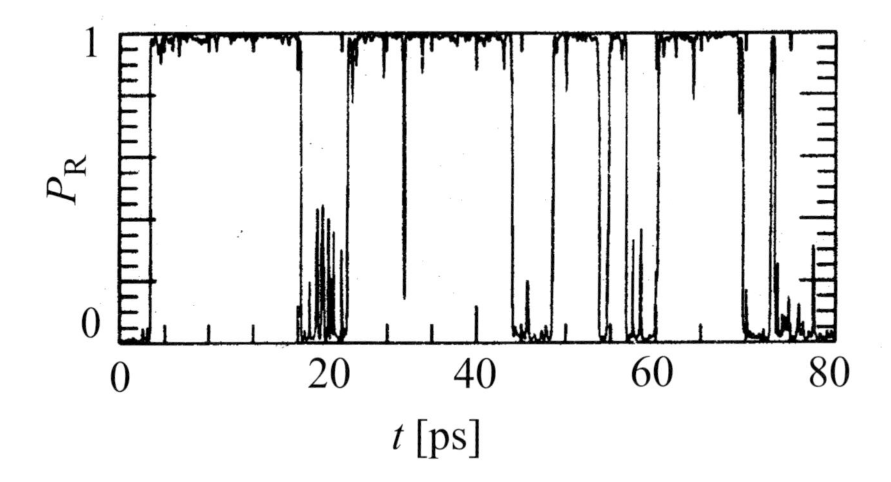


**Figure 7.13:** The probability for the proton to be in the reactant configuration is shown for an adiabatic PT situation. The model system is a strongly bonded $XH^+-X$ complex immersed in a polar aprotic diatomic solvent (reprinted with permission from [Bor92]; copyright (1992) American Chemical Society).

time step according to the canonical equations:

$$\begin{aligned}\frac{\partial Z_k}{\partial t} &= \frac{\partial}{\partial P_k} H_{\mathrm{R}}(Z) \\ \frac{\partial P_k}{\partial t} &= -\frac{\partial}{\partial Z_k} H_{\mathrm{R}}(Z) + F_k \ ,\end{aligned} \tag{7.11}$$

From the new positions obtained in this way, a new interaction potential $V(s, Z)$ is calculated and the stationary Schrödinger equation for the proton wave function is solved again. This procedure is continued until some desired final time. We emphasize that in contrast to the general situation of Section 3.11 the adiabatic limit does not require a simultaneous self–consistent solution of the time–dependent Schrödinger equation and Newton's equations of motion.

How can we use the results of such a simulation to obtain, for instance, reaction rates. Let us consider the situation of a PT system where the position of the barrier along the PT coordinate is at $s = s^*$. Then the probability $P_{\mathrm{R}}$ that the proton is in the reactant configuration can be calculated from the adiabatic ground state proton wave function as follows

$$P_{\mathrm{R}}(Z(t)) = \int_{-\infty}^{s^*} ds \ |\chi_0(s, Z(t))|^2 \ . \tag{7.12}$$

This probability will be a function of time, since the adiabatic proton wave function depends on the actual configuration of the classical coordinates, $Z(t)$. The probability $P_{\mathrm{R}}$ will approach unity in the reactant state and zero after a complete transition to the product state occurred. In Fig. 7.13 we show $P_{\mathrm{R}}$ for a model PT reaction as described in the figure caption. Here the interaction with the solvent is rather strong such that the proton is most of the time stabilized either on the reactant or the product side. Large fluctuations of the solvent dipoles, however, cause occasional transitions between the two configurations, that is, the reaction barrier is crossed.

From the knowledge of the time dependence of the reactant state population one can in principle obtain the transition rate by simple counting the reactive barrier crossings in Fig. 7.13 during a long–time quantum–classical propagation. On the other hand, one could also adapt the definitions of transfer rates given in Section 6.8.2. Alternatively one can use $P_{\rm R}$ to calculate the reactive flux of quantum and classical particles across some surface dividing reactants from products.

## 7.4 Nonadiabatic Proton Transfer

Whenever we have a situation where the hydrogen bond is rather weak, the concepts of adiabatic PT discussed in the previous section can no longer be applied. Here the reaction barrier will be rather high and consequently the splitting between the two lowest eigenstates is small. Thus the different adiabatic states come close to each other and nonadiabatic transitions become rather likely at normal temperatures (cf. upper panel of Fig. 7.7). On the other hand, the transfer time will be long compared with typical relaxation time scales for the environment. We have already seen in Chapter 6 that this situation is most conveniently described using a *diabatic* representation of the Hamiltonian as given by Eq. (7.7). We will focus on a situation of a protonic two–state system. This may be appropriate at temperatures low enough such that the second pair of vibrational states (in an only modestly asymmetric PT potential, cf. Fig. 7.7) is thermally not occupied. The two states will be labelled as $j = (R, P)$.

Since we have assumed that the conditions for nonadiabatic PT are fulfilled we can straightforwardly write down the rate for transitions between the diabatic reactant and product states using the Golden Rule expression of Section 3.59. Suppose the stationary Schrödinger equation for the environmental degrees of freedom

$$[E_j(Z) + H_{\rm R}(Z))]\,\Xi_{j,N}(Z) = E_{j,N}\,\Xi_{j,N}(Z)\,, \quad j = (R, P) \tag{7.13}$$

has been solved, the Golden Rule transition rate reads

$$k_{R\to P} = \frac{2\pi}{\hbar}\sum_M f(E_{R,M})\sum_N |\langle\Xi_{R,M}|V_{RP}|\Xi_{P,N}\rangle|^2\delta(E_{R,M} - E_{P,N})\,. \tag{7.14}$$

Here the $N - \{N_k\}$ comprises the quantum numbers for the environmental degrees of freedom $Z$, in the reactant and product diabatic state. Of course, expression (7.14) is only of limited value since calculating the eigenstates of the environment is in the general case impossible. However, as in the case of electron transfer one can obtain analytical expression for the limit of a harmonic oscillator environment. We are not going to repeat the derivations given in Section 6.4 which can easily be adapted to the present situation. Also the general reasoning which led to the introduction of the dielectric continuum model for the solvent in Section 6.5 applies and the respective expressions can be translated into the present situation.

For PT reactions, however, it may often be necessary to include a coordinate dependence of the diabatic state coupling, that is, to go beyond the Condon approximation, which has been used in the treatment of nonadiabatic electron transfer. This is basically due to the intramolecular promoting modes which may have a drastic influence on the PT rate. Note that this influence will be even more pronounced in the nonadiabatic regime, where the tunnel coupling is rather small. Compared to the dominant effect of possible promoting modes, the dependence of the diabatic coupling on the solvent coordinates is often neglected. For the actual form of this dependence it is reasonable to assume an expression which is similar to the one introduced for electron transfer in Eq. (6.28). Note, however, that the parameter $\beta$ which characterizes the wave function overlap is much larger for PT than for electron transfer, since the protonic wave function will be more localized.

In case of a coordinate dependent state coupling, but also for more general (not harmonic) environments it is necessary to return to the definition of the transfer rate in terms of correlation functions as given in Eq. (3.131). Adopting Eq. (3.131) to the present situation, the PT rate can be written as

$$k_{R\to P} = \frac{1}{2\hbar}\mathrm{Re}\int_0^\infty dt\ \mathrm{tr}_R\left\{\hat{R}_R e^{iH_R^{(0)}t/\hbar}V_{RP}(Z)e^{-iH_P^{(0)}t/\hbar}V_{PR}(Z)\right\}. \tag{7.15}$$

Here we used the shorthand notation $H_{R/P}^{(0)} = E_{R/P}(Z) + H_{\mathrm{R}}(Z)$, $\hat{R}_R$ is the statistical operator for the reactant state, and the trace is also performed with respect to the reactant states. Equation (7.15) can be transformed into a more convenient form by using the operator identity

$$e^{-iH_P^{(0)}t/\hbar} = e^{iH_R^{(0)}t/\hbar}\,\hat{T}\exp\left\{-\frac{i}{\hbar}\int_0^t dt' e^{iH_R^{(0)}t'/\hbar}(H_P^{(0)} - H_R^{(0)})e^{-iH_R^{(0)}t'/\hbar}\right\}. \tag{7.16}$$

Introducing the time–dependent energy gap between reactant and product state configurations as

$$\Delta H^{(\mathrm{I})}(t) = e^{iH_R^{(0)}t/\hbar}\,(H_P^{(0)} - H_R^{(0)})\,e^{-iH_R^{(0)}t/\hbar}, \tag{7.17}$$

we can rewrite Eq. (7.15) as

$$\begin{aligned} k_{R\to P} \;=\; & \frac{1}{2\hbar}\mathrm{Re}\int_0^\infty dt\ \mathrm{tr}_R\Big\{\hat{R}_R\,V_{RP}^{(\mathrm{I})}(Z,t) \\ & \times\hat{T}\exp\left\{-\frac{i}{\hbar}\int_0^t dt'\Delta H^{(\mathrm{I})}(t')\right\}V_{PR}^{(\mathrm{I})}(Z,0)\Big\}, \end{aligned} \tag{7.18}$$

where the interaction representation of $V_{RP}(Z)$ is with respect to $H_R^{(0)}$. In the context of linear optical spectroscopy of molecular systems expressions of the type (7.18) have been shown to be amenable to a classical treatment (cf. Section 5.3.5). This requires to replace the quantum dynamics of the environment, which is introduced via $V_{RP}^{(\mathrm{I})}(Z,t)$ and $\Delta H^{(\mathrm{I})}(t)$, by classical dynamics on the diabatic reactant state potential energy surface, or more specifically $V_{RP}^{(\mathrm{I})}(Z,t)$ is replaced by $V_{RP}(Z(t))$ and $\Delta H^{(\mathrm{I})}(t)$ by $\Delta H(Z(t))$. Here the time dependence of the coordinates is governed by the equations of motion of classical mechanics. In addition the time–ordered exponential in Eq. (7.18) can be replaced by an ordinary exponential in the classical approximation. Finally, the thermal averaging in Eq. (7.18) has to be performed with respect to some classical thermal distribution function for the reactant state as discussed in Section 3.11.

## 7.5 The Intermediate Regime: From Quantum to Quantum–Classical Hybrid Methods

The Golden Rule description in the previous section was based on the assumption of weak hydrogen bonding. In other words, the energetic separation between the two lowest vibrational states of the PT coordinate (tunnel splitting) has to be small. One consequence is that the adiabatic approximation is no longer justified and transitions between different proton states occur. In the previous section this has been described using coupled diabatic proton states.

In Section 7.2 we have already mentioned that the actual barrier height and therefore the tunnel splitting is subject to strong modifications in the presence of a fluctuating environment. Thus unless hydrogen bonding is really strong there may be no clear separation between the adiabatic regime and some intermediate or even the nonadiabatic regime. In this case one has to use an alternative formulation which is suited for all regimes and in particular incorporates transitions between adiabatic proton states.

In principle one could apply the Quantum Master Equation approach of Chapter 3 and treat the quantum dynamics of the relevant system under the influence of the dissipative environment. We have already mentioned, however, that the consideration of only a single relevant coordinate, that is, the proton coordinate, may not be sufficient and it might be necessary to include, for instance, several modes of the environment into the relevant system in order to allow for a perturbative treatment of the system–environment coupling. But, in practice the propagation of reduced density matrices in more than three dimensions requires an immense numerical effort.

In the following we will first discuss a fully quantum mechanical wave packet method in Section 7.5.1 before focusing on a quantum–classical hybrid approach in Section 7.5.2.

### 7.5.1 Multidimensional Wave Packet Dynamics

Suppose it is sufficient to restrict the dynamics to the proton transfer coordinate and a finite number of nuclear coordinates. Let us further assume that the Hamiltonian is available in the reaction surface form given by Eq. (7.1), i.e., with some (intramolecular) oscillator modes $q_\xi$. The simplest possible wave function would have the form of a Hartree product (cf. Eq. (2.28)):

$$\phi(s, q, t) = \chi(s, t) \prod_\xi \Xi_\xi(q_\xi, t) . \tag{7.19}$$

Using the time–dependent Dirac–Frenkel variational principle for finding the wave functions for the different degrees of freedom such that

$$\langle \delta\phi | i\hbar \frac{\partial}{\partial t} - H | \phi \rangle = 0 , \tag{7.20}$$

is fulfilled gives the Schrödinger equation for the reaction coordinate

$$i\hbar \frac{\partial}{\partial t} \chi(s, t) = [T_s + U_{\rm SCF}(s, t)] \, \chi(s, t), \tag{7.21}$$

where the effective potential

$$U_{\rm SCF}(s,t) = U(s) + \sum_\xi \Big[\frac{1}{2}\omega_\xi^2 \langle \Xi_\xi(t)|q_\xi^2|\Xi_\xi(t)\rangle - F_\xi(s)\langle \Xi_\xi(t)|q_\xi|\Xi_\xi(t)\rangle\Big] \quad (7.22)$$

has been introduced. It contains the time-dependent mean–field potential due to the interaction with the oscillator modes. For the latter we obtain the equations of motion

$$i\hbar\frac{\partial}{\partial t}\Xi_\xi(q_\xi,t) = \Big[T_\xi + \frac{1}{2}\omega_\xi^2 q_\xi^2 - F_\xi(t)q_\xi\Big]\Xi_\xi(q_\xi,t)\ . \quad (7.23)$$

Here, we defined the time–dependent linear driving forces for the oscillator dynamics $F_\xi(t) = \langle\chi(t)|F_\xi(s)|\chi(t)\rangle$. This quantity is averaged with respect to the proton coordinate, that is, it contains the mean–field interaction for the oscillator modes. Furthermore, notice that Eq. (7.23) describes a harmonic oscillator with time–dependent driving force. Therefore, if the reservoir is initially in the ground state and described by an uncorrelated Gaussian wave packet, the dynamics which is initiated by the interaction with the proton coordinate is that of a Gaussian wave packet with a time–dependent mean value. Since the dynamics of both subsystems is determined by simultaneous solution of Eqs. (7.21) and (7.23) this approach is called *time–dependent self–consistent field method.*

The approach outlined so far is rather appealing for it allows to treat a fair number of degrees of freedom on a quantum mechanical level. It may provide a reasonable description for hydrogen bond motion in the vicinity of a minimum on the potential energy surface or for strong hydrogen bonds. On the other hand, for proton transfer reactions between reactant and product potential wells it is likely to run into trouble. The reason lies in the mean–field character of the coupling. To illustrate this, suppose that we are interested in the force which acts on some oscillator coordinate if the proton is in its vibrational ground state $\chi_0(s)$. For a symmetric double minimum potential the ground state wave function will obey $\chi_0(s) = \chi_0(-s)$ (cf. Fig. 7.9). Hence, given an antisymmetric coupling like $F_\xi(s) \propto s$ (cf. Fig. 7.8, right panel), the mean force will vanish. Although this is an extreme example it becomes clear that upon PT the force on the oscillator modes may change considerably such that for a rather delocalized proton wave packet details of this coupling are averaged out leading to a qualitatively wrong behavior.

Whenever only a few degrees of freedom have to be considered one can resort to numerically exact methods which allow to account for all relevant correlations. Here, the most versatile approach is the extension of the self–consistent field approach to include a superposition of different Hartree–products, i.e. the multi–configuration time–dependent Hartree method (for an overview, see [Bec00]).

A recipe for including correlations beyond the mean–field approximation even for rather large systems is most easily appreciated if we return to the diabatic picture of some general system–bath Hamiltonian as given by Eq. (7.7).[3] Having defined diabatic proton states for the reactant and product we can use the *coupled–channel* approach introduced in Section 5.7.2. To this end let us expand the time–dependent total wave function in terms of the stationary

[3]It is rather straightforward to map this general Hamiltonian onto the specific reaction surface Hamiltonian for an oscillator reservoir.

*diabatic* proton states as follows (cf. Eq. (7.9))

$$\phi(s,Z;t) = \sum_{j=(j_R,j_P)} \Xi_j(Z;t)\chi_j(s,Z) \ . \tag{7.24}$$

In analogy to Section 5.7.2 one obtains the following coupled–channel equation for the time–dependent wave function of the environment, $\Xi_{j=(j_R,j_P)}(Z;t)$,

$$i\hbar\frac{\partial}{\partial t}\Xi_j(Z;t) = (E_j(Z) + H_\mathrm{R}(Z))\,\Xi_j(Z;t) + \sum_{j'\neq j} V_{jj'}(Z)\,\Xi_{j'}(Z;t) \ . \tag{7.25}$$

Given a diabatic Hamiltonian as in Eq. (7.7) this equation is in principle exact. However, unless the number of environmental degrees of freedom can be restricted to just a few the numerical effort for solving the coupled channel equations is prohibitive. Therefore, it is customary to neglect correlations between different environmental coordinates and assume that the wave function for the different diabatic states of the proton, $\Xi_j(Z;t)$, can be factorized as follows[4]

$$\Xi_j(Z;t) = \prod_k \Xi_j(Z_k;t) \ . \tag{7.26}$$

Using this factorization ansatz and employing again the Dirac–Frenkel time–dependent variational principle (cf. Eq. (7.20)) one obtains the following equation for the wave function $\Xi_j(Z_k;t)$

$$i\hbar\frac{\partial}{\partial t}\Xi_j(Z_k;t) = \sum_{j'} H^{(\mathrm{eff})}_{jj'}(Z_k;t)\,\Xi_{j'}(Z_k;t) \ . \tag{7.27}$$

We can identify this as a *mean–field* approach, i.e., the time evolution of the wave function for the environmental degree of freedom $Z_k$ is determined by the averaged potential of all other degrees of freedom $Z_{k'\neq k}$. The effective time–dependent Hamiltonian entering Eq. (7.27) is given by

$$H^{(\mathrm{eff})}_{jj'}(Z_k;t) = \int d\tilde{Z}\;\Xi^*_j(\tilde{Z};t)\,[\delta_{jj'}\,(E_j(Z) + H_\mathrm{R}(Z)) + (1-\delta_{jj'})V_{jj'}(Z)]\,\Xi_{j'}(\tilde{Z};t) \ . \tag{7.28}$$

Here we introduced $\tilde{Z}$ as the shorthand notation for all coordinates $Z$ except $Z_k$. Further we have used

$$\Xi_j(\tilde{Z};t) = \prod_{k'\neq k} \Xi_j(Z_{k'};t) \ . \tag{7.29}$$

This approximate treatment allows to consider much larger environments but neglects any correlation effects in the dynamics of different environmental degrees of freedom. However, in contrast to the time–dependent self–consistent field approach the force acting on the reservoir particles depends on the diabatic state of the reaction coordinate.

[4]Note that the following treatment is not unique to proton transfer, i.e., it can be applied to the electron–vibrational dynamics as well.

### 7.5.2 Surface Hopping

Due to the complexity of the environment, however, one often wants to retain its classical description. From Eq. (7.24) it is obvious that in general the total system is in a superposition state with respect to the diabatic proton states. This introduces some conceptual difficulty since the classical environment cannot be in such a state, i.e., it cannot experience the forces due to *both* diabatic proton states at the *same* time. One possibility to solve this problem approximately is to average the forces on the classical degrees of freedom with respect to the quantum states. However, this will only be a good approximation if these forces are not very different in the two quantum states which is often not the case.

An alternative and simple classical approach incorporating quantum transitions is given by the *surface hopping* method which had been introduced in Section 3.11.2. Here, the classical propagation of the environmental degrees of freedom is combined with certain prescriptions for quantum transitions in the quantum subsystem.

For the simulation of PT reactions one employs the instantaneous *adiabatic* proton states which have to be determined according to Eq. (7.4) for each time step. The protonic wave function at any time step can then be expanded in terms of this instantaneous adiabatic basis set according to[5]

$$\chi(s,Z;t) = \sum_A c_A(t)\, \chi_A(s,Z(t)) \,. \tag{7.30}$$

Inserting this expression into the time–dependent Schrödinger equation with the Hamiltonian given by Eq. (7.4) one obtains the following set of equations

$$\frac{\partial}{\partial t} c_A = -\frac{i}{\hbar} E_A^{(\mathrm{p})}(Z)\, c_A - \sum_k \frac{\partial Z_k}{\partial t} \sum_{A'} c_{A'} \left\langle \chi_A \left| \frac{\partial}{\partial Z_k} \chi_{A'} \right.\right\rangle \,. \tag{7.31}$$

The last factor on the right–hand side can be identified with the nonadiabatic coupling matrix (cf. Eq. (3.418)). The quantum–classical propagation then proceeds as explained in Section 3.11.2. The surface hopping method gives a means to calculate, for instance, transitions rates without referring to any particular limit of PT.[6] Since the proton coordinate is treated quantum mechanically, effects of tunneling and zero–point motion are naturally included. It should be emphasized again that the incorporation of nonadiabatic transitions relies on some ad hoc stochastic model which, however, uses information about the probability distribution with respect to the proton states.

In the enzymatic catalysis example given in Fig. 7.11, the surface hopping method was used to address the influence of quantum effects for hydride ($H^-$) transfer. Nonadiabatic transitions where found to have only a minor effect on the rate for this process.

---

[5]Note that in general any suitable basis set can be used, but this would require to calculate matrix elements of the protonic Hamiltonian on the right–hand side of Eq. (7.31).

[6]This can be done, for instance, by partitioning the possible values of the PT coordinate into reactant and product side configuration. Based on the expectation value of the PT coordinate for a given classical trajectory, it can be decided whether a reactive transition between the reactant and the product configuration occurred.

# Notes

[Aga02] P. K. Agarwal, S. R. Billeter, and S. Hammes–Schiffer, J. Phys. Chem. B **106**,

[Bec00] M. H. Beck, A. Jäckle, G. A. Worth, and H.-D. Meyer, Phys. Rep. **324**, 1 (2000).

[Bor92] D. Borgis, G. Tarjus, and H. Azzouz, J. Phys. Chem. **96**, 3188 (1992).

[Loc00] S. Lochbrunner and E. Riedle, J. Chem. Phys. **112**, 10699 (2000).

[Mad02] D. Madsen, J. Stenger, J. Dreyer, P. Hamm, E. T. J. Nibbering, and T. Elsaesser, Bull. Chem. Soc. Japan **75**, 909 (2002).

[Ste98] G. Steinberg–Yfrach, J.–L. Rigaud, E. N. Durantini, A. L. Moore, D. Gust, and T. A. Moore, Nature **392**, 479 (1998).

# 8 Exciton Transfer

*The transfer of electronic excitation energy within a molecular aggregate will be considered. Attention is focused on the so-called Frenkel exciton model, where the moving excitation energy is completely built up as an intramolecular excitation and no charge transfer between different molecules occurs. The construction of the Hamiltonian governing Frenkel exciton motion is explained in detail. We will discuss the coupling of electronic excitations between different molecules which causes excitation delocalization and exciton transfer, as well as the interaction with the various types of vibrational degrees of freedom. Exciton transfer in a situation of weak and of strong dissipation is described. The latter case directly leads to the well–established Förster theory for incoherent exciton hopping. It is shown how the Förster transfer rate can be expressed in terms of the emission and absorption spectra of the donor molecule and the acceptor molecule, respectively. The optical absorption spectra of the whole aggregate are also computed. Finally, the exciton–exciton annihilation process is shortly described.*

## 8.1 Introduction

In the following we will discuss the electronic excitation energy transfer (XT) between two molecules according to the general scheme

$$\mathrm{M}_1^* + \mathrm{M}_2 \longrightarrow \mathrm{M}_1 + \mathrm{M}_2^* . \tag{8.1}$$

For clarity each molecule is described by two states, that is, a LUMO and a HOMO. The starting point is a situation where molecule 1 has been excited ($\mathrm{M}_1^*$), for instance, by means of an external laser pulse, and molecule 2 is in its ground state ($M_2$). Then the Coulomb interaction between these molecules leads to a reaction where molecule 1 is de–excited and the electrostatic energy is transferred to molecule 2 which is excited (see Fig. 8.1). Alternatively, the product state can be also reached via an electron exchange between $M_1$ and $M_2$. The electron in the LUMO of $M_1$ moves to the LUMO of $M_2$ and the hole in the HOMO of $M_1$ is filled by an electron of the HOMO of $M_2$. The latter process requires that the wave functions overlap between $M_1$ and $M_2$, while the former process may take place even if both molecules are spatially well separated. The quantum mechanical state consisting of an excited electron and an unoccupied HOMO (a hole) at the same molecule is called *Frenkel exciton*.

Frenkel excitons are encountered in associated and non–covalently bound complexes. Examples are molecular crystals of aromatic compounds such as benzene or naphthalene, for instance, and rare gases in the solid phase. Another important class of Frenkel exciton systems are dye aggregates (for instance, isocyanine or pseudo–isocyanine). Upon aggregation

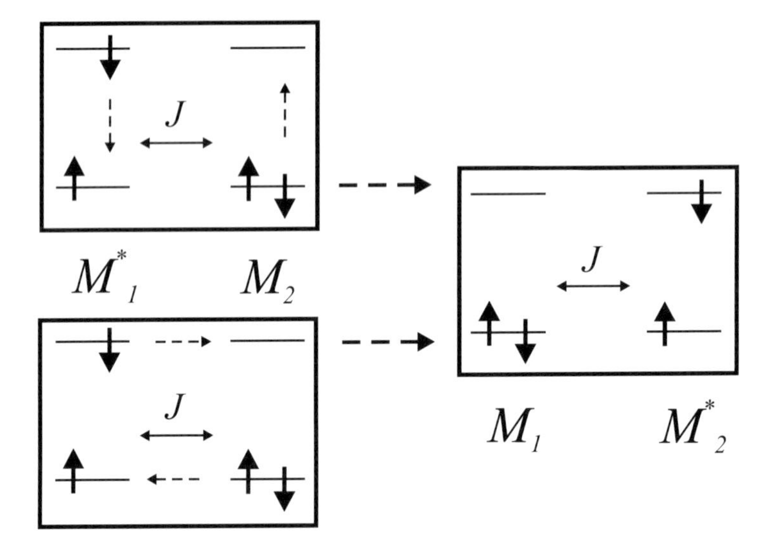


**Figure 8.1:** Excitation energy transfer between molecules $\mathrm{M}_1$ and $\mathrm{M}_2$. $\mathrm{M}_1$ is initially in the excited state in which one electron has been promoted from the HOMO to the LUMO (left panels). In the final state $\mathrm{M}_1$ is in its ground state and $\mathrm{M}_2$ is excited (right panel). In the upper–left scheme the Coulomb interaction $J$ triggers the exchange of excitation energy. In the lower–left scheme the exchange of excitation energy is combined with an exchange of electrons.

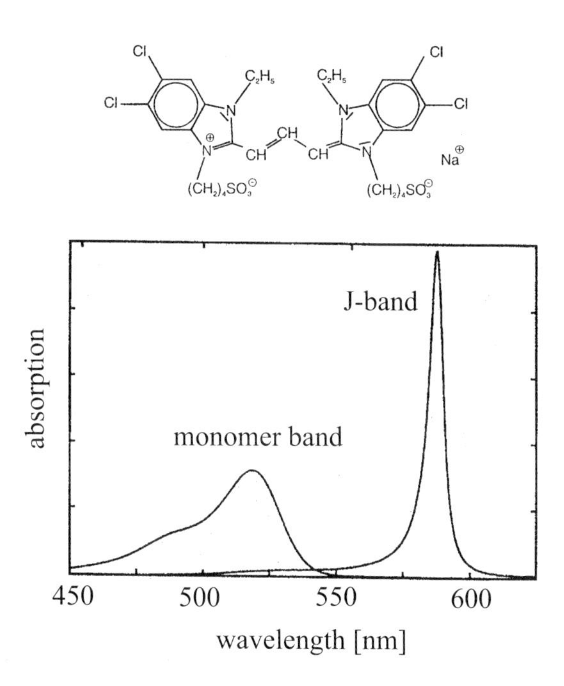


**Figure 8.2:** Molecular structure of the dye TDBC (5,5',6,6'–tetrachloro–1,1'–diethyl–3,3'–di(4–sulfobutyl)– benzimidazolcarbocyanine) forming J–aggregates, together with the room temperature monomer absorption (in methanol) and the J–band (in water). (reprinted with permission from [Mol95]; copyright (1995) American Institute of Physics)

which occurs in solution or in thin solid films, the dyes form rod–like arrangements consisting of several hundreds of molecules (see Figs. 8.2 and 8.3). In the last two decades also biological *chromophore complexes* attracted broad interest. The light–harvesting complex of natural photosynthetic antenna systems represents one of the most fascinating examples where the concept of Frenkel excitons could be applied. Both primary steps of photosynthesis, that is, directed excitation energy transfer in the antenna (solar energy collection) and charge transfer in the reaction center (connected with charge separation), have been subject to considerable efforts over the last decades. Schematic views of two types of antenna systems which form pigment–protein complexes are shown in Figs. 8.4 and 8.5.

An indication for the formation of an aggregate is the change from a broad monomeric absorption band to a comparatively sharp and shifted aggregate absorption band as is shown for TDBC in Fig. 8.2. This narrowing of the absorption is due to the mutual interaction of the monomers in the aggregate. It will be discussed in more detail in Section 8.7 .

The theoretical and experimental investigation of the behavior of excitons in molecular aggregates has a long tradition. Early theoretical contributions by T. Förster and D. L. Dexter were based on an incoherent rate equation approach. The variety of phenomena highlighted in recent discussions ranges from cooperative radiative decay (superradiance) and disorder–induced localization to nonlinear effects like exciton annihilation and two–exciton state for-

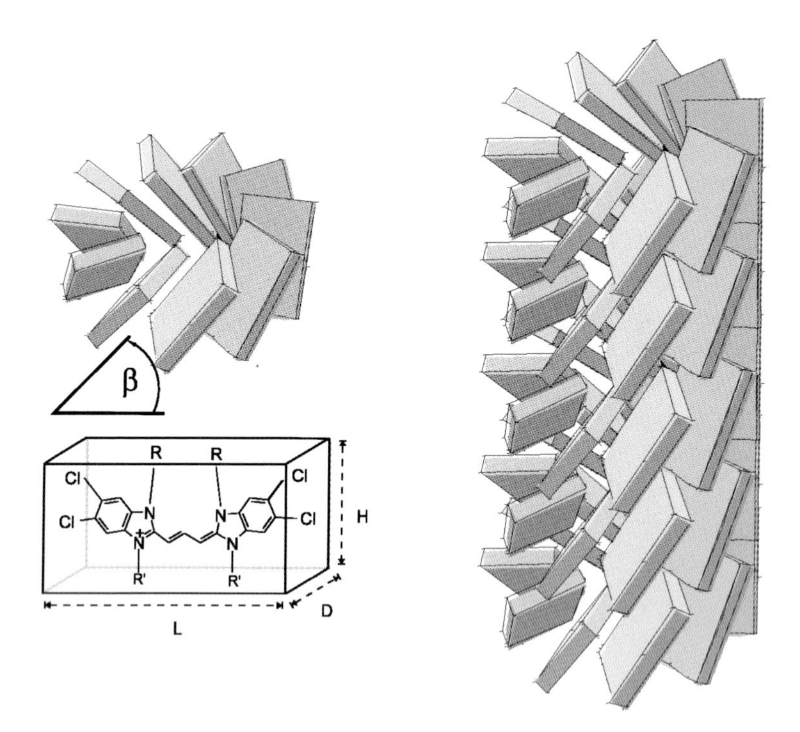


**Figure 8.3:** Cylindrical J–aggregate of an amphiphilic dye. The basic structure is shown in the lower–left part in a box with L = 1.9 nm, D = 0.4 nm, and H = 1.0 nm. The transition dipole moment lies in the direction of the long edge. Upper–left part: single circle built by 10 molecules, right part: tenfold helix formed by five of the shown circles. (reprinted with permission from [Spi02] ; copyright (2002) Elsevier Science B.V.)

mation. There exists also a large number of theoretical investigations focusing on exciton transport in molecular systems beyond the rate limit. Particularly successful in this respect has been the so–called *Haken–Strobl–Reineker* model, which describes the influence of the environment on the exciton motion in terms of a stochastic process.

The case opposite to the Frenkel exciton, where electron and hole are separated by a distance much larger than the spacing between neighboring molecules, is called *Wannier–Mott exciton*. It occurs in systems with strong binding forces between constituent molecules or atoms such as covalently bound semiconductors. Frequently, also an intermediate form, the *charge transfer exciton*, is discussed. Here electrons and holes reside on molecules which are not too far apart. This type of exciton appears if the wave functions of the involved molecules are sufficiently overlapping, as is necessary for an electron transfer reaction (cf. Chapter 6). Charge transfer excitons can be found, for example, in polymeric chains formed by silicon compounds (polysilanes).

This chapter will focus on the description of Frenkel excitons in molecular aggregates.

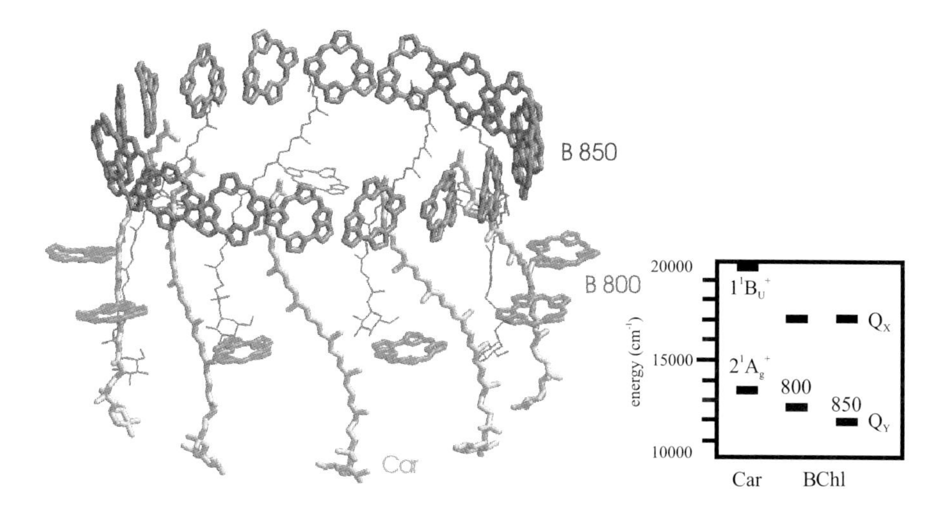


**Figure 8.4:** Schematic view of the so–called LH2 antenna which is typical for a number of photosynthetic bacteria (left panel). The active pigments are bacteriochlorophyll *a* molecules (BChl a), of which only the porphyrine planes are shown. These pigment molecules form two rings interconnected by carotenoids (Car) and stabilized by proteins (not shown). Since the two pigment rings differ by their absorption wavelength (800 nm and 850 nm) they are labelled as B800 and B850. Important excited electronic states of all pigments are displayed in the right panel. (LH2 figure courtesy of J. Herek)

The term "aggregate" is used to characterize a molecular system which consists, at least, of some hundred non–covalently bound molecules. Occasionally, we will also use the term chromophore complex. The electronic excitation energy in an aggregate can move as an exciton over the whole system according to the reaction scheme given in Eq. (8.1).[1] The initial state relevant for the transfer process is often created by means of an external laser pulse resonant to the respective $S_0 \rightarrow S_1$ transitions. In general, this state is a superposition of eigenstates of the molecular system including the mutual Coulomb interaction, i.e., it may contain contributions of all monomers. In terms of the corresponding wave functions this implies a delocalization over the whole aggregate (provided that the wavelength of the exciting light is large compared to the aggregate size). The degree of delocalization and the type of motion initiated by the external field (cf. Fig. 8.6) depends crucially on the interaction between the exciton system with environmental DOF such as intramolecular nuclear motions. As it has been already discussed in connection with our studies on electron transfer reactions (cf. Section 6.3) the ratio between the characteristic times of intramolecular (vibrational) relaxation and intermolecular transitions decides on the particular way the XT proceeds. Two limiting cases are displayed in Fig. 8.6. If the intramolecular relaxation is fast (compared with the intermolecular transitions) then the excitation remains localized and the XT is named *incoherent*. In the contrary case the excitation may move as a delocalized state through the aggregate, i.e. the XT is a *coherent* transfer. A more detailed discussion will be given in Section 8.4

[1]Since in the experiment any regular structure of the aggregate is disturbed by external influences exciton motion is restricted to smaller parts of the whole aggregate (see below).

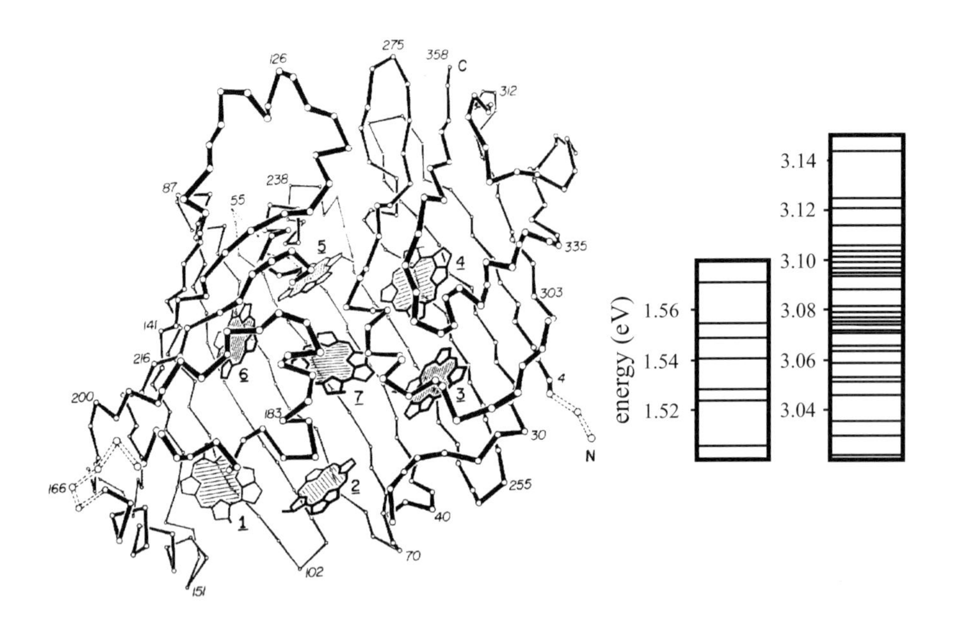


**Figure 8.5:** Schematic view on a single subunit of the Fenna–Mathew–Olsen complex (located in the base plate of the green sulphur bacterium *Prosthecochloris aestuarii*). Shown are the seven bacteriochlorophyll*a* molecules as well as the backbone of the carrier protein. The right panel displays the energetic position of the seven single–exciton levels (around 1.54 eV) and of the 21 two–exciton levels (around 3.8 eV, together with seven higher singlet excitations, cf. Sections 8.2.3 and 8.8.1).

First, we give in Section 8.2 some fundamentals of exciton theory, introducing the single and the two–exciton states, and discussing the coupling to vibrational DOF. Although we introduce the higher excited aggregate states which contain two (or even more) excitations (cf. Figs. 8.7 and 8.16), only the related phenomenon of exciton–exciton annihilation will be discussed in the supplementary Section 8.8.1. Techniques to describe the different regimes of exciton dynamics are presented in the Sections 8.5 and 8.6. Optical properties of different aggregates are described in Section 8.7.

## 8.2 The Exciton Hamiltonian

Let us consider a molecular aggregate consisting of $N_{\rm mol}$ molecules arranged in an arbitrary geometry and with the center of mass of the $m$th molecule located at $\mathbf{X}_m$ with respect to the origin of some coordinate system. The aggregate Hamiltonian $H_{\rm agg}$ can be constructed in similarity to the Hamiltonian $H_{\rm mol}$ of a single molecule (cf. Sections 2.2, 2.3, and 2.7). In the present case, however, a separation into intramolecular and intermolecular contributions is

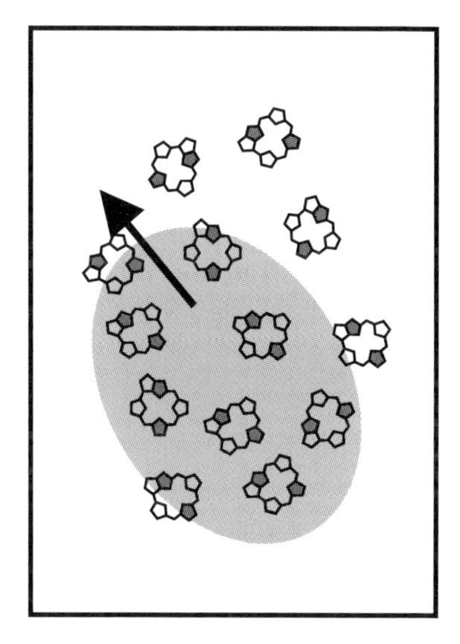
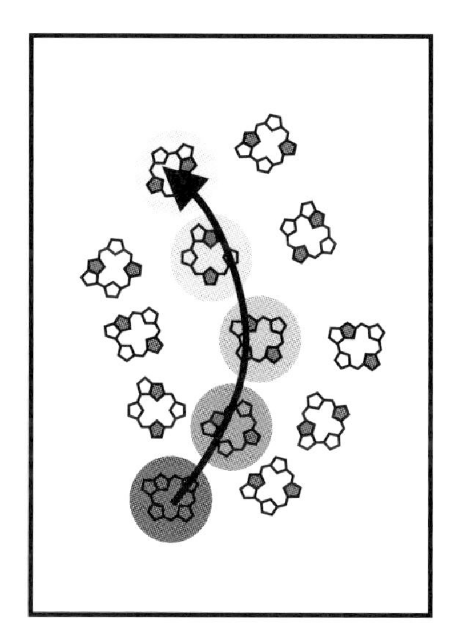

**Figure 8.6:** Schematic illustration of coherent (left) and incoherent (right) exciton motion in a chromophore complex (formed by pheophorbide–a molecules in a layer of behenic acid molecules, after [Kor98]). The shaded area symbolizes the exciton extending over several monomers in the left panel. In the right panel the excitation hops from molecule to molecule (at a certain time the excitation is present at the different molecules with a certain probability corresponding to the grey scale).

advisable:

$$H_{\text{agg}} = \sum_m H_m(R) + \frac{1}{2}\sum_{mn} V_{mn}(R) \,. \tag{8.2}$$

The intramolecular contributions $H_m$ describe individual molecules and are identical with the expression of $H_{\text{mol}}$ in Eq. (2.114). All types of intermolecular Coulomb interaction are comprised in $V_{mn}(R)$, the intermolecular electron–electron interaction $V_{mn}^{(\text{el}-\text{el})}$, the intermolecular coupling among the nuclei $V_{mn}^{(\text{nuc}-\text{nuc})}$, and the electron–nuclei coupling $V_{mn}^{(\text{el}-\text{nuc})}$ (between electrons of molecule $m$ with the nuclei of molecule $n$) as well as the coupling $V_{mn}^{(\text{nuc}-\text{el})}$, where electrons and nuclei have been interchanged. Note that for situations where we can restrict the description to valence electrons only, "nuclei" means nuclei plus core electrons. Since an aggregate formed by single molecules is considered, the nuclear (vibrational) coordinates can be split up into intramolecular coordinates $R_{\text{intra}}$, and intermolecular coordinates $R_{\text{inter}}$. The set $R_{\text{intra}}$ separates into single–molecule contributions, $R_{\text{intra}} = \{R_m^{(\text{intra})}\}$, and the $R_{\text{inter}}$ describe the motion of the molecules relative to one another. However, all environmental degrees of freedom, for example, those of a solvent, will also be included into the set $R_{\text{inter}}$.[2]

[2]Different models of the description of a solvent have been presented in the foregoing chapters. In Chapter 3 we introduced models to handle the coupling of a solvent to intramolecular vibrational processes. If the solvent molecules posses a permanent dipole moment, a description by means of macroscopic electrodynamics for dielectric media becomes possible. This has been presented in Chapter 6.

For further treatment we split up the electronic part $H_m^{(\mathrm{el})}$ of the Hamiltonian $H_m$ in Eq. (8.2). It is formally identical to the Hamiltonian given in Eq. (2.11), but in the present case the set $R$ of *all* nuclear coordinates of the aggregate enters the electronic–nuclear interaction (cf. Eq. (2.6)). Therefore, the electronic spectrum of $H_m^{(\mathrm{el})}$ will differ from that in the gas phase. In order to obtain a classification of transfer processes with respect to intramolecular electronic excitations, we expand the aggregate Hamiltonian Eq. (8.2) in terms of the adiabatic electronic states, $|\varphi_{ma}\rangle$, of the single molecules $m$. The label $a$ counts the actual electronic state ($S_0, S_1$, etc.). These states are defined via the stationary Schrödinger equation for a single molecule (cf. Eq. (2.12), $r_m$ denotes the set of electronic coordinates of the $m$th molecule)

$$H_m^{(\mathrm{el})}(R)\varphi_{ma}(r_m;R) = \epsilon_{ma}(R)\varphi_{ma}(r_m;R)\;. \tag{8.3}$$

Note that states belonging to different molecules are not orthogonal. Further it is important to keep in mind that all quantities in Eq. (8.3) depend on the complete set $R$ of *all* nuclear coordinates.[3]

Next we construct an expansion basis for the electronic states of the total aggregate. This will be done in analogy to the treatment presented in Section 2.4. First, we define the Hartree product ansatz (see Eq. (2.28))

$$\phi_{\{a\}}^{\mathrm{HP}}(\{r_m\};R) = \prod_{m=1}^{N_{\mathrm{mol}}} \varphi_{ma}(r_m;R)\;. \tag{8.4}$$

In a second step we generate an antisymmetric wave function (see Eq. (2.29)):

$$\phi_{\{a\}}(\{r_m\};R) = \frac{1}{\sqrt{N_{\mathrm{p}}!}}\sum_{\mathrm{perm}}(-1)^p\mathcal{P}\left[\phi_{\{a\}}^{\mathrm{HP}}(\{r_m\};R)\right]\;. \tag{8.5}$$

Here, $\mathcal{P}$ generates a permutation of electron coordinates of different molecules in the aggregate, and $p$ counts the number, $N_{\mathrm{p}}$, of permutations.[4] The set $\{a\}$ of single–molecule electronic quantum numbers describes the electronic configuration of the *total* aggregate.

In contrast to the Hartree–Fock procedure of Section 2.4, however, for simplicity the single–molecule state vectors $|\varphi_{ma}\rangle$ are assumed to be known and not the subject to a variational procedure. We note again that the functions Eq. (8.5) are neither orthogonal nor normalized. This becomes particularly obvious upon expanding the Schrödinger equation for the aggregate electronic state $|\psi\rangle$ with respect to the basis (8.5). We write

$$|\psi\rangle = \sum_{\{a\}} C(\{a\})|\phi_{\{a\}}\rangle \tag{8.6}$$

[3]The present use of nonorthogonal single–molecule states is similar to the treatment of bridge–mediated electron transfer in a DA complex discussed in Section 6.2.1.

[4]The number $N_{\mathrm{p}}$ of necessary permutations is obtained as $N_{\mathrm{el}}!/\prod_m(N_{\mathrm{el}}^{(m)}!)$, where $N_{\mathrm{el}}$ denotes the total number of electrons belonging to the different molecules of the aggregate. The number of electrons of the single molecule $m$ is given by $N_{\mathrm{el}}^{(m)}$.

and obtain

$$\sum_{\{b\}} \left( \langle \phi_{\{a\}} | H_{\text{agg}} | \phi_{\{b\}} \rangle - E \langle \phi_{\{a\}} | \phi_{\{b\}} \rangle \right) = 0 \,. \tag{8.7}$$

As an example let us consider a simple aggregate consisting of two molecules (*molecular dimer*). At the moment it suffices to concentrate on the electronic part $V_{12}^{(\text{el-el})}$ of the intermolecular interaction. Then, one recovers matrix elements of the Coulomb interaction which are similar to Eq. (2.138) and Eq. (2.139). They describe the direct and the exchange contributions, respectively. For the dimer we have

$$\begin{aligned} \langle \phi_{a_1 a_2} | V_{12}^{(\text{el-el})} | \phi_{b_1 b_2} \rangle &= \int dr_1 dr_2 \frac{1}{\sqrt{N_\text{p}!}} \sum_{\text{perm}} (-1)^p \mathcal{P} \varphi_{1a_1}^*(r_1; R) \varphi_{2a_2}^*(r_2; R) \\ &\quad \times V_{12}^{(\text{el-el})} \frac{1}{\sqrt{N_\text{p}!}} \sum_{\text{perm}} (-1)^{p'} \mathcal{P}' \varphi_{2b_2}(r_2; R) \varphi_{1b_1}(r_1; R) \\ &\equiv J_{12}^{(\text{el-el})}(a_1 a_2, b_2 b_1) - K_{12}^{(\text{el-el})}(a_1 a_2, b_2 b_1) \,. \end{aligned} \tag{8.8}$$

The direct Coulomb interaction is given by a single term as in Eq. (2.138). But the exchange part $K_{12}^{(\text{el-el})}(a_1 a_2, b_2 b_1)$ contains different contributions depending on the number of electrons which have been interchanged between the two molecules. If only a single electron has been exchanged between certain molecular orbitals, we recover an expression similar to Eq. (2.139). A closer inspection of Eq. (2.139) reveals that the spatial overlap between the two molecular orbitals, which belong to molecule 1 and to molecule 2, is responsible for the exchange contribution (cf. Fig. 8.1). Such a wave function overlap decreases exponentially with increasing intermolecular distance. Usually for distances larger than about 1 nanometer one can neglect the exchange contributions to the interaction energy.

In the following we assume that the intermolecular exchange contribution can be neglected. This means that we can use the Hartree product ansatz (8.4) for the electronic wave function of the aggregate. As a consequence of the neglect of intermolecular wave function overlap, i.e. of the assumption $\langle \varphi_{ma} | \varphi_{nb} \rangle = \delta_{ma,nb}$, the states $|\phi_{\{a\}}^{\text{HP}}\rangle$ form a complete basis. The expansion of the Hamiltonian Eq. (8.2) gives

$$\begin{aligned} H_{\text{agg}} &= \sum_{\{a\},\{b\}} \langle \phi_{\{a\}}^{\text{HP}} | H_{\text{agg}} | \phi_{\{b\}}^{\text{HP}} \rangle \times | \phi_{\{a\}}^{\text{HP}} \rangle \langle \phi_{\{b\}}^{\text{HP}} | \\ &= \sum_m \sum_{ab} H_m(ab) | \varphi_{ma} \rangle \langle \varphi_{mb} | \\ &\quad + \frac{1}{2} \sum_{mn} \sum_{abcd} J_{mn}(ab, cd) | \varphi_{ma} \varphi_{nb} \rangle \langle \varphi_{nc} \varphi_{md} | \,. \end{aligned} \tag{8.9}$$

The expression implies that $H_{\text{agg}}$ acts in the state space spanned by the states $\phi_{\{a\}}^{\text{HP}}$, Eq. (8.4).[5] The quantities $H_m(ab)$ are the matrix elements $\langle \varphi_{ma} | H_m | \varphi_{mb} \rangle$ of the single molecule part of

[5]The given expression avoids a notation where $|\varphi_{ma}\rangle\langle\varphi_{mb}|$ and $|\varphi_{ma}\varphi_{nb}\rangle\langle\varphi_{nc}\varphi_{md}|$ act on the unit operator $\mathbf{1} = \sum_{\{a\},\{b\}} |\phi_{\{a\}}^{\text{HP}}\rangle\langle\phi_{\{b\}}^{\text{HP}}|$ of the electronic state space. However, any use of $H_{\text{agg}}$ has to be understood in this way.

Eq. (8.2). They have off–diagonal contributions due to nonadiabatic couplings. The diagonal parts are given by the eigenvalues $\epsilon_{ma}$ of Eq. (8.3).

The matrix elements of the Coulomb interaction (with the electron–electron coupling reduced to the direct part) read

$$\begin{aligned} J_{mn}(ab,cd) &\equiv \langle \varphi_{ma}\varphi_{nb}|V_{mn}|\varphi_{nc}\varphi_{md}\rangle \\ &= \int dr_m dr_n \; \varphi^*_{ma}(r_m)\varphi^*_{nb}(r_n)V^{(\text{el}-\text{el})}_{mn}(r_m,r_n)\varphi_{nc}(r_n)\varphi_{md}(r_m) \\ &+ \delta_{ad}\delta_{bc}V^{(\text{nuc}-\text{nuc})}_{mn} \\ &+ \delta_{bc}\int dr_m \varphi^*_{ma}(r_m)V^{(\text{el}-\text{nuc})}_{mn}(r_m,R^{(\text{intra})}_n)\varphi_{md}(r_m) \\ &+ \delta_{bc}\int dr_n \varphi^*_{nb}(r_n)V^{(\text{nuc}-\text{el})}_{mn}(R^{(\text{intra})}_m,r_n)\varphi_{nc}(r_n)\,. \end{aligned} \tag{8.10}$$

To be complete we have to mentioned here that all terms in that expression depend on the intermolecular coordinates, too, indicating, for example, a distance modulation among the molecules. A detailed discussion of the processes corresponding to the various types of matrix elements will be given in the next section. So far we have also assumed that the vibrational DOF are fixed. The inclusion of their dynamics, i.e., the vibrational part of the aggregate Hamiltonian Eq. (8.2), is investigated in Section 8.3 .

### 8.2.1 The Two–Level Model

In this section we will specify the Hamiltonian (8.9) to a situation where besides the electronic ground state, $S_0$, only the first excited singlet state, $S_1$, of the different molecules is incorporated in the excitation energy transfer. Such a restriction is possible, for example, if a single $S_1$ state is initially excited, and if the $S_1$ states of all other molecules have approximately the same transition energy. The incorporation of further states such as triplet states or higher excited singlet states is straightforward.

#### Classification of the Coulomb Couplings

We start our discussion of the Hamiltonian, Eq. (8.9), by considering the matrix elements of the Coulomb interaction, $J_{mn}(ab,cd)$, and here first the electron–electron part $J^{(\text{el}-\text{el})}_{mn}(ab,cd)$. According to the two–level assumption all electronic quantum numbers can take only two values corresponding to the ground state $S_0$ ($a = g$), and the excited state $S_1$ ($a = e$). In Table 8.1 we summarize the physical processes contained in the different matrix elements and the combinations of electronic state indices they correspond to. In the first row (I) all matrix elements are listed which describe the electrostatic interaction between electronic charge densities located at molecule $m$ and molecule $n$. These charge densities follow directly from the electronic wave function of the $S_0$ or $S_1$ state. The second row of Table 8.1 (II) contains those matrix elements which are responsible for the interaction of the transition from $g$ to $e$ (or reverse) at molecule $m$ with the charge density of the states $g$ and $e$ at molecule $n$. Next we have those matrix elements which cause the motion of the Frenkel excitons between different molecular sites in the aggregate (III). They describe the transition of molecule $n$ from

| | matrix element | interaction process |
|---|---|---|
| (I) | $J_{mn}^{(\text{el-el})}(gg,gg)$<br>$J_{mn}^{(\text{el-el})}(ee,ee)$<br>$J_{mn}^{(\text{el-el})}(ge,eg)$<br>$J_{mn}^{(\text{el-el})}(eg,ge)$ | between charges<br>at molecules $m$ and $n$ |
| (II) | $J_{mn}^{(\text{el-el})}(eg,gg)$<br>$J_{mn}^{(\text{el-el})}(gg,ge)$<br>$J_{mn}^{(\text{el-el})}(ge,ee)$<br>$J_{mn}^{(\text{el-el})}(ee,eg)$ | between transitions at<br>molecule $m$ with charges at $n$ |
| (III) | $J_{mn}^{(\text{el-el})}(eg,eg)$<br>$J_{mn}^{(\text{el-el})}(ge,ge)$ | between $S_0 \to S_1$ transition at molecule $n$<br>and $S_1 \to S_0$ transition at $m$ |
| (IV) | $J_{mn}^{(\text{el-el})}(ee,gg)$<br>$J_{mn}^{(\text{el-el})}(gg,ee)$ | simultaneous excitation and de–excitation<br>of molecules $m$ and $n$ |

**Table 8.1:** Classification of the Coulomb interaction matrix elements in Eq. (8.10) for electronic two–level systems (note, that $J_{mn}^{(\text{el-el})}$ is symmetric with respect to the site indices).

the ground to the excited state, while the reverse process takes place at molecule $m$. This situation is sketched in Fig. 8.1. Finally, the last row contains the processes of the simultaneous excitation or de–excitation of both molecules (IV).

Noting Eq. (8.10) for the total coupling matrix elements, we see that those electronic matrix elements of Table 8.1 positioned in the first and second row have to be completed by nuclear contributions but those of row (III) and (IV) not. The total matrix elements corresponding to the type (I) describe the electrostatic interactions of two neutral molecules which may be in different electronic states:

$$\begin{aligned} J_{mn}(ab,ba) &= \langle\varphi_{ma}\varphi_{nb}|V_{mn}^{(\text{el-el})}|\varphi_{nb}\varphi_{ma}\rangle \\ &+\langle\varphi_{ma}|V_{mn}^{(\text{el-nuc})}|\varphi_{ma}\rangle + \langle\varphi_{nb}|V_{mn}^{(\text{nuc-el})}|\varphi_{nb}\rangle \\ &+V_{mn}^{(\text{nuc-nuc})} . \end{aligned} \tag{8.11}$$

If the electrostatic interaction energy between all charge carriers of both molecules is well–balanced, the contribution of these matrix elements to Eq. (8.9) is negligible. In the same way we may argue for the types of matrix elements given in row (II) of Table 8.1. We obtain, for example,

$$J_{mn}(ab, bc) = \langle \varphi_{ma}\varphi_{nb}|V_{mn}^{(\text{el-el})}|\varphi_{nb}\varphi_{mc}\rangle + \langle \varphi_{ma}|V_{mn}^{(\text{el-nuc})}|\varphi_{mc}\rangle \ . \tag{8.12}$$

For a further analysis let us define the *transition density* as

$$\rho_{meg}(r_m; R) = \varphi_{me}^*(r_m; R)\varphi_{mg}(r_m; R) \ . \tag{8.13}$$

We can introduce this quantity into $J_{mn}(ab, bc)$ (provided that we identified $a = e, c = g$). Then, the matrix element, Eq. (8.12) determines the interaction energy of the electronic transition density of molecule $m$ with the electronic charge distribution (in state $b$) of molecule $n$ as well as with the charge of its nuclei. Again, we assume that both parts of this energy compensate each other. As a result, the matrix elements Eq. (8.12) do not contribute to Eq. (8.10), and it remains to discuss the pure electronic matrix elements given by the type (III) and (IV) of Table 8.1. However, the latter lead to off–resonant interaction processes which do not conserve energy and thus give negligible contributions. The neglect of these processes is frequently called Heitler–London approximation .

Therefore, the following discussion will be exclusively based on the use of the matrix elements of type (III) which give resonant contributions to the Hamiltonian. They will be denoted as

$$J_{mn} = J_{mn}(e, g, e, g) \ . \tag{8.14}$$

and read after introduction of the transition density, Eq. (8.13)

$$J_{mn} = \int dr_m dr_n \ \rho_{meg}(r_m; R) V_{mn}^{(\text{el-el})} \rho_{neg}^*(r_n; R) \ . \tag{8.15}$$

The transition density gives a measure for the degree of *local* wave function overlap between the electronic ground state and the excited state of molecule $m$. If the transition densities of both molecules have a spatial extension similar to the intermolecular distance one cannot invoke any further approximation at this point. However, if the intermolecular distance is large compared with the transition density extension, there is no need to account for all the details of the transition densities.

### Dipole–Dipole Coupling

We will adopt this point of view and carry out a treatment similar to that of Section (2.5). There it was shown how to remove the short–range part of the intermolecular Coulomb interaction by employing a multipole expansion. To this end the Coulomb interaction is written in terms of the electronic coordinates related to the center of masses, $\mathbf{X}_m$ and $\mathbf{X}_n$, of the considered molecules

$$V_{mn}^{(\text{el-el})} = \sum_{j,k} \frac{e^2}{|\mathbf{X}_{mn} + \mathbf{r}_j(m) - \mathbf{r}_k(n)|} \ . \tag{8.16}$$

Here, we have introduced the intermolecular distance $\mathbf{X}_{mn} = \mathbf{X}_m - \mathbf{X}_n$, and $\mathbf{r}_j(m)$ ($\mathbf{r}_k(n)$) denotes the coordinates of the $j$th ($k$th) electron at molecule $m$. In a next step the multipole expansion in powers of $|\mathbf{r}_j(m) - \mathbf{r}_k(n)|/|\mathbf{X}_{mn}|$ is performed up to the second–order term (cf. Section 2.5). We abbreviate $\mathbf{X}_{mn} = \mathbf{X}$ and $\mathbf{r}_j(m) - \mathbf{r}_k(n) = \mathbf{r}$ and obtain

$$\frac{1}{|\mathbf{X}+\mathbf{r}|} \approx \frac{1}{|\mathbf{X}|} + \mathbf{r}\nabla_{\mathbf{X}} \frac{1}{|\mathbf{X}|} + \frac{1}{2} (\mathbf{r}\nabla_{\mathbf{X}})(\mathbf{r}\nabla_{\mathbf{X}}) \frac{1}{|\mathbf{X}|} \ . \tag{8.17}$$

The two types of derivatives read in detail

$$\mathbf{r}\nabla_{\mathbf{X}} \frac{1}{|\mathbf{X}|} = -\frac{\mathbf{r}\mathbf{X}}{|\mathbf{X}|^3} \ , \tag{8.18}$$

and

$$(\mathbf{r}\nabla_{\mathbf{X}})(\mathbf{r}\nabla_{\mathbf{X}}) \frac{1}{|\mathbf{X}|} = -\frac{\mathbf{r}^2}{|\mathbf{X}|^3} + \frac{3(\mathbf{r}\mathbf{X})^2}{|\mathbf{X}|^5} \ . \tag{8.19}$$

If we insert the obtained approximation for $V_{mn}^{(\mathrm{el-el})}$ into the matrix element, Eq. (8.15), several terms will vanish because of the orthonormality of the electronic wave functions $\varphi_{ma}$ (the transition density $\rho_{meg}$ vanishes if simply integrated with respect to the electronic coordinates). We immediately see that the zeroth– and first–order terms do not contribute. If we replace in the second–order expression $\mathbf{r}$ by $\mathbf{r}_j(m) - \mathbf{r}_k(n)$, we obtain terms where $\mathbf{r}_j(m)$ or $\mathbf{r}_k(n)$ appear twice. Since we have $\int dr_m \rho_{meg} = 0$ these terms are equal to zero. Only those terms contribute which depend on the electronic coordinates of both molecules. They can be collected to give the Coulomb interaction in *dipole–dipole approximation* as follows

$$V_{mn}^{(\mathrm{el-el})} \approx \frac{\hat{\mu}_m \hat{\mu}_n}{|\mathbf{X}_{mn}|^3} - 3\frac{(\mathbf{X}_{mn}\hat{\mu}_m)(\mathbf{X}_{mn}\hat{\mu}_n)}{|\mathbf{X}_{mn}|^5} \ . \tag{8.20}$$

The electronic dipole operator reads

$$\hat{\mu}_m = \sum_j e\mathbf{r}_j(m) \ . \tag{8.21}$$

Introducing Eq. (8.20) into Eq. (8.15) the dipole operators are replaced by transition dipole moments (cf. Eq. (5.7))

$$\mathbf{d}_m = \int dr_m \ \hat{\mu}_m \rho_{meg}(r_m; R) \equiv \langle \varphi_{me} | \hat{\mu}_m | \varphi_{mg} \rangle \ . \tag{8.22}$$

Thus the electronic matrix element of Eq. (8.20) can be cast into the form

$$J_{mn}(\mathbf{X}_{mn}) = \left(1 - \delta_{mn}\right) \kappa_{mn} \frac{|\mathbf{d}_m||\mathbf{d}_n^*|}{|\mathbf{X}_{mn}|^3} \ . \tag{8.23}$$

Note that this notation accounts for the fact that matrix element only exist for $m \neq n$. Furthermore, we introduced an orientational factor defined as

$$\kappa_{mn} = \mathbf{n}_m \mathbf{n}_n - 3(\mathbf{e}_{mn}\mathbf{n}_m)(\mathbf{e}_{mn}\mathbf{n}_n) \ , \tag{8.24}$$

where $\mathbf{n}_n$ and $\mathbf{e}_{mn}$ are the unit vectors pointing in the directions of the transition dipole moment $\mathbf{d}_n$, and the distance vector $\mathbf{X}_{mn}$, respectively. As already stated, this approximate form of the Coulomb interaction is applicable if the spatial extension of both transition densities appearing in Eq. (8.15) is small compared to the intermolecular distance $|\mathbf{X}_{mn}|$. Now, the aggregate Hamiltonian Eq. (8.2) takes the form (to incorporate off–diagonal parts of the single molecule Hamiltonian is postponed to Section 8.8.1)

$$H_{\text{agg}} = \sum_m \sum_{a=g,e} H_{ma} |\varphi_{ma}\rangle\langle\varphi_{ma}| + \sum_{mn} J_{mn} |\varphi_{me}\varphi_{ng}\rangle\langle\varphi_{ne}\varphi_{mg}| \, . \tag{8.25}$$

Remember that this notation includes the property $J_{mm} = 0$. For the following discussion it is useful to separate from the aggregate Hamiltonian the electronic part $H_{\text{el}}$. This is achieved by fixing the nuclear coordinates by their values corresponding to the aggregate ground state ($R \to R_0$). If we replace the $H_{ma}$ in Eq. (8.25) by the electronic energies (at $R_0$) we obtain

$$H_{\text{el}} = \sum_m \sum_{a=g,e} \epsilon_{ma} |\varphi_{ma}\rangle\langle\varphi_{ma}| + \sum_{mn} J_{mn} |\varphi_{me}\varphi_{ng}\rangle\langle\varphi_{ne}\varphi_{mg}| \, . \tag{8.26}$$

Finally, we recall the important approximation already introduced in the first part of Section (8.2). It concerns the neglect of the exchange interaction between the considered molecules. In contrast to the singlet–singlet energy transfer facilitated by the dipole–dipole Coulomb interaction, exchange interaction provides a major mechanism for energy transfer between triplet states. It could also be effective, if singlet–singlet transfer is symmetry–forbidden.

### Second Quantization Notation

It is also customary in exciton theory to introduce creation and annihilation operators as follows

$$B_m^+ = |\varphi_{me}\rangle\langle\varphi_{mg}| \, , \qquad B_m = |\varphi_{mg}\rangle\langle\varphi_{me}| \, . \tag{8.27}$$

This results in

$$H_{\text{el}} = E_0 + \sum_m (\epsilon_{me} - \epsilon_{mg}) B_m^+ B_m + \sum_{mn} J_{mn} B_m^+ B_n \, . \tag{8.28}$$

The first term on the right–hand side denotes the electronic aggregate ground–state energy [6]

$$E_0 = \sum_m \epsilon_{mg} \, . \tag{8.29}$$

The operators are of the Pauli type obeying the commutation relations

$$\left[B_m^+, B_n\right]_+ = \delta_{mn} + (1 - \delta_{mn}) 2 B_m^+ B_n \, , \tag{8.30}$$

[6] The fact that the aggregate ground state energy is a sum of the molecular electronic ground state energies results from the approximations introduced with respect to the intermolecular Coulomb interaction in Section 8.2.1.

and

$$\left[B_m^+, B_n^+\right]_+ = (1 - \delta_{mn})2B_m^+ B_n^+ \ . \tag{8.31}$$

The relation $[B_m^+, B_m]_+ = |\varphi_{mg}\rangle\langle\varphi_{mg}| + |\varphi_{me}\rangle\langle\varphi_{me}|$ can be interpreted as the completeness relation for the electronic state space of the $m$th molecule, i.e., it can be set equal to unity. This relation has been used to derive Eq. (8.28) from Eq. (8.26). For $m = n$ the excitations behave like Fermions and two of them cannot occupy the same molecular state. This local (exciton–exciton) repulsion gives rise to interesting effects in nonlinear optical spectroscopy of aggregates.

## 8.2.2 Single and Double Excitations of the Aggregate

Since we have restricted our discussion to the two–level exciton model including $S_0$ and $S_1$ states only, we are in a position to classify the total wave function according to the number of excited molecules, $N^*$. This is particularly useful when studying optical properties of aggregates where the number of excited molecules can be related to the number of photons absorbed by a single aggregate. The quantum mechanical electronic state of the aggregate $|\phi_{\{a\}}^{\text{HP}}\rangle$ contains the subset of $N^*$ excited molecules and the subset of $N_{\text{mol}} - N^*$ molecules in the ground state (see Fig. 8.16). The superposition of all states with fixed $N^*$ can be used as an ansatz for the $N^*$–exciton eigenstate of the Hamiltonian (8.9).

Multi–exciton states play an important role for the nonlinear optical properties of molecular aggregates. Therefore, we will consider single–exciton states ($N^* = 1$) as well as two–exciton states ($N^* = 2$, cf. Fig. 8.7). The single–exciton state can be reached from the aggregate ground state via an optical excitation process which involves the absorption of a single photon. A subsequent absorption step may lead from the single–exciton to the two–exciton state (excited state absorption). At this point it is important to emphasize that two–exciton states are not only a formal theoretical extension of the single–exciton levels. They have been clearly identified in different experiments [Fid93].

In order to make sure that the aggregate contains not more than two $S_0 \to S_1$ excitations, the completeness relation for the electronic aggregate states has to be rearranged according to the number of molecules which are in the excited state $S_1$:

$$\sum_{\{a\}} = |\phi_{\{a\}}^{\text{HP}}\rangle\langle\phi_{\{a\}}^{\text{HP}}| - |0\rangle\langle 0| + \sum_m |m\rangle\langle m| + \sum_{m,n} |mn\rangle\langle mn| + \dots \ . \tag{8.32}$$

We restrict the expansion to the first three contributions. The first term contains the aggregate ground state wave function

$$|0\rangle = \prod_m |\varphi_{mg}\rangle \ . \tag{8.33}$$

The presence of a single excitation in the aggregate is accounted for by the second term according to (cf. Fig. 8.7, left panel)

$$|m\rangle = |\varphi_{me}\rangle \prod_{n \neq m} |\varphi_{ng}\rangle \ . \tag{8.34}$$

The third term in Eq. (8.32) corresponds to the presence two excitations in a single aggregate (Fig. 8.7, right panel, the expression vanishes for $m = n$)

$$|mn\rangle = |\varphi_{me}\rangle|\varphi_{ne}\rangle \prod_{k\neq(m,n)} |\varphi_{kg}\rangle \ . \tag{8.35}$$

According to the ordering scheme, Eq. (8.32) we approximate the Hamiltonian, Eq. (8.26) as

$$H_{\text{el}} \approx H_{\text{el}}^{(0)} + H_{\text{el}}^{(1)} + H_{\text{el}}^{(2)} \ . \tag{8.36}$$

The three terms on the right–hand side correspond to the case of no, a single, and two excitations in the aggregate; they can be deduced from the matrix elements of Eq. (8.26). The ground state contribution of Eq. (8.36) is given by

$$H_{\text{el}}^{(0)} = E_0|0\rangle\langle 0| \ , \tag{8.37}$$

with the electronic ground state energy $E_0$ introduced in Eq. (8.29). The single–excitation (single–exciton) Hamiltonian is given by

$$H_{\text{el}}^{(1)} = \sum_{mn} \left(\delta_{mn}\{E_0 + \epsilon_{me} - \epsilon_{mg}\} + J_{mn}\right)|m\rangle\langle n| \ . \tag{8.38}$$

We note that the energy of the single excitation at molecule $m$, $\epsilon_{me} - \epsilon_{mg}$, has been added to the ground state energy $E_0$. For the two–excitation (two–exciton) Hamiltonian we obtain

$$\begin{aligned} H_{\text{el}}^{(2)} &= \sum_{mn} \{E_0 + \epsilon_{me} - \epsilon_{mg} + \epsilon_{ne} - \epsilon_{ng}\}|mn\rangle\langle mn| \\ &+ \sum_{mnk} \left(J_{mk}|mn\rangle\langle kn| + J_{kn}|mn\rangle\langle mk|\right) , \end{aligned} \tag{8.39}$$

including in the first term the excitation energy of molecule $m$ and $n$.

### 8.2.3 Delocalized Exciton States

In the present model there is no coupling between the aggregate ground state and the single– and two–exciton states. Therefore, we may solve separate eigenvalue equations for singly and doubly excited states. To remove the unimportant ground state energy we set in the present section $E_0 = 0$. In the case of a single excitation the eigenvalue equation reads

$$H_{\text{el}}^{(1)}|\alpha\rangle = \mathcal{E}_\alpha|\alpha\rangle \ . \tag{8.40}$$

To construct the solutions to this equation we expand the eigenstate $|\alpha\rangle$ with respect to the complete basis of singly excited states:

$$|\alpha\rangle = \sum_{m} c_\alpha(m)|m\rangle \ . \tag{8.41}$$

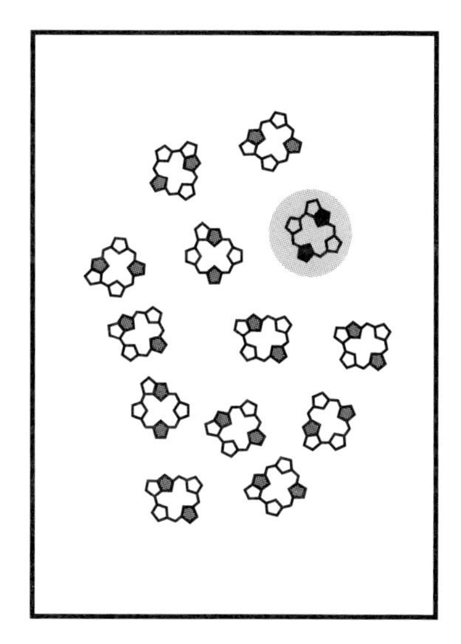

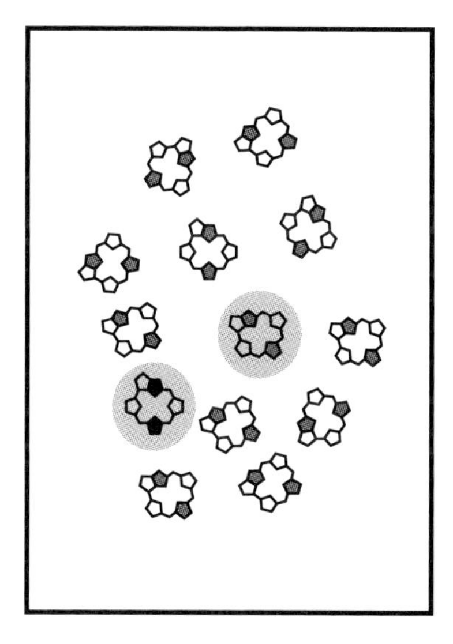

**Figure 8.7:** Schematic illustration of the presence of a singly excited state, Eq. (8.34), (left panel) and of a doubly excited state, Eq. (8.35), (right panel) of the type of chromophore complex introduced in Fig. 8.6. (The grey circle labels the molecule which is in the excited state. To get the one– and the two–exciton states the localized excitations have to be delocalized over the whole complex.)

Introducing this ansatz into Eq. (8.40) and multiplying it by $\langle n|$ from the left yields (note the abbreviation $E_n = \epsilon_{ne} - \epsilon_{ng}$)

$$\mathcal{E}_\alpha c_\alpha(n) = \sum_m \langle n|H_{\rm el}^{(1)}|m\rangle c_\alpha(m) = E_n c_\alpha(n) + \sum_m J_{nm} c_\alpha(m) \ . \tag{8.42}$$

In the following we will call $H_{\rm el}^{(1)}$ the exciton Hamiltonian and write

$$H_{\rm ex} = \sum_\alpha \mathcal{E}_\alpha |\alpha\rangle\langle\alpha| \ . \tag{8.43}$$

To determine the eigenstates corresponding to the presence of two excitations in the considered aggregate we have to solve the eigenvalue equation (note the replacement of $\alpha$ by $\tilde{\alpha}$)

$$H_{\rm el}^{(2)}|\tilde{\alpha}\rangle = \mathcal{E}_{\tilde{\alpha}}|\tilde{\alpha}\rangle \ . \tag{8.44}$$

The *two–exciton* states are defined as

$$|\tilde{\alpha}\rangle = \sum_{mn} c_{\tilde{\alpha}}(mn)|mn\rangle \ , \tag{8.45}$$

with the expansion coefficients following from (remember $E_0 = 0$ and the definition of $E_m$ and $E_n$)

$$\mathcal{E}_{\tilde{\alpha}} c_{\tilde{\alpha}}(mn) = (E_m + E_n) c_\alpha(mn) + \sum_k \left( J_{mk} c_{\tilde{\alpha}}(kn) + J_{kn} c_{\tilde{\alpha}}(mk) \right) \ . \tag{8.46}$$

Note that there are $N_{\rm mol}(N_{\rm mol}-1)/2$ possible two–exciton states. Fig. 8.5 displays the single and two–exciton energies corresponding to the pigment–protein complex also shown in this figure. Before turning to the consideration of the exciton–vibrational coupling two examples of single– and two–exciton spectra are discussed where analytical expressions can be derived.

## The Molecular Heterodimer

As the simplest example we consider the so–called heterodimer which consists of two monomers with different excitation energies $E_1$ and $E_2$ and a coupling $J = J_{12}$. The eigenvalue problem for two coupled two–level molecules has already been solved in Section 2.8.3. The single–exciton eigenvalues are given by

$$\mathcal{E}_{\alpha=\pm} = \frac{E_1+E_2}{2} \pm \frac{1}{2}\sqrt{(E_1-E_2)^2+4|J|^2}\,. \tag{8.47}$$

For the details of the corresponding eigenfunctions we refer to Section 2.8.3. Here we set the arbitrary phase factors appearing in the solution of the two–level problem equal to 1 and write

$$\begin{aligned} |+\rangle &= \frac{1}{\sqrt{1+\eta^2}}\Big(\eta|1\rangle + e^{-i\arg(J)}|2\rangle\Big) \\ |-\rangle &= \frac{1}{\sqrt{1+\eta^2}}\Big(|1\rangle - \eta e^{-i\arg(J)}|2\rangle\Big)\,. \end{aligned} \tag{8.48}$$

The parameter $\eta$ (cf. Eq. (2.167)) is equal to zero for $J=0$ and otherwise given by

$$\eta = \frac{1}{2|J|}\left|E_1 - E_2 + \sqrt{(E_1-E_2)^2+4|J|^2}\right|\,. \tag{8.49}$$

Eq. (8.48) nicely illustrates the delocalization of the wave function over the dimer. In the case of the two exciton state in a molecular dimer there is no way to form a delocalized wave function. Hence it follows $\mathcal{E}_{\tilde{\alpha}} = E_1 + E_2$ and the corresponding eigenstate is $|mn\rangle$.

## The Finite Molecular Chain and the Molecular Ring

First we consider an aggregate which consists of a linear arrangement of $N_{\rm mol}$ identical molecules with $S_0 \to S_1$ excitation energies $E_{\rm exc}$ and nearest–neighbor dipole–dipole coupling $J$. The neglect of long–range dipole–dipole interactions is justified in cases where the distance between molecular entities is not too small (note that according to Eq. (8.23) $J_{m,m+1} = 2^3 J_{m,m+2}$). Such regular structures can be found in systems which show a rod–like arrangement of the molecules after aggregation.

The determination of the energy spectrum of a finite linear chain has already been explained in Section 2.8.4. In the present notation we obtain

$$\mathcal{E}_\alpha = E_{\rm exc} + 2J\cos(\alpha)\,, \tag{8.50}$$

with $\alpha = \pi j/(N_{\rm mol}+1)$, $(j=1,\ldots,N_{\rm mol})$. The wave function expansion coefficients read

$$c_\alpha(m) = \sqrt{\frac{2}{N_{\rm mol}+1}}\sin(\alpha m)\,. \tag{8.51}$$

This result is confronted with that obtained for a regular molecular ring (cf. Fig. 8.4), a system which was discussed already in Section 2.8.4. We again consider identical molecules with excitation energy $E_{\text{exc}}$ and nearest–neighbor dipole–dipole coupling $J$ but now in a ring–like spatial arrangement. It can be easily shown that Eq. (8.50) remains valid but with $\alpha = 2\pi j/N_{\text{mol}}$, $(j = 0, \dots, N_{\text{mol}} - 1)$ and

$$c_\alpha(m) = \frac{1}{\sqrt{N_{\text{mol}}}} e^{i\alpha m} \,. \tag{8.52}$$

Note that due to the periodicity of the aggregate the sine–function in Eq. (8.51) has been replaced by a complex exponential. Moreover, we get the site independent probability distribution $\mid c_\alpha(m) \mid^2 = 1/N_{\text{mol}}$ for the molecular ring, what is different from the probability distribution following for the regular chain.

The two–exciton states for the linear chain can also be constructed [Muk95]. The respective eigenvalues read

$$\mathcal{E}_{\tilde{\alpha}} = 2E_{\text{exc}} + 4J\big(\cos(\alpha) + \cos(\beta)\big) \tag{8.53}$$

with $\alpha = \pi j/N_{\text{mol}}$, $\beta = \pi j'/N_{\text{mol}}$ $(j, j' = 1, 3, \dots, 2N_{\text{mol}} - 1)$ and similar for $\beta$ (note that for this particular case the quantum number $\tilde{\alpha}$ is given by the pair $\alpha$ and $\beta$). The expansion coefficients of the related eigenstates are obtained as:

$$c_{\tilde{\alpha}}(mn) = \frac{\text{sgn}(m-n)}{N_{\text{mol}}} e^{i(\alpha m + \beta n)} \,. \tag{8.54}$$

where $\text{sgn}(m - n) = (1 - \delta_{mn})(m - n)/|m - n|$.

Apparently, for a more complex structure of the aggregate excitonic spectra are only obtainable by numerical means. For example, this is valid for the single and two exciton levels of the biological chromophore complex which is shown in Fig. 8.5 together with the respective energy levels.

## 8.3 Exciton–Vibrational Interaction

Having discussed the electronic problem of an aggregate with the nuclear configurations frozen at $R$, a description which incorporates the nuclear dynamics will be given next. As already explained in Section 8.2 the set $R$ of all vibrational DOF will be decomposed into $R_{\text{intra}} = \{R_m^{(\text{intra})}\}$ where every $R_m^{(\text{intra})}$ describes the intramolecular nuclear coordinates of molecule $m$, and $R_{\text{inter}}$ which stands for all intermolecular and solvent coordinates. Both types of vibrational coordinates enter the aggregate Hamiltonian, Eq. (8.25) in two ways. First, the vibrational Hamiltonian $H_{ma}$ $(a = e, g)$ depend on the set $R$ of all vibrational degrees of freedom. Second, the dipole–dipole coupling $J_{mn}$ depends via Eq. (8.23) on the distance $\mathbf{X}_{mn}$, between molecules $m$ and $n$. Thus it is subject to modulations resulting from vibrational dynamics. Moreover, as can be seen from Eq. (8.23), fluctuations of the mutual orientations of the transition dipole moments may also influence $J_{mn}$. Finally, so–called non–Condon contributions become important, if the transition dipole moments depend on the nuclear coordinates (cf. Section 5.2.2).

In order to account for the coupling to vibrational degrees of freedoms we concentrate on single–exciton states only (described in a two–level model for every molecule). Noting the ordering scheme of the electronic Hamiltonian, Eq. (8.36), the total aggregate Hamiltonian (8.2) will be written as:

$$H_{\text{agg}} = H_{\text{agg}}^{(0)} + H_{\text{agg}}^{(1)} . \tag{8.55}$$

The aggregate ground state Hamiltonian reads

$$H_{\text{agg}}^{(0)} \equiv \sum_m H_{mg}|0\rangle\langle 0| = \big(T_{\text{nuc}} + U_0(R)\big)|0\rangle\langle 0| . \tag{8.56}$$

with PES defined as

$$U_0(R) = E_0(R) + V'_{\text{nuc-nuc}}(R) . \tag{8.57}$$

The ground–state energy (at $R = R_0$) has been introduced in Eq. (8.29), and $V'_{\text{nuc-nuc}}$ is the part of the nuclei–nuclei interaction which follows from the intramolecular contributions $V_m^{(\text{nuc-nuc})}$ (which, however, may depend on all types of vibrations).

The single excitation part of the aggregate Hamiltonian takes the following form:

$$H_{\text{agg}}^{(1)} = \sum_{mn} \big(\delta_{mn}(T_{\text{nuc}} + U_0(R) + \epsilon_{me}(R) - \epsilon_{mg}(R)) + J_{mn}\big)|m\rangle\langle n| . \tag{8.58}$$

The PES for the excited electronic state follows as

$$U_m(R) = U_0(R) + \epsilon_{me}(R) - \epsilon_{mg}(R) . \tag{8.59}$$

Considering for a moment the vibrational coordinates and momenta as classical quantities it is possible to introduce (single) exciton states in analogy to Section 8.2.3. We set

$$|\Phi_\alpha(R)\rangle = \sum_m c_\alpha(m;R)|m\rangle , \tag{8.60}$$

and obtain the aggregate Hamiltonian as

$$H_{\text{agg}}^{(1)} = \sum_\alpha \big(T_{\text{nuc}} + U_\alpha(R)\big)|\Phi_\alpha(R)\rangle\langle\Phi_\alpha(R)| . \tag{8.61}$$

The $U_\alpha(R)$ are the potential energy surfaces for the exciton level with quantum number $\alpha$. It is important to notice that besides the potential energy surfaces also the exciton state vectors depend on the actual nuclear configuration. This type of state is often called the *adiabatic* exciton state. However, its introduction has the disadvantage of a complicated coordinate dependence of the state vectors and potential energy surfaces. In order to introduce delocalized exciton states, therefore, we will proceed in a different way described in Section 8.3.2.

### 8.3.1 Coupling to Intramolecular Vibrations

Let us specify the PES Eqs. (8.57) and (8.59) to the case of an exclusive coupling to (high–frequency) intramolecular coordinates $\{R_m^{(\text{intra})}\}$. First, we note that only the set $R_m^{(\text{intra})}$ of coordinates which belongs to molecule $m$ enters the molecular energy level $\epsilon_{mg}$ and $\epsilon_{me}$. Furthermore, $V_m^{(\text{nuc-nuc})}$ only depends on the $R_m^{(\text{intra})}$. Therefore, we obtain the ground state PES Eq. (8.57) as

$$U_0(R) = \sum_m U_{mg}(R_m^{(\text{intra})}) , \tag{8.62}$$

with the single molecule PES $U_{mg}$. The PES, Eq. (8.59) for the singly excited aggregate state reads:

$$U_m(R) = U_{me}(R_m^{(\text{intra})}) + \sum_{n \neq m} U_{ng}(R_n^{(\text{intra})}) . \tag{8.63}$$

Here, the PES $U_{me}(R_m^{(\text{intra})})$ of the excited state of molecule $m$ follows from $\epsilon_{me} - \epsilon_{mg} + U_{mg}$. If we introduce a similar separation of the total nuclear kinetic energy operator we may introduce a vibrational Hamiltonian for every molecule which may refer to the electronic ground state, $H_{mg}$, or to the excited state $H_{me}$. This description is particularly important for systems where the molecules are well separated one from another and where intermolecular vibrations have only a minor influence on the excitation energy transfer (cf. Section 8.5).

### 8.3.2 Coupling to Aggregate Normal–Mode Vibrations

The following considerations focus on the coupling to (low–frequency) intermolecular and solvent coordinates $R_{\text{inter}}$ (but we will use $R$ instead of $R_{\text{inter}}$). Let us suppose that a restriction to the harmonic approximation for the nuclear motions will be sufficient. This allows us to introduce vibrational normal modes $\{q_\xi\}$. Then, the ground state PES given in Eq. (8.57) becomes

$$U_0(R_{\text{inter}}) \equiv U_0(q) = U_0^{(0)} + \sum_\xi \omega_\xi^2 q_\xi^2/2 . \tag{8.64}$$

Here, the reference energy $U_0^{(0)} \equiv U_0(R_0)$ is related to the nuclear equilibrium configuration $R_0$ of the aggregate ground state. With respect to the normal mode coordinates, the set $R_0$ corresponds to $\{q_\xi = 0\}$. For the following we set $U_0^{(0)}$ plus the zero–point energy of the vibrations equal to zero. The normal mode coordinates can be expressed in terms of oscillator creation and annihilation operators $C_\xi$ and $C_\xi^+$ as $q_\xi = \sqrt{\hbar/2\omega_\xi}\, Q_\xi$ with $Q_\xi = (C_\xi + C_\xi^+)$. Using this notation the vibrational Hamiltonian can be written as

$$T_{\text{nuc}} + U_0(R) = \sum_\xi \hbar\omega_\xi C_\xi^+ C_\xi . \tag{8.65}$$

Next let us consider the single–excitation Hamiltonian Eq. (8.58). Following Section 2.6.1 we assume that the excited electronic state can be described by the same normal mode coordinates as the ground state. Then, the excited state PES, Eq. (8.59) is obtained as

$$U_m(q) = \epsilon_{me}\left(R(q)\right) - \epsilon_{mg}\left(R(q)\right) + \sum_\xi \omega_\xi^2 q_\xi^2/2 \,, \tag{8.66}$$

where we indicated the dependence of the Cartesian nuclear coordinates on the normal mode coordinates introduced for the ground state, $R = R(q)$. An expansion of the excitation energy $\epsilon_{me} - \epsilon_{mg}$ with respect to the various $q_\xi$ gives in the lowest order

$$\begin{aligned}\epsilon_{me}\left(R(q)\right) - \epsilon_{mg}\left(R(q)\right) &\approx \epsilon_{me}\left(R_0\right) - \epsilon_{mg}\left(R_0\right) \\ &\quad + \sum_\xi \left(\frac{\partial\left(\epsilon_{me}\left(q\right) - \epsilon_{mg}\left(q\right)\right)}{\partial q_\xi}\right)_{q_\xi=0} \cdot q_\xi \\ &\equiv E_m + \sum_\xi \hbar\omega_\xi g_m(\xi)(C_\xi + C_\xi^+) \,.\end{aligned} \tag{8.67}$$

Here, we introduced the Franck–Condon transition energy

$$E_m = \epsilon_{me}(R_0) - \epsilon_{mg}(R_0) \,, \tag{8.68}$$

and the dimensionless coupling constants read

$$g_m(\xi) = \left(\frac{\partial\left(\epsilon_{me}\left(q\right) - \epsilon_{mg}\left(q\right)\right)}{\partial q_\xi}\right)_{q_\xi=0} \cdot \frac{1}{\sqrt{2\hbar\omega_\xi^3}} \,. \tag{8.69}$$

Eq. (8.67) enables us to write the excited state PES as

$$U_m(q) = E_m - \sum_\xi \hbar\omega_\xi g_m^2(\xi) + \sum_\xi \frac{\hbar\omega_\xi}{4}\left(Q_\xi + 2g_m(\xi)\right)^2 \,. \tag{8.70}$$

To account for the modification of the dipole–dipole coupling by vibrational motions we introduce the equilibrium value $J_{mn}^{(0)}$ between molecule $m$ and $n$. Expanding the dipole–dipole coupling with respect to $\{q_\xi\}$ around the equilibrium configuration we get in first order

$$J_{mn} \approx J_{mn}^{(0)} + \sum_\xi \left(\frac{\partial J_{mn}}{\partial q_\xi}\right)_{q_\xi=0} \cdot q_\xi = J_{mn}^{(0)} + \sum_\xi \hbar\omega_\xi \tilde{g}_{mn}(\xi) Q_\xi \,, \tag{8.71}$$

where the we defined the coupling matrix $\tilde{g}_{mn}(\xi)$. Alternatively to the inclusion into the definition of the excited state PES Eq. (8.70), the $g_m(\xi)$ can be combined with $\tilde{g}_{mn}(\xi)$ to the *exciton–vibrational coupling matrix*

$$g_{mn}(\xi) = \delta_{mn} g_m(\xi) + (1 - \delta_{mn})\tilde{g}_{mn}(\xi) \,. \tag{8.72}$$

The resulting exciton–vibrational Hamiltonian, which describes the motion of a single excitation in the aggregate under the influence of the nuclear degrees of freedom, reads

$$H_{\text{agg}}^{(1)} = H_{\text{ex}} + H_{\text{vib}} + H_{\text{ex-vib}} \,. \tag{8.73}$$

It contains the electronic (excitonic) part

$$H_{\text{ex}} = \sum_{mn} \left( \delta_{mn} E_m + J_{mn}^{(0)} \right) |m\rangle\langle n| \tag{8.74}$$

which has to be distinguished from Eq. (8.58), the vibrational part [7]

$$H_{\text{vib}} = \sum_{\xi} \hbar\omega_{\xi} C_{\xi}^{+} C_{\xi} \sum_{m} |m\rangle\langle m| \;, \tag{8.75}$$

and for the coupling to the vibrational DOF we obtain

$$H_{\text{ex-vib}} = \sum_{mn} \sum_{\xi} \hbar\omega_{\xi} g_{mn}(\xi) Q_{\xi} |m\rangle\langle n| \;. \tag{8.76}$$

If the intermolecular electronic coupling dominates it is advisable to introduce delocalized exciton states. This will be done in the following section.

### 8.3.3 Exciton–Vibrational Hamiltonian and Excitonic Potential Energy Surfaces

The aggregate Hamiltonian $H_{\text{agg}}^{(1)}$, Eq. (8.73) is rewritten in terms of the eigenstates of the exciton Hamiltonian (8.74), i.e. the *exciton states*, by using the expansion, Eq. (8.41). Assuming the same separation of $H_{\text{agg}}^{(1)}$ as in Eq. (8.73) we obtain the excitonic part $H_{\text{ex}}$ as in Eq. (8.43), and the exciton–vibrational coupling reads

$$H_{\text{ex-vib}} = \sum_{\alpha,\beta} \sum_{\xi} \hbar\omega_{\xi} g_{\alpha\beta}(\xi) \left( C_{\xi} + C_{\xi}^{+}) \right) |\alpha\rangle\langle\beta| \;. \tag{8.77}$$

Here, the exciton–vibrational coupling matrix is given by

$$g_{\alpha\beta}(\xi) = \sum_{m,n} c_{\alpha}^{*}(m) g_{mn}(\xi) c_{\beta}(n) \;. \tag{8.78}$$

The vibrational part remains unaffected, only $\sum_m |m\rangle\langle m|$ is replaced by $\sum_{\alpha} |\alpha\rangle\langle\alpha|$.

This model of linear exciton–vibrational coupling has found wide application in the study of optical properties of excitons in molecular aggregates as well as in the description of exciton dynamics. We will return to this Hamiltonian when describing partly coherent exciton motion in Section 8.6.

If diagonal elements of $g_{\alpha\beta}(\xi)$ are much larger than the off diagonal ones one can introduce a notation of $H_{\text{agg}}^{(1)}$ leading to a certain type of potential energy surfaces (cf. Eq. (8.61)). To this end, we take the potential energy part $\sum_{\xi} \hbar\omega_{\xi} Q_{\xi}^2/4$ of $H_{\text{vib}}$ and combine it with the term $\propto Q_{\xi}$ of Eq. (8.77) to define the (shifted) *excitonic* PES

$$U_{\alpha}(q) = \mathcal{E}_{\alpha} - \sum_{\xi} \hbar\omega_{\xi} g_{\alpha\alpha}^2(\xi) + \sum_{\xi} \frac{\hbar\omega_{\xi}}{4} \left( Q_{\xi} + 2g_{\alpha\alpha}(\xi) \right) \;. \tag{8.79}$$

[7] The projector $\sum_m |m\rangle\langle m|$ ensures that the vibrational Hamiltonian acts in the state space of single excitations of the aggregate.

Then, the exciton representation of the aggregate Hamiltonian is obtained as

$$\begin{aligned} H_{\text{agg}}^{(1)} &= \sum_{\alpha,\beta} \Big( \delta_{\alpha\beta} \{ T_{\text{vib}} + U_{\alpha}(q) \} \\ &\quad + (1 - \delta_{\alpha\beta}) \sum_{\xi} \hbar\omega_{\xi} g_{\alpha\beta}(\xi) Q_{\xi} \Big) |\alpha\rangle\langle\beta| \, . \end{aligned} \tag{8.80}$$

This expression is particularly suitable for carrying out a perturbational expansion with respect to $g_{\alpha\beta}(\xi)$. In this respect it is of interest to clarify the type of the states which are of zeroth–order with respect to the off–diagonal elements. Neglecting their contribution in the aggregate Hamiltonian one easily obtains the zeroth–order states as

$$|\Psi_{\alpha}^{(0)}\rangle = \sum_{m} c_{\alpha}(m)|m\rangle \prod_{\xi} \exp\left( g_{\alpha\alpha}(\xi) C_{\xi} - \text{h.c.} \right) |N_{\xi}\rangle \, . \tag{8.81}$$

This state vector contains the shift of every normal mode oscillator upon excitation of the exciton state $|\alpha\rangle$. The magnitude of this shift is determined by $g_{\alpha\alpha}(\xi)$. It is related to the non–shifted normal mode oscillator states $|N_{\xi}\rangle$ with $N_{\xi}$ excited vibrational quanta. The exciton state has been expressed in terms of local excitations. This type of exciton–vibrational state is usually called Davydov ansatz; it has been widely used (in a time–dependent version) to describe soliton motion in molecular chains [Dav79]. Notice, that it is well–known that this type of state, if understood as an ansatz for a variational determination of the respective energy (with $c(m)$ and $g(\xi)$ to be determined), does not give the best result for the ground state energy of the so–called excitonic polaron. A more general ansatz would contain the $g(\xi)$ in the exponent of the shift operator as a function of the molecule index $m$. Then, the state incorporates a superposition of vibrational displacements which depends on the different sites [Toy61] (see also [Mei97] for a recent application).

## 8.4 Regimes of Exciton Transfer

As already discussed in connection with the description of electron transfer reactions in Chapter 6, the actual type of excitation energy motion is determined by the mutual relation between two time scales. The intramolecular vibrational relaxation time, $\tau_{\text{rel}}$, determines the time which the nuclear vibrations of each molecule need to return to thermal equilibrium after the electronic transition took place. The transfer time, $\tau_{\text{trans}}$, is given by the inverse of the characteristic interaction energy between two molecules. It is the time the exciton needs to move from one molecule to another neglecting any additional perturbations. The different regimes of XT are indicated in Fig. 8.8 (to have a clearer presentation the XT regimes have been drawn along the inverses of the characteristic times, i. e. along the intramolecular and intermolecular interaction strength).

If $\tau_{\text{rel}} \ll \tau_{\text{trans}}$, it is impossible to construct a wave function which involves different molecules. Intramolecular relaxation introduces fast dephasing and we are in the regime of *incoherent transfer* labelled by I in Fig. 8.8. The exciton motion proceeds diffusively, similar to the *random walk* known from statistical physics (see right part of Fig. 8.6). This type of transfer is characterized by an occupation probability $P_m(t)$ for the exciton to be at molecule

$m$ but not by a wave function extending over different molecules. It will be discussed in Section 8.5.

If on the other hand, $\tau_{\text{trans}} \ll \tau_{\text{rel}}$, the exciton can move almost freely from molecule to molecule according to the respective Schrödinger equation. The exciton travels through the aggregate as a spatially confined quantum mechanical wave packet (see left part of Fig. 8.6). Since such a type of motion requires fixed phase relations between exciton wave functions of different molecules, it is called *coherent transfer*. (In Chapter 3 we discussed that this type of motion is typical for closed quantum systems not subject to the influence of environmental forces.) The respective region of coherent motion is indicated as region II in Fig. 8.8, and the related theoretical description will be given in Section 8.6.

Clearly, there are regions of XT between the coherent an the incoherent type ($\tau_{\text{rel}} \approx \tau_{\text{trans}}$). This motion is called *partially coherent* XT (region II in Fig. 8.8), but notice that such a characterization of the intermediate region of XT processes is often not straightforward.

In general, various mixtures of the different types of motion within the same aggregate are possible. For example, if there are two groups of closely packed molecules in the aggregate, XT within any group may appear to be coherent (or partially coherent), but between the groups the XT takes place as a hopping process (cf. Section 8.5.3 and also Fig. 8.4). Moreover, moving in region III of Fig. 8.8 to the upper right corner both basic couplings become large. This region characterizes qualitatively a type of XT motion where the excitation is delocalized but connected with a noticeable displacement of the vibrational DOF (we refer

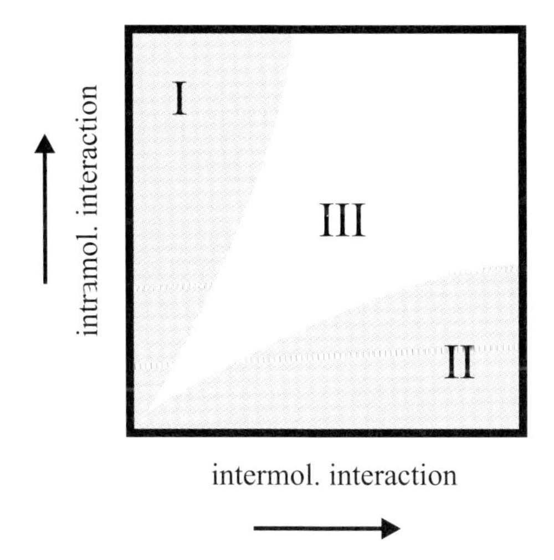


**Figure 8.8:** Schematic representation of different XT regimes. The strength of intermolecular interactions increases along the horizontal axis and that of intramolecular couplings along the vertical axis. Förster transfer as described in Section 8.5 is typical for region I whereas the density matrix description given in Section 8.6 can be applied in Section II. In the intermediate region III delocalized exciton formation and exciton–vibrational coupling have to be dealt with on an equal footing.

to the similarity with the ultrafast ET described in Section 6.8). Its theoretical description has been intensively discussed at the end of the 1980's, although there was little progress in experimental verifications (see, for example [Dav86,Sco92]).

## 8.5 Förster Theory of Incoherent Exciton Transfer

In this section we will be concerned with the regime of incoherent transfer where a localized excitation jumps from molecule to molecule. From our general discussion in Chapter 3 we know that incoherent quantum particle motion is adequately described by the rate equation for the state occupation probabilities, $P_m(t)$. They describe the presence of a single excitation at molecule $m$:

$$P_m(t) = \mathrm{tr}\{\hat{W}(t)|m\rangle\langle m|\} \equiv \langle m|\mathrm{tr}_{\mathrm{vib}}\{\hat{W}(t)\}|m\rangle \; , \tag{8.82}$$

where $\hat{W}(t)$ is the statistical operator characterizing the time–dependent nonequilibrium state of the exciton–vibrational system and $\mathrm{tr}_{\mathrm{vib}}\{...\}$ denotes the trace with respect to the vibrational degrees of freedom. According to Section 3.4.5, $P_m$ follows from the solution of the rate equation

$$\frac{\partial}{\partial t} P_m(t) = -\sum_n k_{m\to n} P_m(t) + \sum_m k_{n\to m} P_n(t) \; . \tag{8.83}$$

Here we will concentrate on the case where the rates $k_{m\to n}$ are proportional to the square of the intermolecular Coulomb interaction leading to a Golden Rule expression. The XT is considered for a system of two molecules which are not necessarily identical (heterodimer). Instead of using the labels $m$ and $n$, and in analogy with the discussion of electron transfer, the heterodimer will be called a donor–acceptor complex.

First, we will concentrate on the case of two *independent* sets of vibrational coordinates, defined either for the donor or acceptor part of the dimer (cf. Section 6.4.2 where a similar situation has been discussed for the electron transfer). This case is of particular importance if intramolecular vibrations of the donor and acceptor molecules dominate the transfer.

### 8.5.1 The Förster Transfer Rate

In order to describe the XT in a donor–acceptor complex, where donor and acceptor are characterized by independent sets of vibrational coordinates, we have to generalize the preceding considerations. Let us introduce (independent) wave functions for both molecules as $\Psi_{maM}(r_m; R_m^{(\mathrm{intra})}) = \varphi_{ma}(r_m; R_m^{(\mathrm{intra})})\chi_{maM}(R_m^{(\mathrm{intra})})$ ($m = D, A$; $a = g, e$ $(S_0, S_1)$; $r_m$ are the electronic coordinates). The $S_0 \to S_1$ transition energies are given by $E_{meM} - E_{mgN}$, and the $R_m^{(\mathrm{intra})}$ comprise the intramolecular nuclear coordinates. If necessary, they also include donor/acceptor specific environmental (solvent) coordinates. We would like to point out that the assumption that every single molecule has its *own* set $R_m^{(\mathrm{intra})}$ is particularly reasonable at large intermolecular distances ($> 10$ Å).

According to Fig. 8.1 the transfer proceeds via de–excitation of the donor and simultaneous excitation of the acceptor. This results in the following general transfer rate:

$$
\begin{aligned}
k_{DA} &= \frac{2\pi}{\hbar} \sum_{M_D,N_D} \sum_{M_A,N_A} f(E_{DeM_D})\, f(E_{AgN_A}) \\
&\times |\langle \Psi_{DeM_D}, \Psi_{AgN_A} | V_{DA} | \Psi_{AeM_A}, \Psi_{DgN_D} \rangle|^2 \\
&\times \delta(E_{DeM_D} + E_{AgN_A} - E_{AeM_A} - E_{DgN_D}) \,. \qquad (8.84)
\end{aligned}
$$

The matrix elements of the Coulomb interaction have already been discussed in Section 8.2.1. The electronic part contains a direct as well as an exchange contribution. Let us assume that the distance between the molecules in the complex exceeds one nanometer. Then, intermolecular wave function overlap is of less importance and we may follow the reasoning given in Section 8.2 and neglect the exchange contributions. Furthermore, we take the coupling matrix elements in the dipole–dipole interaction form, Eq. (8.23). Invoking the Condon approximation, i.e. assuming that $J_{DA}$ does not depend on the nuclear coordinates, we obtain

$$
\begin{aligned}
\langle \Psi_{DeM_D}, \Psi_{AgN_A} | V_{DA} | \Psi_{AeM_A}, \Psi_{DgN_D} \rangle &= J_{DA} \langle \chi_{DeM_D} | \chi_{DgN_D} \rangle \\
&\quad \langle \chi_{AgN_A} | \chi_{AeM_A} \rangle \,. \qquad (8.85)
\end{aligned}
$$

Taking all assumptions together we notice that we have recovered the model which had been derived in the context of electron transfer reactions in Section 6.4.2. There it was shown that in the case of two independent sets of vibrational coordinates the energy conservation for the transfer reaction contained in the $\delta$–function of the Golden Rule formula could be separated into two parts. This separation will also be applied in the present case.

From Fig. 8.1 it is clear that the process of excitation energy transfer can *formally* be viewed as the combined process of optical recombination at the donor and simultaneous optical absorption at the acceptor. The Förster approach is built upon this analogy. Hence the transfer rate shall be expressed in terms of the $S_1 \rightarrow S_0$ donor emission spectrum and the $S_0 \rightarrow S_1$ acceptor absorption spectrum. In order to obtain this appealing form of the transfer rate we rewrite the delta function in Eq. (8.84) as

$$
\begin{aligned}
\delta(E_{DeM_D} + E_{AgN_A} - E_{AeM_A} - E_{DgN_D}) &= \int dE\, \delta(E_{DeM_D} - E_{DgN_D} - E) \\
&\quad \times \delta(E + E_{AgN_A} - E_{AeM_A}) \,. \qquad (8.86)
\end{aligned}
$$

Here, the first delta function on the right–hand side accounts for the donor emission. The energy, $E = \hbar\omega$, which is set free in this process is used to excite the acceptor; it can be considered as the energy of the exciton.

The donor emission spectrum is given by (cf. Section 5.4)

$$
I_D(\omega) = \frac{4\hbar\omega^3}{3c^3} |\mathbf{d}_{eg}^{(D)}|^2 \sum_{M_D N_D} f(E_{DeM_D}) |\langle \chi_{DeM_D} | \chi_{DgN_D} \rangle|^2 \, \delta(E_{DeM_D} - E_{DgN_D} - \hbar\omega) \qquad (8.87)
$$

| donor | acceptor | $R_{\rm F}$ [nm] |
|---|---|---|
| Chl $a$ | Chl $a$ | 8–9 |
| Chl $b$ | Chl $a$ | 10 |
| $\beta$–carotene | Chl $a$ | 5 |

**Table 8.2:** Förster radii for typical biological donor–acceptor systems (data taken from [Gro85]).

and the acceptor absorption coefficient is (cf. Section 5.29)

$$\alpha_A(\omega) = \frac{4\pi^2\omega n_{\rm mol}}{3c}|\mathbf{d}_{eg}^{(A)}|^2 \sum_{M_A N_A} f(E_{AgM_A})|\langle\chi_{AgN_A}|\chi_{AeM_A}\rangle|^2 \times \delta(E_{AgN_A} - E_{AeM_A} + \hbar\omega). \tag{8.88}$$

Using Eqs. (8.86)–(8.88) we obtain the Förster formula which expresses the energy transfer rate in terms of the *spectral overlap* between the monomeric emission and absorption characteristics:

$$k_{DA} = \frac{9c^4\kappa_{DA}^2}{8\pi n_{\rm agg}|\mathbf{X}_{DA}^{(0)}|^6}\int_0^\infty \frac{d\omega}{\omega^4} I_D(\omega)\,\alpha_A(\omega)\,, \tag{8.89}$$

with the orientation factor $\kappa_{DA}$ given in Eq. (8.24).

The rate, Eq. (8.89), decreases like the sixth power with increasing distance between the monomers. The distance, $R_{\rm F}$, for which the transfer rate is equal to the radiative decay rate of the donor,

$$k_{DA}(R_{\rm F}) = \frac{1}{\tau_{\rm rad}^{(D)}} = \int_0^\infty d\omega I_D(\omega) \tag{8.90}$$

is called the *Förster radius*. In Table 8.2 we have listed Förster radii for some typical biological donor–acceptor systems. In terms of the Förster radius the transfer rate is

$$k_{DA} = \frac{1}{\tau_{\rm rad}^{(D)}}\left(\frac{R_{\rm F}}{|\mathbf{X}_{DA}^{(0)}|}\right)^6 . \tag{8.91}$$

The absolute value of the Förster rate is determined by the donor emission rate and the acceptor absorption coefficient.[8] The intuitive and experimentally accessible form of the transfer

[8]Although the idea of the combination of an optical emission and absorption process has been used to derive the Förster rate, the transfer does *not* involve the exchange of a photon. The interaction Eq. (8.8) or (8.10) is of pure Coulombic type. The term photon can only be used if the coupling between donor and acceptor molecule includes retarded (transverse) contributions of the radiation field (see also the discussion in the first part of Section 6.5.2 and the literature list).

rate has led to a wide use of Förster theory. It should be noted, however, that Eq. (8.89) is strictly valid only for homogeneously broadened spectra. Moreover, molecular systems where the dipole–dipole coupling is of the order or even larger than the homogeneous line width, cannot be described using the incoherent Förster approach which is based on the Pauli Master equation. This situation requires the solution of the density matrix equation taking into account coherent exciton dynamics. Before discussing this in Section 8.6 some variants of the excitation transfer processes discussed so far will be considered.

### 8.5.2 Nonequilibrium Quantum Statistical Description of Förster Transfer

It is of general interest to apply the technique introduced in Section 3.9 to derive rate equations for excitation energy transfer. If we aim at a nonequilibrium quantum statistical description of Förster transfer we are interested in transfer rates of the type $k_{m\to n}$. These are derived via a perturbation series with respect to the intermolecular coupling $J_{mn}$, whereby the exciton–vibrational coupling is treated exactly.

We can directly translate the treatment of Section 3.9 if we neglect the nuclear coordinate dependence of the dipole–dipole coupling in the single–exciton Hamiltonian, Eq. (8.73). The single–exciton aggregate Hamiltonian $H_{\rm agg}^{(1)}$ has to be split up according to Section 3.9 into a zero–order part $H_0$ and into a perturbation $\hat{V}$. To describe the Förster transfer the zero order Hamiltonian is given by

$$H_0 = \sum_m (T_{\rm nuc} + U_m(q))|m\rangle\langle m| \;, \tag{8.92}$$

and the inter–site coupling defines the perturbation

$$\hat{V} = \sum_{mn} J_{mn}|m\rangle\langle n| \;. \tag{8.93}$$

The PES $U_m(q)$ introduced in $H_0$ have already been given in Eq. (8.70); $T_{\rm nuc}$ is the kinetic energy operator of the vibrational degrees of freedom.

According to Section 3.9.2 we obtain a rate equation like Eq. (8.83) for the probability $P_m$, Eq. (8.82) that the $m$th molecule has been excited. The respective transition rate (of second order with respect to $J_{mn}$) can be deduced from the general result Eq. (3.391) as:

$$k_{m\to n} = \frac{2\pi}{\hbar}|J_{mn}|^2\mathcal{D}_{mn}(\omega_{mn}) \;. \tag{8.94}$$

Here, we introduced the combined density of states $\mathcal{D}_{mn}$ for a transition from molecule $m$ to molecule $n$ (cf. e.g. Eq. (6.90))

$$\mathcal{D}_{mn}(\omega_{mn}) = \frac{1}{2\pi\hbar}\int dt\; {\rm tr}_{\rm vib}\{\hat{R}_m\hat{U}_m^+(t)\hat{U}_n(t)\} \;, \tag{8.95}$$

The transition frequencies are defined via the minima of the PES introduced in Eq. (8.92), i.e. $\omega_{mn} = (U_m^{(0)} - U_n^{(0)})/\hbar$. Moreover, the $\hat{U}_m(t)$ denote the time–evolution operator describing nuclear motions at molecule $m$ (note the inclusion of the minima of the PES into $\hat{U}_m(t)$; $\hat{R}_m$ is the respective statistical operator).[9]

The Förster rate can be recovered after introducing two sets of independent exciton–vibrational states as in Section 8.5.1. But in the following we shortly present a computation where the (combined) density of states for the intramolecular transitions are used (cf. Section 5.2.3). As demonstrated in Section 8.3.1 the complete vibrational Hamiltonian can be decomposed into local vibrational Hamiltonian corresponding either to the ground or to the excited electronic level. Accordingly, the vibrational equilibrium statistical operator in Eq. (8.94) takes the following form $\hat{R}_m = \hat{R}_{me} \prod_{n \neq m} \hat{R}_{ng}$. In a similar manner the time–evolution operator is obtained as $\hat{U}_m(t) = \hat{U}_{me}(t) \prod_{n \neq m} \hat{U}_{ng}(t)$. As a result the trace expression in Eq. (8.95) reads

$$\begin{aligned} \mathrm{tr}_{\mathrm{vib}}\{\hat{R}_m \hat{U}_m^+(t) \hat{U}_n(t)\} &= \mathrm{tr}_m\{\hat{R}_{me} \hat{U}_{me}^+(t) \hat{U}_{mg}(t)\} \mathrm{tr}_n\{\hat{R}_{ng} \hat{U}_{ng}^+(t) \hat{U}_{ne}(t)\} \\ &\times \prod_{k \neq m,n} \mathrm{tr}_k\{\hat{R}_{kg} \hat{U}_{kg}^+(t) \hat{U}_{kg}(t)\} . \end{aligned} \tag{8.96}$$

Since the total set of vibrational DOF separate into sets of site–local modes the complete trace could be factorized into site–local traces, and the last product exclusively referring to the electronic ground state equals to one. To proceed we introduce the density of states for individual molecules into Eq. (8.96). The de–excitation of molecule $m$ can be characterized by $\mathcal{D}_m^{(\mathrm{em})}(\omega - \omega_{meg})$, Eq. (5.110), and the excitation of molecule $n$ is described by $\mathcal{D}_n^{(\mathrm{abs})}(\omega - \omega_{neg})$, Eq. (5.35) ($\omega_{meg}$ and $\omega_{neg}$ denote the local transition frequencies corresponding to the difference between the minima of the excited state and the ground state PES, cf. Eq. (5.32)). Then Eq. (8.95) becomes (note the specification of the transition frequency $\omega_{mn}$)

$$\mathcal{D}_{mn}(\omega_{meg} - \omega_{neg}) = \hbar \int d\omega \; \mathcal{D}_m^{(\mathrm{em})}(\omega - \omega_{meg}) \mathcal{D}_n^{(\mathrm{abs})}(\omega - \omega_{neg}) . \tag{8.97}$$

In order to illustrate this results let us focus on the case of harmonic intramolecular vibrations, and in particular on the high–temperature case where both densities of states can be described by the expression (cf. Eq. (5.99))

$$\mathcal{D}(\omega, S) = \frac{1}{\sqrt{2\pi k_{\mathrm{B}} T S}} \exp\left\{ - \frac{(\hbar\omega - S/2)^2}{2 k_{\mathrm{B}} T S} \right\} . \tag{8.98}$$

The Stokes shift $S$ was introduced in Eq. (5.43), here it has to be defined separately for molecule $m$ and $n$. Setting $\mathcal{D}_m^{(\mathrm{em})}(\omega - \omega_{meg}) = \mathcal{D}(-\omega + \omega_{meg}, S_m)$ as well as $\mathcal{D}_n^{(\mathrm{abs})}(\omega - \omega_{neg}) = \mathcal{D}(\omega - \omega_{neg}, S_m)$, and carrying out the integration in Eq. (8.97) gives $\mathcal{D}_{mn} = \mathcal{D}(\omega_{meg} - \omega_{neg}, S_m + S_n)$. The total Förster rate follows as

$$\begin{aligned} k_{m \to n} &= |\, J_{mn} \,|^2 \left( \frac{2\pi}{\hbar^2 k_{\mathrm{B}} T (S_m + S_n)} \right)^{1/2} \\ &\times \exp\left\{ - \frac{\left(\hbar(\omega_{meg} - \omega_{neg}) + (S_m + S_n)/2\right)^2}{2 k_{\mathrm{B}} T (S_m + S_n)} \right\} . \end{aligned} \tag{8.99}$$

In Fig. 8.9 we plotted $k_{m \to n}$ in dependence on the energetic detuning between the electronic transition frequencies, $\omega_m - \omega_n$, for different temperatures.

[9]To avoid confusion with the PES we write $\hat{U}_m(t)$ instead of $U_m(t)$.

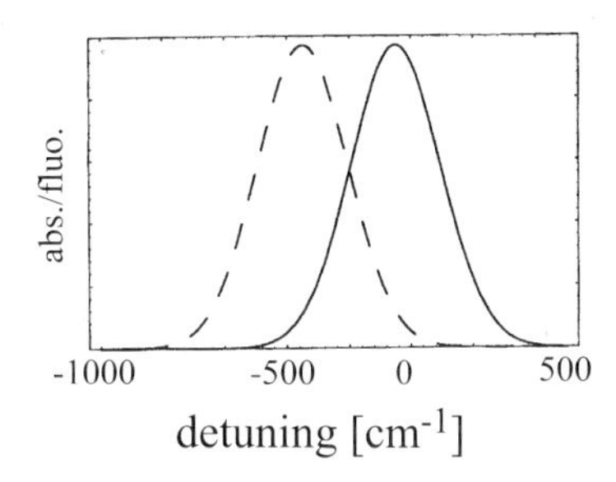


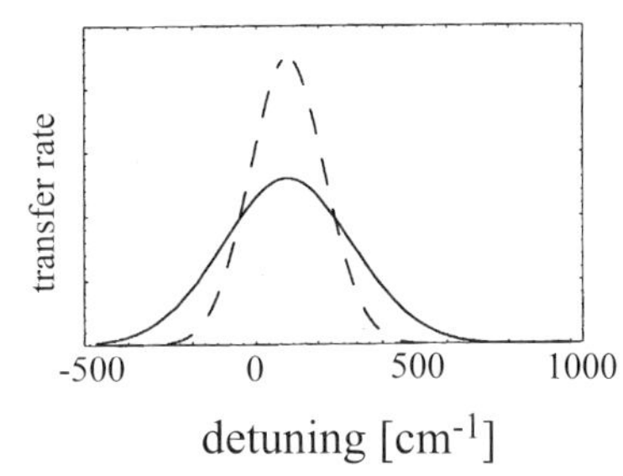


**Figure 8.9:** Left panel: fluorescence spectrum of molecule $m$ (dotted) and absorption spectrum of molecule $n$ (solid); the overlap between both determines the Förster transfer rate according to Eq. (8.99) ($\hbar(\omega_m - \omega_n) = 400\text{cm}^{-1}$, T=300 K, $S_m = S_n = 100\text{cm}^{-1}$). Right panel: $k_{m\to n}$ in dependence on $\hbar(\omega_m - \omega_n)$ for T=300 K (solid) and T=100 K (dotted).

### 8.5.3 Energy Transfer Between Delocalized States

As demonstrated in the previous section Förster theory is based on the assumption of the motion of excitations which are localized at a single molecule of the aggregate. In the following we shortly comment on the calculation of transition rates if such a localization on a single molecule does not exist and the electronic excitation has to be described by a delocalized exciton (cf. Fig. 8.6). First, we compute the rate for transitions between different exciton levels of an aggregate. Afterwards, the formulation is generalized to the case of excitation energy transfer between two aggregates or molecular complexes where the internal Coulomb coupling is strong but the coupling between both aggregates is weak.

An appropriate model would be the adiabatic exciton Hamiltonian, Eq. (8.61). Here one has to carry out an expansion of the PES up to the second order with respect to the deviation from the equilibrium position. Then the introduction of normal mode vibrations leads to excitonic potential energy surfaces as introduced in Eq. (8.79). Nonadiabatic couplings $\Theta_{\alpha\beta}$ will appear which induce transitions between the different PES. Alternatively, one could start with $H_{\text{agg}}^{(1)}$, Eq. (8.80), where the off–diagonal part of the exciton–vibrational coupling, Eq. (8.77), would be responsible for such transitions. In any case one obtains the lowest–order transition rate between different exciton levels as

$$k_{\alpha\to\beta} = \frac{2\pi}{\hbar} \sum_{M,N} f(\mathcal{E}_{\alpha M}) \mid \langle\chi_{\alpha M}|\Theta_{\alpha\beta}|\chi_{\beta N}\rangle \mid^2 \delta(\mathcal{E}_{\alpha M} - \mathcal{E}_{\beta N}) \;, \tag{8.100}$$

where the excitonic energies include the set of vibrational quantum numbers $M$ or $N$. One should emphasize that this rate expression Eq. (8.100) can be understood as a rate for an internal conversion process between delocalized excitonic aggregate states. In Section 8.6.2 a similar transition rate between delocalized states will be discussed (however in the framework of a perturbational description of the exciton–vibrational coupling). Therefore, we do not further comment on Eq. (8.100).

Another situation is realized in the case of two weakly coupled aggregates which have, however, strong internal Coulomb couplings. For the derivation of the transfer rate one can closely follow the argument of Section 8.5.1. However, the initial and the final state of the transitions are not those of a single molecule, but of the whole aggregate. To see how this modifies the rate let us first consider the coupling matrix (cf. Eq. (8.15)). We neglect as in Eq. (8.85) the dependence of the electronic matrix elements on the vibrational coordinates and discuss the transition from the exciton levels $|D\alpha\rangle$ of the donor aggregate to the exciton levels $|A\beta\rangle$ of the acceptor aggregate. If we expand the exciton states according to Eq. (8.41) we obtain

$$\begin{aligned} J_{D\alpha,A\beta} &= \langle D\alpha, A0|V_{DA}^{(\text{el-el})}|A\beta, D0\rangle \\ &= \sum_{m,n} c^*_{D\alpha}(m)\langle Dm, A0|V_{mn}^{(\text{el-el})}|An, D0\rangle c_{A\beta}(n) \\ &= \sum_{m,n} c^*_{D\alpha}(m)c_{A\beta}(n)J_{Dm,An} \,. \end{aligned} \tag{8.101}$$

Invoking the dipole–dipole approximation $J_{Dm,An}$ describes the dipole–dipole coupling between a molecule $m$ of the donor aggregate and a molecule $n$ of the acceptor aggregate. Since both expansion coefficients are proportional to $1/\sqrt{N_{\text{mol}}}$, where $N_{\text{mol}}$ is the number of molecules in every aggregate, a rough estimate of the coupling matrix elements gives $N_{\text{mol}}\bar{J}$. Here, $\bar{J}$ is a representative mean value of the various $J_{mn}$. Hence we have to expect a certain enhancement of the excitation energy transfer rate in relation to the ordinary Förster transfer, if the transitions take place between delocalized states of two separated aggregates (provided that the spectral overlap does not become to small).

The transition rate can be written as

$$\begin{aligned} k_{D\alpha\to A\beta} &= \frac{2\pi}{\hbar} \mid J_{D\alpha,A\beta} \mid^2 \\ &\times \sum_{M,N} f(\mathcal{E}_{D\alpha M}) \mid \langle \chi_{D\alpha M}|\chi_{A\beta N}\rangle \mid^2 \delta(\mathcal{E}_{D\alpha M} - \mathcal{E}_{A\beta N}) \,. \end{aligned} \tag{8.102}$$

Obviously, a similar expression like Eq. (8.89) for the total rate $k_{DA} = \sum_{\alpha,\beta} k_{D\alpha\to A\beta}$ can be derived, where the emission and absorption spectrum now belongs to the whole donor and acceptor aggregate, respectively. In principle one can also go beyond the dipole–dipole coupling and take all multipole contributions into account. If one in addition describes the donor emission and acceptor absorption without invoking the dipole approximation the electronic contribution the rate becomes proportional to the square of the exact Coulomb interaction matrix element, Eq. (8.15), (see, also [Dam99,Scho91]).

This type of XT mechanism has been extensively discussed in relation to the excitation energy transfer in the photosynthetic light harvesting complex shown in Fig. 8.4. Inspecting this figure it becomes obvious that there are chromophores which are closely packed (the B850 bacteriochlorophyll molecules) and chromophores where this is not the case (the carotenoids and the B800 bacteriochlorophylls). Table 8.3 presents the Coulombic interaction energy $J_{mn}$ (without invoking the dipole approximation) and the transfer rate according to the generalization, Eq. (8.102), of the Förster rate for different chromophores and exciton levels. Apparently, the effect of an enhanced electronic coupling on the transfer rate may be reduced if the spectral overlap part remains small.

| Type of Transfer | $\vert J_{DA}\vert$ in eV | $k_{D\to A}$ in 1/s |
|---|---|---|
| B800 $\to$ B850 | $1.6 \cdot 10^{-2}$ | $2.9 \cdot 10^{12}$ |
| B800 $\to$ $\mathcal{E}_1$ | $2.7 \cdot 10^{-3}$ | $8.2 \cdot 10^{9}$ |
| B800 $\to$ $\mathcal{E}_{2,3}$ | $8.7 \cdot 10^{-3}$ | $8.6 \cdot 10^{11}$ |
| B800 $\to$ $\mathcal{E}_{4,5}$ | $1.0 \cdot 10^{-2}$ | $8.8 \cdot 10^{12}$ |
| B800 $\to$ B850-exciton | | $1.4 \cdot 10^{13}$ |
| Car $\to$ B850 | $5.4 \cdot 10^{-3}$ $(1.4 \cdot 10^{-10})$ | $4.4 \cdot 10^{11}$ |
| Car $\to$ $\mathcal{E}_1$ | $1.8 \cdot 10^{-2}$ | $2.3 \cdot 10^{7}$ |
| Car $\to$ $\mathcal{E}_9$ | $6.4 \cdot 10^{-3}$ | $3.4 \cdot 10^{9}$ |

**Table 8.3:** Coulomb interaction and excitation energy transfer rates for a biological antenna similar to that shown in Fig. 8.4 (purple bacterium *Rs. molischianum* with 24 instead of 27 chromophores). B800 labels one bacteriochlorophyll molecule from the lower ring and B850 one of the upper ring (both in the lowest excited so–called $Q_y$–state ) whereas Car stands for the second excited state of the carotenoid. $\mathcal{E}_{1,2,..}$ are the various exciton states of the B850 ring, and "B850–exciton" labels the contribution from all exciton levels of the B850 ring. The couplings and transition rates between single molecules are related to those which are positioned close together. Calculations are based on Eq. (8.10) for the Coulomb interaction (without invoking the dipole–dipole coupling approximation) and Eqs. (8.84) and Eq. (8.102) for the transfer rate (the number in the bracket gives the related exchange coupling, data taken from [Dam99]).

## 8.6 Transfer Dynamics in the Case of Weak Exciton–Vibrational Coupling

The Förster theory of XT as discussed in the previous section corresponds to the case of fast intramolecular relaxation as compared to the transfer time. It has to be considered as the incoherent transport of excitation energy which is localized at single molecules. In the present section we will discuss the case where the exciton–vibrational coupling is weak compared to the dipole–dipole interaction. The appropriate theoretical tool for discussing this limit is density matrix theory introduced in Section (3.5.6). We identify the electronic excitations (limited to the singly excited electronic states of the aggregate) with the *relevant* system in the sense of Chapter 3. The vibrational DOF of the aggregate are considered as the reservoir (heat bath), responsible for electronic energy dissipation and dephasing of the coherent exciton motion. The adequate description of the exciton dynamics is given by the Quantum Master

Equation approach of Section 3.7.1.

In principle there are two possibilities for the representation of the single–exciton reduced density matrix (cf. Section 3.8.7) which follows from the reduced statistical operator, $\hat{\rho}(t) = \text{tr}_{\text{vib}}\{\hat{W}(t)\}$. First, one may choose a site representation in terms of the localized basis set $|m\rangle$ defined in Eq. (8.34). This gives the reduced density matrix

$$\rho_{mn}(t) = \langle m|\hat{\rho}(t)|n\rangle \ . \tag{8.103}$$

Since the Förster transfer is realized as a hopping process between different sites it is reasonable to assume that a site representation is well–suited for establishing the link between Förster theory and the density matrix approach. However, we have repeatedly pointed out that a non–eigenstate representation like Eq. (8.103) requires some caution when calculating relaxation and dephasing rates (cf. Section 3.8.7). In order to simplify the discussion of the site representation we will assume that there is a considerable detuning between different monomer transition energies, $|\hbar\omega_{mn}| > |J_{mn}|$, in order to justify the neglect of the dipole–dipole interaction in the calculation of the relaxation matrix (cf. Eq. (8.107) below).

The alternative to this treatment is provided by an eigenstate (energy) representation of the single–exciton reduced density matrix:

$$\rho_{\alpha\beta}(t) = \langle\alpha|\hat{\rho}(t)|\beta\rangle \ . \tag{8.104}$$

Here the equations of motion are based on the exciton Hamiltonian introduced in Section 8.3 (Eq. (8.73)). Of course, it is straightforward to calculate from the density matrix in the representation of delocalized exciton states, for example, the probability $P_m(t)$ the molecule $m$ is excited at time $t$. Using the expansion coefficients $c_\alpha(m)$ introduced in Eq. (8.41) it follows that

$$P_m(t) = \sum_{\alpha,\beta} c_\alpha(m)c^*_\beta(m)\rho_{\alpha\beta}(t) \ . \tag{8.105}$$

There is a further aspect of the present density matrix approach on XT dynamics which particularly favors it for the study of optical properties. This statement is related to the fact that the definition Eq. (8.103) or Eq. (8.104) can be easily extended to include the ground state $|0\rangle$ as well as the states $|mn\rangle$, Eq. (8.35), with two excitations in the aggregate (or the two–exciton state, Eq. (8.45)). Accordingly, one may compute off–diagonal density matrix elements like $\langle\alpha|\hat{\rho}(t)|0\rangle$. The latter is proportional to the polarization between the ground state and the single exciton state $|\alpha\rangle$ and allows to directly include the coupling to the radiation field (written like in Eq. (5.9)) into the density matrix equation. Then, as described in Section 5.6, there is no need to introduce any linear or nonlinear response function characterizing the exciton system. Instead the density matrix obtained at a finite strength of the radiation field offers a direct access to nonlinear spectra (cf., e.g., [Ren01]). However, we will not further touch this issue and refer to the application of this concept in Section 8.7.3.

## 8.6.1 Site Representation

To study exciton motion in space we choose the local (site) representation given by the single–exciton states $|m\rangle$ of Eq. (8.34); the density matrix is defined by Eq. (8.103). In order to write down the equations of motion for the single–exciton reduced density matrix, we start by identifying the different contributions to the Hamiltonian defined in Eq. (8.73). As already mentioned the basic assumption for the following is that the elements of the exciton–vibrational coupling matrix $g_{mn}(\xi)$ are weak enough to justify the second–order perturbation expansion which leads to the relaxation matrix (Eq. (3.284)). Accordingly, $H_{\rm ex}$ can be identified with the system Hamiltonian $H_{\rm S}$ as used in Chapter 3 to characterize the active (relevant) system. The reservoir Hamiltonian $H_{\rm R}$ is given by the vibrational Hamiltonian $H_{\rm vib}$, Eq. (8.65). Finally, the system–reservoir coupling $H_{\rm S-R}$ corresponds to the exciton–vibrational coupling $H_{\rm ex-vib}$ (Eq. (8.76)). To use the general formulas of Section 3.8.2 we have to rewrite $H_{\rm ex-vib}$ in the form of Eq. (3.146). In the present case the summation in Eq. (3.146) is carried out with respect to the index $u = (m, n)$ combining the two possible site indices. Therefore, the system part of the interaction Hamiltonian is $K_u = |m\rangle\langle n|$ and the bath part is $\Phi_u = \sum_\xi \hbar\omega_\xi g_{mn}(\xi)(C_\xi^\dagger + C_\xi)$.

This identification enables us to write down the reservoir correlation function Eq. (3.186) as

$$\begin{aligned} C_{kl,mn}(t) &= \sum_\xi \omega_\xi^2 g_{kl}(\xi) g_{mn}(\xi)\, \mathrm{tr}_{\rm vib}\{\hat{R}_{\rm eq}(C_\xi^\dagger(t) + C_\xi(t))\,(C_\xi^\dagger + C_\xi)\} \\ &= \sum_\xi \omega_\xi^2 g_{kl}(\xi) g_{mn}(\xi) \left[(1 + n(\omega_\xi))\, e^{-i\omega_\xi t} + n(\omega_\xi) e^{i\omega_\xi t}\right] . \end{aligned} \tag{8.106}$$

(The equilibrium correlation function of the dimensionless normal mode coordinates has been calculated before in Section (3.6.2).) The damping matrix, Eq. (3.281), entering the expression (3.284) for the relaxation matrix (Redfield tensor) can now be written as

$$\Gamma_{kl,mn}(\omega) = \pi\omega^2 \,(1 + n(\omega))\,(J_{kl,mn}(\omega) - J_{kl,mn}(-\omega)) . \tag{8.107}$$

Here we introduced the spectral density of the vibrational modes

$$J_{kl,mn}(\omega) = \sum_\xi g_{kl}(\xi) g_{mn}(\xi)\, \delta(\omega - \omega_\xi) . \tag{8.108}$$

In contrast to Section 3.6.4 it contains *correlations* between four molecular sites.

Let us discuss the equations of motion for the single–exciton reduced density matrix in view of the present model for exciton–vibrational coupling. As outlined in Section 3.8.2 the relaxation matrix $\Gamma_{kl,mn}(\omega)$ contains population relaxation and dephasing rates. Population relaxation corresponds to energy relaxation accompanied by the transfer of excitons from one molecule to another. The respective (Golden Rule type ) rate for a transition from state $|m\rangle$ to state $|n\rangle$ reads

$$\begin{aligned} k_{m\to n} &= 2\Gamma_{mn,nm}(\omega_{mn}) \\ &= 2\pi\omega_{mn}^2 (1 + n(\omega_{mn}))\,(J_{mn,nm}(\omega_{mn}) - J_{mn,nm}(\omega_{nm})) . \end{aligned} \tag{8.109}$$

Obviously, $k_{m\to n}$ is only different from zero, if the coupling to the vibrational modes is *off–diagonal* in the site index $m$. This requires that $g_{mn}(\xi) \neq 0$ for $m \neq n$. Since the off–diagonal elements of $g_{mn}(\xi)$ derive from the modulation of the dipole–dipole interaction (cf. Eq. (8.71)), this is in accordance with the mentioned transfer character of energy relaxation.

The energy relaxation rates can now be used to express the dephasing rate of the exciton state $|m\rangle$ (cf. Eq. (3.291)) as

$$\gamma_m = \sum_n \Gamma_{mn,nm}(\omega_{mn}) = \frac{1}{2}\sum_n k_{m\to n} \,. \tag{8.110}$$

In the following we will assume the validity of the Bloch model and neglect all elements of the relaxation matrix which cannot be written in terms of energy relaxation and dephasing rates (cf. Section 3.8.3). This decouples the equation of motion for the population and coherence type density matrix elements and we obtain

$$\begin{aligned}\frac{\partial}{\partial t}\rho_{mn} &= -i\omega_{mn}\rho_{mn} - \frac{i}{\hbar}\sum_l (J_{ml}\rho_{ln} - J_{ln}\rho_{ml}) \\ &\quad -\delta_{mn}\sum_l \left(k_{m\to l}\rho_{mm} - k_{l\to m}\rho_{ll}\right) - (1-\delta_{mn})(\gamma_m + \gamma_n)\rho_{mn} \,.\end{aligned} \tag{8.111}$$

First, let us discuss the coherent contribution to the right–hand side which derives from the matrix elements of $H_{\text{ex}}$ (first line). Apparently, the motion of a single–exciton in the aggregate is enforced by the dipole–dipole coupling. This means that an exciton initially localized at a single molecule or at a small number of molecules, will move through the aggregate like a wave packet. This motion is reversible and results from a nonperturbative consideration of the dipole–dipole coupling.

The dissipative part on the right–hand side of Eq. (8.111) (second line) is responsible for irreversibility. In particular we have energy relaxation which affects the occupation probabilities, $\rho_{mm}(t) = P_m(t)$, and dephasing of the single–exciton coherences described by $\rho_{mn}(t)$. It has already been discussed in Section 3.8.7 that for a definition of the relaxation rates in the site representation, the asymptotic state is given by a thermal distribution over the localized molecular states according to

$$\rho_{mn}(\infty) = \delta_{mn}\frac{e^{-E_m/k_{\text{B}}T}}{\sum_{m'} e^{-E_{m'}/k_{\text{B}}T}} \,. \tag{8.112}$$

For an aggregate having identical monomer transition energies, such as the regular chain discussed in Section 8.2.3, the occupation probabilities at thermal equilibrium should be equal for all monomers. However, an initial preparation of an exciton at a particular molecule will result in a coherent motion over the whole aggregate since $\omega_{mn} = 0$ leads to vanishing energy relaxation rates (cf. Eq. (8.109)). This contradiction is due to the restriction to a *linear* exciton–vibrational interaction. In particular the incorporation of pure dephasing contributions (cf. Eq (3.292)) would result in a proper equilibration. Nevertheless, the theory in its present form is appropriate for situations where some irregularity of the monomeric $S_0 \to S_1$ transitions is present. But, this is already required to justify the neglect of the dipole–dipole interaction when calculating the relaxation rates.

In order to illustrate the dynamics according to Eq. (8.111), we consider a linear chain model as shown in Fig. 8.10. For simplicity we restrict ourselves to situations where a factorization of the exciton–vibrational coupling matrix according to $g_{mn}(\xi) = g_{mn}g(\xi)$ is justified

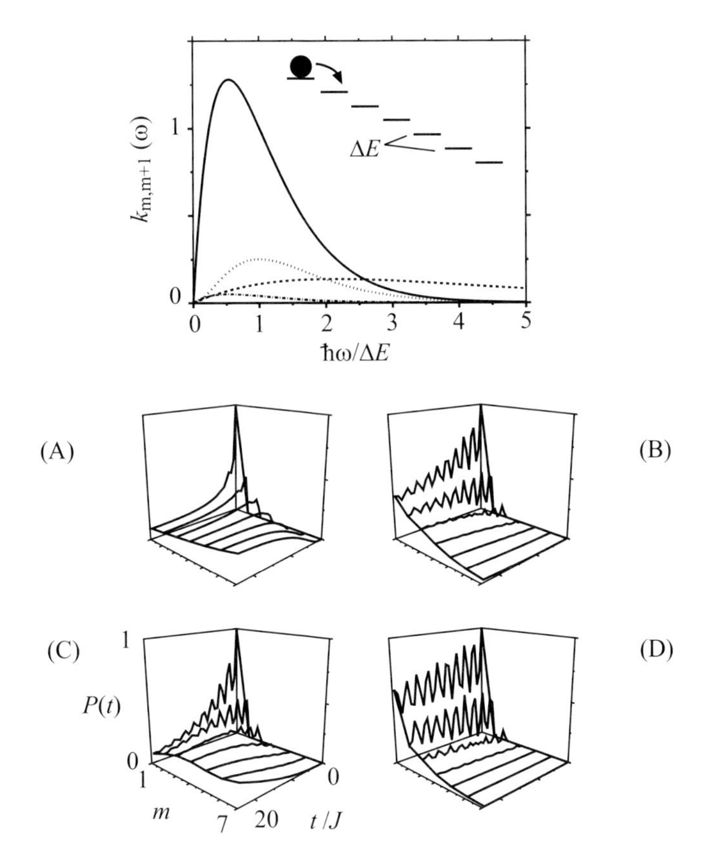


**Figure 8.10:** Dissipative exciton dynamics according to Eq. (8.111) for a linear aggregate shown in the inset of the upper panel ($P_m(t=0) = \delta_{m0}, \Delta E = 2J_{mn} = 2\delta_{n,m\pm1}J$). The upper panel shows the rates (8.114) corresponding to the different situations in the lower panel. The basic parameters are: solid line (A) – $g_{mn} = 0.5J_{mn}, T = 300$ K, $\hbar\omega_c = \Delta E/2$. They are then modified as follows: dashed line (B) – $\hbar\omega_c = 1.7\Delta E$, dotted line (C) – $T = 4$ K, and dash–dot line (D) – $g_{mn} = 0.5J_{mn}$.

(cf. Section 3.6.4). In this case the dissipative influence of the aggregates's vibrational modes can be described by a single, exciton state independent, spectral density

$$J(\omega) = \sum_{\xi} g^2(\xi)\delta(\omega - \omega_\xi) . \tag{8.113}$$

and the energy relaxation rates become

$$k_{m\to n} = 2\pi|g_{mn}|^2\omega_{mn}^2\left[1 + n(\omega_{mn})\right]\left[J(\omega_{mn}) - J(\omega_{nm})\right] . \tag{8.114}$$

It has already been pointed out in Section 3.6.4 that the microscopic calculation of spectral densities is a rather complicated task for realistic systems. Quite often one resorts to model functions which, for instance, reproduce experimental spectra reasonably well. In our example we used the type $j(\omega) = \Theta(\omega)\exp\{-\omega/\omega_c\}/2\omega_c^3$. Here $\omega_c$ is a cut–off frequency (cf. discussion in Section 3.6.4).

In Fig. 8.10 we have plotted the population dynamics for a linear aggregate consisting of seven molecules. The behavior of $P_m(t)$ apparently reflects the magnitude of the rate $k_{m\to n}$ at the relevant transition frequency ($\hbar\omega = \Delta E$, see inset of upper panel). The resulting exciton motion covers different dynamic regimes ranging from the almost incoherent motion (A), and the partly coherent motion (B,C), to the almost coherent motion (D).

In order to examine in which manner the present density matrix theory includes hopping–like Förster transfer as a limiting case, we will derive the respective hopping transfer rate $k_{m\to n}$. We notice that Förster theory implies a weak Coulomb interaction and the transfer dynamics can be categorized as being in the *nonadiabatic* limit according to the terminology introduced for electron transfer in Section 6.3. Stressing the similarity to the case of electron transfer we can adopt the results of Section 6.8.2, where the nonadiabatic electron transfer rate has been derived from density matrix theory (Eq. (6.211)). In the present case the nonadiabatic rate for exciton (hopping) transfer is given by

$$k_{m\to n} = \frac{2\pi}{\hbar^2}|J_{mn}|^2 \frac{\gamma_m + \gamma_n}{\pi\left(\omega_{mn}^2 + (\gamma_m + \gamma_n)^2\right)} . \tag{8.115}$$

This formula contains a broadening of the transition energy $\omega_{mn}$ which is of Lorentzian type. Any vibrational substructure which is accounted for in the overlap integral of Förster's rate formula Eq. (8.89) is absent. This is a consequence of the treatment of vibrational modes within lowest order of perturbation theory.

### 8.6.2 Energy Representation

In this section we extend the foregoing discussion by including the effect of the dipole–dipole interaction into the definition of the relaxation rates. Therefore, the results presented below will be valid for strongly coupled aggregates with and without intrinsic energetic disorder. By construction the stationary limit of the equations of motion for the single–exciton reduced density matrix will be given by

$$\rho_{\alpha\beta}(\infty) = \delta_{\alpha\beta}\frac{e^{-E_\alpha/k_B T}}{\sum_{\alpha'} e^{-E_{\alpha'}/k_B T}} , \tag{8.116}$$

instead of Eq. (8.112) which was obtained in the site representation. (The reduced density matrix was defined in Eq. (8.104).)

In order to describe aggregates having appreciable dipole–dipole interaction energies, which is reflected in a rather large shift of the absorption band upon aggregation (cf. Fig. 8.2), we formulate the equations of motion for the density matrix in the eigenstate representation, Eq. (8.104). The derivation proceeds along the same lines as for the site representation.

First, we take $H_{ex}$ from Eq. (8.43) as the system Hamiltonian $H_S$. The reservoir Hamiltonian $H_R$ again is given by the vibrational Hamiltonian $H_{vib}$, Eq. (8.65). Finally, the system–reservoir coupling $H_{S-R}$ is defined by the exciton–vibrational coupling, Eq. (8.77). Since we have performed a simple basis set transformation only, the reservoir correlation function

Eq. (3.186) (and thus the relaxation matrix) has the same structure as in the previous section:

$$C_{\alpha\beta,\gamma\delta}(t) = \sum_{\xi} \omega_{\xi}^2 g_{\alpha\beta}(\xi) g_{\gamma\delta}(\xi) \left[[1+n(\omega_\xi)]e^{-i\omega_\xi t} + n(\omega_\xi)e^{i\omega_\xi t}\right] . \tag{8.117}$$

Following Chapter 3 and Section 8.6.1 we obtain for the reduced density matrix $\rho_{\alpha\beta}$ the following equations of motion in the Bloch approximation

$$\begin{aligned} \frac{\partial}{\partial t}\rho_{\alpha\beta} &= -i\omega_{\alpha\beta}\rho_{\alpha\beta} \\ &\quad -\delta_{\alpha\beta}\sum_{\kappa}(k_{\alpha\to\kappa}\rho_{\alpha\alpha} - k_{\kappa\to\alpha}\rho_{\kappa\kappa}) - (1-\delta_{\alpha\beta})(\gamma_\alpha + \gamma_\beta)\rho_{\alpha\beta} . \end{aligned} \tag{8.118}$$

Since the basis $|\alpha\rangle$ diagonalizes the single–exciton Hamiltonian $H_{\rm ex}$, the coherent part on the right–hand side contains only the transition frequencies between exciton eigenstates, $\omega_{\alpha\beta} = (\mathcal{E}_\alpha - \mathcal{E}_\beta)/\hbar$. The transition rates are similar to Eq. (8.109), but now formulated in the delocalized exciton states

$$k_{\alpha\to\beta} = 2\Gamma_{\alpha\beta,\beta\alpha}(\omega_{\alpha\beta}) = 2\pi\omega_{\alpha\beta}^2\big(1+n(\omega_{\alpha\beta})\big)\big(J_{\alpha\beta,\beta\alpha}(\omega_{\alpha\beta}) - J_{\alpha\beta,\beta\alpha}(\omega_{\beta\alpha})\big) . \tag{8.119}$$

They define the dephasing rates as in Eq. (8.110):

$$\gamma_\alpha = \frac{1}{2}\sum_{\beta} k_{\alpha\to\beta} . \tag{8.120}$$

Assuming again that the exciton–vibrational coupling matrix factorizes, $g_{\alpha\beta}(\xi) = g_{\alpha\beta}g(\xi)$, the relaxation rates are given by

$$k_{\alpha\to\beta} = 2\Gamma_{\alpha\beta\beta\alpha}(\omega_{\alpha\beta}) = 2\pi|g_{\alpha\beta}|^2\omega_{\alpha\beta}^2\left[1+n(\omega_{\alpha\beta})\right]\left[J(\omega_{\alpha\beta}) - J(\omega_{\beta\alpha})\right] . \tag{8.121}$$

Even though Eqs. (8.118)–(8.120) look formally identical with the respective equations in the site representation (Eqs. (8.109)–(8.111)), there is a fundamental difference since the dipole–dipole interaction has been fully accounted for in the determination of the relaxation rates in Eq. (8.121). This can be seen from the argument of the spectral density and the distribution function (see related discussion in Section 3.8.7). It guarantees that the energy relaxation rates (8.121) fulfill detailed balance with respect to the exciton eigenstates $k_{\alpha\to\beta}/k_{\beta\to\alpha} = \exp\{-\hbar\omega_{\alpha\beta}/k_{\rm B}T\}$. Accordingly, the system will relax to the correct equilibrium distribution (Eq. (8.116)) after initial preparation in a nonequilibrium state. We note in passing that Eq. (8.118) also highlights that applying the limit of the Bloch model in the site and the eigenstate representation has a different physical meaning in terms of the exciton dynamics.

Finally, it should be mentioned that the assumption of a linear exciton–vibrational coupling which neglects pure dephasing, for instance, leads to a relaxation matrix according to which coherences between degenerate eigenstates ($\omega_{\alpha\beta} = 0$) are not subject to dephasing processes. The equations of motion (8.118) might therefore not be appropriate for the description of highly symmetric aggregates having degenerate exciton eigenenergies such as regular molecular rings. In realistic systems, however, static distributions of monomer transition energies

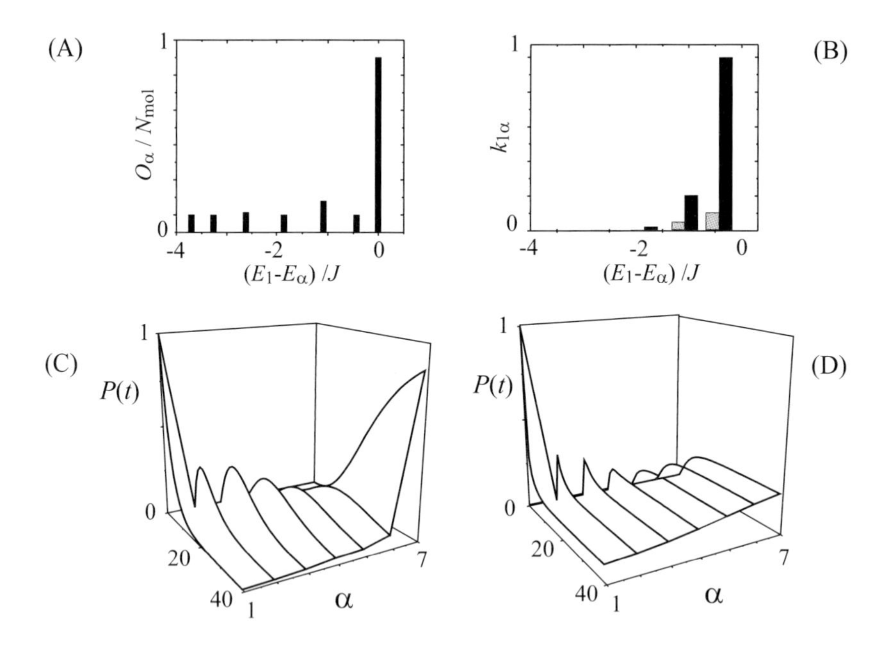


**Figure 8.11:** Dissipative dynamics in a regular chain of seven molecules: (a) transition amplitudes at the corresponding single–exciton eigenenergies, (b) relaxation matrix $w_{1\alpha}$ for $g_{mn} = 0.77J$ and $\hbar\omega_c = 0.25J$ (black T=300 K, grey T=4 K). Panels (c) and (d): $\rho_{\alpha\alpha}(t) = P_\alpha(t)$ for T=300 K (c) and T=4 K (d).

and dipole–dipole interactions are likely to remove any degeneracy thus justifying the use of Eq. (8.118).

To illustrate the dynamics in the eigenstate representation we show in Fig. 8.11 the numerical solution of Eq. (8.118) for a regular chain with seven sites, using the same model spectral density as in the example of Fig. 8.10. First, we plotted in panel (a) the transition amplitudes defined in Eq. (8.127). Being in H–aggregate configuration, the energetically highest exciton state has by far the largest transition amplitude. This allows us to assume that an external field can prepare the system in this particular state. With the highest state being initially excited with probability one, the subsequent dynamics shows no oscillations but a relaxation towards the equilibrium distribution (8.116). The latter will be different for different temperatures (panels (c) and (d)). The relaxation proceeds via emission and absorption of single vibrational quanta. The relaxation rates relevant for the initial excited state are shown in panel (b).

As a final remark we like to point out that in analogy to the creation of vibrational wave packets which was discussed in Section 5.5, a spatially localized *excitonic wave packet* can in principle be prepared by an ultrafast optical excitation. This requires the pulse to be spectrally broad enough to excite a coherent superposition of exciton states $|\alpha\rangle$. However, unless the exciton energies are equidistant the wave packet is likely to become rapidly delocalized due to wave packet dispersion.

## 8.7 The Aggregate Absorption Coefficient

Having considered dissipative exciton dynamics we now discuss the stationary linear absorption (cf. Fig. 8.12) coefficient of the aggregate and its relation to the energy spectrum of the single–exciton Hamiltonian, Eq.(8.74). Following the discussions given in Chapter 5 and in particular in Section 5.2.2, we expect an expression for the absorption coefficient which is similar to Eq. (5.29), that is, there should be an absorption line for every possible transition from the ground to some excited exciton–vibrational state. To be more precise, the initial state is defined by the Hamiltonian $H_0$, Eq. (8.65). The final states of the transition are obtained from eigenstates of the single–exciton Hamiltonian Eq. (8.73). Unlike for the initial states, a general expression for the final states cannot be derived in a simple way. This is due to the fact that one would have to account for arbitrarily strong exciton–vibrational and Coulomb coupling at the same time. If there are many vibrational modes this is practically impossible.

Therefore, we concentrate on a situation similar to that of Section 8.6 where the exciton–vibrational coupling is sufficiently small compared to the Coulomb coupling, $\sum_\xi \hbar\omega_\xi g_{mn}^2(\xi) \ll J_{mn}$. If we neglect any contribution of the exciton–vibrational coupling to the absorption spectrum we obtain the absorption as sharp lines at the different exciton energies $\mathcal{E}_\alpha$. This will provide a reference for further discussions. Including the exciton–vibrational coupling a broadening of the sharp lines will occur as already discussed in Section 5.3.4. We start by giving a short derivation of the absorption coefficient $\alpha(\omega)$ for an ensemble of aggregates using the general expression Eq. (5.72) which defines $\alpha(\omega)$ via the dipole–dipole correlation function $C_{\rm d-d}(t)$, Eq. (5.70).

In the present case the dipole operator comprises the contributions $\hat{\mu}_m$, Eq. (8.21) of all molecules in the aggregate according to $\hat{\mu} = \sum_m \hat{\mu}_m$. This expression is valid for optical transitions into all excited states as well as for transitions between different excited states. For the present purposes, however, it is sufficient to restrict the model to transitions into the single exciton state (corresponding to a $S_0$–$S_1$ transition). Therefore, the aggregate dipole operator which is used in the following reads

$$\hat{\mu} = \sum_m \hat{\mu}_m \equiv \sum_m \mathbf{d}_m |m\rangle\langle 0| + \text{h.c.} \,, \tag{8.122}$$

where $\mathbf{d}_m$ is the transition matrix element of the two–level model of Section 8.2.1. Further, we note that when computing $\alpha(\omega)$ the volume density of molecules appearing in Eq. (5.72) has to be replaced by that of the aggregates $n_{\rm agg}$. Let us first study the coherent case where exciton–vibrational coupling is neglected.

### 8.7.1 Absence of Exciton–Vibrational Coupling

We start by introducing the exciton representation of the dipole operator as

$$\hat{\mu} = \sum_\alpha \mathbf{d}_\alpha |\alpha\rangle\langle 0| + \text{h.c.} \,, \tag{8.123}$$

with the transition matrix elements given by

$$\mathbf{d}_\alpha = \langle \alpha | \sum_m \hat{\mu}_m | 0 \rangle = \sum_m c_\alpha^*(m) \mathbf{d}_m \,. \tag{8.124}$$

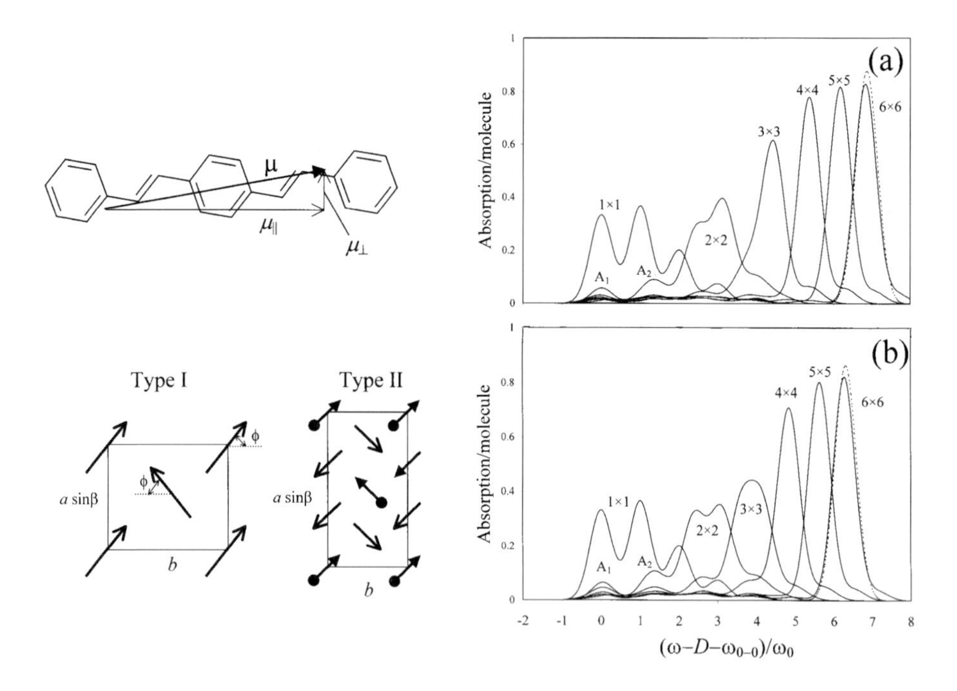


**Figure 8.12:** Absorption spectra of $p$–distyrylbenzene nanoaggregates. Upper–left part: single molecule with the transition dipole moment shown. Lower–left part: two types of nanoaggregates studied (arrows represent the transition dipole moments projected along the short molecular axis, the configurations correspond to known crystal structures of related molecules). Right panel: change of the absorption spectra with aggregate size (part a: aggregate type I, part b: aggregate type II). (reprinted with permission from [Spa02]; copyright (2002) American Institute of Physics).

Then, the time–dependent dipole operator entering the dipole–dipole correlation function $C_{\mathrm{d-d}}(t)$, Eq. (5.70) simply reads

$$\begin{aligned}\hat{\mu}(t) &= \sum_{\alpha} \mathbf{d}_\alpha e^{iH^{(1)}_{\mathrm{agg}}t/\hbar}|\alpha\rangle\langle 0|e^{-iH^{(0)}_{\mathrm{agg}}t/\hbar} + \mathrm{h.c.} \\ &\approx \sum_{\alpha} \mathbf{d}_\alpha e^{i\mathcal{E}_\alpha t/\hbar}|\alpha\rangle\langle 0| + \mathrm{h.c.}\ . \end{aligned} \tag{8.125}$$

The second line follows since $H_{\mathrm{ex-vib}}$ is set equal to zero. Using the same approximation the trace in $C_{\mathrm{d-d}}(t)$ reduces to a trace with respect to the electronic states $|0\rangle$ and $|m\rangle$. Furthermore, the equilibrium statistical operator gives a projection onto the electronic ground state of the aggregate, $\hat{W}_{\mathrm{eq}} = |0\rangle\langle 0|$. Inserting this into Eq. (5.72) and neglecting antiresonant contributions gives (the prefactor $1/3$ follows from the orientational averaging, cf. Chapter 5)

$$\alpha(\omega) = \frac{4\pi^2 \omega n_{\mathrm{agg}}}{3c} \sum_{\alpha} |\mathbf{d}_\alpha|^2 \delta(\mathcal{E}_\alpha - \hbar\omega)\ . \tag{8.126}$$

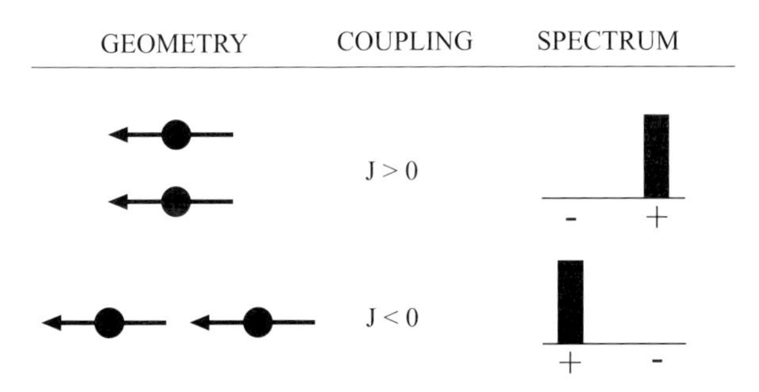


**Figure 8.13:** Dependence of the transition amplitudes of the eigenstates of a molecular dimer on the mutual arrangement of the monomeric transition dipole moments depicted by the arrows on the left–hand side. Changing the sign of one dipole leads to excitation of the antisymmetric state without changing the spectrum.

The strength for transitions from the ground state into the single–exciton state $|\alpha\rangle$ is determined by the respective transition dipole moment, Eq. (8.124), where the expansion coefficients $c_\alpha(m)$ give the contribution of the $m$th molecule to the single–exciton eigenstate $|\alpha\rangle$. In order to characterize this quantity we compute the oscillator strength. For a collection of molecules with identical transition dipole moments (same magnitude and same spatial orientation), $\mathbf{d}_m = \mathbf{d}$, it reads

$$O_\alpha = \frac{|\mathbf{d}_\alpha|^2}{|\mathbf{d}|^2} = |\sum_m c_\alpha(m)|^2 . \tag{8.127}$$

Note that as a consequence of the neglect of environmental influences the external field can in principle excite exciton states which are delocalized over the whole aggregate according to Eq. (8.124).

To illustrate the derived formulas let us first consider the molecular heterodimer which energy levels have been already introduced in Section 8.2.3. The oscillator strength defined in Eq. (8.127) for transitions into the symmetric and antisymmetric eigenstates is (for $\eta$ see Eq. 8.49)

$$O_\pm = \frac{|1 \pm \eta \exp(\pm i \arg(J))|^2}{1+\eta^2} . \tag{8.128}$$

In Fig. 8.13 we show the distribution of oscillator strength in the dimer absorption spectrum for the degenerate case, $E_1 = E_2 = E_0$ ($\mathcal{E}_\pm = E_0 \pm |J|$), in dependence on the geometry of the transition dipole moments. The sign of the dipole–dipole coupling can be positive or negative, as indicated for two extreme cases in Fig. 8.13. For dipoles pointing in the same direction, this implies that the energy shift $\mathcal{E}_+ - E_0$ corresponding to the symmetric eigenstate $|+\rangle$ carrying all oscillator strength can be positive, $\mathcal{E}_+ = E_0 + J$, or negative, $\mathcal{E}_+ = E_0 - |J|$. In molecular aggregates this energy shift can be observed upon aggregation. Depending on whether the absorption band shifts to the red or to the blue, aggregates are classified as J– or H–aggregates, respectively. An example for a J–aggregate was TDBC shown in Fig. 8.2.

If there is some detuning between the monomer transition energies ($E_1 \neq E_2$), which can be caused by different local environments for two otherwise identical molecules (static disorder), both eigenstates will carry oscillator strength. In the limit that $|E_1 - E_2| \gg |J|$, the absorption spectrum becomes monomeric and the eigenstates are localized at the respective molecules.

Next let us consider an aggregate consisting of a linear arrangement of $N_{\text{mol}}$ identical molecules as also introduced in Section 8.2.3. If we consider the absorption spectrum we note that the single–exciton state $\mathcal{E}_0 = E_0 + 2J$ will have the lowest (highest) energy for a J– (H–) aggregate. It also has the largest transition amplitude for optical absorption. The respective oscillator strengths, Eq. (8.127), are given by

$$O_\alpha = \frac{1-(-1)^j}{2} \cot^2\left(\frac{\pi}{2}\frac{j}{N_{\text{mol}}+1}\right) . \tag{8.129}$$

The expression for $O_\alpha$ shows that nearly all oscillator strength is contained in a single exciton state ($j = 1$, cf. part of Fig. 8.11). As an example of a linear chain type aggregate we show the absorption spectrum of TDBC in Fig. 8.2. It is a J–aggregate and the interaction induces a red–shift of the band upon aggregation.

## 8.7.2 Static Disorder

An important factor determining the width of absorption lines of artificially prepared or naturally occurring aggregates is static disorder. In this section we will outline an approach which takes the effect of energetic and structural disorder into account. The formulation is rather general and can be applied to the much simpler case of single molecules in solution as well.

As has already been discussed in Section 5.2, a change of the energy level structure, for example, from aggregate to aggregate leads to an additional broadening of the absorption which is measured on a sample containing a large number of aggregates. In general one can characterize such a behavior by a set of parameters $y \equiv \{y_j\}$ which enter the Hamiltonian and describe a specific energetic and structural situation in the aggregate. The parameters $y$ will be additionally labelled by $A$, which counts all aggregates contained in the sample volume $V$. This should indicate that set $y$ varies from aggregate to aggregate. Accordingly, every aggregate will have its own absorption cross section $\sigma = \sigma(\omega; y_A)$. The cross section follows from the absorption coefficient as $\sigma = \alpha/n_{\text{agg}}$ and we may write:

$$\alpha_{\text{inh}}(\omega) = \frac{1}{V} \sum_{A \in V} \sigma(\omega; y_A) . \tag{8.130}$$

The inhomogeneous broadening can be described as an averaging with respect to different realizations of the aggregate's structure and energy spectrum. This is called a *configurational average*. If there exist a large number of different realizations one can change from the summation to the integration with respect to the different parameters $y_j$

$$\alpha_{\text{inh}}(\omega) = \int dy\ \mathcal{F}(y)\sigma(\omega; y) . \tag{8.131}$$

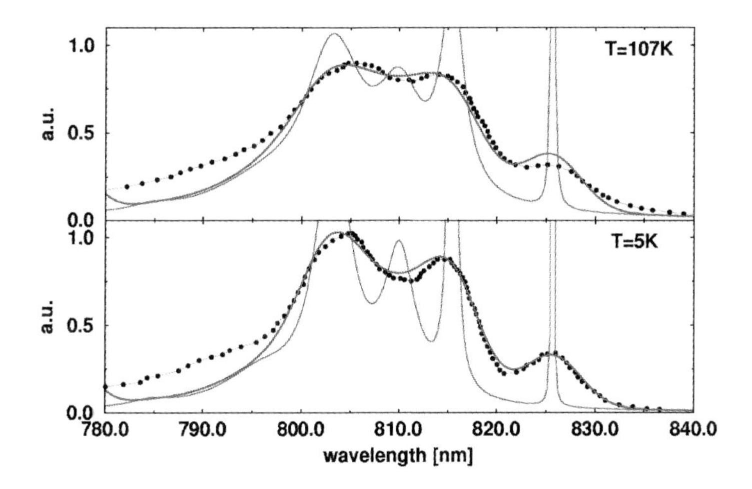


**Figure 8.14:** Linear absorption of the pigment–protein complex shown in Fig. 8.5 at two different temperatures. The calculated spectra including inhomogeneous broadening are drawn as thick lines. Thin lines show the related homogeneous spectra, Eq. (8.137) (i.e., before convoluting with a Gaussian distribution function). The points represent the experimental values measured in [Fre97].

The integration extends over the whole set of parameters. The appropriate normalized distribution function $\mathcal{F}(y)$ can formally be introduced as

$$\mathcal{F}(y) = \frac{1}{V} \sum_{A \in V} \prod_j \delta(y_j - y_{Aj}) \ . \qquad (8.132)$$

For specific applications $\mathcal{F}(y)$ is taken to be a continuous function of the parameters $y_j$.

In the following we consider the simple case, where disorder can be described by Gaussian distributions of the various exciton levels around certain mean values $\overline{\mathcal{E}}_\alpha$. In fact the Gaussian form of the distribution function can be justified from the central limit theorem of probability theory. We set

$$\mathcal{F}(y) \to \mathcal{F}(y \equiv \{\mathcal{E}_\alpha\}) = n_{\text{agg}} \prod_\alpha \mathcal{F}_\alpha(\mathcal{E}_\alpha - \overline{\mathcal{E}}_\alpha) \ , \qquad (8.133)$$

with

$$\mathcal{F}_\alpha(E) = \frac{1}{\sqrt{2\pi \Delta_\alpha}} \exp\left\{ -\frac{E^2}{2\Delta_\alpha^2} \right\} \ . \qquad (8.134)$$

Here $\Delta_\alpha$ is the width of the Gaussian distribution for the state $|\alpha\rangle$. Taking the cross section according to Eq. (8.126), the inhomogeneously broadened absorption spectrum is obtained as

$$\alpha_{\text{inh}}(\omega) = \int d\mathcal{E} \ \mathcal{F}(\mathcal{E}) \sigma(\omega; \mathcal{E}) = \frac{4\pi^2 \omega n_{\text{agg}}}{3c} \sum_\alpha |\mathbf{d}_\alpha|^2 \ \mathcal{F}_\alpha(\hbar\omega - \overline{\mathcal{E}}_\alpha) \ . \qquad (8.135)$$

In this simple case the distribution of microscopic parameters directly determines the line-shape of the inhomogeneously broadened spectrum. If the disorder should be related in a more direct manner to the actual situation, for example, to fluctuations of the intermolecular distance and thus to $J_{mn}$, a numerical calculation of the configurational average becomes necessary. Fig. 8.14 compares the absorption spectra of the biological pigment–protein complex shown in Fig. 8.5 with and without static disorder. In the absence of disorder four absorption lines can be identified which are homogeneously broadened. The way to calculate these spectra is briefly explained in the next section.

### 8.7.3 Limit of Weak Exciton–Vibrational Coupling

If the exciton–vibrational coupling is sufficient weak we may compute the absorption coefficient following the procedure of Section 5.3.4. There, the coupling of a single molecule to a thermal environment has been taken into account perturbatively. This situation is similar to the case of weak exciton vibrational coupling as already discussed in Section 8.6. The following computation is based on the arguments given, in particular, in Section 8.6.2.

To calculate the absorption coefficient we start with Eq. (5.88) where the dipole–dipole correlation function $C_{\mathrm{d-d}}(t)$ is determined by a density operator propagation. If we translate the notation of Section 8.6 to the present case we obtain Eq. (5.89) in the following form:

$$C_{\mathrm{d-d}}(t) = \sum_{\alpha} \left( \mathbf{d}_{\alpha}^{*} \langle \alpha | \hat{\sigma}(t) | 0 \rangle + \mathbf{d}_{\alpha} \langle 0 | \hat{\sigma}(t) | \alpha \rangle \right) . \tag{8.136}$$

Instead of a single excited state as in Eq. (5.89) we have here the set of exciton levels, but the trace with respect to the vibrational states does not appear. The density operator $\hat{\sigma}(t)$ follows from the propagation of the initial state $\hat{\sigma}(0) = [\hat{\mu}, |0\rangle\langle 0|]_{-}$. Taking into account the density matrix equations introduced in Section 8.6.2 (but generalized here to the off–diagonal type of functions $\rho_{\alpha 0}$) and the dephasing rates Eq. (8.120) we obtain the absorption spectrum in analogy to Eq. (5.92) as

$$\alpha(\omega) = \frac{4\pi\omega n_{\mathrm{agg}}}{3\hbar c} \sum_{\alpha} |\mathbf{d}_{\alpha}|^2 \frac{\gamma_{\alpha}}{(\omega - \omega_{\alpha,0})^2 + \gamma_{\alpha}^2} . \tag{8.137}$$

This formula is apparently a generalization of Eq. (8.126) with the transitions into the exciton states broadened by the dephasing rates $\gamma_{\alpha}$. An application one may find in Fig. 8.14. It displays the absorption spectrum of the biological pigment–protein complex presented in Fig. 8.5. There, the four exciton levels which mainly contribute to the absorption are broadened to a different extend. This is mainly caused by the spectral density entering Eq. (8.121). The different values of $J(\omega_{\alpha\beta})$ at different transition frequencies finally lead to different values of the dephasing rates (cf. also Fig. 3.5). Notice that the homogeneously broadened spectra become relatively structureless if inhomogeneous broadening is introduced.

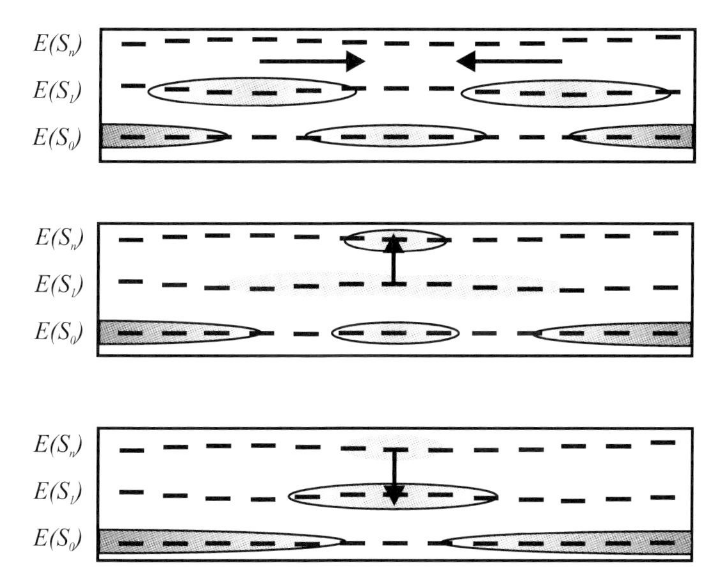


**Figure 8.15:** Scheme of exciton–exciton annihilation in a linear chain of three–level molecules (cf. Fig. 8.16). Upper panel: two partly delocalized excitations of the $S_1$–state moving toward each other. Middle panel: transformation of two $S_1$–state excitations into a single excitation of the $S_n$–state. Lower panel: internal conversion of the $S_n$–state excitation into an excitation of the $S_1$–state.

# 8.8 Supplement

## 8.8.1 Exciton–Exciton Annihilation

The Förster transfer considered in Section 8.5 concentrates on the description of a single excitation in the aggregate. This is appropriate whenever the light intensity used to excite the aggregate is low enough to justify the restriction on a singly excited state. However, upon increasing the light intensities one may study states where different molecules of the aggregate are excited simultaneously. This opens new relaxation channels as will be discussed in the following. Such experiments have originally been focused on dye aggregates (see for example [Sun88,Sti89]). But there is also some recent work on different photosynthetic antenna systems. In all these experiments the most dominant relaxation mechanism is the process of *exciton–exciton annihilation*.

Exciton–exciton annihilation is usually characterized as a two step process (cf. Fig. 8.15). First, two excitations being in the $S_1$–state of the molecules have to move close to each other so that their excitation energy can be used to create a higher excited $S_n$–state ($n > 1$) at one molecule. This step leaves behind the other molecule in the $S_0$ ground state and is usually named exciton fusion. In a second step a probably ultrafast internal conversion process brings back the molecule which is just in the higher excited $S_n$–state to the $S_1$–state. The whole process can be represented by the following scheme:

$$M_1(S_1) + M_2(S_1) \stackrel{\text{fusion}}{\longrightarrow} M_1(S_n) \stackrel{\text{IC}}{\longrightarrow} M_1(S_1)\,. \tag{8.138}$$

So far, exciton–exciton annihilation has been often described by the rate equation $\dot{n}(\mathbf{r},t) = -\gamma n(\mathbf{r},t)^2$, with the exciton density $n(\mathbf{r},t)$ at the spatial position $\mathbf{r}$ and the annihilation rate constant $\gamma$ [Ame02]). Besides such a macroscopic description valid for larger aggregates (and organic semiconductors) where exciton diffusion may take place, various microscopic theories have been presented [Sun70,Ryz01]. A consequent microscopic description has to consider the dynamics of the process of exciton–exciton annihilation as shown in Fig. 8.15. Therefore, a theoretical formulation has to use a three–level model for every molecule of the aggregate and has to account for, at least, two–exciton states. The process of internal conversion has to be considered, too.

If the excitations which undergo the annihilation process are not completely localized the description has to be done using delocalized single– and two–exciton states. In the opposite case it will be sufficient to calculate the annihilation rate for the transition from localized states. Both cases will be considered in Section 8.8.1. In the next section we shortly comment on a model for two–exciton states if a double excitation of the single molecules into a $S_n$–state has been incorporated.

### Three–Level Description of the Molecules in the Aggregate

Instead of the two–level model used so far for the description of the aggregate's molecules we will additionally incorporate a third state $|\varphi_{mf}\rangle$ which corresponds to a higher $S_n$–level (cf. Fig. 8.15). The related energy level is denoted by $\epsilon_{mf}$ with the energetic position determined by the relation $\epsilon_{mf} - \epsilon_{me} \approx \epsilon_{me} - \epsilon_{mg}$. Moreover, it is assumed that there exists a nonvanishing transition dipole matrix element $\tilde{\mathbf{d}}_m = \langle\varphi_{mf}|\hat{\mu}|\varphi_{me}\rangle$ which connects the $S_1$ state with the higher excited state. But we set the matrix elements $\langle\varphi_{mf}|\hat{\mu}|\varphi_{mg}\rangle$ equal to zero.

Resulting from this a new class of Coulomb coupling matrix elements arises extending those introduced in Section 8.2.1. Since we will deal with aggregates where the molecules are sufficiently separated in space, we can restrict ourselves to the direct Coulomb interaction taken in dipole–dipole approximation. Having only the coupling matrix elements $J_{mn}(eg,eg)$ in the two–level model, Eq. (8.10) tells us that the new types $J_{mn}(fe,fe)$ and $J_{mn}(fg,ee)$ (as well as the complex conjugated expressions) must be considered additionally (see also Fig. 8.16). The first describes excitation energy transfer between the $S_1$ and the $S_n$ state (molecule $m$ undergoes the transition $S_1 \rightarrow S_n$ while the revers process takes place in molecule $n$). The second type of matrix element characterizes the excitation of molecule $m$ and the de–excitation of molecule $n$, both being initially in the $S_1$–state (see also Fig. 8.16). Therefore, the general Hamiltonian, Eq. (8.9) valid for a multi–level description of every molecule in the aggregate (including the intermolecular nuclear repulsion term) reduces to

$$\begin{aligned}
H_{\text{agg}} &= \sum_m \sum_{a=g,e,f} H_{ma}|\varphi_{ma}\rangle\langle\varphi_{ma}| \\
&+ \sum_{mn} \Big( J_{mn}(eg,eg)|\varphi_{me}\varphi_{ng}\rangle\langle\varphi_{ne}\varphi_{mg}| + J_{mn}(fe,fe)|\varphi_{mf}\varphi_{ne}\rangle\langle\varphi_{nf}\varphi_{me}| \\
&+ \big[J_{mn}(fg,ee)|\varphi_{mf}\varphi_{ng}\rangle\langle\varphi_{ne}\varphi_{me}| + \text{h.c.}\big]\Big) . \qquad (8.139)
\end{aligned}$$

The off–diagonal part of the single molecule Hamiltonian describing nonadiabatic transitions has been separated. We introduce

$$H_{\text{na}} = \sum_m \Theta_m(ef)|\varphi_{me}\rangle\langle\varphi_{mf}| + \text{h.c.} \; . \tag{8.140}$$

what describes nonadiabatic coupling between the $S_n$ and the $S_1$–state (cf. Eqs. (2.114), (5.139)). A similar expression had already been used in Section 5.7 to describe the internal conversion process.

In a next step we introduce the two–exciton state by extending the derivations given in Section 8.2.3. (Obviously there is no need to define the single exciton states anew.) Instead of Eq. (8.46) the two–exciton state is now written as (the quantum numbers $\tilde{\alpha}$ exclusively refer of the two–exciton states):

$$|\tilde{\alpha}\rangle = \sum_{m,n} c_{\tilde{\alpha}}(m,n)|me, ne\rangle + \sum_m c_{\tilde{\alpha}}(m)|mf\rangle \; . \tag{8.141}$$

This two–exciton state covers two $S_1$–excitations at molecule $m$ and $n$ (note the additional label "$e$") as well as higher excitation at the $m$th molecule (note the label "$f$" here). The extensions introduced in this section will be used to discuss different types of an exciton–exciton annihilation rate in the following section.

### The Rate of Exciton–Exciton Annihilation

Let us start with the consideration of exciton–exciton annihilation in the limit of delocalized exciton states. In this case one can directly utilize the results obtained for internal conversion in Section 5.7. But now, the reactant state is given by the two–exciton state $|\tilde{\alpha}\rangle$ and the product state by the single–exciton state $|\beta\rangle$. The rate follows as

$$k_{\tilde{\alpha}\to\beta} = \frac{2\pi}{\hbar} \mid \Theta(\tilde{\alpha},\beta) \mid^2 \mathcal{D}(\tilde{\alpha},\beta;(\mathcal{E}_{\tilde{\alpha}} - \mathcal{E}_\beta)/\hbar) \; , \tag{8.142}$$

where the density of states can be defined in similarity to Eq. (5.141) or Eq. (5.143), but based on the PES of the single– and two– exciton states. The coupling matrix elements $\Theta(\tilde{\alpha},\beta)$ are given by the exciton representation of the nonadiabatic coupling, Eq. (8.140):

$$\Theta(\tilde{\alpha},\beta) = \langle\tilde{\alpha}|H_{\text{na}}|\beta\rangle = \sum_m \Theta_m(fe)c^*_{\tilde{\alpha}}(mf)c_\beta(me) \; . \tag{8.143}$$

Let us assume that all molecules in the aggregate are identical and characterized by the same nonadiabatic coupling $\Theta_m(fe)$. Furthermore, we replace the density of states in Eq. (8.142) by a quantity referring to the local internal conversion processes. Then we obtain

$$k_{\tilde{\alpha}\to\beta} = \mid \sum_m c^*_{\tilde{\alpha}}(mf)c_\beta(me) \mid^2 \; k^{(\text{IC})}_{f\to e} \; . \tag{8.144}$$

Here, $k^{(\text{IC})}_{f\to e}$ is rate of internal conversion which according to our assumption is identical for all molecules of the aggregate. The exciton–exciton annihilation, therefore, can be described

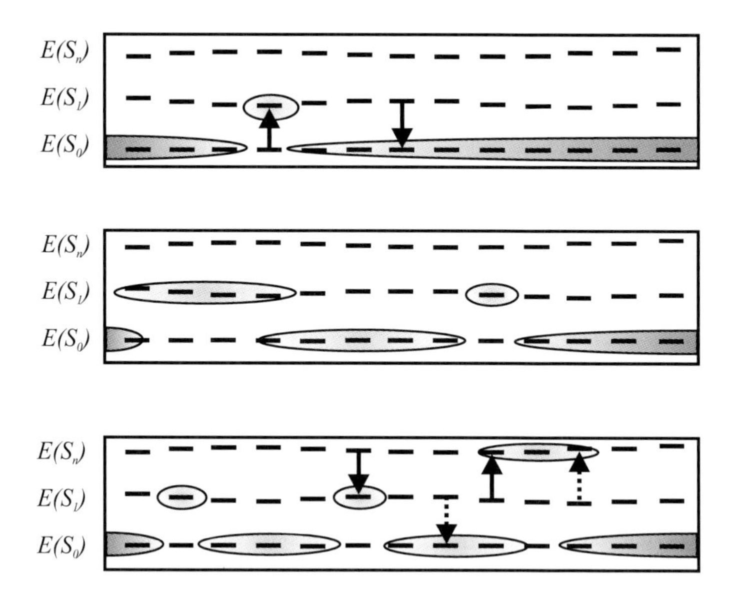


**Figure 8.16:** Different excitations in a linear chain of three–level molecules (with the $S_0$ ground state, the first excited singlet state $S_1$, and a higher excited singlet state $S_n$; plotted are the energy levels of the molecules versus their spatial position). Upper panel: First excited state of the chain with the fourth molecule in the $S_1$–state (the arrows symbolize the dipole–dipole coupling between the $S_0$ and $S_1$–state). Middle panel: second excited state of the chain with two excitations in the $S_1$–state (the left excitation is assumed to be delocalized across four molecules). Lower panel: third excited state of the chain with two localized excitations in the $S_1$–state and one partially delocalized excitation in the $S_n$–state (arrows with full line symbolize the dipole–dipole coupling between the $S_1$ and $S_n$–state and those with the broken lines describe coupling responsible for the transition of two excitations from the $S_1$–state to one excitation of the $S_n$ and one of the $S_0$–state).

by this local internal conversion rate, however, weighted by the square of an overlap expression. This expressions incorporates the overlap of the probability amplitudes $c^*_{\tilde{\alpha}}(mf)$ and $c_\beta(me)$ for having a double and single excitation, respectively, at site $m$. In this description the first step of exciton–exciton annihilation, namely the exciton fusion is masked by the two–exciton state in particular by the nonvanishing expansion coefficient $c_{\tilde{\alpha}}(mf)$ measuring the probability to have a double excitation at a single molecule.

If the annihilation process proceeds via localized states as indicated in scheme (8.138) one has to start with the doubly excited state $|me, ne\rangle$. It is transferred to the intermediate state $|mf\rangle$ of a higher excited single molecule, and the product is simply given by the single excited state $|me\rangle$ at molecule $m$. This scheme reminds on bridge–mediated electron transfer reactions as discussed in Section 6.6. There, the transfer from the initial donor state through the intermediate bridge states into the final acceptor state could take place as a direct transition (superexchange transfer) or as a step–wise process going from the donor to the bridge and then to the acceptor (sequential transfer). The latter appears if vibrational relaxation in the intermediate state (the bridge states) interrupts the direct transfer from the donor to the acceptor. One can expect similar conditions in the case of exciton–exciton annihilation.

Let us consider, for instance, the two–step annihilation process. It is characterized by the rate $k_{me,ne\to mf}$ describing the creation of the higher excited state at molecule $m$, and by the rate $k_{mf\to me}$ characterizing the internal conversion at molecule $m$. The first rate is computed with the Coulomb matrix element $J_{mn}(fg,ee)$ as the disturbance, and the second rate is simply the rate of internal conversion $k^{(\mathrm{IC})}_{f\to e}$. Both should enter rate equations for the various state populations with the solution characterizing the two–step annihilation process. But like the introduction of the superexchange process in 6.6 one may also describe the annihilation as a process without intermediate state relaxation.

---

# Notes


[Ame00] H. van Amerongen, L. Valkunas, and R. van Grondelle, *Photosynthetic Excitons* (World Scientific, Singapore, 2000).

[Dam99] A. Damjanovic, Th. Ritz, and K. Schulten, Phys. Rev. E **59**, 3293 (1999).

[Davy79] A. S. Davydov, Phys. Scr. **20**, 387 (1979).

[Dav86] A.S. Davydov, Annalen der Physik **43**, 93 (1986).

[Fid93] H. Fidder, J. Knoester, and D. A. Wiersma, J. Chem. Phys. **98**, 6564 (1993).

[Fre97] A. Freiberg, S. Lin, K. Timpmann, and R. E. Blankenship, J. Phys. Chem. B **101**, 7211 (1997).

[Gro85] R. van Grondelle, Biochem. Biophys. Acta **811**, 147 (1985)

[Kor98] O. Korth, Th. Hanke, and B. Röder, Thin Solid Films **320**, 305 (1998).

[Mei97] T. Meier, Y. Zhao, V. Chernyak, and S. Mukamel, J. Chem. Phys. **107**, 3876 (1997).

[Mol95] J. Moll, S. Daehne, J. R. Durrant, and D. A. Wiersma, J. Chem. Phys. **102**, 6362 (1995).

[Muk95] S. Mukamel, *Principles of Nonlinear Optical Spectroscopy*, Oxford University Press, 1995.

[Ren01] Th. Renger, V. May, and O. Kühn, Phys. Rep. **343**, 137 (2001).

[Ryz01] I. V. Ryzhov, G. G. Kozlov, V. A. Malyshev, J. Knoester, J. Chem. Phys. **114**, 5322 (2001).

[Sch01] G. D. Scholes, X. J. Jordanides, and G. R. Fleming, J. Chem. Phys. **105**, 1640 (2001).

[Sco92] A. Scott, Phys. Rep. **217**, 1 (1992).

[Spa02] S. Spano, J. Chem. Phys. **116**, 5877 (2002).

[Spi02] C. Spitz, J. Knoester, A. Ourat, and S. Daehne, Chem. Phys. **275**, 271 (2002).

[Sti88] H. Stiel, S. Daehne, K. Teuchner, J. Lumin.**39**, 351 (1988).

[Sun70] A. Suna, Phys. Rev. B **1**, 1716 (1970).

[Sun88] V. Sundström, T. Gillbro, R. A. Gadonas, and A. Piskarskas, J. Chem. Phys. **89**, 2754 (1988).

[Toy61] Y. Toyozawa, Progr. Theor. Phys. **26**, 29 (1961).

# 9 Laser Control of Charge and Energy Transfer Dynamics

*The microscopic picture for the real–time motion of particles and quasi–particles which emerged from the combination of theoretical efforts and advances in ultrafast spectroscopies calls for going one step further, that is, not only to analyze but to control this dynamics. Elementary charge and energy transfer processes might be guided from a reactant to desired product states without a priori specification of the reaction pathway. On the other hand, the potential energy surface including nonadiabatic couplings but also the details of the system–bath interaction might be explored in unprecedented detail. In order to accomplish these goals molecular quantum dynamics has to be influenced by making use of its inherent interference features.*
*Various strategies for the theoretical design of laser fields exist, the most successful one being optimal control theory. It can be formulated for coherent and dissipative time evolution. Experimentally, adaptive closed–loop control has been demonstrated to be a versatile tool for generating ultrashort laser pulses of almost arbitrary temporal shapes and spectral contents, tailored to achieve specific goals.*
*Optimal control theory is briefly introduced putting emphasis on the dissipative control kernel which follows from maximization of an appropriate control functional. Instead of covering the broad range of applications in theory and experiment we follow the lines of the previous chapters and discuss exemplary results for the control of ultrafast photoinduced electron transfer and infrared laser driven proton transfer processes.*

## 9.1 Introduction

In Section 5.5 it had been demonstrated that the shape of a nuclear wave packet which is prepared by an external field in an excited electronic state depends on the properties of that field. If the pulse duration is shorter than the time scale for vibrational motion, it was found that the vibrational wave function of the electronic ground state is almost instantaneously transferred to the electronically excited state. As a result, vibrational wave packet motion in the excited state occurred. In the opposite case of a very long pulse, only a single vibrational level of the excited electronic state is populated provided that the laser frequency fits the respective energy difference (and the excited state is not dissociative). Having this behavior in mind, one may ask whether it is possible to design specific laser fields which *manipulate* the coupled electron–vibrational dynamics in such a way that the wave packet created in the excited electronic state has some well–defined shape or travels into a particular exit channel as shown in Fig. 9.1.

The question posed belongs to the general type of problems which are related to the *active*

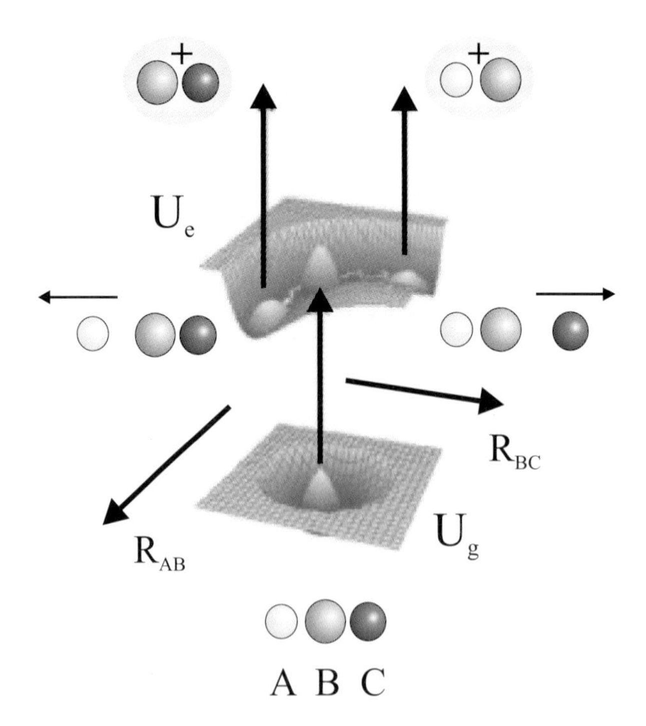


**Figure 9.1:** Ground and excited state PES of an ABC type molecule drawn as a function of the A–BC and AB–C bond lengths. The control objective could be to design an optical laser field which excites the shown ground state vibrational wave packet onto the excited state surface such that it is guided into a particular exit channel which corresponds to breaking either the A–BC (right) or the AB–C (left) bond. The reaction product could be detected, for instance, by photoionization.

*control* of molecular dynamics by means of external fields. Of course, traditional chemistry already involves to a large extent the control of reactions, however, in the statistical sense. By this we mean that during the last 200 years chemists have been successful in modifying the outcome of chemical reactions by adjusting external conditions such as temperature or concentrations of reactants. With the help of catalysts, on the other hand, the speed of reactions can be influenced. During the last decade emerging flexible laser systems offered the possibility to step out of the statistical realm and influence the microscopic dynamics of molecular systems. Why this should be of any interest? It is certainly more than an intriguing challenge to theory and experiment. Consider, for example, the scenario of a reaction where certain exit channels would be closed under normal (statistical) conditions. Sequences of laser pulses, for instance, can be used as an *active* means for opening such channels and yield products which could not have been obtained otherwise. This idea is not restricted to chemical reaction dynamics, but it is of potential importance, for instance, for the emerging field of molecular electronics. Here ultrafast optical switches could be designed whose function depends, e.g., on details of the laser spectrum.

In principle the conception to use lasers for controlling molecular dynamics is as old as the laser itself. However, early attempts, for instance, to break particular bonds simply by exciting a certain (local) vibrational mode with a narrow bandwidth laser, did not hold the promises. Intramolecular vibrational energy flow out of the excited mode often prevented an efficient bond breaking. However, the selective preparation of vibrationally excited states in relatively small molecules has been shown to provide an effective means for controlling subsequent reactions such as photodissociation or bimolecular atom transfer. Here, the key is to prepare a vibration which has a large overlap with the reaction coordinate. For example, it has been shown for HOD that selective bond breaking , e.g., into H+OD, is possible if the OH vibration is first prepared in a high overtone state in the electronic ground state and then photoexcited (see, e.g, [Cri96]).

It was only in the 1980s that it was theoretically demonstrated that utilization of quantum mechanical interference effects is very important for a successful laser control (for reviews see [Ric00,Sha03]). This holds irrespective of the mode of laser field operation, that is, in the time as well as in the frequency domain. In terms of the simple example shown in Fig. 9.1 this could imply that the evolution of the electron–vibrational states, $|\chi_{aM}\rangle|\varphi_a\rangle$, forming the coherent superposition

$$|\Psi(t)\rangle = \sum_{a=g,e}\sum_{M} C_{aM}(t)|\chi_{aM}\rangle|\varphi_a\rangle \tag{9.1}$$

is manipulated individually by the laser pulse such that the resulting wave packet shows the desired behavior. Compared with the monochromatic excitation, for instance, in vibrationally controlled photodissociation, such a scheme offers greater flexibility also for tackling more complex molecular systems. In a sense the coherent nature of the laser field is transferred to the evolving molecular quantum wave packet. Since in general a wave packet will have a multitude of spectral components, its manipulation is most conveniently done with ultrafast laser pulses offering a broad spectrum of frequencies whose occurrence in the field even might be a function of time itself.

The importance of having different frequency components is also apparent when considering infrared laser–driven ground state dissociation. As a consequence of the anharmonicity of

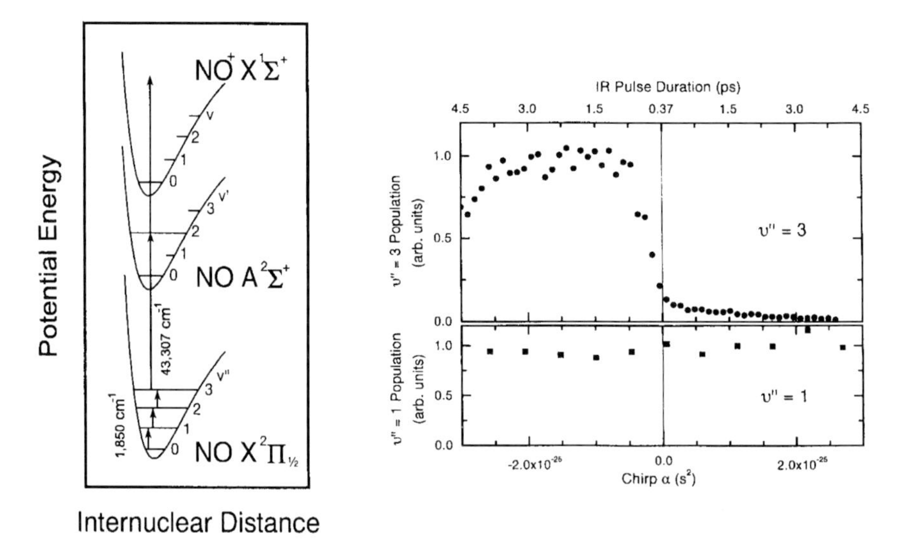


**Figure 9.2:** Chirped pulse vibrational ladder climbing in nitric oxide. Left panel: Potential energy curves showing the anharmonic vibrational levels in the electronic ground state as well in the two electronically excited states which are used to monitor the ladder climbing via resonance enhanced multiphoton ionization. Right panel: Populations of the first and the third vibrationally excited state in dependence on the chirp rate $\alpha$ which is the second frequency derivative of the phase of the pulse. Notice that the pulse duration changes upon chirping. (reprinted with permission from [Maa98], copyright (1998) Elsevier Science B.V.)

the PES the frequency for transitions between states decreases with increasing energy. Thus climbing the vibrational ladder will be particularly efficient if photons of decreasing energy are offered to the molecule during the interaction with the laser field. This is illustrated in Fig. 9.2 for the preparation of NO in the third vibrationally excited state by means of a laser field whose center frequency changes during the pulse. If this so–called *pulse chirp* is negative such that the frequency is reduced during the pulse, the population of the target state is dramatically increased as compared with the case of a positive chirp. Notice that the bandwidth of the ultrashort laser pulses is sufficient to excite the target state already in the chirp–free case. An example for chirp–free ground state dissociation control which also includes competing intramolecular processes is shown in Fig. 9.3.

The first successful theoretical and experimental demonstrations of laser control have been for rather small molecules under gas phase conditions. For a theoretical modelling this amounts to solving the time–dependent Schrödinger equation[1]

$$i\hbar\frac{\partial}{\partial t}|\Psi(t)\rangle = \Big(H - \hat{\mu}E(t)\Big)|\Psi(t)\rangle\,, \qquad |\Psi(t=0)\rangle = |\Psi_i\rangle. \tag{9.2}$$

[1] In the following we assume that the external field is aligned with the dipole moment of the molecule. In passing we note that there are also strategies for using strong laser fields to align molecules (see, e.g., [Sha03]).

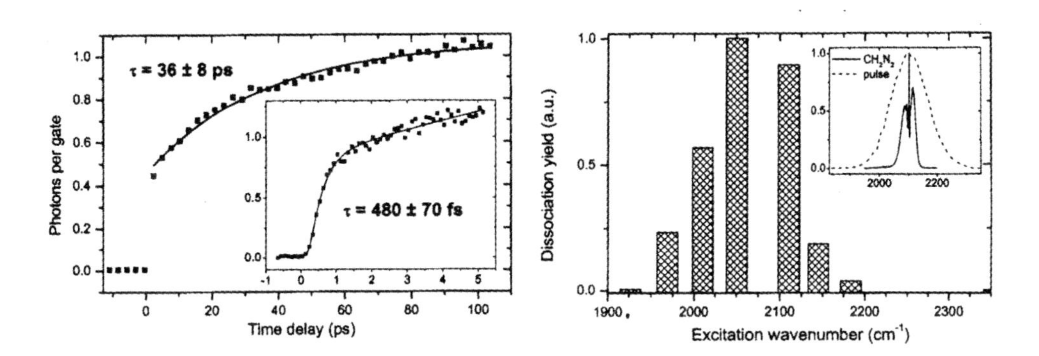


**Figure 9.3:** Ultrafast electronic ground state dissociation of the C–N bond in gaseous $CH_2N_2$ using a mid–infrared laser with 140 $cm^{-1}$ bandwidth; (See inset of the right panel where also the IR absorption spectrum is shown.). In the right panel the dissociation yield for different laser excitation wavelengths is shown. In the left panel the dissociation process is monitored by detecting laser–induced fluorescence from the nascent $^1CH_2$ . The fast time scale corresponds to the direct dissociation due to the driving of the reaction coordinate, whereas the longer time scale can be attributed to the statistical dissociation after vibrational energy redistribution. (reprinted with permission from [Win03], copyright (2003) American Institute of Physics)

Then the task is to find an external field $E(t)$ which drives the molecular dynamics towards a specific *target state*, $|\Psi_{\rm tar}\rangle$, at a certain time $t_f$, that is, we demand that $|\Psi(t_f)\rangle = |\Psi_{\rm tar}\rangle$. In the example of Fig. 9.1 the target state could be the outgoing wave packet; for bound state problems it may correspond to a particular quantum state such as a high overtone excitation of some vibrational mode. Obviously, this is not an unequivocal task and a solution is not guaranteed to exist especially in situations involving unbound potentials and many degrees of freedom.

The simplest approach to finding an appropriate field is to use a specific parameterized analytical pulse such as

$$E(t) = \sum_i \Theta(\tau_i + t_i - t)\Theta(t - t_i)\, A_i \sin^2\left(\frac{(t-t_i)\pi}{\tau_i}\right) \cos(\Omega_i t + \eta_i) \, . \tag{9.3}$$

This is a sequence of pulses; the unit–step functions guarantee that each pulse has a finite amplitude only for $t_i \leq t \leq \tau_i$. For each pulse there are five parameters at our disposal: the time $t_i$ where the pulse is switched on, the pulse duration $\tau_i$, the pulse frequency $\Omega_i$, its amplitude $A_i$, and the phase $\eta_i$. Given a target state these parameters can be optimized by available numerical methods through repeatedly solving the appropriate equation of motion. An example for the excitation of a specific vibrational state of HOD is shown in Fig. 9.4.

Although the simplicity of the pulse form is appealing, this strategy might fail for more complex situations where control is likely to require rather complicated pulse shapes. A general theoretical tool for pulse design is provided by *optimal control theory*. Here the subject is mapped onto an optimization problem for an appropriate functional as outlined in more detail in Section 9.2.

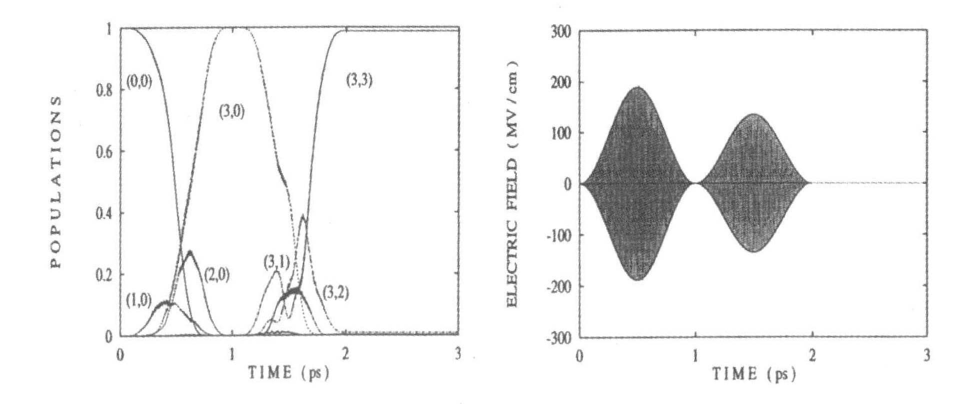


**Figure 9.4:** Selective preparation of the state ($\nu_{\rm OH} = 3, \nu_{\rm OD} = 3$) in a two–dimensional model of coupled OH and OD bond stretching vibrations in HOD by a sequence of two pulses of the form given in Eq. (9.3) (right panel). The left panel shows the population dynamics of the different quantum states starting from the ground state $(0,0)$. Apparently, the excitation mostly proceeds via the intermediate state $(3,0)$. The laser frequencies are tuned such that both steps involve the absorption of three photons. (reprinted with permission from [Kor96], copyright (1996) American Chemical Society)

So far we have discussed the control of molecular dynamics on the basis of the time–dependent Schrödinger equation which implies a gas phase situation. Switching to the condensed phase, however, or considering large polyatomic molecules, the control of selected degrees of freedom will have to compete with concurring vibrational energy redistribution and relaxation. Let us consider the example of an excited state preparation in a multi–level system. Here, the process of state preparation is accompanied by energy dissipation such that any intermediate state occupied during the action of the external field will have a finite lifetime. Therefore, one would expect only an incomplete or transient control compared with the gas phase. In principle, for a low–dimensional relevant system optimal control theory might provide us with a control scenario which compensates for the counter–productive influence of the interaction with the reservoir (cf. Fig. 9.12 below). This, of course, only holds as long as the lifetime of the target state is not finite. In this case the control objective can be reached only temporarily. Nevertheless, this target state might act as an intermediate state in a more complex control scheme.

The theoretical methods for predicting pulse shapes outlined so far have been well ahead of the experimental possibilities for quite some time. It has only been recently that it became possible to generate ultrashort laser pulses of almost arbitrary form by using pulse shapers as shown in Fig. 9.5. However, the theoretically predicted pulse forms inevitably depend on the specific choice of the Hamilton operator which will always contain uncertainties in particular for condensed phase situations. Therefore, it is not surprising that recent progress in the control of complex systems has been achieved by a different strategy, that is, the so–called *adaptive closed–loop control* which relies on experimental data and does not require the knowledge of the Hamiltonian. The principle of closed–loop control is sketched in Fig. 9.5.

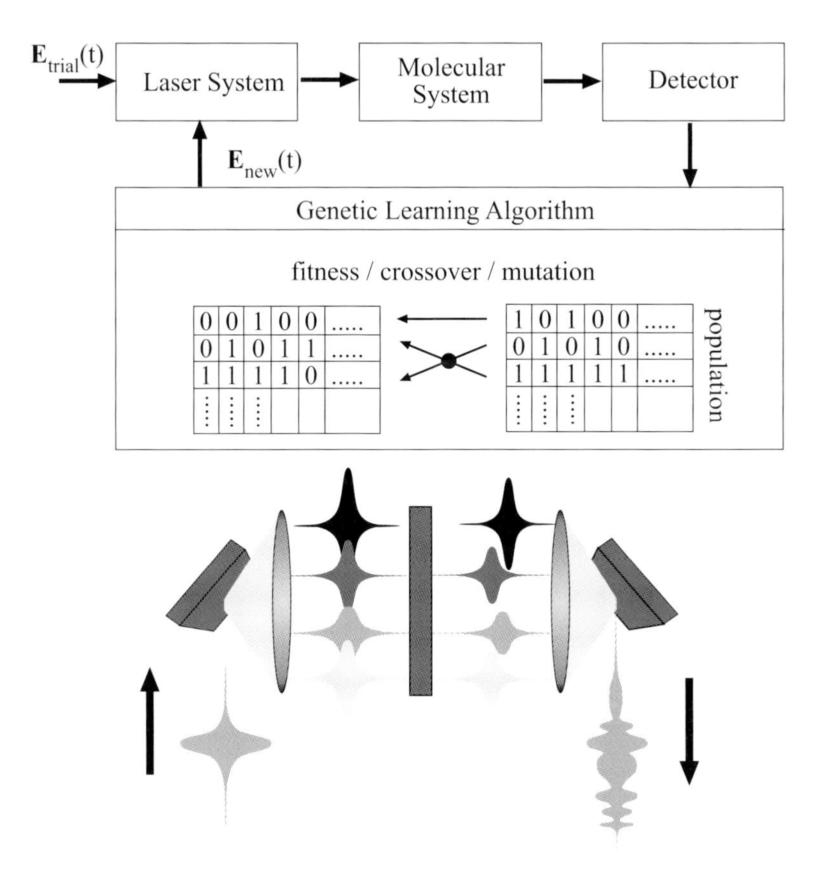


**Figure 9.5:** Principle of adaptive closed-loop control: A trial field $\mathbf{E}_{\mathrm{trial}}$ is optimized to fulfill a certain objective by learning about the system's response. For the learning algorithm evolutionary strategies are frequently used, where the properties of the field are encoded in a gene (e.g., in binary form) which is part of a population. After a population is tested the fitness of each gene is evaluated. The fittest genes (here the first one) move directly into the next population generation. Pairs of the remaining genes are combined (crossover) and enter the next generation. This crossover proceeds in a way that, e.g., the first 3 bits of gene 2 are combined with the last bits of gene 3 as shown here. To enhance the searching capabilities mutations (bit flips) are introduced (here in gene 1). The genes are then used to address the elements of a pulse shaper shown in the lower part. Here, an incoming laser pulse is spectrally decomposed and the phases and amplitudes of the individual components are manipulated. This is achieved by using a multipixel liquid–crystal display (LCD). Finally, the modified components are combined to the desired pulse shape.

Given an initial guess for the laser field which is applied to the molecular system the response of the latter is analyzed and compared with some objective such as the reaction yield. Upon controlled changing of the pulse form in a pulse shaper this objective is optimized by going through the loop many times. By this way of trial, error, and feedback, the laser field learns about the system dynamics. More specifically this requires that the laser field is parameterized,

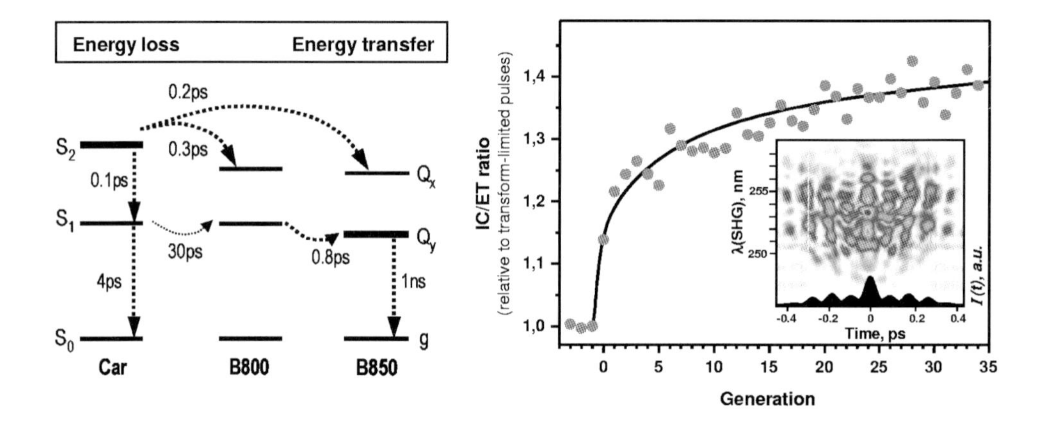


**Figure 9.6:** Closed–loop control of the light–harvesting efficiency of carotenoids (Car) in the peripheral antenna of the *Rps. acidophila* (left, cf. Fig. 8.4). Upon excitation of the $S_0 \rightarrow S_2$ Car transition with a transform–limited pulse internal conversion is as efficient as energy transfer (ET) to the bacteriochlorophyll acceptors of the B800/B850 system. Upon pulse shape optimization the ratio between IC and ET can be changed considerably as shown for the different generations of a genetic algorithm. The pulse shape is characterized by its autocorrelation function and its wavelength resolution in the inset. (figure courtesy of J. Herek, for more details see also [Her02])

for instance, by the relative phases of its spectral components. An efficient and systematic way of optimizing these parameters is provided by a genetic learning algorithm (cf. Fig. 9.5). The overall performance greatly benefits from the high duty cycle of present laser system which allow to perform thousands of experiments with different pulse shapes in a few minutes. An impressive example where the competing energy transfer and internal conversion dynamics in a light-harvesting photosynthetic complex has been controlled is discussed in Fig. 9.6. Other applications include, for instance, the chemical reaction dynamics in polyatomics (see, e.g., [Bri03,Dan03]) or the high–harmonic generation form the scattering of intense laser fields with atoms [Bar00].

From the theoretical point of view this experimentally very successful approach raises several challenges. Foremost is the question to understand the often rather complicated pulse shapes in terms of the underlying molecular dynamics. Ideally this would involve inverting the experimental data to reconstruct the actual Hamiltonian. In practice this ambitious task can be approached by means of simplified model Hamiltonians whose properties can be straightforwardly related to specific features of the optimized laser field. Here, it is often helpful to scrutinize the robustness of the field by introducing additional constraints which simplify the pulse form. Moreover, the control objectives can be modified such that, for instance, well–defined parts of the potential energy surface are explored. Finally, it should be emphasized that this type of learning algorithm is of limited value for theoretical pulse optimization in view of the fact that a single numerical propagation with a given field may be rather expensive.

Despite the experimental success of adaptive closed–loop control, optimal control theory still bears the potential to understand fundamental issues of laser–driven molecular dynamics. In the following Section 9.2.1 we will give an outline of the working equations of optimal control theory. In Section 9.3 we will discuss we complete the discussion of the previous chapters and present applications to electron and proton transfer processes. Notice that there are numerous applications of control theory to other problems in molecular dynamics and spectroscopy which are covered in excellent books to be mentioned in the Suggested Reading section below.

## 9.2 Optimal Control Theory

### 9.2.1 The Control Functional and the Control Field

The simplest and most intuitive strategy for designing a laser pulse which drives a system from an initial state $|\Psi_0\rangle$ at $t = t_0$ to a target state $|\Psi_{\text{tar}}\rangle$ at $t = t_f$ certainly rests on parameterized analytical pulses as given in Eq. (9.3) (cf. Fig. 9.4). However, this approach often assumes that the reaction path is known beforehand. More flexibility is certainly provided if the whole temporal shape of the laser pulse is subject to optimization. To put this into mathematical terms one has to introduce a functional $J$ of the laser field which reaches its extremum if the laser pulse drives the system in the required manner. Apparently, the functional $J$ can be defined as the overlap between the target state $|\Psi_{\text{tar}}\rangle$ and the wave function $|\Psi(t_f)\rangle$ of the driven system at $t = t_f$. Setting $J = |\langle\Psi_{\text{tar}}|\Psi(t_f)\rangle|$, the expression equals 1 if the control task has been solved. Otherwise, one gets $J < 1$. Additional constraints can be included via the method of Lagrange multipliers. An obvious constraint would be an upper limitation for the total field energy, $\mathcal{E}_{\text{field}}$, of the laser pulse which should control the molecular system. To introduce this quantity we assume that the control field propagates as a linear polarized plane wave with polarization unit vector $\mathbf{e}$ and wave vector $\mathbf{k}$:

$$\mathbf{E}(\mathbf{r}, t) = \mathbf{e}E(t)e^{i\mathbf{kr}} + \text{c.c.} \,. \tag{9.4}$$

According to Maxwells's equations we obtain $\mathcal{E}_{\text{field}}$ from the vector of energy flow density of the field (Poynting vector), if integrated over a surface $A_0$ of the probe and over the time-interval $[t_0, t_f]$:

$$\mathcal{E}_{\text{field}}(E, t_0, t_f) = \frac{iA_0}{4\pi \mid \mathbf{k} \mid} \int\limits_{t_0}^{t_f} dt \, \left(E(t)\frac{\partial}{\partial t}E^*(t) + \text{c.c.}\right) . \tag{9.5}$$

In praxis, however, one takes the simpler constraint

$$S_{\text{field}}(E; t_0, t_f) = \int\limits_{t_0}^{t_f} dt \mid E(t) \mid^2 , \tag{9.6}$$

stating that the square of the field strength should remain finite during the whole pulse.[2] And, instead of using the overlap between the target state and the wave function propagated up to

[2]This ambiguity in the choice of the constraint (and the concrete form of $J$) indicates that the control theory is not governed by laws of nature but by certain objectives defined such as to manipulate the molecular system.

$t = t_f$ we can also choose the square of the overlap and get as the functional (including the constraint via a Lagrange multiplier $\lambda$):[3]

$$J(E; t_0, t_f) = |\langle \Psi_{\text{tar}} | \Psi(t_f; E) \rangle|^2 - \lambda \big( S_{\text{field}}(E; t_0, t_f) - S_0 \big) . \tag{9.7}$$

Here, $S_0$ is the desired value of the expression $S_{\text{field}}(E; t_0, t_f)$.

Comparing different pulses $E(t)$ the functional $J$ should become extremal if the pulse is found which drives the system into the target state at time $t = t_f$. This pulse will be called *optimal pulse*, and the requirement for the extremum can be written as $\delta J = J(E + \delta E) - J(E) = 0$, where $\delta E$ is a small variation of the pulse field. We briefly demonstrate how to calculate $\delta J$ (a more involved derivation will be given in the subsequent section). As it is the case whenever an extremum is determined, $\delta J$ has to be calculated in the limit $\delta E \to 0$, so a linearization with respect to $\delta E$ suffices. It is easy to linearize the part of $J(E + \delta E)$ which accounts for the constraint. For reasons of a clear presentation we additionally assume that $E(t)$ is a real function. Then the expression $\int_{t_0}^{t_f} dt \; 2E(t)\delta E(t)$ can be derived. To get the contribution of the first part of $J(E + \delta E)$ we introduce the following separation of the total system Hamiltonian $H(E(t)) = H_{\text{mol}} + H_{\text{field}}(E(t))$ which determines the wave function propagation in the presence of the driving field. For the slightly modified field $E + \delta E$ we set $H(E + \delta E) = H(E) + H_{\text{field}}(\delta E)$. In a next step, the time–evolution operator $U(t_f, t_0; E + \delta E)$, driving the wave function $|\Psi_0\rangle$ (initially prepared in the system) into the final state, is separated in the same way. Noting the introduction of the $S$–operator in Section 3.2.2 we may write $U(t_f, t_0; E + \delta E) = U(t_f, t_0; E) S(t_f, t_0; E, \delta E)$. Here, the $S$–operator is defined by $H_{\text{field}}(\delta E)$ (of course, $S$ depends on $E$ via $U(t_f, t_0; E)$). Linearizing $S$ with respect to $\delta E$ (what is possible by an expansion of the $S$–operator, cf. also Section 5.3.1) we obtain

$$\begin{aligned} \langle \Psi_{\text{tar}} | \Psi(t_f; E + \delta E) \rangle &= \langle \Psi_{\text{tar}} | U(t_f, t_0; E) S(t_f, t_0; E, \delta E) | \Psi_0 \rangle \\ &\approx \langle \Psi_{\text{tar}} | U(t_f, t_0; E) | \Psi_0 \rangle \\ &\quad + \langle \Psi_{\text{tar}} | U(t_f, t_0; E) \frac{i}{\hbar} \int_{t_0}^{t_f} dt \; U^+(t, t_0; E) \\ &\quad \times \hat{\mu} \delta E(t) U(t, t_0; E) | \Psi_0 \rangle . \end{aligned} \tag{9.8}$$

Since the square of the overlap expression has been incorporated into $J$ the conjugated complex version of the upper expression has to be discussed, too.

Collecting all terms, the requirement that $\delta J = 0$ is written as $\int_{t_0}^{t_f} dt \; K(t)\delta E(t) = 0$. Since $\delta E(t)$ is arbitrary we get $K(t) = 0$. This equation reads in more detail

$$2\lambda E(t) = \frac{i}{\hbar} \Big( \langle \Psi(t_f) | \Psi_{\text{tar}} \rangle \; \langle \Psi_{\text{tar}} | U(t_f, t; E) \hat{\mu} U(t, t_0; E) | \Psi_0 \rangle - \text{c.c.} \Big) . \tag{9.9}$$

It represents a (functional) equation for the determination of the optimal pulse. Because of its nonlinear character Eq. (9.9) can be solved by iterative methods as will be outlined in

[3] Often the additional constraint that the state vector is propagated according to the Schrödinger equation, Eq. (9.2) is included via a Lagrange multiplier state vector. This renders the optimization of the functional $J$ to become unconstraint. However, the equation of motion can also be included via the calculation of the expectation value of the target operator as will be shown in the following section.

the following section. Any solution of Eq. (9.9) depends parametrically on the Lagrange multiplier $\lambda$. Setting $S_{\text{field}} = S_0$ gives the Lagrange multiplier $\lambda$ as a function of $S_0$, and consequently the optimal field as a function of $S_0$. The outlined method of setting up the functional, Eq. (9.7) including the constraint, Eq. (9.6), and of solving Eq. (9.9) to get the optimal pulse for a concrete control task is known as *Optimal Control Theory* [Rab88,Kos89].

The formulation of optimal control theory given so far is valid for any field strength. Nevertheless, one may restrict the search for the optimal pulse to weak field excitation, where the time–evolution operator $U(t, t_0; E)$ as well as Eq. (9.9) can be linearized in $E$. To do this we first note that $U(t, t_0; E) = U_{\text{mol}}(t - t_0)S(t, t_0; E)$, where $U_{\text{mol}}$ is exclusively determined by $H_{\text{mol}}$, and $S$ denotes the $S$-operator defined by the coupling Hamiltonian, $H_{\text{field}}$. The linearization of $S$ with respect to $E$, i.e. $1 + S^{(1)}$, gives the required approximation. If inserted into Eq. (9.9), $U_{\text{mol}}\ (1 + S^{(1)})$ appears three times in this expression.

To derive a simpler expression we further assume that $\langle\Psi_{\text{tar}}|\Psi_0\rangle = 0$, that is, the initial state is orthogonal to the target state. This property would be fulfilled if a system of two electronic PES is considered as in Fig. 9.1 (cf. Section 5.2.2), where the initial state is the vibrational ground state of the electronic ground state $|\Psi_0\rangle = |\chi_{g0}\rangle|\phi_g\rangle$ and the target state is a certain vibrational wave packet at the excited electronic state, $|\Psi_{\text{tar}}\rangle = |\chi_{\text{tar}}\rangle|\phi_e\rangle$. The factor $\langle\Psi_{\text{tar}}|\Psi(t_f)\rangle^*$ of Eq. (9.9) becomes $\langle\Psi_{\text{tar}}|U_{\text{mol}}(t_f - t_0)\ S^{(1)}(t_f, t_0)|\Psi_0\rangle^*$ after linearization in the field strength. The other field–dependent contributions to the right–hand side of Eq. (9.9) have to be neglected and one obtains:

$$\begin{aligned} 2\lambda E(t) &= \frac{i}{\hbar}\Big(\langle\Psi_{\text{tar}}|U_{\text{mol}}(t_f - t_0)S^{(1)}(t_f, t_0)|\Psi_0\rangle^* \\ &\quad \times\langle\Psi_{\text{tar}}|U_{\text{mol}}(t_f - t)\hat{\mu}U_{\text{mol}}(t - t_0)|\Psi_0\rangle - \text{c.c.}\Big) . \end{aligned} \tag{9.10}$$

We assume that $\hat{\mu} = d|\phi_e\rangle\langle\phi_g| + \text{h.c.}$, and take the electronic matrix elements of all expressions ($\langle\phi_a|U_{\text{mol}}|\phi_b\rangle = \delta_{ab}U_a$) to finally get:

$$\lambda E(t) = \frac{|d|^2}{\hbar^2}\int_{t_0}^{t_f} d\tau\ \text{Re}\big(\Omega(t)\Omega^*(\tau)\big)E(\tau) . \tag{9.11}$$

Here we introduced the overlap function

$$\Omega(t) = \langle\chi_{\text{tar}}|U_e(t_f - t)U_g(t - t_0)|\chi_{g0}\rangle . \tag{9.12}$$

which describes the overlap between the target vibrational state and the initial vibrational state $\chi_{g0}$, but propagated from $t$ to $t_f$ on the excited PES (its propagation from $t_0$ to $t$ with $U_g$ introduces only an unimportant phase factor). An alternative interpretation would be to view it as an overlap between $\chi_{g0}$ and $\chi_{\text{tar}}$, but with the latter propagated backwards in time from $t_f$ to $t$. Thus, for the condition $\delta J = 0$ to hold it is required here that at each time $t$ the optimal field incorporates information about the complex amplitudes of the forward *and* backward propagations.

The overlap functions $\Omega(t)$ and $\Omega^*(\tau)$ at different time–arguments define the kernel of a homogeneous Fredholm integral equation for the determination of the field. This type of equation can be transformed into an eigenvalue equation after the time integration has been discretized. Optimal fields are then given by the eigenvectors with the corresponding eigenvalues $\lambda$ being the relative yields with respect to applied pulse intensity. Thus the eigenvector corresponding to the largest $\lambda$ is the optimal field (for a more complete account see also [Yan93]). In the next section we will elaborate optimal control theory further. Specifically, we will obtain the generalization of Eq. (9.9) to the case of dissipative dynamics.

### 9.2.2 Mixed–State and Dissipative Dynamics

The formulation of optimal control theory given so far has some shortcomings concerning the definition of the control task objective as well as the type of molecular dynamics which can be used for the simulations. We will turn to the latter problem below, where we describe the modifications of the optimal control theory which are necessary when changing from closed to open system dynamics. First, however, we concentrate on some generalizations of the quantity $|\langle\Psi_{\rm tar}|\Psi(t_f)\rangle|^2$ whose extremum we are looking for. Let us introduce the notation $\langle\Psi(t_f)|\hat{O}_{\rm tar}|\Psi(t_f)\rangle$ which is identical with the previous one, if we set $\hat{O}_{\rm tar} = |\Psi_{\rm tar}\rangle\langle\Psi_{\rm tar}|$. The quantity $\hat{O}_{\rm tar}$ is named *target operator* and projects here simply on the earlier introduced target state $\Psi_{\rm tar}$. However, numerous other forms can be introduced.[4]

In a further step let us split up the time evolution operator $U(t_f, t_0; E)$ in the expectation value of the target operator to write:

$$\begin{aligned}\langle\Psi(t_f)|\hat{O}_{\rm tar}|\Psi(t_f)\rangle &= \langle\Psi_0|U^+(t_f,t_0;E)\hat{O}_{\rm tar}U(t_f,t_0;E)|\Psi_0\rangle \\ &= {\rm tr}\{\hat{O}_{\rm tar}\hat{W}(t_f)\} \end{aligned} \tag{9.13}$$

The latter step simply includes the identification $\hat{W}(t) = U(t_f, t_0; E)|\Psi_0\rangle\langle\Psi_0|U^+(t_f, t_0; E)$. It is the great advantage of this notation that we can extend optimal control theory to the propagation of mixed state dynamics. For example, we may set $\hat{W}(t_0) = \hat{R}_{\rm eq}$, where $\hat{R}_{\rm eq}$ denotes the equilibrium statistical operator of the considered system, instead of $\hat{W}(t_0) = |\Psi_0\rangle\langle\Psi_0|$. And if the control of dissipative dynamics of an open quantum system is considered we may even set for the expectation value of the target operator, Eq. (9.13):

$$O_{\rm tar}(t_f) = {\rm tr}_S\{\hat{O}_{\rm tar}\hat{\rho}(t_f)\} \tag{9.14}$$

with $\hat{\rho}$ being the reduced density operator.

In the following we will derive the equations for the determination of the optimal laser field if $O_{\rm tar}(t_f)$ is defined according to Eq. (9.14), i.e. if a condensed phase situations has to be considered. In such a case the dynamics may be deduced from the propagation of the reduced statistical operator by means of a Quantum Master Equation (cf. Eq. (3.261)):

$$\frac{\partial}{\partial t}\hat{\rho}(t) = -i\mathcal{L}_{\rm mol}\hat{\rho}(t) - i\mathcal{L}_{\rm F}(t)\hat{\rho}(t) - \mathcal{D}\hat{\rho}(t) \ . \tag{9.15}$$

---

[4]For example, we may introduce a more general target operator which might be a superposition of two different states $|\Psi_{\rm tar}\rangle = |\Psi^{(1)}_{\rm tar}\rangle + |\Psi^{(2)}_{\rm tar}\rangle$. But one can also introduce a mixture of projectors as $\hat{O}_{\rm tar} = \sum_\alpha w_\alpha |\Psi_{\rm tar}(\alpha)\rangle\langle\Psi_{\rm tar}(\alpha)|$. This definition may incorporate a complete excited electronic state as the target regardless of the concrete vibrational state. However, it is not yet understood if for such general target operators (as well as for more sophisticated molecular dynamics involving, for instance, bond breaking) a unambiguous solution of the optimal control problem exists.

Here $\mathcal{L}_{\text{mol}}$ and $\mathcal{L}_{\text{F}}(t)$ are the Liouville superoperators corresponding to the molecular Hamiltonian and relevant system–field interaction Hamiltonian, respectively. $\mathcal{D}$ is the relaxation superoperator (see, e.g., Eq. (3.261)). Eq. (9.15) can be rewritten after introducing a time evolution superoperator, i.e.

$$\begin{aligned}\hat{\rho}(t) &= \mathcal{U}(t, t_0; E)\hat{\rho}(t_0) \\ &= \hat{T} \exp\Big\{ -i \int_{t_0}^{t} d\tau \, \big(\mathcal{L}_{\text{mol}} + \mathcal{L}_{\text{F}}(\tau) - i\mathcal{D}\big)\Big\}\hat{\rho}(t_0)\end{aligned} \tag{9.16}$$

where the dependence on the external field has been indicated explicitly. Note that in general $\mathcal{D}$ will depend on the external field as well. This may play a role when the system–field interaction is strong enough to modify the spectrum of eigenvalues of the relevant system. In this case the relaxation will proceed according to the time–dependent spectrum. However, if the system–bath coupling is weak such that the related relaxation times are much longer than the laser pulse duration this effect is unlikely to be significant.

The control objective will be defined in terms of a target operator whose expectation value which is given in Eq. (9.14) should take a maximum value at a certain time $t = t_f$. An example for a target operator has been given already in the previous section where it was the projection operator onto a specific target state. Of course, the target operator should act in the space of the states of the relevant system. In analogy to Eq. (9.7) we write the control functional as (recall that $E(t)$ is taken to be real)

$$J = O_{\text{tar}}(t_f) - \frac{\alpha}{2} \int_{t_0}^{t_f} d\tau \; E^2(\tau) \, . \tag{9.17}$$

Notice, that the field intensity constraint has been modified compared with Eq. (9.7). Here it enters via a penalty which is given for high intensities. Its importance is weighted by the positive parameter $\alpha$.[5] Since the variation of the constant term $\propto S_0$ in Eq. (9.6) gives zero, the resulting equations for the optimal field are not influenced. However, whereas the Lagrange multiplier $\lambda$ is fixed by the assumed field energy $S_0$, the penalty factor $\alpha$ is a parameter. As mentioned in Section 9.2.1 the equation of motion (9.15) enters via the dependence of the target on the evolution of $\hat{\rho}(t)$ which in turn depends on the field $E(t)$.

The functional Eq. (9.17) is maximized allowing variations in the field $E(t)$, that is, we have to find an expression for $\delta J$, where we already performed the variation for the second term in Section 9.2.1. For the first term it is required to calculate the functional derivative of $\hat{\rho}(t_f)$, that is,

$$\begin{aligned}\frac{\delta \hat{\rho}(t_f)}{\delta E(t)} &= \frac{\delta}{\delta E(t)} \mathcal{U}(t_f, t_0; E)\hat{\rho}(t_0) \\ &= -i \int_{t_0}^{t_f} d\tau \, \mathcal{U}(t_f, \tau; E) \, \frac{\delta \mathcal{L}_{\text{F}}(\tau)}{\delta E(t)} \, \mathcal{U}(\tau, t_0; E) \, .\end{aligned} \tag{9.18}$$

Noting that

$$\frac{\delta}{\delta E(t)} \mathcal{L}_{\text{F}}(\tau) \bullet = -\delta(\tau - t) \frac{1}{\hbar} [\hat{\mu}, \bullet]_- \, , \tag{9.19}$$

[5]This parameter could even be chosen as a time–dependent function under the integral in order to enforce a certain overall shape of the field envelope, e.g., to ensure smooth switching–on and off behavior.

one obtains the functional derivative of $O_{\rm tar}$ as

$$\frac{\delta O_{\rm tar}(t_f)}{\delta E(t)} = K(t_f, t; E) , \tag{9.20}$$

where the field–dependent dissipative *control kernel* has been introduced as

$$K(t_f, t; E) = \frac{i}{\hbar} {\rm tr_S}\{\hat{O}_{\rm tar} \mathcal{U}(t_f, t; E) \, [\hat{\mu}, \mathcal{U}(t, t_0; E)\hat{\rho}(t_0)]_-\} . \tag{9.21}$$

Inserting this result into the equation for $\delta J = 0$ (see previous section) gives an expression for the optimal field:

$$E(t) = \frac{1}{\alpha} \, K(t_f, t; E) . \tag{9.22}$$

The control kernel is obtained by propagating the reduced density operator in the presence of the external field from the initial time $t_0$ until some time $t \leq t_f$ where the commutator with respect to the dipole operator is calculated. Then the result is propagated from $t$ to the final time $t_f$ where the operator $\hat{O}_{\rm tar}$ acts. According to Eq. (9.22) the control kernel has to be calculated such that it coincides (up to the prefactor $1/\alpha$) with the field.

### 9.2.3 Iterative Determination of the Optimal Pulse

Eqs. (9.21) and (9.22) constitute a highly nonlinear set of equations for the determination of the optimal field. These equations can be solved by iteration, for instance, according to the following procedure (see also [Oht99]): First, the control kernel is rewritten as

$$K(t_f, t; E) = \frac{i}{\hbar} {\rm tr_S}\{\hat{\sigma}(t; E)[\hat{\mu}, \hat{\rho}(t; E)]_-\} , \tag{9.23}$$

where the two time–dependent operators $\hat{\sigma}(t; E)$ and $\hat{\rho}(t; E)$ are propagated separately up to the intermediate time $t$. However, the new auxiliary operator $\hat{\sigma}(t; E)$ has to be propagated *backwards* in time, starting from $\hat{O}_{\rm tar}$ at $t = t_f$, that is,

$$\hat{\sigma}(t; E) = \hat{O}_{\rm tar} \mathcal{U}(t_f, t; E) = \mathcal{U}^+(t_f, t; E) \, \hat{O}_{\rm tar} . \tag{9.24}$$

As shown in the supplementary Section 9.4.1 the following coupled sets of equations of motion can be derived

$$\frac{\partial}{\partial t}\hat{\rho} = -i\mathcal{L}_{\rm mol}\hat{\rho} - \mathcal{D}\hat{\rho} - \frac{1}{\alpha\hbar^2}{\rm tr_S}\{\hat{\sigma}(t)[\hat{\mu}, \hat{\rho}]_-\} \, [\hat{\mu}, \hat{\rho}]_- \tag{9.25}$$

and

$$\frac{\partial}{\partial t}\hat{\sigma} = -i\mathcal{L}_{\rm mol}\hat{\sigma} + \tilde{\mathcal{D}}\hat{\sigma} - \frac{1}{\alpha\hbar^2}{\rm tr_S}\{\hat{\sigma}[\hat{\mu}, \hat{\rho}]_-\} \, [\hat{\mu}, \hat{\sigma}]_- . \tag{9.26}$$

where the dissipative superoperator for backward propagation is denoted $\tilde{\mathcal{D}}$, see Eq. (9.37). Here, nonlinearities which mediate the feedback in the evolution of $\hat{\rho}$ and $\hat{\sigma}$ via the field have been made explicit by using Eq. (9.22).

The separation of the control kernel into the propagation of two operators, Eqs. (9.25) and (9.26), can be used to devise an iterative scheme for obtaining the control kernel and therefore the optimal field. To this end we start from the zeroth order approximation for $\hat{\rho}(t)$

$$\frac{\partial}{\partial t}\hat{\rho}^{(0)} = -i\mathcal{L}_{\rm mol}\hat{\rho}^{(0)} - \mathcal{D}\hat{\rho}^{(0)} + i\mathcal{L}_{\rm F}^{(0)}(t)\hat{\rho}^{(0)} , \tag{9.27}$$

where $\mathcal{L}_{\rm F}^{(0)}(t)$ contains an initial guess $E^{(0)}(t)$ for the control field. This allows to start an iteration where the $n$th order auxiliary operator is given by

$$\frac{\partial}{\partial t}\hat{\sigma}^{(n)} = -i\mathcal{L}_{\rm mol}\hat{\sigma}^{(n)} + \tilde{\mathcal{D}}\hat{\sigma}^{(n)} - \frac{1}{\alpha\hbar^2}{\rm tr_S}\{\hat{\sigma}^{(n)}[\hat{\mu},\hat{\rho}^{(n)}]_-\}\,[\hat{\mu},\hat{\sigma}^{(n)}]_- \tag{9.28}$$

and the $n$th order reduced density operator follows from

$$\frac{\partial}{\partial t}\hat{\rho}^{(n)} = -i\mathcal{L}_{\rm mol}\hat{\rho}^{(n)} - \mathcal{D}\hat{\rho}^{(n)} - \frac{1}{\alpha\hbar^2}{\rm tr_S}\{\hat{\sigma}^{(n-1)}[\hat{\mu},\hat{\rho}^{(n)}]_-\}\,[\hat{\mu},\hat{\rho}^{(n)}]_- \,. \tag{9.29}$$

Notice that the coupling between the evolution of the two operators enters on the right hand side via $\hat{\sigma}^{(n-1)}$. The optimal field for the $n$th iteration step is given by

$$E^{(n)}(t) = \frac{i}{\alpha\hbar}{\rm tr_S}\{\hat{\sigma}^{(n-1)}(t)[\hat{\mu},\hat{\rho}^{(n)}(t)]_-\} \,. \tag{9.30}$$

Apparently, the method for calculating optimal control fields outlined so far can also be applied to mixed and pure states in the dissipation–free case.

## 9.3 Laser Pulse Control of Particle Transfer

The successful demonstration of controllable photoinduced reactions in gas and condensed phase is doubtless a milestone in ultrafast dynamics research. Among an increasing number of applications are those where branching ratios of competing photodissociation channels are controlled (for an overview, see [Bri03]). On the other hand, guiding the motion of particles with the help of tailored laser fields is of fundamental importance because it provides a means for controlled charge flow in molecular systems. This could be useful, for instance, to design ultrafast molecular switches for the emerging field of molecule–based electronics. On the other hand, the occurring change of the charge distribution and thus of the associated multipole moments, might be used to probe details of the interaction with, e.g., a polar solvent.

The principal problem one faces when turning such ideas into praxis is related to the complex dynamics of coupled electronic and nuclear DOF which is inherent to particle transfer reactions. In the following we will illustrate some general aspects of optimal control theory as it can be applied to proton and electron transfer reaction. In doing so we focus on simple model systems such as PT in a two–dimensional double minimum potential and the ET in a donor–bridge–acceptor (DBA) complex.

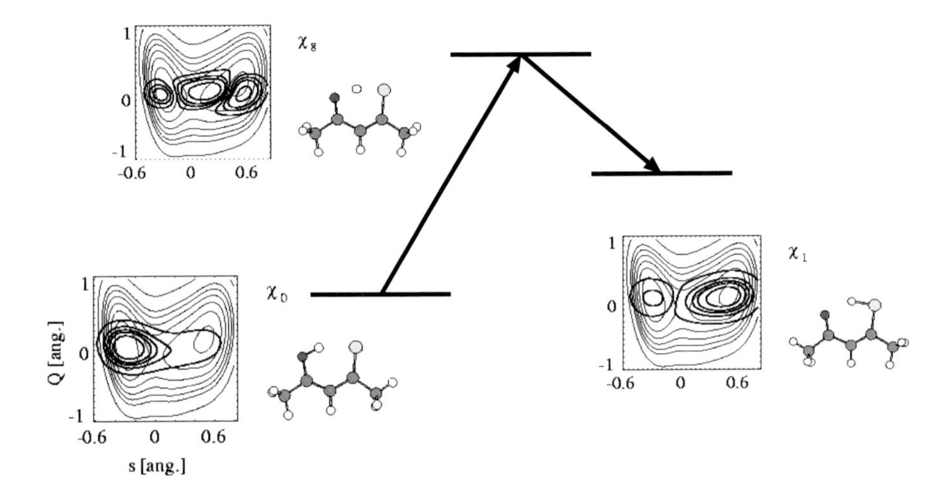


**Figure 9.7:** PT between an enol (left) and a enethiol (right) configuration in thioacetylacetone. In the upper part the transition state is shown. The barrier height is 0.22 eV and the asymmetry between the potential minima is 0.07 eV. Also shown are the corresponding probability distributions according to the eigenfunctions of the ground ($|\chi_0\rangle$), the first ($|\chi_1\rangle$) and the 8th ($|\chi_8\rangle$) excited state of a model Hamiltonian taking into account the reaction coordinate for PT, $s$, and the coupling to a vibrational mode $Q$ which modulates the strength of the hydrogen bond (for more details see [Dos99]). The arrows in the level scheme indicate a pathway for laser–driven PT which is discussed in Fig. 9.8.

### 9.3.1 Infrared Laser–Pulse Control of Proton Transfer

In view of the importance of hydrogen bonds and proton transfer reactions (cf. Chapter 7) it is not surprising that this type of elementary transfer reaction is rather frequently discussed in the context of laser control. Driving the proton motion across a hydrogen bond is not only fundamentally important but this capability could have interesting applications such as to initiate subsequent reactions. Moreover, the complicated potential energy surface could be explored by combining laser driving with spectroscopic detection in real time. This also concerns systems in the condensed phase where, e.g., phase and energy relaxation processes specific for the different local configurations could be explored.

In the following we will discuss a simple example which serves to illustrate possible strategies for proton transfer control in the electronic ground state. In addition it will highlight the power of optimal control theory to predict reaction pathways. The model comprises two relevant coordinates, that is, a proton transfer coordinate and a harmonic coordinate describing the modulation of the proton transfer distance (cf. the Hamiltonian Eq. (7.1)). For simplicity the interaction with some environment has been neglected. The potential for the PT is asymmetric as a consequence of different local environments for the two tautomers. This also necessitates to include higher order terms in the coupling function $F(s)$ in Eq. (7.1). The mentioned asymmetry is, together with an appreciable change in the dipole moment, a prerequisite for implementation of control schemes which require distinct initial and final states. The specific

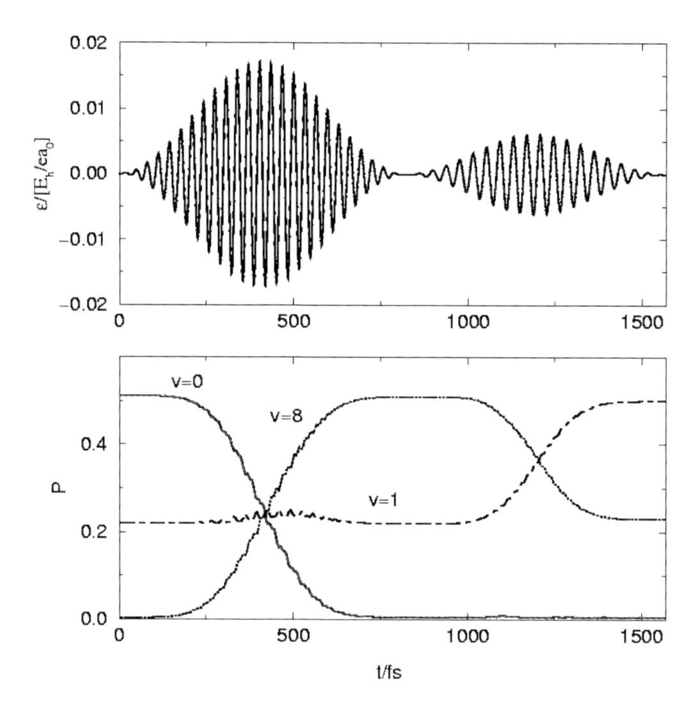


**Figure 9.8:** Pump–dump control of the system of Fig. 9.7 with a two–pulse sequence as shown in the upper panel. The respective population dynamics reveals that the transfer proceeds from the localized initial states ($v = 0$ and $v = 1$, the system starts from a thermal initial population) to the final state ($v = 1$) via a delocalized intermediate state ($v = 8$). This pathway is indicated in Fig. 9.7. (For more details see [Dos99].)

parameterization of the potential in Eq. (7.1) has been chosen to mimic quantum chemistry results for thioacetylacetone. In Fig. 9.7 some eigenfunctions, $\chi_v(s, Q)$, of the Hamiltonian are shown which reveal rather localized states on the reactant and product site.

Looking at Fig. 9.7 it is tempting to consider the straightforward reaction path, that is, the direct excitation $|\chi_0\rangle \to |\chi_1\rangle$ by a single pulse. This is, of course, only possible if there is a finite transition dipole matrix element $d_{01} = \langle\chi_0|\hat{\mu}|\chi_1\rangle$, which is typically the case for asymmetric potentials with a rather low barrier as shown in Fig. 9.7. Using the pulse form given by Eq. (9.3) we have several parameters which can be adjusted to give the optimum pulse say for complete population inversion. If the dynamics can be restricted to the two lowest vibrational states (with energies $E_0$ and $E_1$), one can use a result known from optics which states that in case of resonant driving, that is, for $\Omega_1 = (E_1 - E_0)/\hbar$, complete population inversion in the interval $[t_0 : t_f]$ can be obtained, if the integrated pulse envelope is equal to $\pi$

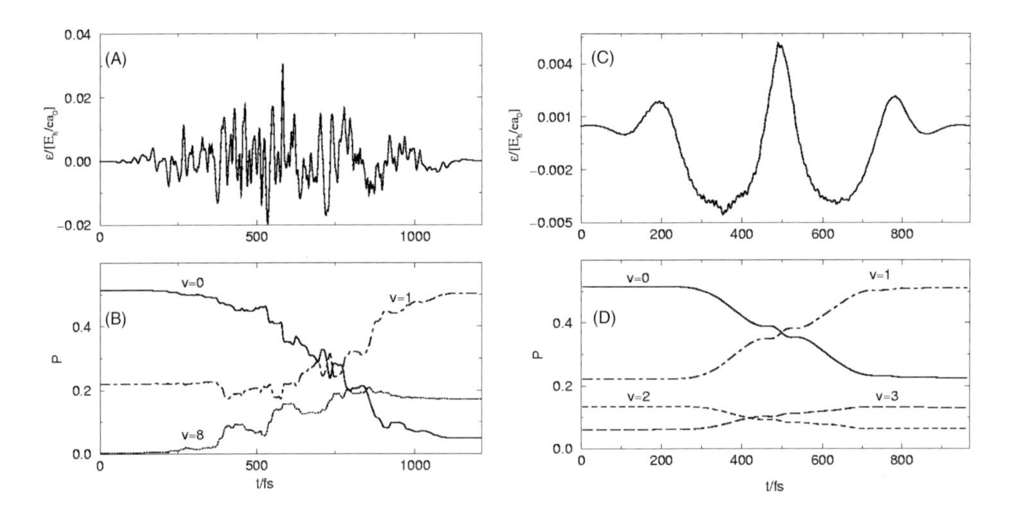


**Figure 9.9:** Optimal control theory results for the system of Fig. 9.7. Panels A and B correspond to the case of a small penalty factor $\alpha$, i.e. for a high field intensity. Here, the reaction proceeds in analogy to the pump–dump mechanism of Fig. 9.8. In panels C and D results are shown for a high penalty. Now, a reaction path "through" the barrier is taken, which can be rationalized in terms of the populations which are exchanged between pairs of states like $|\chi_0\rangle, |\chi_1\rangle$ and $|\chi_2\rangle, |\chi_3\rangle$. These pairs would be degenerated in the case of vanishing potential asymmetry. As in Fig. 9.8 the system starts from a thermalized initial population. (For more details see also [Dos99].)

(so–called $\pi$–pulse):

$$\frac{|d_{01}|}{\hbar} \int_{t_0}^{t_f} dt A \sin^2\left(\frac{t\pi}{\tau}\right) = \pi \ . \tag{9.31}$$

This relation makes it easy to find suitable pulse parameters. It should be noted, however, that the transition frequency $(E_1 - E_0)/\hbar$ may become very small (in particular for weaker hydrogen bonds) which makes the experimental realization of this type of ultrashort (infrared) pulse very difficult.

In this situation the following so-called pump-dump scheme may be more appropriate: Here one uses a sequence of two laser pulses as shown in 9.8A. The first (pump) pulse excites the system from the localized reactant ground state to some delocalized intermediate which is energetically above the reaction barrier (here state $|\chi_8\rangle$ shown in the middle panel of Fig. 9.7). A subsequent second (dump) pulse then de-excites the system to the localized product state. Also in this case, optimization of the pulse parameters may give a field $E(t)$ which almost completely switches the system from the reactant into the product state. The respective population dynamics of the three states is shown in 9.8B.

So far the reaction has been driven along a predetermined path. Optimal control theory does not have this kind of bias and the reaction path is determined by the balance between the target state optimization and the penalty for high field intensities as given by Eq. (9.7). If this penalty is small, the predicted pulse form is rather complicated, although the reaction path is

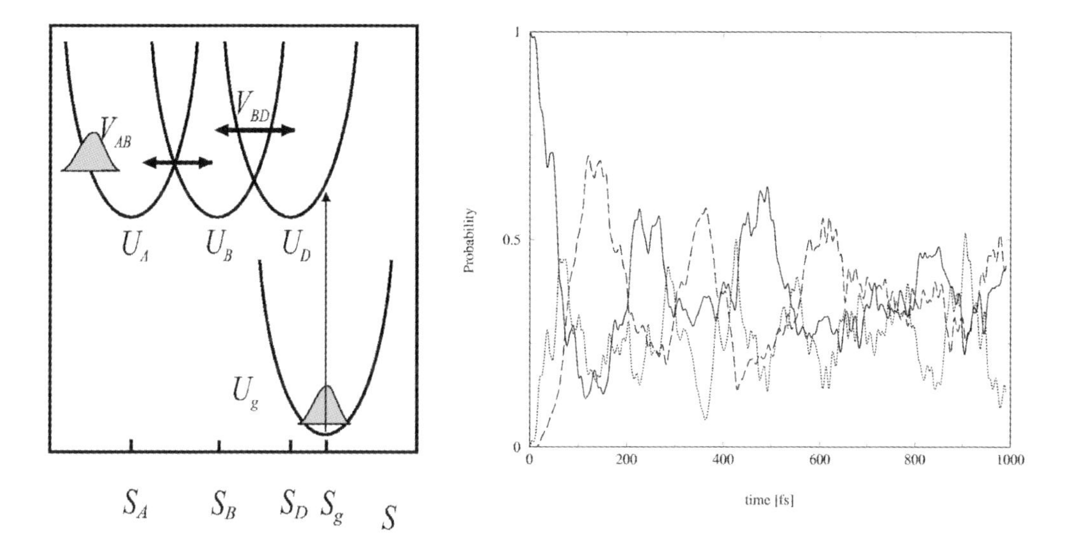


**Figure 9.10:** Left: DBA model system of diabatic PES along a reaction coordinate $s$ for describing photoinduced ET starting from the ground states PES $U_g$. The vibrational frequency of all states is 0.1 eV and the interstate coupling is 0.03 eV. The wave packet in the acceptor potential corresponds to the target state in Fig. 9.11. Right: ET dynamics after impulsive excitation of the donor state as characterized by the population of the donor (solid), bridge (dotted), and acceptor (dashed) states.

very similar to the pump–dump scheme as seen by comparing Fig. 9.8 with Figs. 9.9A and B. However, if the penalty factor $\alpha$ is increased the reaction path over the barrier becomes too costly and a path "through" the barrier is taken. The predicted pulse form is shown in Fig. 9.9C; the frequency of the field does not correspond to any transition frequency of the system. Thus the mechanism has to be understood as a deformation of the potential energy surface such that the asymmetry between the two wells is compensated and tunneling becomes an effective reaction channel. This argument is supported by the population dynamics of the system's eigenstates shown in Fig. 9.9D. Here, the driven PT corresponds to a population exchange between pairs of states like $|\chi_0\rangle$ and $|\chi_1\rangle$ which would be degenerated in case of vanishing potential asymmetry.

Finally, we note that since all driving schemes are very sensitive to the vibrational energy level structure the pulses will be isotope selective.

In the discussed schemes of infrared laser control of proton transfer the reactions have been restricted to the adiabatic electronic ground state. In the next section we will focus on electronically excited states where the coupling between different diabatic states will introduce the new aspect of coupled electron–vibrational dynamics.

## 9.3.2 Controlling Photoinduced Electron Transfer

The following discussion, however, will focus on the application of optimal control theory to *photoinduced* ET. In passing we note that there has been a large effort to understand the effect

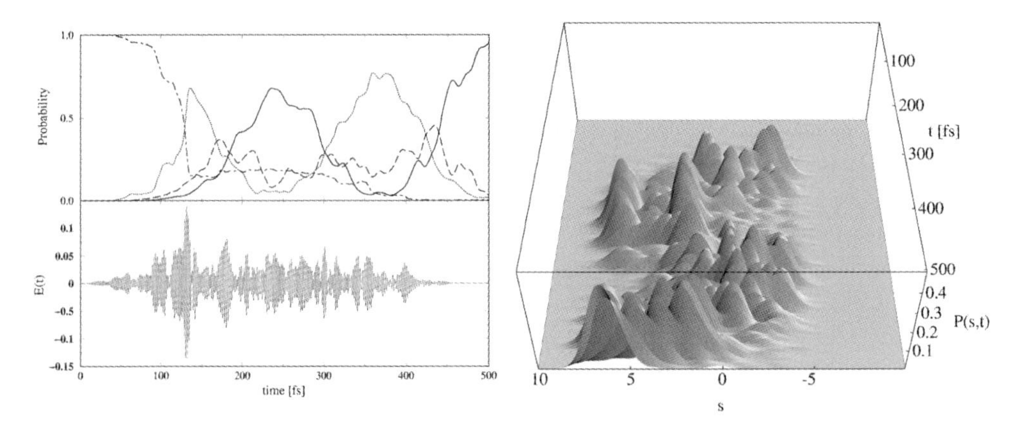


**Figure 9.11:** Optimal control of the wave packet dynamics in the DBA system shown in Fig. 9.10. The target state has been the ground state of the acceptor PES displaced from $s^{(A)} = 4$ to $s^{(A)} = 6.4$ (right panel, cf. Fig. 9.10). In the left panel we show the diabatic state populations for the ground (dash-dotted), the donor (dotted), the bridge (dashed), and the acceptor (solid) as we as the optimal field (in units of $10^7$ V/cm). (For more details, see [Man01]).

of monochromatic field driving on the particle motion in a DA type complex being modelled in terms of the spin–boson model (for a review, see [Gri98]). In contrast, in the present section it will be of importance to consider the scheme of photoinduced ET (including optical excitation from a donor ground state, cf. Fig. 6.6) as well as ultrafast field control of electron–vibrational dynamics.

For the investigation of photoinduced ET, let us first consider the model DBA system shown in Fig. 9.10. The ET will be described in terms of four diabatic state PES along some reaction coordinate $s$. The ground state $U_g$ is coupled to the donor state only via a constant transition dipole matrix element. The DBA system consists of three identical but shifted PES with a constant coupling between D and B and B and A. In order to have a reference we first discuss the case of impulsive excitation (cf. Section 5.5, Eq. (5.124)). The diabatic state populations are shown in Fig. 9.10 (right panel). Initially, only the donor PES is populated but the subsequent dynamics shows oscillations which are typical for coupled PES in the coherent dissipation-free limit. Viewed in terms of the corresponding wave packet motion, this behavior corresponds to a delocalization of the initially prepared ground state wave packet over the available part of the reaction coordinate axis.

Turning our attention to the controllability of this type of model electron–vibrational dynamics we first have to address the question of time scales. In optimal control theory it is required to specify a final time $t_f$ at which the target state has to be realized. On the other hand, ET reactions have an intrinsic time scale $\tau_{\rm ET}$ resulting from the coupling of the shifted PES. Obviously, $t_f > \tau_{\rm ET}$ has to be fulfilled in order to control the evolving wave packet. However, given the complicated coherent oscillations in Fig. 9.10 as a result of quantum mechanical interferences it is apparent that even for $t_f > \tau_{\rm ET}$, the control field and also the respective yield may strongly depend on the final time $t_f$. In a sense the control field has to be synchronized with the oscillating wave packet motion. In Fig. 9.11 we have set the target such

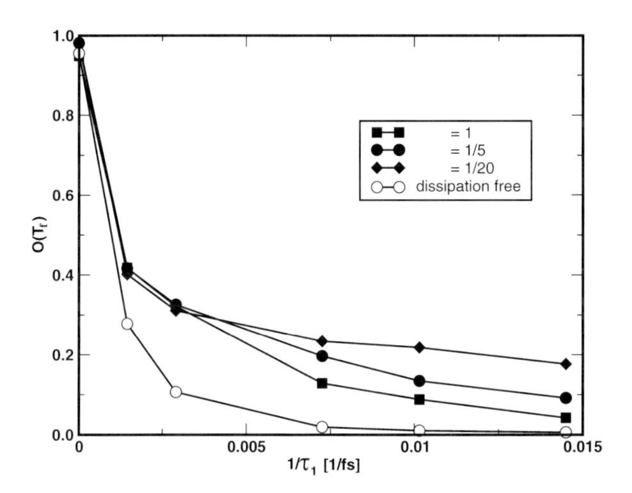


**Figure 9.12:** ET control efficiency for a DA (no bridge) system in dependence on the inverse life time of the target state which was set to be the first excited vibrational state of the acceptor PES. (For more details, see [Man02].)

that the final state should correspond to a superposition state, that is, a compact wave packet given by the acceptor PES vibrational ground state but displaced along the reaction coordinate (cf. Fig. 9.10). The final state and the way it is reached is shown in the right panel of Fig. 9.11. The control field as well as the dynamics of the diabatic state populations are shown in the left panel of Fig. 9.11.The interpretation of the pulse form is not straightforward. Generally, one expects that the field–driving of the populations has to be synchronized with the time scales for the field–free evolution which is due to the state couplings. This argument is supported by the fact that, for instance, the population of the acceptor state in the field– free evolution (Fig. 9.10) has passed two times a maximum when it reaches the target time $t_f = 500$fs. This is similar to the evolution in the control field as shown in Fig. 9.11.

So far we have not considered the influence of the interaction with some environment on the controllability of ET reactions. In general it is assumed that relaxation processes will act against the control objective if, for instance, the target state has a finite lifetime. Therefore, the question arises whether the relaxation process can be at least partially compensated by the laser–driving. In Fig. 9.12 we show for a simple DA system the efficiency for populating the first excited vibrational state in the acceptor PES starting from the ground state PES. Comparing the case of a field which has been optimized for the dissipation–free case with fields whose optimization took into account the finite lifetime of the target state, we clearly see that laser–driving can compete with relaxation although the overall yield at the target time is diminished. However, the latter can be increased at the expense of using higher field intensities represented by the decreasing penalty factor $\alpha$.

## 9.4 Supplement

### 9.4.1 Dissipative Backward Time Evolution

The determination of the dissipative control kernel in Section 9.2.2 required to propagate the auxiliary density operator $\hat{\sigma}$ backward in time according to Eq. (9.26). In the following we will derive the respective dissipation superoperator $\tilde{\mathcal{D}}$ as well as the respective equation of motion for $\hat{\sigma}$ . We start from the representation of dissipative time evolution given in Eq. (3.445). Since for the present application a time dependent field enters the time evolution we have to generalize Eq. (3.445) to

$$\begin{aligned} \mathcal{U}(t,t_0)\hat{\rho}(t_0) &= \hat{A}(t,t_0)\hat{\rho}(t_0) + \hat{\rho}(t_0)\hat{A}^+(t,t_0) \\ &+ \sum_j \int_{t_0}^{t} d\tau\; \hat{C}_j(\tau;t,t_0)\hat{\rho}(t_0)\hat{C}_j^+(\tau;t,t_0)\,. \end{aligned} \tag{9.32}$$

Inserting this relation into Eq. (9.21) gives

$$\begin{aligned} K(t_f,t;E) &= \frac{i}{\hbar}\mathrm{tr_S}\{\hat{O}_{\mathrm{tar}}\Big(\hat{A}(t_f,t)[\hat{\mu},\hat{\rho}(t)]_- + [\hat{\mu},\hat{\rho}(t)]_-\hat{A}^+(t_f,t) \\ &+ \sum_j \int_{t_0}^{t_f} d\tau\; \hat{C}_j(\tau;t_f,t)[\hat{\mu},\hat{\rho}(t)]_-\hat{C}_j^+(\tau;t_f,t)\Big)\}\,. \end{aligned} \tag{9.33}$$

Rearranging the different terms in the trace leads to

$$\begin{aligned} K(t_f,t;E) &= \frac{i}{\hbar}\mathrm{tr_S}\{\Big(\hat{A}^+(t_f,t)\hat{O}_{\mathrm{tar}} + \hat{O}_{\mathrm{tar}}\hat{A}(t_f,t) \\ &+ \sum_j \int_{t_0}^{t_f} d\tau\; \hat{C}_j^+(\tau;t_f,t)\hat{O}\hat{C}_j(\tau;t_f,t)\Big)[\hat{\mu},\hat{\rho}(t)]_-\}\,. \end{aligned} \tag{9.34}$$

This expression confirms the principal existence of some $\tilde{\mathcal{U}}$ in Eq. (9.24). Given the form Eq. (9.32) for $\mathcal{U}$ one can obtain $\mathcal{U}^+$ or equivalently the equations of motion for $\hat{\sigma}$. To do this we use Eq. (9.16) and get

$$\frac{\partial}{\partial t}\mathcal{U}(t_f,t) = \mathcal{U}(t_f,t)\,\big(i\mathcal{L}_{\mathrm{mol}} + i\mathcal{L}_{\mathrm{F}}(t) + \mathcal{D}\big)\,. \tag{9.35}$$

In a next step the control kernel, Eq. (9.23), is generalized such that it depends via $\hat{\rho}$ on a second independent time–argument $\bar{t}$. Then the time derivative with respect to $t$ yields

$$\frac{\partial}{\partial t}K(t_f,t,\bar{t};E) = \frac{i}{\hbar}\mathrm{tr_S}\{\hat{\sigma}(t)\big(i\mathcal{L}_{\mathrm{mol}} + i\mathcal{L}_{\mathrm{F}}(t) + \mathcal{D}\big)[\hat{\mu},\hat{\rho}(\bar{t})]_-\}\,. \tag{9.36}$$

A rearrangement of the terms in the bracket leads to the required equation of motion for $\hat{\sigma}$, (9.26), with the dissipative superoperator $\tilde{\mathcal{D}}$ being defined via

$$\tilde{\mathcal{D}}\hat{\sigma}(t) = \sum_u \Big(\Lambda_u^+ K_u\hat{\sigma}(t) + \hat{\sigma}(t)K_u\Lambda_u - \Lambda_u^+\hat{\sigma}(t)K_u - K_u\hat{\sigma}(t)\Lambda_u\Big)\,. \tag{9.37}$$

Notice that this form valid for backward time evolution differs from the forward evolution superoperator $\mathcal{D}$, Eq. (3.261).

---

# Notes


[Bar00] R. Bartels, S. Backus, E. Zeek, L. Misoguti, G. Vdovin, I. P. Christof, M. M. Murnane, and H. C. Kapteyn, Nature **406**, 164 (2000).

[Bri03] T. Brixner and G. Gerber, Chem. Phys. Chem. **4**, 418 (2003).

[Cri96] F. F. Crim, J. Phys. Chem. **100**, 12725 (1996).

[Dan03] C. Daniel, J. Full, L. González, C. Lupulescu, J. Manz, A. Merli, S. Vajda, and L. Wöste, Science **299**, 536 (2003).

[Dos99] N. Došlić, K. Sundermann, L. González, O. Mó, J. Giraud-Girard, and O. Kühn, Phys. Chem. Chem. Phys. **1**, 1249 (1999).

[Gri98] M. Grifoni and P. Hänggi, Phys. Rep. **304**, 229 (1998).

[Her02] J. L. Herek, W. Wohlleben, R. J. Cogdell, D. Zeidler, and M. Motzkus, Nature **417**, 533 (2002).

[Kor96] M. V. Korolkov, J. Manz, and G. K. Paramonov, J. Phys. Chem. **100**, 13927 (1996).

[Kos89] R. Kosloff, S. A. Rice, P. Gaspard, S. Tersigni, D. J. Tannor, Chem. Phys. **139**, 202 (1989).

[Maa98] D. J. Maas, D. I. Duncan, R. B. Vrijen, W. J. van der Zande, and L. D. Noordam, Chem. Phys. Lett. **290**, 75 (1998).

[Man01] T. Mančal and V. May, Eur. Phys. J. D **14**, 173 (2001).

[Man02] T. Mančal, U. Kleinekathöfer, and V. May, J. Chem. Phys. **117**, 636 (2002).

[Oht99] Y. Ohtsuki, W. Zhu, and H. Rabitz, J. Chem. Phys. **110**, 9825 (1999).

[Pie88] A. P. Pierce, M. A. Dahleh, and H. Rabitz, Phys. Rev. A **37**, 4950 (1988).

[Ric00] S. A. Rice and M. Zhao, *Optical Control of Molecular Dynamics*, Wiley, New York, 2000.

[Sha03] M. Shapiro and P. Brumer, *Principles of the Quantum Control of Molecular Processes*, Wiley and Sons, Hoboken, 2003.

[Win03] L. Windhorn, J. S. Yeston, T. Witte, W. Fuß, M. Motzkus, D. Proch, and K.-L. Kompa, J. Chem. Phys. **119**, 641 (2003).

[Yan93] Y. J. Yan, R. E. Gillilan, R. M. Whitnell, K. R. Wilson, and S. Mukamel, J. Phys. Chem. **97**, 2320 (1993).

# 10 Suggested Reading

## Chapter 1

- for an introduction into theoretical concepts and basic experiments see:
  H. Haken and H. C. Wolf, *Molecular Physics and Quantum Chemistry*, (Springer Verlag, Berlin, 1999).

## Chapter 2

- for an excellent account on electronic structure theory see:
  A. Szabo and N. S. Ostlund, *Modern Quantum Chemistry*, (Dover, New York, 1996).
- for the most recent overview on computational chemistry (including a discussion of reaction dynamics) see:
  P. von Ragué Schleyer (ed.), *Encyclopedia of Computational Chemistry*, (Wiley, Chichester, 1998).
- an introduction into Density Functional Theory is given in:
  W. Koch and M. C. Holthausen, *A Chemist's Guide to Density Functional Theory*, (Wiley–VCH, Weinheim, 2000).

## Chapter 3

- general relaxation theory
  - a introduction is given in:
    K. Blum, *Density Matrix Theory and Applications*, (Plenum Press, New York, 1996).
  - a standard textbook for nonequilibrium dynamics is:
    R. Kubo, M Toda, and N. Hashitsume, *Statistical Physics II: Nonequilibrium Statistical Mechanics*, (Springer, Berlin, 1985).
  - the original paper of A. G. Redfield can be found in:
    Advances of Magnetic Research **1**, 1 (1965).
- further developments
  - for the superoperator approach to dissipative dynamics see:
    E. Fick and G. Sauermann, *The Quantum Statistics of Dynamic Processes*, (Springer, Berlin, 1990).

- for recent advances in numerical methods see:
  J. Broeckhove and L. Lathouwers (eds.), *Time–Dependent Molecular Dynamics*, (Plenum Press, New York, 1992).
- for a recent review on the use of Redfield theory see:
  W. T. Pollard, A. K. Felts, R. A. Friesner, Advances in Chemical Physics **93**, 77 (1996).
- the relation of Redfield theory and phase space dynamics is detailed in:
  D. Kohen and D. Tannor, Advances in Chemical Physics **111**, 219 (1999).
- for recent topics in the theory of dissipative quantum dynamics see:
  U. Weiss, *Quantum Dissipative Systems*, (World Scientific, Singapore, 1993);
  T. Dittrich, P. Hänggi, G.–L. Ingold, B. Kramer, G. Schön, and W. Zwerger, *Quantum Transport and Dissipation*, (Wiley–VCH, Weinheim, 1998).
- for a review on numerical path integral methodology see:
  N. Makri, Journal of Physical Chemistry A **102**, 4414 (1998).
- recent developments:
  Chemical Physics, special issue *Quantum Dynamics of Open Systems* (eds. P. Pechukas and U. Weiss), **268** (2001).
- the multiconfiguration self–consistent field method for wave packet propagation is described in:
  M. H. Beck, A. Jäckle, G. A. Worth, and H.-D. Meyer, Physics Reports **324**, 1 (2000).

## Chapter 4

- for recents review on vibrational energy flow see:
  T. Uzer, Physics Reports **199**, 73 (1991);
  M. Gruebele and R. Bigwood, International Reviews in Physical Chemistry **17**, 91 (1998);
  M. Gruebele, Advances in Chemical Physics **114**, 193 (2000).

- for the effect of multi–phonon transitions see:
  V. M. Kenkre, A. Tokmakoff, and M. D. Fayer, Journal of Chemical Physics **101**, 10618 (1994).
- for an introduction into instantaneous normal mode theory and its relation to relaxation see:
  G. Goodyear and R. M. Stratt, Journal of Chemical Physics **105**, 10050 (1996).

## Chapter 5

- optical spectra of diatomic molecules including external perturbations are discussed in:
  H. Lefebvre–Brion and R. W. Field, *Perturbations in the Spectra of Diatomic Molecules*, (Academic Press, New York, 1986).

- examples for more complex systems can be found in:
  J. Michl and V. Bonačić–Koutecký, *Electronic Aspects of Organic Photochemistry*, (Wiley, New York, 1990).

- for the relation between wave packet motion and linear spectroscopy see:
  R. Schinke, *Photodissociation Dynamics*, (Cambridge University Press, Cambridge, 1993).

- for an introduction into the Liouville space description of nonlinear optical spectroscopy see:
  S. Mukamel, *Principles of Nonlinear Optical Spectroscopy*, (Oxford University Press, Oxford, 1995).

- for a review on nonadiabatic electron–vibrational dynamics see:
  W. Domcke and G. Stock, Advances in Chemical Physics **100**, 1 (1997).

## Chapter 6

- the classical work on Marcus theory is:
  R. A. Marcus, Journal of Chemical Physics **24**, 966 (1956).

- for applications of Marcus theory to electron transfer in proteins:
  R. A. Marcus and N. Sutin, Biochimica Biophysica Acta **200**, 811 (1985).

- for a density matrix approach to ultrafast photoinduced electron transfer see:
  V. May and M. Schreiber, Chemical Physics Letters **181**, 267 (1991);
  J. M. Jean, R. A. Friesner, G. R. Fleming, Journal of Chemical Physics **96**, 5827 (1992).

- review articles and books on electron transfer:

  - E. G. Petrov, *Physics of Charge Transfer in Biosystems*, (Naukova Dumka, Kiev, 1984).

    J. Jortner and B. Pullman (eds.), *Tunneling*, (Reidel Publishing Company, 1986).
  - J. Jortner and B. Pullman (eds.), *Perspectives in Photosynthesis*, (Kluwer Academic Publishers, 1990).
  - J. N. Onuchic, D. N. Beratan, and J. J. Hopfield, Journal of Physical Chemistry **90**, 3707 (1986).
  - P. F. Barbara, Th. J. Meyer, and M. A. Ratner, Journal of Physical Chemistry **100**, 13148 (1996).
  - A. M. Kusnezov and J. Ulstrup, *Electron Transfer in Chemistry and Biology*, (Wiley, Chichester, 1998).
  - recent developments: Chemical Physics, special issue *Processes in Molecular wires* (eds. P. Pechukas and U. Weiss), **281** (2002).

## Chapter 7

- on overview on the different aspects of proton transfer can be found in:
  T. Bountis (ed.), *Proton Transfer in Hydrogen–Bonded Systems*, (NATO ASI Series B: Physics Vol. 291, Plenum Press, 1992);
  H.–H. Limbach and J. Manz (eds.), *Hydrogen Transfer: Experiment and Theory*, Berichte der Bunsengesellschaft Physikalische Chemie **102**, No. 3 (1998);
  T. Elsaesser and H. J. Bakker (eds.), *Ultrafast Hydrogen Bonding Dynamics and Proton Transfer Processes in the Condensed Phase*, (Kluwer Academic, Dordrecht, 2002).

- for an introduction into hydrogen bonding see:
  D. Hadži (ed.), *Theoretical Treatments of Hydrogen Bonding*, (Wiley, Chichester, 1997).

- a general treatment of nonadiabatic proton transfer rates is given in:
  D. Borgis and J. T. Hynes, Chemical Physics **170**, 315 (1993).

- ultrafast excited state proton transfer is reviewed in:
  S. Lochbrunner and E. Riedle, Recent Research Developments in Chemical Physics **4**, 31 (2003).

- the surface hopping method described in the text has been originally developed in:
  J. C. Tully, Journal of Chemical Physics **93**, 1061 (1990).

  - an application to proton transfer as well as technical details can be found in:
    S. Hammes–Schiffer and J. C. Tully, Journal of Chemical Physics **101**, 4657 (1994).
  - an extension to correlated multiple proton transfer is given in:
    S. Hammes–Schiffer, Journal of Chemical Physics **105**, 2236 (1996).

## Chapter 8

- a classical monograph is:
  A. S. Davydov, *Theory of Molecular Excitons*, (Plenum, New York, 1971).

- for a more recent introduction see:
  V. M. Agranovich and M. D. Galanin, *Electronic Excitation Energy Transfer in Condensed Matter*, in Modern Problems in Condensed Matter Sciences, V. M. Agranovich and A. A. Maradudin (eds.) (North–Holland, Amsterdam, 1982).

- for a density matrix approach to exciton transfer dynamics see:
  V. M. Kenkre and P. Reineker, *Springer Tracts of Modern Physics* **94** (Springer, Berlin, 1982).

- for a review on exciton–vibrational coupling in photosynthetic antenna complexes see:
  O. Kühn, Th. Renger, V. May, J. Voigt, T. Pullerits and V. Sundström, Trends in Photochemistry and Photobiology **4**, 213 (1997).

- recent developments:
T. Kobayashi (ed.), *J–aggregates*, (World Scientific, Singapure, 1997);
V. Chernyak, N. Wang, and S. Mukamel, Physics Reports **263**, 213 (1995);
Chemical Physics, special issue 213 (1997), D. Markovitsi and P. Trommsdorf (eds.) *Photoprocesses in Multichromophoric Molecular Assembles* **275** (2002);
Th. Renger, V. May, and O. Kühn *Ultrafast Excitation Energy Transfer Dynamics in Photosynthetic Pigment–Protein Complexes*, Physics Reports **343**, 137 (2001).

## Chapter 9

- textbooks on laser pulse control of molecular dynamics:
S. A. Rice and M. Zhao, *Optical Control of Molecular Dynamics*, (Wiley, New York, 2000);
M. Shapiro and P. Brumer, *Principles of the Quantum Control of Molecular Processes*, (Wiley, New York, 2003).

- for a general perspective see:
I. Walmsley and H. Rabitz, Physics Today, August 2003, p. 43.

- an overview on feedback control in the experiment is given in:
T. Brixner and G. Gerber, ChemPhysChem **4**, 418 (2003).

- a number of special issues of different journals is also available:
Kent Wilson *Festschrift*, Journal of Physical Chemistry A **103** (1999);
Special issue on laser pulse control, Chemical Physics **267**, (2001).

- an overview on laser pulse control as well as ultrafast spectroscopy and ultrafast phenomena in molecular systems can be found in the proceedings of the *Femtochemistry Conferences*:
J. Manz and A. Welford Castleman, Jr. (eds.), Journal of Physical Chemistry **97**, No. 48 (1993);
M. Chergui, *Femtochemistry*, (World Scientific, Singapore, 1996);
A. Welford Castleman, Jr. (ed.), Journal of Physical Chemistry A **102**, No. 23 (1998),
and in the following books:
A. H. Zewail, *Femtochemistry – Ultrafast Dynamics of the Chemical Bond*, (Vol I and II, World Scientific, Singapore, 1994);
J. Manz and L. Wöste (eds.), *Femtosecond Chemistry*, (Verlag Chemie, Weinheim, 1995);
V. Sundström (ed.), *Femtochemistry and Femtobiology*, (World Scientific, Singapore, 1997);
A. Douhal, J. Santamaria (eds.) *Femtochemistry and Femtobiology*, (World Scientific, Singapore, 2002)

- the control of molecular systems is discussed in:
  W. Domcke, P. Hänggi, and D. Tannor (eds.), *Dynamics of Driven Quantum Systems*, special issue, Chemical Physics **217** No. 2 and 3 (1997);
  P. Gaspard and I. Burghardt (eds.), *Chemcial Reactions and their Control on the Femtosecond Time Scale*, Advances in Chemical Physics **101** (1997).

# Index